PATTERNS ON THE EARTH

PATTERNS ON THE EARTH:
An Introduction to Geography

FOURTH EDITION

Rhoads Murphey
University of Michigan

Rand McNally
College Publishing Company,
Chicago

Cover photograph:
Death Valley National Monument, California,
by David Muench.

78 79 80 10 9 8 7 6 5 4 3 2 1

Printed in U.S.A. by Rand McNally & Company
Library of Congress Catalog Card Number 77-77728

for Eleanor

Preface to the fourth edition

This is really a new book, and appropriately, therefore, has a new title, *Patterns on the Earth*. Its predecessor, *An Introduction to Geography*, was originally conceived as an introduction to the discipline of geography as a whole, and included regional chapters only as further detailed illustrations of geographical concepts.

The Scope of Geography, first published in 1971, incorporated the first chapters from the original text. The second edition of *Scope*, published in 1973, further rounded out and brought this general survey of the discipline and its principles up to date in a relatively brief paperback format.

The original text has continued to be used largely for its coverage of the regions of the world, and it seemed appropriate to plan this new version, in effect, this new book, even more completely around the regional theme. The regional material is still used, as before, to explain and illustrate a succession of geographic concepts, a formula that has proved successful in many courses over many years, and that I first evolved in my own introductory course. The material on the discipline as a whole, which formed the first eleven chapters of the original text (now available in *Scope*), has been completely rewritten for this book and shortened, forming the first three chapters. This much of an introduction to what geography is all about seems essential even for a primarily regional text, and has been designed to lead into and make more useful the regional chapters that follow.

The regional material has been basically rewritten as well as carefully updated, not only in factual terms, but in keeping with changing perspectives. The information and discussion should remain valuable well into the 1980s. Equally important for many users, the entire text has been carefully edited to improve its readability. Some users had reported occasional reading-level problems, and this has been taken very seriously. Each sentence has been examined critically, and the majority of them rewritten. The result is a book that now reads more simply and clearly and should pose no problems for beginning students.

Equally careful attention has been given to the maps. Virtually all of them are new, and all are greatly improved in clarity and effectiveness. Of course, maps are basic to any geography book, and I am grateful to Rand McNally for the care and willingness they have shown to bring the maps up to the present high standard. The same applies to the photographs, many of which are new, but all of which have been subjected to the same painstaking quality control. The Statistical Appendix has been completely redone (as opposed to simple updating), and it is now even more useful than before. Statistical tables and other references in the text have similarly been redone. This has been a very large and difficult undertaking, and I am especially grateful to James Chan for his careful work with the statistical compilations.

Other friends and colleagues have helped me greatly through critical readings of all or parts of the new manuscript where they have special expertise: Everette Bannister, Basheer K. Nijim, and William Rosenberg. Richard Dastyck, Fullerton College; James R. McDonald, Eastern Michigan University; Elliot McIntire, California State University, Northridge; Kennard W. Rummage, State University of New York, Brockport; and Marilyn Silberfein, Temple University, provided me with excellent critiques and suggestions for improvement. James Chan read the whole manuscript and has given me abundant critical help. Others still, unknown to me by name, have sent comments, suggestions, or corrections based on their reading or use of the third edition, and to them also I am grateful. Each chapter has been read by at least one and usually by several geographers and others with special knowledge of the subject, and accordingly, the text has been further improved in accuracy, completeness, and effective presentation.

I owe a special debt to Stella Jenks of Rand McNally College Publishing Company, who has done most of the new reading-level editing. Even more importantly, she has shepherded this project as a whole from original manuscript to a handsome finished book with remarkable conscientiousness, patience, and high standards; she has become almost as much of its parent as I am. I am sure she would agree that for all the hard work that has gone into it, this is supposed to be a book to *enjoy*. May all of its readers do so!

R. M.
Ann Arbor, November, 1977

Contents

Preface vii

Maps xvi

Figures xviii

Tables xviii

Introduction for the student 1
What is geography? 2
Geographic analysis 2
About this book 3

1 Spatial form and spatial interaction 5
Spatial relations 6
Cities and spatial interaction 8
Change and spatial relations 10
Maps and the spatial frame 11
Spatial systems 16
Spatial inequalities 20

2 The environmental factor 21
People remake the earth 21
Determinant or influence? 23
The meaning of resources 25
Future resources 29
The ecological balance 31
Change and process 32
Can humanity survive? 32

3 The region 35
Areal differences 35
Lines, barriers, and transition zones 36
Distinction, identity, and regionalism 36
The regional hierarchy 37
Bases of distinctiveness 38
Regional cores 39
Space, time, and change 40
Single-factor and multiple-factor regions 41
Study of regional wholes 42

World maps 45

4 Europe: a general geographic introduction 69
The sea, rivers, and the plain 69
Industrial resource base 73
Trade, manufacturing, and politics 77
Europe's future 78

5 The British Isles: Britain and Ireland 80
Britain: a prized location 80
Celtic areas 90
Wales and the coal revolution 92
Barren Scotland and emigrant Scots 92
Proudly distinctive Ireland 95
The United Kingdom 96

6 Scandinavia and the Low Countries 97
Norway 97
Sweden 101
Finland 103
Denmark 105
The Low Countries 106
The Netherlands 108
Belgium 112
Problems of smallness: Benelux 114

7 France and southern Europe 115
France 115
The Mediterranean world 122
Spain and Greece 129
Italy 130

8 Germany and central Europe 136
Germany 136
Switzerland 146
The shatter zone 148
Poland and its resource base 151
The Baltic states: open and unstable 152
Geography and the Czech achievement 152
The Balkans: mountains and competing states 154

The Turkish influence 156
Austria and its former empire 156

9 The U.S.S.R.: the gift and the problem of space 159

Empty distances and excess space 162
Agricultural land 163
Mineral resources and manufacturing 164
Transport problems 169
Rivers and the rise of the Russian state 171
The agricultural crisis 172
The transformation of nomadism 176
Russian authoritarianism 178

10 People and land in monsoon Asia 180

Monsoon Asia 181
Physical base 182
Agrarian civilizations 183
Irrigation and its significance 185
The peasant society 185
Village and field 188
Limited spatial mobility and exchange 189
Tradition versus change 190
Traditional Asian superiority 192
Influence of Western traders 193
Population explosion in Asia 194
Western investments and economic stimulation 196
Agriculture and industrialization 197

11 The Indian subcontinent: between two worlds 199

The Indus valley civilization and the early invasions 201
Disunity and regional variety 203
Political fragmentation and the role of Hindustan 204
The Muslim contribution 205
Major regional divisions 207
Larger north-south distinctions 209
Language and regionalism 210
Force of outside influences 211

12 The subcontinent decides and plans 213
The Muslim east 213
The Muslim northwest 214
Indian industrialization 217
Problems of industrial development 218
Small-scale and handicraft industry 221
Economic growth and agriculture 221
Traditional Indian agriculture 224
Sri Lanka (Ceylon) 227
Between two worlds—the British impact 228

13 China, the middle kingdom: splendid isolation 231
Relations with other cultures 231
The beginnings of China 234
China's northern cradle 235
Settlement of the south 237
Expansion of political control 237
The field and the steppe 239
Manchuria: conflict and development 242
China Proper 243

14 Changes on the good earth of China 252
Extensions of cultivated area 253
Agricultural yields 253
Mobilization and collectivization 255
Mineral resources 257
The industrial pattern 260
China chooses 261

15 Japanese achievements and the problem of Korea 263
Physical framework 265
The spread of Japanese settlement 267
Economic achievement 269
Mineral resources 269
Agricultural development 272
Fishing 275
Food and population 276
Industrialization 276

Patterns of manufacturing and cities 279
Foreign trade 282
Korea: the tragic region 283

16 Southeast Asia: tropical islands, deltas, and the sea **288**
Soils and population patterns 289
Shifting cultivators 293
Pluralism and politics 297
The Indian and Chinese impact 297
Colonial-commercial cities and plural economies 298
The openness of southeast Asia 301
Modern economic development 302
Resource base and resource use 302
Tin and petroleum 303
Population, resource use, and colonial economies 305
Agriculture and population densities 308
Pluralism compounded 310
The future of an independent southeast Asia 311

17 At the ends of the earth: Australasia and the Pacific world **313**
Patterns of resources and barriers 316
Human interactions 317
Australia and New Zealand 319
New Zealand 324
Melanesia 326
Micronesia and Polynesia 327

18 The Dry World: civilization and conflict **330**
Regional lines 331
Bases of unity 331
Nomad versus farmer 334
The desert 337
The oasis economy 339
The spread of Islam 342
Climatic factors and human changes 343
The beginnings of civilization 345
Mesopotamia and the Nile 346
The succession of empires 347

Turkish identity and the persistence of Iran 349
The paradox of Dry World history 350

19 Modern states and issues in the Dry World **353**
Colonialism 353
Independence 354
Petroleum and politics 355
North Africa and its new states 357
Egypt 359
Desert east Africa 364
Turkey 365
Arab Asia 368
Arabia 370
Israel 371
Iran 374
Afghanistan 376

20 Sub-Saharan Africa and its modern development **378**
Ownership and use of resources 382
African development 382
Environment, location, and African cultures 385
European settlers 386
Colonialism in Black Africa 388

21 African regions **391**
West Africa 391
Central Africa 399
South Africa 403
East Africa 409

22 Latin America: a blend of Indian and European **413**
The Spanish and Portuguese rulers 414
Racial and cultural mixtures 417
Population patterns and political fragmentation 417
Ancient Indian civilizations 420
Negro mixtures 423
Resource use and the Latin American system 423
Population growth and economic development 425

23 Latin American regions 429
The West Indies 429
Mexico and Middle America 435
Northern South America and the west coast 438
Brazil 443
The pampa-Chaco region 447
Chile and the tip of South America 450
Latin American cultural unity 453

24 North America: a modern prodigy 454
Europe and America 454
Resources 457
Patterns of resource use 457
The role of transport 459
The agricultural dilemma 463
Labor and resource use 464
Urbanization 466
Broader spatial patterns 470
The North American essence 476

25 Regions and contemporary problems of North America 478
Regionalization 481
A summing up 494

Glossary 496
Statistical Appendix 498
Index 537

Maps

1	Maps of spatial patterns	13
2	Strasburg, Va.	15
3	Growth of the Detroit urbanized area	17
4	Climatic regions	46–47
5	Natural vegetation	48–49
6	Population density	50–51
7	Predominant economies	52–53
8	Languages	54
9	Religions	54
10	Energy production	55
11	Energy consumption	55
12	Mineral fuels	56–57
13	Iron ore and ferroalloys	58–59
14	Europe	60–61
15	Asia	62
16	Africa	63
17	Australia and New Zealand	64–65
18	North America	66
19	South America	67
20	Centrality of Europe	70
21	Minerals in Europe	71
22	Railways and waterways in Europe	72
23	Precipitation in Europe	73
24	Wheat production in Europe	75
25	Rye acreage in Europe	76
26	Potato acreage in Europe	77
27	Population of Europe	78
28	The British Isles	82
29	Cities and coal fields in the United Kingdom	86
30	Land use in Scandinavia and the Low Countries	98
31	The Low Countries	107
32	Reclamation in the Low Countries	110
33	Land use in France	116
34	The Paris basin	119
35	Landforms of the Mediterranean basin	124–25
36	Citrus production in the Mediterranean basin	128
37	Olive production in the Mediterranean basin	128
38	Grape production in the Mediterranean basin	128
39	North Italy	131
40	Landforms of central Europe	138–39
41	Boundary changes in Germany	142
42	Peoples of the shatter zone	149
43	Boundary changes in central Europe	150
44	Landforms and transportation in the Soviet Union	160–61
45	Population of the Soviet Union	162
46	Minerals and economic patterns in the Soviet Union	164
47	Climate in the Soviet Union	166
48	Vegetation and landscape zones in the Soviet Union	174
49	Precipitation in monsoon Asia	182
50	Population of monsoon Asia	184
51	Colonial empires in Asia about 1900	195
52	Landforms and states of the Indian subcontinent	200
53	Ancient and modern India	202
54	Major languages of the Indian subcontinent	211
55	Population of the Indian subcontinent	214
56	Tea and wheat production in the Indian subcontinent	216
57	Rice production in the Indian subcontinent	216
58	Jowar and bajra production in the Indian subcontinent	216

59 Coal and iron ore in the Indian subcontinent 216
60 Average annual rainfall in Sri Lanka 226
61 Tea and rubber acreage in Sri Lanka 226
62 Landforms of China 232–33
63 Population of China 236
64 Precipitation in China 238
65 Main railroads in China 254
66 Main waterways in China 257
67 Fuels and power in China 258
68 Landforms of Japan and Korea 264
69 Population of Japan 266
70 Land use in Japan and Korea 268
71 Population of Korea 285
72 Landforms of southeast Asia 290–91
73 Population of southeast Asia 292
74 Rainfall in southeast Asia 294
75 Minerals and economic patterns in southeast Asia 296
76 Landforms of Australasia and Oceania 314–15
77 Precipitation in Australia 320
78 Vegetation in Australia 320
79 Population of Australia 320
80 Minerals and economic patterns in Australia 320
81 Landforms of the Dry World 332
82 Population and rainfall in the Dry World 338
83 Dry World empires 348
84 Oil in the central Dry World 356
85 Irrigation in Egypt, Sudan, and Israel 360
86 Precipitation in Africa 380
87 Vegetation in Africa 380
88 Population of Africa 381
89 Minerals and economic patterns in Africa 381
90 Tsetse fly infested areas 385
91 Regions of Black Africa 392
92 Central and south Africa 398
93 Precipitation in Central America 415
94 Vegetation in Central America 415
95 Population of Central America 415
96 Minerals and economic patterns in Central America 415
97 Precipitation in South America 418
98 Vegetation in South America 418
99 Population of South America 419
100 Minerals and economic patterns in South America 419
101 Landforms of Central America 430–31
102 Peninsular South America 448
103 Railways in North America 456
104 Generalized farming types in North America 458
105 Major minerals in North America 462
106 Value added by manufacture in North America 466
107 Precipitation in North America 468
108 Vegetation in North America 468
109 Population of North America 469
110 Core areas of North America 474
111 Regions of North America 480

Figures

1 Hypothetical spatial patterns of city hinterlands 8

2 City hinterlands in a hypothetical landscape 8

3 Village and agricultural landscape on a south Indian plain 19

4 Standard Asian village pattern 19

5 World population expressed in areal terms 26

6 Sample regionalizations of the United States 41

7 Climographs of representative cities in Europe 74

8 Climographs of representative cities in the Mediterranean basin 126

9 Climographs of representative cities in the Soviet Union 167

10 Climographs of representative cities in monsoon Asia 183

11 Climographs of representative cities in China 239

12 Climographs of representative cities in southeast Asia 295

13 Climographs of representative cities in the Dry World 339

14 Climographs of representative cities in Africa 382

15 Climographs of representative cities in South America 420

16 Basic Andean landforms of Peru and Bolivia 442

17 Climographs of representative cities in North America 469

Tables

1 Population of England and Wales, 1700–1975 85

2 Countries of the Dry World and their approximate population 333

3 Countries of Black Africa and their approximate population 397

PATTERNS ON THE EARTH

ntroduction for the student

Present-day students are faced with a dismaying problem. They live in an age when the boundaries of knowledge are still expanding. Often they despair of absorbing the flood of new ideas or new facts that modern learning has produced. Yet the problems of the modern world make it necessary to swim in this flood and to clutch at whatever solid material can be found. Given this urgency, people too frequently tend to bypass ideas that require an effort to understand, and concentrate instead on the simple acquisition of information that they hope will equip them adequately. Americans, plunged so suddenly and so recently into the complex affairs of the world, suffer especially from this urgency and often tend to confuse information with knowledge, or to prefer hard facts to thoughtful analysis. Neither facts nor ideas are of adequate use unless one makes an effort to think about them and to apply them intelligently. Indeed, facts without ideas, or both without individual thought and intelligent application, may be worse than nothing and may do real harm. In the simplest possible terms, one must be able to put two and two together if one is to understand anything. It is not safe or useful to depend on someone else to do it.

The disciplines of geography and history, which have much in common, also share a basic role in general education, that of widening intellectual horizons. If we know only the time and place in which we live, we understand dangerously little about the world, or about our immediate time and area. We cannot measure and assess our own circumstances or make fully valid judgments about our own affairs unless we acquire some knowledge of other times. Nor can we understand the circumstances and decisions of others with whom we are increasingly closely involved unless we know more of other societies. It is difficult to exaggerate the importance of such knowledge. Indeed, it is more urgently needed than knowledge of the technical and scientific advances that frequently command more attention in our own times. We must learn to understand and manage ourselves, and to understand and live with the varied world community, before scientific and technical advances can be made fully useful. History and geography attempt to lay this essential foundation. Together they widen our basic perspectives about ourselves by examining our accumulated historical experience and its variations worldwide.

What is geography?

Geography comes from a Greek word whose literal meaning is "description of the earth." But modern geography is concerned with people as well as with the earth, and with relationships and analysis as well as with description. Geographers analyze the physical world and examine relations between places in order to throw light on the patterns and nature of human society. They investigate the interrelationship that exists between people and their physical environment. They examine regional differences, and attempt to account for them. Geographers pick out regional patterns, and try to draw regional lines and identify regional relationships. The earth and its spatial framework, and the pattern of distribution of people and elements on the earth's surface are studied by geographers to develop a better understanding of the human world. They set people in the framework of the earth they inhabit.

The broad scope necessary to analyze spatial patterns over the world and people's varying use of the earth involves the geographer in the process of generalization. Generalization, or abstraction, is seldom easy, because it tries to go beyond the individual uniqueness of facts or single happenings to discern the broader common features, relationships, or causes. Language itself is an abstraction. Practically all words in any language represent generalizations that ignore individual variations, for example, "table" or "man." Few generalizations are safe from attack unless they are so general as to be useless, but understanding suffers unless we attempt to construct a larger frame of common features into which separate facts may be fitted. To generalize about broad patterns or about cause-and-effect relationships in human society is especially difficult because the problem is highly complex. Generalization can be misleading, as it glosses over exceptions and contradictions. The individual learns by generalizing from accumulated experience. The scholar must learn to establish patterns and reach conclusions based on the variety of accumulated facts about the world. The dangers of generalization are clear, and even beginning students are under obligation to make a careful and reasoned use of facts. But facts must be used, not gathered as if they were ends in themselves. Facts never "speak for themselves"; they have meaning only when they can be related or grouped into some sort of system.

This book attempts to guide students toward the use of certain facts by means of the ideas employed in geography. Since any one book is limited, and since this one attempts to introduce students to a large field of knowledge, it necessarily deals in a great number of broad generalizations. Many of these can be challenged by pointing out exceptions or contradictions, although it is hoped all of them have enough validity to stand in general terms. It is usually not possible to write a general introduction as a scholarly treatise, nor is it necessarily desirable. Once the dangers and limitations of abstraction are made clear, students must be encouraged to develop a conceptual framework and a pattern of ideas to help them analyze, relate, and interpret certain facts. To begin with, they can usually do this most usefully in broad terms. Clarity and understanding are usually enhanced by attempting to see the pattern of the forest as well as the individual trees.

Geographic analysis

It is probably true that each intellectual or academic discipline is primarily an attitude of mind. Each of the social sciences represents one approach among many to the common objective of knowledge about human society. The distinctiveness of these approaches is more a matter of what each is looking for than of a set of mutually exclusive facts, tools, or methods. Geographers look for particular factors and particular relationships to which they are attracted by a particular attitude. The same material may also be examined by the historian, for example, or the economist, just as the geographer may make use of historical or economic data. Broadly, the geographer is looking for spatial form and spatial relations, or, in other words, for patterns of distribution and interaction. The other disciplines are less concerned with these patterns and, therefore, are less likely to discover them, although they may find the geographer's conclusions useful for their own purposes.

There is a necessary division of labor among allied disciplines, but it should not be assumed that the need for collaboration can be avoided thereby. Allied disciplines often overlap, but this does not destroy their respective identities or responsibilities. Effective analysis depends on broad collaboration as well as on distinct functions. The distinct areas of analysis explored by geography may be summarized under three main headings:

1. The distribution and relationships of people over the earth, and the spatial aspects of human settlement and use of the earth (Chapter 1).
2. The interrelationship between human society and the physical environment as part of the study of areal differences (Chapter 2).
3. The regional framework and the analysis of specific regions (Chapter 3).

You may complain that *space* is a very broad or even an ambiguous word. Indeed, all three of these headings might be lumped together as spatial study. The division of the earth into regions and the analysis of specific regions are both concerned with spatial form and distribution. Examination of differences in human society from place to place as they may be related to differences in the physical environment, and of differing landscapes as they reflect a changing human state, are also concerned with earth space. Environmental factors such as mountains, rivers, or deserts affect relations between one place and another, and also affect the nature and distribution of human activities. In turn, human activities may transform or severely damage the physical environment. Geography is basically concerned with the patterns of human distribution over and use of the earth, why people use it differently in different places or at different times, and how this is related to the whole of their cultures and economies. These are broad, complex questions. Many approaches in various fields may be used toward answering them. Geographers must depend not only on their own analysis but also on the approaches and findings of others. They choose to investigate spatial distribution, the physical environment and landscape, or the region, which lend themselves to geographic analysis. In this sphere they have the principal responsibility for assessing the many lines of evidence and suggesting conclusions.

About this book

A textbook should perform two main functions. It should provide selected information and ideas that will be useful and reliable beginning guides toward understanding, and it should encourage a maximum of individual thinking and application. To do both jobs effectively, the text should somehow whet the curiosity of the reader. Curiosity is a sound and universal basis of learning. Most people are naturally curious about geography, if only about places and maps. This book is written to stimulate its readers to think. It would be no service to do all of their thinking for them, although some selection of material and ideas must be made. The process of learning is the process of forming opinions, as well founded in fact and reason as one can make it, but still one's own.

There are probably as many notions about how to approach the field of geography for introductory purposes as there are geographers. If the objectives of an introduction, as outlined above, can be carried out, the particular facts treated are of secondary importance. But they must be accurate and relevant. No book can include more than a tiny sample of existing knowledge or hope to present more than a few facts. The information should be selected to contribute to the understandings that are this book's main aim. The pieces should hang together in some coherent way rather than being picked at random; and they should have some essential interest and importance. This book's objective is only secondarily information; primarily, it is the development of geographic ideas and the geographic approach to the study of human society. Especially for this reason, it can be only a beginning, and it will have inevitable gaps. Something must be left to the individual instructor and the reader as they supplement any book, necessarily incomplete, with their own material and ideas.

The first section of the book explores the nature of geography, the kinds of problems it investigates, the methods it uses, and the subtle implications of humanity's changing interrelationship with the earth it inhabits. The second and larger section provides a series of regional studies in which the evolution of settlement, culture,[1] economy, and political forms in each area is treated in historical perspective, as well as in a current context. These samples of regional analysis attempt to convey an understanding of the various aspects of each area, and to apply the ideas developed in the first section to the core of geography—the total study of areas in their spatial frame, or their pattern of arrangement on the earth. Not all of the world is treated, nor can any of the regional chapters deal completely with a regional totality, given the obvious limitations of space. Each attempts to *introduce* the reader to a regional whole, rather than to provide a factual handbook. Each regional chapter also focuses on one or a few leading characteristics of the area it treats, or on one or two major ideas, used as threads to tie the interpretation of the region together.

Although the arrangement of chapters in the larger section of the book is regional, rather than topical, each chapter is concerned more with the illustration and development of geographic ideas and analysis than with regional information per se. To understand sea power or a maritime economy, for instance, one turns to the example of Britain or Japan or Norway. In Chapter 7 certain geographic ideas conveniently illustrated by France are stressed, but not everything of geographic significance about France is discussed. Each of the regional chapters, however, is designed to include essential factual coverage, not as an end in itself, but as a means to the intelligent understanding of each area. The book's aim is to introduce beginning students to geography in all of its aspects and to provide them with selected samples of geographic study. The bulk of the book is appropriately contained in the regional chapters, for the study of actual areas as people have used them is the heart of geography.

The book is designed to be used with an atlas. Atlas maps, with their colors and their larger size, as well as their quantity and variety, can show more detail more clearly than is possible in a textbook, except at prohibitive expense. For most place names, physical features, boundaries, and the distribution of elements like climate, commodity production, or language, an atlas is an indispensable companion to this book, or to any book about geography. Each chapter should be read with an open atlas, referring to regional and topical maps as necessary to illustrate what the book discusses. Most of the chapters have maps designed to illustrate special features or problems with which the text is concerned. But these maps are no subsitute for an atlas, and they cannot include all the relevant mapable information: not only place names, but the great variety of other data on altas maps that will be helpful, if not essential.

Spellings for the names of many cities, rivers, and mountains may vary from country to country, and even from one series of maps to another. Where there is more than one spelling or name in general usage, the English name is given first and used in the text, followed by the alternate spelling where the name appears for the first time in the text.

At the end of most chapters there are "Questions for Further Study and Discussion." In most cases they are not simple questions of fact answered in the preceding text. There seems little value in posing such questions. Often the information necessary to answer these questions is not given in the chapter. "Further study" means what it says, and the questions provided are only samples of the kind of additional inquiry it is hoped the book will provoke. Any book, by itself, can be only the smallest beginning. It is well said that "a little learning is a dangerous thing." One can seldom discuss facts, only relations between facts. In doing so, one evolves and applies ideas. This may mean, among other things, that there is no one right answer, as with most important questions involving more than simple facts. In nearly all of social analysis we are concerned with probabilities or possibilities rather than with certainties. The discussion questions are designed to encourage students to use their own minds and to reach their own conclusions. That is also the purpose of this book.

[1]The word *culture* in its general sense, as used by anthropologists and throughout this book, means the total way of life of a people, including their system of values, system of social organization, language, material structures, clothing, diet, techniques, and art forms.

1 Spatial form and spatial interaction

The Earth from space. *(NASA photo.)*

The first function of geography listed in the Introduction is a study of "the distribution and relationships of people over the earth, and the spatial aspects of human settlement and use of the earth." The geographer wants to know how and why people are distributed as they are. What accounts for the great clusters of population and for the empty areas, for the cleared and uncleared land, the cities, the villages, and the roads? What spatial patterns can be observed? What are the interactions between them?

If the roughly 4 billion people of the world were spread evenly over the earth's land surface, there would be an average of about 70 people per square mile (about 28 per square kilometer). If they were gathered together as closely as an audience in a theater, 300 square miles (about 550 square kilometers) would hold them all. As it is, about half of the earth's land area supports fewer than 2 people per square mile. Some densely populated rural areas support as many as 3 thousand per square mile (about 1,100 per square kilometer), and population density in the centers of large cities may reach or exceed 100 thousand per square mile (38 thousand per square kilometer). Settlements of various kinds are scattered widely over the earth, but only a few have become big cities. In some rural areas of the world, farmers live in villages (nucleated settlements), and in others they live in isolated farmhouses (dispersed settlements).

Mineral ores and fuels occur in scattered locations over much of the earth, but heavy metal manufacturing has developed on a large scale in only a few places. There are approximately 3,000 different languages in the world. Some languages are spoken only by a few people, and some by hundreds of millions; some are limited to a single, large or small area, and others have spread over widely separated areas. What are the implications, causes, and interrelations of these varied spatial patterns?

Geography is concerned whenever the fact or the concept of *place* or the question *where* is involved. Places and spatial patterns, as in the examples above, are not static events or simple facts, but are different sets of relationships. It is only of incidental importance where Timbuktu is (or China, or the wet tropics, or the area where Swahili is spoken). But knowledge of the location helps to understand the relations between Timbuktu and the rest of the world and the effects these relations have had on Timbuktu, and vice versa. No place and no pattern of distribution in the world is exactly like any other place or pattern. There are also marked similarities among groups of places and spatial patterns. In each case, the nature of relations with other places may be fully as important as local conditions in accounting for the differences or the similarities. Geographers attempt to analyze human society in its spatial frame. They examine not only where people and their activities are found and their groupings, but why they are there, and how their location, nature, and development are related to the influences exerted by other places. This is the meaning of the term *spatial interaction*. Spatial form is significant only when it can be related to spatial interaction.

A theory that can help to explain real spatial patterns can often be developed by imagining an unreal, featureless earth as smooth as a billiard ball on which people and resources are evenly distributed. By adding the varied elements of complex reality to this simplified image one at a time, the effects of each element may be estimated, and some of the aspects of experimental science simulated. Location theory, in particular, makes use of this technique in developing the theory of the *central place* discussed later. The understanding of complex reality requires the kind of generalization that theory makes possible and out of which theory is constructed. But generalization, and the necessary oversimplification on which it rests, must be tested continually against reality.

Spatial relations

Timbuktu

Timbuktu (Tombouctou) is a real city, despite its unreal overtones. Why is it where it is, and why has the settlement on this site grown? (Consult an atlas map of North Africa.) Location on the Niger River is clearly relevant, for Timbuktu lies near the margin of the Sahara Desert. In the absence of adequate rainfall and groundwater for wells, settlement in this area must be limited to sites where water is available. *Site* refers to the actual ground on which a city is located. It is often paired or contrasted with *situation*, which refers to spatial relations with other places, principally accessibility rather than simple distance. Situation is thus both a relative and a changeable matter.

Timbuktu's situation—its relations with other places—has been much more important than the local conditions of its site in accounting for the city's growth and decline. Northward is the desert. Southward the land becomes more humid, therefore more densely populated, with a productive economy different from that of the desert. Settlements that lie in the zone of transition between one type of physical or economic condition and another may be convenient centers for the exchange of the different kinds of goods each area produces. Timbuktu has prospered as a trade center in part for this reason. The city also has the advantage of easy access to the moister and more productive area to the south by means of the Niger River, which flows southward to the coast.

Towns and cities are specialized, concentrated settlements that have grown primarily because they have access to wider areas around them, and can perform economic, political, or social functions for these wider areas. These functions, such as government, marketing, manufacturing, or the servicing of trade routes, are most conveniently carried on in a central place. The differences in the physical nature of the areas north and south of Timbuktu promoted a *complementary* (filling mutual lacks) trade relationship between them, since these areas produced and needed to exchange different kinds of products. Regional differences do not always result in exchange. The amount of exchange depends on the effective demand in one region for the commodities produced for sale in another region, and on the cost of transport between the two places.

Timbuktu is a convenient center through which the complementary exchange between the different areas around it can take place. This exchange is due not only to its central location, but because transport carriers are likely to change there from land to river, from desert transport to humid-area transport. In fact, Timbuktu has been called "the place where camel meets canoe." Places where carriers change, or *break-in-bulk points*, have economic opportunities for their populations. This applies not only to loading and unloading, storage, or servicing of transport carriers, but also to the financing and insuring of trade and to the processing or manufacturing of goods passing through. The goods must be unloaded at the break-in-bulk point. Transport and loading-unloading costs for these goods are minimized at that point, and a variety of routes are likely to focus there, funneling goods through the city. This trade is important for any manufacturing or service industry dependent on the assembly of raw materials or the distribution of finished goods. The trade through Timbuktu has never been very great by modern standards, because the tributary areas the city served were sparsely populated and unproductive (the desert), or economically organized largely on a subsistence rather than a commercial exchange basis (the moister areas to the south). Hence Timbuktu is not a very big city, but it is the biggest and almost the only city in this part of northwest Africa.

Timbuktu would be very much smaller, however, if only the desert were to the north. The desert produced almost nothing for exchange, but beyond it is the moister Mediterranean coast. This area did produce something different from the areas south of Timbuktu. This coastal region also exchanged goods with humid central Africa over the most practicable route, which went through Timbuktu. Some of these goods were redistributed from north Africa by cheap sea transportation to wider markets in Europe and elsewhere. There, because of different physical and cultural conditions, the same goods, such as ivory, gold, spices, or slaves, were not produced. These wider relationships were reflected in the size and functions of Timbuktu, hundreds or even thousands of miles away.

The goods that moved along the route north of Timbuktu, however, had to be of certain kinds. Here there was no river to lessen transport costs, and it was a very long haul to the Mediterranean coast. The area was largely empty of population and provided few opportunities for exchange. The carriers (camels) are very expensive, measured by the unit cost of transport, or the cost per ton-mile (for carrying one ton one mile), mainly because their individual capacities are small. Transport costs depend largely on the capacity of the carrier, which is the principal reason why water transport is so cheap. The length of the route between Timbuktu and the Mediterranean coast increased the expense, as did the lack of trading centers along the way. Such centers are important in making a transport route profitable. Therefore this route could carry only goods that were very high in value by weight, such as gold, ivory, slaves (who provided their own transport), or salt (which is very high in value in most preindustrial economies).

With the development of mechanized sea transport and the widening of Africa's commercial relations with the rest of the world, the goods that used to move through Timbuktu to wider markets began to move instead through west African seaports. Timbuktu has become a much smaller place and more dependent on local trade and services. Change, especially technological change, is continually altering spatial relations and spatial interaction patterns. At its height, about the fifteenth century, Timbuktu was a commercial metropolis. Because of its size and its excellent access, it was also a famous cultural center for much of the large western African Sudan, and supported a well-known Muslim university. With the change in trade routes that fed it and their interruption by political disorders, Timbuktu's glories faded. By the end of the nineteenth century, the French found it a largely ruined town. Thus, both political and technical change were responsible for Timbuktu's decline, as they affected the city's relations with other places, its pattern of spatial interaction.

Other examples of spatial interaction

Britain's economy and society have been profoundly influenced by its access to and from the rest of the world. Similar physical conditions exist in many other parts of the world, but the particular type and degree of accessibility that Britain enjoys cannot be duplicated. The country's human characteristics may be related to its unique location. Other factors having little or nothing to do with location are also important, but a study of the locational or spatial factors can provide considerable insight toward understanding Britain.

California's agriculture now specializes in growing fruit and vegetables partly because of a favorable local climate, but also because California has developed cheap, easy, and quick access to the large market in the eastern United States, where the climate is different. If the east were a different and less complementary place, or if there were no railroads, California's agriculture would be a smaller and very different enterprise.

The effects of weak spatial relations may be equally important and obvious in other situations. Isolation was as responsible as environment for the unique culture of the Australian aborigines, the North American Indians, and of the various groups native to the rain forests of Africa and South America.

The relations between places can be analyzed to throw light on every problem geographers investigate. Part of their principal function is the study of space and place. They must attempt to see a region, or the world, as a whole. Comparison is also an essential part of geographic analysis and interpretation. The two great poles of world civilization, Europe and the Orient, cannot be fully understood until they have been compared. Comparison helps to reveal the distinctive qualities of each, and may suggest some of the dynamics that are taken for granted or overlooked without the comparison. One's own region or country can be understood better when it is compared with other regions or countries, and seen in some perspective. Knowledge of only one area or society may also be seriously misleading if it is applied to the interpretation or evaluation of another area or society. Use of a single comparative criterion without knowledge of the wide differences elsewhere (and without realizing that there is no single cultural norm) distorts the purposes and benefits of comparison. This is regrettably common among Americans, whose knowledge of most of the rest of the world is dangerously limited. The comparative method, of course, is no special property of geography. Anthropologists and sociologists use it as the basis of much of their work; it is as important in history as in geography. The geographer merely applies the comparative method to spatial form, spatial relations, and regional differences.

Figure 1. Hypothetical spatial patterns of city hinterlands. Modified after Christaller and Brush.

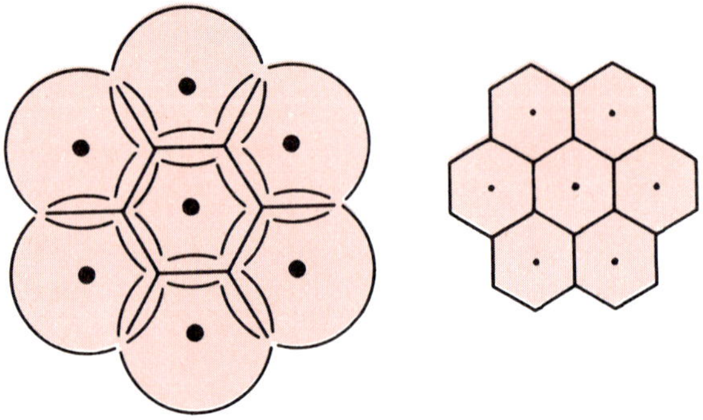

Figure 2. City hinterlands in a hypothetical landscape.

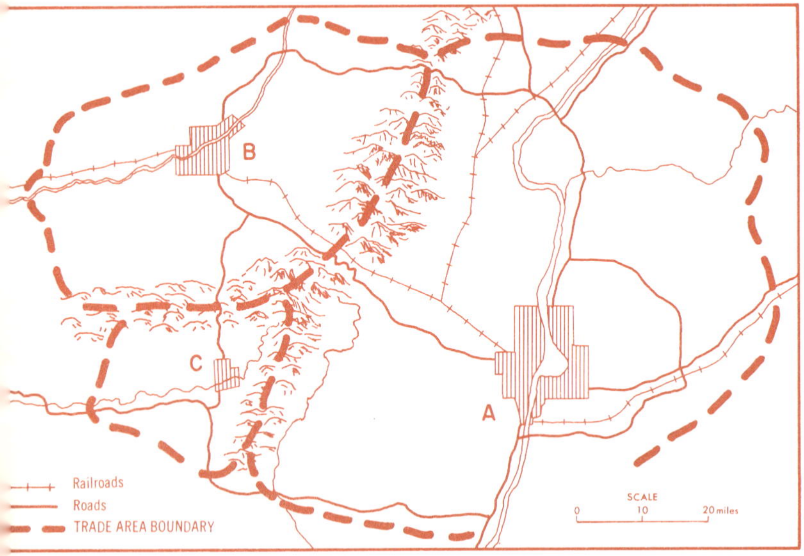

Cities and spatial interaction

The city is perhaps the simplest and clearest example of spatial form and spatial interaction. To explain the size, function, and spatial form of a city, one turns first to its relation with other places. The city is as big as it is and specialized in its particular way because it provides services for a particular area dependent on it, the city's *hinterland.* The size of this hinterland is determined by the ease of access to the city, plus the competition of other cities that may also serve the same area (see Figures 1 and 2, page 8). The total productivity of the hinterland and the size of its market are matters over which the city can exercise little influence, except to improve its own access. It is the hinterland that determines the city's size and specializations. How an urban center grows is primarily a reflection of access, of spatial factors over a very wide area, and of the interrelations that take place within the area.

The size and productivity of the American Middle West and the easy routes that lead from it down the Mohawk and Hudson rivers are responsible in large part for the size and prosperity of New York City. Provincial residents of Manhattan may find it hard to accept the Middle West as a major basis of their existence, but this shortsightedness is characteristic of most city dwellers. In most big cities the butchers, bakers, and candlestick makers for the city's own population employ more people than are employed in the so-called *basic* enterprises that produce goods or services to be sold outside. It is these basic enterprises that make it necessary to have butchers and bakers, cab drivers and street cleaners. It has been estimated that one person employed in a basic enterprise, such as wholesaling, manufacturing, or insurance for areas outside the city, will support approximately two other people employed in the manufacturing and service enterprises supplying basic workers and their families with what they need: food, schooling, department stores, housing, and so on. This roughly one-to-two relationship between employment in production for external sale and employment in production for internal consumption is called the *basic-service ratio*. But we cannot all live by taking in each other's washing, even though most city people may do so. The concept of basic enterprises re-emphasizes that cities exist in order to perform certain specialized functions for a wider area. A big city like New York, in fact, will overlap the hinterlands of other cities because it has highly specialized goods or services to offer that can be obtained nowhere else. For less specialized goods and services, a smaller and nearer city will be the center. Specialization of any kind, even in a village general store, may reflect particular local advantages, or may be simply the result of bigness. But neither of these will exist unless the village or city has effective connections with a tributary hinterland.

Various attempts have been made to explain the distribution, location, and size of cities and towns, beginning with the assumption that urban settlements arise to serve a more or less well-defined hinterland. Attempts have also been made to find consistent spatial patterns of city location and size on the assumption that there might be a regular order of distribution. From some of these investigations the theory of the *central place* has been developed. It argues in part that cities may be spaced at more or less regular intervals, each serving a hinterland around it whose limits are fixed primarily by distance. Where distance is the only factor, hinterlands theoretically would assume the shape of a circle whose center is occupied by a city. Or, since several competing cities actually occupy the landscape, the hinterlands would be straight on all sides, and would have the shape of regular polygons, as shown in Figure 1 (page 8).

But simple distance is rarely, if ever, the only factor affecting access. In reality, city hinterlands are grossly irregular in both size and shape. The cities themselves differ widely in size, and the pattern of city distribution is far from regular, as the physical and human landscapes they serve are not regular or uniform. Mountains, rivers, plains, seas, deserts, and other features of the physical landscape affect access or spatial interaction. Human factors, differing widely from place to place, and especially technological factors as they influence transportation and production, also affect access. The degree of complementary relationships between different regions, on both physical and human grounds, is further reflected in differing degrees of spatial interaction. Political boundaries, cultural differences, or linguistic barriers, among many other factors, affect the flow of goods, services, and ideas, and also affect the distribution of cities. Among regions, productivity differs, demand levels differ, and each region tends to require a different arrangement of different-sized cities to serve it. A large semiarid area with a sparse population and low productivity, or a large area occupied by a dense but mainly subsistence or self-sufficient population cannot support as many or as large cities as a smaller but more productive and more commercialized area.

The theory of regularly spaced cities and hinterlands is also disproved in reality by the clearly observable hierarchy among urban places. A large city with good, long-range access, New York, for example, may include in its hinterland most of the other cities in the country, to differing degrees. For large capital loans, exclusive dress fashions, or specialized publishing, New York is the only supplier in many cases. For these, even Chicago is tributary to New York. Most of the Middle West and the larger area served by Sears Roebuck, for example, is tributary to Chicago. Cities of middle rank may include other smaller cities in their hinterlands. At the bottom of the size scale, a rancher in Wyoming, for example, will obtain the bulk of his goods and services from the nearest or most accessible small town in whose immediate hinterland he lives. But he will also be tributary to Cheyenne (the state capital), to New York for cut diamonds or books, and to Chicago for barbed wire and sheep or cattle brokerage. If his wife wants a new hat, she will probably

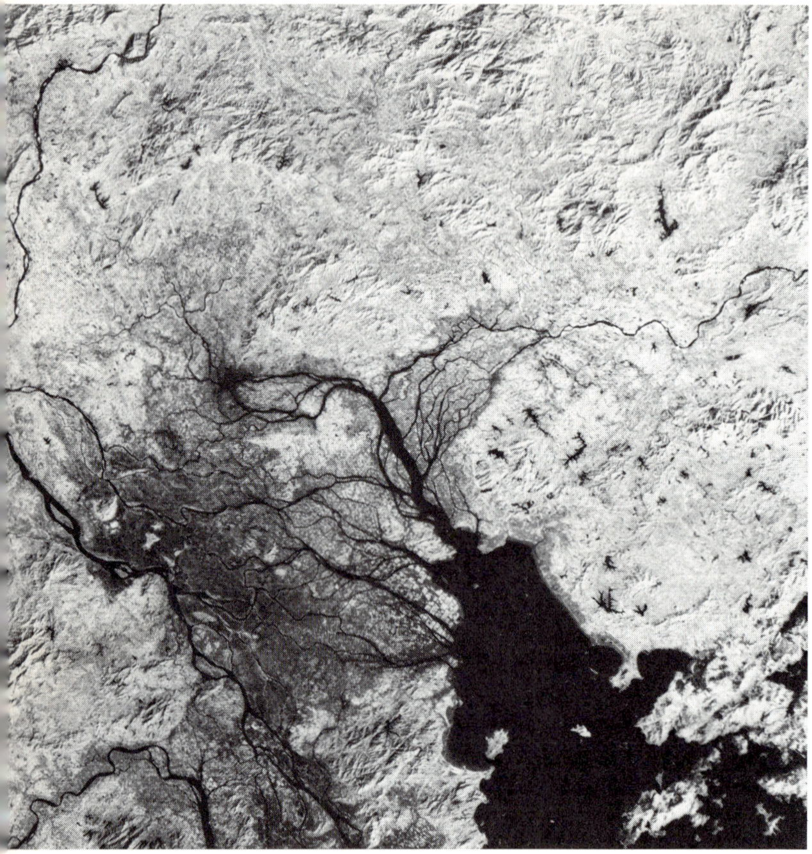

The Canton delta. Seen from 568 miles (914 km) up, the city of Canton shows as a dark blob a little left of center, at the head of the Pearl River, which empties into the large bay. Even from this height, Canton is clearly more compact than Houston (see picture on page 11). Most cities in the developing world, where the private automobile has made less impact, are this way. The Hsi (Si) River and its braided channels wander over the intensively cultivated delta in the lower left, while the hilly landscape of south China marks the edge of the delta at the top of the picture. Hong Kong and its small mainland territories are included in the lower right. *(NASA satellite photo.)*

not shop for it in the small town, but in the nearest or most accessible small city, whose hinterland for more specialized goods and services overlaps and incorporates that of the small town.

When a city grows big because it serves a productive hinterland, it acquires more and more specialized functions that further increase the city's size as they enlarge its hinterland. The biggest cities thus tend to have a self-perpetuating advantage. But for any village, town, or city, tributary hinterlands tend to be different for each type of commodity sold or service performed. Medical services will cover one area, food retailing another, wholesaling another, and sale of heavy manufactured goods another.

Where manufacturing becomes highly specialized, even a relatively small city may sell its products over a very wide area. For example, Hershey, Pennsylvania, and Hollywood, California, include large parts of the world in their respective hinterlands in this specialized sense. The same may be said of the tiny island of Harris off the coast of Scotland, which is world famous for tweed, even though it is poor and sparsely populated. In all of these cases, urban location and growth reflect spatial interaction. Relations between places are affected not only by distance but by differing degrees of access. The complementary relationships resulting from regional differences of all kinds, and the simple factor of the size and consequent specialization of urban places are, in turn, a reflection of access.

Change and spatial relations

Human society is not constant nor are its geographic relationships. Physical distances and the environment may not change, but their effects on people are changing all the time. The most important change is in human knowledge and technical powers. People learn to do more and different things with the earth. Thus its resources and its distances affect them differently. Less tangible matters, such as human wants, tastes, or cultural and ideological standards, also change. Movement of people helps to spread change and to alter land use and other aspects of cultures or economies. The earth does not mean the same thing to people at different periods, nor does it have the same effect on them, even though some of the major outlines of the relationship may remain relatively constant in a given area.

Probably the most dramatic reminder of the dynamic quality of people's relation to the earth and of the impact of changing spatial relations is the changes in North America since 1492. A land that supported perhaps 1 million Stone Age people in 1492 now supports about 245 million of one of the most technically and economically advanced people in the world. California in 1492 was the home of what appear to have been some of the most primitive of all the American Indians, although its sunshine was just as warm and its oil reserves even larger than they are now.

Any society occupies a certain stage in the development of its geographic base and its place in the spatial frame. Its development may progress up or down relative to the rest of the world. Civilizations decay as well as grow. In fact, the earliest city civilizations, so far as we know, occupied sites in or on the margins of the great desert of Eurasia. At that time these sites were highly advantageous, but now they support societies that are relatively poor and lacking in advanced economic development. It is not space or the environment that have changed, but people's use of them. People have made more productive use of other areas whose advantages they now feel are greater in light of their new powers and wants.

A man-made environment—Houston, Texas. Taken from 60,000 feet (18,300 m), this picture shows clearly the urban sprawl characteristic of most North American cities. Note how the original downtown center clusters near the river, and how later, outward growth in the automobile age has followed a radial pattern along major highways. *(NASA photo.)*

Maps and the spatial frame

Maps are the distinctive tool of the geographer, for only with maps can we best see and analyze spatial relations and spatial form. Direct observation of any area, even over a long period, can pick out only individual details or small pieces of landscape. Larger patterns can be understood only when larger areas are seen as a whole. They must be reduced to an observable scale, and many of the almost infinite details of reality eliminated in order to concentrate attention on a few. Any map selects only a few patterns to present; for example, elevations above sea level or distribution of population. An aerial photograph shows much directly observable concrete reality at a reduced scale in which some large patterns can be seen. But the profusion of detail and the fact that many spatial patterns are not directly observable or do not stand out in an aerial photograph make it much less clear than a selective map.

A map can also present areal traits, such as language distribution, density of population, average annual temperature, illiteracy, flow of traffic, or distribution of incomes, that cannot be seen directly. Most of the aspects of the earth and of human society can be mapped. Representing them on a map makes it possible to understand them better or from a different point of view. In a map they are placed within the frame of space and can be related to other elements that have spatial form. There are several mapping techniques to represent different types of spatial form. Map 1 (page 13) shows a few examples. Most aspects of human society have spatial form. Any atlas uses several different mapping techniques, and others are used elsewhere in this book.

Essentially, maps perform two kinds of jobs.

They show spatial distribution or occurrence (spatial form), and they show relationships. Of course, the two are interconnected, since the nature and degree of relationships depend in part on the amount and nature of the intervening space: how far it is from *A* to *B* and whether mountains, plains, seas, forests, farming, manufacturing, and dense or sparse populations lie between them. An atlas is designed to show as many of these factors as possible in a series of maps—no one map can show them all without being illegible. Any map is thus an oversimplified and selective symbol of reality. Some distributions are best shown by a pattern of dots, some by lines, some by colors or shades, some by other symbols. Relationships may be indicated by lines, such as transport routes whose thickness varies with the volume of flow, or by superimposing distribution patterns to show how they overlap. Roads or travel routes may be indicated by line symbols. Most relations, however, are indicated more implicitly than explicitly on maps, and must be deduced or extracted.

A map is also a convenient inventory of selected information available for immediate visual inspection. It shows exact locations and spatial patterns that indicate a variety of relationships. Often such relationships may be best understood by making or interpreting a map, and insights revealed that were only guessed at earlier (for example, see the discussion of China as a region in Chapter 3). The map is a powerful tool of analysis, not merely a collection of data. It can suggest explanations and interrelations through its selective simplification of complex reality. It can direct attention to questions whose existence or importance were not seen before, thus advancing both theory and understanding. Anyone who uses maps should realize the varied dimensions of their usefulness, and should look for more than simple locational data. Selected spatial patterns in cartographic form can raise and suggest answers to a great variety of basic questions about the composition and interrelations of both the human and the physical landscape.

Scale

Patterns of spatial form and spatial interaction can be portrayed at widely varying levels of detail. Any map must eliminate irrelevant or confusing detail and include only what is directly helpful. A map is not only a selective and oversimplified symbol of reality, but a greatly reduced image. The degree of reduction is indicated by the scale of the map. For example, if one unit on the map represents 1 million units on the earth, the scale is 1 to 1,000,000. A map of the United States alone at this scale would nearly fill a small room. Thus, maps of large countries or continents are usually drawn at smaller scales like 1 to 20,000,000 or more, and maps of the world at 1 to 50,000,000 or less. ("Small" refers to the fractions; 1 to 50,000,000 is a smaller scale than 1 to 1,000,000, although it may show a much larger area.) Naturally, the smaller the scale, the less detail that can be shown. A map with a large scale—like 1 to 62,500—can show individual buildings. (This scale is often used in the United States because it is an easily divisible fraction nearly the same as 1 inch to 1 mile, or 1 to 63,360.) For detailed analysis of small areas, such large-scale maps are essential, but most atlas maps are drawn at smaller scales. Look at the variety and progression of scales used in an atlas. The bar scale is also commonly used. It is a line representing distance in miles or kilometers on the map, and is a great convenience in measuring distances.

Distortion

Maps must be used with caution because they are distorted representations of the earth, as well as greatly reduced and oversimplified selective symbols. The earth is spherical and three-dimensional; a map is flat and two-dimensional. Looking at flat maps tends to make one forget that it is impossible to represent a spherical object or any part of its surface on a flat plane without distortion. The simplest proof of this is to flatten a whole orange skin, or the two halves of a rubber ball. It cannot be done without tearing or stretching. A map distorts reality in shape, distance, direction, or total area. Most maps do mainly one or mainly another, depending on the method used to project the round earth onto flat paper. Most atlases include some projections that distort shape and some that distort area. Each is useful for different purposes. Of course, distortion is maximized in maps of the world as a whole (which cannot really be seen as a whole at all, since one cannot see around it). It is barely noticeable in maps of areas the size of the United States, and negligible for smaller areas. In small-scale maps distances or directions are often shown inaccurately. Use care in measuring distances, especially on projections that do not show the true size of areas. The only approximately true representation of both shape and area is a

MAP 1
MAPS OF SPATIAL PATTERNS

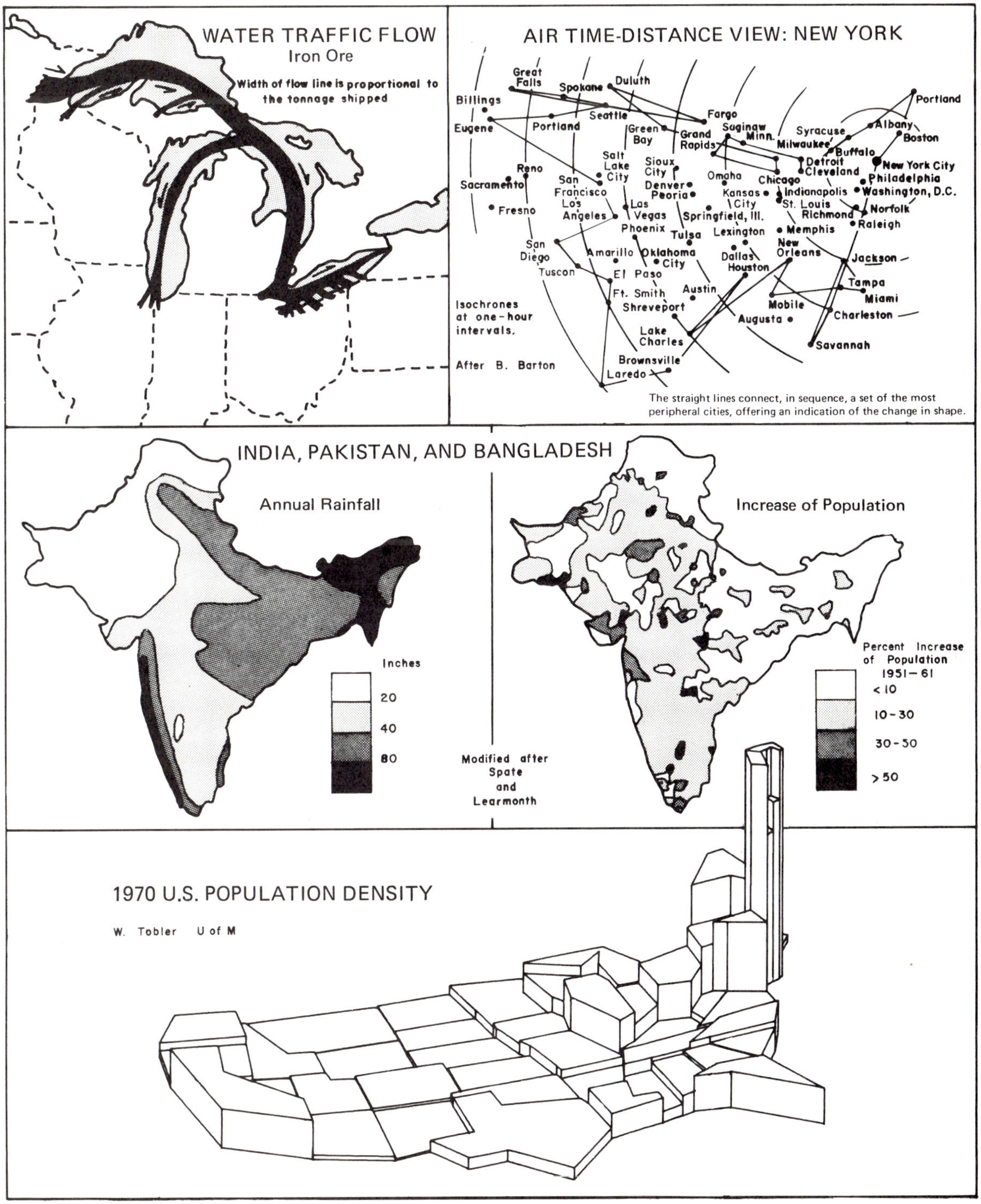

globe, and flat maps should be used with that model in mind.

Latitude and longitude

To indicate exact locations, a grid of lines is shown on most maps. The parallel lines extending east and west are called *parallels of latitude,* both north and south of the equator (which is midway between the north and south poles). These parallels of latitude are intersected by lines extending north and south, called *meridians of longitude.* Meridians are not parallel, because each originates and terminates at the poles. Therefore they are most widely separated at the equator and converge toward the poles. One of these meridians of longitude is arbitrarily chosen as the base point of reference or the prime meridian. For convenience and by established practice, the meridian that runs through the Royal Observatory at Greenwich near London is generally recognized as the prime meridian, although any other one could just as well have been used. The fact that Greenwich has been generally accepted reflects the prominence of Britain in the modern world and the pioneer work of British scientists, explorers, and map makers. Longitude is measured in terms of distance east or west of the prime meridian of Greenwich, and latitude in distance north or south of the equator. The earth is divided for these purposes into 90° of latitude north and south of the equator, and 180° of longitude east and west of Greenwich. The degrees are further subdivided into minutes and seconds for greater accuracy. Thus any location may be stated as so many degrees, minutes, and seconds north or south latitude and so many degrees, minutes, and seconds east or west longitude. Thus, it is possible to fix locations very accurately and to pinpoint a house or a ship, for example.

Of course, 180° east of Greenwich and 180° west of Greenwich are the same thing. Here another line separates east and west as the equator separates north and south. It also marks the change in time from one day to another, according to the rotation of the earth. Moving eastward from one time zone to another, clocks are advanced one hour, but moving eastward across 180° longitude, one enters the preceding day. Travel toward the rising sun has progressed so far by 180° that it has caught up with and overtaken the day that arbitrarily began there twenty-four hours earlier. Moving in a westerly direction, the reverse is true; clocks are set back from one time zone to another westward. But at 180° one enters the following day, which arbitrarily begins there. To avoid confusion, remember the day begins and ends at 180°, the International Date Line. Crossing this line changes the date on the calendar.

Mapping relief

Landforms, arbitrarily but conveniently, may be divided into four categories: plains, hills, plateaus, and mountains. Hills are not only lower than mountains but are usually more rounded and more gently sloping. Plateaus are simply elevated plains. Elevation, however, is less important than *local relief.* Local relief includes the degree of slope of the ground, how much difference there is between the highest and the lowest parts of an area, and how much of the area as a whole is level or sloping. A high plateau may be more usable than a rough, low-lying area. Many areas labeled mountains may actually contain extensive sections of low relief, and many plains may be interrupted by high relief.

Most colored relief maps of large areas have altitude tints to show elevation above sea level, often using green for elevations under a thousand feet and darkening shades of brown, purple, or red for the higher elevations. Altitude tints may obscure large variations in local relief. They may also suggest topographic barriers where none exist, as in the North American Great Plains. Map makers must differentiate between altitudes at some more or less arbitrary elevations. In mapping a large area with widely varying elevations, each elevation tint must cover hundreds or even thousands of feet within which there may be considerable local relief that cannot be shown. A map of India, for example, must include both the Himalayas and the Ganges valley. Such a map of normal page size would be very difficult to read if it used more than eight or ten altitude tints.

More accurate relief maps of smaller areas are made using multiple *contour lines,* as in Map 2 (page 15). (A contour line connects points of equal elevation.) The smaller the *contour interval* (the difference in elevation between two contour lines), the more accurate the map, and the more difficult it is to read. Where small local variations in relief are especially important, such as in irrigated areas or the centers of cities, a contour interval of 5 feet (about 1.5 m) or less may be used. For larger areas or areas with heavy

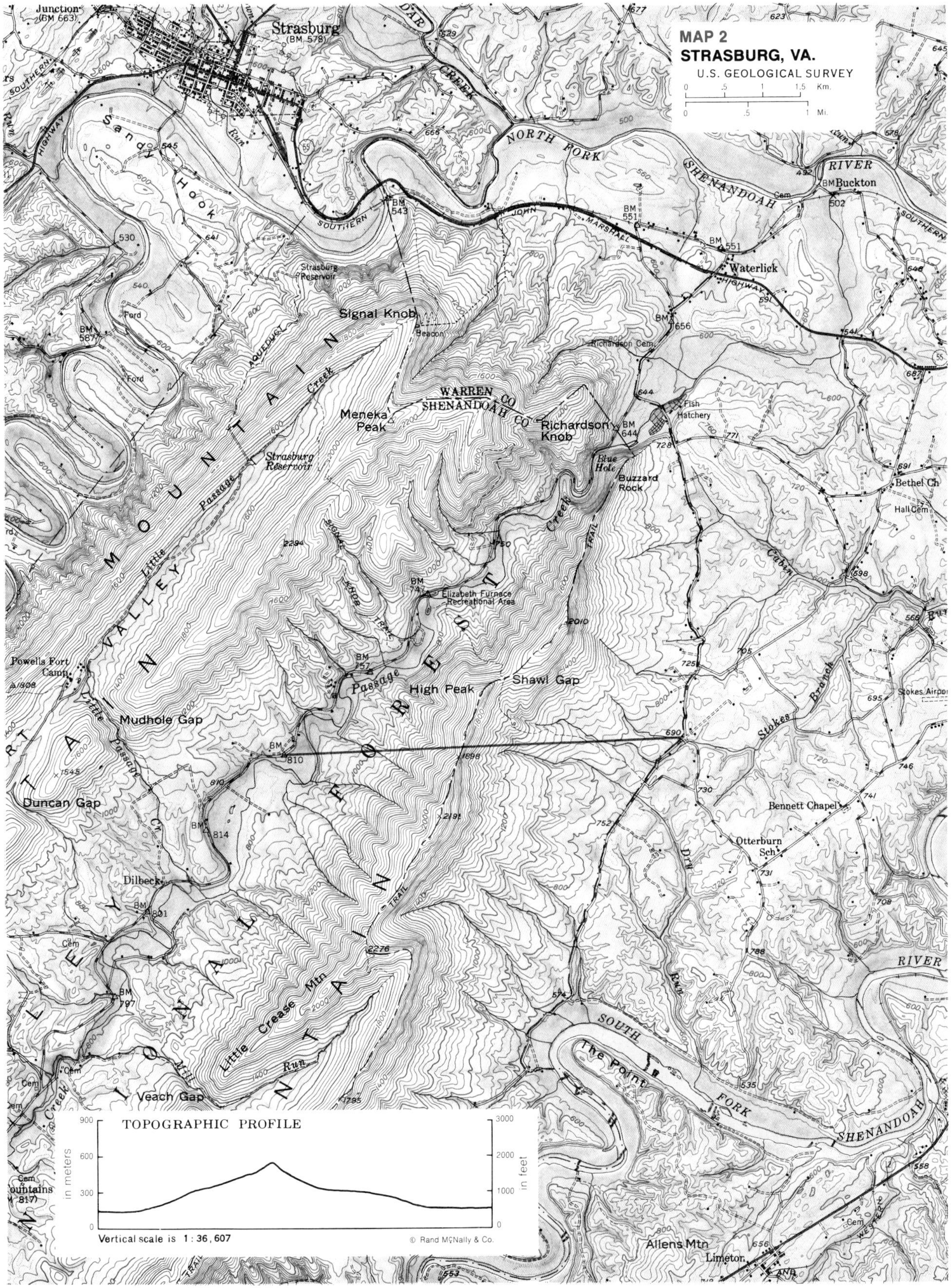
MAP 2
STRASBURG, VA.
U.S. GEOLOGICAL SURVEY
0 .5 1 1.5 Km.
0 .5 1 Mi.
Strasburg
(BM 578)
Junction
(BM 663)
Sandy Hook
NORTH FORK
SHENANDOAH
RIVER
Buckton
Waterlick
Strasburg Reservoir
Signal Knob
Beacon
Richardson Cem.
WARREN CO
SHENANDOAH CO
Meneka Peak
Richardson Knob
Fish Hatchery
Blue Hole
Buzzard Rock
Bethel Ch
Hall Cem
MASSANUTTEN MOUNTAIN
Little Passage Creek
FORT VALLEY
Elizabeth Furnace Recreational Area
Powells Fort Camp
Mudhole Gap
High Peak
Shawl Gap
Stokes Branch
Stokes Airport
Duncan Gap
Bennett Chapel
Otterburn Sch
Dilbeck
Little Crease Mtn
Veach Gap
Dry Run
SOUTH FORK
SHENANDOAH
The Point
Allens Mtn
Limeton
TOPOGRAPHIC PROFILE
in meters
in feet
900
600
300
0
3000
2000
1000
0
Vertical scale is 1 : 36, 607
© Rand McNally & Co.

local relief, an interval of 50 or 100 feet (about 15 to 30 m) may be more appropriate. A *vertical profile* may also be constructed on the basis of a contour map, as illustrated on Map 2. For a large area, small-scale contour maps are almost illegible or unworkable unless generalized so much that altitude tints might as well be used. The best small-scale relief maps of large areas make use of a combination of altitude tints and *plastic shading*, now used in most atlas maps of landforms. This suggests the landforms visually without losing any accuracy. But no small-scale relief map can be accurate in detail. It can give only a highly generalized picture.

Spatial systems

Geographic analysis assumes that spatial patterns are distributed over the earth or any smaller part of it, according to one or more systems of arrangement that are potentially identifiable. Many of these spatial systems are quite obvious: political units, such as nations, states, or counties; patterned systems of land use, such as rubber plantations or market-garden areas around cities; the placing of houses along streets; and the arrangement of rooms and furniture within each house. Some are less obvious, such as the distribution of income, the incidence of particular diseases, or the spread of ideas. But, whether obvious and directly observable or not, spatial patterns can be fully understood only when they can be shown as individual or collective systems resulting from factors that affect spatial behavior. Some patterns of spatial distribution are random or haphazard, but this is quite rare. Nearly all distributions are non-random; they are shaped by particular factors, although the relevant factors may not always be easy to distinguish. The random–non-random distinction is particularly important in any study of spatial systems.

Behavioral space and spatial perception

Spatial arrangement is evident in the entire range of human activities. Each culture has its biases, and its own perceptions of space. Most individuals and most culture groups have a spatial perception that is both warped and incomplete. You may have seen examples of cartoon maps, or cartograms, illustrating this point—for example, the New Yorker's view of the United States or the American tourist's view of Europe. In these, viewers recognize the existence, location, importance, or size of a city or country according to their perception of what is important. Culture affects individual perceptions of space in consistent ways because it tends to define what is important. It is the principal medium through which individuals learn or acquire information. North American culture is relatively self-contained, isolated, and self-satisfied. Beyond their own national space, North Americans are most aware of Western Europe, although their information is scanty and often inaccurate. Much of the rest of the world is perceived even more vaguely, if at all, as many studies and questionnaires have demonstrated. Such cultural biases are perfectly natural and universal. However, as travel time shrinks, all parts of the world become increasingly interdependent. If conflicts threaten as a result of imperfect knowledge or understanding, warped and incomplete perceptions of global reality become dangerous. There is less excuse for geographic ignorance among North Americans, or for the dangerous biases their culture tends to produce, than for similar attitudes on the part of less-wealthy populations in the past.

Living space

Over time, perceptions of space vary from place to place or culture to culture and among individuals. But patterns and spatial flows are arranged according to identifiable systems. It is part of the geographer's job to identify and analyze them. Settlement patterns make a good illustration, and are one of the most widely studied forms of spatial systems. Cities, towns, and villages grow up at certain locations and perform certain functions primarily as a result of spatial interaction, as already discussed. Modern Western cities have increased in size during the past century in particular, as means of transport have multiplied. Better transport has made it possible to move larger amounts of goods more cheaply over greater distances and at greater speeds. Thus, the city can perform more functions on behalf of larger areas, bringing in the food and other supplies it needs to maintain its own population, and also drawing its own work force from greater distances. The urbanized area (including commuter suburbs) has

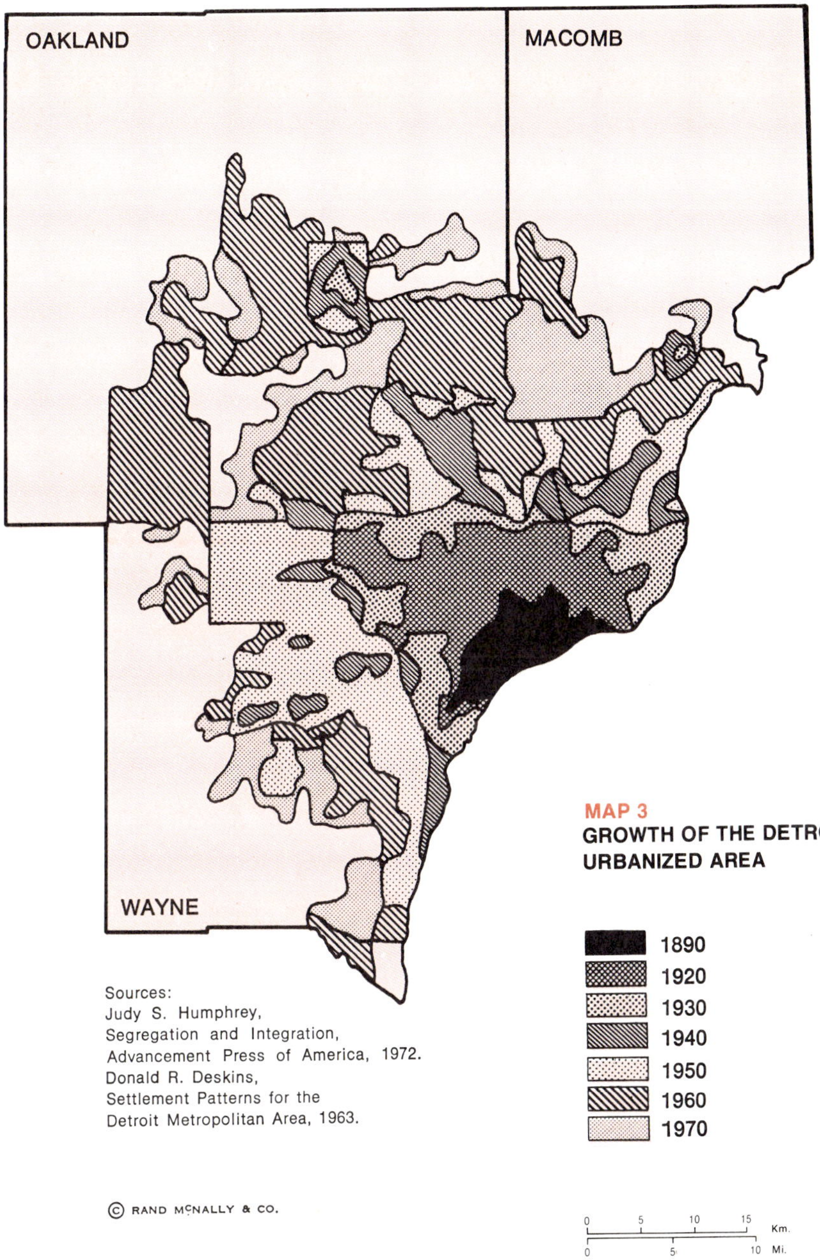

MAP 3
GROWTH OF THE DETROIT URBANIZED AREA

Sources:
Judy S. Humphrey, Segregation and Integration, Advancement Press of America, 1972.
Donald R. Deskins, Settlement Patterns for the Detroit Metropolitan Area, 1963.

rapidly expanded as cars, buses, and trains bring larger areas within bearable travel time from the city (see Map 3, page 17).

The arrangement of agricultural villages reflects the same kinds of spatial constraints or circumstances, even in areas where mechanized transport is absent. Where movement is by foot, animal power, or water, the spacing of large settlements will reflect most importantly the routes along which bulk goods can be moved cheaply, primarily rivers, lakes, or seacoasts. Smaller settlements tend to be spaced according to local agricultural productivity, and are closest to each other where productivity is highest. This pattern occurs in most of monsoon Asia, for example, where the villages on an agricultural plain look like Figure 3 (page 19), which is based on an aerial photograph of a small part of south India.

Where villages are close, productivity per acre is high and fields are small. The fields are close to the village, and the farm workers from each household walk to their plots every day. The size of the village is effectively limited by the number of households the area can support on the agricultural land within easy walking distance. It is determined also by the productivity of the land. As both Figures 3 and 4 (page 19) show, an agriculturally productive landscape tends to be broken up into groups of small fields clustered around villages that are relatively close together. The outer borders of the fields of one village are adjacent to those of the next village. In a less productive setting, village spacing would be wider. Like most spatial systems, these patterns also reflect the principle of least effort: things are arranged to minimize movement, labor, or expense.

Spatial arrangements within each village reflect their function, and their spatial constraints, as shown in Figure 4 (page 19). Houses are closely clustered along a central or single path, and fields radiate out from the village on all sides. People necessarily live close together, and various forms of cooperative effort are involved. Relations with other places and other people tend to be minimal, but within the village, relations and interdependence are close. These circumstances, in turn, are reflected in the traditional social structure characteristic of monsoon Asia: an exceptionally close-knit family system, firm intra-village connections, little scope for individual as opposed to group interests or values, and a generally closed or self-protective face to the rest of the world. All of these characteristics have clearly been influenced by the spatial system, the particular frame in which village Asia existed in the past, and, to a large extent, still exists.

The contrasting spatial arrangements of the

Detroit-Toledo from the air. Another sample of North American urban sprawl showing the capital of the private automobile. The radial pattern of Detroit's growth from its early center on the St. Clair River stands out clearly. The letters identify: *A*—Lake St. Clair, *B*—Lake Erie, *C*—Detroit, *D* and *E*—silt from the mouths of the Maumee and Sandusky rivers, *F*—farmlands that show as black because of surface water from a recent rain, *G*—lakes that dot this glaciated landscape, *H*—Ann Arbor, *I*—Pontiac, *J*—Toledo, *K*—Bowling Green, and *L*—Sandusky. This photograph was taken June 7, 1973 from an altitude of 560 miles (905 km). *(NASA satellite photo.)*

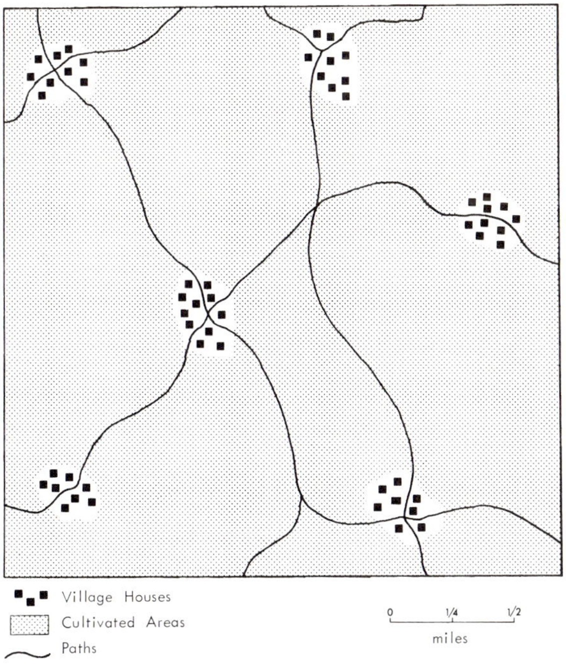

Figure 3. Village and agricultural landscape on a south Indian plain.

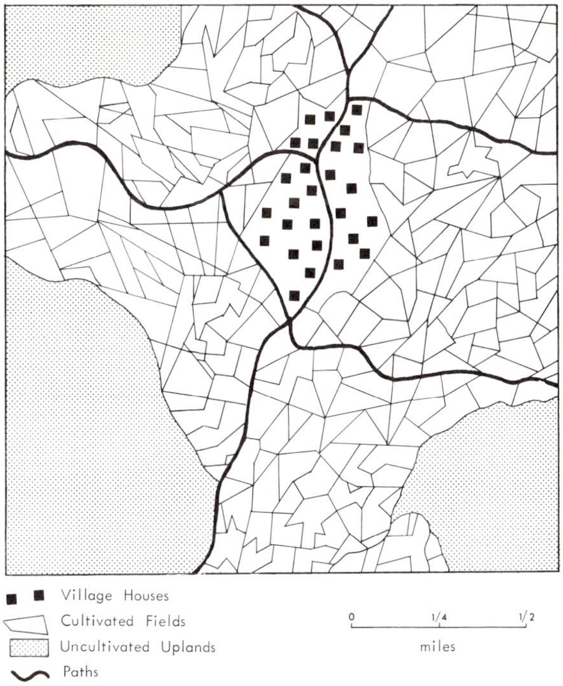

Figure 4. Standard Asian village pattern.

modern Western city are clearly related to its different social and cultural characteristics, and to many of its contemporary problems: individualism, alienation, rootlessness, mobility, and rapid change. Modern Western cities are also primarily commercial, and their internal spatial patterns and forms are shaped principally by their commercial-industrial functions and by the commercial value of urban land. Their layout and architecture contrast with cities whose chief functions are ceremonial, symbolic, or administrative, including most cities built before the nineteenth century.[1] In most contemporary American cities, land values tend to reach a peak in or close to the city's central business district, but to fall off away from the center in an irregular pattern. Large space-users (railways, most factories) must obviously seek cheaper land. Functions whose success depends on easy access in high volume, especially department stores and other retailers, must seek the center, or outlying centers where transport access is maximized. Retailers tend to cluster near points where pedestrian traffic is greatest or where individual customers can most easily reach the area by car or public transportation.

With the increased use of the private car and the consequent decay of public transportation in many cities, department stores and other retailers, formerly concentrated where public transport lines converged (but where the private car was at a disadvantage) have had to move or to establish stores in outlying locations. There car access is better and ample, lower-cost space is available for huge parking lots, many times the size of the store itself. It is obvious that land values are reflected also in residence patterns, including segregation by race, cultural origin, and occupation, since these factors are related to in-

[1]An excellent book on this subject is Paul Wheatley's *Pivot of the Four Quarters* (Chicago, 1971). Chapters 3, 4, and 5 deal in great detail with early cities in Asia, Europe, Africa, and the Western Hemisphere as ceremonial centers.

come. Black residential patterns make the clearest case, at least in most American cities, because blacks are the largest, easily identified, separate, racial and cultural group, often restricted to low-income occupations.

Spatial inequities

Whatever the changes and tendencies toward uniformity brought about by the spatial diffusion of innovation, the world remains enormously varied, both in its regional range of physical diversity and in the way in which people live. This is not merely a matter of regional culture, preferences, and attitudes, but involves great differences and inequities in the material bases of life. Two-thirds or more of the world's people, sometimes euphemistically called the "developing world," live and die close to and sometimes below the margin of subsistence, while the remaining one-third enjoy ever-increasing affluence. People have not yet learned how to diffuse the economic abundance their new technology makes possible. Most of that new technology is used instead to strengthen the power and wealth of those who already have it, and to support international rivalries and conflicts aimed at the same objective. Relatively speaking, the poor grow poorer and the rich richer as the gap between the "developed" and "developing" sectors continues to widen. To these inequities are added a pre-existing pattern of racial and cultural differences. Circumstances of very recent origin, during the past two centuries—a brief moment in the course of human history—have brought wealth and power mainly to the people of Europe and North America (plus Japan). Most of these people share a common racial and cultural grouping; the rich of the world are predominantly white. It is not easy for the rest of the world to avoid resentment, and to use pre-existing racial, religious, linguistic, and other cultural differences as symbols of rivalry or as divisive rallying points.

These spatial patterns and the factors that account for them are clearly the most important of all. The geographer's efforts to identify and analyze the spatial dimensions of human society need to be concentrated on those that divide people in destructive ways. Their search for answers can help to ensure that differences, which will always exist, do not lead to injustice or to conflict. We are beginning to learn a good deal about spatial patterns of difference and interaction. That developing knowledge needs to be used.

Questions for further study and discussion

1. In what ways is urbanization dependent on spatial interaction?

2. Are there any aspects or characteristics of human society that do not have spatial form or that cannot be mapped? What does the answer suggest?

3. How is any agricultural system affected by spatial interaction? How and why might such effects differ among different types of agricultural systems?

4. State the latitude and longitude for the following: Berlin, Peking (Peiching), Alma-Ata, Denver, Honolulu, Manila, Cape Town, Buenos Aires, Tonga Islands.

5. Why does the International Date Line not follow a straight north-and-south course along the 180° line? What are the reasons for the deviations? What practical advantages are there for the location of the International Date Line in using Greenwich as the prime meridian?

6. New Zealand and Great Britain are roughly the same size, are both islands, and have similar physical environments. How can you explain their very different development, including the sizes and economic specializations of their respective populations that come predominantly from the same racial and cultural stock?

2 The environmental factor

Saudi Arabia. *(Photo from Arabian American Oil Company.)*

The second function of geography has been listed as a study of "the interrelationship between human society and the physical environment." To the geographer, the physical environment means everything except people and their works: climate, soils, vegetation, wild-animal life, landforms, the atmosphere, minerals, and water. Geographers call these the environment, or literally, *the surround,* because they are concerned with people. The word *environment* has meaning only in reference to some living unit existing within it for which it provides the setting. Obviously, the physical environment may have a profound effect on people and on how they organize their activities. But it is appropriate to speak of an interrelationship because people, in turn, act on their environment, and alter both it and its effects on them. Thus, the environmental influence is far from being a constant, as people themselves are not constant.

People remake the earth

Humans change their environment in a great variety of ways and have fundamentally remade the landscape of every area of the world they have occupied in large numbers. In many cases succeeding periods of human occupance of the same area have each brought about a different change in the landscape: drained or irrigated it, cut the trees or planted it, enclosed it or left it as one great common. People usually remove the forest first, although even this has not yet been accomplished in the taiga forests of northern Asia and North America or in most of the equatorial rain forest. Removing the trees probably has more widespread effects on the rest of the environment and, in turn, on people than anything else. The agriculture that usually follows removal of the trees changes the soil or may seriously deplete it, and again revolutionizes the appearance of the landscape. Depending on the nature of the agricultural system, woods and clearings are replaced by hedgerows and plowed fields, fences of varying kinds in each region, terraced rice fields, orchards, plantations, irrigated basins, vineyards, truck farms, pastures, open ranges, artificial lakes, dams, wells, ponds, or fields of grain. People also build a more or less dense network of roads, and later, railroads and

airfields. They dig canals and dredge harbors, dig mines and oil wells, use or exhaust mineral deposits, and reap a harvest from the sea, changing it as well. All these changes in the environment react in turn upon people and influence their settlements and activities. Because people have altered the environment and learned to use it differently, they have come under the influence of a variety of new factors—some positive, such as new resources; some negative or limiting, such as soil erosion— that were unimportant to them before or were important in different ways. Finally, villages and towns grow into cities that now may cover many square miles with acre after acre of paved streets, factories, shops, and dwellings.

Where "nature" has been left, it has almost always been modified, at least in the environmentally favorable parts of the world where there are more people. Even the undestroyed temperate forests are not left alone, but are systematically managed or replanted. Few parts of the modern earth outside the polar areas or the equatorial tropics would be recognizable to Neolithic, or even to Medieval people. Animals kept or encouraged by people also play a part in modifying the environment. This is especially true of the grazing animals, and particularly of sheep and goats, whose close cropping prevents regrowth of trees or bushes and sometimes even of grass. Overgrazing leads to serious erosion and may even produce a change in the effective climate. Many desertlike areas, in fact, are the work of goats. People have seriously upset the complex interdependence of the biotic elements (animals, plants, birds, fish, insects) and the rest of the environment.

The strong influence of environmental factors on society has led some investigators to give them undue weight. As usual with such approaches, there

The permissive climate of monsoon Asia—irrigated paddies in Taiwan. Large investments of labor and organized management are necessary to build, maintain, and cultivate irrigated fields like these. The stooping figure is transplanting rice seedlings. This back-breaking job is performed each year by millions of Asian rice farmers in paddies whose water level is carefully controlled. The warm, moist climate helps to ensure that such heavy investments in the land will be repaid by high yields. In turn, high yields can support dense populations and highly organized water-control systems. *(From Joint Commission for Rural Reconstruction.)*

is a core of validity in the environmental approach. This idea has great usefulness, if it is used intelligently. The physical environment is one of many factors that influence society. Political considerations, such as subsidies, tariffs, central planning of the economy, or differences in market size between different political units, may also be important or even predominant influences on agricultural systems, manufacturing patterns, or trade flows. Cultural factors, such as taste preferences, traditional systems of agriculture or grazing, or systems of social organization, may be more important than the physical environment in determining land use or the nature of the economy. Just as importantly, human technical abilities and wants are continually changing. For all of these and other non-environmental reasons, human use of earth varies enormously from place to place and from century to century.

Similar environments are used differently, different environments are used similarly, at different times and in different places. Similar cultures may make similar use of different environments. Western Europe, southeastern Australia, and eastern North America tend to look alike. Despite the differences in physical environments, the people have used and modified their landscapes similarly, because these areas were settled by people with similar cultures and techniques. By contrast, southeastern China and southeastern United States look very different, despite the close similarity of their physical environments. The great cultural and technical differences between their respective inhabitants are reflected in their use of the land. Chinese cultural traits and their associated patterns of land use have also spread into other parts of Asia, and have created there what might be called a Chinese landscape. When the Europeans went into the Far East, they made a different use of the land. They evolved a distinctive agricultural system in the form of plantations. They built cities that are European in appearance. Even in the desert, there are pronounced differences in culture and economy between arid areas that are physically very similar. For example, the deserts of central Asia, Australia, South Africa, and southwestern United States all support different people with widely differing cultures and land use. Different people look for different things in the same environment and produce quite different results.

Determinant or influence?

These obvious facts and the continuing changes in human use of the environment in any one area discredit the theory of environmental determinism. It is a tempting theory because it tries to explain the complexities and variations in human society in terms of a nonhuman constant. The physical environment may be relatively constant, as opposed to human society, but its effects on mankind are far from constant, and often far from clear or direct. Historians have remarked that the only thing that does not change is change itself. However, even this statement is not entirely true, since the process, nature, speed, and direction of change frequently differ widely. To a large extent environmental changes are reflections of the overall growth of or change in human society. Such an analysis applies also to changes in society as a result of environmental influences.

Nevertheless, environmental influence is profound. Few aspects of human society are free from it, although its effects may be disguised and are often not clear. There is a risk that in discarding the theory of environmental determinism, the baby may be thrown out with the bath water. Many scholars abandoned all attempts to consider environmental influence, or even to consider causality in general. They confined themselves to a simple description of spatial patterns and regional units without attempting to interpret what they described. This was a serious loss. The purpose of study is to understand and explain, or at least to ask why, not merely to accumulate data. Avoidance of the environmental factor also neglected a major function of geography, which is to set people within the frame of the earth they inhabit. This necessarily involves critical examination of human interrelations with their physical base, as well as the other factors influencing them and contributing to regional differences. Fortunately, the discipline of geography has increasingly realized the central importance of that task.

"Determinism" as an approach has been replaced by "possibilism." It assumes that people have a large degree of choice even though the environment may set certain limits on what they can do profitably. Of the many things environmental conditions allow, people choose to do certain things, not only because of their own inclinations, but as a result of other influences or factors. These other factors help to shape their civilization: political pressures, traditional forms, cultural values, or the development of particular techniques. Environmental conditions may make one or a few choices more attractive or more successful. But choices may differ widely among different societies and at different times. The rejection of determinism is not merely a belated recognition of free will. More importantly, it is a realization that human society and its choices are extremely complex, that they are continually changing, that no one factor or influence can hope to explain this complex reality, and that what is observed is the result of a mass of interacting factors in which the human side is fully as important as the physical environment. This is not to say that environmental influence can safely be ignored. Its effects can be traced, sometimes with striking results, sometimes as a subdued note in the influences over most aspects of human society and in its differences over time and over space.

In some ways, however, the use of the word *influence* is misleading. It may suggest, or even promote, an unconscious belief that the environment is a direct force, that it acts on human society. This, of course, is not strictly true except in a very limited sense, as, for example, the destruction of property by floods or earthquakes. It is primarily people who act and people who choose. The physical environment merely exists. Although people may adjust to it or may follow certain courses of development in part because of the nature of environmental conditions, their actions are not influenced by the environment in any direct sense as if it were an active force. One speaks of environmental influences mainly because it is awkward and difficult to speak of the relationship in any other way without excessive words. But it is important to remember that the environment is predominantly passive. People are the main active agents although they may be either preservers or destroyers.

When one selects a single element of the environment, such as climate, and attempts to base complete explanations on it, the risk of overemphasis is greatest. Climate may be largely responsible for the Eskimo's way of life and for the lack of population in the desert. It is a temptation to use the same simple key on the more complicated human use of the land in more temperate parts of the earth.

Arid margins of the steppe—stock around a water hole near Happy, Texas. Climate here offers a relatively limited range of choice and a meager basis for livelihood in terms of the number of inhabitants who can maintain themselves. This area is too dry for farming, and its sparse vegetation is barely adequate to support hardy cattle on open ranges where local water is available. Towns or ranch houses, like those in the background, are few, small, and widely spaced. The area is not productive enough or requires more to support large investment in land improvements or extensive transport networks. *(From Standard Oil of New Jersey Collection, University of Louisville.)*

Obviously such a limiting or conditioning factor as climate that varies so greatly from place to place has a discernible influence on human society. But this does not mean that people who live in certain climates think or live in a certain fixed way. It does not mean that the climate of some parts of the world gives the inhabitants a distinctly "progressive" or "backward" character, as many climatic propagandists would have us believe. The environment usually operates as a whole. To isolate one factor and ignore the others is irresponsible and inaccurate where there are many non-environmental factors operating as well, and where technological change is also taking place. Human efforts are fully as important as the physical characteristics of the earth on which civilization is built. A man standing on a hill is there because the hill supports him, but equally because he has climbed there himself.[1] The physical environment is important as one of many influences acting together on society, not as a determinant. It is both difficult and misleading to examine the environmental influence without equal attention to other influences.

The meaning of resources

Resources are things on which people depend at any given time. These are primarily nonhuman sources provided by the environment, but human skills and capacities, of course, may also become resources. Since human civilization began, economic, cultural, and technological changes have continued to alter the pattern and extent of resources. Because such change is not uniform over the world, at any given time there are pronounced areal differences in what constitutes a resource. A resource is less a physical fact than a cultural, technical, or economic appraisal. This applies not only to minerals, but to all of the things on which a society may depend, including location, climate, soils, animals, and every usable nonhuman advantage or material. Japan's or Great Britain's location, California's or Cuba's climate did not become resources until the technical and economic abilities of their respective occupants and the changed spatial relations between these places and other places made their use possible. Once the nonhuman resources were put to use, population increased, created new human resources, and the income level rose even more. Nonhuman resources are part of the physical environment. People change or even eliminate them by using them. Their influence on society changes as people increase their technical powers and use resources more fully, exhaust or destroy them, or pollute and degrade their environment. Human labor and skills, varying from place to place and from time to time, may also be considered resources, and may be an area's chief resource. This is so obvious that it should be unnecessary to stress it, but there is a logical and necessary distinction between people on the one hand, and their nonhuman environment on the other.

Population densities and resource use

Population densities are understandable only if resources are considered in flexible terms. For example, it has been estimated that the population of Britain in 10,000 B.C. was about 1,000. The same area now supports well over 55 million people on an enormously higher level of per capita income and consumption. Today more of the potential resources of Britain have been put to work by a technically more advanced as well as a more numerous population. During the Industrial Revolution Europe's population increased seven times, increasing its per capita income at the same time. Europe's latent resources and physical environment changed very little, but they became much more effectively used. The same process took place even more strikingly in the United States. Its population increased thirty-one times between 1790 and 1940 as new resources were developed. Obviously, measures of population density or overpopulation mean nothing unless they can be related to usable resources. Significant measures involve more than the number of people per square mile. The often-used term *man-land ratio* may be misleading. The huge area of Baffin Island (in northern Canada) now supports about 300 people at a very low standard of living. Belgium, a much smaller area, supports over 10 million people at a very high standard. The *man-resource ratio* is a clumsier term and may also be misleading, unless resources and their use are understood as changeable.

[1]This illustration was often used by the British geographer Sir Halford Mackinder in attempting to explain the dual nature of geographical analysis.

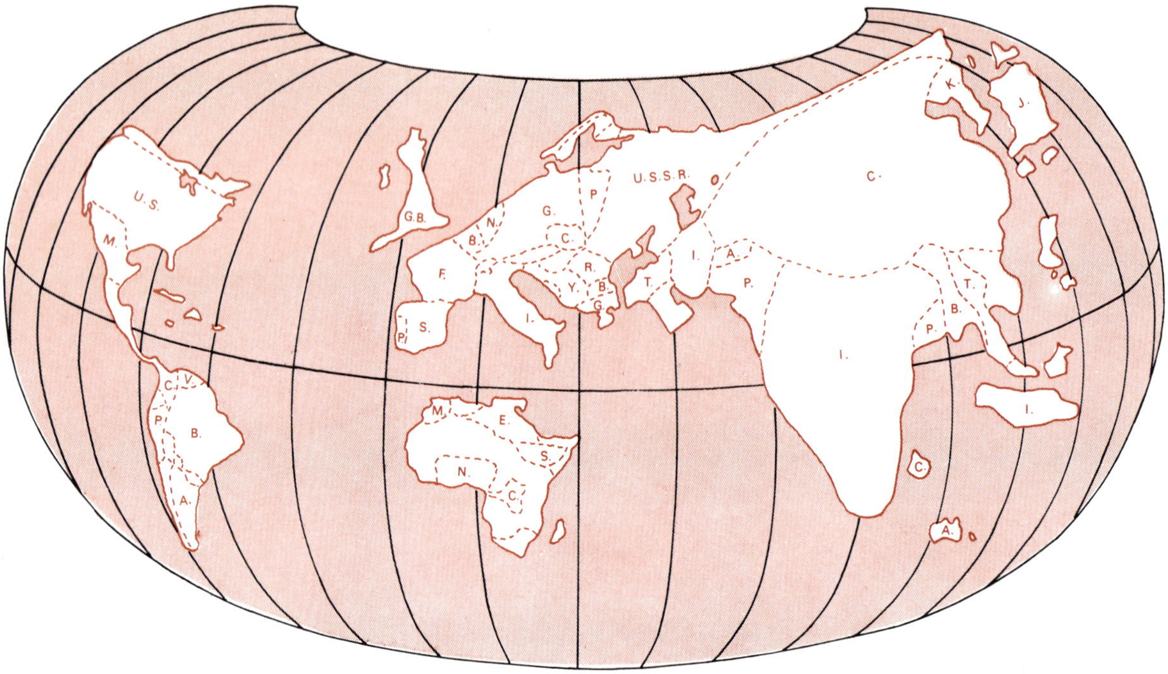

Figure 5. World population expressed in areal terms. Each continent and each country appears on a scale in proportion to its population. Updated after Woytinsky and based on 1977 population figures.

But properly understood, it is a better measure of the population balance.

Baffin Island is probably no less overpopulated, for example, than Java (Djawa) or India with their hundreds of millions of people. Java and India will remain overpopulated, in terms of average per capita income or consumption (the best measures of overpopulation), as long as they make inadequate use of their present and latent resources. The resources in these countries are potentially capable of supporting even more people on a very much higher standard of living than now. Europe also had relatively low living standards and inadequate use of potential resources two hundred years ago as compared with its present position. The West broke out of this situation primarily through an economic and technological revolution, first in trade and agriculture, then in manufacturing. The changes made it possible to use resources that had been previously neglected or poorly used.

Resources and the location of manufacturing

Of the many thousand raw materials on which an industrial economy is dependent, a few are of paramount importance because they help to determine where industrialization may occur. Modern manufacturing uses large amounts of them, and normally cannot afford to haul them great distances. The mineral alloys necessary to make steel hard or resistant to heat or corrosion are just as necessary as iron ore for making that steel. But such relatively small amounts of these alloys are needed for every ton of steel that they can economically be brought from the other side of the world if necessary. Some resources that are used on a large scale occur or can be produced only in parts of the world where there are no other advantages for manufacturing. Therefore, they do move considerable distances to reach the established manufacturing centers. High-value tropical crops like rubber, chocolate, or vegetable oils do this. Mineral deposits occur in widely scattered locations, and only by chance are they found with favorable climate, landforms, population centers, or other factors likely to encourage manufacturing.

When their locations are not favorable, the minerals may not be usable, if they are bulky or heavy in proportion to their value.

There are only a limited number of major heavy manufacturing areas in the world: northwest Europe, the Ukraine-Urals area in the Soviet Union, northeast United States, central and western Japan, north China–Manchuria, and the Damodar region (Bihar–West Bengal) of India. But there are plenty of mineral deposits in other places. Heavy manufacturing, in which raw materials and transport costs are a large proportion of total production costs, must concentrate where assembly costs for its raw materials and distribution costs for its finished products are at their lowest. Heavy manufacturing must be particularly sensitive to its economic distances from the raw materials used in the largest amounts. If these economic distances are small, and if distances to market are also small, heavy manufacturing can ignore distances from other materials needed in lesser amounts. Large deposits of copper, aluminum, manganese, antimony, tungsten, or nickel by themselves are unlikely to attract much manufacturing locally, aside from the immediate processing of the ores, because they are used in relatively small amounts in manufacturing as a whole. If an ore contains only a small proportion of pure metal, it may not pay to ship what is mostly waste material for long distances. Such ore is usually somewhat refined on the spot. Copper ore, for example, is often smelted or concentrated into *blister copper* at the mines. The copper sent to market or to copper-using industries from smelters is about 99 percent pure. But heavy or general manufacturing would find little advantage in locating at a copper or tungsten mine if the mine were far from sources of fuel and power, from other bulky raw materials, from labor, and from the market.

Heavy manufacturing and its role

Heavy manufacturing, which includes most importantly iron and steel, construction materials, transport equipment, chemicals and refining, and heavy machinery, is fundamental to industrialization because it provides the essential means of producing and exchanging more goods by power-driven machinery. Without these industrial essentials, or *capital goods* (sometimes called *producers' goods*), it is difficult or impossible to produce or exchange *consumer goods* like textiles, furniture, or food for wider markets, except in small quantities or at a relatively high cost. Thus all industrialization in the modern sense depends on heavy industries. The supply and cheapness of fresh, canned, or frozen foods, ready-made clothing, houses, or beds depend on steel, machinery, and transport equipment. Most consumer-goods industries are market-oriented, since raw materials for them are a relatively small proportion of total production costs. They are often attracted to centers of heavy manufacturing and their attendant markets, power, services, capital, labor, and transport facilities. Consumer goods are made nearly everywhere in some form and to some degree, and have been for thousands of years. But they are made more cheaply and in large quantities only where they can depend on heavy industries to supply them with the means of mechanical production and exchange.

The big three: coal, iron, and oil

Coal, iron ore, and (more recently) petroleum form a triad of heavily used resources to which most manufacturing must be especially sensitive. Fresh water might also be included, since it is used in even larger amounts by most manufacturing plants. But fresh water is much more widely available, and generally influences the location of manufacturing on a smaller or more local scale.

1. *Coal*. Of the big three, coal was long the most important resource because it was the cheapest available material for a great variety of uses and was consumed in huge quantities. Coal is often important enough to draw to it whatever other raw materials are needed, and thus to minimize total transport or assembly costs. But the factors of location, market, technology, capital, labor, and transport must be favorable in the area where the coal occurs. There are very few large industrial areas that are not near to coal, and most of the largest ones are very close to coal, at least in economic distance. Coal is used not only in the metallurgical industries as coke,[2] but as the largest single source of electric power (where the power is generated by steam-driven turbines). It is a

[2]Coal is reduced to coke by heating it in an airless oven to drive out the water, and is used for making iron and steel. Not all coal is suitable for making metallurgical coke. The coal must have the right chemical and structural properties, ash content, heating value, volatility, and other characteristics.

leading raw material in many chemical industries, where it is especially important for production of synthetic substances, such as dyes and plastics. As petroleum, hydroelectricity, natural gas, and fissionable materials have become cheaper and increasingly important sources of power, coal has lost some of its former importance. For example, coal is no longer the chief fuel for home heating or for railways or ships. But coal will remain the best or cheapest material for many years for a variety of manufacturing needs, and it continues to be the most-used industrial raw material, after fresh water.

It is possible that coal's relative importance may increase in the future, since world petroleum supplies will almost certainly be exhausted long before world coal deposits. Oil can be obtained from oil shales, which are relatively plentiful, but the cost in the future may be higher than obtaining it from coal. Petroleum or its products can be made from coal, as is done already on a small scale in certain areas where petroleum is scarce. The coal-rich areas will probably continue to have major economic advantages, as they have since the eighteenth century. Much of the enormous and rapid industrial development of the United States was due to it being "the land of cheap coal," with about half of the world's total workable coal reserves. In addition, the United States has a large share of the world's petroleum, iron ore, and other valuable minerals. Extraction costs for coal are relatively low in the United States, due to the favorable nature of most of the deposits. But the United States is rapidly consuming its minerals, and is already a large importer. With the fortunate exception of coal, we may shortly be a net importer of most of the heavily used minerals. Americans use minerals at many times the rate of the rest of the world, and the rate is still rising.

2. *Iron Ore.* Since iron ore is used in somewhat smaller total amounts than coal, it is usually transported to places where coal is cheap. But manufacturing uses enough iron ore that its location is profoundly affected by ore availability, as illustrated by the movement of iron and steel manufacturing in the United States toward the east coast in response to changed supplies of iron ore. Technical economies in the use of fuel have decreased the proportion of coal necessary for iron and steel making (and for other manufacturing processes also) so slightly more iron

ore than coal by weight is now used in making a ton of pig iron. But coal has many other uses. Steel manufacturing also makes large use of scrap metal as a raw material, mixing it with pig iron to make steel, thus replacing iron ore to some extent. Scrap is usually cheapest and most plentiful in large market centers. These centers are often associated with coal production and import, and they consume large amounts of metal products. Thus coal's locational dominance remains, as well as its association with the dense population necessary for manufacturing. Comparison of the distribution of coal deposits and population in Europe provides a good illustration (see Maps 6 and 12, pages 50–51 and 56–57).

Markets, the transport facilities that serve them, and their pools of labor, capital, special services (engineering, banking, or insurance), and con-

The Industrial Revolution—traditional blacksmith's forge in Lacock, England, and modern steel-rolling mill in South Wales. The obvious contrast in productivity per man-hour and effectiveness of resource use illustrates the essence of the Industrial Revolution. *(From British Information Services.)*

sumers are a major attraction for nearly every kind of manufacturing, wherever transport is cheap and adequate enough to bring the necessary raw materials to the market at a bearable cost. Coal is closely associated with markets and dense populations in so many cases because coal is too bulky and low in value to be transported far in the huge quantities needed. So the markets and populations have frequently grown up close to the coal. As transport facilities improve and transport costs for bulky raw materials decrease, markets become an even more desirable or profitable location for manufacturing. This process continued throughout the Industrial Revolution as sites of raw material and power sources have increasingly lost their locational attraction, and the markets have gained. Since so many of the markets grew up close to coal sources, and since coal is still used in such large amounts, this development has not affected coal as much as it has affected other raw materials.

If water transport is available, iron ore may move considerable distances to market centers where coal, scrap metal, other necessary raw materials, and power, labor, capital, and transport facilities are present. Specialized carriers also help make the movement of iron ore relatively cheap despite its bulk and weight.

3. *Oil.* Although petroleum and natural gas are used in huge quantities, they are much more cheaply transportable, by pipeline overland (almost as cheap as ocean transport) and by tanker at sea. They can be loaded and unloaded quickly, mechanically, and thus cheaply. Special carriers built to haul only petroleum and loaded to capacity with no waste space can offer very low haulage rates. The relatively low transport cost for petroleum is fortunate, since so much of the world's petroleum lies in the Middle East, the East Indies, Mexico, and Venezuela, far from major market centers. A modern industrial area must have low-cost access to petroleum, even though petroleum alone does not create industrialization where the deposits occur. Oil can be moved too easily to places that have greater total advantages for manufacturing. The oil of Iraq or Saudi Arabia, for example, does not stimulate local industrialization, except to provide capital to be invested in manufacturing. Other requisites for industrialization, such as other raw materials, technical skills, adequate political leadership, markets, and skilled labor are largely lacking, and oil alone cannot attract or create them.

Future resources

No one can be certain how far or how rapidly technological change or other factors can alter resource use, although the record of the past hundred years suggests that the rate and degree of change have been underestimated consistently. One can be sure only that the meaning and use of resources are highly changeable, and that the future cannot be measured accurately in terms of the present. We may well be on the threshold of vast new developments in resource use that will add new resources to the list

Results of overgrazing—New Mexico. The land to the left of the fence has been protected from wandering, grazing animals. To the right, unrestricted grazing has left the soil dangerously exposed to erosion. In this semiarid climate, vegetation so severely damaged will not recover rapidly. *(From U.S.D.A., Soil Conservation Service.)*

and change the meaning of many present ones. Because a material or resource is technically usable does not necessarily mean that it is economically usable, if total costs are too high, or it is ecologically unsound. It is also generally unwise to assume that when new resources are developed the present locational pattern of economic activity will be greatly distorted. It might be in certain cases, as clearly happened with the increased use of coal, the single most important mineral resource of modern world. But because central Africa has much uranium, for example, does not mean that it will become a great manufacturing or population center on the basis of nuclear power. As with resources in the past and present, assembly and distribution costs will continue to influence market and manufacturing locations.

The broad, varied bases of settlement, manufacturing, transport, markets, and capital will continue to be more important than specific mineral deposits in themselves, even though all rocks or all soils become usable sources of minerals. General technological change is more likely to benefit established centers than presently underdeveloped areas. The economic gap between the developed and underdeveloped areas is still widening rather than narrowing. This trend cannot be reversed merely by the discovery or use of new minerals or power sources in the underdeveloped areas. The industrial or "developed" areas of the West are still increasing their per capita production and consumption faster than most of the rest of the world. Many areas will continue to be "underdeveloped" even if they are able to increase their per capita production. One important factor is the rapid increase in the population of most of the underdeveloped areas, as their economic growth proceeds. Population increases threaten to consume most or all of the new production, so per capita consumption may rise very little, if at all. It will be difficult for the presently underdeveloped areas to reach the point where their per capita production increases as or more rapidly than Europe's at present.

Nuclear power may be economically and safely usable on a large scale in competition with coal and oil in the near future, but the advantages of nuclear power are likely to benefit the existing industrial areas. There all the other necessary factors are concentrated, especially the major markets with their high, constant demand for power that helps to keep power costs low. An even more promising development in the long run is the direct use of energy from the sun as heat, or in some other form as plants use it. This still seems to be a long way from being economically possible on a large scale, but if and when it can be achieved, the Sahara Desert will not become the industrial capital of the world.

Additions to the list of usable resources, including many less dramatic than nuclear or solar energy, may continue to increase the standard of living, and, one may hope, also the common ground between nations. In a world of relative plenty where goods are better distributed and resources more widely used, there may be less occasion for discord. Such possibilities are important for all of humanity. But resources are not exclusively physical facts as much as they are reflections and creations of humanity. This is apparent in the tremendous regional diversity of the world and in the particular patterns of resource use that have grown up in its different areas. The materials or advantages on which civilizations and economies depend vary enormously. Resource appraisals vary as places, people, and societies vary. One of the greatest challenges to modern civilization is whether it can increase and spread resource use widely enough that, by varying local combinations of resources and wants, economic abundance is available to all.

The ecological balance

Humans are distinguished from other animals by their ability to manipulate the environment, first (perhaps half a million years ago) by the use of fire, then by progressively more effective hand and mechanical tools, and finally by a great variety of machines and chemicals. The last hundred years have seen this process grow tremendously, as people rapidly increased their abilities to manipulate the environment through industrialization. They have probably made a more destructive impact on the environment during the past century than during the whole preceding period of human history, an impact that is still continuing.

Ecology is the study of the mutual interplay among and between organisms and their environment. Originally only a biological concept, it has recently been expanded to include, and now to emphasize, the role of the human organism and human interrelationships with those parts of the natural environment that are important to mankind. In this context, the nature of the environment as interlocking systems in balance with one another is especially clear. But this balance is easily disrupted by aggressive human powers, and part of the ecological system has grown out of control.

The natural environment is a fragile system; it is dangerously vulnerable to the destructive ways people have manipulated it. Other species have adapted to their environment or perished. Humans survived a major environmental change in the last glacial period with the help of new weapons, such as fire. But some newer weapons threaten the very base that sustains them. On a human time scale, the environment changes very slowly. People have become an external force acting on the environment to produce rapid and sweeping changes. They have divorced themselves from their environment, and may have cut the cord that nourished them. Without this new external pressure, the environment would heal itself in time, but the pressure of modern human technology is unrelenting and still increasing. It continues to disrupt the comparatively slower and more fragile processes by which the various elements of the environment remain in a dynamic balance.

The long-run effects of the internal combustion engine; the use of chemicals in agriculture, food processing, and industrial products; nuclear weaponry or testing; and the burning of fossil fuels (coal, oil, and gas) may tip the balance sharply to the negative or even the fatal side. Awareness of these risks is still so recent that we do not yet know their dimensions, nor how best to guard against them, short of abandoning the whole structure of industrial technology. Such a solution, of course, is unrealistic, even if it were appropriate. But each year it becomes clearer that a drastic change in human behavior must take place, and as rapidly as possible, if we are to avoid altering our environment to the point where we can no longer survive in it. We cannot eliminate the environment, but we can eliminate, disrupt, or destroy those parts of it that sustain us.

The burden of human impact on the environment has been greatly increased, not only by industrialization and the revolution in power use, but by the rise in human population resulting from the increase in productivity. People have upset the ecological balance by multiplying their own numbers and by spreading their occupance and use of the environment over almost the whole world. At the same time they have increased the number of demands and the impact of their demands on the ecological system. Although increased human survival and greater numbers may be or may seem a strongly positive matter, they clearly have negative consequences. They strain or subvert a total ecological system on which people are vitally dependent, especially as they continue to weaken it. Finally, one may raise the more philosophical question of whether humanity can be preserved at all unless it can learn not to destroy itself, either indirectly by destroying its environment, or directly through war and hate. Humanity, the supposed conqueror of nature, has much to learn. Perhaps it is the beginning of wisdom that we have at last come to recognize some of our own limitations.

Change and process

Fortunately, neither people nor their environment are static. The ecological system is best seen as a *process* rather than as a fixed system. As part of this system, people also change and will continue to change, hopefully in a more effective response to their ecological setting. The environment as a whole, and each part of it, is continually changing and adapting in a mutually linked system. Daily, monthly, and seasonal alterations are obvious, but on a much larger time scale, the environment is dynamic. Soils and vegetation, for example, alter slowly over time in a mutually responsive fashion, and also in response to changes in climate. It is possible to reconstruct these changes throughout most of the earth's history by the fossilized evidence of major geologic epochs, and to measure the magnitude of the change. People appeared very recently on such a time scale, and their long-run place in the ecological system is still unclear. No doubt much will depend on how well they are able to adapt.

During the evolution of the human species, mankind has had to adapt to drastic changes in climate and the other elements of the environment. The emergence of *homo sapiens,* about fifty thousand years ago, coincided with the last advance of the ice. Humans began to evolve toward agriculture, domestication of animals, and ultimately civilization during the equally drastic changes accompanying the retreat of the ice sheets and the return of warmer climate. So far the human species has proved itself eminently adjustable. It has taken maximum advantage of both major and minor environmental fluctuations rather than being limited or eliminated by them, as with many other species. The greatest challenge, however, is of human making: having reduced the power of the environment, can people continue to manipulate it without destroying its ability to sustain them? In other words, can they change now to adapt to the new problems they have created by upsetting the ecological balance? Or will they be eliminated, as so many other species have been, by their inability to change, to reverse their present disastrous course?

Much of the answer will depend on how we perceive our environment: as obstacle, as enemy, as tool, or as the basis of our existence that requires nurturing care. Culture forms a lens through which people perceive reality. Since culture changes over time, and is different in different places at the same time, the distortion of this lens is not constant. In other words, it is perfectly possible for us to alter our perception of the environment in the face of this new challenge, and to adapt our behavior accordingly. We may already be seeing the beginning of such a change as people in a variety of contemporary cultures become more sharply aware of relatively long-standing problems, such as atmospheric pollution, to appreciate their urgency, and to be ready to do something about them. This country's standards for good or acceptable or clean air changed radically and rapidly about 1900, again about 1950, and again in 1970 as the atmosphere became progressively degraded. The problem has forced itself to the attention of more and more people. The same changes have been repeated in Western Europe, the Soviet Union, and Japan as industrialization and the internal combustion engine have spread. Tokyo, for example, has probably the most polluted air of any big city in the world, but it may also be the most pollution-conscious. In other words, its citizens perceive atmospheric pollution especially sharply. At major street junctions in the downtown area automatic pollution sensors flash continuous readings of the changing count on huge illuminated billboards.

Can humanity survive?

In a world where there is nearly global exchange of information and awareness of differences in experience and culture, each area may be able to benefit from the perspectives, insights, and mistakes of others. All people share a planet whose environment operates as a single system. What is done to it in one area is likely to affect it elsewhere. Nuclear testing in the south Pacific or central Asia puts radioactive particles into global atmospheric circulation and into the circulation system of the oceans, so all parts of the world are covered. Industrial pollution in any area also enters the global circulation system. The effects of biological and chemical warfare in Indochina cannot be confined to that unhappy area. We are all in the same boat, even if some of the most obvious leaks are only in certain parts of it. The more industrialized nations necessarily have the major responsibility, partly because they are the chief polluters and the chief users of finite resources, and partly because their degrading effects on the environment are inherited by everyone else. The poor two-thirds of mankind, whose economic margin is much smaller, can less well tolerate the consequences of a deteriorating environmental base. The United States alone, with about 5 percent of world population, accounts for over one-third of world consumption of fossil fuels. It generates nearly half of the world's mechanical energy, producing nearly half of the world's atmospheric and water pollution in the process. This is both irresponsible and shortsighted and is increasingly resented. Change is urgently overdue. All societies are interdependent, just as people are interdependent with their environment.

We still have much to learn technologically about the consequences of and the means for controlling environmental degradation. New technology is continually being evolved; its directions are determined in large part by felt needs: what a given society considers important or urgent, which in turn is dependent on that society's perceptions. The West, especially the United States, has tended to emphasize individual freedom—"rugged individualism." In

"Hillsides bare, rivers gorged" (Chinese proverb)—siltation in the San Pablo Reservoir, California. Removal of most of the trees on the surrounding hills has brought heavy erosion, washing silt into this reservoir and clearly threatening its function—a common problem. *(From U.S.D.A.)*

the frontier past, when space, resources, and economic opportunity were ample, such an attitude was understandable. In a more crowded world where individual actions profoundly affect public welfare, this attitude may be less understandable. In any case, environmental degradation can be checked only by concerted public action. Perhaps the West can alter its perceptions and begin to reverse the disastrous course industrialization has set. We live in a closed ecological system. It can nourish or extinguish us; the choice is ours.

Questions for further study and discussion

1. What examples can you find, in addition to those cited, of similar environments that are used differently? Of different environments that are used similarly? How can each case be accounted for?

2. What other examples can you find of the similar use of similar environments that help to support the idea of the physical environment influencing mankind in a more or less consistent way? To begin with, consult atlas maps dealing with world agricultural and crop patterns, but do not limit your investigation to agriculture.

3. How may environment influence culture in particular cases? How may culture act on environment?

4. What environmental factors are relevant in the site of the city where you live? In its economic specializations? In its cultural characteristics? How have spatial relations been important in all of these matters?

5. Many ocean-going ships can reach Albany on the Hudson River. Why is New York the major break-in-bulk point and by far the bigger city?

6. What is the national population density (people per square mile or square kilometer) of Norway, Argentina, Holland, Java, Tibet, Yugoslavia, Australia, Japan, Canada, Afghanistan, Malta, and Puerto Rico? Which of these countries are overpopulated? What does overpopulation mean?

7. What articles or goods in daily use in the United States have no connection with steel? With refining? How does the supply and cheapness of fresh fruit or ready-made clothing depend on steel?

8. Resource surveys and inventories have been made in great detail for the United States. Is it known precisely what the resources of the United States are or what their quantitative total is? Is the answer subject to change?

9. What changes in resource use and what new resources can you distinguish in the contemporary world? In the contemporary United States?

3 The region

Springtime in the Colorado Rockies.

The third function of geography has been listed as a study of "the regional framework and the analysis of specific regions." Examination of spatial form or of the spatial dimensions of human society must consider distinct areal units, so geography is concerned with the concept of the region. The world may be divided into continents, or countries, or counties according to more or less arbitrary existing lines. But there is a more important and more complex set of areal divisions to be made on the basis of units that contain distinct and internally consistent patterns of physical features or of human development.

Areal differences

Areal divisions are often difficult to make. There are almost endless criteria for defining a given region. An area may be distinctive and internally consistent because it coincides with distribution of particular climates, landforms, or soils. Or it may be distinctive because it is characterized by a particular economic activity, occupational structure, land use, or language. Areal distinctions may also be drawn on the basis of administrative or commercial services, or trade relations in a particular territory. One may describe and mark off on the map manufacturing regions, agricultural regions, climatic regions, landform regions, religious regions, or political regions, to mention only a few of the more obvious ones. Some of these may coincide exactly, although in many cases the various factors or characteristics that make an area distinct may differ in their spatial distribution. The city, as it is discussed in Chapter 1, is a good illustration of this problem. A city's newspapers will cover one area, its trade another (or a variety of others). The plain or river valley in which the city lies covers another area, the city's specialized services, such as bus routes or library circulation, another. And its political control will cover still another, this one seldom coinciding with the compact urban area of the city.

For a variety of reasons it is essential to pick out some degree of regional distinction and regional coherence from the jumbled mass of distribution and areal differences. This is necessary to understand any one area adequately, to distinguish it from other areas, and to see the basic spatial framework within which people live and which they have helped to create. Planning for the management and development of any area requires an understanding of the limits and the distinctive components of the regional unit involved. Regional planning, in which development is planned for an area as a whole in all of its interrelated aspects, is now accepted as an essential method for promoting economic growth or areal development. It requires a knowledge not only of regional differences and their limits, but of the interrelated factors, from climate to transport patterns to historical trends, that compose any regional unit. Business firms and political administrations also need to establish and to understand areas for the efficient operation of their activities. Trade areas and political boundaries are often matters of dispute, reflecting the complex problems involved. So the geographer attempts to provide some guidance. The study of political boundaries: what makes a good one or a bad one, where the boundary should be, and what impact a political boundary may have on the area through which it runs, is a whole field in itself, political geography. But it is also part of the geographer's more general, regional study.

Lines, barriers, and transition zones

To make regional divisions intelligently, first one must understand as well as possible what gives the region its coherence and distinction, what its essential unifying or common qualities are. Second, one must determine the points or zones where these qualities diminish significantly, and are replaced by other qualities belonging presumably to another region. Usually there is a wide zone of transition, and the line is difficult to draw. Prominent physical features, such as mountain ranges, often coincide areally with the zone of transition from one region to another. Mountains may act as barriers to economic and cultural movement or interchange. They may also be responsible for a transition in the physical aspects of the landscape. But the effect on people of even pronounced physical features of the environment is seldom constant. If such features are used as regional lines, the line may become obsolete or awkward as human occupance changes on both sides of such a boundary or meets across it.

Sharp cultural gradients on the map are rare, and even in physical terms, it is seldom possible to draw a definite line between regions. Regional characteristics change more or less gradually over space, and the transition area between one region and another may be a wide zone. This may be true even where a major mountain range is used as a regional line. The sea makes a clearer boundary, but its barrier character has been tremendously lessened by the modern revolution in communications and transport. The North Atlantic Treaty Organization, for example, now exhibits more genuine unity among its members than exists among the countries of Europe as a whole, although the Atlantic is as wide and deep as ever. Rivers may be physical barriers to an extent, but where they are navigable they are usually unifiers. Settlements on both banks are likely to be dependent on river-borne trade, hence, they will be part of a single economic and cultural community. Unnavigable rivers may support a more or less uniform pattern of agricultural land use in their valleys, which may be dependent on river water for irrigation. River basins are usually physical and human regional units. Attempts to divide them or to use rivers as political boundaries, as is commonly done, are generally not satisfactory.

Distinction, identity, and regionalism

Since regions tend to merge with each other over a wide zone of transition despite physical barriers, any single line may be arbitrary and seriously misleading, as well as awkward. For many purposes and especially for political boundaries, a single line must be used. Physical features, especially rivers, provide convenient lines, and are often used despite the disadvantages. For regional analysis, however, the basic problem is the determination of what gives any region its essential distinctive quality. It is only on that basis that satisfactory regional lines can be drawn. To draw them, we must be concerned with the physical framework, with economic, political, and social characteristics and functions, and also with the regional beliefs and loyalties of the inhabitants. We are dealing in many ways with what psychologists call a *gestalt,* a pattern of factors operating as a unit that is more than the simple sum of its parts, as water is more than the total of its hydrogen and oxygen components. Each aspect of an area exists in context, is interrelated with other aspects, and cannot be understood completely unless it is seen in its total context. Nor can the area as a whole be understood unless all of its interrelated aspects are seen. This greater whole, this areal pattern of characteristics in the minds of the inhabitants is called *regionalism*, or in the case of a national state, *nationalism*. Separatism, or the desire for political autonomy, is pronounced regionalism that is similar to nationalism, but that may not insist on complete independence. Outside threats, pressures, or competition, real or imaginary, commonly strengthen regional consciousness and cohesion. But regionalism in basic terms results simply from the awareness of areal differences. People identify themselves with their own area and with its characteristics, including diets, dress, house styles, art forms, political attitudes, landscapes, seasons, and a host of other attributes that differ from region to region. Parents teach regional cultural traits to their children, take pride in their regional identity, and feel themselves different from and usually better than the people of other regions.

Such attitudes provide one of the clearest clues to the existence of regional entities, and are particularly helpful when it comes to drawing regional lines. This may be done in some situations merely by asking the inhabitants to which cultural region they feel they belong. Regionalism is often especially pronounced in isolated or mountainous areas. There the physical distinction from surrounding areas is strong and the culture and economy are likely to be distinctive also. Switzerland is a good example, and there are many others. But regionalism is not the particular property of the mountains, and every area that can be distinguished from a neighboring area feels it, in differing degrees. In the United States this feeling was strong enough in the southern states to support a movement for independence as a separate sovereignty. It remains strong, partly because of sentiment and loyalties, partly because of the geographical differences that fostered the distinctive feelings in the first place.

Language is often used as a measure of regionalism and as a basis for determining regions. Of all aspects of culture, language may be the most important mark of identity as well as one of the clearest. It often tends to coincide areally with most other aspects of a regional *gestalt*. Although it may be a badge of regional identity whose importance is emotional as well as practical, language is not a necessary or sufficient cause for regionalism or an adequate measure of a regional entity. The American Confederacy suggests that regionalism may be very strong and that a true region may exist without a regional language. Such examples could be multiplied. Switzerland, on the other hand, suggests that regionalism on grounds other than language may be strong enough to thrive despite language differences within the region. Until about the eighteenth century, language differences were seldom an issue or a cause for separatism, and multilingual states were the rule. People were not closely involved with one another over large distances, so language was less important, and the literacy level was relatively low. In general, only with the rise of modern nationalism and the national state on an enlarged areal base has language become an emotional part of regional identity. In some cases, regionalist or separatist or nationalist movements have actually created or revived a regional language to emphasize identity. Where the regional language was primarily a spoken one and had little or no literature, a literature has been manufactured in some cases. This suggests that factors other than language are important in promoting regionalism and in creating a regional entity. Language has become a more important part of regional identity than it used to be, but it has not been primarily responsible for creating or establishing that identity. Improvements in communication and the increase of interaction in the last several centuries have also been accompanied by an increase, rather than a decrease, in regional consciousness and rivalry. As regions are involved more closely with each other, perhaps differences are more apparent.

It should be made clear, however, that the concept of the region is primarily an intellectual device, useful as a tool for the better understanding and analysis of human society. The regional concept picks out certain spatial patterns. But political boundaries or other regional distinctions we may draw are products of our own minds rather than complete descriptions of reality, or of any "natural" order. In this sense, there are no "natural" regions, only the human-perceived classifications of the welter of areal differences.

The regional hierarchy

The differentiation of regions may begin very low on the size scale. It is important to establish the order of magnitude in which one is dealing in a particular case. Differences one may call regional exist, for instance, between different parts of a city. At the other end of the scale, they are apparent between continents and hemispheres. Two neighborhoods or two cities, however, will have much in common as part of a larger urban, provincial, national, or even continental region. Thus, there is an overlapping hierarchy of regions, depending on the criteria used. One may look at the world or at any part of it through the big end of the telescope to see gross, general spatial patterns, or through the small end to see more detail and more minute subdivisions. Both devices are useful.

Areal differences are universal, and are expressed in details of culture, from diet to dress, house types, language, prejudices, tastes, occupations, skills, and social and political systems. The United States is composed not only of a hierarchy of political subregions, such as wards, cities, counties, and states, but of a great variety of more broadly based subregions involving areal differences in land use, employment, investment, and production, voting habits, income, marketing patterns, cultural attitudes, and speech. These and a host of other characteristics have spatial form and are discernible as spatial patterns. Some of these broader spatial patterns are commonly recognized in geographical expressions as "the South" or "New England." But each of these regions may be further subdivided to recognize more complex variations from one part of the South or New England to another. At the other end of the size scale, it is clear that both of these American regions as a whole belong within the larger region of the United States and, in turn, of North America. At that level, or on a world scale, their similarities become more important than their differences.

Bases of distinctiveness

Regional differences may be related in part to differences in the physical environment, but only in part. Human society is a complex affair in which a great variety of factors operate. The regional analyst must consider all of these factors. An area covered by a given climate or soil may support a more or less consistent type of land use or culture, related in part to climate and soil. But land use or culture are also related to non-environmental factors, and the same patterns may often occur in physically different areas. A region may derive much of its distinctiveness from its interconnections with other areas. The Netherlands, for example, is a distinctive region partly because of its widespread trade relations with the rest of the world. The Rhine (Rhein) River, which is partly responsible for the commercial importance of the Netherlands as a break-in-bulk point at the river's mouth, is also responsible for considerable economic unity in the Rhine valley. Although political boundaries cross the Rhine and use it as a political line in some stretches, a heavy flow of trade follows the river. It is possible to argue that the "Dutch region" or the "Rhine region" actually extends beyond the limits of the conventional political or physical lines and includes the area with which the region trades. This kind of analysis, however, brings us back to the essential problem that few aspects of a region are likely to cover exactly the same area, and that the application of a regional label to any area is ambiguous until the criteria used are made clear.

The example of China

For example, China as a political state covers a vast territory, containing a great variety of physical and human differences. About half of this area is occupied predominantly by people who are ethnically not Chinese, speak a non-Chinese language, and have a non-Chinese culture and economy. Over half of it is so dry, so cold, or so mountainous that it contrasts sharply with the relatively well-watered, warm plains where most of the Chinese people live. On the other hand, important aspects of Chinese culture and large blocks of Chinese people are found in several places in east Asia beyond China's political frontiers: the Chinese written language and cultural forms in Japan, Indochina, and Korea; and heavy Chinese settlement in Thailand, Malaysia, and Java. Where is the true China region? The answer depends on the criteria chosen, or on those to which the greatest weight is given. Generalized lines may be drawn on the basis of single factors—around the area where Chinese-speaking people are in the majority, or where agriculture is the predominant occupation, or where the rainfall exceeds 20 inches

(50 cm) a year. This may be a useful method, but it avoids the major question. What makes China China, and where does it cease to be China?

One convenient solution is to plot the areal distribution of all factors that could be considered relevant as measures of regional identity: consistent language, ethnic type, economy, climate, landforms, social and political institutions, political control, cultural forms, historical alignments and changes, and so on through any region's characteristics, although they will be different for different regions. The basic minimum area indisputably China would be the area where all of the factors designated as consistent Chinese characteristics were found. But none of these characteristics covers exactly the same total area, and somehow the margins must be disposed of. A decision must be made about which of the characteristics are most important. Areas where they are all absent can be eliminated; areas where only some of them are absent will have to be treated on the basis of the priority given to each characteristic as a measure of Chineseness. By such means one can ultimately arrive at a reasonable regional differentiation, but only after first deciding what makes the region distinct. Plotting the areas on a map will usually help suggest the critical factors. It will also be necessary to analyze thoroughly the regional *gestalt* to avoid an arbitrary decision and to understand adequately the relative importance of the factors to be evaluated. This is obviously not a simple problem, for all the ease with which lines are often drawn or the glibness with which people talk about "natural" regions. Chinese civilization has traditionally rested on agriculture. Where agriculture and the farmer stop on the map, "China" has tended to weaken or disappear. In this case most of the essential regional characteristics coincide with agriculture in their areal distribution. In the excessively dry, cold, or mountainous parts of political China, agriculture is greatly reduced or eliminated. There many of the other "China" characteristics are replaced by clearly non-Chinese characteristics, although the Communist government has fostered much cultural uniformity since 1950.

It is logical then to ask why Chinese political control extended so far beyond the limits reached by the other factors, and why political control was not extended over the areas in other parts of east Asia where Chinese people or Chinese culture did spread. This question is not raised to answer it, for that would be beyond the scope of this brief example. But it emphasizes the implications of the regional concept and the insight it may provide.

Regional cores

The essential characteristics of most regions gradually decline as the regional margins approach, and regional cores or nuclei are another common feature. Usually spatial patterns of human development tend to focus or to settle on a hub or axis. Such a pattern may center on a physical phenomenon, such as a fertile plain, a river valley, a seacoast, or a crossroads of easy routes. Or it may be a cultural development, such as a city, the center of a particular type of regionally dominant economy, or a center of political or military control. Most clearly defined regional cores have all of these characteristics. They coincide with the largest area of dense population, include the largest share of the region's economic activity and the political capital, and tend to set the region's cultural pattern. Usually this core area is the first, in point of time, to develop the cultural characteristics of the region as a whole, and it spreads its dominant influence over the surrounding area as far as the original physical and cultural conditions allow. At some point these conditions become different enough or far enough away so they are no longer dominated by influences exerted from the core. This distant area will lie in a zone of transition. Or it may be influenced more by another less distant and more congenial center and belong to that center's region. Most regions gain much of their coherence from comparatively well-defined cores around which their areal traits tend to cluster.

Two prominent examples of regional cores are the London basin and the Paris basin. Both are the chief centers of population in their respective national regions. Both are the centers of political control and major market centers. Both acted historically as nuclei for the development and spread of their respective regional cultures, and both are still the dominant cores in this respect. Transport lines and circulation patterns focus on both to a high degree. The pattern of circulation and the density and degree of interaction within the region are often a leading part of or clue to regional identity and distinction. Regional lines may be drawn on the basis of orientation of circulation patterns toward a particular regional core, or on the basis of common interaction. Beyond such a line, orientation toward the core of another region may predominate, even though it may be distant, or toward a different set of interaction partners. There may be a zone of transition or indifference lying between competing cores. London and Paris are both strong cores whose dominance extends throughout and, to some extent, beyond their respective national areas. But in both countries there are also subsidiary cores, most importantly the centers of heavy manufacturing that developed much later than London and Paris. These manufacturing centers are areally separate, and associated with mineral resources largely lacking in the London and Paris basins. These lesser centers are subsidiary or tributary to London or Paris in part because they are mainly single-factor cores rather than multiple-factor or generic in nature. In other regions or countries, more evenly balanced competing cores may develop as, for example, in Germany (Berlin and the Ruhr), and to some degree in the United States (the middle Atlantic coast and the Great Lakes–Chicago district). As with the hierarchy of regions, however, there is a hierarchy of cores that may be distinguished at different orders of magnitude. Strictly speaking, a core performs some service or exerts some influence over the whole of the larger region to which it belongs. Manchester and the English Midlands are subsidiary to London, and Chicago and the Great Lakes district to New York, but these subsidiary cores play a prominent national role in the provision of certain major goods and services. St. Louis or Atlanta, however, although they are centers of settlement cores within small subregions of the United States, do not play a marked national role, so they are not cores in the larger regional sense. They are merely centers of population clusters and of certain local functions or local regional characteristics.

Space, time, and change

Geographers undertake the determination and analysis of regions because they are concerned with the relationships between regions, with the association of areas, and with spatial interaction. The patterns they map are not static, nor are they self-sufficient. They are part of an integrated whole that must be fitted together as well as taken apart. Before one can examine relationships within the whole, it is necessary to distinguish the parts. This involves not only separate forces and factors but the interplay of the complex factors making up each region and the areal associations that result.

In drawing regional boundaries, analyzing regional units, and considering the association of areas, the time dimension is of critical importance. Regional boundaries are seldom constant, since they reflect a particular stage of development. As the distinctive aspects of a society change, the degree and nature of its spatial expression change also. This results not only from technological change, such as improvements in communication extending cultural unity, but changes over time may alter the basic nature of the community. Wholly new factors or new emphases may develop in the economic or political system, for example. They may produce a different core, the exclusion of former areas, or the inclusion of new ones. Invasions, conquests, or migrations may distort or alter regions basically. Cultures change, or may be powerfully influenced by other areas. New forms of production may dominate the economy or revolutionize its nature.

However, there tends to be a certain generic harmony of the components of any region that may reassert itself, perhaps in a somewhat different form, even as the components change. Only over relatively long periods or under special circumstances do regions vanish or wholly new ones arise. The physical environment is relatively constant; even though its influences change as man changes, it maintains a certain amount of regional continuity over time. Spatial relations are also manipulated and altered as man changes, but physical distances and absolute location remain constant. While the region called "China" has remained relatively constant areally during the past two thousand years (despite recent political expansion into Manchuria and Inner Mongolia), tremendous changes resulting from outside influences in Europe and North America have destroyed most of the regions that existed two thousand years ago and have produced new regions in their places. The same temporal processes are at work, slowly or rapidly, in

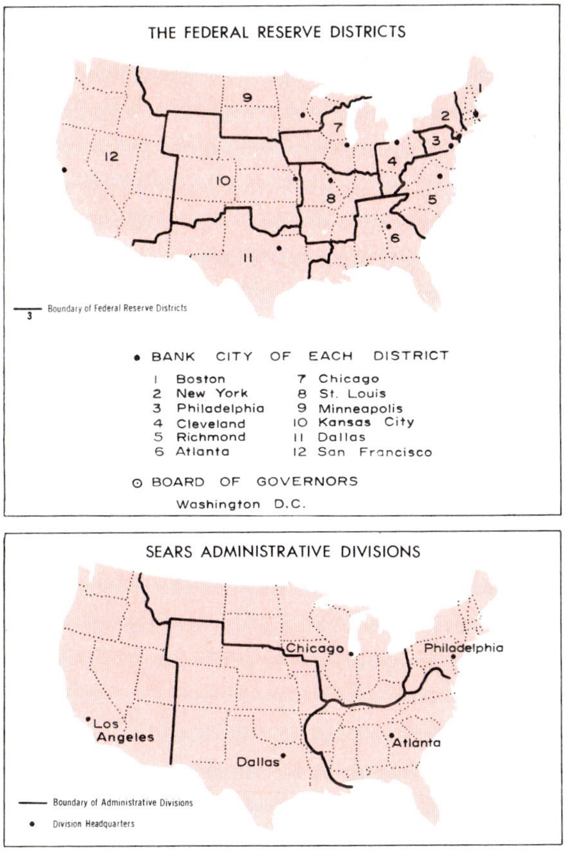

Figure 6. Sample regionalizations of the United States.

any situation, continually changing the essential quality of every region and affecting the region's spatial limits.

Any fact of human geography, or any region, can be examined and explained adequately only by considering its historical evolution. The location, size, and nature of a settlement, for example, may be explained statically in terms of particular present advantages of local site or relations with other places. But it must also be explained dynamically in terms of historical growth and the changing factors that have affected that growth—cultural and technical changes, varying political influences or decisions, the rise and fall of new states, or changes in spatial interaction. Any or all of these may be essential for understanding the present status of the settlement in question.

As the separate regions change, so the associations between regions change. Indeed, alterations may come about within a given region wholly or largely as a result of changes in its associations with other regions. Contemporary China is a good example. Many of China's characteristics are being rapidly altered as a result of new and closer relations with the West since the eighteenth century, and more recently with the Soviet Union. Mexico and Canada are profoundly affected in their regional development by their closeness to the United States; changes within them reflect changes in their huge neighbor. But Mexico and Canada are also associated with other more distant areas, with Spain and Great Britain most obviously, and in differing degrees with the rest of the world. The nature and force of these associations is intimately bound up with the human development of the two countries. The association of areas in its broadest sense is relevant at every stage of regional differentiation. A single region, whatever its size, whether based on a single functional criterion or on more general characteristics, is simply an association of smaller areas that have enough in common to be distinguished from other associations of areas. This region then is involved in a broader pattern of association with other similarly constituted regions.

Single-factor and multiple-factor regions

Although we may speak in general terms of China or Mexico or Canada as regions, such a label, of course, grossly oversimplifies the complex of culture and environment involved in such large areas. We can more accurately speak of *single-factor regions,* such as political, linguistic, or climatic areal units, or we can outline *functional regions,* such as trade or administrative areal units. For example, the United States is divided by the Federal Reserve System into twelve districts, each a recognizable and viable commercial-financial region grouped around one dominant urban financial center (see Figure 6 on page 41). Sears, Roebuck & Co. divides the country into five major and eleven minor merchandising districts, each with its own urban center for warehousing and servicing the district as a whole. Each of these districts offers slightly different goods, reflecting differences in regional needs or tastes and consumer preferences. The division of the United States and Canada into states and provinces is an obvious example of political administrative regions. Broad patterns of climatic differences are recognized in climatic regions. Single-factor regions such as these can be sharply drawn with a single line, of course. A political boundary, the boundaries of a Federal Reserve District, or of a precisely classified climate can be defined easily, and *one* consistent spatial pattern extracted from reality can be outlined simply and clearly.

There is unity implied and apparent in broader geographic expressions like "the South," "New England," "the Corn Belt," Mexico, China, or Canada. These expressions are applied to much more varied regions including a great number of different characteristics. It may be difficult to define these elements, aside from the obvious political boundaries. Each of the single factors involved almost certainly covers a different area, overlapping each other near the center, but leaving much of the "region" ambiguous. The American South, the Corn Belt, or New England, for example, are all multiple-factor regions. Their identity depends on a variety of individual characteristics more strongly present in some parts (Alabama, Iowa, Vermont) than in others (Kentucky, Ohio, Connecticut). Areas lying spatially close to the edge of any region may be influenced by regions adjacent to them. But the idea of the region assumes the existence of some spatial unit that exhibits common traits. This broader concept of the region might be stated as *a definable area possessing a wide range of more or less homogeneously distributed and interrelated characteristics.* The region is a complex of homogeneous factors that are interdependent and that operate together as part of a larger whole or *gestalt.* The idea of the region has often been criticized because it is ambiguous and oversimplified. The distinction between single-factor or functional regions on the one hand, and broader, multiple-factor generic regions on the other, helps to avoid much of this criticism. The general concept of distinct areal homogeneity is extremely useful, provided it is based on an analysis of the many factors that make an area distinct.

Study of regional wholes

Regional information and understanding are objectives of geographers as well as the classification of regions. A large part of geographic work consists of the accumulation of regional material. Here also geography undertakes a task generally not done as an integrated whole by the other disciplines. Economists are interested in production, finance, or the market system in any region; sociologists, in kinship groups, social structures, or institutions; and so on. Geographers attempt to examine and understand the region as a whole, in all of its physical and human aspects. They try to fit these aspects together, and to relate the region to other regions. This goal is demanding and may never be completely met. But someone must make the attempt to fit the regional mosaic together and to see regional wholes. So much has to be examined and understood to do this adequately that geographers must depend on the specialized studies of others as well as their own studies. They frequently function to some extent as editors or integrators. They must also compile and analyze a great bulk of regional information. This collecting and editing is an essential but elementary first step, a means to an end. Other students of society need the information that geographers can provide. Regional planners, the general public, and those who make policy decisions of many kinds need regional information too. But the information alone is not fully useful until it can be fitted into the regional whole. It can also be dull to deal only with factual detail, to see only the individual trees and miss the pattern of the forest. Sometimes the "pack-rat mind," which collects endless bits of information for their own sake, is apparent in the craze for facts and the confusion of information with knowledge that handicaps some regional study. Information can seldom be gathered intelligently unless an overall pattern is seen. Buildings are constructed of bricks, but the bricklayer cannot, or should not, choose and lay bricks without a blueprint or image of the building. Knowing the number of miles or kilometers of railroads in Chile, for example, is less important than knowing the patterns of their distribution. The interrelations the railroads reflect and stimulate in Chile, the role of railroads in its regional economy, and their part in the whole Chilean regional unit must be known. The simple information is necessary first, and the need for that information is fully apparent only when the larger regional context is understood.

Regional analysis tries to show the particular use

people have made of the earth in a given region, and the patterns and effects of spatial relations within and between each region and the rest of the world. Such analysis usually centers on: (1) population and its distribution, density, and growth; (2) the nature of the settlements; (3) the agricultural system; (4) the use of other resources; (5) the pattern and nature of manufacturing, trade, and transportation or circulation; and (6) the nature and distribution of the regional culture as a whole. Consideration is also given to historical factors and changes, and to other significant factors having spatial form or spatial distribution. Regional analysts must depend on the work of particular specialists in many cases. Climate is important to them, for example, but they may turn to climatological specialists for the detailed facts and processes. They will also turn to the work of soils scientists, agricultural economists, anthropologists, or historians. Geographers must fit these varied data together within the regional framework it is their business to provide. The regional whole must be understood as the context in which all of its parts are to be examined, since all of its parts are interrelated. A specialist's approach to agriculture, for example, might make sense of the Russian agricultural system that a more general regional approach might miss. But it also might be misinterpreted because the specialist did not adequately evaluate the influence of politics, tradition, or social organization on agriculture in this particular regional situation. Each region must be treated as a single areal unit with an integrated culture.

This kind of regional understanding is the basis for regional comparisons, whether of a single aspect of the region or the whole complex. Comparison is obviously necessary in making regional distinctions. Historical comparison is necessary, too, for any region is a dynamic entity whose nature is continually changing.

One cannot speak geographically without speaking of places, and one cannot speak effectively of places without considering the regional concept and the patterns of areal relationships. Special analysis of industrial resources, transportation, agricultural types, settlements, or political factors is not only incomplete but liable to serious error until it can be applied within a variety of regional frameworks and regional contexts. This means more than superimposing data or ideas on a landscape, or extracting them from it. The regional whole must be understood. Regional models must be constructed against which any theory or study or discovery with regional relevance can be measured.

The progress of knowledge is not served by insulating the boundaries of convenience that separate its various branches. No student of society is self-sufficient, and all of the branches of study are interdependent. Within geography, any sound analysis must make thorough use of both the topical and the regional approaches as well as contributions from other disciplines. But it is the nature of geography that at the heart of all its work is the concept of the region.

Questions for further study and discussion

1. Find several other examples of regional cores. How do they differ from each other? Can you find regions, or political states, without clearly defined cores? With two or more cores of approximately equal prominence? With one dominant core and other subsidiary ones? What conclusions do you draw from these examples?

2. Apply a three-level regional division to your own state.

3. How would you describe and outline the core area or areas of the United States? To what extent are there subsidiary or evenly balanced separate cores? How would you apply the same analysis to Germany, and to Japan? Why is a single core less dominant in Germany and Japan than in Britain or France?

4. Select and discuss two examples of settlements whose present status must be explained primarily in terms of historical or dynamic rather than static factors.

World maps

Maps showing world patterns of distribution are essential for geographic study of many problems, including individual regions as well as particular topics of multi-regional relevance. The patterns that follow have been selected for their importance and usefulness in geographic study, but they are only a selection, and in no sense intended as a substitute for an atlas.

Each of the maps necessarily deals in gross generalizations. Each shows major world patterns according to broad arbitrary categories rather than accurate local details since it is not possible to show both on any one map in most cases. It is also very difficult to show gradations or transitions on small-scale maps of the world. On Maps 4 through 9 the various regional patterns are separated by sharp lines because it makes multicolor printing easier. The reader must remember, however, that on an actual landscape, zones of gradation and transition separate regional patterns of nearly every kind. Maps 10 and 11 focus on the critical problem of energy in its present, most commonly used forms. Here too the generalizations are broad. But note how the United States stands out as both a producer and a consumer of energy (especially natural gas), despite its small share of the world's population.

Maps 12 and 13 show the distribution and production of the most important mineral fuels and metals, as currently known. Deposits too small to be economically significant are not shown.

There is an inevitable time lag in publishing reliable statistics dealing with the world as a whole. Such statistics are compiled by a great variety of organizations, so it is rarely possible to include in any book accurate or representative figures that are less than three or four years old, except on a few specific matters from a single country. These maps are designed to show world patterns; the data on which they are based must be broadly representative of the world as a whole. There is a greater time lag for statistics from some areas than from others, and a complete series of figures no more than five years old represents a considerable achievement. It is also much more valuable, and much less misleading, to use averages of three or four consecutive years than to use only the figures for the latest year available, which is often seriously unrepresentative. Maps 10 through 13 and the bar graphs accompanying each map are based on the latest available reliable and representative data for the world as a whole.

Maps 14 through 19 are basic locational and physiographic maps of each continent. Their full page size and the use of color greatly enhance their usefulness. They should be used by the student in connection with the relevant chapters.

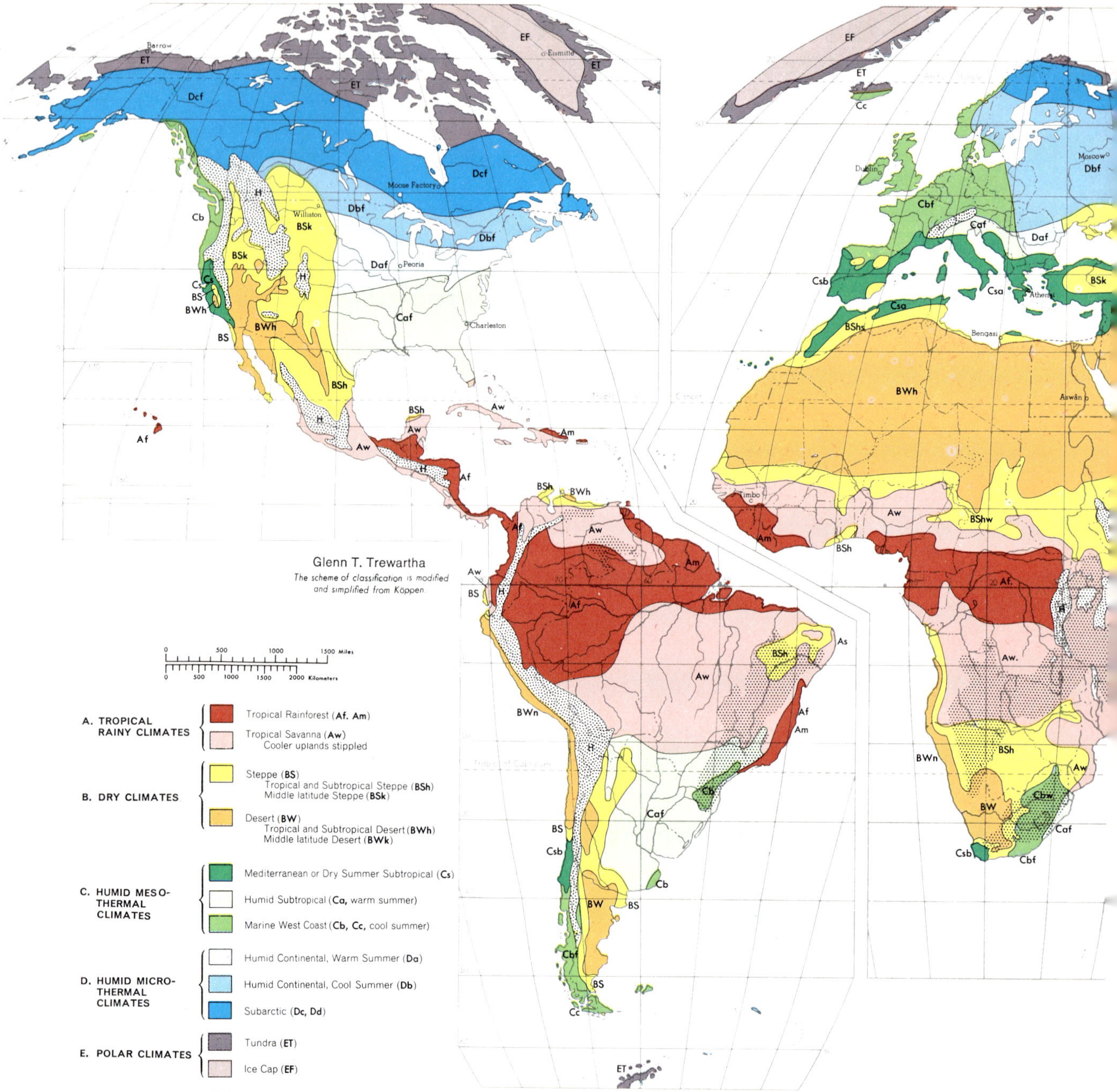

Glenn T. Trewartha

The scheme of classification is modified and simplified from Köppen

0 500 1000 1500 Miles

0 500 1000 1500 2000 Kilometers

A. TROPICAL RAINY CLIMATES
- Tropical Rainforest (**Af**, **Am**)
- Tropical Savanna (**Aw**) Cooler uplands stippled

B. DRY CLIMATES
- Steppe (**BS**)
  - Tropical and Subtropical Steppe (**BSh**)
  - Middle latitude Steppe (**BSk**)
- Desert (**BW**)
  - Tropical and Subtropical Desert (**BWh**)
  - Middle latitude Desert (**BWk**)

C. HUMID MESO-THERMAL CLIMATES
- Mediterranean or Dry Summer Subtropical (**Cs**)
- Humid Subtropical (**Ca**, warm summer)
- Marine West Coast (**Cb**, **Cc**, cool summer)

D. HUMID MICRO-THERMAL CLIMATES
- Humid Continental, Warm Summer (**Da**)
- Humid Continental, Cool Summer (**Db**)
- Subarctic (**Dc**, **Dd**)

E. POLAR CLIMATES
- Tundra (**ET**)
- Ice Cap (**EF**)

H. UNDIFFERENTIATED HIGHLANDS

EXTENSIVE UPLANDS

The various alphabetical formulas designating climates on the map are explained on the opposite page. Each formula constitutes a short description of the chief characteristics of a climate.

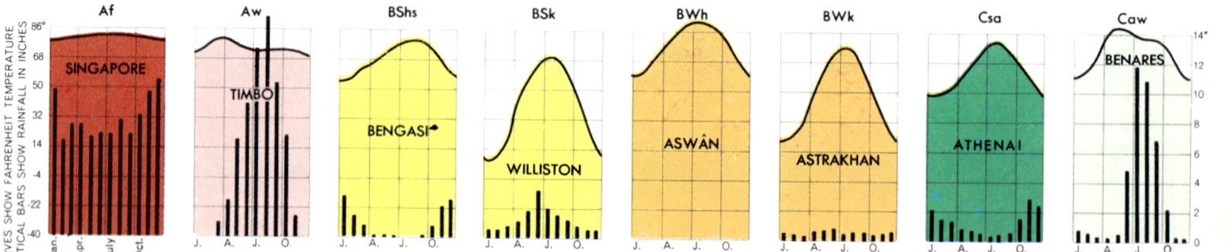

Tropical rain-forest climate | Tropical savanna climate; with wet and dry seasons | Tropical and sub-tropical steppe climate | Middle latitude steppe climate | Tropical and sub-tropical desert climate | Middle latitude desert climate | Mild climate; summer drouth and winter rain | Subtropical climate; winter drouth and summer rain

Reprinted by permission "Elements of Physical Geography" Copyrighted 1957 by Glenn T. Trewartha. Published by the McGraw-Hill Book Company, Inc.
B-510000-66 -2-2-4 D

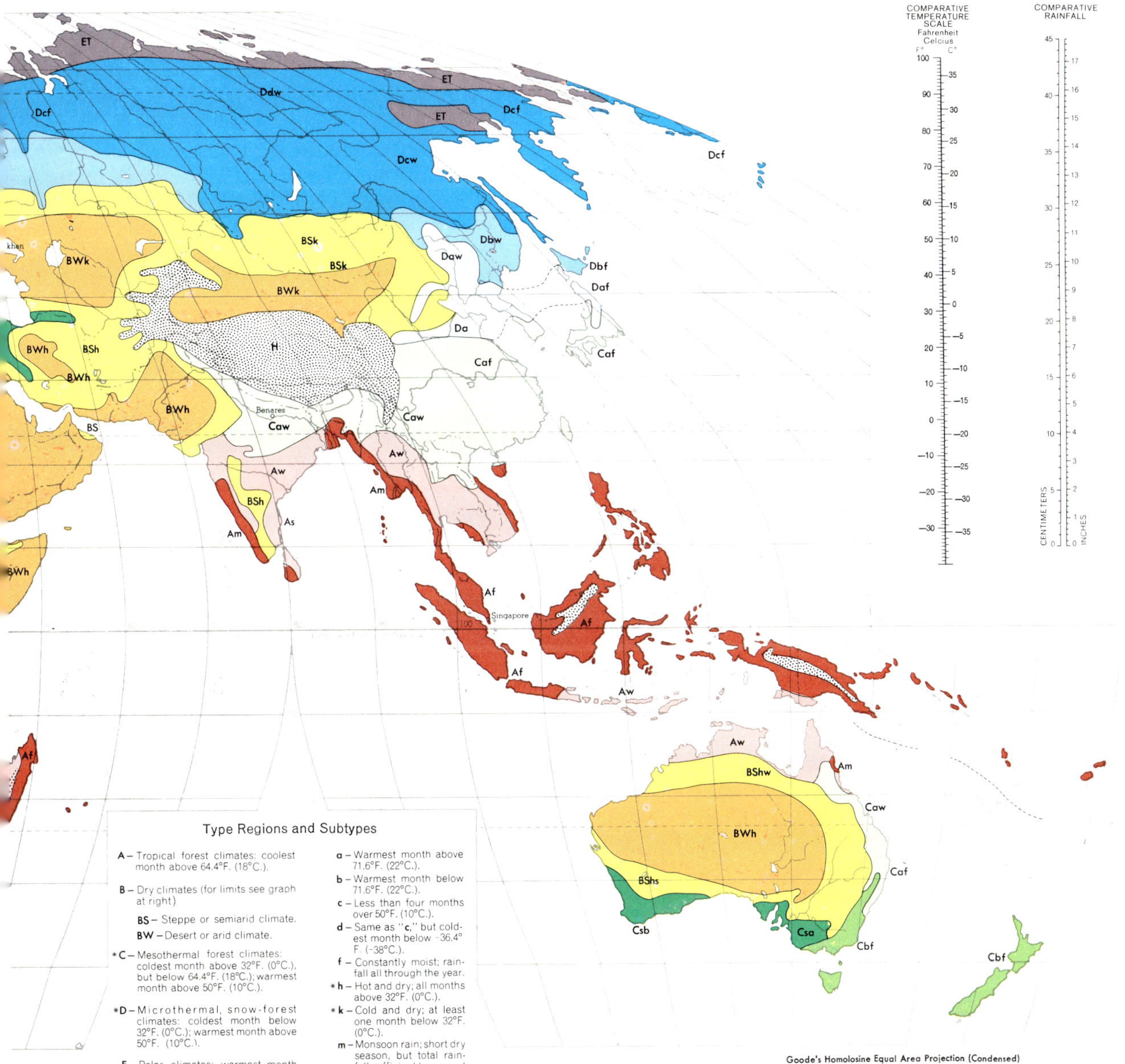

Type Regions and Subtypes

A – Tropical forest climates: coolest month above 64.4°F. (18°C.).

B – Dry climates (for limits see graph at right)

BS – Steppe or semiarid climate.

BW – Desert or arid climate.

*C – Mesothermal forest climates: coldest month above 32°F. (0°C.), but below 64.4°F. (18°C.); warmest month above 50°F. (10°C.).

*D – Microthermal, snow-forest climates: coldest month below 32°F. (0°C.); warmest month above 50°F. (10°C.).

E – Polar climates: warmest month below 50°F. (10°C.).

ET – Tundra climate: warmest month below 50°F. (10°C.) but above 32°F. (0°C.).

EF – Perpetual frost: all months below 32°F. (0°C.).

a – Warmest month above 71.6°F. (22°C.).

b – Warmest month below 71.6°F. (22°C.).

c – Less than four months over 50°F. (10°C.).

d – Same as "c," but coldest month below -36.4° F. (-38°C.).

f – Constantly moist; rainfall all through the year.

*h – Hot and dry; all months above 32°F. (0°C.).

*k – Cold and dry; at least one month below 32°F. (0°C.).

m – Monsoon rain; short dry season, but total rainfall sufficient to support rainforest.

n – Frequent fog.

n' – Infrequent fog, but high humidity and low rainfall.

s – Dry season in summer

w – Dry season in winter.

** Modification of Köppen definition*

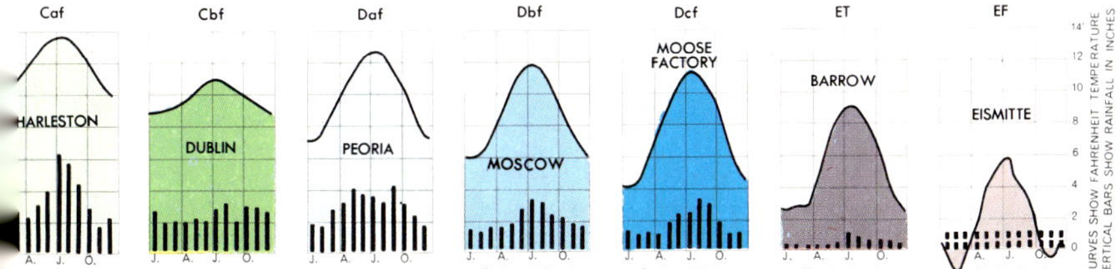

Moderate continental forest climate; mild winters

Moderate marine forest climate; mild winters

Continental forest climate; warm summer

Continental forest climate; cool summer

Continental taiga climate; very severe winters

Tundra climate

Glacial climate (Data incomplete)

Limits of the Regions of Dry Climate

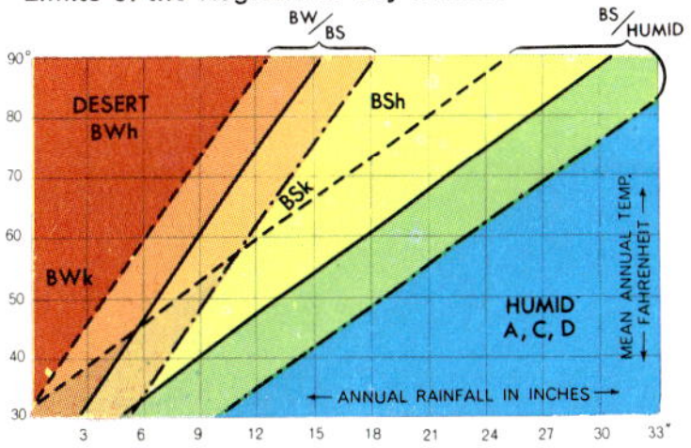

– – – – Winter concentration of precipitation
——— Precipitation evenly distributed throughout the year
– · – · – Summer concentration of precipitation

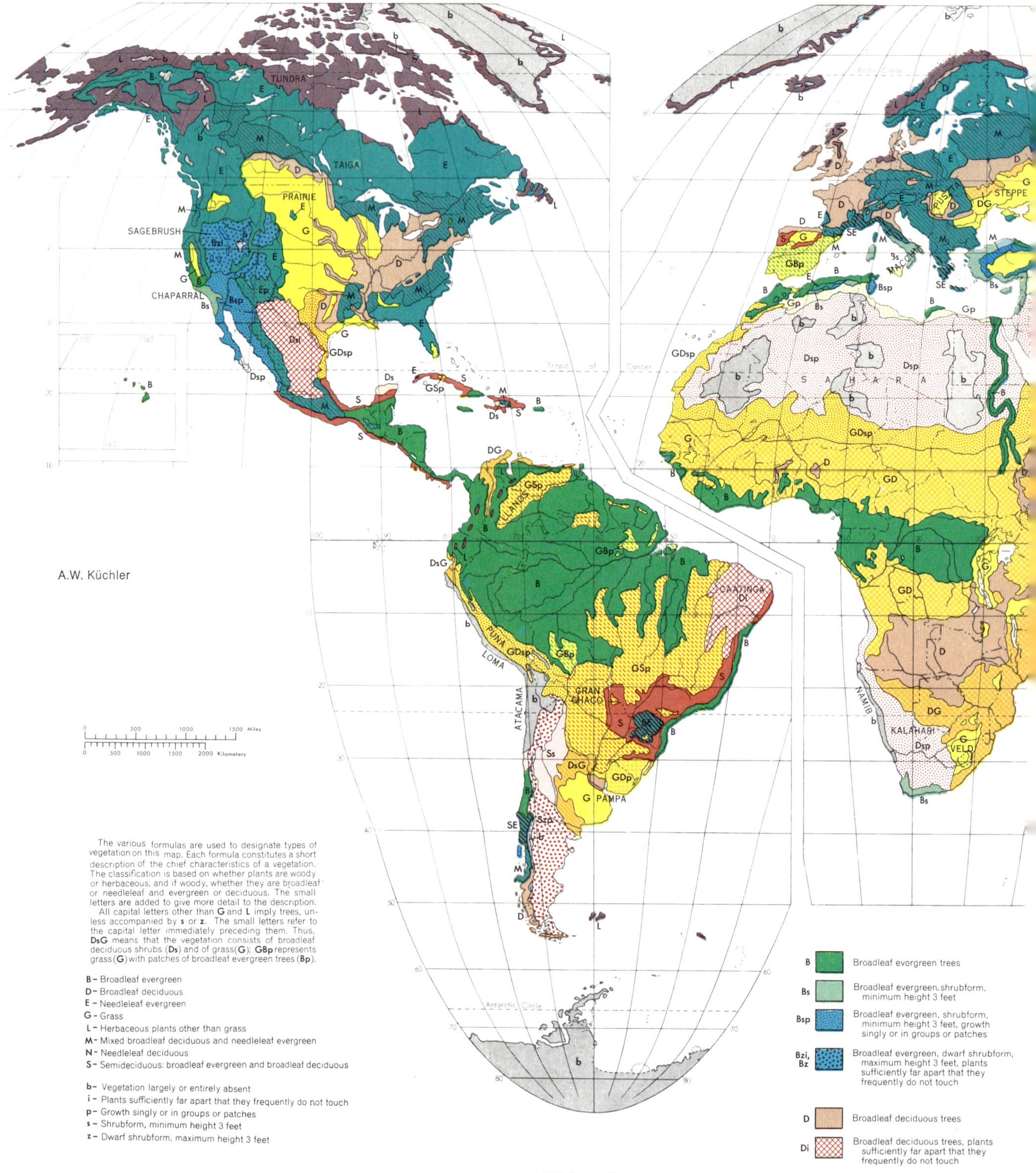

A.W. Küchler

The various formulas are used to designate types of vegetation on this map. Each formula constitutes a short description of the chief characteristics of a vegetation. The classification is based on whether plants are woody or herbaceous, and if woody, whether they are broadleaf or needleleaf and evergreen or deciduous. The small letters are added to give more detail to the description. All capital letters other than **G** and **L** imply trees, unless accompanied by **s** or **z**. The small letters refer to the capital letter immediately preceding them. Thus, **DsG** means that the vegetation consists of broadleaf deciduous shrubs (**Ds**) and of grass (**G**); **GBp** represents grass (**G**) with patches of broadleaf evergreen trees (**Bp**).

- **B** – Broadleaf evergreen
- **D** – Broadleaf deciduous
- **E** – Needleleaf evergreen
- **G** – Grass
- **L** – Herbaceous plants other than grass
- **M** – Mixed broadleaf deciduous and needleleaf evergreen
- **N** – Needleleaf deciduous
- **S** – Semideciduous: broadleaf evergreen and broadleaf deciduous

- **b** – Vegetation largely or entirely absent
- **i** – Plants sufficiently far apart that they frequently do not touch
- **p** – Growth singly or in groups or patches
- **s** – Shrubform, minimum height 3 feet
- **z** – Dwarf shrubform, maximum height 3 feet

B-510000-86-3 3-3.5 D

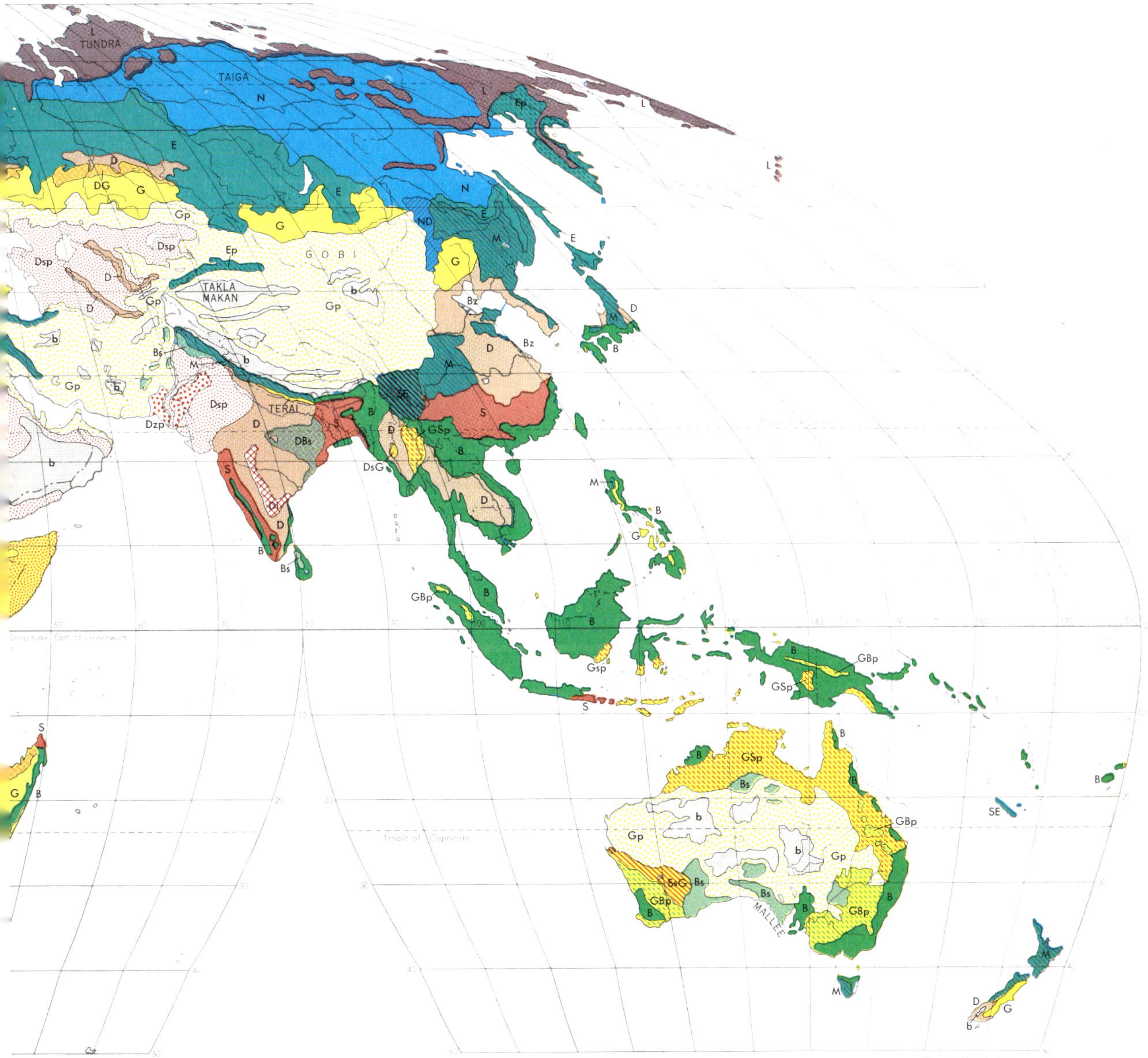

Goode's Homolosine Equal Area Projection (Condensed)

- Broadleaf deciduous, shrubform, minimum height 3 feet
- Broadleaf deciduous, shrubform, minimum height 3 feet, plants sufficiently far apart that they frequently do not touch
- Broadleaf deciduous, shrubform, minimum height 3 feet, growth singly or in groups or patches
- Broadleaf deciduous, dwarf shrubform, maximum height 3 feet, growth singly or in groups or patches
- Broadleaf deciduous, shrubform, minimum height 3 feet; Grass and other herbaceous plants
- Broadleaf deciduous trees; Grass and other herbaceous plants
- Broadleaf deciduous trees; Broadleaf evergreen, shrubform, minimum height 3 feet
- E — Needleleaf evergreen trees
- Ep — Needleleaf evergreen trees, growth singly or in groups or patches
- G — Grass and other herbaceous plants
- Gp — Grass and other herbaceous plants, growth singly or in groups or patches
- GBp — Grass and other herbaceous plants; Broadleaf evergreen trees, growth singly or in groups or patches
- GD — Grass and other herbaceous plants; Broadleaf deciduous trees
- GDp — Grass and other herbaceous plants; Broadleaf deciduous trees, growth singly or in groups or patches
- GDsp — Grass and other herbaceous plants; Broadleaf deciduous, shrubform, minimum height 3 feet, growth singly or in groups or patches
- GSp — Grass and other herbaceous plants; Semideciduous: broadleaf evergreen and broadleaf deciduous trees, growth singly or in groups or patches
- L — Herbaceous plants other than grass
- M — Mixed: broadleaf deciduous and needleleaf evergreen trees
- N — Needleleaf deciduous trees
- ND — Needleleaf deciduous trees; Broadleaf deciduous trees
- S — Semideciduous: broadleaf evergreen and broadleaf deciduous trees
- Ss — Semideciduous: broadleaf evergreen and broadleaf deciduous, shrubform, minimum height 3 feet
- SsG — Semideciduous: broadleaf evergreen and broadleaf deciduous, shrubform, minimum height 3 feet; Grass and other herbaceous plants
- Szp — Semideciduous: broadleaf evergreen and broadleaf deciduous, dwarf shrubform; maximum height 3 feet, growth singly or in groups or patches
- SE — Semideciduous: broadleaf evergreen and broadleaf deciduous trees; Needleleaf evergreen trees
- b — Vegetation largely or entirely absent

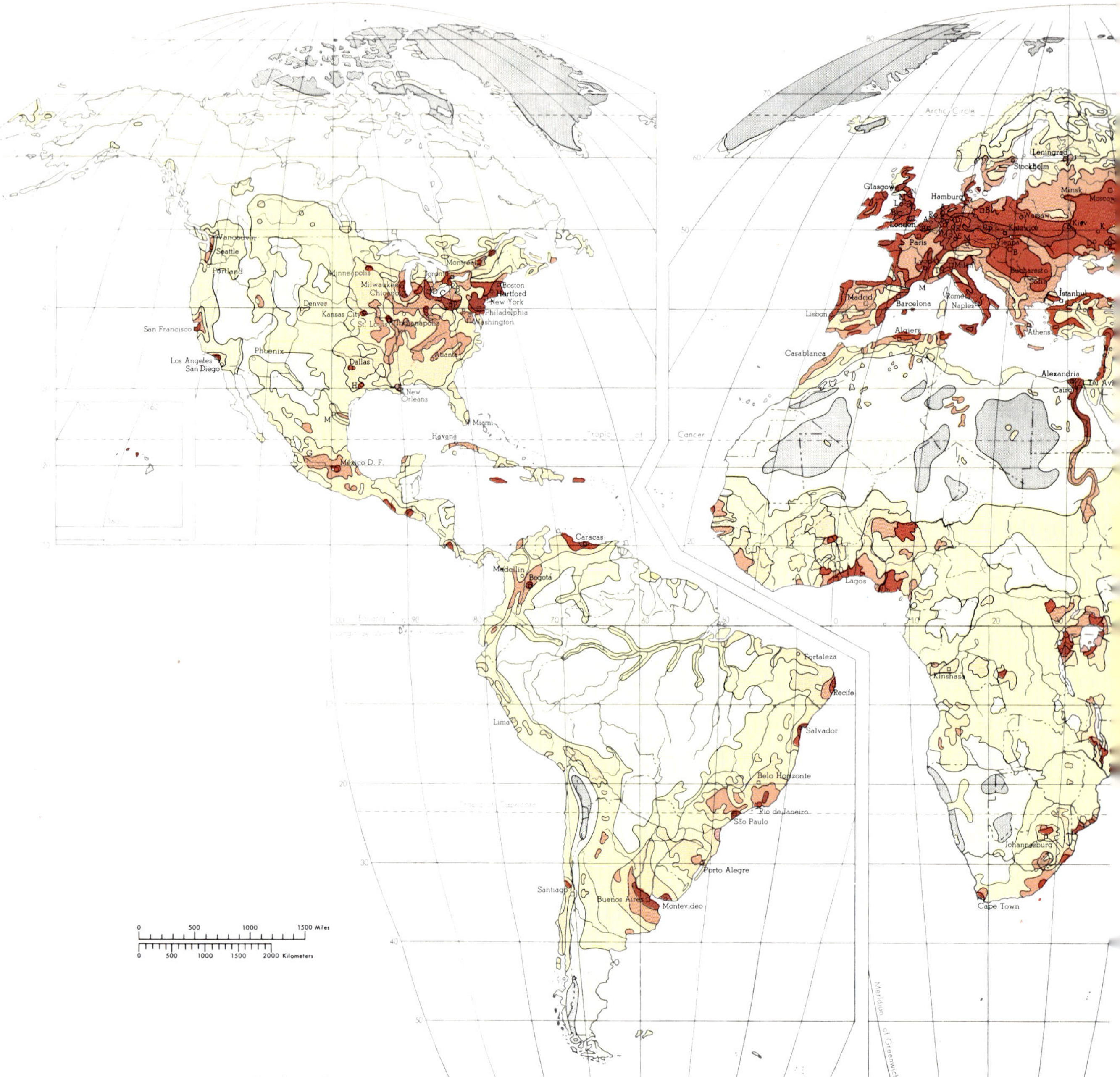

Arctic Circle
Tropic of Cancer
Tropic of Capricorn
Antarctic Circle
Meridian of Greenwich
Vancouver
Seattle
Portland
San Francisco
Los Angeles
San Diego
Phoenix
Denver
Minneapolis
Milwaukee
Chicago
Kansas City
St. Louis
Indianapolis
Dallas
Atlanta
New Orleans
Miami
Toronto
Montreal
Boston
Hartford
New York
Philadelphia
Washington
Havana
México D. F.
Caracas
Medellín
Bogotá
Lima
Fortaleza
Recife
Salvador
Belo Horizonte
Rio de Janeiro
São Paulo
Porto Alegre
Santiago
Buenos Aires
Montevideo
Leningrad
Stockholm
Glasgow
Hamburg
London
Paris
Minsk
Moscow
Warsaw
Kiev
Katowice
Vienna
Milan
Bucharest
Sofia
Madrid
Barcelona
Lisbon
Rome
Naples
Istanbul
Athens
Algiers
Casablanca
Alexandria
Cairo
Tel Aviv
Lagos
Kinshasa
Johannesburg
Cape Town
0 500 1000 1500 Miles
0 500 1000 1500 2000 Kilometers

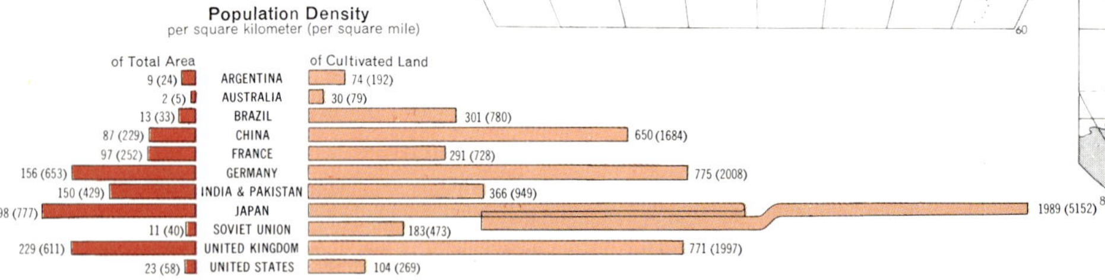

Population Density
per square kilometer (per square mile)
of Total Area
of Cultivated Land
9 (24) ARGENTINA 74 (192)
2 (5) AUSTRALIA 30 (79)
13 (33) BRAZIL 301 (780)
87 (229) CHINA 650 (1684)
97 (252) FRANCE 291 (728)
156 (653) GERMANY 775 (2008)
150 (429) INDIA & PAKISTAN 366 (949)
298 (777) JAPAN 1989 (5152)
11 (40) SOVIET UNION 183(473)
229 (611) UNITED KINGDOM 771 (1997)
23 (58) UNITED STATES 104 (269)

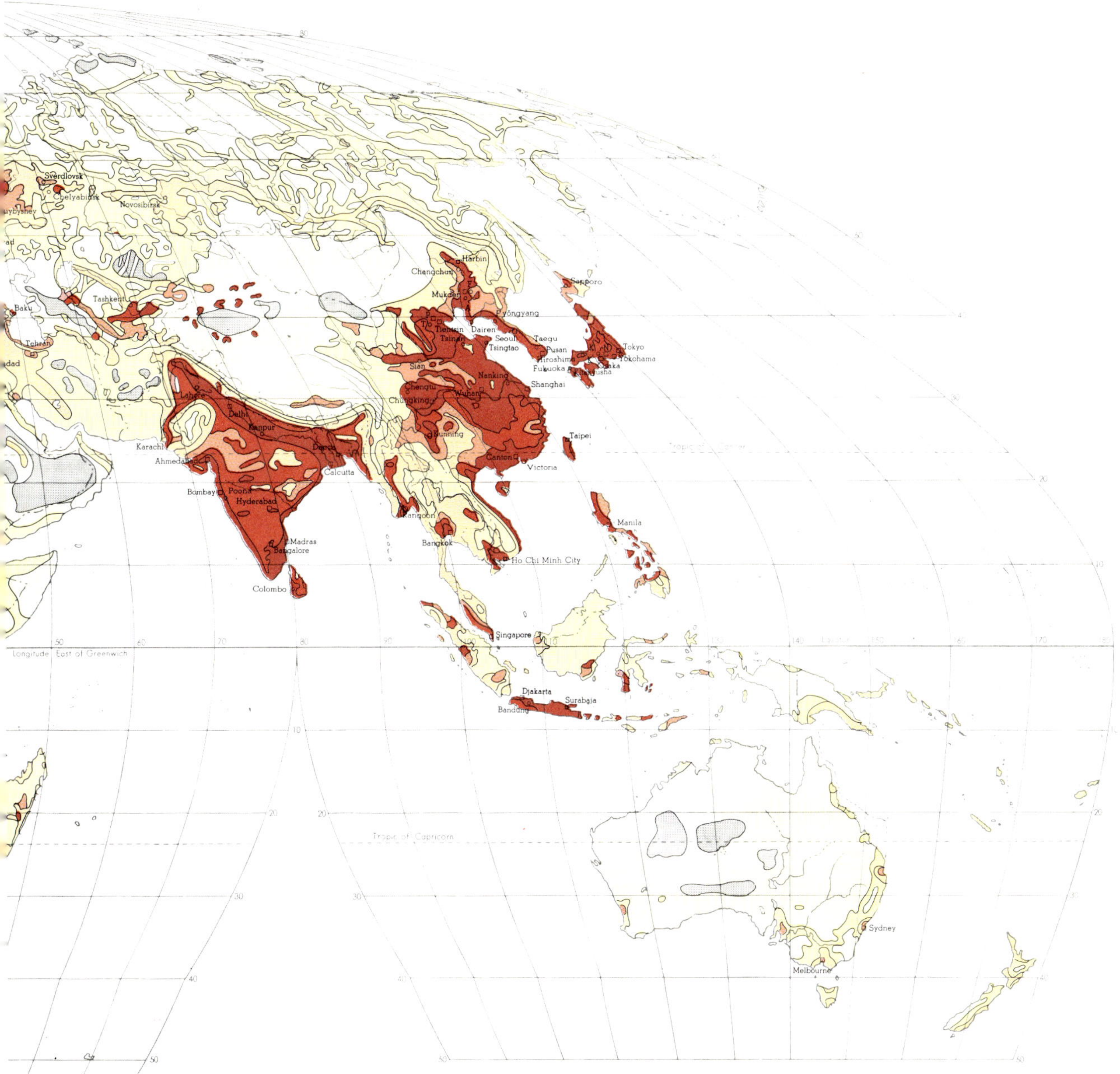

Goode's Homolosine Equal Area Projection (Condensed)

Per Sq. Km.	Per Sq. Mile
Uninhabited	Uninhabited
Under 1	Under 2
1-10	2-25
10-25	25-60
25-50	60-125
50-100	125-250
Over 100	Over 250

□ Metropolitan areas over 2,000,000 population
○ Metropolitan areas 1,000,000 to 2,000,000 population

Not all cities are named and some are identified by initial letter only.

Rural/Urban Population Ratios

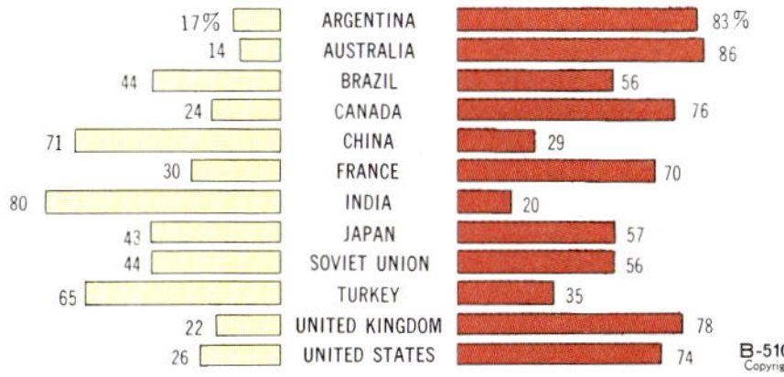

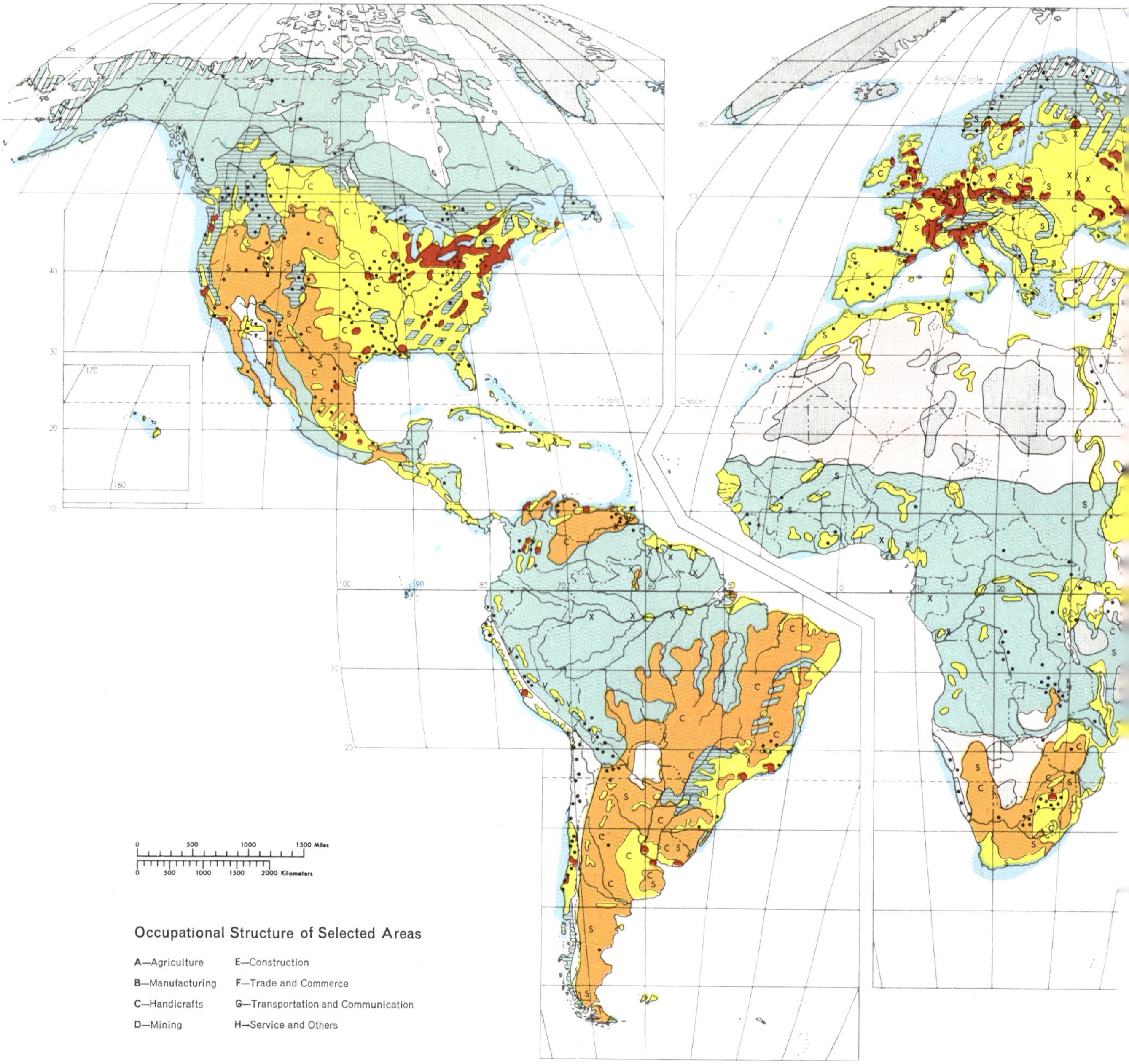

Occupational Structure of Selected Areas

A—Agriculture
B—Manufacturing
C—Handicrafts
D—Mining
E—Construction
F—Trade and Commerce
G—Transportation and Communication
H—Service and Others

UNITED KINGDOM
24,709,000 gainfully employed—1975

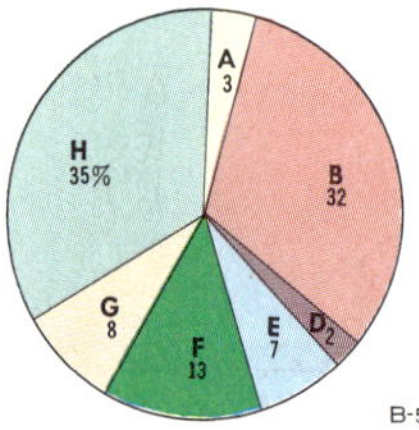

UNITED STATES
84,783,000 gainfully employed—1975

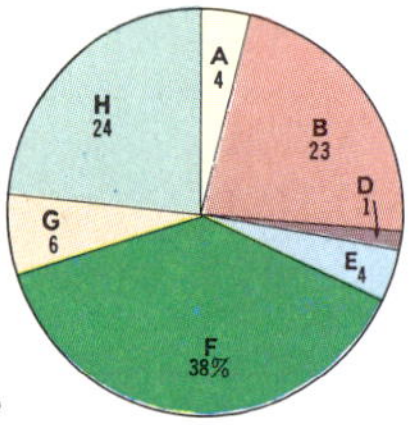

CANADA
9,364,000 gainfully employed—1975

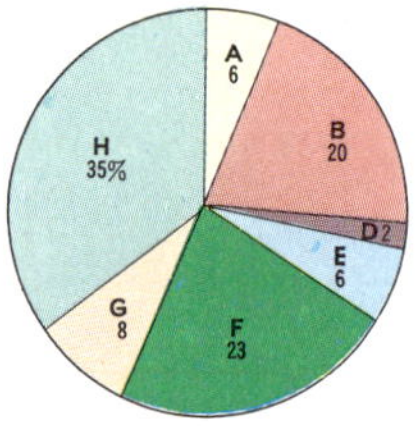

WESTERN EUROPE
76,414,300 gainfully employed—1975

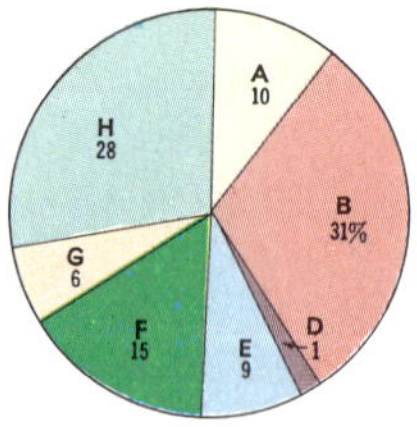

B-510000-36-4-A-3-4D

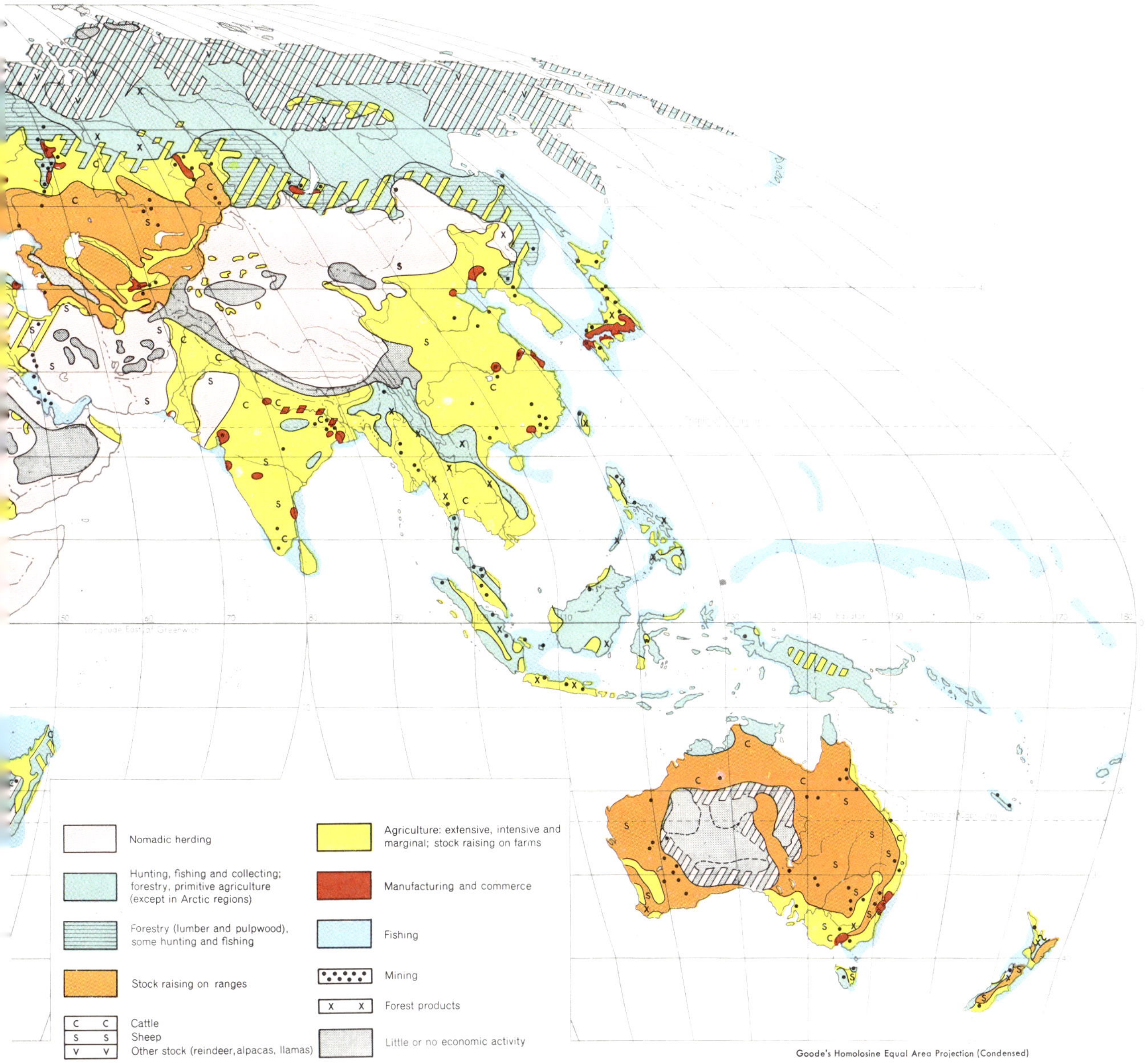

Goode's Homolosine Equal Area Projection (Condensed)

SOVIET UNION
,094,000 gainfully employed—1975

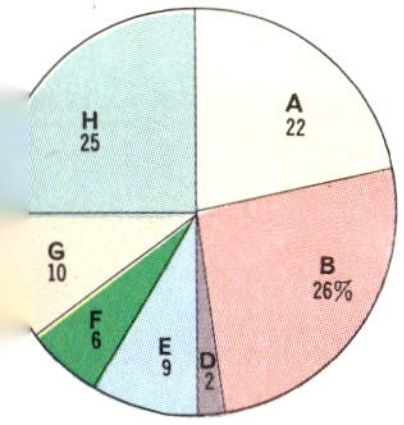

BRAZIL
29,545,400 gainfully employed—1970

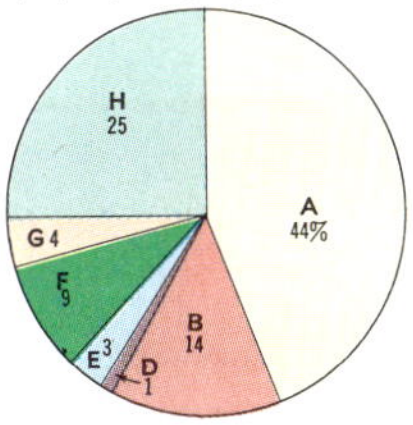

INDIA
180,373,000 gainfully employed—1973

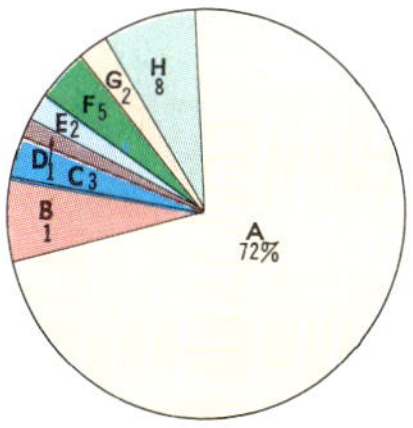

CHINA
270,000,000 gainfully employed—Est. 1970

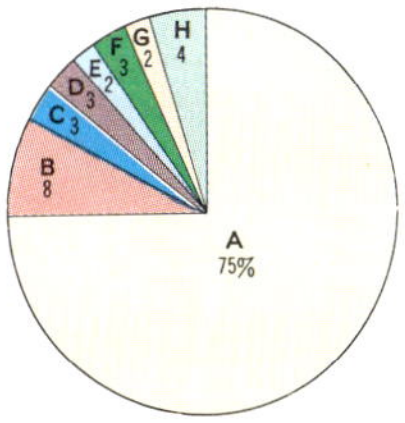

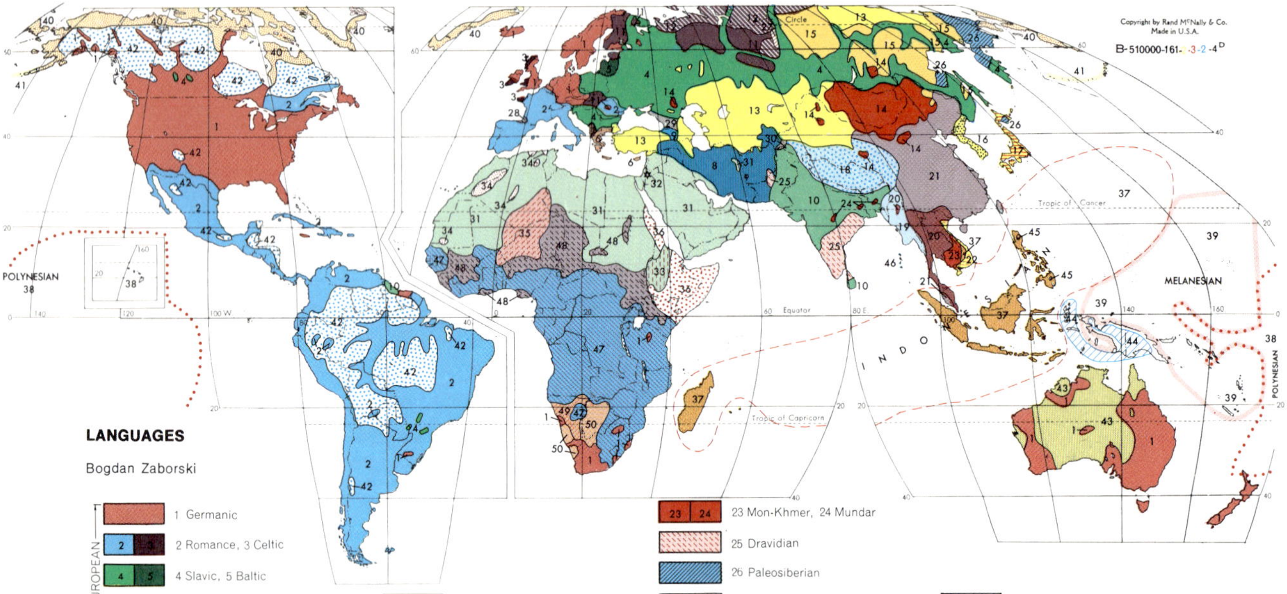

LANGUAGES

Bogdan Zaborski

INDO-EUROPEAN
- 1 Germanic
- 2 Romance, 3 Celtic
- 4 Slavic, 5 Baltic
- 6 Greek, 7 Albanian
- 8 Iranian, 9 Armenian
- 10 Indo-Aryan

- URALIAN: 11 Finno-Ugric, 12 Samoyed

ALTAIC
- 13 Turkic
- 14 Mongolic
- 15 Tungus-Manchu
- 16 Korean
- 17 Japanese

- 18 Tibetan, 19 Burmese
- 20 Thai, 21 Chinese
- 22 Vietnamese
- 23 Mon-Khmer, 24 Mundar
- 25 Dravidian
- 26 Paleosiberian
- 27 Ket
- ASIANITIC: 28 Basque, 29 Caucasian, 30 Burushaski
- SEMITIC: 31 Arabic, 32 Hebrew, 33 Amharic
- HAMITIC: 34 Berberic, 35 Tuareg, 36 Kyshitic
- AUSTRONESIAN: 37 Indonesian, 38 Polynesian, 39 Melanesian
- 40 Eskimo, 41 Aleut
- 42 American Indian
- 43 Australian Aborigine
- 44 Papua, 45 Negrito, 46 Andaman
- 47 Bantu
- 48 Language of intermediate zone between Bantu and Hamit-Semit
- 49 Hottentot, 50 Bushmen

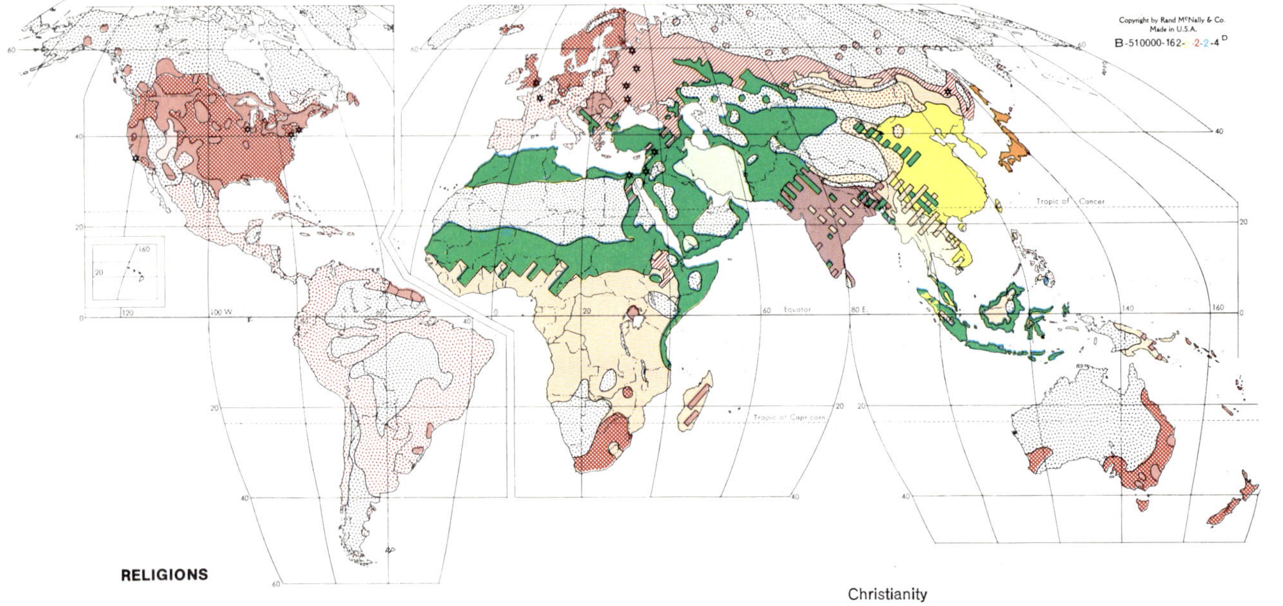

RELIGIONS

- Southern Buddhism
- Chinese Religion *(Confucianism, Taoism and Buddhism, superimposed and more or less fused)*
- Japanese Religion *(Shinto and Buddhism superimposed)*
- Islam: Sunni Moslems
- Islam: Shiah Moslems
- Lamaism *(Northern Buddhism)*
- Hinduism
- Tribal Religions
- Judaism *(Important minorities, chiefly in cities)*
- Religions Undifferentiated

Christianity

- Roman Catholic *(Western Rite)*
- Eastern Churches *(Orthodox, Armenian, Copt, Jacobite, Nestorian and Roman Catholic of Eastern Rites)*
- Protestantism
- Mormonism
- Christianity, sect not distinguished

After a map compiled by the Office of the Geographer, Department of State

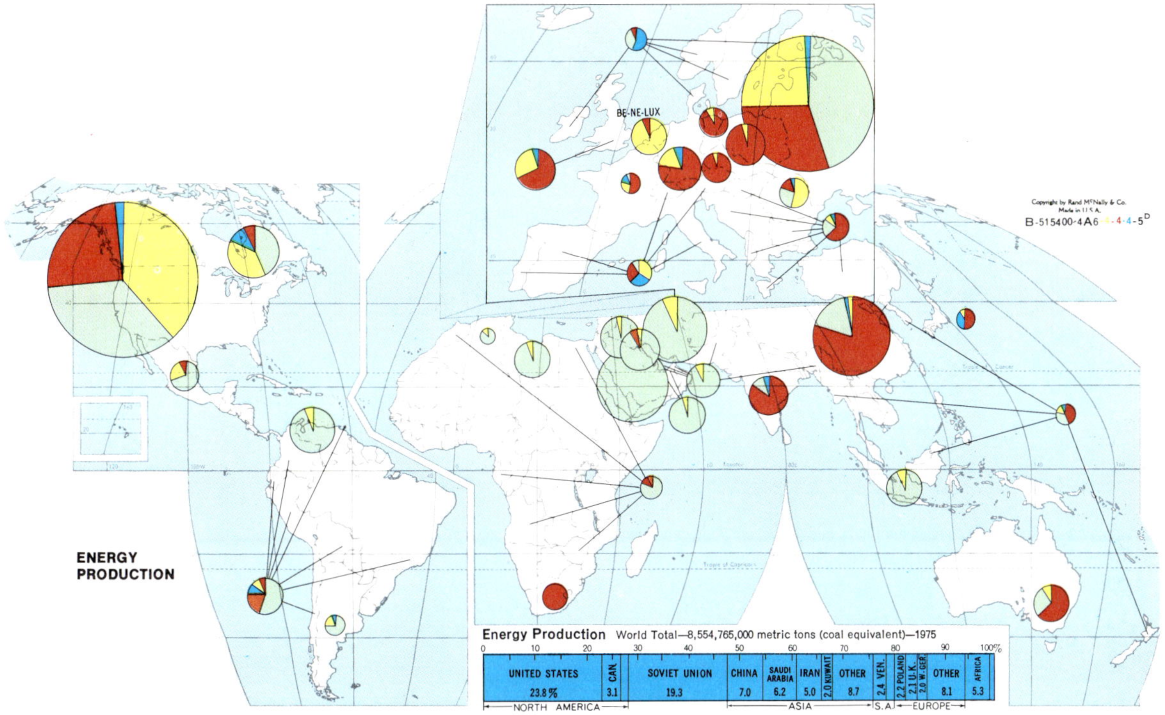
BE-NE-LUX
ENERGY PRODUCTION
Energy Production World Total—8,554,765,000 metric tons (coal equivalent)—1975
0 10 20 30 40 50 60 70 80 90 100%
UNITED STATES 23.8%
CAN. 3.1
SOVIET UNION 19.3
CHINA 7.0
SAUDI ARABIA 6.2
IRAN 5.0
2.0 KUWAIT
OTHER 8.7
2.4 VEN.
2.2 POLAND
2.1 U.K.
2.0 W. GER.
OTHER 8.1
AFRICA 5.3
NORTH AMERICA
ASIA
S.A.
EUROPE

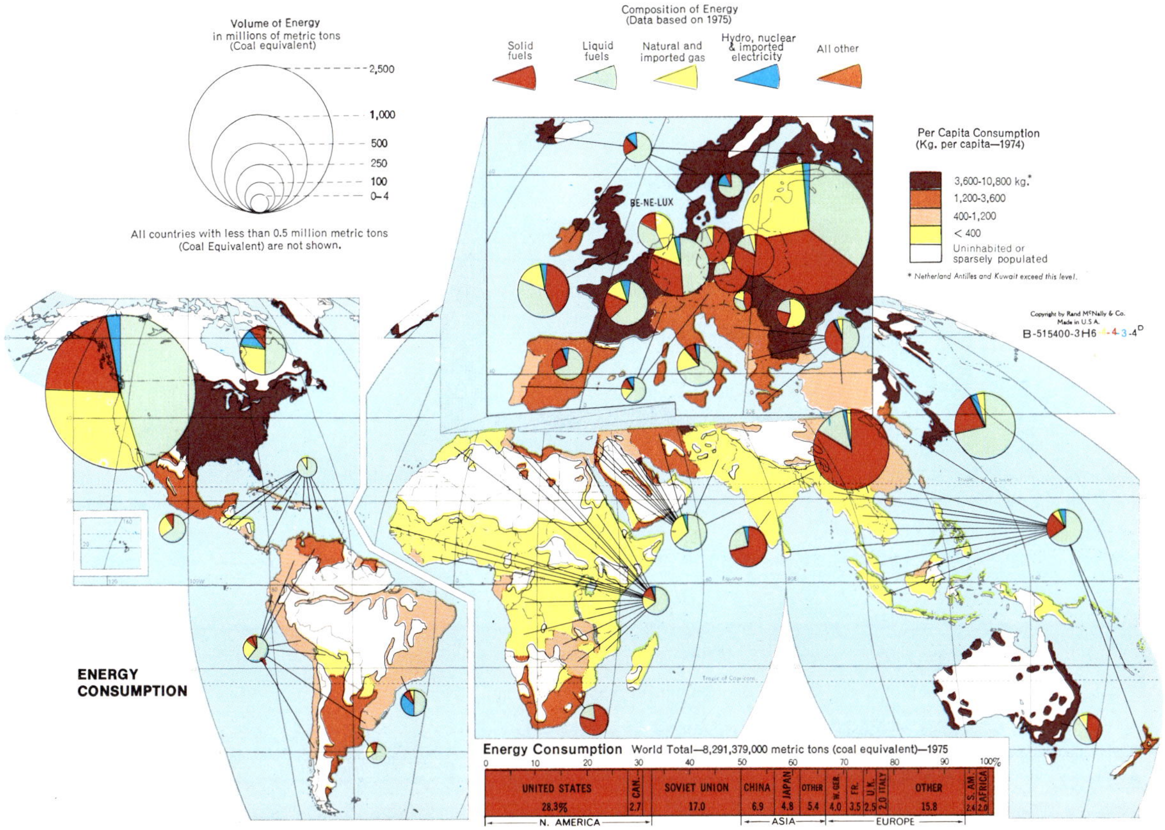
Volume of Energy
in millions of metric tons
(Coal equivalent)
2,500
1,000
500
250
100
0–4
All countries with less than 0.5 million metric tons
(Coal Equivalent) are not shown.
Composition of Energy
(Data based on 1975)
Solid fuels
Liquid fuels
Natural and imported gas
Hydro, nuclear & imported electricity
All other
BE-NE-LUX
Per Capita Consumption
(Kg. per capita—1974)
3,600-10,800 kg.*
1,200-3,600
400-1,200
< 400
Uninhabited or sparsely populated
* Netherland Antilles and Kuwait exceed this level.
ENERGY CONSUMPTION
Energy Consumption World Total—8,291,379,000 metric tons (coal equivalent)—1975
0 10 20 30 40 50 60 70 80 90 100%
UNITED STATES 28.3%
CAN. 2.7
SOVIET UNION 17.0
CHINA 6.9
JAPAN 4.8
OTHER 5.4
W. GER. 4.0
FR. 3.5
U.K. 2.5
2.0 ITALY
OTHER 15.8
S. AM. 2.4
AFRICA 2.0
N. AMERICA
ASIA
EUROPE

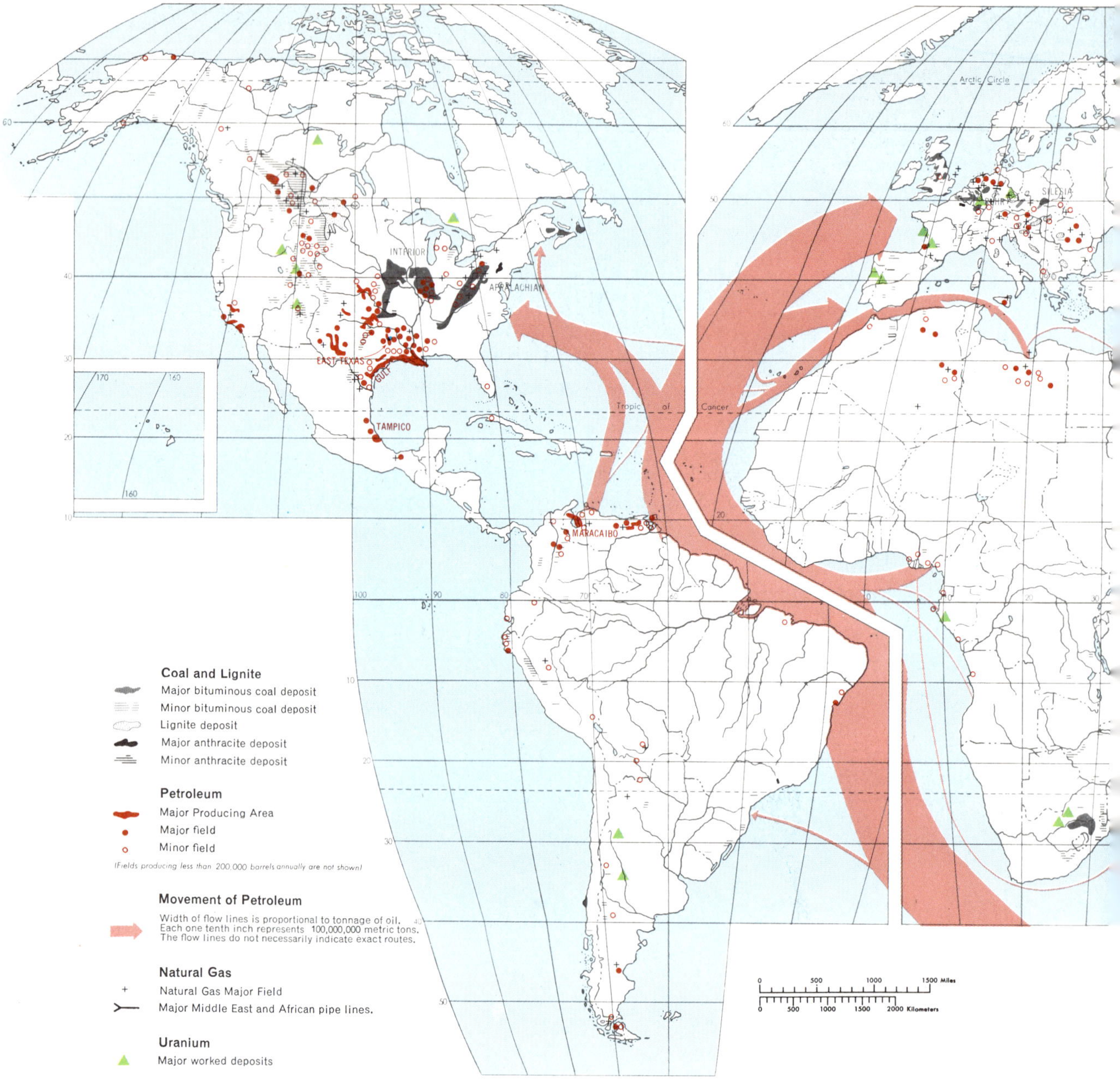

Coal Production* World Total—2,227,270,000 metric tons—1974

UNITED STATES	SOVIET UNION	CHINA	INDIA	OTHER	POLAND	UNITED KINGDOM	W. GERMANY	OTHER	S. AFRICA		OCEANIA
24.2%	21.3	19.3	3.7	3.6	7.3	4.9	4.5	3.8	2.9		2.7
N. AMERICA		ASIA	ASIA	ASIA	EUROPE	EUROPE	EUROPE	EUROPE	AFR.		

*Bituminous and Anthracite Lignite World Total—749,975,000 metric tons, 19% of which was produced in the Soviet Union.

Coal Reserves** World Total—1,417,048,000,000 metric tons—1975

UNITED STATES	CHINA	OTHER	SOVIET UNION	W. GER.	UNITED KINGDOM	POLAND	E. GER.	OTHER	AUSTRALIA		AFRICA
25.6%	21.1	2.8	19.3	7.1	7.0	2.7	2.1	3.5	5.2		2.1
N. AMERICA	ASIA	ASIA		EUROPE	EUROPE	EUROPE	EUROPE	EUROPE	OCE.		

Petroleum Production World Total—2,799,000,000 metric tons (20,578,000,000 barrels)—1976

SOVIET UNION	SAUDI ARABIA	IRAN	IRAQ	KUWAIT	U.A.E.	INDONESIA	CHINA	OTHER	UNITED STATES	VENEZUELA	CANADA	OTHER	NIG.	LIBYA	OTHER	EUROPE	
18.2%	14.7	10.5	3.7	3.0	2.4	2.8	2.7	3.8	14.4	4.1	2.3	3.5	3.6	3.4	3.1	2.3	
	ASIA	ASIA	ASIA	ASIA	ASIA	ASIA	ASIA	ASIA	W. HEMISPHERE	W. HEMISPHERE	W. HEMISPHERE	W. HEMISPHERE	AFRICA	AFRICA	AFRICA		

Petroleum Reserves World Total—75,530,000 metric tons (555,368,000,000 barrels)—Av. 1974

SAUDI ARABIA	KUWAIT	IRAN	IRAQ	U.A.E.	CHINA	INDO.	OTHER	SOVIET UNION	UNITED STATES	VEN.	OTHER	NIG.	OTHER	U.K.		
19.6%	13.7	12.3	6.2	4.6	2.8	2.1	3.2	8.7	6.1	2.8	3.2	3.5	6.7	2.2		
ASIA	ASIA	ASIA	ASIA	ASIA	ASIA	ASIA	ASIA		W.HEMIS.	W.HEMIS.	W.HEMIS.	AFRICA	AFRICA	EUR.		

**Including lignite and coke

B-510000-462-5-5-4 6D

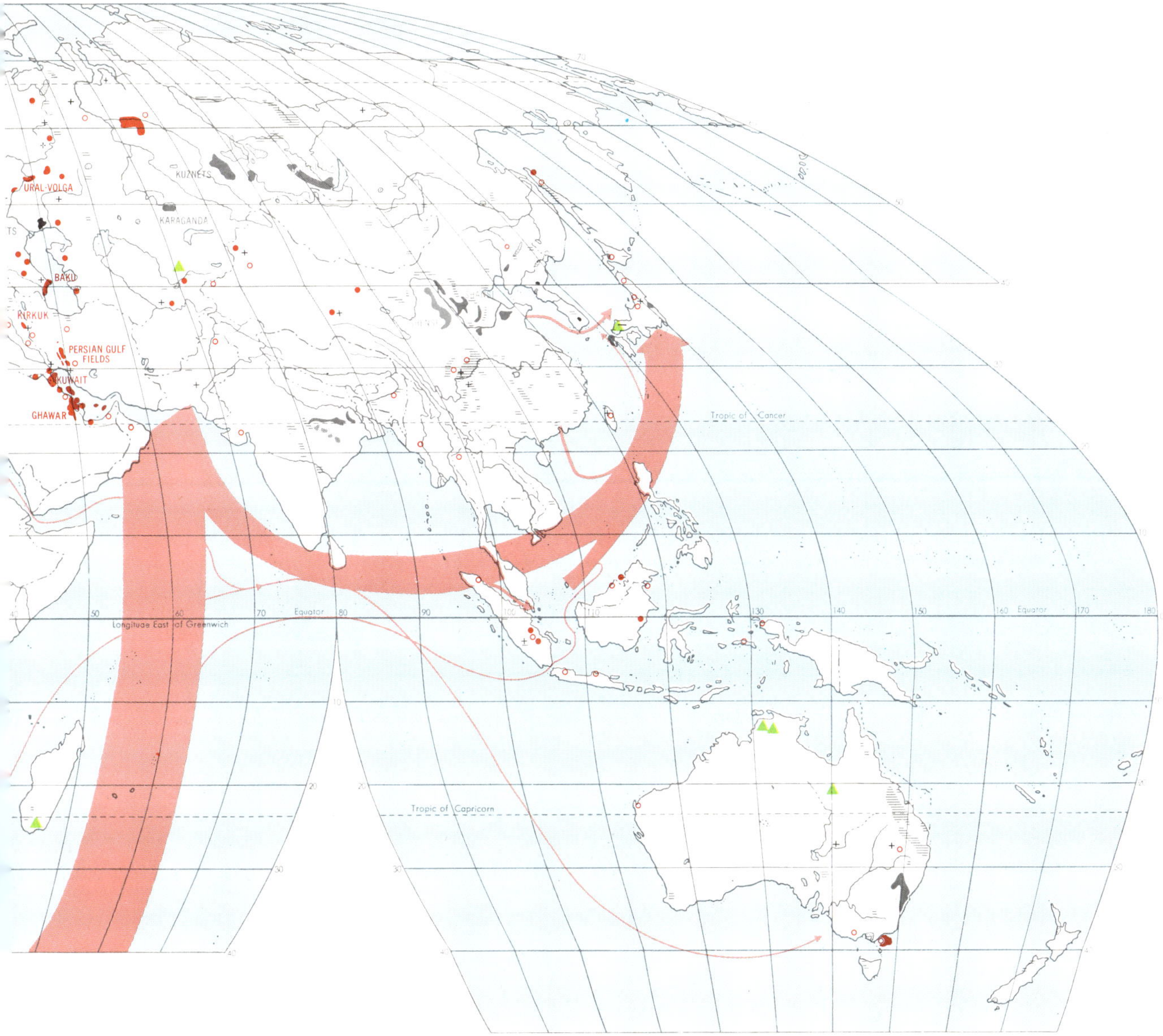

Goode's Homolosine Equal Area Projection (Condensed)

...tural Gas Production

World Total—1,423,371,000,000 cubic meters—1976

UNITED STATES	CAN.	OTHER	SOVIET UNION	NETH.	U.K.	OTHER	IRAN	CHINA	OTHER	AFRICA	ALL OTHER
39.7%	6.2	4.1	22.5	7.1	2.3	4.7	3.4	2.8	2.0	3.2	2.0
W. HEMISPHERE				EUROPE			ASIA				

...tural Gas Reserves

World Total—59,195,000,000,000 cubic meters—1974

SOVIET UNION	IRAN	SAUDI ARABIA	OTHER	UNITED STATES	CAN.	VEN.	OTHER	ALGERIA	NIG.	OTHER	NETH.	OTHER
33.5%	17.9	2.9	8.5	11.3	2.7	2.1	3.7	4.7	2.4	2.2	3.7	4.4
	ASIA			W. HEMIS.				AFRICA			EUROPE	

Uranium Production

World total— 27,800 metric tons—1976

UNITED STATES	CANADA	SOUTH AFRICA	NIGER	GABON	FRANCE	OTHER
44.0%	24.8	10.5	6.5	3.9	7.2	3.1
NORTH AMERICA		AFRICA			EUR.	

Uranium Reserves

World Total—1,080,000 metric tons—1975

UNITED STATES	CANADA	AUSTRALIA	SOUTH AFRICA	NIGER	ALGERIA	OTHER	FRANCE	OTHER	ALL OTHER
29.6%	13.3	22.5	17.2	3.7	2.6	2.8	3.4	2.0	2.9
N. AMERICA			AFRICA				EUR.		

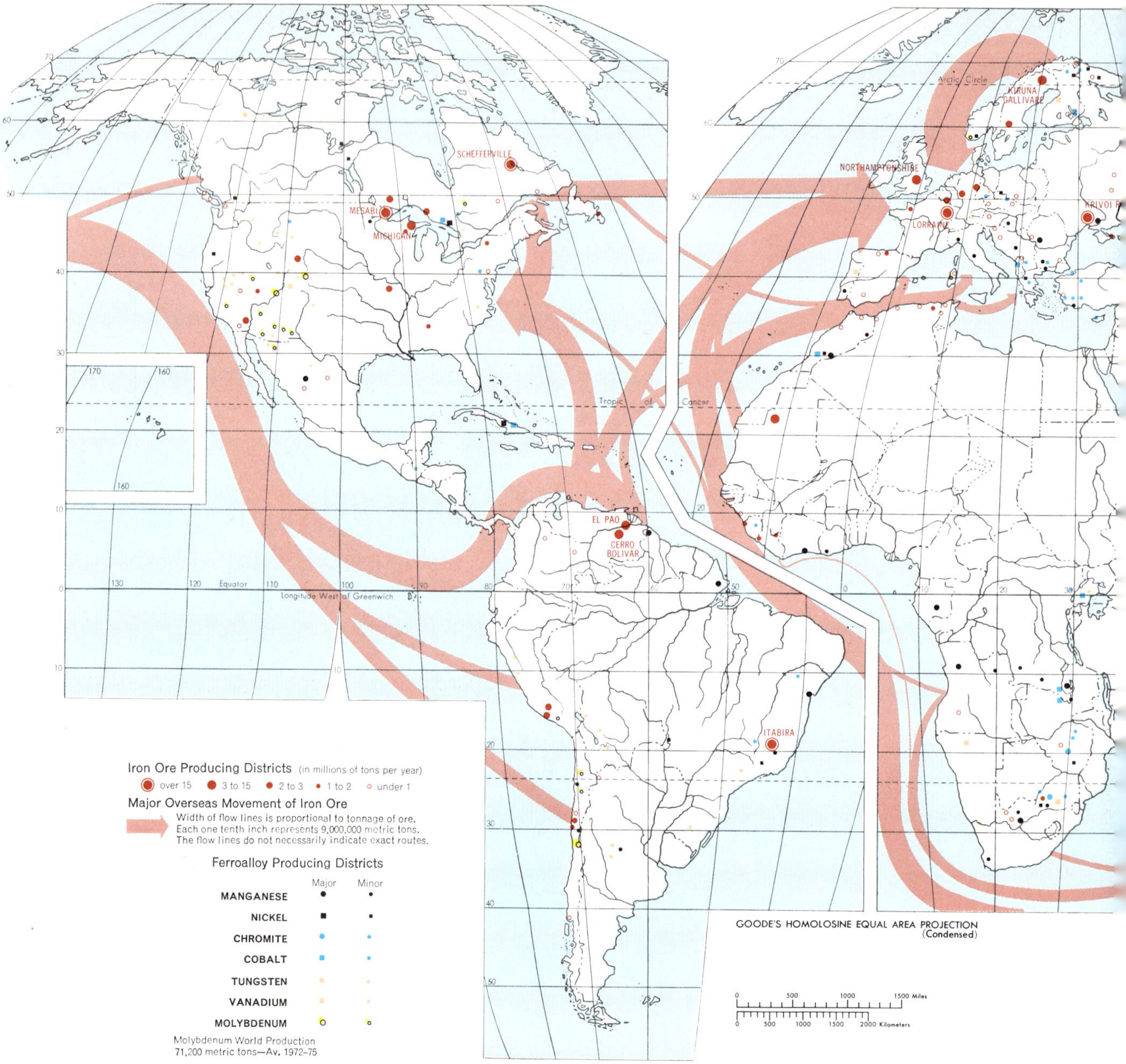

Manganese World Production—22,834,000 metric tons (metal content)—1974

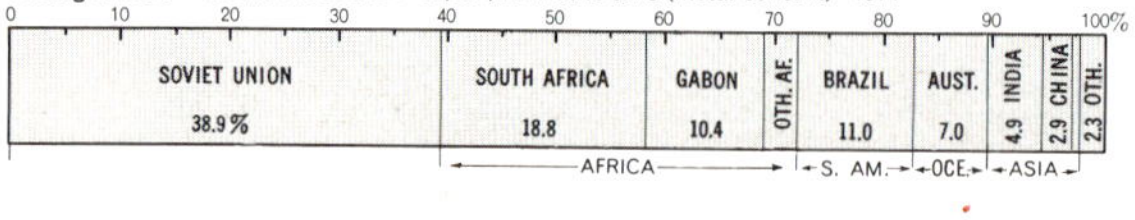

Tungsten World Production—36,200 metric tons (60% WO_3)—1974

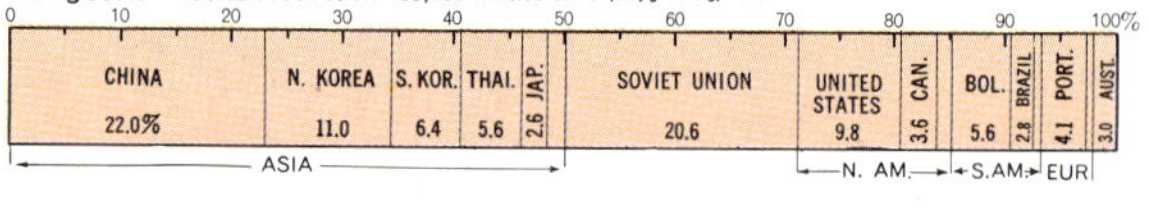

Nickel World Production—723,000 metric tons (metal content)—1975

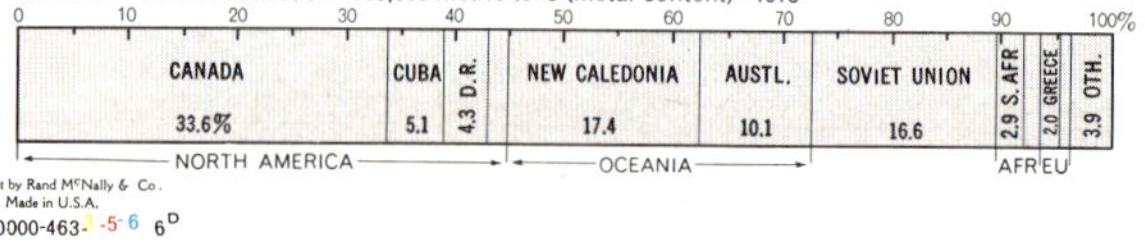

Vanadium World Production—19,700 metric tons—1974

B-510000-463- -5 6 6D

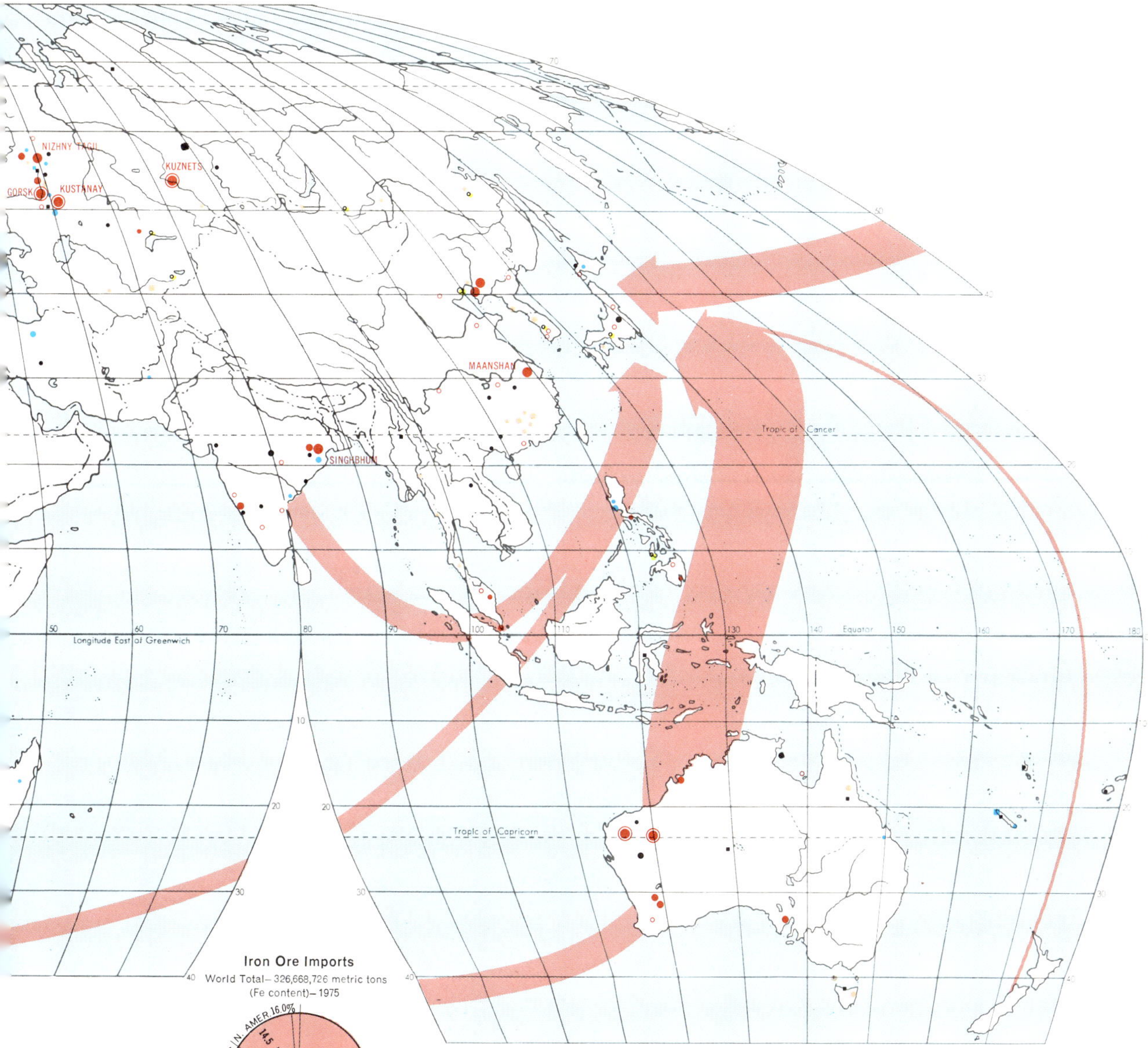

Iron Ore Imports

World Total—326,668,726 metric tons (Fe content)—1975

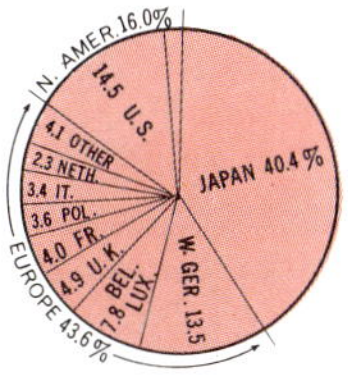

...me Ore World Production—7,197,000 metric tons (Cr_2O_3 content)—1974

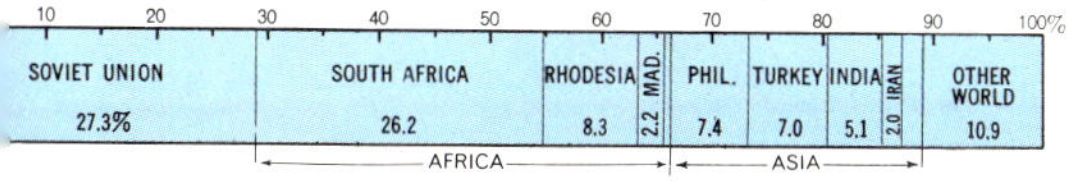

Iron Ore World Production—900,000,000 metric tons—1975

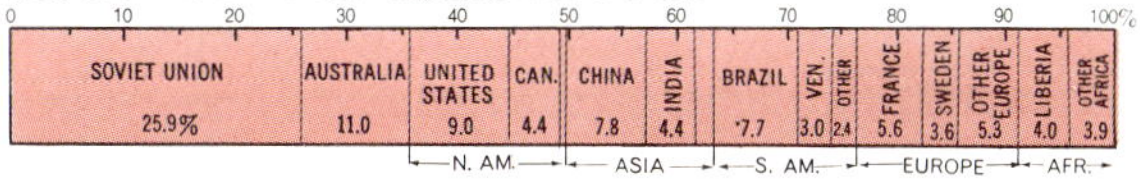

...lt World Production—30,500 metric tons (metal content)—1974

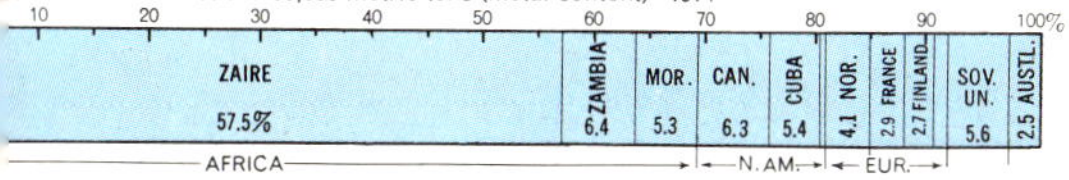

Iron Ore Reserves World Total—251,456,000,000 metric tons (Fe content)—Est. 1975

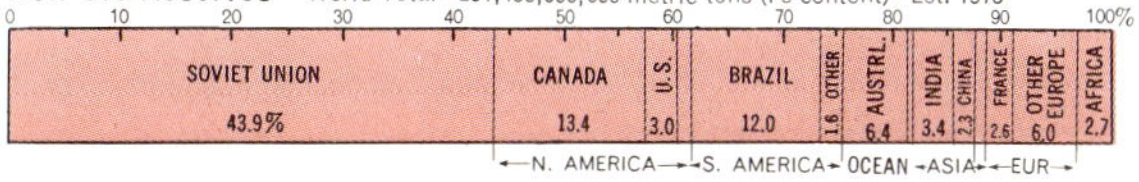

Conic Projection
Elevations and depressions are given in feet

UNION OF SOVIET SOCIALIST REPUBLICS
RUSSIAN SOVIET FEDERATED SOCIALIST REPUBLIC
KAZAKH S.S.R.
UZBEK S.S.R.
TURKMEN S.S.R.
KIRGHIZ STEPPE
TURKEY
IRAN (PERSIA)
SYRIA
IRAQ
LEBANON
CYPRUS
AFGHANISTAN
BLACK SEA
CASPIAN SEA
Surface—92 Ft. Below Sea Level
ARAL'SKOYE MORE (Aral Sea)
AZOVSKOYE MORE (Sea of Azov)
CASPIAN DEPRESSION
MANYCH DEPRESSION
PLATO UST'-URT
P-OV MANGYSHLAK
PESKI KYZYLKUM (DESERT)
PESKI KARAKUMY (DESERT)
DASHT-E-KAVIR (DESERT)
KAVIR-E-LUT (DESERT)
ELBURZ MTS.
ZAGROS MTS.
CAUCASUS MOUNTAINS
ASIA MINOR
KURDISTAN
URAL MTS.
MOSCOW (Moskva)
Arkhangelsk (Archangel)
Petrozavodsk
Vologda
Cherepovets
Kostroma
Yaroslavl'
Rybinsk
Kineshma
Ivanovo
Kovrov
Vladimir
Kalinin
Dzerzhinsk
GORKI
Noginsk
Kolomna
Serpukhov
Ryazan'
Kaluga
Tula
Novomoskovsk
Smolensk
Bryansk
Orël
Yelets
Lipetsk
Tambov
Michurinsk
Kursk
Voronezh
Chernigov
Nezhin
Sumy
KHAR'KOV
Poltava
DNEPROPETROVSK
DONETSK
Makeyevka
Gorlovka
Voroshilovgrad
Shakhty
Novocherkassk
Rostov-na-Donu
Taganrog
Zhdanov
Berd'ansk
Zaporozh'e
Nikopol'
Krivoy Rog
Nikolayev
Melitopol'
Kherson
Odessa
Simferopol'
Sevastopol'
Yalta
Kerch'
Novorossiysk
Krasnodar
Armavir
Pyatigorsk
Stavropol'
Maykop
Groznyy
Makhachkala
Derbent
Ordzhonikidze
Sochi
Sukhumi
Kutaisi
Poti
Batumi
TBILISI
Kirovabad
BAKU
Yerevan
Leninakan
Kars
VOLGOGRAD
Kamyshin
Saratov
Engel's
Vol'sk
Balashov
Borisoglebsk
Penza
Syzran'
Ul'yanovsk
KUYBYSHEV
Chapayevsk
Orenburg
Orsk
Aktyubinsk
Ural'sk
Gur'yev
Astrakhan
Kazan'
Cheboksary
Yoshkar-Ola
Kirov
Kotlas
Syktyvkar
Izhevsk
Votkinsk
Sarapul
Ufa
Sterlitamak
Magnitogorsk
Beloretsk
Zlatoust
CHELYABINSK
SVERDLOVSK
Perm'
Krasnokamsk
Lys'va
Chusovoy
Berezniki
Solikamsk
Serov
Nizh. Tagil
Kamensk-Ural'skiy
Kurgan
Tyumen'
Tobol'sk
Petropavlovsk
Omsk
NOVOSIBIRSK
Barnaul
Biysk
Novokuznetsk
Leninsk-Kuznetskiy
Slavgorod
Pavlodar
Semipalatinsk
Kokchetav
Tselinograd (Akmolinsk)
Karaganda
Temir-Tau
Balkhash
Kustanay
Troitsk
Turgay
Irgiz
Aral'sk
Kzyl-Orda
Dzhambul
Chimkent
TASHKENT
Samarkand
Bukhara
Chardzhou
Mary
Ashkhabad
Kizyl-Arvat
Krasnovodsk
Nebit-Dag
Urgench
Khiva
Tashauz
Kungrad
Chimbay
Turkestan
Novo-Kazalinsk
Baykonur
Herat
Mashhad
Gorgan
Babol
Rasht
Tabriz
TEHRAN
Kashan
Hamadan
Esfahan
Yazd
Kerman
Shiraz
Abadan
Khorramshahr
Basra
Baghdad
Babylon (Ruins)
Kirkuk
Al Mawsil
Nineveh (Ruins)
Persepolis (Ruins)
Zahedan
Farah
Aleppo
Homs
Palmyre
Beirut
Nicosia
Iskenderun
Adana
Mersin
Antalya
Konya
Gaziantep
Diyarbakir
Kayseri
Sivas
Erzurum
Trabzon (Trebizond)
Samsun
Sinop
Zonguldak
Ankara (Angora)
Afyonkarahisar
Euphrates
Tigris
Volga
Ural
Don
Ob'
Irtysh
Syr-Dar'ya
Amu-Dar'ya
C-519697-76-
COPYRIGHT BY RAND McNALLY & COMPANY MADE IN U.S.A.

Lambert's Azimuthal, Equal Area Projection
Elevations and depressions are given in feet

Lambert's Azimuthal, Equal Area Projection

Elevations and depressions are given in feet

Lambert's Azimuthal, Equal Area Projection
Elevations and depressions are given in feet

PAPUA NEW GUINEA
Mt. Albert Edward 13 100
Buna
Mt. Victoria 13 363
Port Moresby
OWEN STANLEY RA.
Torres Strait
SOUTH CAPE
Samarai
TROBRIAND IS.
WOODLARK
D'ENTRECASTEAUX ISLANDS
LOUISIADE ARCHIPELAGO
TAGULA
ROSSEL
CHOISEUL
VELLA LAVELLA
NEW GEORGIA
RENDOVA
SANTA ISABEL
MALAITA
FLORIDA
RUSSELL IS.
TULAGI
Honiara
GUADALCANAL
SOLOMON ISLANDS
(British)
SAN CRISTÓBAL
RENNELL
SANTA CRUZ ISLANDS
TORRES IS.
BANKS ISLANDS
ESPÍRITU SANTO
MAEWO
PENTECOST
MALEKULA
AMBRIM
EPI
NEW HEBRIDES
(British and French Condominium)
EFATE
Vila
ERROMANGA
TANA
ANEITYUM
CAPE YORK
CAPE YORK PENINSULA
OSPREY REEF
CAPE MELVILLE
CORAL SEA
Laura
Cooktown
Palmerville
ATHERTON
Cairns
Mungana
Croydon
Mt. Bartle Frere 5287
PLATEAU
Forsayth
Ingham
GREAT BARRIER REEF
HOLMES REEFS
WILLIS IS.
FLINDERS REEFS
LIHOU REEFS
TREGROSSE IS.
HINCHINBROOK
Townsville
MARION REEF
Charters Towers
Bowen
WHITSUNDAY
CUMBERLAND IS.
Richmond
Hughenden
GREGORY RANGE
GREAT DIVIDING RANGE
Mackay
Mt. Dalrymple 4190
Kynuna
Winton
NORTHUMBERLAND IS.
SWAIN REEFS
WRECK REEFS
ÎLES CHESTERFIELD (Fr.)
ÎLES BÉLEP
OUVÉA
LIFOU
NEW CALEDONIA (Fr.)
ÎLES LOYAUTÉ (French)
MARÉ
Nouméa
ÎLE DES PINS
Tropic of Capricorn
PACIFIC OCEAN
QUEENSLAND
GREAT ARTESIAN BASIN
Barcaldine
Clermont
Emerald
Longreach
Jericho
Rockhampton
Mount Morgan
CURTIS
Gladstone
Dingo
Blackall
Yaraka
Tambo
BUCKLAND TABLELAND
Windorah
Bundaberg
SANDY CAPE
FRASER
Maryborough
Gympie
Quilpie
Charleville
Roma
Thargomindah
Cunnamulla
St. George
DARLING DOWNS
Dalby
Toowoomba
Ipswich
Brisbane
N. STRADBROKE I.
Southport
Warwick
Dirranbandi
Hungerford
Mungindi
Lismore
Tenterfield
Moree
Brewarrina
Walgett
Glen Innes
NEW ENGLAND RANGE
Grafton
Bourke
Inverell
Narrabri
Armidale
The Round Mountain
Coonamble
Wilcannia
Cobar
Tamworth
WARRUMBUNGLE RA.
Kempsey
Port Macquarie
Nyngan
LIVERPOOL RANGE
Dubbo
NEW SOUTH WALES
LORD HOWE (NEW S. WALES)
Maitland
Newcastle
Cessnock
Orange
Bathurst
Lithgow
BLUE MTS.
Forbes
West Wyalong
SYDNEY
Botany Bay
Wollongong
Wentworth
RIVERINA
Hay
Narrandera
Goulburn
Wagga Wagga
Canberra
Jervis Bay
AUST. CAP. TER.
Deniliquin
Kerang
Echuca
Albury
Mt. Kosciusko 7316
SNOWY MTS.
Cooma
Bendigo
Benalla
VICTORIA
Ballarat
Maryborough
Bega
Bombala
Geelong
MELBOURNE
CAPE HOWE
Bairnsdale
NINETY MILE BEACH
Warrnambool
Wonthaggi
CAPE OTWAY
WILSON'S PROMONTORY
Bass Strait
KING
FLINDERS
FURNEAUX GROUP
CAPE BARREN
HUNTER IS.
TASMANIA
Burnie
Ulverstone
Devonport
Launceston
Mt. Ossa 5305
Strahan
New Norfolk
Hobart
Risdon
BRUNY
SOUTH EAST CAPE
TASMAN SEA
0 50 100 200 300 400 500 Miles
0 100 200 400 600 800 Kilometers
NORTH CAPE
Kaitaia
Russell
Hauraki Gulf
GREAT BARRIER
Devonport
Auckland
NORTH ISLAND
Hamilton
Bay of Plenty
EAST CAPE
NEW ZEALAND
North Taranaki Bight
New Plymouth
C. EGMONT
South Taranaki Bight
Ruapehu
Vol. 9175
Gisborne
Hawke Bay
Napier
Hastings
Wanganui
Palmerston North
CAPE FAREWELL
Tasman Bay
Nelson
Karamea Bight
Lower Hutt
Wellington
Cook Strait
CAPE FOULWIND
Greymouth
Hokitika
SOUTH ISLAND
SOUTHERN ALPS
Mt. Cook 12 349
Pegasus Bay
Christchurch
CASCADE PT.
Canterbury Bight
Timaru
RESOLUTION ISLAND
Dunedin
CAPE SAUNDERS
Invercargill
Foveaux Strait
STEWART ISLAND
SOUTHWEST CAPE
©RMcN.
Same scale as main map

0 200 400 600 800 1000 Miles

0 400 800 1200 1600 Kilometers

Lambert's Azimuthal, Equal Area Projection

Elevations and depressions are given in feet

Lambert's Azimuthal, Equal Area Projection

Elevations and depressions are given in feet

4 Europe: a general geographic introduction

The Rhine River in Germany. *(Photo by G. Willert for Van Cleve.)*

Europe has had the strongest modern claim to the title of hub of the world. The most important part of this claim is that Europeans originated the commercial and industrial revolutions, and were principally responsible for their spread to the rest of the world. Europe is still the world's primary manufacturing and trading economy, measured in terms of population and total output, and set much of the pattern for the modern economic development of all other areas. Perhaps of equal importance is the development in Europe of distinctive cultural and political institutions that have spread widely to create replicas of themselves or to alter existing societies. This was especially true for representative government and its connected ideas of individual liberties, individual enterprise, and socialism. Modern nationalism and national states also began first in Europe, and these ideas were exported to or imposed upon others. In many respects the civilization and techniques that grew up in Europe, if not Europeans themselves, have conquered most of the globe. Modern Europe has been a center of change, a dynamic creator of new institutions and new ideas. The concepts of change and progress were themselves a vital part of Europe's exports.

The sea, rivers, and the plain

In effect Europe is a great peninsular extension of the major land mass of Eurasia; the sea virtually surrounds it. In Europe west of Russia, only a small part (eastern Czechoslovakia, eastern Hungary, southeastern Poland, and western Rumania) is more than 300 miles (480 km) from the sea. Further, the great European peninsula is composed of a series of subpeninsulas or close-lying islands: Scandinavia, Denmark, Great Britain, France, Spain and Portugal, Italy, and the Balkans. Europe has the longest coastline of any continent in proportion to total land area. (Europe is conventionally called a continent even though it is merely a peninsula of Asia.) The surrounding sea and Europe's many harbors provide the advantage of cheap, easy circulation. This was a strong asset in the development of a commercial-industrial economy. Maritime climate, influenced by the sea, dominates most of the continent, especially west of Russia. This climate favors agriculture and contributes to the number and usability of Europe's rivers, adding significantly to internal circulation. Water routes provide cheap exchange among the parts of Europe and between it and the rest of the world. Exchange and economic development are closely interrelated.

The coastline itself has virtues other than its length. It is highly indented and contains many excellent natural harbors. Several of them lie toward the landward end of penetrating sea fingers and are well located with respect to markets and population con-

MAP 20
CENTRALITY OF EUROPE

MAP 21
MINERALS IN EUROPE

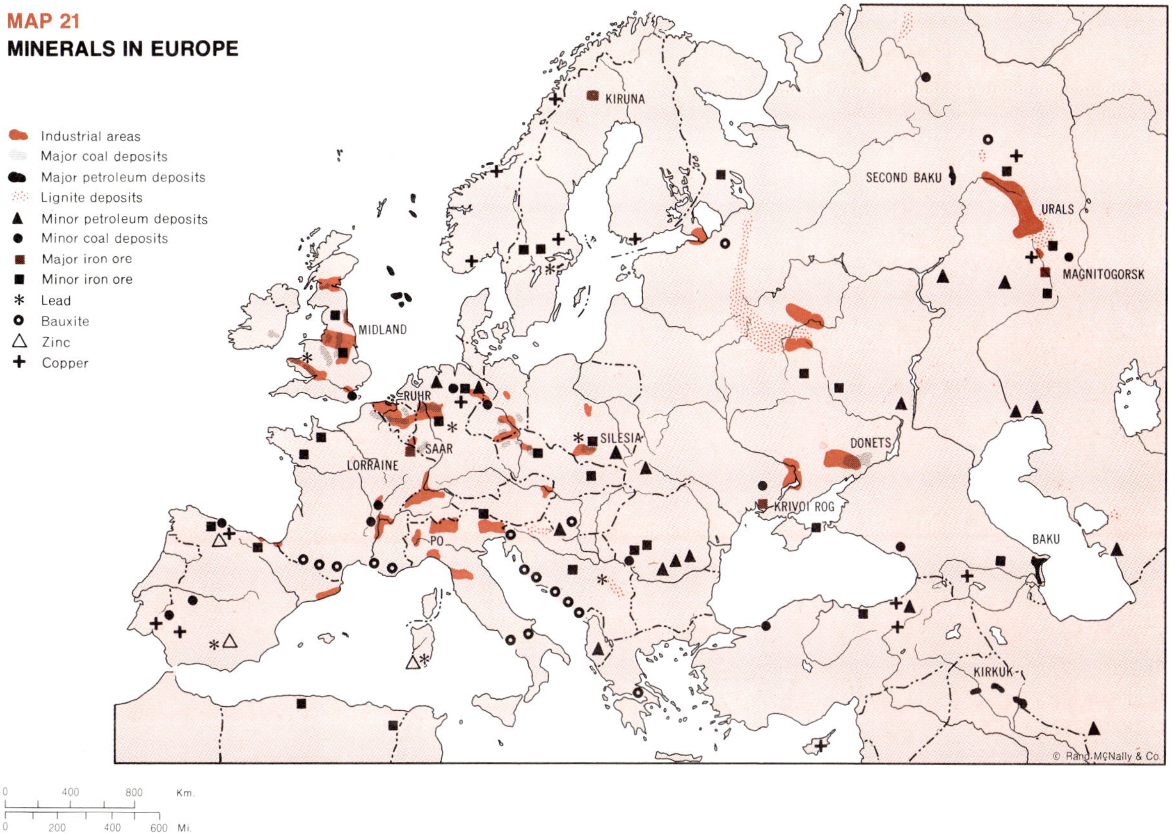

centrations. In other stretches of the coast where good natural harbors are less available, navigable rivers give access to productive hinterlands. The broad coastal plain, such a prominent part of the European landscape, is an area of dense population, characterized by easy circulation by land, by navigable rivers flowing across the level plain, or by the surrounding sea. Of all the continents, Europe has the lowest average elevation and the greatest proportion of plains.

Europe enjoys high mobility, reflected not only in exchange, trade, and manufacturing, but also in cultural interaction. The peninsular character of Europe, plus regional differences in topography and climate, allowed the growth of cultural, economic, and political centers that were distinct from each other. Physically Europe is more diverse than the strong influence of the sea and the prominence of the coastal plain might suggest. Britain, Denmark and Scandinavia, France, Spain, Italy, and Greece, for example, are logically separate units, on both physical and human grounds. But the ease of circulation so pronounced in Europe brought these different regions into close contact with each other. Culturally there was a continual cross-fertilization, and economically, a varied and mutually beneficial exchange. Each area was able to take advantage of local conditions contributing to specialization, and also to profit from the specialization and creativeness of other areas. Europe contains a wide range of environmental differences and a number of complementary areas. Trade and cultural flows between the separate regions are aided by the ease and cheapness of circulation.

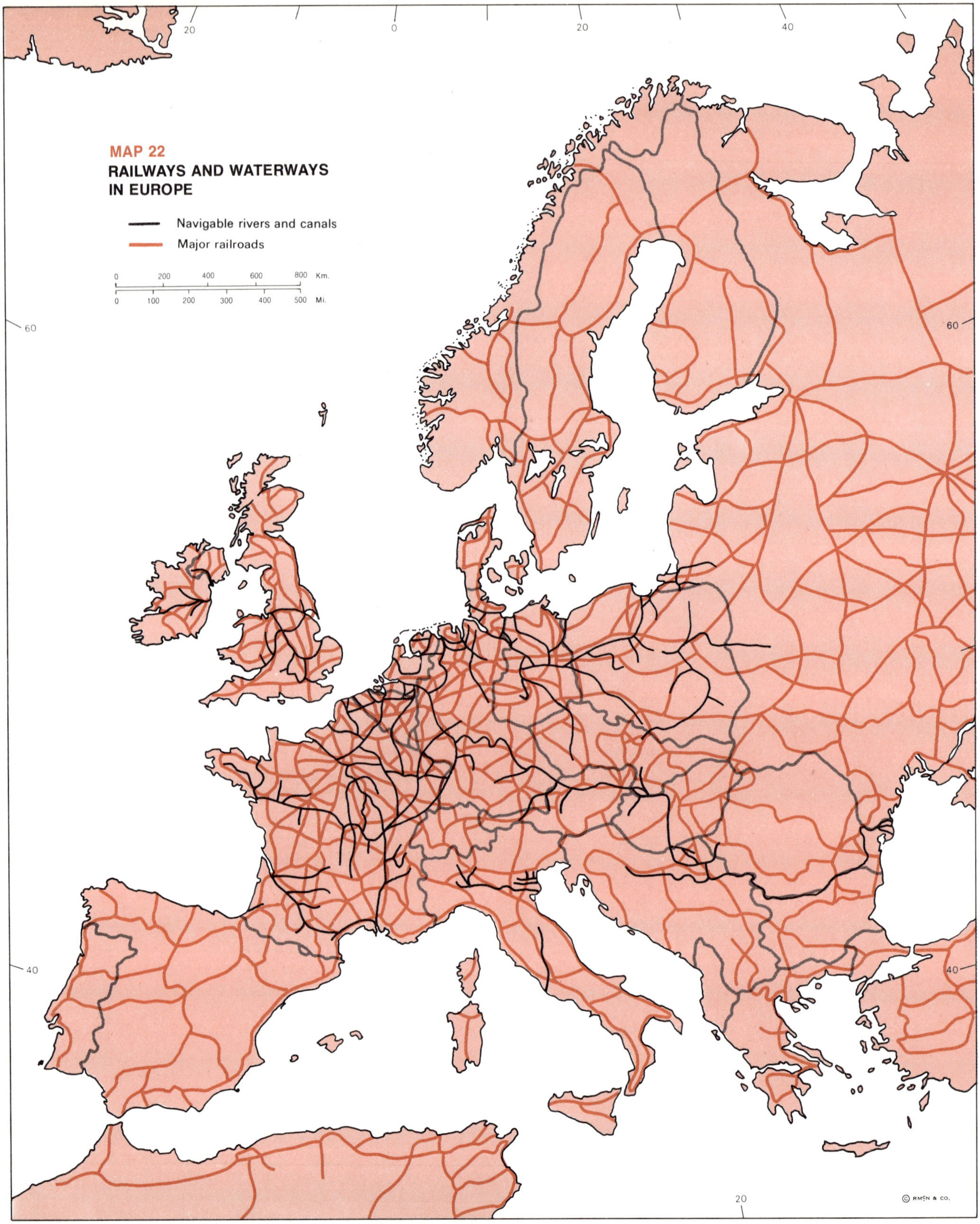
MAP 22
RAILWAYS AND WATERWAYS IN EUROPE
Navigable rivers and canals
Major railroads
0 200 400 600 800 Km.
0 100 200 300 400 500 Mi.
20
0
20
40
60
60
40
40
20
© RM©N & CO.

External access

Mobility within Europe is matched by its access to and from the rest of the world. Map 20 (page 70) emphasizes this advantage. Europe lies at the center of the world's land masses where the major modern sea routes converge between London and Paris. Of all areas in the world, Europe was placed best to spread its control or influences on a worldwide basis, and to profit from close relations with other areas in modern times. Centrality of position is no more important than the advantages of this position for the use of the sea—the cheap highway of trade, of expansion, and of ideas.

Europe is equally accessible by land. It is located along the tip of the great triangle formed by Eurasia as a whole. The European plain lies unbroken along the northern side of this triangle, and widens out into the steppes of central Asia. The southern side, leading eastward from Europe, has also been a route of contact with the economies and cultures of the Arab-Turkish world and with India. Influences starting almost anywhere within these areas ultimately were likely to make their impact on Europe; it is the outlet of the land funnel. Europe is open on virtually all sides, open to receive stimuli and trade from the rest of the world and to send out its own culture.

Industrial resource base

By fortunate accident Europe also has rich mineral resources, especially in coal, natural gas, oil (around and under the North Sea), and iron ore. As usual, the total reserves, although they are large, are no more important than the location of the deposits. Most of them, including the largest (see Maps 21 and 22, pages 71 and 72), are on navigable waterways or the sea, and in or near the lowland centers of the greatest

MAP 23
PRECIPITATION IN EUROPE

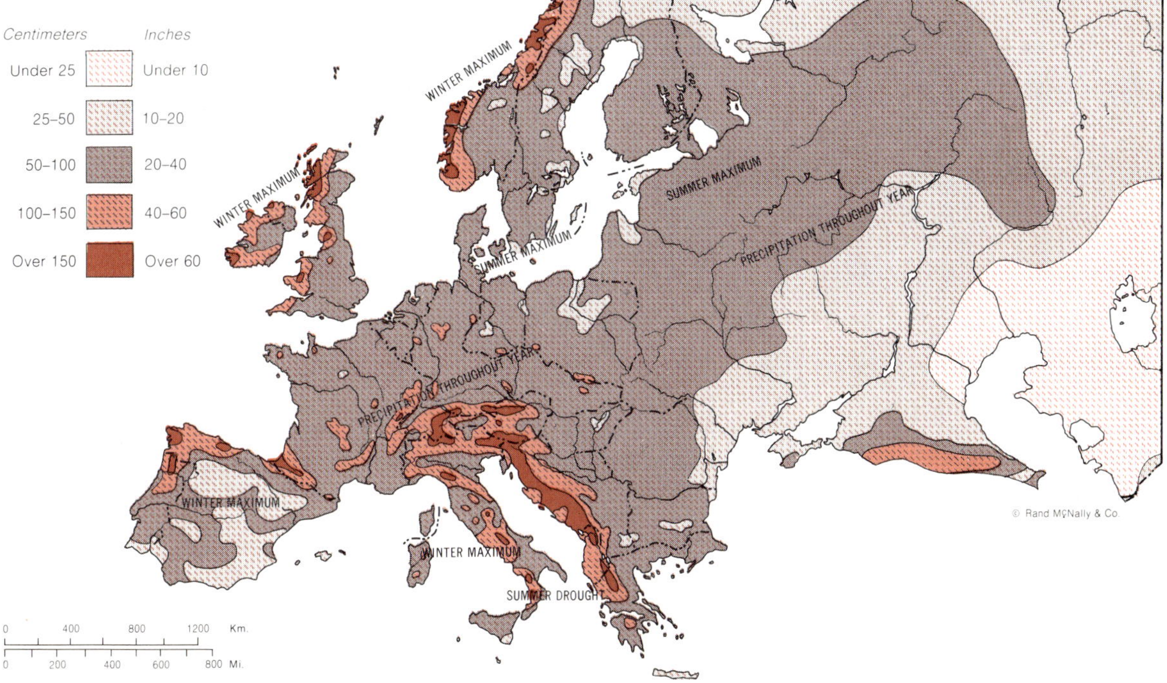

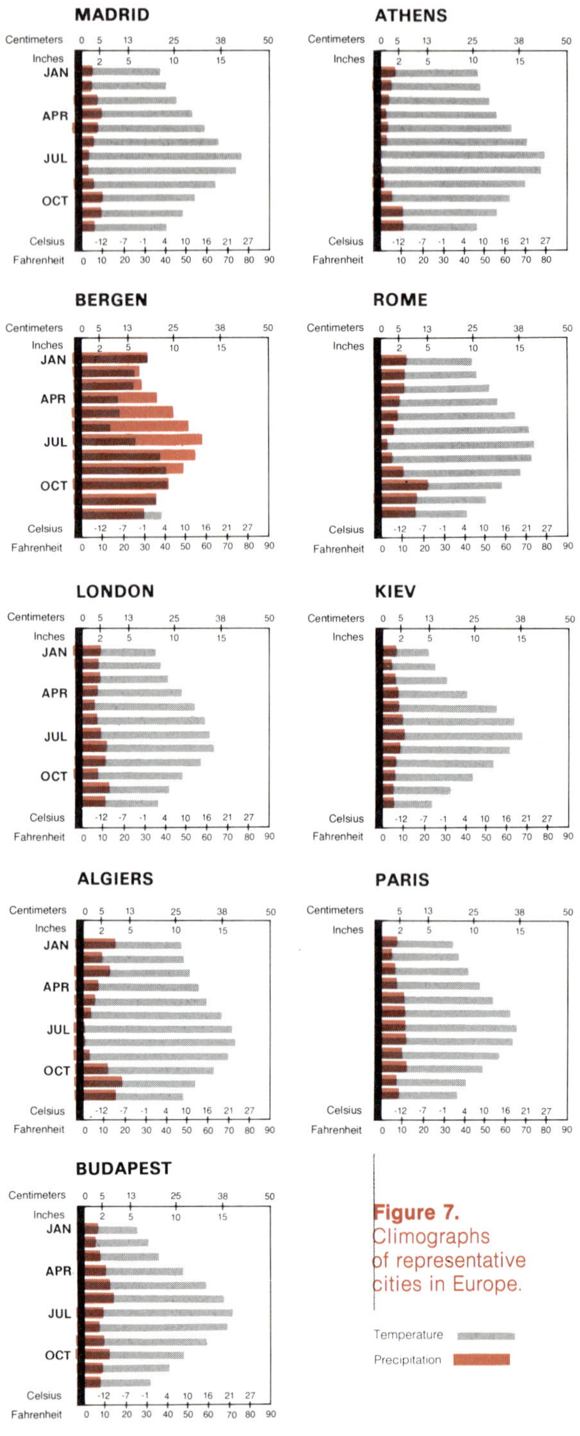

Figure 7. Climographs of representative cities in Europe.

markets. Coal and iron have helped to create these markets, but the fact that minerals occur close to productive areas where population and market growth can increase is an important advantage. Industrialization is largely concentrated in a ragged zone across north-central Europe, turning north into the British Midlands and southern Scotland. It coincides largely with the major workable coal deposits, located between the north European plain and the hilly or mountainous areas farther south in the heart of the continent. Industrial developments in northern Italy and in the Barcelona district of Spain depend in large part on local hydro power. Neither of these areas is as important as the main industrial belt running through north Europe that is closely associated with coal and iron ore.

Natural waterways have been added to extensively by a dense system of canals, built and used in moist lowland Europe on a greater scale than in any area of comparable size in the world. Map 22 (page 72), showing canals and navigable waterways, illustrates this advantage of cheap transport and the benefits of investing in canal building even in a railway age. New canals are still being built or extended, further linking Europe's major navigable rivers and connecting complementary regions, like the Mediterranean basin and the Rhine valley through the Rhône corridor, or the Danube basin and the Rhine through the Rhine-Danube (Ludwig) canal. The transport bills that Russia or the United States have to pay as part of the manufacturing process, for example, are proportionately higher, since railroad haulage plays a larger part in assembling the raw materials, many of which are also remote from population centers. Water transport was particularly important before and during the early stages of industrialization, when land haulage was much more expensive than it is now. If it had not been for the pioneer industrial development in water-rich Europe, industrialization in the United States and elsewhere would probably have been delayed. These larger areas with their greater agricultural and mineral resources and the help of railroads can now profit from economies of scale to achieve lower production costs than Europe's in some kinds of manufacturing. Tropical commodities like rubber, cacao, or vegetable oils are largely lacking in all the world's major industrial areas. But Europe's sea trade with former colonial holdings in the tropics ensures a steady flow of these goods.

MAP 24
WHEAT PRODUCTION IN EUROPE

Each dot represents 100,000 bushels

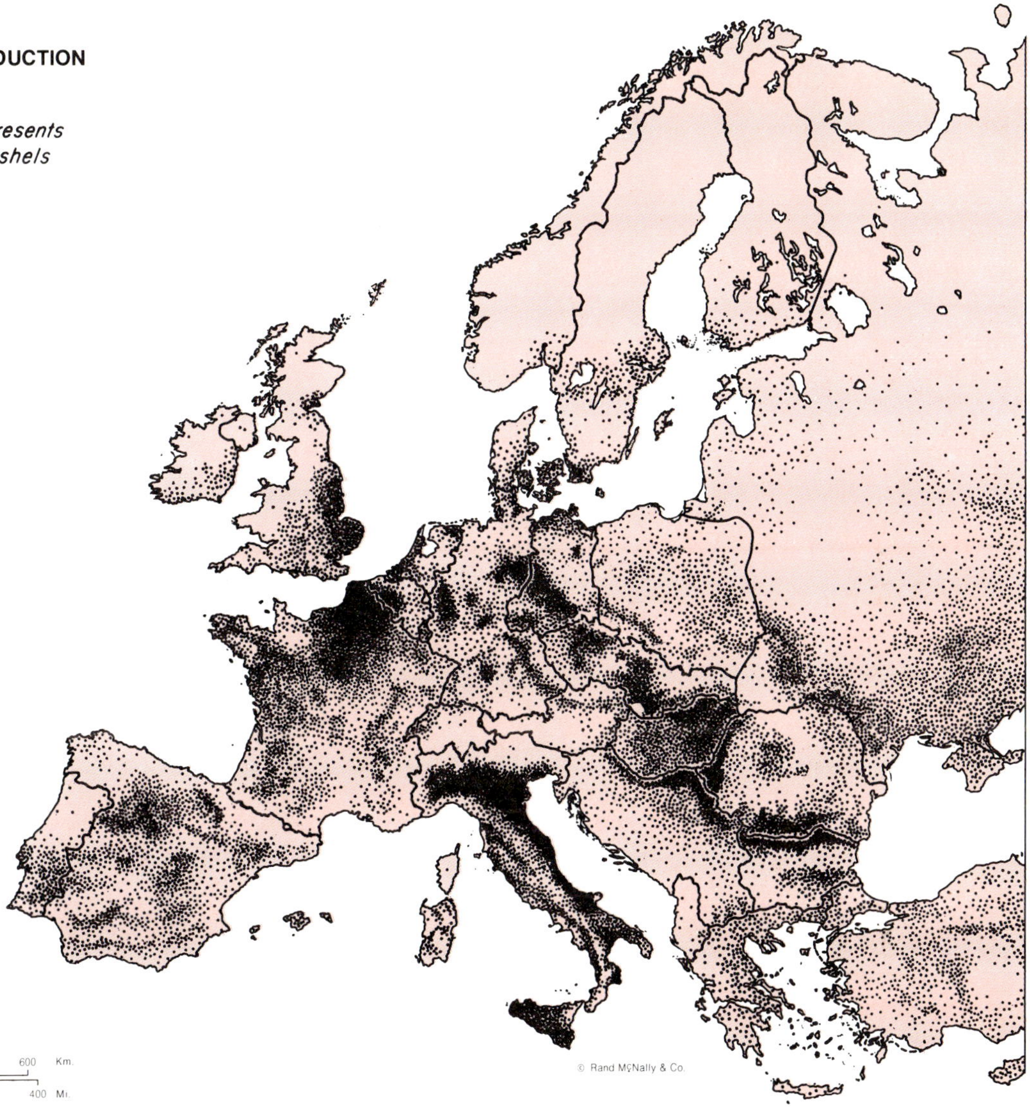

0 200 400 600 Km.
0 200 400 Mi.

These tropical areas are also major sources of petroleum, tin, copper, and other minerals, providing a further basis for profitable investment overseas by Europeans.

With the end of the Middle Ages, Europeans began to use resources of location, raw materials, and power that had not been as important before, and the European environment began to affect human activity differently. Northwest Europe had been the center of the world only in a physical sense before this time. It was a relatively remote and backward place compared with the Mediterranean basin or monsoon Asia until the Renaissance. As north European knowledge of and ability to use the sea and the resources of the land increased, northwest Europe became a central trading area, and began to realize the great potential of its total environment and location.

The advantages of mobility, climate, centrality, access, and mineral deposits did not cause the development of modern Europe, however. That development can be understood better in the light of Europe's advantages, but it cannot be *explained* in those terms. The advantages remained unused to a large degree for most of Europe's premodern history. Other areas elsewhere, eastern North America, for example, had similar advantages that did not result in industrial development until Europeans settled there. There was nothing inevitable about what happened in Europe after the sixteenth century. It is important to guard against the assumption that a particular result was bound to occur from a particular set of

MAP 25
RYE ACREAGE IN EUROPE

Each dot represents 5,000 acres

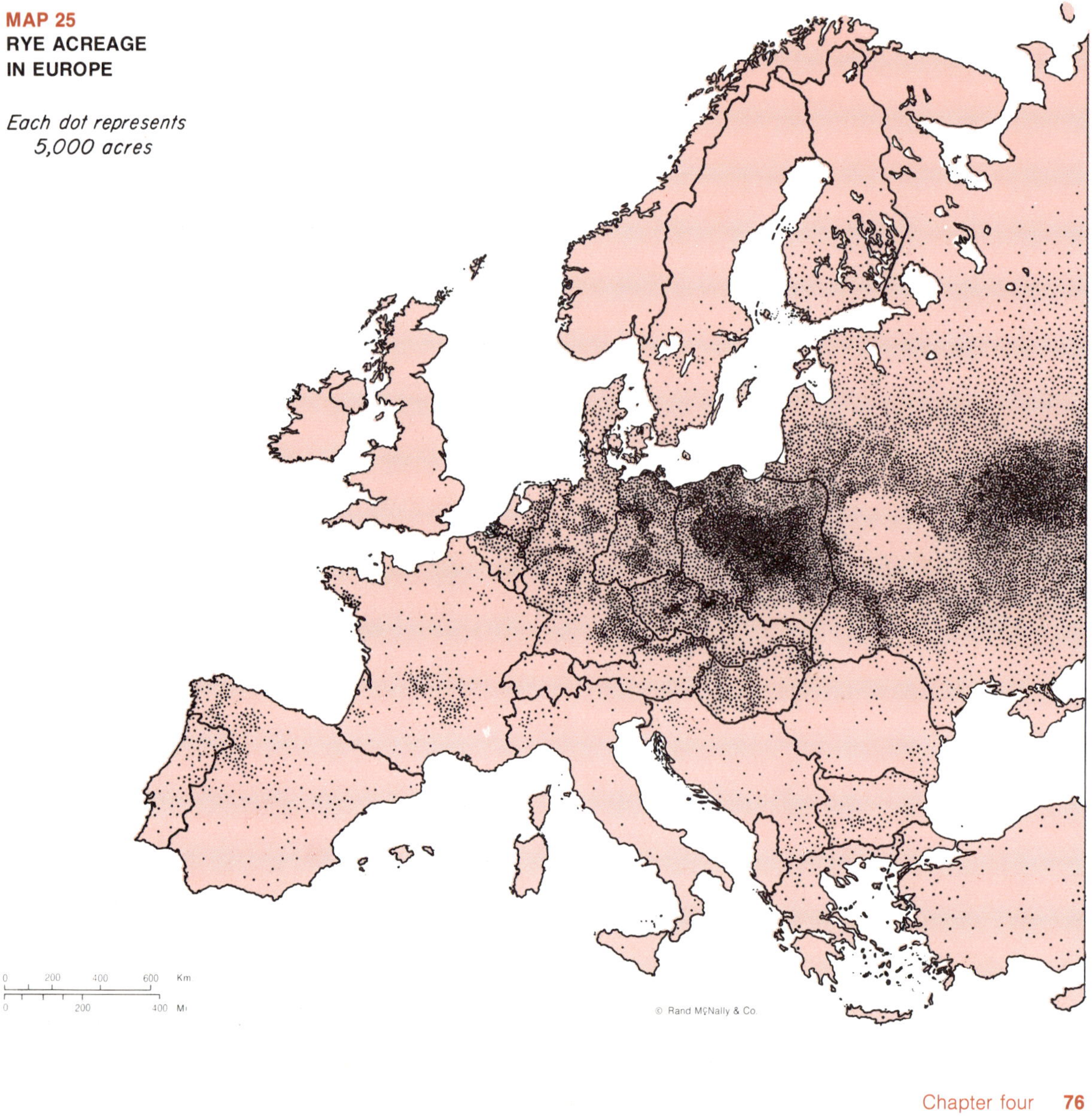

circumstances, and to assume that the circumstances were causal factors merely because we can observe the result. The reasons behind the development of modern Europe are almost infinitely complex, and development might well have been different. But it would not have taken place as it did if the particular advantages outlined above had been absent. While they did not produce the development, they allowed it to take place, once the varied complex of other factors became favorable.

Trade, manufacturing, and politics

Commercial and industrial development in Europe also contributed to a society where political institu-

MAP 26
POTATO ACREAGE IN EUROPE

Each dot represents 2,500 acres

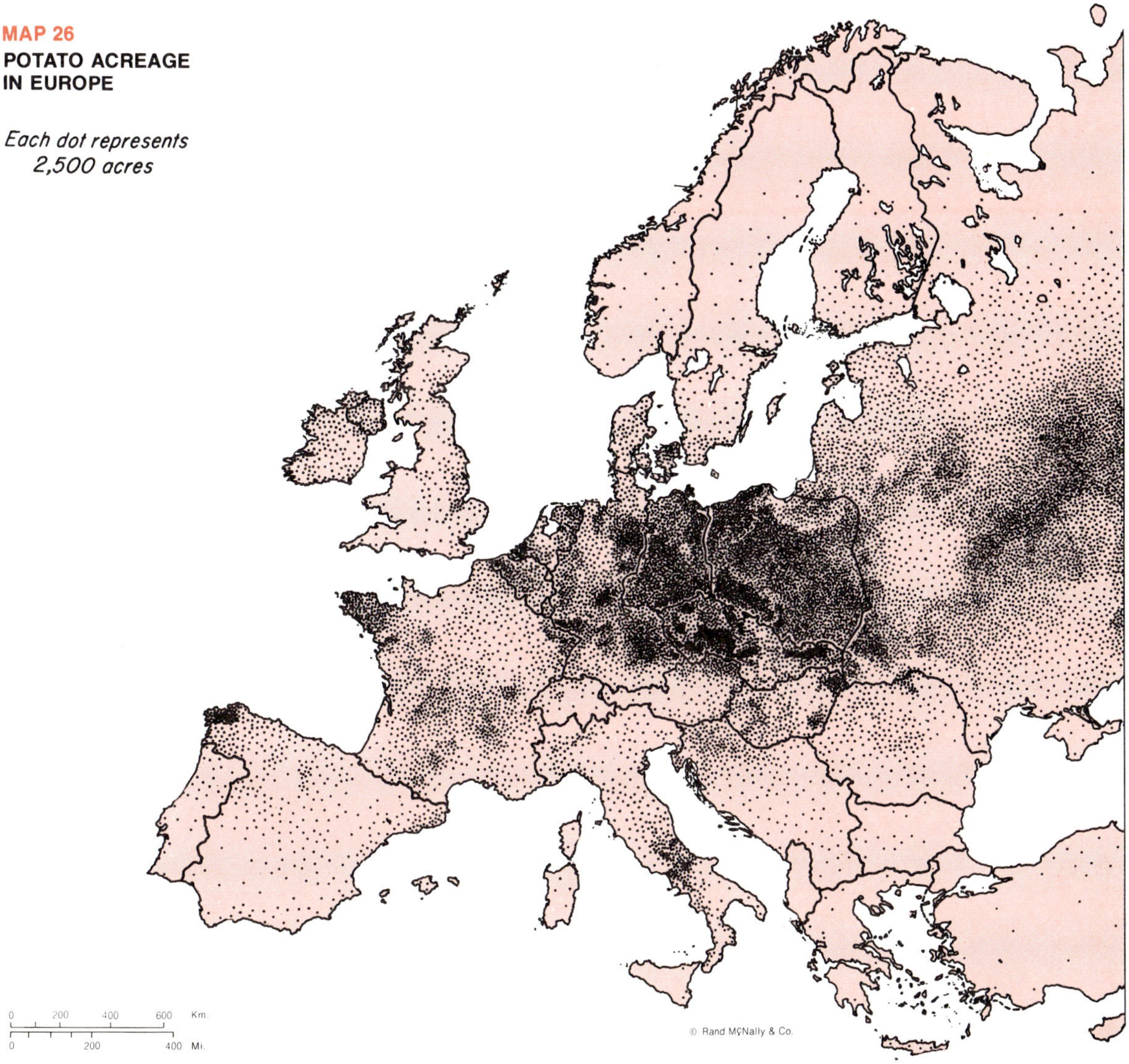

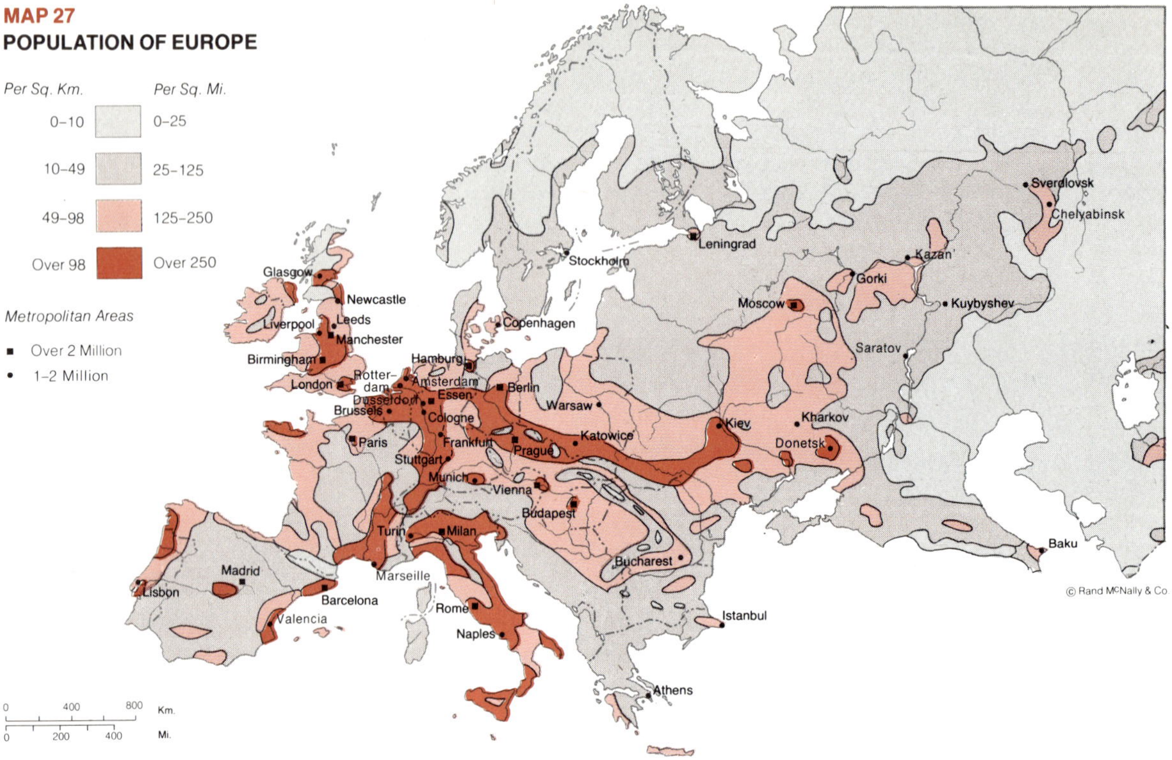

MAP 27
POPULATION OF EUROPE

tions became broadly representative of the people as a whole. The industrial and commercial revolutions started with a small group of urban-based merchants and entrepreneurs. Their newly won economic power helped them achieve growing independence from the traditional society and its representatives, the king and the landed nobility. This was perhaps the most important origin of modern constitutional representative government, and later of socialism. Parliaments and political parties were, to a significant extent, the instruments of an urban, commercial middle class anxious to enhance its position and rights against royal or aristocratic tyranny. The tremendous increases in trade and manufacturing raised the economic level of the population as a whole, as well as adding to its numbers. Economic changes provided the basis for a mass democracy that had never existed before and that increasingly demanded political recognition. Individual economic opportunity, actual or potential, grew rapidly and to a completely unprecedented degree. From at least the seventeenth century, Europe provided a congenial atmosphere for ideas about individual enterprise, progress, and human rights that accompanied the Industrial Revolution and the rise of political democracy.

Europe's future

There are great human differences between regions on a world scale. Even within Europe, cultural unity is limited and stops far short of political uniformity. Nearly one-fourth of the world's population lives in Europe. Since the Second World War, western Europe has increasingly realized the advantages of greater unity, at least economically. The large number of European states inherited from a preindustrial age prevents the best use of Europe's resources under modern conditions. The European

Coal and Steel Community of the early 1950s has been enlarged into the European Common Market, officially created on January 1, 1959. It links in differing degrees most of the countries west of Poland and Czechoslovakia (except Spain) in a growing customs union. Resources are being pooled, production increasingly integrated, currencies made more freely exchangeable, and tariff duties progressively smaller.

By 1990 there may be almost as much economic unity within western Europe as within the United States. If so, Europe can begin to enjoy more economies of scale and the benefits of wider and fuller exchange of goods. Defense planning has long been integrated, and the Council of Europe with headquarters at Strasbourg, still an advisory body, may ultimately grow into a more substantial political union. The unparalleled devastation of two world wars greatly accelerated the relative decline of Europe's wealth, power, and influence in comparison with the rapid growth of North America, Japan, the Soviet Union, and, more recently, China and India. Europe has recovered with remarkable speed from the ruin of 1939–1945; production and trade exceeded prewar levels as early as 1952. For the past three centuries Europe has led the world and pioneered modern economic and political development. European civilization and its genius for innovation are still vigorous. The present century may see a united Europe firmly established as the senior member of the world community.

Questions for further study and discussion

1. What parts of the world have been settled largely from Europe? In which of them is a European type of culture dominant: in language, political institutions, and economic system?

2. What parts of the world have been largely free of European influences? Why?

3. Why are canals proportionately more important in northwest Europe than in eastern United States? Why has it apparently paid to continue investing in canals in Europe, but not in the United States, where a great many old canals have been abandoned?

4. Why was the formation of the Common Market difficult, and its inception greeted with disturbances reflecting economic problems, as, for example, in Belgium? What policies has France followed toward the Market, and why? Why was the British application to join the Market in 1961 turned down? What changes led to Britain becoming a member in 1973?

5 The British Isles: Britain and Ireland

Big Ben and the Houses of Parliament, London. *(From Pan American Airways.)*

Britain: a prized location

Britain[1] can be divided best into lowland and highland areas. Lowland Britain is dominated by England, where trade and manufacturing have helped to form the kind of society discussed in the preceding chapter. Highland Britain is physically and culturally different and supports a different people. Here and there on its lowland margins, it has been industrialized into an English likeness. But the United Kingdom of Great Britain and Northern Ireland is a union of distinct cultural strains and the distinct physical areas they inhabit: plains and mountains.

England has an outstanding location to use the sea. This situational advantage led to a structure of trade, manufacturing, and empire as a pioneer of the modern European revolution. England shares Europe's favorable climate, soils, and landforms, but improves on Europe's locational assets. Insularity favored the early development of English nationalism as an awareness of regional distinctiveness. Nationalism and the national state stimulated the expansion of trade and empire, giving England an advantage over rivals on the Continent, where nationalist feelings were developed only later. Insularity also gave vital military protection to the modern English state. And, as an island, Britain's use of the sea led to its development as a maritime power.

The British Isles lie off the most productive part of the European mainland, where the northern and southern arms of the European plain come together to meet the sea. Here modern Europe's population is most concentrated, and here is its largest market (see Map 27, page 78). Through the heart of this area runs the Rhine River, Europe's principal trade artery, with its mouth directly opposite the mouth of the Thames and the city of London. On both sides of the Rhine are other major rivers draining into the English Channel at England's doorstep. In this area also are some of Europe's major deposits of mineral resources and the extensive manufacturing functions that have developed near them (see Map 21, page 71). The Channel is just the right width: wide enough to deter invasion (even in the Second World War, the Royal Navy and the Royal Air Force could patrol it), but narrow enough not to deter close cultural and

---

[1]*Britain* includes Scotland, Wales, and England.

commercial contact between England and the Continent. If it chose, England could maintain politically "splendid isolation" (as its own nineteenth-century politicians called it), but it could continue to profit to the full from cheap, easy access to the mainland.

On a broader scale, the British Isles virtually surround one side of what is almost a closed sea, the North Sea, around which the economic heart of modern Europe lies. Movement of goods by cheap sea routes between this major area of Europe and the rest of the world has had to go by way of England or the Channel in most cases. London, and to a somewhat lesser extent other British ports, became prominent break-in-bulk points for this trade, distributing goods from overseas to Europe and collecting European goods for further shipment.

Islands lying close to a productive mainland perform this *entrepôt* function in many parts of the world. Hong Kong, Trinidad, and Curaçao are other examples. An entrepôt (literally, "between ports") is a collecting and distributing center for the trade of several other ports that lie around it. It may produce little itself except port, storage, and trade services. A central location and water routes to a wide productive trade area are its major requirements—hence, most entrepôts are on offshore islands along a commercial coast. A further advantage is the physical security of insularity for goods in storage and for general commercial functions. In some cases the island is different politically from the mainland, and the trade security on which the island's economy depends is thus increased.

British manufacturing developed relatively early and continued to profit from the transport advantages of entrepôt trade. Europe's large market was easily accessible. The shallow North Sea was also important as a major fishing ground. Fishing was an important part of the national economy, and sailors for the navy were recruited from the fishing fleets.

The island, like the Channel, approached an ideal size in terms of Britain's historical development. It was large enough and contained adequate plains areas to be economically and politically self-sufficient, but small enough to allow early achievement of national unity and an integrated economy. Its size and productivity were especially important in the competition with the other countries of Europe for commercial, political, and colonial leadership. The island also contained a wealth and variety of mineral resources, well located near sea or river. These were important factors in the Industrial Revolution that began in England.

External influences and cultural mixture

Population and economic activity were concentrated and cultural influences were most varied at the western end of the European triangle. Here Roman and Germanic Europe met, and England benefited from both, ultimately receiving most of the influences generated within Europe or coming to it from its landward approaches. A hybrid society and culture grew up in Britain, the inheritor of the variety of continental patterns. The westward migration of the Celts across Europe ended in Britain. Celts still live in mountain corners of Britain, having once ruled the whole of the island. Before the Celts, archeological evidence shows a great variety of immigrations and invasions from Europe, and continuous close cultural contact between Britain and the mainland. Later, the Romans conquered the island, ruled it for four hundred years (about 43–425 A.D.), and left a permanent legacy of Latin civilization. Germanic Angles and Saxons followed the Romans, invaded and settled in Britain, and brought their culture and language with them. Through several centuries, A.D. 500–1000, Danes and Norwegians raided and settled not only the coasts but large inland areas, too, leaving a further mixture of peoples and cultures.

The last successful invasion of England, by the Normans under William the Conqueror, occurred in 1066. The Normans represented a Latin culture, like that of the Romans. *Norman* is a corruption of *Northmen*. These people were originally Scandinavian Viking raiders. But about a hundred and fifty years before their conquest of England, they had settled in France on an exposed peninsula in the English Channel. They conquered it by a seaborne invasion and called it Normandy (Normandie). They had absorbed French culture, and came to England as carriers of a superior civilization to a backward outpost. The modern English language shows the mixture of cultures making up modern Britain. This variety of influences led to the vigorous cultural development often associated with such a situation.

Sea power

Sea power helped Britain, less well developed economically than France at the beginning of their

MAP 28
THE BRITISH ISLES

rivalry, win the prizes overseas and play an important role in European affairs as well. The strength of sea power is its flexibility, which gives it the advantage of surprise. A fleet can strike with concentrated force without warning at a time and place of its choosing, and can withdraw its forces to redeploy them at will. The number of men and amount of capital necessary for operations of this kind are much smaller than those necessary to support a land army. And a navy is much more cost-effective. With warships Britain could dispense with large armies and concentrate its resources in a fleet. It was able not only to defend itself, its trade, and its empire, but to exercise a large degree of control over the European mainland, so neatly blocked by the British Isles. British economic existence depended on free relations with an independent Europe, so England consistently resisted French, and later German, attempts to weld Europe together forcefully under their own domination. Had this happened, it would have left England in an untenable minority position politically rather than allowing it to act as a power among numerous equals.

Tower Bridge and the Tower of London. The old Norman tower, built in the eleventh century, still guards the lowest easy crossing point and head of ocean navigation on the Thames River. *(From Pan American.)*

The air age and the development of missile warfare have reduced the importance of both navies and armies. Britain is no longer secure. World security in the nuclear bomb era can come only from the disappearance of war as a gruesome mistake. But British sea power and its advantages have not been limited to military considerations. England's union with the sea has been the main basis of its economic prosperity and national strength, and a recurring influence on many of its most important institutions. Sea power made insularity effective by establishing independence and security. From the time of the Norman conquest, England was able to resist invasions from the mainland and to exercise some choice over what came into the country. Development was independent, although it was continually stimulated from the Continent. This security allowed England to develop without the crushing burden of heavy military investment to defend land frontiers, a problem common to all the powers of the mainland. As Shakespeare put it in Act II, Scene I of *Richard II:*

> This other Eden, demi paradise;
> This fortress built by Nature for herself
> Against infection and the hand of war; . . .
> This precious stone set in a silver sea
> Which serves it in the office of a wall . . .
> Against the envy of less happier lands . . .
> England, bound in with the triumphant sea.

This remained a precious gift in the English mind: the tradition of freedom and of a society of free men, whose protection by the sea and the fleet allowed them to forgo tight control in return for the blessings of liberty.[2] It was mirrored in their legal and political institutions, and transplanted by them to North America, where similar geographic conditions of protection helped it to flourish.

Mercantile and parliamentary England

England's location and sea power were put to use by mercantile England. By 1640 only half of the total population of England remained directly employed in agriculture. This marked the beginning of a division of labor that did not take place elsewhere in Europe (with the exception of the Netherlands) for more than a hundred years. The new urban, commercial middle class demanded appropriate political representation and an effective voice in making national policy that was so important to them financially. Especially they wanted to preserve their independence from royal infringement. They found a way to accomplish these ends in the House of Commons. Parliamentary government developed gradually and was seldom characterized by the violence that accompanied similar changes later on the Continent. Britain's secure position was certainly involved in this. Protected by the Channel and the fleet, the state could accept compromises in domestic affairs with less risk than the Continent could. In the French Revolution, a century and a half after Charles I was beheaded in London, the Terror (as violence and guillotining were called) was closely related to the fact that France was surrounded by hostile powers. Their announced aim was the overthrow of the revolutionary government, and their expeditionary forces came perilously close to Paris. With rare exceptions the English have been able to deal with their problems less urgently and have even poked fun at authority and prestige within their society. It is said that the English count heads rather than breaking them.

The strains and changes in the English economy and society brought about by the increase in commerce after the fifteenth and sixteenth centuries were paralleled, but exceeded, by the changes in industrialization after the eighteenth. Broadly, there was a tremendous strengthening of the nonagricultural, urban sector, and a consequent change in population distribution. The agricultural area of the English plain lost its economic dominance to the new industrial area of the Midlands, where coal and iron ore helped to create the major manufacturing district. This district runs roughly in a wide band from Birmingham-Coventry northward through Leicester, Nottingham, Sheffield, and Bradford-Leeds and northwest through Stoke-on-Trent, Liverpool, and Manchester (see Maps 21 and 29, pages 71 and 86). Other lesser industrial areas, also associated with coal, developed in south Wales, on the northeast coast near Newcastle-on-Tyne, and along the Clyde River in Scotland. The population of these areas boomed; many cities were new, and had inadequate or no representation in Parliament under the old electoral laws. Apart from district representation, industrialization brought mounting pressure for economic and

---

[2]Insularity is also sometimes accompanied by provincialism. A headline in the London *Times* in recent years read: "Thick Fog in Channel. Continent Isolated."

Industrial England—Stoke-on-Trent. This is part of the "Black Country" of the English industrial belt—grimy towns, smoking chimneys, and uniform workers' housing. Stoke-on-Trent is still a major pottery center. *(From Aerofilms Ltd.)*

institutional change. The dominant political and social position of the rural gentry and aristocracy was no longer acceptable. Nor were their outmoded laws and administrative machinery adequate for an industrial-commercial age.

Attention was also centered on tariffs imposed on imports, especially food grains (the Corn Laws), and on the restrictions and dues on foreign shipping entering British ports or carrying British goods (the Navigation Acts). Both had originally been designed to increase government revenues and to protect domestic farmers and shippers from foreign competition. But commercial-industrial England, bringing in food and raw materials and selling finished goods abroad, and acting as Europe's entrepôt, had far more to gain from complete freedom of trade. The Corn Laws and Navigation Acts were repealed. Free trade became the cornerstone of British policy and the basis of the country's prosperity. British society, also reacting to these changes, was transformed from its old, aristocratic, rural base to today's urban, commercial democratic structure.

Power and machines: the Industrial Revolution

The power and machine revolution, which began in eighteenth-century England, was the result of many interacting factors. It is difficult to say which were most important, and even difficult to be sure what all the factors were. Britain, however, did have certain clear advantages. Coal and iron ore, the two most important industrial raw materials, were relatively plentiful in Britain, and both were favorably located (see Maps 21, 22, and 29, pages 71, 72, and 86). Most of the major coal fields were near the sea or

Table 1. Population of England and Wales, 1700–1975

1700	6,000,000
1750	6,500,000
1801	8,892,536
1811	10,164,256
1821	12,000,236
1831	13,896,797
1841	15,914,148
1851	17,927,609
1861	20,066,224
1871	22,712,266
1881	25,974,439
1891	29,002,525
1901	32,527,843
1911	36,070,492
1921	37,885,242
1931	39,947,031
1941	41,614,248
1951	43,757,889
1960	46,071,604
1970	48,310,060
1975	49,218,700

MAP 29

CITIES AND COAL FIELDS IN THE UNITED KINGDOM

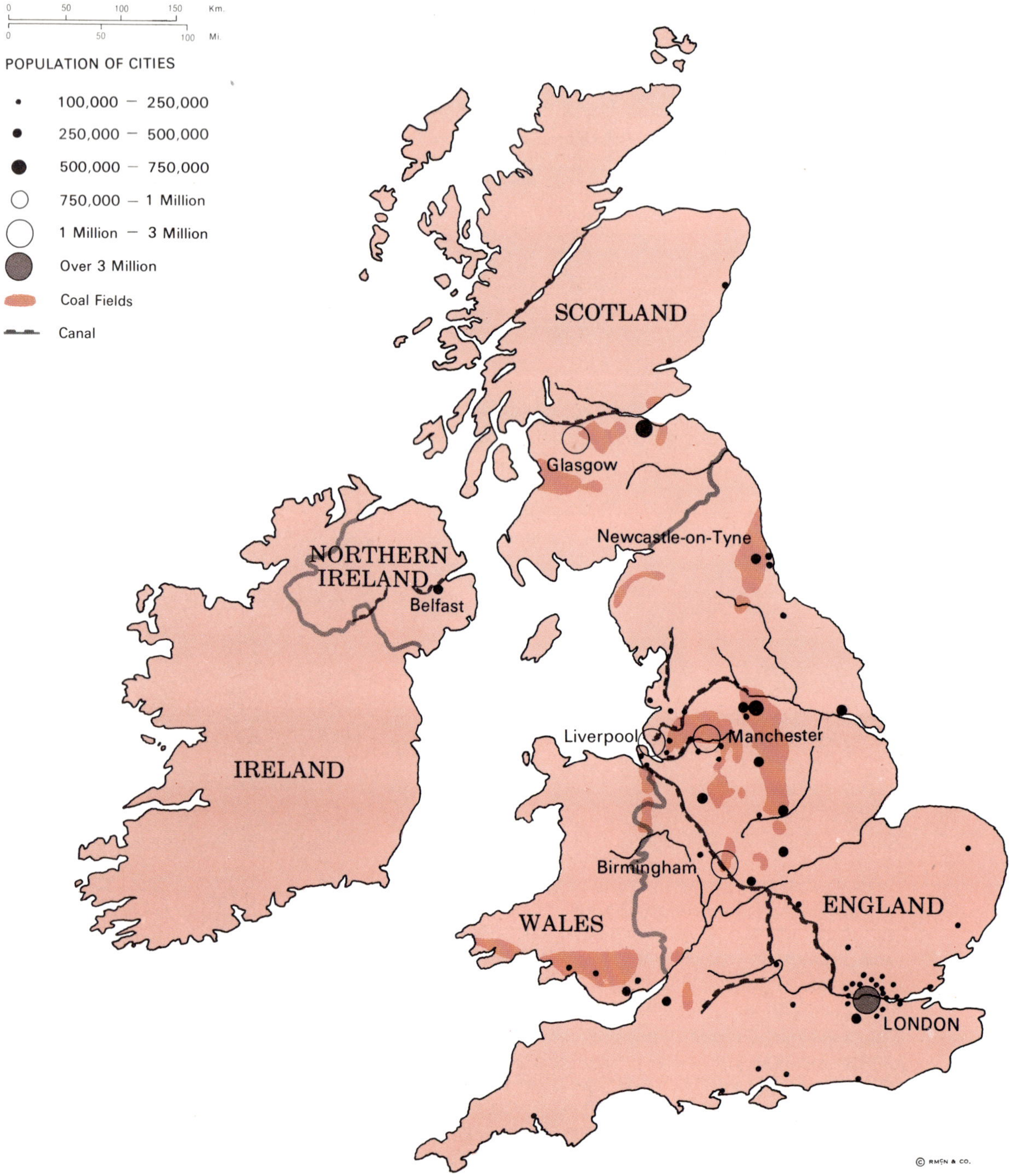

navigable rivers: southern Wales; the Manchester and Stoke-on-Trent areas served by the Mersey and Trent rivers, by canals, and by the sea; the Birmingham area; the eastern side of the Pennine Chain (a range of hills running north and south through north-central England); the northeast coast near Newcastle-on-Tyne and along the River Tyne (hence the famous redundant expression, "coals to Newcastle"); and Scotland along the Clyde. By fortunate accident iron ore was found remarkably close to coal: right next to it at Birmingham, south Wales, the Pennines, and the Clyde, and not far away from Newcastle or Manchester. Their location was a tremendous asset, especially in the early stages of industrialization before railways were built. Then bulky low-value commodities like coal or iron could not be shipped any great distance overland without prohibitive expense. Where they occurred together naturally, heavy manufacturing was cheaper, and, where water transport was available, raw materials and finished products could be moved cheaply.

In a more general sense, Britain's economic structure, already based to a large extent on maritime trade, was favorable to the development of manufacturing. An investment in trade provided an incentive for increases in production and a generally receptive attitude toward economic innovations. Increased production was a stimulus to trade, and was eagerly sought. Technological inventions like Thomas Newcomen's steam pump and James Watt's improvement on it as a steam engine, or James Hargreave's spinning jenny, might have been made elsewhere or at an earlier time. But inventions were more likely to be adopted by the economy of late eighteenth-century Britain, and perhaps even to be stimulated by the general attitude and interests of the time and place. It was an inquiring age, and its inquiries were most often directed to the improvement and increase of production and commerce. New machines were invented elsewhere in the eighteenth century, notably in France. Others had been invented earlier, in the Italian Renaissance, in China during the Han dynasty two thousand years before, and in other parts of the world. But in eighteenth-century England, the time and the place were right. Merchant capital, earned in trade and available for investment in manufacturing, was a further stimulus to industrialization. Certain local advantages, such as mechanical water power on the slopes of the Pennine Chain (soon displaced by power from coal) or good pottery clay in the vicinity of Stoke-on-Trent, were minor factors that would have produced little industrial development by themselves. But England's locational advantage permitting use of the sea made it a center of exchange in the modern period, an excellent situation for manufacturing.

The modern agricultural revolution also began in England in the seventeenth and eighteenth centuries. New crops, rotation systems, improved varieties of plants and breeds of animals, better methods of fertilization and cultivation, and better organization of land use were developed and applied, and later spread to the Continent. The old, small medieval fields, divided by hedges or by uncultivated ridges, were consolidated into larger fields. Agricultural reforms could be applied more quickly and effectively, and larger and more economic units of land were created. The introduction of crop rotation made it possible to reduce the fallow area, which had been such a prominent part of medieval agriculture in the old, three-field system. This effectively increased the total crop land. The consequent rise in food production and the freeing of labor from the farms were important and necessary bases for the industrialization that followed.

Trading ships

Industrialization also increased Britain's profitable interest in the shipping trade. This became an important source of income, for British ships carried not only imports of food and raw materials for the home market and exports of finished goods, but the commerce of much of the rest of the world. As a major entrepôt, Britain was a profitable base for a merchant fleet. In turn, this stimulated the growth of a shipbuilding industry, especially along the Clyde and the Tyne rivers, where heavy manufacturing had cost advantages from cheap local coal and iron, plus access to deep water. British shipyards supplied the country's merchant and naval fleets, and additionally about a third of the seagoing ships for the world as a whole, including Britain's trade competitors. During the nineteenth century, and, until recently, with a few breaks due to wars, over half of the world's merchant ships were British, and there were many more British-built ships sailing under other flags. Profits from trade and transport services were reinvested in industrial development at home, and were

Rural England—Cotswold Hills. This main street of a small English town in the Cotswold Hills area shows how little life in rural England has changed. Rolling farmland and fields with grazing cattle and sheep are found close to the village. *(Photo by Adam Woolfitt for Woodfin Camp & Associates.)*

loaned out to the less-fortunate rest of the world at profitable rates of interest. British capital built railroads in North and South America, factories in Europe, plantations in Africa and the East, and fortunes in London.

Modern cities

For at least the past hundred years, urbanization, or the proportion of people living in cities, has been high in England (close to 80 percent since about 1900 —see Map 29, page 86), and the urbanized area has covered a large part of the country. Urbanization accompanies and serves commercialization and industrialization. In some respects most of south-central England became a vast urban or semiurban landscape, crisscrossed by a dense network of railways and roads, inhabited more by commuters than by farmers, and devoted to essentially urban uses such as market gardening or recreation. London became the giant city, and includes the many large cities of the Midlands and the north in its tributary hinterland. Although London is not the center of the industrial area, since the plain of southeast England has almost no coal, it grew as the financial and commercial center for the economic development of Britain and Europe, and for the trade and finance of the world as a whole. Its function as the political capital of a vast empire also contributed to its size.

There is considerable manufacturing carried on in London. Service industries, publishing, and other production take advantage of the huge urban and nearby market and of transport connections. But manufacturing has never been the city's principal business. Heavy manufacturing gravitated to coal: to the Midlands and Birmingham, to Sheffield, Cardiff, Stoke-on-Trent, Leeds, Glasgow. Textile industries developed their major center in Lancashire County and the West Riding of Yorkshire, especially in Manchester (with Liverpool for its port). There they could bring in raw cotton from overseas, wool from Australia and the British highlands, and have access to cheap coal nearby. But London, although never part of the main manufacturing district, remained the major base and focus of the shipping trade, as important as any single manufacturing industry. Southampton became important with the increase in the size of passenger ships, as London's principal outport for passengers. London also came to insure as well as finance a large share of international trade, even though much of the goods never physically came to England, and became banker to the world.

Britain's decline

Having pioneered so much of modern industrialization, Britain entered the First World War already losing ground to the new industrial giants: Germany, the United States, Japan, and Russia. The First and Second World Wars, in which Britain bore a disproportionately heavy burden, further accelerated its relative decline. But more than this was happening. The structure of free world trade that Britain helped to build and on which it was so dependent was largely destroyed by rising nationalism, imperialism, and tariffs in newly industrialized areas. Although the British Commonwealth retaliated by erecting its own tariffs, a large net loss resulted since it worked against Britain's locational advantage for world trade. The more restricted trade became, the smaller share of it Britain could keep on a national rather than a freely competitive basis. To make matters worse, Britain's merchant fleet was nearly destroyed by war twice, and took time and money to rebuild. Each war was followed by a tightening of trade restrictions to Britain's detriment. Profitable investments British capital had found overseas had to be liquidated to pay for the wars, and this important part of the economy was greatly decreased.

The terms of trade on which Britain bought and sold in the rest of the world changed to its disadvantage. The country had to pay more for what it bought, and received less for what it sold. As industrialization spread from England to the rest of the world, competition for food and raw materials increased. In some cases local nationalism or local economic development raised prices with the same effect. At the same time, other newly industrialized areas sold their finished goods in competition with Britain. Some of these new industrial areas had cost advantages over Britain beyond their domestic markets. Cheap coal and the economies of scale were the most important advantages. In the twentieth century, much British coal became relatively expensive, because the best and cheapest domestic reserves were used up. Other areas, which had lacked the early advantage of being near the sea, could now depend on cheap, volume rail transport to exploit their larger and richer inland reserves. The British coal industry, where railways were born, had to work a greater proportion of small, bent, deeply buried seams, while newer coal-rich areas could mine their larger, more accessible deposits. Since the Second World War, Britain has become a net importer of coal.

The same relative situation occurred with iron ore. Britain had profited from the economies of scale when larger units could not be integrated effectively. But, as railroads were built, large continental areas like the United States, Russia, and Germany began to grow as a result of the advantages provided by the size of their domestic markets, protected by tariffs, and the larger size and newer equipment of their manufacturing plants. With the accelerated development of the west European economy, especially West Germany, Rotterdam, at the mouth of the Rhine, by about 1960 had surpassed London in volume of port business and had become the leading European entrepôt.

In political terms, the contemporary world is dominated by two super powers, as they have been called—the Soviet Union and the United States. In the nineteenth century both were areally too big to be effectively integrated; industrialization and railroads have made them a more effective size than any other national units. In time China and India, and perhaps Brazil, may join their ranks, but the smaller countries must either establish a close economic and political union among themselves, or inevitably fall behind.

While Britain's relative position has deteriorated since the two World Wars, its future is not necessarily one of continuing relative decline. Britain has managed to keep an important place in world affairs and world trade despite the effects of two wars, whose costs were really beyond its reasonable powers to support as it did. By 1953 losses in the merchant fleet had been largely replaced, a reminder of the vital importance of maritime trade to Britain, and of Britain's unchanged locational advantages for it. But world economic shortages, heavy and unproductive military expenditures, sharp inflation, and severe trade restrictions have been especially burdensome for the British. With the change in the terms of trade, it has become more profitable and more necessary for Britain to produce food at home, especially meat and dairy products. The country now provides over half of its own food supply from an agricultural system competing in cost and efficiency with overseas imports. Imports must be paid for by exports or services, and, with the changed balance of advantage, domestic agriculture now has a larger place economically. New or disused agricultural land is being cultivated, and the use of all farm land is being intensified.

Extensive efforts are being made to modernize and retool British manufacturing in many areas. British plants had fallen behind the newer, more efficient, and highly mechanized plants in Europe and North America. Important regional shifts have also taken place. The north and northwest have been in a prolonged decline in both general economic and specific industrial terms, despite government planning efforts to attract new industry to the north. The south and southeast areas in the immediate hinterland of London have become increasingly important as industrial centers. Such shifts reflect the rapidly declining importance of coal and the advantages of newer and more efficient plants that have developed closer to the center of the market. There they can readily import necessary raw materials and draw on cheap and abundant supplies of power, capital, and skilled labor. With their large interest in world trade, the British hesitated to align themselves exclusively with a European economic bloc, but their application to join the Common Market in 1961 was greeted by a Paris paper with the headline "Britain Is No Longer an Island." British membership in 1973, reconfirmed by a plebiscite in 1975, may help further to stimulate industrial modernization and extend the market for British goods.

Surveys off the east coasts of England and Scotland in the shallow North Sea have also revealed extensive fields of both petroleum and natural gas that are becoming a major new source of cheap fuel. The first oil from these fields came ashore in 1975. In the 1980s Britain will be self-sufficient in petroleum, and is already exporting it.

Britain is still at the commercial center of the world and is still a major entrepôt for world trade. Its prosperity depends on the prosperity of the global market it serves. Manufacturing will continue to find a profitable base in Britain. It benefits from transfer functions and cheap water transport serving a highly developed domestic market of over 56 million, a nearby European market of over 300 million for which Britain has prominent locational advantages, and an overseas market where specialized British manufactures have a good position. As a people experienced with the modern revolution to which they contributed so largely, the British seem certain to maintain an important role in world affairs.

Celtic areas

Wales, Scotland, and Ireland are the principal outposts of Europe where the Celts have successfully maintained themselves and their culture. These areas, with mountains, unfavorable climate, and remoteness, shelter a people who have been numerically and economically weaker than their lowland neighbors for a long time. The Celts are found in areas of poor living: cold, wet, mountainous, remote. The more desirable crops will not tolerate their climates and soils, so potatoes, oats, rye, turnips, and hardy animals take their place. Agricultural England profits from the position of the Welsh and Scottish mountains acting as barriers to the moisture-bearing winds off the Atlantic. The area between 30° and 60° latitude lies in a global belt of predominantly westerly winds. But wherever moisture-laden winds meet a topographic barrier, such as the Welsh and Scottish highlands, and are forced to rise, they tend to lose much of their moisture since they are cooled as they rise. The windward sides of such mountains tend to have much rainfall, and areas on the leeward side tend to be much drier. The leeward areas are said to lie in a *rain shadow* cast by the mountains. Thus most of England is drier than Wales, Scotland, or Ireland. In this part of the world with its low evaporation rates, lack of high temperatures or strong sunshine, and slow, gentle showers, a little rainfall goes a long way. England is better off climatically than the Celtic highlands. English summers are warmer and the growing season is longer than in the highlands that border England on the west and north. England's lowland soils and broad stretches of level or rolling land also contrast with the barren, stony slopes and narrow valleys of the highlands. These advantages, plus mobility, cheap exchange, and, later, industrialization, have provided a much greater range and amount of economic opportunity in England. In numbers and economic strength the highland Celts could not rival the lowland English, so they have been absorbed into the English orbit. The highlands are also cut up into valleys, each tending to support its own individualistic group or clan in the past, hampering the development of effective unity. Physical and economic poverty, plus the usual and understandable human clannishness in the mountain areas, hampered the growth of strong states and furthered the eventual union of the Celtic areas with populous, productive England, the political center of gravity.

Traditional livelihood

Originally these Celtic upland areas had only the subsistence economy of most mountains. Agriculture was precarious and minimal, not only because of slope, but because of the miserably cold, wet climate, even during the growing season. Sheep were the most profitable and common means of support, since they could maintain themselves under these poor conditions on the sparse, stony mountain pastures and could grow a heavy coat of fine-quality wool. Hardy types of highland cattle were also raised (see the photograph on page 91), but sheep produced the most famous specialty. As mountain areas both physically and economically remote from markets, the highlands needed a high-value product for export. Place names give an unmistakable clue: Cardigan (although the sweaters were actually named for the Earl of Cardigan), the Cheviot Hills between Scotland and England, the River Tweed, Braemar (west of Aberdeen on the River Dee), Argyll (a highland county north of Glasgow), Paisley, Harris and Lewis islands, Inverness. Weaving was such a profitable specialty that it is still an important source of income for the poorer areas. They have little else to depend on and are too far from markets to trade in low unit-value agricultural goods such as raw wool, or perishables like meat or milk. The textiles associated with these places, of course, are exceptionally high in value; they must be to be profitable. Most of them are made by hand labor, often in small lots by individual cottage producers. Other places are better suited for machine manufacture, and the artisan touch, although expensive, adds to the market price in an industrial world.

Some agriculture is possible in the highlands. Minor cereals like oats and rye, which will grow in the climate and soils, provide most of the local food supply. But these bulky, low-value commodities could not be sent to market over the mountains profitably; they had to be converted to a more concentrated and valuable form. Hence, Scotch and Irish whiskey was produced, and, like tweed cloth, they became even more profitable as new markets developed in the rest of the industrial world. Whiskey has remained an important specialty. It is also high in quality and brings a high price. The word itself is Celtic in origin—*uisge beatha,* "water of life." Tradition and hand labor are still involved, including the use of peat as fuel in the curing process over open fires, which helps give these whiskeys their characteristic smoky taste.

A final mountain specialty is tourism, nearly as important in Celtic Britain as in Switzerland. Barren hills, moors, and seacoasts that are otherwise unproductive draw the urban inhabitants of crowded England. They are near enough to be within reach of most of England's people. Almost every Celtic town

Highland cattle—Argyllshire, western Scotland. These hardy and well-protected animals are suited to conditions in the Scottish highlands. The loch (lake) and mountains in the background are typical of the highland landscape and attract tourists, especially in the summer. *(From British Information Service.)*

and rural district enjoys income from tourists, and in some areas this is the principal economic base.

Wales and the coal revolution

Coal was discovered in Wales even before the Industrial Revolution. By the end of the eighteenth century, Welsh coal was one of the major ingredients in the industrialization of England. The great virtue of Welsh coal was not so much its quality, although this was excellent, as its location close to water transport (see Map 29, page 86). On the north side of the Bristol Channel in south Wales is an extensive coal field, especially between Swansea and Cardiff, from which coal could be distributed very cheaply. Coal mining and the attendant local industrial development became important features of Welsh life, and brought south Wales into the English economy. The Cardiff area became the principal center of this new development, attracting rural Welsh labor on a large scale. Those who stayed on farms and in the valleys felt these people had sold their souls to the English. This led to the contemptuous term "Cardiff Welshman" for one who had supposedly forsaken his Celtic birthright and deserted to the grimy mines and factories.

North of Wales, the development of steamships in the nineteenth century and the massive industrialization in Lancashire County created the need for a deep-water port. Liverpool, just over the border from Wales, was a suitable site with the addition of artificial harbor facilities. Industrial-commercial England thus grew up on both sides of Wales. Coal was also discovered near Liverpool on the Welsh side of the border, and its discovery further accelerated the industrialization process. "Liverpool Welshman" was added to the vocabulary as this city, too, drained labor from the hill and rural districts. Birmingham and the industrial Midlands completed the circuit on the east. Wales was ringed by industrialization.

Ironically, coal helped make Wales a chronically depressed area, bound economically to an industry characterized by disastrous fluctuations of demand and price. Added to the seasonal fluctuations, Welsh coal grew relatively more expensive than coal from newer areas overseas. There was too little other local industrial opportunity, so Wales played a satellite role as the supplier of raw material to England and the world. It had few profitable alternatives, especially when the coal industry languished. The economic health of south Wales came to depend on coal, and too often its health was poor. A best-selling novel of the 1940s, *How Green Was My Valley,* tells much of this unhappy story. It recounts the ugly transformation of a beautiful countryside by a coal-mining industry that brought Wales too few material benefits in return for what it destroyed.

The manufacturing of iron and steel, the smelting of imported tin ores, and the making of tin plate are highly developed in south Wales, close to the coal. Iron and steel manufacturing has grown in importance, especially since 1945, and has helped to remove the economic depression often associated with south Wales in the past. The largest iron and steel works in western Europe, a plant opened in the early 1950s, is near Port Talbot, just east of Swansea. The British government has tried to locate new industrial development in previously depressed areas, including the Clyde lowland in Scotland as well as south Wales.

Interior Wales is still remarkably isolated and rural, although it is less than a day's drive from London. The long Welsh coastline on the Atlantic has brought little advantage. There are almost no adequate harbors except for fishing boats. More important, the people, the manufacturing, and the trade are elsewhere. There are numerous tourist centers along the coast, but otherwise the sea that made England great brings only storms and unwanted rain to Wales.

Barren Scotland and emigrant Scots

Scotland's development has followed a similar pattern. The country may easily be divided into the highlands and the lowlands, with the line between them extending through the southern foothills of the Grampians, north of the Clyde. The Clyde–Firth of Forth lowland is obviously a separate area. South of it hills are lower than they are to the north. Anglo-Saxon and modern English influences have been strong south of the Clyde for centuries. The traditional Scotsman is a highlander, or at least one who fully subscribes to the sentiment "Hail Caledonia, stern and wild" (Caledonia is an old name for Scotland).

Rural Wales—Merionethshire. Gentler slopes than are characteristic of the Scottish highlands and much more easily accessible from lowland England have hindered Welsh efforts to maintain their political independence in the past, but are a boon to the modern tourist trade. *(From British Information Service.)*

Northern Scotland has the least favorable physical environment in all the British Isles, although its hills and its bleakness are attractive to tourists. The land is largely covered with mountains nearly bare of workable soil. Only scattered patches of farm land are available, and the predominant vegetation is gorse or heather. Climate is highly unfavorable to agriculture; rain is almost continuous, often in the thin showers euphemistically called "Scotch mists." The growing season is damp and cold. Such conditions are suitable for sheep and for some dairying, but this highland country understandably has been a consistent supplier of emigrants, who have gone all over the world as perennial pioneers. Canada and New Zealand still have a noticeable Scottish air. Scots came early to settle in North America; the emigrant Welsh gravitated to the coal-mining areas of Pennsylvania and elsewhere. One of the most famous and typical Scottish emigrants to North America, Andrew Carnegie, supposedly said from his wealth of experience that "pioneering doesn't pay." However true this is, many of the Scots had few alternatives. Not many were as successful as Carnegie, who came to America as a poor boy. The most prosperous ones avoided the frontier and went into business during the expanding economic conditions of the nineteenth century.

Scots ran the largest trading firms in India and China, and are still prominent in the economic and financial life of London.

Industrial Revolution in Scotland

Before the Industrial Revolution most of Scotland was inhabited by poor herders and weavers who shared a fierce national feeling that enabled them to resist English control for centuries. But coal parallels the Clyde–Firth of Forth lowland, and the Clyde estuary is one of the finest deep-water harbors in all Europe, artificially deepened as far inland as Glasgow. This combination of factors in the end proved more effective than Roman or English arms in bringing Scotland within the fold. The Scottish lowlands became a great industrial and commercial center. This area did not concentrate as heavily on coal mining for the support of other areas as Wales, since it could take advantage of its own lowlands and ports to establish basic manufacturing. Shipbuilding became a leading specialty, since iron and steel could be made cheaply and other necessary materials assembled easily beside the Clyde. The lowlands added marine engineering to the list of Scottish accomplishments, another skill Scottish emigrants took with them throughout the world.

Glasgow is the principal center of this industrial development, although Edinburgh has remained the chief administrative and cultural city. Most of the best coal is on the Glasgow side of Scotland. Since Glasgow faces the Atlantic instead of the North Sea, as Edinburgh does, it gained an early and continuing trade advantage. Glasgow also had local iron ore that literally lay on top of the coal seams, so unusually low extraction costs as well as transport advantages contributed to Glasgow's rise as a center of heavy manufacturing. Local iron ores were exhausted relatively quickly, but the Scottish industrial complex was able to draw ores to it from England and from overseas, assisted by Glasgow's location. Glasgow's cheap fuel, its nearness to wool-producing areas in the highlands, and its locational advantages for bringing in cotton from America also provided the basis for considerable textile manufacturing.

Scottish industrialization developed mainly on the lowlands. While it transformed Glasgow into the grimy industrial likeness of the English Midlands, it left the highlands largely alone. Almost as many people in lowland Scotland now live in cities as in England, while in the highlands, which represent nearly three-fourths of Scotland's area and less than one-fifth of its population, most people are still rural. A newly important resource in the highlands, hydroelectric power, has already begun to make some impact on the old patterns. As in parts of Wales, the adjacent industrial lowlands are reaching out for power sources, damming streams and lakes, and setting up a power grid to feed into their electrical systems. Both Scotland and Wales are well situated for this, being mountainous areas of heavy rainfall well within economic range of a lowland industrial market with a high constant load factor. But highland Scotland, like highland Wales, still finds tourists one of its most profitable sources of income. However, the North Sea oil fields are off the Scottish coast, and are transforming ports there with refineries and operations bases. Oil is becoming a major part of the Scottish economy.

Irish landscape—County Antrim, Northern Ireland. The typically green and gently rolling Irish countryside is shown in the foreground, combining field crops with dairying and grazing. In the background are the low Antrim Mountains. *(From British Information Services.)*

Proudly distinctive Ireland

Ireland still stands aloof. Without coal (except for some deposits in the south that are mainly too low grade to be economically workable) and without other resources to attract industrialization until recently, Ireland stayed in a rural, agrarian world. Insularity has doubtless contributed to the strong Irish nationalism and sense of distinctiveness, even as far back as the Dark Ages when Ireland was a center of monastic learning. But Irish distinctiveness and ultimate independence from England have been related also to the lack of common economic development, particularly since the seventeenth century. Without manufacturing or trade on an English scale, Ireland remained a different kind of place. Repeated English devastation of Ireland to stamp out rebellions, and the unfortunate system of English absentee landowners further embittered the Irish and strengthened their resistance to union with England. Ireland has fewer mountain barriers than Wales or Scotland. There is more productive land as a conqueror's prize and more opportunity for wholesale military success. But English victories stimulated Irish nationalism rather than destroying it. In the end, Ireland was granted independence because the cost of denying it, in lives and continued conflict, was too great and outweighed the questionable benefits.

The Republic of Ireland (Eire), independent of all ties with England since 1922, consciously emphasizes its own distinctiveness. It supports Gaelic as well as English in its school systems and in its official policies. The Republic has cultivated a rural attitude and fought against becoming a cultural or economic subsidiary of England. To some degree it has cut off its economic nose to spite its political face. As an almost exclusively agricultural country, it would normally (and did before independence) sell almost all of its exports to food-importing England, and buy there almost all of what it needed to import. The Irish government reduced this dependence somewhat, but at an economic cost. Despite efforts at distinctiveness, Dublin, the capital and largest city, remains a British city in many respects, speaking English, living in an English cultural world, and sending a stream of emigrants to England in search of better opportunity. The list of Dubliners and Irishmen who have won high places for themselves in English affairs is long and distinguished, including Jonathan Swift, Edmund Burke, the Duke of Wellington, George Bernard Shaw, James Joyce, and many others.

Since about 1960, official relations with England have improved considerably, and the memory of a bitter past has begun to recede with a new generation. The Irish labor pool, cheap land, and relatively low tax levels have attracted German and other European capital for light industrial investment. Manufacturing is rapidly growing in importance in the economy, and the emigration of young Irish workers has decreased greatly. Urbanization and income levels are rising fast. Ireland is changing, becoming less distinctive and more a part of an expanding European industrial-commercial economy.

Partition

Northern Ireland, or the region of Ulster, remains under British control. There has been heavy British acculturation there, and considerable settlement by both Scots and English. Northern Ireland is more accessible from Britain, especially from the mouth of the Clyde. For this and other reasons, it was more attractive to the Scottish and English settlers. The Mountains of Donegal act as a partial shield, casting a modified rain shadow over most of Ulster and helping to bring higher summer temperatures. The border around Northern Ireland excludes these mountains. Ulster is also better drained than most of the rest of Ireland, more productive agriculturally, and more like England or lowland Scotland physically. Belfast has a major deep-water harbor conveniently located to serve the productive part of Northern Ireland and the west coast of England and Scotland. An extensive shipbuilding industry has grown up here, taking advantage of this location for cheap assembly of heavy raw materials and distribution of finished products in close proximity to the Scottish shipbuilding centers. The Irish population of Ulster, as well as the English and Scots, have been mainly Protestant rather than Catholic ever since the Reformation. This further underlines its distinction from southern Ireland. But tensions between Catholic and Protestant groups in Northern Ireland have led to violence and terrorism in Belfast and elsewhere. They have become chronic since 1969, despite British efforts at military and political control. Many Ulster Protestants resist the idea of union with the Irish Republic, but that may be the most promising long-run solution.

Irish agriculture

Although Ireland is more favored agriculturally than Wales or Scotland, this does not make it a good agricultural area. There are extensive lowlands, but there is also excessive rainfall. The Emerald Isle, as Ireland is fancifully called from its green, grassy landscape, is usually grey, receiving the full force of rain-bearing winds off the Atlantic. Topographically Ireland is roughly similar to a shallow saucer, with most of the mountains in an exterior fringe and most of the lowlands in the central plain, where drainage is frequently poor. In this sodden environment, with its poor soils and inadequate growing season, most of the major cereals cannot be grown profitably, and their place is taken by potatoes (see Map 26, page 77). Potatoes tolerate poor soils and damp, cool climates. Even under these conditions, they may yield as much food per acre as a cereal crop under good conditions. When they were introduced from the Americas in the seventeenth century, Ireland's population boomed, but its agriculture was based on an almost one-crop system. The great potato famine of the middle 1840s, followed by mass Irish emigration overseas, particularly to America, resulted from unusually wet weather that rotted the potatoes in the fields. Ireland's population decreased by half in the remainder of the nineteenth century while the rest of Europe and most of the rest of the world continued to grow rapidly. Ireland's losses were mainly through emigration rather than famine, but both point to poverty and an overcrowded agriculture with limited domestic alternatives. Dairying, poultry, pigs, horses (for hunting, draft, and meat), and sheep have since become more important, and are reasonable uses of Ireland's fine upland and lowland pastures. Nearby markets in industrial urban England have also encouraged this development. Ireland supplies much of England's butter, eggs, bacon, cheese, milk, and meat.

The United Kingdom

Wales was the first Celtic area incorporated with England at the end of the thirteenth century. Scotland, after losing and regaining a measure of independence several times in successive struggles, joined with England under a common king in 1603. The final Act of Union in 1707 joined Scots and English in a common parliament at Westminster, and opened to Scotsmen full political and economic privileges within the union. Regional feeling still flourishes in all the Celtic areas, whatever their political status. Scotland and Wales support nationalist parties. They continue to press for full independence, and have recently won a much larger degree of local administrative autonomy. Although local sympathies flourish, most people are willing to regard the political union as settled, in the main, barring the sectional grievances that are part of every multi-cultural state. For good or for ill, Wales and Scotland share a common destiny with England, primarily as a result of the Industrial Revolution. This does not necessarily mean that Scottish and Welsh individuality is doomed; indeed it seems extremely vigorous. As long as the Welshman sings or the Scot plays his pipes, no one will mistake him for an Englishman. Should anyone unaccountably make such an error about a citizen of the Irish Republic, he is likely to regret it.

Questions for further study and discussion

1. Why is it so difficult to determine the population of a city? With the aid of large-scale maps of the New York, Tokyo, and London areas, what conclusions can you reach? How would you draw significant lines around your own city? Around Boston, Mass.?
2. Why did London become an insurance and financial center?
3. The English Channel, geologically speaking, is quite recent; not many thousands of years ago, it was only a large river. In what ways would modern Britain be different if the Channel had remained that size?
4. Why has eastern England (around Norwich and Ipswich) changed so much less than the rest of the country since the Middle Ages?
5. Trace the boundary between Scotland and England. How do you account for its course? Does it seem a suitable one, and does it accurately divide two distinct areas? How does it compare in these respects with the boundary between England and Wales?
6. Why has no large city except Cardiff developed in Wales? Why do most of the existing urban areas cluster so closely in the extreme southeast?
7. Why should many Irish soils tend to be poor?

6 Scandinavia and the Low Countries

Early morning in Copenhagen. *(Photo from Standard Oil Company of New Jersey.)*

Scandinavia is an example of varying adjustments to an originally poor land. Most of Norway, Sweden, Finland, and Denmark were relatively ill supplied with usable resources by general standards. Yet they are highly developed, progressive states with an enviable record of accomplishment. Physical poverty appears in mountainous terrain, in a generally marginal climate, in poor soils, and in the absence of extensive industrial raw materials.[1] Each of the three major countries, Norway, Sweden, and Denmark, has a different set of advantages and disadvantages. But they cannot be considered physically favorable by comparison with most of the rest of northwest Europe. Nevertheless, each has attained a high level of cultural and economic development. Sweden and Denmark have long had some of the highest per capita incomes in Europe. Each country has adjusted to a particular geographic situation. Comparing them will bring out their distinctive adjustments.

Norway

Norway has a classic maritime economy, and consistently maintains the third- or fourth-largest merchant marine in the world. With a population of around 4 million, the size of their merchant marine is amazing. Norway is important as a European commercial fishing nation; most of the population hugs the sea. Norway's heavy dependence on the sea is closely related to the lack of profitable alternatives. Other maritime powers depend on advantages of location, a commercial base, a strategic site for break-in-bulk functions, a large foreign trade, or at least a large home market. England, Holland, Japan, Denmark, Sweden, the United States, Germany, and even Italy (the other main maritime powers) especially contrast with Norway in this respect. Norway has a few advantages: a long seacoast with good natural harbors, nearby fishing grounds in the North Sea, and a position on the Atlantic. But Norway's location is off to the side, not convenient to Europe, and most of Norway is far from the main shipping lanes. Its tiny population and generally unproductive area are neither a significant trade generator nor an important commercial base. There is relatively little manu-

[1]Norway's share of the recently discovered oil and gas fields under the North Sea has greatly changed the Norwegian resource base, but its land areas are not well endowed.

facturing to stimulate foreign trade in raw materials or finished products.

Norway has almost no really good ports, as distinct from harbors. The fjorded coast does contain many fine harbors. But a good shelter for ships is not the same as a center of land-sea exchange. Often the two do not coincide. A port is a center of land-sea exchange because it has access to a productive hinterland and provides a place where ships can dock. A country that produces little or nothing cannot support a port. The harbors may be good, but there is nothing to be exchanged. A productive country seeking outlets for its products or needing to import goods will create an artificial harbor if necessary, or

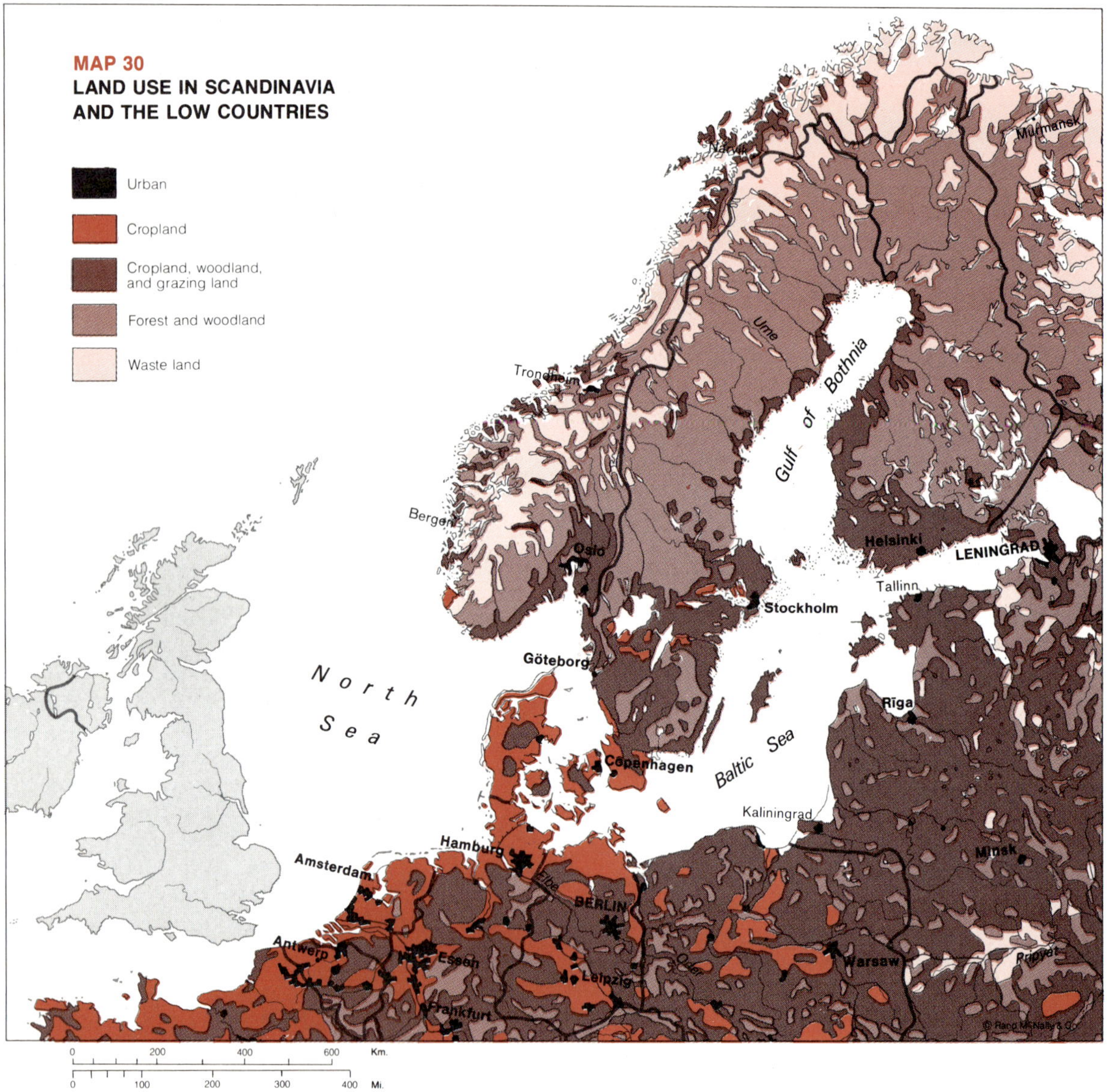

make use of a site that is physically unsuitable but economically convenient. Many of the world's largest ports, especially those in river deltas that can tap a large hinterland, have poor harbors. *Port* is an economic concept, *harbor* a physical one.

The glaciers, covering Scandinavia until relatively recently as measured in geologic time, left behind after their retreat a mountain country scraped nearly bare of soil, deeply gouged into steep valleys, and fringed on the seaward side by fjords like the one in the photograph on page 99. Fjords are trenches, deepened by the glaciers where they fell into the sea, then further flooded as the continental ice sheets melted and the sea level rose. The fjords are deep, well-protected waterways for shipping, although there is sometimes a shallow sill at the sea entrance that must be dredged. Inside the sill the extreme depth of the water with its sheer rock bottom may create anchoring problems, but in general, fjords are excellent harbors. In North America roughly the same sort of fjorded coast occurs northward from Puget Sound through Alaska, at latitudes comparable with Norway's. Since both lie on the westward edge of a large continental land mass in the zone of the prevailing westerly winds, the climate is not nearly as cold as at the same latitude on the east coasts. Norway further profits from the North Atlantic current (Gulf Stream), which sweeps warm waters up the entire coast; thus the fjords are seldom frozen. But the glaciated mountains that lie behind most of the Norwegian fjords are very sparsely settled, and produce little for trade. Access to the inland areas is also difficult and expensive, because of the nature of the fjords as well as the mountains.

Norwegian cities and ports

All Norwegian cities, as distinct from market towns, are on the sea at the heads of fjords. Oslo, with a population of nearly half a million, is the only one of large size, because it is the only one that has access to a more or less productive hinterland. Bergen, Norway's second city, with a population of about 180 thousand, has a harbor as good as Oslo's, but Bergen's tributary hinterland is much less productive. Until the present century and the completion of a railroad along the difficult mountain route to Oslo, Bergen was actually connected more easily and closely, both commercially and culturally, with Britain, Denmark, and southern Norway by sea. The land

Fjord and farm—coastal Norway. This is the upper reach of a fjord showing the glaciated mountains of the *fjeld*, and a typical mountain farm, most of it good only for rough pasture. *(From Norwegian National Travel Office.)*

area that lies behind Bergen is still difficult to reach and of very slight commercial importance.

Aside from Oslo and Bergen, both at the southern end of Norway and both profiting to different degrees from nearby or accessible lowland, Norway has no other large cities. Trondheim, the third largest, has a population of about 135 thousand. It, too, has a lowland around it, but it is quite far north. With these exceptions, there are no productive lowlands in Norway large enough to support more than a small town, and no trade functions to make effective use of most of Norway's excellent fjord harbors. Narvik, in the far north, really serves a Swedish hinterland, and is discussed under Sweden.

The Norwegian region

Norway includes the barrenest, barest, rockiest part of Europe. By comparison Sweden is much better off. It occupies the gentler eastern slopes of the Scandinavian mountain chain that merge into the coastal plain on the Swedish side well before they reach the sea. Sweden has much level or rolling agricultural land and more accessible timber. The political distinction between the two countries is relatively recent. Norway was united with Sweden from 1814 to 1905, having been joined previously with Denmark for several centuries. The separation from Sweden, when it finally came, was a recognition of strong Norwegian nationalism, understandable considering the obvious physical differences between the two countries. The mountains and the sea in Norway developed a very different culture and society than the plains of eastward-facing Sweden.

Norway's domestic resources, aside from the sea, are limited. There is no significant coal and only a little iron ore in central Norway. Oil and gas from the North Sea fields are a major resource now beginning to make a big impact on Norway. Most of these resources will probably be exported, given the very small size of the domestic market, but they will greatly increase the country's wealth and living standards. Adjusting to this new wealth may now be Norway's major problem.

Agriculture in this mountainous, poor-soil country is limited and uncertain. The fjorded west coast offers only tiny patches of level land. On the central plateau, or *fjeld,* soil is thin or nearly absent as a result of glaciation, and even trees are lacking over large areas. Only in southeastern Norway and in a small strip of the southwestern coast called Jaeren, south of Stavanger, do mountains and nearly bare rock give way to more rolling hills and deeper, though still not strikingly fertile soils. In the south winters are fairly mild, but summers are wet and cool. Turnips, barley, oats, potatoes, cabbages, and similar crops must often take the place of the more desirable cereals like wheat or even rye, which cannot be grown profitably in most of Norway.

Animal husbandry is the largest single sector of the agricultural economy. The animals make use of land that is marginal agriculturally. The acute shortage of good land also leads to the use of grazing areas that are marginal even for animals. Dairying is important in the Norwegian economy. Much of the Norwegian cheese is made from goat's milk, since this nimble animal eats a wide variety of food and can maintain itself even under Norwegian conditions. Cattle are concentrated in the limited lowland areas, or on the nearby gentler slopes, where their milk and meat are accessible to urban markets. Sheep, also pastured mainly on the uplands, are far more numerous than cattle.

Large parts of Norway are covered with coniferous (evergreen) forests, but this does not mean that all of the timber is a usable resource. Most of it cannot be cut and brought out at a reasonable cost. Mountains, remoteness from population centers, the general lack of railways (there is too little population or business to support them) and of navigable rivers, plus the physical access problems of the fjord harbors, all contribute to this. Logging is expensive on slopes, and a product like wood, relatively low in unit value, cannot be moved long distances economically except by very low-cost carriers. Where the forests are more accessible or nearer the sea, they are used extensively. For a long time timber and pulp were Norway's leading exports, moving to markets all over the world by cheap sea transport. Most of the wood comes not from the mountains but from the lowland and hill country in the southeast around Oslo. The mountains and the *fjeld* of central Norway, with their thin soils and high altitude, are not extensively forested, in addition to having poor access.

Potential hydroelectric power is plentiful in Norway, but its use is limited by many of the same disadvantages that hamper use of Norway's forests. Mountains and a moist climate may mean potential water power, but possible power sites for much of

Norway are too far from large markets to make their development profitable, even though there are no other domestic power sources available. Hydro power is heavily used where it is accessible, particularly for much of Norway's limited manufacturing. Despite the problems of hydro development, Norway has the highest per capita production of hydroelectricity in the world. This calls attention to the lack of other resources until recently, making heavy dependence on hydro power necessary, and keeping the size of the population down. By west European standards, industrial development is relatively small.

The familiar mountain specialty of tourism is important in Norway, cultivated and advertised by the government and railways, and an important source of foreign exchange. Folk art is also commercially important. Emigration, especially to the frontier areas of North America, has taken place on a large scale over a long period of time. Norwegian achievements in polar exploration are understandably famous. The other achievements of Norwegian development—a creative culture that has contributed far more than its share to European arts, learning, and diplomacy—are even greater evidence of how this small nation has risen above its material problems in the past. Now it must contend with the problems as well as the benefits that its new oil wealth may bring.

Skåne farmland. These gently rolling plains and good soils in southwestern Sweden are in sharp contrast with Norway. *(From Swedish National Tourist Office.)*

Sweden

The northern mountains between Norway and Sweden are high and rugged. Their crests mark a physical as well as a political boundary between Norway's maritime region and continental Europe. Sweden lies in a noticeable rain shadow that clearly distinguishes it from most of maritime Europe (see Map 23, page 73). It tends to have colder winters and warmer summers in common with the continental areas. While Sweden has strong ties with the Atlantic world and a vigorously Western culture, geographical position limits Sweden's freedom of action. Norway joined the North Atlantic Treaty Organization as a matter of obvious self-interest, but Sweden has held aloof wherever possible from all power alignments. It cannot afford to alienate the states and economies of central and eastern Europe.

The Baltic Sea is no substitute for the Atlantic. In this largely continental climate, the sea and its harbors freeze in winter for varying periods depending on the latitude. There is ice at Stockholm for about two months, although icebreakers normally keep the port open all winter. The Baltic is a small sea and has only a narrow, easily blocked outlet. Land power usually dominates the Baltic, and naval might is relatively ineffective. Sweden was the dominant Baltic power at various periods until the eighteenth century, when the rise of the great Russian state and the subsequent development of Germany ended Swedish imperial ambitions permanently. These much larger and more productive countries had the further advantage of massive industrialization with which Sweden was unable to compete.

Domestic resource base

Sweden has physical advantages that Norway lacks; dependence on the sea is less necessary. There is more relatively good agricultural land in central Sweden on the edge of the wide coastal plain and also at the southern tip of the country. The somewhat drier climate is an asset, and the summer growing season is warmer and longer. Wheat is grown profitably as far north as the Stockholm area, and is especially important in southwestern Sweden and in the southernmost tip of the country, called Skåne (see Map 24, page 75). Skåne is significantly farther south than any part of Norway, and no important mountain barriers lie between it and the westerly winds off the Atlantic. Barley, rye, and oats are also grown widely in Sweden.

Sweden's extensive coniferous forests are concentrated mainly in the north, away from the major centers of population. Nevertheless, they are far more accessible than Norway's forests. They lie on the gentler eastern slopes of the mountains, where there are also a large number of rivers that can be used to float the timber to market. The northern rivers freeze in winter, but enough logs can be floated down in the spring and early summer to supply the pulp and sawmills at the river mouths for the whole year. The trees are cut mainly in winter, when movement and transport by sleds and skis within the forest is easier, and labor can be spared more easily. The largest timber production, however, comes from southern Sweden on the poor glacial soils of the lowlands north of Skåne. There agriculture is unprofitable or impossible, but higher temperatures than those in the north mean faster growth for timber.

Hydro power, heavily used for the support of Sweden's urban centers and for the demands of manufacturing, is well developed. Usable minerals are present, most importantly iron ore, although there is very little coal. The iron ore is exceptionally rich, averaging between 60 and 70 percent pure iron. It is one of the largest reserves of high-grade ore in the world. Its quality makes it suitable for the manufacture of special steels, apart from the richness of the ores. The most important mines are inconveniently located near the northern tip of the country at Kiruna (see Map 21, page 71), but there are others in central Sweden, not far from Stockholm. One of the earliest iron industries in Europe developed here, and still remains important.

Kiruna is inconvenient not only because of remoteness but because the Baltic freezes during three or four winter months this far north. The ores move by rail across the mountains to Narvik on the Atlantic coast of Norway. Narvik is closer to Kiruna than any Baltic port, and it is never frozen; ore shipments are Narvik's major function, aside from its small role as a fishing base.

Lacking coal, Sweden is not in a good position to make iron or steel from its ores on a competitive basis for export. Domestic iron and steel production in central and northern Sweden, using imported coal, is large enough to supply the Swedish market. Most of the ores are exported for the manufacture of high-quality steel for cutlery, Swiss watches, and other products needing steel with special properties. Cheap hydro power does enable the Swedes to smelt and refine some of the ores, and to make very fine steel by the electric furnace process, much of which is exported. This steel is more expensive than steel made with coal, regardless of the ore used, so despite its high quality, it has only a limited market. Razor blades, ball bearings, precision parts, and products that require unusual resistance to heat, wear, or corrosion are characteristic Swedish specialties.

Nevertheless, Sweden, especially in contrast with Norway, has been a country of domestic opportunity. The Swedes had little part in the Viking raids carried on by Norwegians and Danes. In more recent times, Sweden's political expansion was limited to the lands immediately around the Baltic, in what may be called a power-vacuum situation. Thus, Finland and most of the southern Baltic shores were included in Sweden for several centuries.

Swedish cities

Göteborg is the country's major maritime center. It lies on what is really the best Atlantic coastline in all Scandinavia. This area was part of Norway or Denmark until the seventeenth century, and its development in Swedish hands as a maritime trading center is quite recent. Göteborg is now the base of Sweden's large merchant marine, and is also a prominent shipbuilding and automobile-manufacturing center. Shipbuilding is probably Sweden's leading heavy industry. Production is considerably greater than in the United States, much of it for foreign customers. Göteborg has a population of over half a million, while Stockholm (including suburbs) has about 1.3 million, or about one-sixth of the national total of slightly over 8 million. Göteborg is the country's largest port in shipping volume, and handles most of Sweden's imports and exports of finished goods. Malmö, the third city, located in southern Skåne opposite Copenhagen, has a little over 400 thousand inhabitants, and is mainly a service center for this most productive part of Sweden's farmland.

Since the late nineteenth century in particular, Norland (northern Sweden) has become important as a source of iron ore and timber, so Stockholm is now less on the fringe of the Swedish core than it was in the years following the loss of Finland and the addition of Skåne. Stockholm's central position within the Baltic area was also an important asset in the earlier period of Swedish dominance in the Baltic. Stockholm is now largely an administrative and cultural center. By far the largest Swedish city, it also serves as the dominant central place for all of Sweden, and contains the headquarters of most Swedish business firms and other organizations. Its lagoon harbor is not well suited to most deep-draft shipping, so it handles little of Sweden's large foreign trade. But its financial services are important. Simple size, as usual, is a great advantage, and Stockholm also profits from its function as the national capital.

Sweden buys some of its food abroad, but has little difficulty maintaining a healthy trade balance because of its valuable exports. Sweden has one of the highest per capita levels of foreign trade in Europe, and income from its merchant marine in the shipping trade is also very large. The greatest economic problem is fuels. The lack of domestic coal and oil is especially awkward in wartime, and has stimulated large Swedish investment in domestic nuclear power plants. Like most of Scandinavia, the Swedish land is mostly physically poor, as the relatively small population in such a large area may suggest. The human achievements of Sweden are remarkable, politically, culturally, and economically. Geographical location does not affect the vigor of Swedish democracy. Neutrality in this case does not mean impartiality, but intelligent discretion by a country that cannot hope to upset the balance of power and that would be ruined quickly in any war with Russia or Germany. Certainly Western naval power could be of little help. Neutrality has brought solid rewards, but the record of Swedish accomplishments is no less impressive because the Swedes have chosen not to subject the country to the destruction of war.

Finland

Despite the central Asian origin of the Finnish people and their language in the remote past, Finland is culturally an integral part of Scandinavia, and is especially closely related to Sweden. For many centuries it was part of Sweden politically, and its cul-

Finnish summer landscape. This scene is in what the Finns call the eastern lake district. Most of Finland is thickly dotted with lakes in this recently glaciated area. The land is covered with the coniferous and birch forest shown here. *(From Embassy of Finland.)*

ture still retains a strong Swedish element. Most of Finland is a low, glacially scoured plain, thickly dotted with lakes, especially in the southern half—a landscape similar to that of central and northern Minnesota and with a similar climate. The recently deposited glacial soils mainly have low fertility and, in many areas, poor drainage. The subarctic climate adds a further disadvantage for agriculture, especially the critically short growing season. Barley, oats, wheat, hay, and potatoes are the chief crops, but cultivated land occupies only about 8 percent of the total area.

Finnish land use is dominated by forestry, and forests are its chief resource. Pulp, paper, and lumber account for about two-thirds of its exports. Unfortunately, coal and oil are lacking, there is very little iron ore, and few other significant mineral resources. But, as in Norway and Sweden, there is extensive development of hydroelectric power. Finnish industrial structure is dominated by the processing of forest products. It is concentrated in the southern edge of the country where all of the major cities are also clustered. Helsinki, the capital and largest city, lies on the shores of the Gulf of Finland, an arm of the Baltic. In this location, it serves the bulk of the population, most of which is concentrated in the southern fourth of the country.

Northern Finland is very thinly settled, a largely unbroken expanse of birch, pine, and fir forest, snow-covered for nearly ten months of the year. The far north is occupied by small, nomadic groups of Lapps, a people and culture related to the Eskimos, but dependent on herds of reindeer, which graze on the tundra vegetation and are hardy enough to endure the Arctic winters. Lapp occupance extends over all of northernmost Scandinavia, including parts of Norway and Sweden. But like most nomadic groups elsewhere, their culture and numbers are declining because of growing contact with more highly developed groups and national states, where opportunity is greater. The Lapp population dwindles as those in each successive generation find more opportunity in the larger Finnish, Swedish, or Norwegian worlds.

Finland itself has suffered from even greater pressures from its overpowering Russian neighbor. As a consequence of the Second World War, it lost a strategic piece of its most valuable territory in the southeast around Lake Ladoga (Ladozhskoye Ozero), including the Karelian Isthmus, which the Russians wanted as a protection for the approaches to Leningrad. However, Finland has successfully retained its Scandinavian character in cultural and political terms, including the distinctive Democratic Socialism that it shares with Sweden, Norway, and Denmark. This political structure has played a major

Copenhagen. The statue of the little mermaid on the rock is a Copenhagen landmark, but the city's chief business is its port, part of which shows here. *(From Danish National Travel Office.)*

role in enabling all these countries to create a society in which, despite limited domestic resources, public welfare and free expression have been preserved and enhanced, winning them wide admiration.

Denmark

Denmark is a morainic (glacial) spit of land dominated by a single city. Greater Copenhagen (København) contains well over one-fourth of Denmark's total population of about 5 million. The other urban settlements are towns rather than cities, and agricultural market and service centers. Copenhagen, like Stockholm, is a strong example of a *primate city*, one that is at least twice the size of its nearest domestic rival. Most national states contain a primate city, but, in some, the overwhelming advantages of its urban location have led to unusual dominance. London, Berlin, Paris, Tokyo, Oslo, Bangkok (Krung Thep), Rangoon, Djakarta, Colombo, Montevideo, and Buenos Aires are other examples. In most of these cases, size appears to have a self-perpetuating quality; once a city has become the largest in its domestic unit, more and more advantages of specialization and market attraction come to it. Copenhagen dominates Denmark to a degree unequalled in any other country but Uruguay, although Denmark is small in both area and population and cannot really support more than one large city. Copenhagen is not on the main peninsula, called *Jutland* (Jylland), sticking out into the North Sea like a thumb, but on the island of Sjaelland (see Map 30, page 98). Like the rest of eastern Denmark, Sjaelland is geologically different and is more closely related physically to Sweden. The spits, bars, and beaches along the northwest coast of Jutland, marking a sandy coastline of *emergence* characteristically having shallow water and few good harbors, are much less evident around rocky Copenhagen, which has an excellent natural harbor, the only one of any consequence in Denmark.

Sjaelland effectively blocks the entrance and exit of the Baltic, and Copenhagen (literally "merchant's harbor") lives on the Baltic trade. It is a major break-in-bulk point between the Atlantic and the Baltic, assembling and distributing goods from and to both. Deeper-draft Atlantic shipping finds Baltic harbors too shallow, while Baltic shipping is often an uneconomic size for the Atlantic. The Baltic ports do not generate enough trade to support many or frequent large shipments. Copenhagen serves as an entrepôt; smaller ships that collect and distribute goods among a number of smaller ports within the Baltic region connect at the entrepôt with larger ships hauling greater distances to and from larger ports.

Agricultural development

The Danish participation in the Viking raids of the Dark Ages reflected an earlier poverty. The commercial development of Copenhagen from the Middle Ages onward changed the situation. With the revival of trade in Europe, there was also a prosperous agricultural development in the good clay soil areas of eastern Jutland and the Danish islands, especially the raising and export of cattle. But as late as the 1880s, the sandy, glacial plain covering most of Jutland was still a relatively poor, backward place.

The development of large, urban markets in nearby Britain and Germany provided a new opportunity. The agricultural system was revitalized with a nationwide network of cooperatives, and became one of the most efficient in the world, producing high-quality dairy, poultry, truck, and specialized meat products. The sandy soils of western Jutland are not highly fertile, but good soils are not necessary for poultry raising or dairying. Pigs, like poultry, fed largely on imported grain, were another profitable use of poor soil areas. They also consume skim milk, a by-product of dairying, and are often raised in dairying areas for this reason. In Denmark, sausages, bacon, and ham were added to the list of high-value exports. Located at a more southern latitude than most of Scandinavia, open to the Atlantic, and with its peninsular character and the moderating influences of the sea, the climate of this flat country is mild. This is an advantage for the raising of poultry, swine, and dairy animals, since their housing costs are low, and more winter grazing is available than in colder places.

The better soils can be fertilized for field crops, and their light and well-drained structure is an advantage. They warm up quickly and, with the maritime temperatures, are able to produce vegetable crops early in the spring, a valuable asset in truck farming. Cooperative organizations and government planning have established high standards and an excellent reputation for Danish animal and agricultural

Sjaelland landscape. This productive agricultural land in a mild climate is well tended in typically Danish fashion. *(From Danish National Tourist Office.)*

goods. They bring high prices from the nearby food-importing, industrial regions. Danish prosperity often reflects the state of the British and German markets. The Danes have an intensive agricultural system of predominantly small farms, partly because of the nearness of the market and the consequent concentration on high-value or perishable products.

The Scandinavian countries are a challenging example to the rest of the world. Economically they show how much can be done by hard work, careful national planning, and skill with a physical base that is much less favorable than many economically underdeveloped areas enjoy. Socially, culturally, and politically, they show that achievements in the art of living and contributions to civilization do not depend on political power or numbers. The challenge is as appropriate for the great powers as for the economically underdeveloped parts of the world. Scandinavian economic, social, and political democracy, their constructive role in international affairs, their cultural accomplishments, and the success with which the people and governments of Scandinavia have met the problems of the modern world in their own countries provide an impressive model. Their accomplishments are especially impressive when measured against the relative poverty of their land and the small size of their population. These less than 20 million people (Sweden has fewer people than Tokyo, or New York, or Paris) have set a mark that few others in the world have equalled.

The Low Countries

The coastal plain along the North Sea between the French and German borders, roughly bisected by the mouths of the Rhine,[2] has had a complex political and cultural history. In the late Middle Ages the revival of European trade and the beginning of textile manufacturing was centered mainly in Flanders, the area occupying the northern half of what is now Belgium and part of the southern Netherlands (see Map 31, page 107). For several centuries Flanders and its Germanic people dominated the economy and culture of the Low Countries, and the flourishing commercial-industrial cities of Ghent (Gent), Ieper (Ypres), and Bruges (Brugge) led European development. Private merchants and weavers and their guilds (associations) controlled the government of these and other Flemish towns and briefly won their independence from feudal or royal authority. But in the fifteenth century Flanders and most of the Low Countries came under the domination of Spain. Protest movements against Spanish rule culminated in a long war of independence that lasted from the middle of the sixteenth century until 1648. The Dutch, whose trading activities at the Rhine mouths had begun to take the commercial lead away from Flanders, were the political and military leaders throughout the long struggle for independence. With the successful conclusion of the war, which had repeatedly ravaged Flanders, the north emerged as the chief center of the Low Countries.

During the war, the Dutch Republic extended its control southward into parts of the largely Catholic provinces of Flanders, Brabant, and Limburg. These areas still have a strong Dutch culture, and are still Catholic in religion. The rest of the Low Countries, what is now Belgium, remained under Spanish, Austrian, or French control until Napoleon's defeat in 1815. Then, Belgium was incorporated with Holland.[3]

[2]The Rhine (Rijn) River divides into the Waal River, the Lek River, and the IJssel River, all of which flow into the North Sea in the Netherlands.

[3]*Holland,* strictly speaking, is the name of one of the western provinces of the Netherlands along the coast, although the name is generally applied to the whole of the Netherlands. *Netherlands* means simply "low lands."

MAP 31
THE LOW COUNTRIES

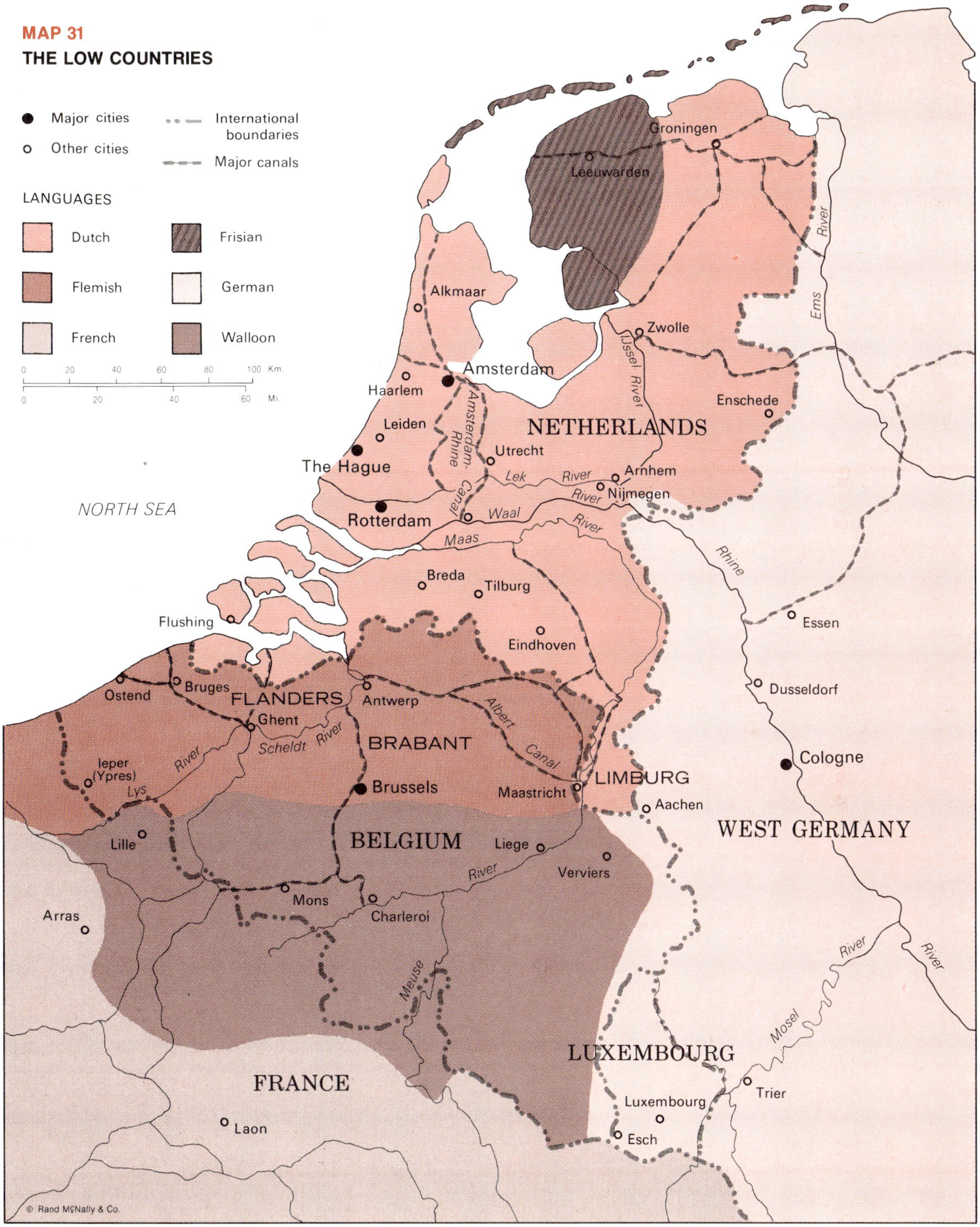

The Low Countries enjoyed a brief period of political unity unlike any they had known since the Middle Ages. But cultural differences and regional feelings were strong, and the union was not successful. With the creation of an independent Belgian state in 1839, the present political pattern was established.

The Netherlands

Protection and distinction for the Dutch had come earlier from the deltaic marshes that originally covered part of their country; they helped to keep the Romans out. After the seventeenth century especially, extensive drainage operations replaced most of the marshes and swampy lagoons behind the curving sand bars along the coast with dry, dike-encircled *polders* (see Map 32, page 110) drained by windmill-driven pumps. Crops might be raised or cattle grazed on this land. An intricate system of drainage canals made this possible. The canals also had some defensive value; they could be used to flood the land and restore something of its original barrier character.

But marsh defenses are hardly adequate to explain the development of the Netherlands as they have become almost entirely ineffective in the age of modern warfare. Canals and dikes were cut in the Second World War, but the speed of the German invasion by mechanized troops and the strength of the German air force made this a largely futile gesture. Most of the Netherlands is in easy range of foreign naval or land artillery and has been for many years. Earlier, the size and strength of surrounding states could not have been resisted only with the help of marshes and canals. In any case, most of the country was never marshy. It is covered by glacial sand deposits, moors, and relatively well-drained alluvium in the valleys and flood plains of the Meuse (Maas) and Rhine rivers.

Most national units depend on some source of economic strength for their existence and growth. Control of the mouths of the Rhine has been a consistent basis of strength for the Dutch state since the Middle Ages. The Rhine is Europe's greatest trade highway, and the ports near its mouths are the mainland's most important break-in-bulk points. The Dutch have never been dislodged from their position here despite the smallness of their country, the feebleness of its physical protection, and the determined efforts of surrounding powerful enemies to take this prize for themselves. Control of the Rhine mouths enabled the Dutch to build a prosperous state that could maintain itself economically against competition.

The European balance of power has also been involved. None of the great states was willing to see the Netherlands controlled by any of the other powers because of its strategic location within Europe. Every attempt at conquest of the Low Countries has been met by a coalition opposition. England, although frequently at war with the Dutch itself, sprang to Holland's defense against Spain, against France, and against Germany. Germany, with the massive industrial might of the Ruhr just across the Dutch border, met united European resistance in its attempt to incorporate the Netherlands, just as in earlier days Prussia had defended Holland against France. The survival of most small states must necessarily depend to some degree on such great power rivalries.

But Dutch survival is also due to the economic advantages of location. The Netherlands developed trade as a specialty even before Britain did. England exported raw wool to textile-manufacturing centers in Flanders long before it began to manufacture cloth. The Low Countries became a prototype of the urban commercial society of modern Europe, a development in which the Dutch shared increasingly after the decline of Flanders. Vermeer, Rembrandt, the van Eycks, Pieter de Hooch, and other Netherlands painters record on their sixteenth- and seventeenth-century canvases groups of Dutch burghers (literally, city dwellers), who might be contemporary members of a board of directors or solid citizens of a modern city.

The Dutch Republic had its formal modern origins in the Treaty of Westphalia in 1648, which ended the long war with Spain. But the federative union of all of the Netherlands provinces developed considerable representative government well before the seventeenth century. The estates, or parliament, were similar to the English House of Commons, and through all the period of its modern growth, Holland has remained a vigorous representative democracy.[4]

[4]The addition of a monarch since 1814 has not been a contradiction.

The close connection between this form of government and the prominence of merchants in Dutch life appears for the same reasons it does in England. The Low Countries encouraged economic and political individualism and enterprise, and the political forms that accompanied them.

Land from the sea and Dutch agriculture

Perhaps the most remarkable thing about Holland is the Dutch achievement in draining and reclaiming from the sea, sand dune, and deltaic marsh the productive agricultural land that now covers the landscape. Over a third of Holland's present land is below the level of mean high tide in the North Sea, protected by massive dikes, and kept dry by pumps and canals. Much of the rest of the country is artificially drained. This work has been in progress for centuries. The largest recent project is the reclamation of land from IJsselmeer (see Map 32, page 110), originally lost to the sea in the Middle Ages. The present *delta plan* is designed to protect Holland against catastrophic storms from the sea. Strong dikes have closed off all the tidal channels between the islands along the southwest coast, except for the mouth of the Scheldt (Schelde) River leading to Antwerp and the artificial waterway serving as a mouth for the Lek River and as a commercial outlet for the port of Rotterdam.

Intensive land use in the Low Countries. The location is Flanders and the crop is flax, heaped to dry. Cultivation is intensive and yields are high. But the industrial and urban part of the Low Countries' economy is also shown in the background. This is a typical Belgian or Dutch scene. *(From Belgian Linen Association.)*

Reclamation and drainage on such a scale are expensive, and Dutch land produces high-value crops. Holland is famous for perhaps the most valuable crop all, flower bulbs. Special varieties may bring as much as a dollar a bulb, and bulb land is worked very intensively with heavy fertilization. Truck crops for the huge nearby European market and dairying in the less well-drained or poorer pasture areas make up most of the rest of the agricultural system. Fluid milk is profitable in this location, but butter and cheese are also made in an attempt to wring still more return from dearly bought land. Cheese, of course, can be exported overseas.

Holland does not attempt to feed itself completely in cereals. Just as England found it more profitable to specialize in manufacturing, Holland imports grains and uses its own limited land as much as possible for more lucrative crops. Fortunately there is much alluvial soil, brought down by the Rhine and its tributaries that is highly productive with proper drainage. Soil fertility helps to justify the heavy investment in reclamation and drainage. Holland's commercial location at the mouths of the Rhine and near the center of the European market further increases land values and makes reclamation rational.

Extensive areas of sandy soil in the Netherlands, especially in the east and south, have had to be heavily fertilized. But the natural productivity of most Dutch soil and the related nature of the agricultural system contrast with Denmark, despite their joint advantages of location. Danish bacon, eggs, poultry, and sausages suggest poor soil; Dutch bulbs and veg-

MAP 32
RECLAMATION IN THE LOW COUNTRIES
THE IJSSELMEER POLDERS (ZUIDER ZEE)

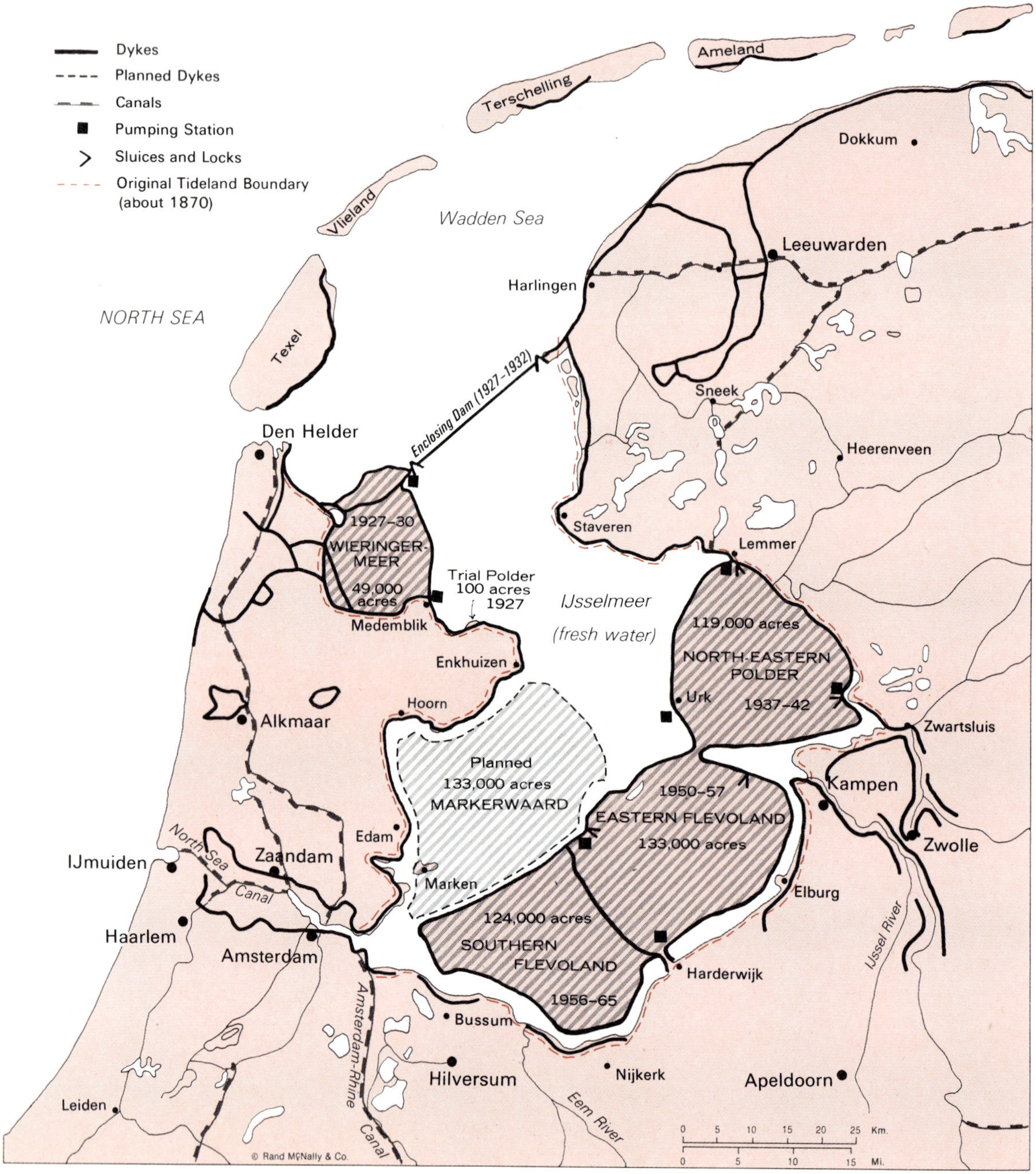

etables, good soil. The best Dutch soil is in the west, where Rhine alluvium and sea clay have been deposited, including many of the areas reclaimed from the sea. Eastern Netherlands is generally more hilly and less fertile, especially where sand and gravel have been deposited in glacial moraines near the edge of the former continental ice sheet. The east is more dependent on dairying and other animals; most of the truck farming is in the west. Holland may be described as Denmark plus the Rhine, but the Rhine is an important addition, both for soils and for trade. It is the Rhine that helps to support Holland's more than 13.5 million people, as against Denmark's 5 million in its larger area. The Low Countries as a whole are among the most densely populated areas in the world, for economic opportunity is high.

Manufacturing and cities

Despite the intensiveness and productivity of Dutch agriculture and the importance of Dutch trade, manufacturing is still more important and employs about one-third of the labor force. Many of the manufacturing industries produce for the domestic market or process agricultural commodities, either from Holland, or from former Dutch and other colonial areas overseas: cheese, milk, and butter from home, chocolate and vegetable oils from abroad. But much of the agricultural production is exported, especially tinned milk and other dairy products. There is a large amount of manufacturing for the European market done in Dutch ports. They are good break-in-bulk points for much of the Continent and bring in the necessary raw materials from overseas for European distribution. Cotton textiles made in Holland are also exported. There has been a rapid development of specialized electrical industries for export, especially radio equipment. In these industries raw materials are less important than skill, capital, and market conveniences. Holland is also fairly well supplied with mineral resources. There is some coal in the south (see Map 21, page 71) that supports a significant amount of heavy manufacturing, and generally supplies Holland's total coal demands. Some coal is also exported. With the additional advantage of location, there is an important shipbuilding industry, primarily in Rotterdam, the leading port.

Petroleum wells in the north have been important for several years, but the discovery of North Sea oil off the Frisian Islands (see Map 21) has already begun to revolutionize Holland's fuel situation. Natural gas production from wells in Dutch territory is now extensive. Natural gas is being exported, as well as displacing, at lower cost, a great deal of domestic coal and oil consumption.

Amsterdam, the old commercial and financial center, is still the largest Dutch city, with over 1 million inhabitants (including the suburbs, nearly 2 million). A canal links Amsterdam with the Rhine mouths (the Lek and Waal rivers) and another with the North Sea, but the city cannot rival Rotterdam in maritime trade. Amsterdam is still the major financial center, and still supports significant heavy manufacturing, including chemical industries. There are only two Dutch cities of large size other than Amsterdam. Rotterdam (over 1 million) is mistress of the Rhine mouths and the chief port of Europe. Hamburg, at the mouth of the Elbe River in Germany, is its nearest rival on the European mainland. Rotterdam's port business is even larger than London's, although it must compete with several Dutch and Belgian rivals, most of whom are building canals to improve their Rhine connections. Rotterdam is, in effect, the ocean port for much of the heavily industrialized Rhineland of Germany, including the Ruhr district. The growth of the port since 1850 has closely reflected development in western Germany. The Hague ('s Gravenhage) is the seat of the Dutch government, and administration is its principal function. It has a population of about 800 thousand. Although there are relatively few large cities in the Netherlands, over a third of the population lives in cities of 200 thousand or more. There are eight cities this size in addition to Amsterdam, Rotterdam, and The Hague: Utrecht, Heerlen, Groningen, Eindhoven, Arnhem, Tilburg, Nijmegen, and Enschede. Small and generally pleasant cities, lined with canals and often preserving fine samples of Medieval and Renaissance architecture, are as typical of the Dutch landscape as are the polders, pastures, and tilled fields of the agricultural system (see the photograph on page 112).

The Netherlands is a crowded country with an extremely dense rural population associated with an intensive agricultural system. There are a large number of trading-manufacturing cities and towns as well. The Dutch population has grown rapidly since about 1850, so it has become increasingly difficult for this tiny country, productive though it is, to

Amsterdam canal. Most towns and cities in the Low Countries are crisscrossed by canals lined with traditional houses built during the period of late-Medieval wealth. *(Photo by R. D. Westerwelle for Van Cleve.)*

support more people. The death rate in Holland has been one of the lowest in Europe for many years. Although the high birth rate has declined, it did so more slowly than the death rate. The success of the Netherlands in maintaining about 14 million people at a relatively high standard of living is striking. With its standard of living rising faster than the population, Holland constitutes a model of efficient use of resources.

Belgium

The Belgian state, little more than a century old, has developed on the open coastal plain of Europe, although within the country the coastal lowlands are narrowed by encroaching hills. People and armies have often moved through Belgium. The hills in the eastern part of the country offer little protection, since there is access from both north and south. The country has been an invasion highway since the beginning of history, and the narrow plain (which spills over into northern France) is pockmarked with the sites of famous battles: Fontenoy, Passchendaele, Agincourt (Azincourt), Malplaquet, Ypres, Mons, Dunkerque, and Waterloo. Belgium's location on the narrow waist of the English Channel has also made it a center of rivalry and conflict. Many European powers have tried to control this area. Within it neither Teutonic north nor Latin south has prevailed permanently. Belgium lies across the cultural line separating German and its dialects from French, and formerly barbarian from formerly Roman-controlled territory. The Germanic language and culture of Flanders and Brabant, the northern half of Belgium, are apparent from the place names; Flemish is a variant of Dutch (see Map 31, page 107). The area around Brussels (Bruxelles) also belongs with Flanders, although the capital city itself speaks French. The language line, however, is to the south of Brussels, as the place names show. Only in religion is the cultural line inconsistent, since southern Netherlands is largely Roman Catholic.

As part of the Low Countries, Belgium was controlled successively by the Frankish Empire under Charlemagne, Burgundy, Austria, Spain, France, and then incorporated with Holland from 1815 to 1839. Despite the obstacles, a definite regional consciousness grew in Belgium through this period. Flanders, the country of Bruges, Ghent, and Antwerp (Antwerpen), strove to distinguish itself from Holland. The French-speaking people of southern Belgium, the Walloons, also protested against what they regarded as alien Dutch influence or control. But there was no opportunity for formal independence until the nineteenth century. With the defeat of Napoleon and the relative decline of French power following the Industrial Revolution, Britain's power on the Continent grew. British policy traditionally feared control of the Low Countries by any other major power. Their strategic location on the Channel close to the heart of England must be kept in friendly hands. A nationalist revolution in Belgium in the 1830s against the union with Holland provided an occasion for British support, with the aim of creating a new state under tacit British guardianship to forestall French or German ambitions. Thus, the balance of power rescued Belgium as it had helped to rescue

Holland. Belgian claims of distinction from the Dutch, their complaints against Dutch discrimination, rivalry between Antwerp and Rotterdam, differences of language and religion, all were translated into independence because far more than Belgian interests were involved. But within Belgium, unfortunately, there is still continual friction between the two cultural halves of the country, the French south and the Germanic north.

Belgium's economic base

In the 1830s Belgium's future advantages for manufacturing could not be foreseen. Flanders, the chief center of Medieval economic development, was considered the most important part of the new Belgian state, and a canal was dug linking the Flemish port of Antwerp with the Rhine. Antwerp has never been able to compete with Rotterdam; in fact it transships a large part of its trade there through the canal and the Rhine. Antwerp's connections with the Rhine and with the sea are not good, and are inadequate for most deep-draft ocean shipping. But almost immediately after winning independence, major and rapid industrial development took place in southern Belgium, and put Flanders in a secondary position economically.

Coal and iron ore, and the manufacturing associated with them, have given Belgium new economic coherence and national strength (see Maps 21 and 22, pages 71 and 72). Deposits of these essential raw materials, the iron close to the coal, occur in the southern part of the country, and Belgium has become a densely industrialized state. Its location near the center of the European market, to which Belgium has access by cheap water and rail transport, is a further advantage. Belgium is now one of Europe's primary heavy manufacturing centers, concentrating on the production of iron and steel, chemicals, heavy machinery, and vehicles. There is also much specialized light manufacturing for the surrounding market: textiles, metallurgical goods, scientific equipment, soap, margarine. One of the specialties is glass, illustrating the general tendency of this industry to gravitate to cheap fuel, as in Pittsburgh. Sand is more universally available than fuel and represents a smaller proportion of total costs in glass manufacture, so it is seldom the determining factor for locating glass making.

The original Belgian iron ores were limited and are now largely exhausted, but local coal and location and transportation advantages have made it profitable for Belgium to import ores. Belgium also imports other necessary raw materials (including cotton, wool, and vegetable oils), and produces for the European market, as one of its cheapest assembly and distribution points. Some of this manufacturing is in the Brussels-Antwerp area, convenient to the sea and to the Rhine. Most of the heavy industry is in the south, and associated with coal, as at Liège, Namur, Charleroi, and Mons. Recently developed coal fields in the Kempen area, east of Antwerp and along the Albert Canal, however, are increasingly important and have stimulated heavy manufacturing in the Antwerp area. The Scheldt River, Antwerp's link with the sea, has been deepened and supplemented by a wide network of canals, making Brussels a port, too. The centrality of Brussels, both within Belgium and between its Walloon and Flemish halves, makes it a better capital than Antwerp, although Brussels proper is still smaller than its older rival. Antwerp is virtually the only maritime port, except for small passenger steamers crossing the Channel that use Ostend (Oostende). Antwerp's position for Rhine trade is poor, however, and it deals mostly in Belgian and French business.

Belgian agriculture

Belgium has made intensive agricultural use of its land, as have Holland and Denmark, although agriculture is somewhat less important in Belgium. Soil is fairly good, especially in the southern part of the country where glacial soils merge with the loam of France. There is a generally mild maritime climate. But the principal factor influencing the Belgian agricultural system is location. Truck crops are dominant, and Belgium offers an example of the intensive use of land close to the market. Brussels sprouts suggest the importance this type of agriculture has in Belgium for feeding both the dense domestic population and the large surrounding population. Dairying, meat animals, and poultry in the areas of poorer soil also supply the local and surrounding urban markets. Most of Belgium's people live in cities. The combination of dense industrialization, urban-based trade, and intensive agriculture in the rural districts makes Belgium the most thickly populated national state in the world, although densities in England alone or Java alone exceed it. In an area only a little larger than

Maryland or Vermont, Belgium supports about 10 million people on a relatively high average standard of living.

Problems of smallness: Benelux

The problems of small size have accelerated with the industrialization that has been so profitable to Belgium. States as small as Holland or Belgium have become out of date from a military or political point of view. And economic disadvantages are also pressing. Economies of scale apply to almost every aspect of modern economic life. Since the Second World War these two countries have followed their mutual interest by establishing the union of Benelux between Belgium, the Netherlands, and Luxembourg, somewhat reversing the separation of the 1830s.

Luxembourg is an odd survival of one of the many duchies that used to make the political map of Europe even more complex than it is now. As with Belgium, the balance of power and the local presence of iron ore have preserved Luxembourg's independence. The country is an important iron and steel manufacturing center. As such, and because of its position and postage-stamp size, Luxembourg is an appropriate member of Benelux. The Benelux community involves a pooling of military resources and a limited customs union minimizing restrictions on trade among the three countries. This makes excellent sense, and it was followed by the wider economic and political union among other European states in the Common Market. The Dutch, in fact, have long supported the idea of a single supranational European state. But Benelux has its problems, and warns that European unification will require a long period of compromise and adjustment. Dutch manufacturing suffers from the competition of stronger Belgian and Luxembourg manufacturing. Belgian trade suffers at the hands of Holland. Two areas that have grown up as independent economies will seldom fit together neatly without considerable sacrifice on both sides. If this is true for Benelux, how much more true for Europe. For Benelux the benefits outweigh the problems. Although they would do so in the long run for Europe as well, and although most Europeans are painfully aware of the need for fuller union, it cannot come quickly or easily.

Questions for further study and discussion

1. Where in the United States are there parallels to Danish agriculture, in physical base and in specialization? Why?
2. Why do Sweden and Denmark have two of the highest per capita foreign-trade levels in Europe?
3. Where else in the world are there excellent natural harbors but few large ports? Large ports and poor harbors?
4. Finland has a slightly larger population than Norway, despite a much less favorable orientation to the sea and a harsher climate. Why?
5. Why are truck crops so important in both Holland and Belgium? Is this consistent with the general locational pattern of truck agriculture?
6. With the industrialization of the Ruhr, why were the Low Countries not incorporated into the German national economy, especially since raw materials to feed the Ruhr pass through the Low Countries?
7. How would you summarize the locational factors in the growth of Rotterdam? What are the city's site problems?

7 France and southern Europe

Eiffel Tower, Paris. *(Trans World Airlines photo.)*

France

France is continental Europe's oldest national state in the modern sense. Its origins may be traced back before Charlemagne to the Frankish kingdom of Clovis that developed out of the confusion of Rome's collapse. The modern French state began to grow in the Paris basin in the early Middle Ages, and France became the chief center of European culture during the seventeenth century. The French language has been spoken all over Europe among educated groups from the late sixteenth or early seventeenth century almost to the present. To a degree, French language and culture were common to the educated world as a whole. France was politically and militarily the dominant power in Europe until about 1870. Its central location within maritime northwest Europe with easy access to all of its areas, repeated and varied external stimuli, and a large and productive home base contributed to this leadership. Industrialization and the rise of larger and stronger states upset the traditional European balance to France's disadvantage, but France still retains much of its position of cultural prominence.

The regions of France

The achievement of France's national unity and cultural leadership within Europe has taken place despite its regionalism, which can still be seen. The three major subdivisions are Atlantic France on the coastal plain facing the Channel, Mediterranean France (the Midi) on the plain facing the Mediterranean Sea, and highland France in the Massif Central, the Alps, the Jura, and the Vosges. The rocky peninsula of Brittany (Bretagne) and the large plains area facing the Bay of Biscay on the west, traditionally called Aquitaine, are additional regions. Each is physically distinctive; each has developed a distinctive culture. The Midi still speaks its own regional French dialect, and in some areas preserves the old Romance language of Provencal. The Midi was the first part of Gaul to come under Roman influence. Even before Roman times the Greeks established colonies along this coast, which still moves in a Mediterranean world to a large extent. The famous climate of the Côte d'Azur, the south coast of France from Marseilles (Marseille) to the Italian border, has warm, usually sunny winters. Its clear blue skies and the sparkling sea, celebrated in French art and dis-

MAP 33
LAND USE IN FRANCE

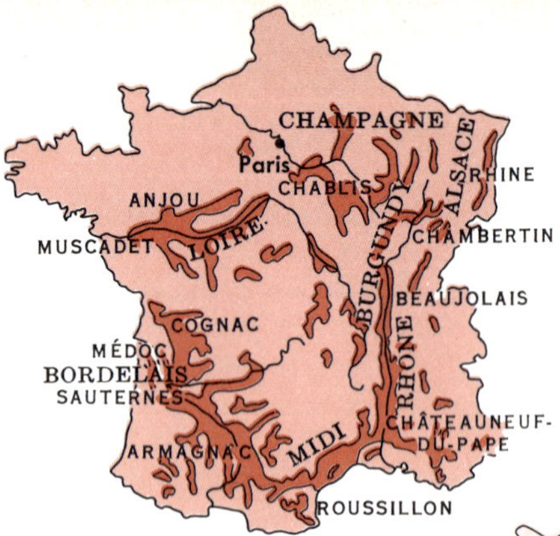

PRODUCTS OF THE GRAPES OF FRANCE

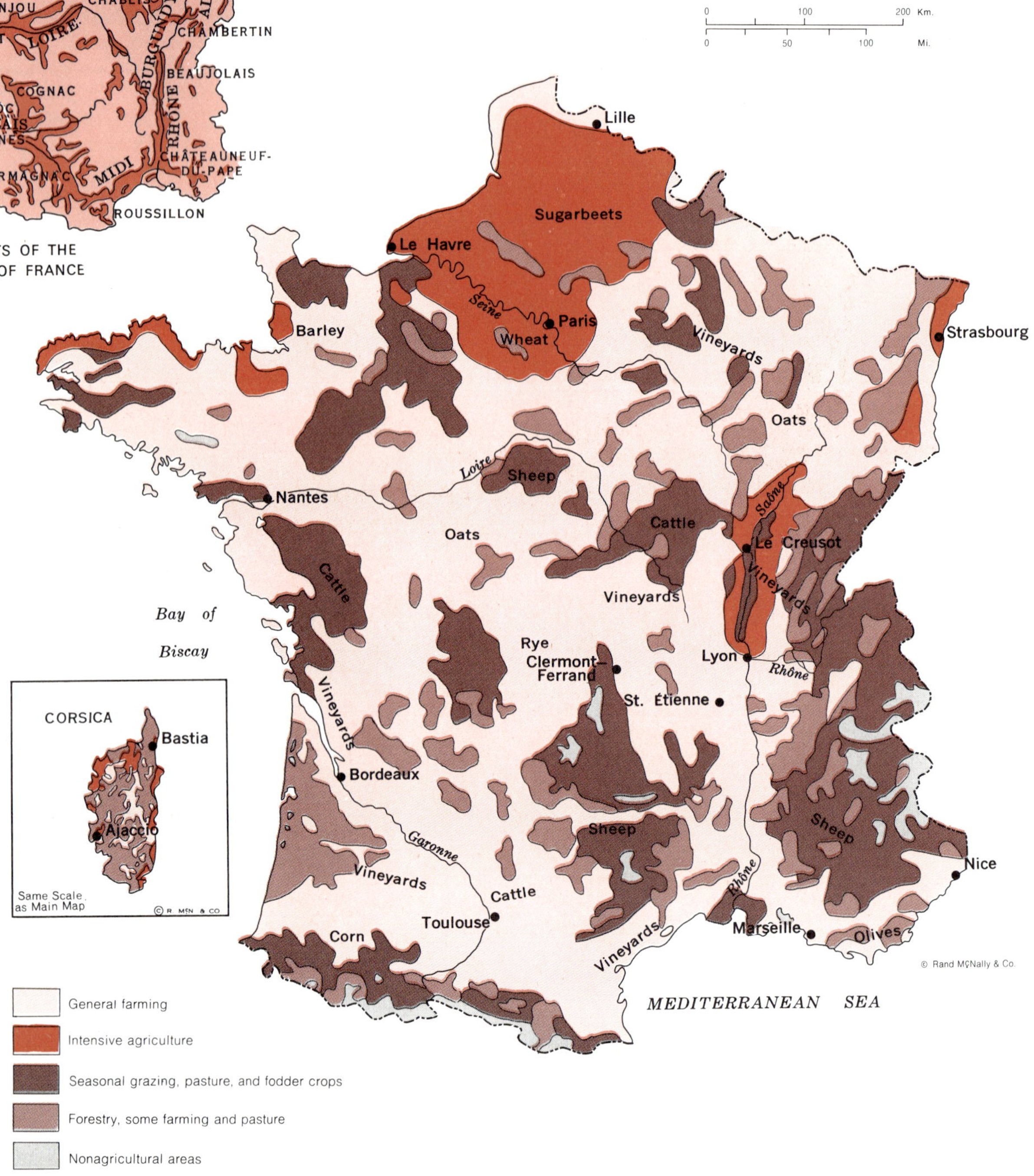

tinctive within France, foster a regional feeling and make the Midi an important international tourist area.

Aquitaine has never been important in economic, political, or cultural terms in proportion to its size. Poor, sandy soil in its southern and southwestern parts (the Landes), marshes in the north, the shortage of good harbors on this emergent coastline with its silted estuaries, and the storminess of the Bay of Biscay combined to restrict Aquitaine's development. Bordeaux, the chief port, and Toulouse, the inland mart and fort at the lowland approaches to passes across the Pyrenees, have remained regional rather than national centers. Bordeaux achieved more importance at the height of the French empire in the New World from export trade in wine (see Map 33, page 116). Like the other French regions, Aquitaine asserted its political separatism and was even under English control for a time during the Hundred Years' War. But by the fifteenth century, it was brought under the control of Paris.

Highland France, the last region to be drawn into the wider national or European orbit, has many of the characteristics of other highland areas. It has traditionally been a refuge for dissident groups, including religious heretics like the Albigenses in the Middle Ages. But industrialization made inroads on local coal and iron deposits, especially in the Massif Central. Later, hydro-power sites were developed. Highland France has been incorporated into the national whole to a large extent in somewhat the same way that Scotland was brought into the English orbit.

Atlantic France has become the dominant core of the modern state. Its advantages have been strong enough to offset the divisive tendencies of regionalism and have made France firmly a part of Atlantic Europe. The benefits of a location on the Channel and near the greatest centers of European population are supplemented by the concentration in northern Atlantic France of most of the country's naturally navigable waterways and most of its productive agricultural land. These resources are common to the European plain. In addition, a large share of France's usable mineral deposits lie on this coastal plain or are easily accessible from it (see Maps 21, 22, and 33, pages 71, 72, and 116). Most of France's manufacturing and people are concentrated in the Atlantic region.

The Midi—Roman theaters at Arles. Evidences of classical Greek and Roman settlement are plentiful in the Midi. The more recent buildings with their red tile roofs and dark rows of cypress trees are also typical of the south of France, and of the European sections of the Mediterranean basin as a whole. *(From French Government Tourist Office and YAN.)*

Paris, the focus

Regional regrouping and concentration under the leadership of the north have been hastened by Paris. Paris is the epitome of an emotional capital, and its hold on men's minds extends well beyond the limits of France. In some respects Paris is still the cultural and intellectual capital of Europe; it has been said that "every man has two cities, his own and Paris." But Paris became prominent because it had certain practical advantages: it was the urban center of Atlantic France and became the nucleus of the modern French state, with a corresponding role in European affairs.

An almost exactly central position within the productive agricultural plain of Atlantic France was

improved when the Industrial Revolution used the coal and iron to the north, east, and south. Paris itself did not become a major heavy manufacturing center, but many cities in its tributary area did. Paris continued to be the chief central place for this most important third of the country.

Paris has prospered in part because of its centrality within the most populous and productive part of France, and in part because of its accessibility. The navigable Seine is joined within 40 miles (64 km) of Paris by most of its major tributaries, including the Yonne, the Aube, the Aisne-Oise, and the Marne, the most important tributary, that joins the Seine just above the city. This river system, plus an extensive network of canals over the relatively level plain, joins the Seine and its tributaries with the Loire, the Rhône, the Rhine, and with Belgium. These waterways ensure for Paris excellent access to Europe's centers of manufacturing, trade, and population, in addition to its central position on the French lowland (see Map 22, page 72). The route from the Mediterranean up the Rhône-Saône valley, now supplemented by canals leading to the Paris basin, helps Paris exert its influence over the Midi as well. The Loire makes similar canal connections between south-central France and the hub of routes near Paris.

French urbanization

The French railroad system helped to strengthen the dominance of Paris, as lines were built centering on the city like the spokes of a wheel (see Map 22, page 72). This is partly for strategic reasons and partly because most of the railroads were built outward from the capital. Greater Paris contains over 9 million people out of a national total of over 53 million. Lyons (Lyon), at the major confluence of the Rhône and Saône; Marseilles, the economic hub of the Midi; and Lille, the major heavy industrial center of northern France, are all close to 1 million. Toulon (almost a suburb of Marseilles) is about 900 thousand. There are a fair number of smaller cities over 300 thousand: Rouen, Bordeaux, Nantes, Nice, Strasbourg, St.

The Cradle of Paris—Îsle de la Cité. Like a ship with its prow pointed downstream, the Îsle de la Cité has been the symbol of Paris and of France for centuries. The Cathedral of Notre Dame stands in the middle background at the upstream end of the island.

Etienne, Toulouse, and Grenoble, each with its own local advantages—but France does not have a highly urbanized economy. Paris is the only really big city, which is a little surprising in a west European state of this size.

The principal reason for the relatively low urbanization, by comparison with Britain, the Low Countries, or Germany, is the continued importance of the agricultural sector in France. France has solid advantages for agriculture, and is less well suited than Britain, Belgium, or Germany for manufacturing or trade. France contains the largest area of productive soils and mild maritime climate of any of the European nations. This was the country's greatest strength in the preindustrial era and contributed much to its position of leadership. France is still largely self-sufficient agriculturally, and in some commodities, notably wines, has a surplus for export. But population and power in the modern world depend far more on industrialization and trade, in which France has had a smaller share.

French agriculture

The French agricultural system is diversified nearly everywhere, as is most agriculture in northwest Europe and in humid North America. Both areas have a widely permissive base of soils and climate. Except for the areas limited only to vine cultivation (for wine), every French region grows many crops in rotation and in mixed cultivation, and also raises animals. Animal manure helps to maintain soil fertility. Animals can also be supported on pasture land that is not well suited for field crops or that can be rotated between pasture and field crops. This combination of different field crops and animals, often called *mixed farming,* is typical of most of northwest Europe and most of eastern United States and Canada. It spreads the risk of loss or gain among a number of farm products. Such a varied system also makes optimum use of the differing types of soil and relief even on small farms. Labor demands are also spread more evenly through the agricultural year. Grains are often rotated with root crops or fodder crops, each making

MAP 34
THE PARIS BASIN

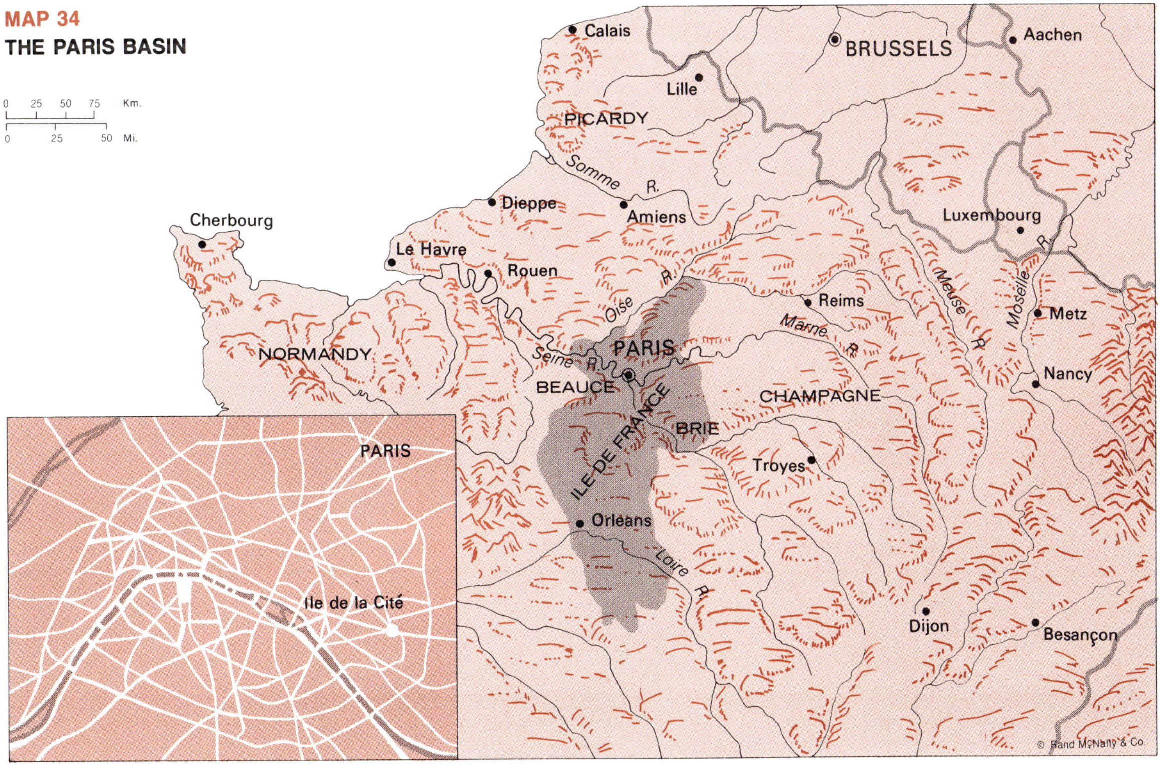

different demands on the soil, and some vegetable or truck crops may also be grown.

On a national basis, France's leading cultivated crop in acreage is wheat. Hay and other coarse fodder crops are areally dominant in nearly all mixed-farming systems, but they are fed to animals. Where relief, soil, and climate allow, wheat is usually the largest crop in most European agricultural systems, except on land close to the big urban markets; there truck crops predominate. Truck crops may also replace wheat in certain other special areas, as shown on Map 33 (page 116). France is rich in good wheat land, and the importance of wheat in its agricultural system emphasizes the high quality and amount of its agricultural resources.

After wheat in cultivated acreage come barley and oats, grown on the less good soils that they tolerate, and also rotated to some extent with wheat. Considerable corn is grown, mostly cut for silage. Grapes are important in many districts, a specialty reflecting traditional skill and a large home and export market for wine. Good wine, however, must be made from grapes grown under special and rigorously observed conditions of soil and climate. Thus, the major vine areas are sharply localized in several small districts, most importantly in the Midi, the Rhône valley, the lower Garonne valley (Bordeaux district), the famous Champagne district not far from Paris, and other localities as shown in the inset on Map 33. Sugar beets are also raised in many areas. They are often grown in rotation with wheat, since both are soil exhausting and both demand good soil. The Atlantic plain of northern France is the leading agricultural region, and specifically the leading wheat area. The rest of France has relatively less level land, less good or less well-drained soil, and, except for the Midi, a less equable climate.

Industrial resources and manufacturing

Domestic supplies of coal are inadequate to support extensive heavy manufacturing on a German or British scale, and have tended to be high in cost. This is partly a result of relatively low-grade deposits, partly of extraction costs, partly of the low national total of reserves, and partly of awkward location (see Map 21, page 71). The coal of the north, along the Belgian border, is fairly well situated, close to major population centers, transport routes, and iron ores. But the deposits in the Massif Central and the French Alps are clearly not well located and require higher transport charges. A large proportion of France's hydroelectric power potential is developed, especially near the Alps and Pyrenees, and its growth has been particularly rapid since 1945. But hydro power cannot make up either for the shortage of coal or for its relatively high costs. Nowhere in France is there anything approaching the easily accessible coal deposits of Great Britain or those of the Ruhr-Rhine valley in Germany. In these areas easy extraction, good grade, and the cheapest possible transportation are combined with dense nearby markets and population concentrations. With such a relatively unfavorable fuel and power base, France has found it

Agricultural France. The village of Le Petit Andely on the Seine River is supported by the productive agricultural land that surrounds it. *(Photo by Ace Williams for Black Star.)*

hard to compete in manufacturing requiring proportionately large amounts of either.

France does have notable deposits of iron ore, especially in Lorraine, which led in part to the past Franco-German disputes over control of this area. There are other, smaller iron ore deposits along the Spanish border in the Pyrenees (an expensive location), near Bordeaux, on the edges of the Massif Central, and in the north near the Belgian border (see Map 21, page 71). A good deal of iron and steel is made in France, mostly in the north, where transport, coal, and markets provide optimum conditions. France also imports German coal extensively since it is found close to Lorraine. Ore and coal flows in western Europe and the localization of manufacturing will be increasingly affected and rationalized under the Common Market, which continues to benefit France.

French manufacturing is highly developed in terms of quality and specialization. For the most part it concentrates on consumer goods or on goods where raw materials are a small proportion of total cost and where skill and value added by manufacture are important. Precision machinery, high-quality consumer goods, metal industries, and art and luxury goods account for most of the industrial production, aside from the processing of agricultural commodities. The industrial structure of Paris, the largest single manufacturing center, is highly diversified, although the heavy industries such as iron and steel are largely lacking. One of its largest components is dress fashions, a good example of a product in which France can best compete because of her long history of cultural and style leadership in Europe. Fine china and pottery, electric machinery, silk, perfume, automobiles, cheese, and wines are other characteristic French specialties.

Population and modern France's decline

The relatively small place of trade and manufacturing in the French economy has brought with it other problems. Europe's population has grown with industrialization and commercialization. This growth occurred on a larger scale in Britain, Germany, and elsewhere than in France; their populations boomed while France's grew more slowly. France fell farther and farther behind in population until by 1900 it was the smallest instead of the largest of the great powers in terms of numbers.

The continued importance of agriculture in France also tended to restrict population growth. The French peasant or the more modern commercial farmer on a small landholding did not have strong incentives for a large family. Land cannot go on being divided among many sons, and France had fewer nonagricultural economic opportunities than most of the rest of western Europe. France's population stopped increasing between the two world wars and began to decline slightly. This was probably also related to the worldwide economic depression that had similar effects on population growth in the rest of western Europe and the United States. After 1945 the French population began to rise again, and the birth rate was one of the highest in Europe for a time. It fell again as Western birth rates generally declined after about 1965. Urbanization and the relative importance of manufacturing have been increasing since about 1960, but the present population is a little over 53 million, still leaving France the smallest of the major European nations numerically.

France's declining power within Europe has been more bitter because of the long and proud record it established in European affairs in the past. The France of Louis XIV or Napoleon was the France of numbers, power, and wealth. A peasant agrarian base was no disadvantage then, and helped to ensure a large and competent army. France has always been predominantly a land power, taking advantage both of its large productive base and its central position within western Europe. Another asset was the *shape* and *size* of the country. France approaches the ideal shape of a national unit: compact and nearly round or square. Administration, defense, economic organization, and national control are much easier and more efficient in a country with this shape than in a country the shape of Chile or of populated Canada, for example. The most densely settled parts of France are also fairly central within the country, and there are relatively few remote, sparsely settled, or offside areas to create administrative, economic, or military difficulties. France also approached the preindustrial ideal in size. Now it is too small for an age of railroads and the modern economies of scale.

"Natural" frontiers

For some time France was obsessed with what it considered to be another asset, France's *natural frontiers,* or *les limites naturelles.* This was part of the eighteenth-century preoccupation with what it called *natural law,* especially as applied to society. France does profit from a number of barrier defenses: on the northwest, west, and south it has the sea; on the southwest, the Pyrenees (nearly unbroken from coast to coast); and on the east the succession of Alps, Jura, Vosges, and Ardennes. These barriers have all been of value to the French state, but natural frontiers do not exist. There are topographic or water or desert or forest barriers, and there are man-made political lines, which by definition are not natural, although they may make use of natural features of the landscape. In the case of France the confusion arose when it was assumed that all landscape features near the spatial limits of French culture would make good political lines, and that these lines were thereby endowed with sanctity or justice. The best example of this is the Rhine. For centuries it has been included in the list of *les limites naturelles de la France.* But the Rhine is primarily a German and a Dutch river. The place names make it clear that French settlement does not reach the Rhine. French control is exerted in Alsace-Lorraine over a population with a German linguistic majority, although the Alsatians do seem to prefer French sovereignty to German. German settlement has spread over both banks of the Rhine in a wide zone. On the French side, within French political territory, the eastern slopes of the Vosges tend to coincide with the line between German and French settlement, as the place names again indicate.

Except for a brief period under Napoleon, the French state has never been able to realize its dream of a realm bounded by the sea, the Pyrenees, the Alps, and the Rhine. Far from being a barrier, the Rhine, like all navigable rivers, has helped to make its basin an economic region. The river has never made a good political line in its lowland sections. Access from the German side helped to make this region a German one, while French influence was hampered by the barrier of the Vosges. As an artery of trade, the Rhine stimulated common rather than divergent cultures and economies along its banks.

French attempts to dominate the Low Countries have been even more ineffectual than the country's precarious foothold on the upper Rhine. Thus, the French state has been left with a gaping hole in its barrier defenses, the open plain along the Channel. The long land frontiers of France had to be guarded carefully. They helped to involve the country in frequent wars, as French claims or ambitions were pursued. The fighting contributed to the development of a central state that was stronger and more authoritarian than those in most of the rest of western Europe until the nineteenth century. The great French political figures of the past were dictators or absolute monarchs. It was the heirs of the French revolution of the 1790s who reversed the traditional pattern by establishing representative government for a remarkably individualistic people.

France's future, like that of most European countries, depends in large degree on establishing close economic and political ties with its continental neighbors and partners. United with Europe, and with its imperial and nationalistic ambitions laid to rest, France's role as a center of European civilization and as its oldest mature member may yet equal, in different ways, the role it played in the days of its imperial greatness.

The Mediterranean world

The Atlantic Ocean's modern importance has pushed southern Europe from the center of the world stage. For many centuries the shores of the Mediterranean were the cultural and economic center of the world and contained most of its population. Many important concepts of modern civilization began in classical Greece and Italy, including the ideas, techniques, and institutions of parliamentary democracy, commerce, banking, diplomacy, universities, classical and Renaissance art forms, and Christianity. Even in science and technology, the Greeks, Romans, and Italians laid important groundwork on which we are still building. The first Western cities in southern Europe were different from the cities that developed as early or earlier in China, India, central Asia, Mesopotamia, and the Nile valley. The pre-Roman Mediterranean cities were primarily centers of trade rather than the central places for agrarian areas or the administrative centers more typical elsewhere. Much of the character of Mediterranean and Western civilization is directly related to the prominence of trade in Mediterranean life. The easily usable waters of the Mediterranean connected a variety of physically different and complementary regions. Each produced different products, exchanged by sea, and each also supported a different culture. The Mediterranean

A Mediterranean town—Villefranche-sur-Mer in southern France. Villefranche-sur-Mer, like so many Mediterranean towns, has been built around a small bay where the mountains come right down to the sea and the houses must climb the hills. Note the remains of the old fort that guarded the small harbor against pirates, also a common feature of most Mediterranean coastal towns. *(Photo by Ciccione from Photo Researchers, Inc.)*

basin became a cosmopolitan place with a variety of cultural cross-fertilization that contributed to the vigor of Mediterranean civilization.

The Mediterranean has been politically unified only once, by the Roman Empire. At the same time it unified most of the known Western world under its control. Despite its easy interconnections and its long-standing flow of trade and ideas, the basin is not a consistent or uniform region. In fact it contains a greater variety of cultures than almost any area of similar size. The distribution of population (Map 27, page 78) also suggests extreme regional differentiation in settlement patterns. The fact that three physically different continents, Europe, Asia, and Africa, come together in the Mediterranean has been a stimulus to trade. But trade alone does not create cultural homogeneity when the trading partners are so different, and when each place is also part of a huge, separate land mass with its own distinctive physical and human characteristics and its own set of spatial relations.

Political disunity

In addition to physical and cultural differences in the Mediterranean region, there is little political unity. The mountains in most of southern Europe are close to the sea (see Map 35, pages 124–25). Few plains have been large enough or strong enough to support a state that could dominate this basin. Rome was the exception, but its dominance was due more to other factors, including its central position in Italy near the sea lanes. The mountains close to the sea and the lack of large plains areas encouraged the growth of many small, independent centers. Each developed its own cultural traits and institutions, and each maintained its political independence. Even though most of these small states had close and extensive trading relations with each other, they maintained their regional identities.

Both before and after the Roman Empire, the characteristic regional and political form in the Mediterranean area was the city-state. Each state generally developed around a confined agricultural base ringed by mountains. It was usually on or near the seacoast so it was able to grow on the basis of sea-

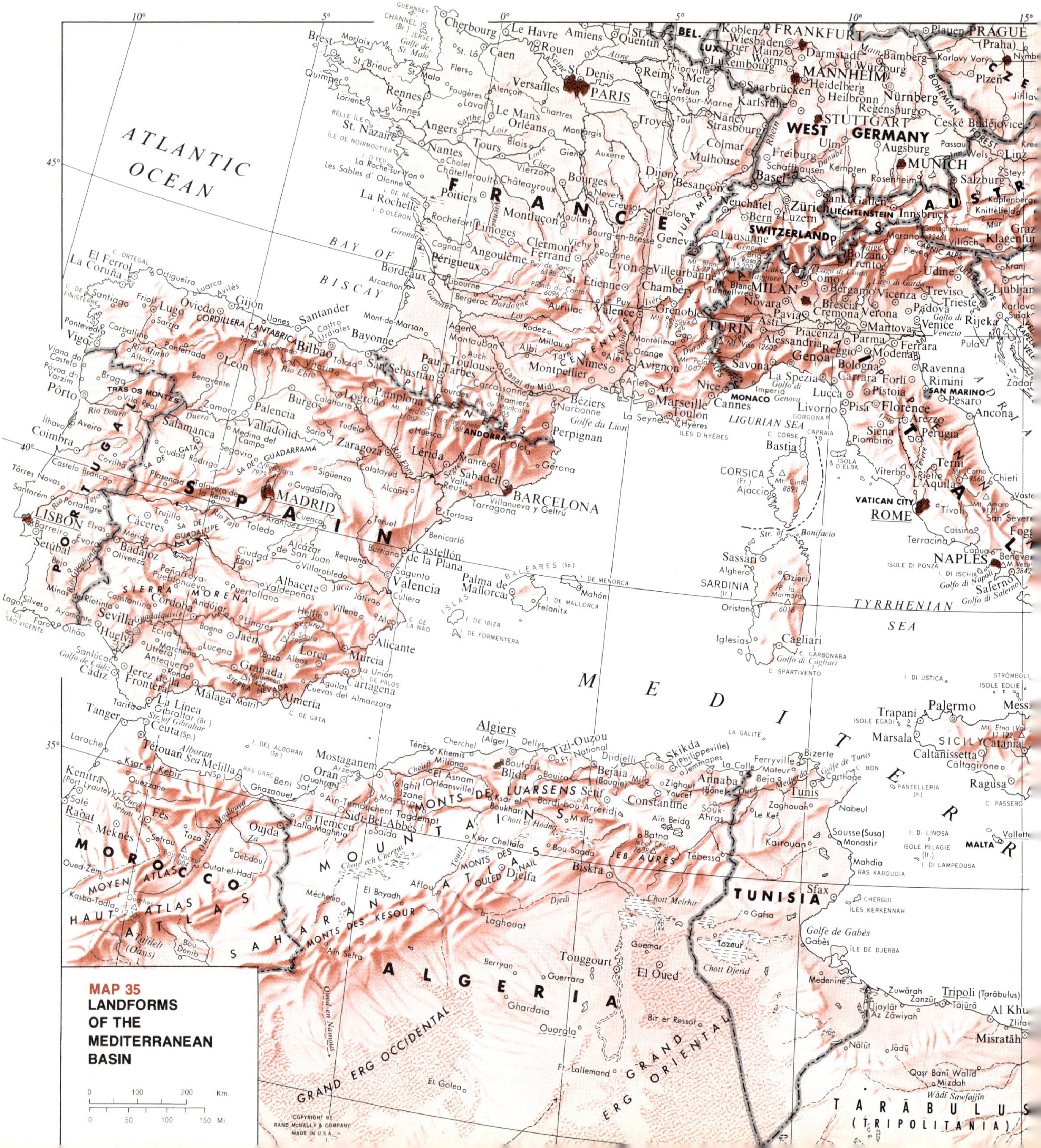

MAP 35
LANDFORMS OF THE MEDITERRANEAN BASIN

20°
25°
30°
35°
40°
45°
40°
35°
30°
KATOWICE
Zabrze
Rzeszów
Kraków
Tarnów
Ostrava
POLAND
Přerov
Žilina
Prešov
Uzhgorod
Košice
Miskolc
Nyíregyháza
VIENNA
Bratislava
Nové Zámky
Wiener-Neustadt
Győr
Pápa
BUDAPEST
Gyöngyös
Debrecen
Hajdúszoboszló
Székesfehérvár
Kecskemét
Szombathely
HUNGARY
Nagykanizsa
Szekszárd
Szeged
Pécs
Subotica
Sombor
Senta
Zagreb
Osijek
Novi Sad
Vršac
Sisak
Našice
Sl. Brod
Zemun
Pančevo
Belgrade
(Beograd)
Šabac
Tuzla
Loznica
Kragujevac
Paraćin
Kraljevo
Niš
Pirot
Sarajevo
Mostar
Livno
Šibenik
Novi Pazar
Leskovac
Priština
Peć
Dubrovnik
Cetinje
Prizren
Skopje
Shkodër
Tetovo
Ulcinj
Tiranë
Durrës
Elbasan
ALBANIA
Korçë
Vlorë
(Valona)
Gjinokastër
Prilep
Bitola
Strumica
YUGOSLAVIA
Lvov
Ternopol'
Ivano-Frankovsk
Kamenets-Podol'skiy
Kolomyya
Khotin
Chernovtsy
Mukachevo
Khust
Satu-Mare
Baia-Mare
Oradea
Cluj
Tîrgu-Mureş
Bistrita
Piatra-Neamţ
Botoşani
Iaşi
Bacău
Bîrlad
Sibiu
Braşov
Arad
Timişoara
Lugoj
Deva
Alba Iulia
ROMANIA
Galaţi
Brăila
Buzău
Ploeşti
Tîrgovişte
Piteşti
Craiova
Turnu-Severin
BUCHAREST
Călăraşi
Constanţa
Tulcea
Izmail
Sulina
Vilkovo
Bolgrad
Kishinëv
MOLDAVIAN S.S.R.
Orgeyev
Bel'tsy
Bendery
Tiraspol'
Odessa
Belgorod-Dnestrovsky
Vinnitsa
Zhmerinka
Khmel'nitskiy
Tul'chin
Mogilëv-Podol'skiy
Soroki
Balta
Uman'
Kirovograd
Pervomaysk
Voznesensk
Nikolayev
Kherson
Krivoy Rog
Smela
Kremenchug
Dneprodzerzhinsk
DNEPROPETROVSK
Pavlograd
Zaporozh'e
Nikopol'
Melitopol'
Osipenko
Zhdanov
Taganrog
DONETSK
Novocherkassk
Rostov-na-Donu
Yeysk
Tikhoretsk
Kropotkin
Stavropol'
Armavir
Labinsk
Maykop
Krasnodar
Timashevskaya
Novorossiysk
Tuapse
Sochi
Sukhumi
Anapa
Kerch'
Feodosiya
Simferopol'
Yalta
Sevastopol'
Yevpatoriya
KRYMSKIY P-OV (CRIMEA)
Dzhankoy
Genichesk
Perekop
AZOVSKOYE MORE (Sea of Azov)
UKRAINIAN S.S.R.
R.S.F.S.R.
U.S.S.R.
BLACK SEA
Sofia
(Sofiya)
BULGARIA
Pleven
Sevlievo
Lovech
Vratsa
Lom
Vidin
Ruse
Tutrakan
Silistra
Razgrad
Tolbukhin
Varna (Stalin)
Svishtov
Nikopol
Tŭrnovo
Kolarovgrad
Sliven
Stara Zagora
Jambol
Burgas
Pomorie
Chirpan
Plovdiv
Khaskovo
Pazardzhik
Kyustendil
Dimitrovo
Kirklareli
Edirne
Midye
STARA PLANINA (BALKAN MTS.)
RHODOPE MTS.
Drama
Xanthe
Komotini
Kavalla
Sérrai
Thessaloniki
Édhessa
Flórina
Kastoria
Kozáni
Ioánnina
Árta
Lárisa
Trikkala
Vólos
Lamía
Khalkis
Agrinion
Pátrai
Kórinthos
Tripolis
Spárti
Kalámai
Pírgos
Kiparissía
ATHENS
(Athinai)
Piraiévs
Thívai
(Thebes)
GREECE
AEGEAN SEA
Mitilíni
Khíos
CRETE
Khaniá
Iráklion
Ródhos
KARPATHOS
İSTANBUL
İzmit
Tekirdağ
Gelibolu
Bandırma
Çanakkale
Bursa
Gemlik
Balıkesir
Edremit
Ayvalık
Bergama
Akhisar
Manisa
İzmir
Ödemis
Aydın
Tire
Söke
Milâs
Muğla
Denizli
Burdur
Isparta
Antalya
(Adalia)
Fethiye
Alanya
Kütahya
Afyonkarahisar
Uşak
Konya
Aksaray
Niğde
Karaman
Ereğli
Mersin
Tarsus
Adana
İskenderun
Antakya
Gaziantep
Kilis
Urfa
Maraş
Malatya
Elâzığ
Kayseri
Sivas
Tokat
Amasya
Zile
Çorum
Yozgat
Kırşehir
Ankara
(Angora)
Eskişehir
Bolu
Adapazarı
Zonguldak
Ereğli
Bartın
Kastamonu
Çankırı
Sinop
Samsun
Ordu
Giresun
Trabzon
Erzincan
TURKEY
ASIA MINOR
TOROS DAĞLARI
PONTIC MTS.
Aleppo
Al Lādhiqīyah (Latakia)
SYRIA
Hama
Homs
Tartous
T'arabulus
(Tripoli)
LEBANON
Beirut
Zahlé
Damascus
(Dimashq)
Saida (Sidon)
Sour (Tyre)
Acre
Haifa
Nazareth
Nablus
ISRAEL
Tel Aviv-Yafo
Jerusalem
Bethlehem
Hebron
DEAD SEA
-1286
'Ammān
JORDAN
Gaza
Beersheba
Rafah
Ma'ān
Al 'Aqabah
SAUDI ARABIA
Areas occupied by Israel since June 1967
CYPRUS
Nicosia
Famagusta
Larnaca
Limassol
ALEXANDRIA
(Al Iskandarīyah)
Rashid
Damietta
Port Said
(Būr Sa'īd)
Ismailia
Tanta
Damanhūr
CAIRO (Al Qahirah)
Suez (As Suways)
Al Jīzah
Hulwan
Al Fayyūm
Bani Suwayf
EGYPT
SINAI PEN.
Gulf of Suez
G. of Aqaba
LIBYAN PLATEAU
MUNKHAFAD AL QATTARAH
Matrūh
Sidi Barrānī
As Sallūm
Tubruq (Tobruk)
Darnah
Shahhāt
Tulmaythah
Tukrah
Bengasi
JABAL AL AKHDAR
LIBYA
BARQAH
(CIRENAICA)
Sulūq
Ajdābiyah
Khalīj Surt
Surt
An Nawfalīyah
MEDITERRANEAN SEA
IONIAN SEA
Bari
Barletta
Molfetta
Altamura
Taranto
Brindisi
Lecce
Otranto
Golfo di Taranto
Rossano
Nicastro
Catanzaro
Reggio di Calabria
Monte Sant'Angelo

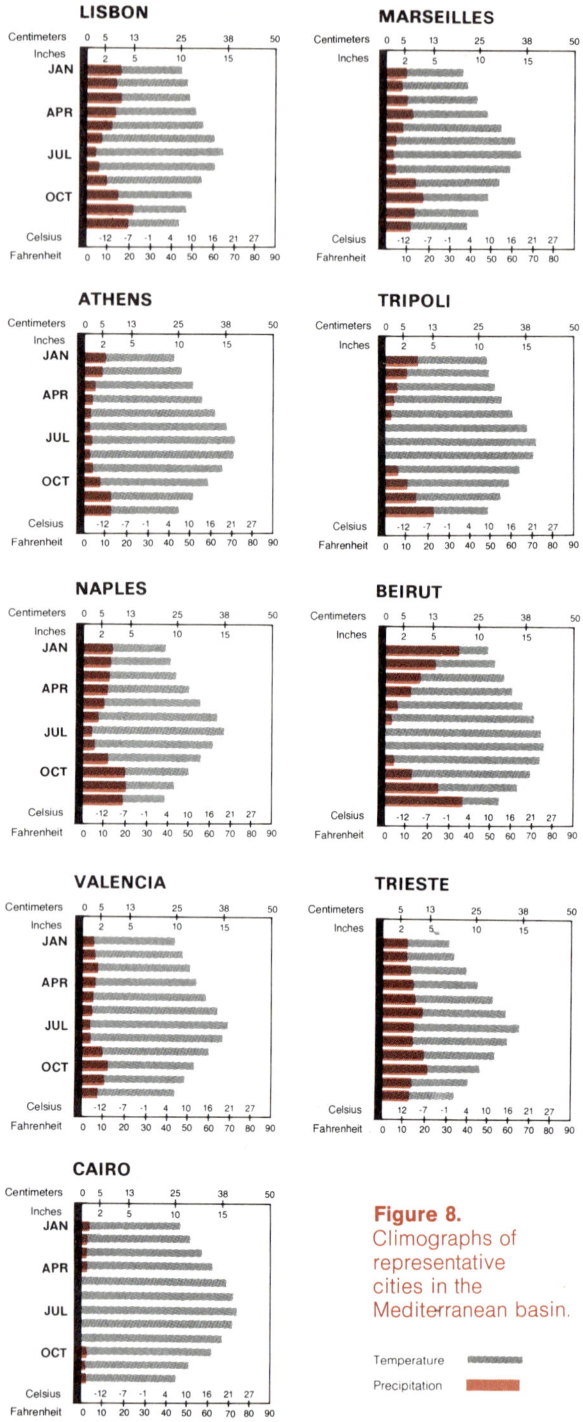

Figure 8. Climographs of representative cities in the Mediterranean basin.

borne trade. Athens (Athinai), Corinth (Korinthos), Marseilles, Syracuse (Siracusa), Palermo, Genoa (Genova), Naples (Napoli), and Pisa are all typical of this pattern. Each was the capital of a larger political unit around the city region. Italy itself was not politically united for fourteen hundred years after the collapse of the Roman Empire. The French Midi was a separate region politically until relatively recently. Spain's political unification dates only from the time of Columbus; Portugal remains separate. And Greece was not freed from the Ottoman Empire until 1831.

Resource poverty

The poor resources of southern Europe will not support industrial-commercial development or large populations. The prevalence of mountains, the shortage of good agricultural land, and the inadequate amount or distribution of rainfall (see Map 35, pages 124–25) are important disadvantages for competition with northwest Europe. The Mediterranean area also largely lacks most of the mineral raw materials essential for modern, large-scale manufacturing. Coal is almost totally absent, a critical handicap. Petroleum is scarce in Mediterranean Europe, although it occurs on the north African side. Recent discoveries of sizable petroleum and natural gas reserves in Sicily and in northern Italy can ease Italy's problems to some extent, but southern Europe as a whole is not well supplied with mineral fuels.

Iron ore occurs in the basin in several places, including Spain, north coastal Algeria and Morocco, Corsica and Sardinia, and Mediterranean Yugoslavia (see Map 21, page 71). Other ores found with several of these deposits, especially in Spain and Yugoslavia, include aluminum, copper, lead, zinc, and ferroalloys. Most are not used to support local or even Mediterranean manufacturing; they are shipped to the industrial centers of northwest Europe. In the absence of cheap coal for smelting, of local markets, or of other interdependent forms of manufacturing, the ores move to points where they are concentrated, and where their processing and ultimate use is cheapest and most convenient.

There is some hydro-power potential in the basin, for it is an area of generally heavy relief. But most of the potential water power is not used outside of Italy. The Mediterranean climate is characterized by a concentration of rainfall in winter (which is also a disadvantage for agriculture), and an annual average that is not very high, making obvious problems for power production (see Map 23, page 73). But at most of the sites where physical factors would allow power production, economic conditions are poor, and the markets are too small or too far from power sites to provide the high constant load factor necessary for profitable power production. Northern Italy is the exception and is discussed later.

Mediterranean land use

Mediterranean agriculture has adjusted to its limited good level land, many thin stony slopes, and a shortage of rainfall with a winter concentration. However, its temperature and sunshine are much more favorable than north Europe's. The most important Mediterranean crop is wheat (see Map 24, page 75). It is consumed locally and well suited to a winter concentration of rainfall and a dry summer. Wheat occupies the best and most level land because it will not grow in poor soils or on slopes as well as other crops, especially trees or vines. Wheat can be planted in late autumn or early winter, develop its root system and grain heads during the wet winter and spring, and ripen in the strong summer sun without suffering from the long dry spells. Bread, macaroni, and spaghetti are important in the Mediterranean diet, and wheat is by far the leading Mediterranean crop in acreage. The other prominent crops of Mediterranean agriculture are all tree or vine crops, sun- and heat-loving, but tolerant of slopes, poorer soils, and drought: citrus and other fruits, nuts, olives, figs, and wine grapes (see Maps 36, 37, and 38, page 128). This list is typical of agriculture in California, too, parts of which are also mountainous or hilly and have a Mediterranean climate. Near large urban centers or in specially fertile level areas, truck crops replace wheat or tree crops, as in every agricultural system.

In specially arid or specially hilly areas, grazing is the most prevalent use for the land. Mediterranean slopes, including those in Turkey, Israel, Jordan, Lebanon, and Syria, may have been grazed on a large scale longer than any others in the world. The tree cover on most of them was removed long ago, although some trees remain in the less accessible mountains. Scanty rainfall discourages regrowth of the forest, but the rainfall is enough to produce serious erosion and flood damage. The persistent overgrazing of slopes in the basin has had disastrous consequences. Cattle cannot be supported profitably in any numbers on the steeper and more sparsely covered slopes. Sheep and goats will tolerate these conditions, but they have done the greatest harm because they crop much closer than cattle and even destroy young trees. Cheese, often made from goats' milk, is important in the Mediterranean diet, from Greek feta to Italian parmesan or provolone. Especially in the last two hundred years, population pressures in an area inadequately supplied with good agricultural land has led to increasing destruction of vegetation and consequent erosion of the slopes. But Mediterranean agriculture as a whole, especially in Italy, uses the land efficiently, making the most of its far from favorable physical base.

Tourism and the Mediterranean landscape

Tourism is a large and welcome source of income. The warm, sunny Mediterranean climate contrasts with grey, chilly north Europe. The Mediterranean landscape is attractive by almost any standards, partly because it has been intensively occupied for so long. Nearly every hill or hillside in the European section of the basin is crowned with a castle, an ancient walled town, a Medieval or Renaissance church, or sometimes a classical ruin. Hillsides are terraced for fruit or nut trees or grapes; the valleys or small plains are carefully tended in rippling fields of wheat and red poppies. The roads are lined with the typical dark-green Mediterranean trees—cypress or the so-called Lombardy poplars (whose hard, glossy leaves restrict loss of moisture).

"Tourists" of an earlier age were less welcome after the fall of Rome. They came as invaders from the north, drawn by the productivity of Italy in particular. Italy was a persistent attraction for the Goths, Vandals, Germans, Austrians, French, and their armies. North Europeans now tour Italy, Spain, and Greece, but their ancestors were largely responsible for the hillside or hilltop sites of so many Mediterranean settlements and for the prevalence of walls and castles. Defensibility was a prime consideration in selecting the site for a settlement. This has been true in a great many parts of the world, wherever or whenever the local order was precarious. Hillsides are also common town sites in the Mediterranean because level agricultural land is at a premium in many areas. A town can tolerate slopes better than wheat

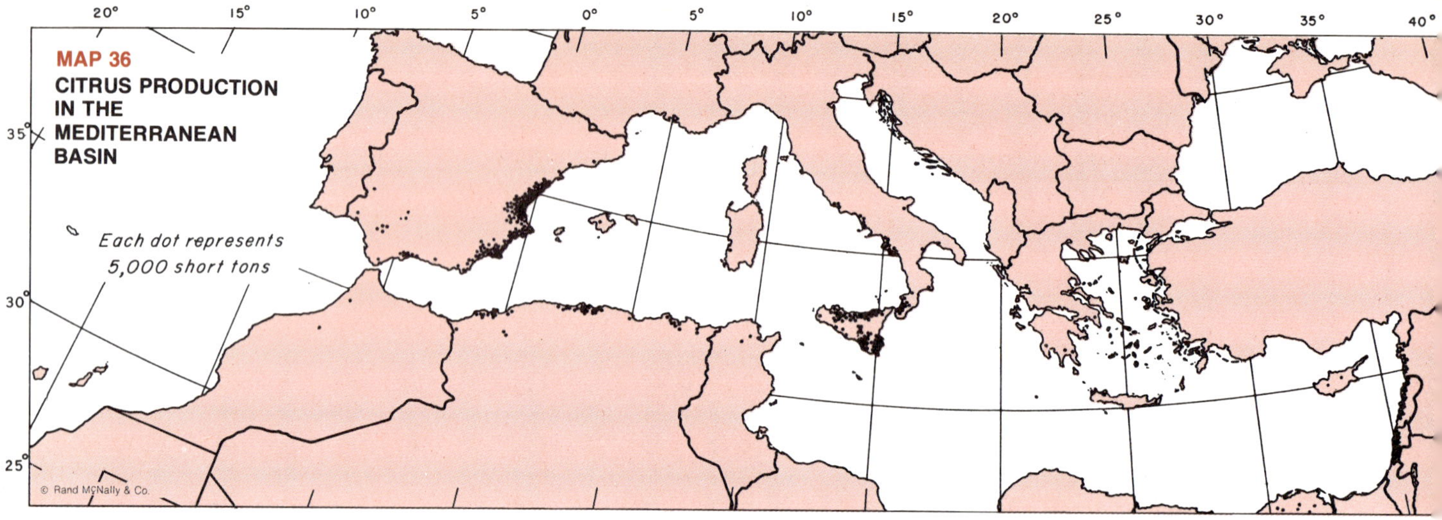
MAP 36
CITRUS PRODUCTION IN THE MEDITERRANEAN BASIN
Each dot represents 5,000 short tons
20°
15°
10°
5°
0°
5°
10°
15°
20°
25°
30°
35°
40°
35°
30°
25°
© Rand McNally & Co.

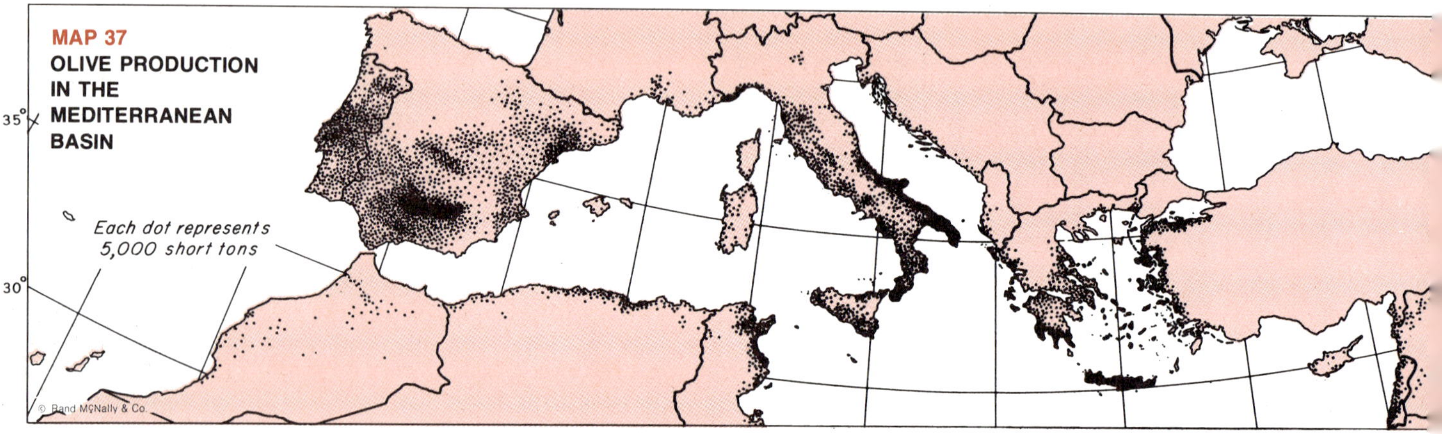
MAP 37
OLIVE PRODUCTION IN THE MEDITERRANEAN BASIN
Each dot represents 5,000 short tons
35°
30°
© Rand McNally & Co.

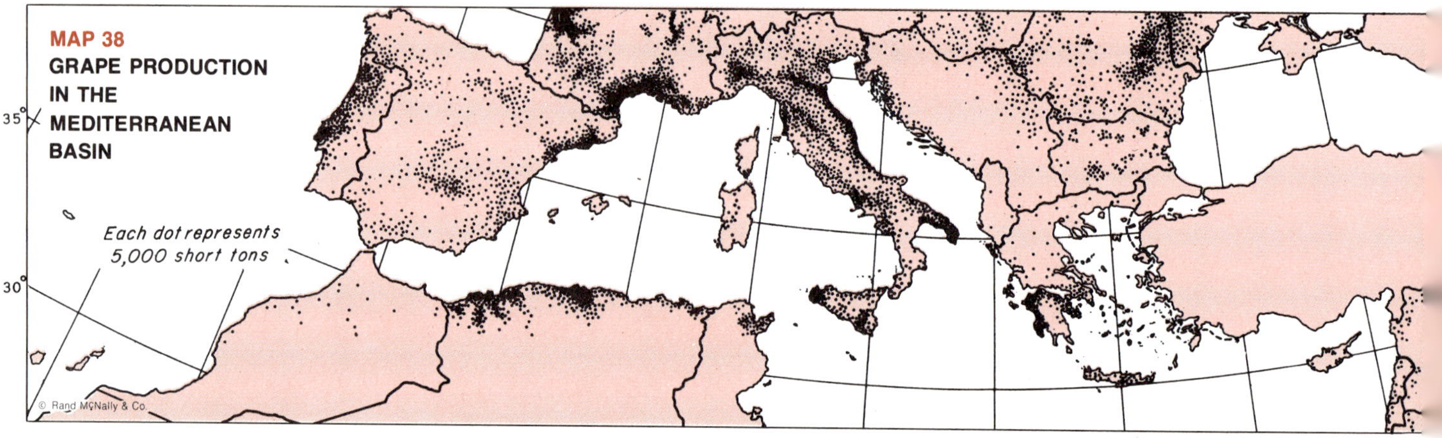
MAP 38
GRAPE PRODUCTION IN THE MEDITERRANEAN BASIN
Each dot represents 5,000 short tons
35°
30°
© Rand McNally & Co.

can. Where grapes, figs, nuts, olives, and fruits are important crops, a hillside town may also be nearer to its work than a town in the valley or plain. A further factor was malaria. After its spread over Italy and the rest of the basin during the latter days of the Roman Empire, many lowland settlements were shifted to hills at a time when security was also rapidly decreasing.

Spain and Greece

Italy is probably the most important Mediterranean country. Spain, perhaps Italy's closest rival in importance, for most of its history has been a poor area economically, a cultural backwater, and seldom a strong influence within the Mediterranean basin. Spain has been isolated from the main stream of European development. Its isolation comes in part from the barrier of the Pyrenees. Its climate is excessively dry, navigable rivers are few, and productive agricultural lowlands are severely limited. Most of Spain is covered by mountains, hills, or a rough plateau, and local regionalism is still very strong. Trade and cultural exchange between Italy and north Europe have flowed across the Alps in much greater volume than has crossed the Pyrenees. Italy has also been much more involved than Spain in Mediterranean trade and cultural development. The domestic Italian base is physically more favorable and supports a much larger population.

For centuries Spain was controlled and occupied by the Islamic Moors from Africa (roughly from the eighth to the fifteenth centuries). With the help of an advanced system of irrigation, the Moors built a culturally and economically impressive civilization, superior in many respects to anything since. But it was largely destroyed when the Spanish finally drove them out.

Spain was a major colonial power in the early period of European expansion overseas, but its importance was relatively short lived. The silver and gold coming from the New World were not enough to

Spanish shepherd. This was taken in the Pyrenees, but illustrates well the arid, rocky, and mountainous landscape characteristic of most of Spain. *(Photo from Yan and Photo Researchers, Inc.)*

breed a more productive Spanish economy at home. Even the presence of significant mineral resources has not stimulated much modern industrial development in the country, except in the Barcelona area.

Greece once had the most highly developed Western civilization, but it did not last. After the death of Alexander the Great, Greece declined. As a small, mountainous, semiarid country, it has not advanced above the general Mediterranean level of relative poverty.

Italy

Physical and regional diversity

Although Italy's peninsular shape is set off at its landward end by high, sharp mountains and is a distinctive physical region, it is not uniform either physically or in its human occupance. Like most Mediterranean countries, it is divided internally by mountains. Only in the north is there a large unbroken plain, the valley of the Po River in the province of Lombardy (Lombardia). The major internal Italian mountain chain, the Apennines (Appenninos), runs primarily north and south, parallel to the sea. South of Rome there are fewer plains than in the northern half of the country, so the north as a whole, even outside the Po plain, is better favored physically than the south (see Map 35, pages 124–25). In the north the Apennines run almost east and west before turning south. This barrier further helps to make Lombardy and Piedmont (Piemonte), along the French border, a distinct subregion.

Rainfall decreases progressively southward and is increasingly concentrated in winter. Sunshine has agricultural as well as tourist advantages. But the shortage of water in the south, added to inadequate level land and the poor soils characteristic of most slopes, severely handicaps this area. Some of the slopes of south Italy and Sicily are deeply covered with volcanic ash and lava. It weathers rapidly and forms a very productive soil whose fertility is almost inexhaustible. This helps to explain the relatively intensive agricultural use of areas lacking in level land and moist climate. But south Italy and Sicily are poor areas on the whole, not as productive agriculturally as the north. For that reason, Sicily and southern Italy are not as good bases for trade or as important trade-generating centers as the coastal cities of north Italy and their agricultural hinterlands.

Lombardy's climate is in many ways more like northwest Europe's than like that of the rest of the Mediterranean, especially in its larger annual rainfall. This is also a major advantage for agriculture. The rainfall is slightly concentrated in winter, but there is considerable summer rain as well. Agriculture in north Italy is based more on grains and forage crops and less on tree crops than in the south. Rice is an important crop in the river valleys or in formerly marshy areas where it can be irrigated easily.

Italy lacks coal, has only small reserves of iron ore (including deposits in Sardinia, which belongs politically to Italy), and has only recently discovered fairly sizable domestic petroleum and natural gas deposits in the Po valley and in Sicily. But the Po valley is favorably situated for the development of hydro power. The Po plain lies between two mountain ranges, the Alps and the Apennines, and rivers rising in each flow down to the plain and into the Po. Rivers from the Alps have their major flow in summer, when the Alpine snows are melting. Rivers from the Apennines are fullest in winter, when most of the rainfall comes. This situation eases one of the common problems in the development of hydro power, particularly common in the Mediterranean, by supplying a more steady, year-round flow of water. Streams from the Apennines do not evenly balance the summer flow of Alpine streams, but they help. Falling water of any sort is of little use for electric power, however, unless there is a large, constant market for it within practical range. In this important respect, too, the Po valley meets the necessary conditions, as few areas in the Mediterranean basin do. Productive agricultural land supports a dense population with relatively high incomes by Mediterranean standards, and Italy's manufacturing centers are also heavily concentrated in the north.

Genoa and Venice

Genoa (Genova) and Venice (Venezia) both deserve separate comment. Genoa is not in Lombardy but in the province of Liguria, separated from Lombardy by the Apennines, and largely covered with mountains or steep hills even in its coastal areas. But Genoa would be an insignificant place without Lombardy, for the Po plain is its major hinterland, and mountainous Liguria alone could not support a major port.

MAP 39

NORTH ITALY

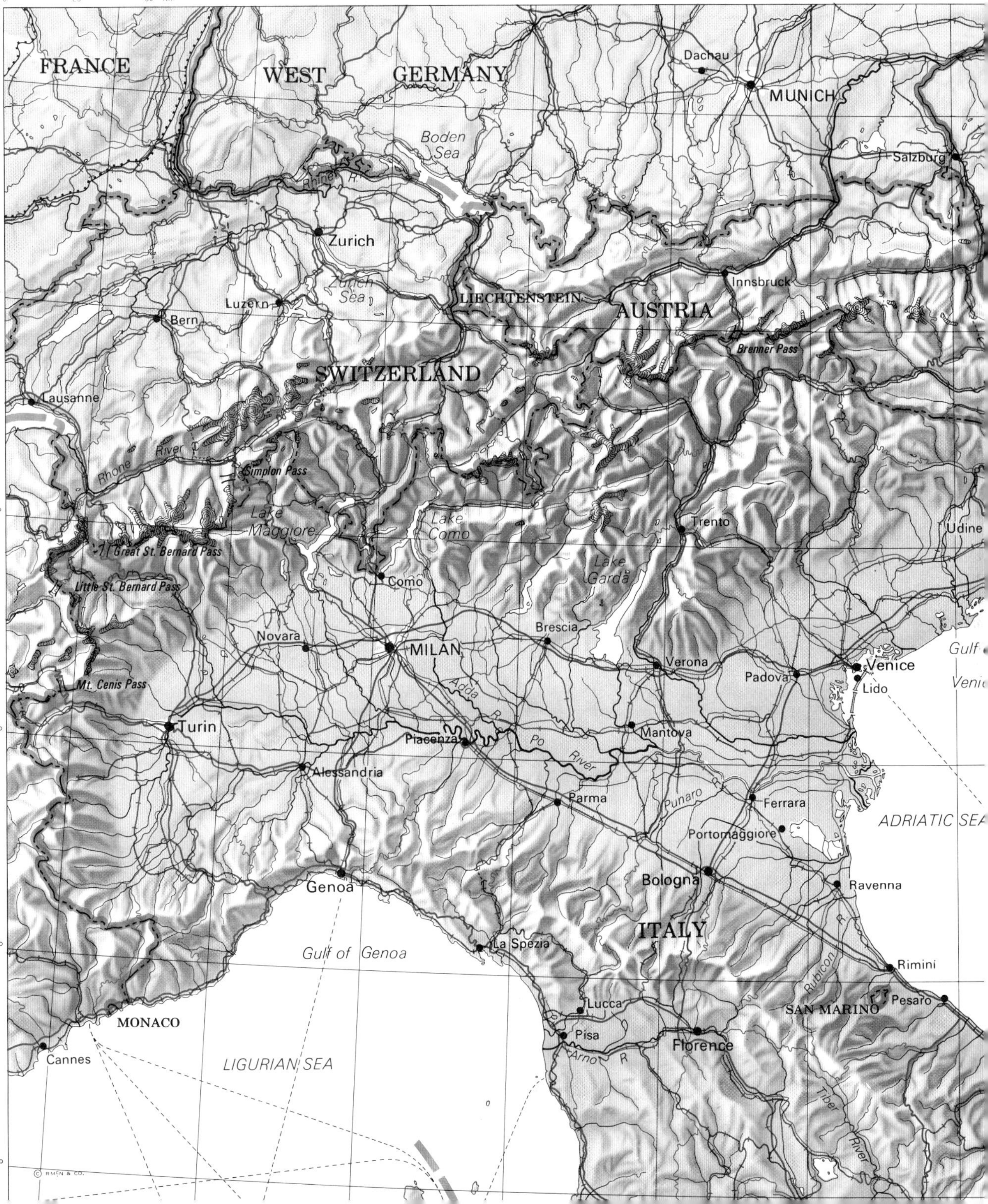

In fact Genoa is the largest port in Italy in the volume of its trade. It serves the most populous and productive part of Italy. Since Italy must import so many essential raw materials due to its specialty of manufacturing, it is particularly dependent on foreign trade. Although the mountains that lie between Genoa and its Po plain hinterland present a problem, a small tributary of the Po, originating in the hills immediately behind Genoa, has cut a valley providing an easy route into the plain (see Map 39, page 131).

Venice has no mountain barrier shutting it off from Lombardy, and it is also near the mouth of the Po. On both grounds it should have monopolized the maritime trade of north Italy long ago. Venice's shallow deltaic harbor is inferior to Genoa's, which occupies a drowned coastline where mountains reach the sea. But this has not been an important factor in keeping down the volume of trade at Venice. Harbors are built where trade requires them, and good ones are neglected where trade does not seek entry. It seeks entry at Genoa and forces its way over the mountains (as the railroads do across the Appalachians or across the Alps) because Genoa is at the western end of the Po plain and Italy's and Lombardy's trade is predominantly with the Atlantic world.

Venice and Genoa were bitter rivals until the sixteenth century, and were forever fighting with each other. Generally Venice had the upper hand, at least in trade, and built a territorial and commercial empire in the eastern Mediterranean. But the increase of shipping in the Atlantic is overwhelmingly in Genoa's favor. Venice is not an insignificant port now; it could hardly be otherwise with the Po plain at its back. But its days of greatness are past. Perhaps this is best for the preservation of its magnificent artistic monuments whose Renaissance and Baroque glory may be second only to Rome's.

Rome from the top of St. Peter's. Looking east toward the Tiber River, we can see the Piazza S. Pietro and look down the Via delle Conciliazione. The Piazza, like St. Peter's, is in high-Renaissance style, but Rome is a mix of ancient and modern. The city has flourished for over two thousand years through many rulers and styles. *(Photo by Mary Baber from Van Cleve.)*

Venice has also profited from its easy access to Alpine passes from Austria and to routes from central Europe (see Map 39, page 131). The site of Venice, north of the Po rather than at its mouth, is related to the routes leading to the Brenner and other Alpine passes. The Po is not very navigable; its gradient is slight and siltation is heavy. Venice is still a city built quite literally on the water and on a group of islands. Its location at the head of the Adriatic placed it closer to north Europe than any of the other Mediterranean or Italian ports, as well as nearer to the Levant (the east coast of the Mediterranean) and routes to Asia. This double locational advantage supported Venice's near monopoly as the major center for trade between Europe as a whole and the East until the sixteenth century. No other city was able to compete with it.

The rise of Rome

Rome lies in the transition zone between north and south Italy near the center of a broken coastal plain that extends from Pisa to Naples. More important, it is near the center of the only extensive lowland outside of the Po valley, the plain of Latium (Lazio). Rome's central position within Italy makes it a good modern capital, as it was a good capital of the Roman Republic and Empire. However, it is significant that the modern drive for Italian unification did not come from Rome but from the north, from the region of Piedmont and from Genoa. They were both far more in touch with the nationalistic and other currents of modern Europe than was antiquarian, Mediterranean Rome. Until about 1930 Milan (Milano) was bigger than Rome, and Florence (Firenze) and Venice were more lively artistic centers. But Rome remained the epitome of all capitals, for Italy and for much of the world.

Since 1870, as the center of national administration, Rome has grown fast, and is now the largest Italian city. Like all big cities, it has acquired some trade and manufacturing, primarily service industries. Its major business is government, plus education, tourism, religious administration in the tiny enclave of the Vatican City, and the support of several nonreligious organizational headquarters, including United Nations agencies. Rome's urban development continues to center on the Tiber (Tévere), which played a major part in fixing the city's site nearly three thousand years ago. Most of the capital cities of Europe have developed on a similar river-centered pattern: London, Paris, Vienna (Wien), Budapest, Berlin, Prague (Praha), Warsaw (Warszawa), Belgrade (Beograd), Riga, Moscow (Moskva), and even Dublin. In nearly all of them the river still runs through the heart of the modern city, so they have a sameness in this respect. Outside Europe many other cities have grown around a river that still bisects them, including Rangoon, Bangkok, Calcutta, Shanghai, Baghdad, Pittsburgh, and Philadelphia.

Italian manufacturing

The great bulk of Italy's modern manufacturing, especially the heavy industries, is in Lombardy and Piedmont. Shipbuilding is concentrated in Genoa. Italy manufactures many of the same products as France, including automobiles, consistently producing the finest racing cars in the world. The automobile industry as a whole is centered in Turin (Torino). Like France, Italy lacks many advantages for heavy manufacturing. Raw materials are expensive, so Italy specializes in high-quality production. Its manufactured goods require special skill or represent a large proportion of value added by manufacturing. Industries using large amounts of electric power have also prospered. Until recently electric power has been Italy's significant industrial resource together with a large, skilled labor force. But hydro power cannot economically displace coal and oil, and Italy cannot make iron and steel at a competitive cost, except for certain uses where special qualities are essential. Domestic shortages of essential raw materials have been an important handicap. Although there is a large domestic Italian market of over 56 million that can support much production, Italy's industrialization is proportionately smaller than that of most of the countries of northwest Europe or North America. Consumer goods dominate factory production. Textiles and clothing together form the largest single industry. It is closely followed by metal products, reflecting in part the automobile and precision-metal industries and the influence of hydro power.

The great city-states of the Italian Renaissance in Tuscany (Toscana), Emilia (Emilia-Romagna), and Umbria—Florence, Pisa, Bologna, Lucca, Siena, Arezzo, Perugia—have been very little involved in modern changes. Rich and prominent as trade centers in a prerailway, preindustrial, and presteamship age, they are no longer accessible, and their seaports have been outgrown by increasingly large ocean carriers. Only Genoa, and to a lesser extent Venice, retain their trade positions, because they are more closely related to Italy's modern economic center in the Po plain.

Southern Italy and Sicily

Southern Italy remains poor except for Naples, which has an agricultural area tributary to it on the only extensive plain in south Italy, the province of Campania. Mountains and low rainfall limit agricultural productivity. Irrigation is less well developed in the south, where it is needed more, than in the north. Water sources are fewer, and the expense of and engineering for large irrigation projects have been too much for the local people. The modern novel, *Christ Stopped at Eboli,* may overstate south Italy's poverty. (This is also a local Campanian proverb—Eboli is about 50 miles [80 km] south of Naples in the hills beyond the southern edge of the Campanian plain.) The area may not be godforsaken, but it has certainly been forsaken by the advantages that make for prosperity in the modern world. It has been, and still is, the principal source of Italian emigrants, many of whom have come to the United States.

The people of Sicily and Sardinia are distinct from those of peninsular Italy. Sicily was colonized by Greeks, Byzantines, and Arabs, and later conquered by Normans, French, and Spanish. As a result partly of this continual political disorder and partly of the area's economic poverty, Sicilians have a well-earned reputation for violence and terrorism. Their secret societies spread to the United States with Sicilian immigrants and contributed substantially to the American gangsters. In Sicily and in the provinces of Calabria and Apulia (Puglia) (the toe and heel of Italy), even today there are almost no villages. The areas are predominantly agricultural, but the people are clustered into towns and cities for protection against the pirate attacks and local disorders that have prevailed for centuries in this exposed and impoverished area. But Italian immigrants as a whole have made rich and valuable contributions to American life. The United States is indebted to the heritage of Italy's practical and cultural achievements.

Italian hillside town—Perugia. Wheat grows in the productive Tiber valley, and olives, grapes, tree fruits, and nuts grow on the hillsides below the town. This combination of agriculture is often found around Italian towns in the Apennines. *(Photo from ROLOC.)*

Regional disunity and the rise of the modern state

Italy's unification came late in the modern era. It reflects in part the poverty of its economic and resource base by modern standards, in part its relative isolation from the mainstream of modern European development, and in part its excessive internal fragmentation because of mountain barriers and other strong regional differences. Until nearly the end of the nineteenth century, local regionalism flourished in practically every province. Most provinces cherished traditions of independence and cultural distinction.

The Apennines act as a barrier between the east and west coasts; even today there are relatively few roads across these mountains. There is less productive level land on the east side. A large number of small streams from the Apennines flow to the coast. The even, gradual eastern coastline of emergence is poorly drained in many areas, and includes few good harbors. There are a few bays on the west coast, including the beautiful Bay of Naples (Gulfo di Napoli) with its fine harbor overlooked by smoking Vesuvius. But the west coast, too, has suffered from a shortage of good harbors, especially inconvenient for a would-be maritime power. The sense of common Italian nationality appeared relatively late. North and south, east and west, plains and mountains, and small basins cut off from each other understandably developed differently, despite the peninsular shape of the country and its barrier boundaries in the Alps.

A big effort has been made since the Second World War to develop the south. Public and foreign capital is being invested there on a large scale in an effort to increase agricultural and industrial production so the south may have an equal place within the national whole. One of the largest steel mills in Italy was located at Taranto on the southern coast of Apulia in the heel of the peninsula. It uses Sardinian ore. The oil and gas of Sicily will also help to raise the economic level of the south significantly. Much can be and is being done in this neglected corner of Italy, but it seems doubtful if the south can ever rival the economic dominance of the north.

Despite the relative poverty of its domestic resources, Italy has long dominated the Mediterranean and played a major role in the growth of Western civilization. The centrality of the Italian peninsula was an advantage. But, even with the apparent barrier of the Alps, it seems likely that Italian development resulted in large part from the benefits and repeated stimuli of close interaction with Europe north of the Mediterranean. This distinguishes Italy from Spain or Greece. Italy's connections with the rest of Europe remained vigorous, and made possible the extension northward of Italian influence and an extensive and continuing racial and cultural mixture within Italy. Spain's long isolation contrasts sharply with Italy's vigor. Greece is both too small and too disadvantaged by the Balkan mountain barrier to compete with Italy as the chief center of economic and cultural development. The Italian achievement is especially laudable for its persistent humanism in every field. Although this is clearly related to the diversity and individuality of the several parts of the peninsula, Italy's success within the Mediterranean world owes much to its wider pattern of spatial interaction.

Questions for further study and discussion

1. Why should a mainly agricultural economy be characterized by a low degree of urbanization? What kinds of cities would such an economy support? In what pattern would they be located?

2. What is the function of Marseilles? Why is it there, and why has it grown so big? Apply the same analysis to Bordeaux, and to Lyons.

3. How do you account for the fact that Nice has been the fastest-growing city in France recently? Can you suggest a parallel to this situation in the United States?

4. Why did the Alps not isolate Lombardy or Italy from the rest of Europe in historical or cultural development, political interaction, or trade? Compare the Pyrenees, which are no greater barriers physically than the Alps.

5. Could Spain profitably develop more of its potential hydro power to smelt its ores rather than exporting them in a raw state? Why or why not? Compare it with Scandinavia, and compare the market areas to which Spanish ores move.

6. Were there other early cradles of navigation besides the Mediterranean? Where, and why?

7. California has a Mediterranean climate in part. Although its agricultural system has much in common with agriculture in the Mediterranean basin, there are also important differences. What are they, and why?

8 Germany and central Europe

Castle of Rheinstein on the Rhine River, near Bingen, Germany. *(Photo from Pan American Airways.)*

Germany

Germany[1] straddles the area in which many of the characteristics of maritime Europe begin to change into those of continental Europe. The creation of a modern German state despite geographic disunity is a measure of the strength of German nationalism. But this achievement bears the marks of a prolonged struggle against geography. That eastern and western Germany, divided most easily (although not necessarily most accurately) along the line of the Elbe River, are basically different should not need stressing after the events of the last hundred years. Broadly speaking, the east has been dominated by authoritarian government and society, by the Prussian state. Poor peasants and the small group of hereditary aristocratic landowners with their large estates dominated the agricultural system. For most of the east, manufacturing and trade were of secondary or minor importance, and change was slow. West Germany has had industrial and trading cities, a growing middle class, and German experiments with democratic ideas, both socially and politically. The west was the Germany of Beethoven, Goethe, Karl Schurz, the early German Socialists, the Ruhr, the ill-fated Weimar Republic, and most recently of the Bonn Government. But German nationalism thrived wherever Germans lived, and its force was great enough to override such differences, as it may one day override the present unhappy division of the German state. Where nationalism works against obstacles, its force may be increased rather than weakened. This has certainly been the case with Germany. Beethoven and Goethe were no more and no less German than Bismarck.

In addition to the basic differences between east and west, Germany may be divided into three distinct regions from north to south: (1) the north German plain (part of the great European plain) from the North Sea and Baltic coasts south to approximately the Harz Mountains; (2) the wide zone of hills in the middle of the country reaching about as far south as the Danube; and (3) the highland or mountain zone of Germany in the southern third of the country. The

---

[1]Throughout this chapter *Germany* refers to the entire area occupied by the present Federal Republic of Germany (West Germany) and the German Democratic Republic (East Germany).

Germans have sometimes called these regions low, middle, and high Germany. They recognize almost as many physical and human differences between them as between east and west. Low German is a dialect distinct from High German, and the people and cultures of the two areas are almost equally different.

In France and England, although regional differences were originally as great as in Germany, a national state developed much earlier. This was partly because of the growth of a strong central core in the Paris and London basins, each of which extended its influence over the rest of the country. In Germany no such clearly dominant central core ever developed. Most of Germany's big cities, its ocean ports, and most of its manufacturing, including the Rhineland (Rheinland-Pfalz), still lie on the north German plain or its southern margins. But the core role of the northern plain has never been as strong as the unifying roles played by the London and Paris basins. The distinctiveness of the north has done as much to set it off from the rest of Germany as to enable it to draw the rest of the country into its own orbit.

The rise of modern Germany

Until the middle of the nineteenth century Germany was divided into many separate states, some large and powerful like the kingdoms of Bavaria (Bayern) or Prussia, some tiny principalities, but all representing pronounced regionalism. German language and German culture spread over a wide area within which there was very little other unity. The almost complete lack of barrier boundaries except in the south, and its location on the great European plain made Germany's security an important issue. During its long history, it has been surrounded by different and usually hostile people and states. Its location on a major European corridor led to many invasions, and its political boundaries have changed many times (see Map 41, page 142).

Elements of economic strength: coal and the Ruhr

Germany persisted as an idea through centuries of frustration and compromise until the ultimate creation of a national state. Although it is difficult to overestimate the role of emotional nationalism in building a strong state, something more concrete or substantial must hold the state together or give it necessary momentum in the modern world. Modern Germany's great economic strength has been coal (see Map 21, page 71). In the Rhineland (most importantly in the valley of the Ruhr River, a tributary of the Rhine [Rhein]), in the Saar basin, in Silesia (now part of Poland), in Saxony (Sachsen), and scattered elsewhere in the northern half of the country are some of the world's major coal deposits.

The Ruhr, the Saar, and Silesia all contain very large deposits of excellent bituminous coal well suited for iron and steel manufacture (coking coal). Especially in the Ruhr and the Saar, the seams are thick and extraction costs low. The coal is well located near the populous heart of the European plain, surrounded by a huge and highly developed market, and served by some of Europe's major navigable waterways. These waterways include not only the Rhine and the Ruhr, but several other Rhine tributaries, plus an intricate network of canals joining the river systems together. The Ruhr district is connected by water with most of the European market. Railroads have also focused their lines on this key location where economic activity has concentrated and haulage volume is high.

Iron ore is available from nearby Lorraine or from more distant sources by sea. The Ruhr is principally a center of heavy industry, especially iron and steel. Many other types of manufacturing have been attracted to it, including chemicals, electrical equipment, vehicles, textiles, and the variety of production nearly always associated with a major iron and steel district. Cheap coal, power, transport, and market advantages have wide industrial appeal.

Other industrial resources

Large-scale industrial development in Silesia came somewhat later than in the Ruhr, partly because the German position in Silesia developed later and was less secure, and partly because Silesia's transport advantages did not compare with those of the Ruhr. In the present century, however, Silesia has become one of the major heavy-manufacturing districts of Europe, serving a largely central and eastern European market. One of the bitterest results of the Second World War for the Germans was the redrawing of their eastern boundary to follow the Oder-Neisse Line. This left not only Silesia but parts of Brandenburg and most of Pomerania and East Prussia in Polish hands. Poland gave up a roughly equivalent amount of territory to Russia in the east, but

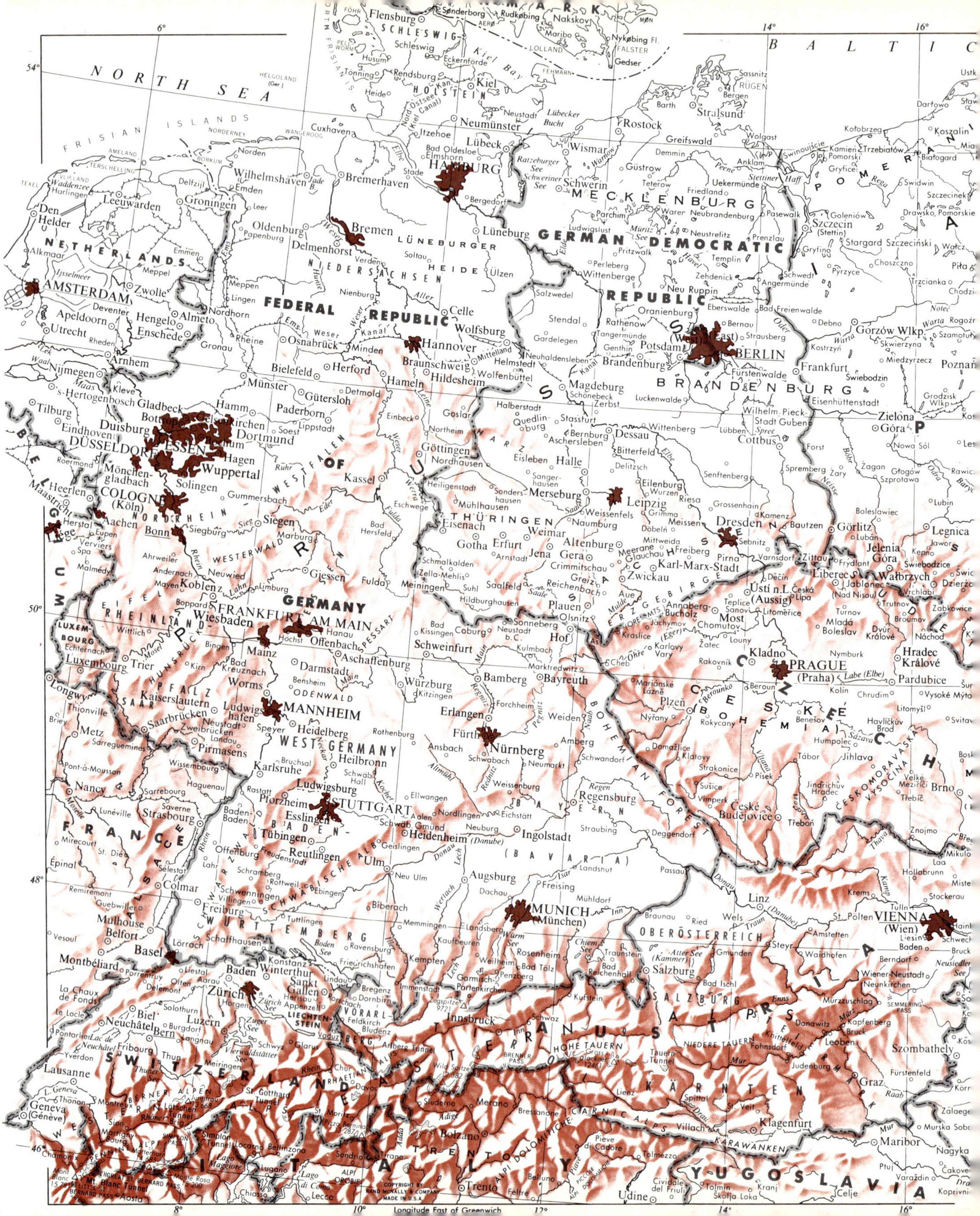
NORTH SEA
BALTIC
FRISIAN ISLANDS
DENMARK
SCHLESWIG
HOLSTEIN
Flensburg
Kiel
Kiel Bay
Neumünster
Lübeck
Hamburg
Bremerhaven
Bremen
Wilhelmshaven
Emden
Oldenburg
Cuxhaven
Rostock
Stralsund
Greifswald
Wismar
Schwerin
MECKLENBURG
GERMAN DEMOCRATIC REPUBLIC
FEDERAL REPUBLIC OF GERMANY
NETHERLANDS
AMSTERDAM
Groningen
Leeuwarden
Zwolle
Utrecht
Arnhem
Nijmegen
Hannover
Braunschweig
Wolfsburg
Magdeburg
Potsdam
BERLIN
BRANDENBURG
Frankfurt
Szczecin (Stettin)
POMERANIA
Lüneburg
LÜNEBURGER HEIDE
NIEDERSACHSEN
Osnabrück
Bielefeld
Münster
Dortmund
Duisburg
ESSEN
DÜSSELDORF
Wuppertal
COLOGNE (Köln)
Aachen
Bonn
Siegen
Kassel
Göttingen
Halle
Leipzig
Dresden
Karl-Marx-Stadt
Erfurt
Weimar
Jena
Gera
Zwickau
THÜRINGEN
HARZ
WESTFALEN
NORDRHEIN
WESTERWALD
EIFEL
RHEINLAND-PFALZ
Koblenz
Wiesbaden
FRANKFURT AM MAIN
Mainz
Darmstadt
Würzburg
Bamberg
Bayreuth
Nürnberg
Erlangen
Regensburg
Ingolstadt
Augsburg
MUNICH (München)
MANNHEIM
Heidelberg
Karlsruhe
STUTTGART
Ulm
Saarbrücken
Kaiserslautern
Trier
LUXEMBOURG
Luxembourg
BELGIUM
Liège
Metz
Nancy
Strasbourg
FRANCE
ALSACE
Colmar
Mulhouse
Basel
Zürich
SWITZERLAND
Bern
Lausanne
Geneva (Génève)
Luzern
LIECHTENSTEIN
AUSTRIA
Innsbruck
Salzburg
Linz
VIENNA (Wien)
Graz
Klagenfurt
HOHE TAUERN
KÄRNTEN
OBERÖSTERREICH
STEIERMARK
CZECHOSLOVAKIA
PRAGUE (Praha)
Plzeň
BOHEMIA
BOHEMIAN FOREST
ERZGEBIRGE
Brno
Liberec
POLAND
Zielona Góra
Poznań
ITALY
Bolzano
Trento
Merano
YUGOSLAVIA
Maribor
CARNIC ALPS
KARAWANKEN
ALPI DOLOMITICHE
Lago Maggiore
Lugano
Lago di Como
Longitude East of Greenwich
COPYRIGHT BY RAND McNALLY & COMPANY MADE IN U.S.A.

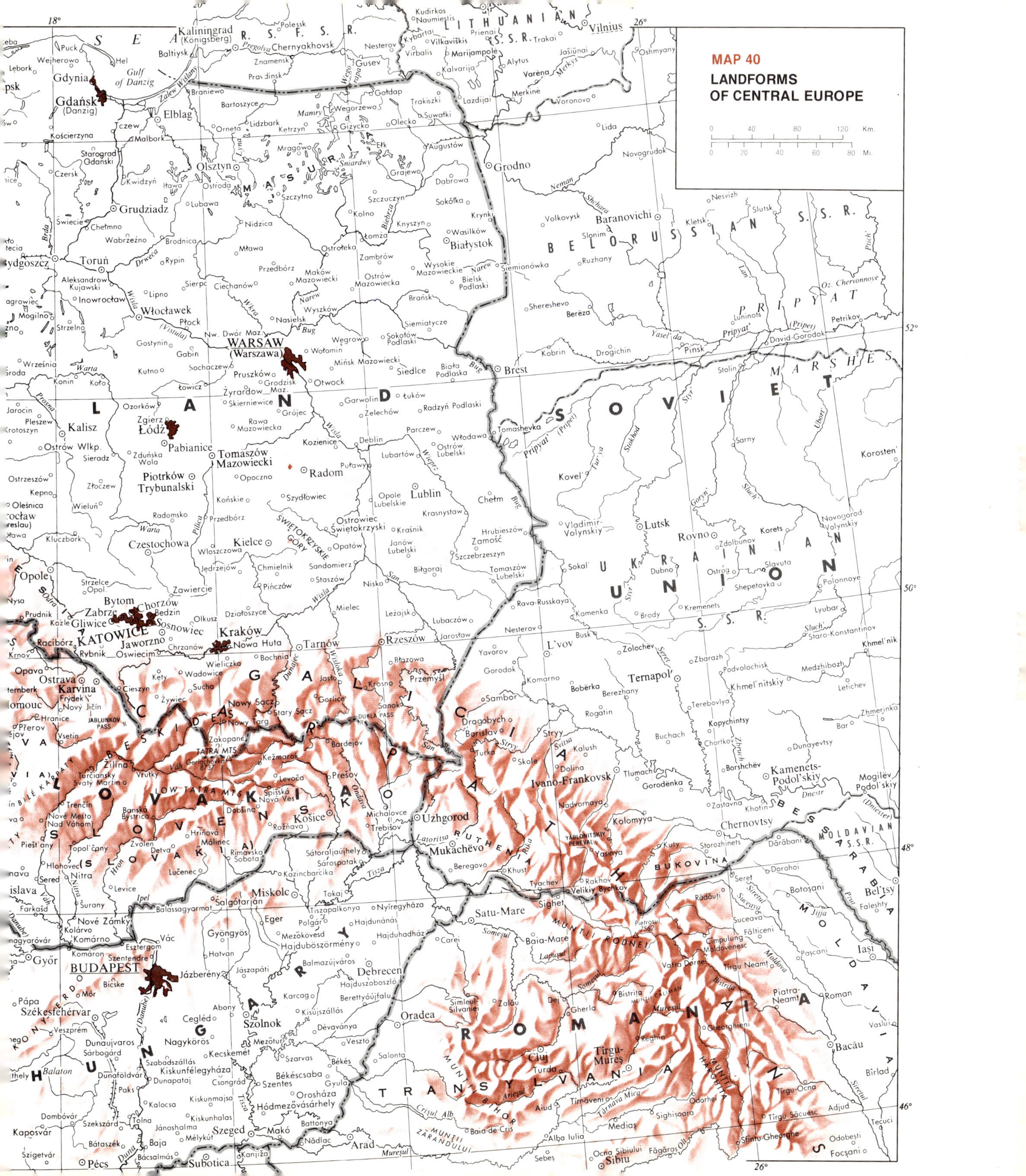

MAP 40
LANDFORMS OF CENTRAL EUROPE

this is cold comfort for the Germans. The new frontier is within 35 miles (56 km) of Berlin, and the city itself exists as a divided enclave within the Communist state of East Germany.

The location of coal in the Ruhr and the advantages it brought for local economic development tended to accentuate further the differences between east and west Germany. Industrialization drew the Rhineland increasingly into the main cultural stream of western Europe, to which it was also bound by close commercial ties and traditional orientation. External stimuli from western Europe did not affect the east nearly so much, and its distinctiveness as a culturally and economically backward appendage to the west was thrown into sharper relief.

One other resource has contributed substantially to the advantage of the modern German state: chemical deposits. These include the world's largest reserves of potash and several other chemicals that cheap coal has made especially easy to use. Coal is itself a chemical raw material of the first importance. German discoveries and processes, and coal's occurrence with a variety of chemical salts laid the basis for the development of the large German chemical industry. There has also been a substantial development of hydroelectric power in middle and high Germany where humid mountains are well within range of large markets.

In other respects Germany is not well supplied with industrial raw materials. Even in wood it is not self-sufficient, despite careful attention to reforestation and selective cutting, and draws heavily on Scandinavian or Russian timber. In the postwar period, rising demand in West Germany and the loss of Silesian and other eastern deposits have made some coal imports necessary. But Germany became an industrial giant with the help of ample cheap coal. This advantage was great enough to offset other lacks, given the technical skill and concerted effort of the German people.

Cities

Not all of German manufacturing was located in the Ruhr and Silesia, but between them these two centers contained most of the heavy manufacturing, with the Ruhr by far the larger of the two. Breslau (renamed Wroclaw by the Poles) was the main Silesian center. In the Ruhr there is no single dominant city, although Duisburg, Essen, Gelsenkirchen, Dortmund, and Düsseldorf all lie within 20 miles (32 km) of each other. However, Düsseldorf became the largest. Local coal seams and the earlier development of many of these cities as trade or administrative centers worked against a single dominant city. With their close original spacing and mushroom growth since 1870, they now make in effect a single mammoth urban industrial area.

Cologne (Köln)[2] is a much older trading and administrative city, only incidentally involved with the industrial structure of the Ruhr. It was originally important because of its location near the downstream entrance to the Rhine gorges (which stretch from Bonn to Bingen), feared by Medieval users of the river. Downstream from Cologne the populous and productive lowlands of the Rhineland widen out, and the city is in a good position to serve them. Upstream are Bonn, Koblenz (at the Moselle [Mosel] junction), Mainz (at the Main junction), Frankfurt (on the Main River before it enters the Rhine), and Mannheim (at the Neckar junction), all regional, political, and commercial centers long before the Industrial Revolution. They have since become centers of diversified manufacturing as well. Elsewhere in Germany, their parallels in origin and function and as central places include Munich (München), Stuttgart, Nuremberg (Nürnberg), Hannover, Leipzig, Magdeburg, and Berlin.

Berlin is an artificial creation to some extent. From a small Prussian capital town to one of the largest cities in the world, its growth was the result of conscious policy. The decision was made to locate the capital of united Germany in what was then the least-developed part of the country and to focus industrial development and many railway lines there. Berlin's communications across the north German plain by land and by river and canal are excellent.

Its location is thus not a great handicap. Hamburg and Bremen, near the heads of estuaries leading to the open sea, are the major German port cities, and trade is their principal business. Like Berlin and all of the regional centers listed above, however, Hamburg and Bremen engage in some manufacturing to supply their own hinterlands and make use of the advantage of their break-in-bulk functions. Ham-

[2]Fom the Latin *Colonia;* this was a Roman-founded colony, a trade and control point on the major north European route.

The Ruhr—Duisburg. Part of this premier industrial complex of Europe and the main basis of the German economy is shown here. *(From Bundesbildstelle, Bonn.)*

burg is also a major shipbuilding center. Hamburg's commercial hinterland, especially the Elbe basin, is extensive, and the city proper is second only to Berlin in population.

Chemnitz (Karl-Marx-Stadt) and Dresden, both in the northwestern foothills of the Erzgebirge (Ore Mountains) in Saxony, were specialized manufacturing centers long before the Industrial Revolution, and have retained a strong industrial emphasis. Cheap, nearby coal and local deposits of fine china clay helped to establish Dresden very early as a world-famous center of ceramic production, a specialty it still pursues. Chemnitz became a textile center, despite its distance from sources of raw cotton, largely because of cheap local fuel. Local chemicals are a further element in its industrial structure; it is one of Germany's leading centers for chemical manufacture. Leipzig, the regional capital of Saxony, is somewhat farther away on the plain where it has transport advantages and market access as well as administrative convenience. It is an even larger industrial city with a great variety of production, including heavy goods. Taken as a whole, the industrial structure of Germany is by far the largest in non-Soviet Europe, despite its domestic resource gaps.

German agriculture

Germany has a relatively poor agricultural base for the support of nearly 80 million people, most of whom are not engaged in agriculture. Productive level land is limited by the heavy relief of middle and high Germany. Soils of the north German plain tend to be sandy, and, like most northern forest soils, acidic. In the eastern half, as maritime conditions yield to continental, winters are cold and growing seasons unfavorably short. Much of north Germany is a glacial outwash plain, deposited recently as geologists measure time. For the the most part, the soils are not particularly fertile, and there are many interruptions in the normal pattern of stream and subsurface drainage as a result of glacial action. With heavy use of chemical fertilizers and careful, hard work on the part of the farmers, Germany manages to grow about 80 percent of its food requirements. The potato, brought over from the New World (see Map 26, page 77) became a staple German food. Much of Germany is potato, turnip, and cabbage country. Had it not been for these crops that tolerate poor soils, drainage, and climate (especially the potato with its high caloric yields), Germany's food deficit would be much larger. The distinctive German cereal crops are barley and rye, also tolerant of these poor growing conditions (see Map 25, page 76). Like potatoes, both play a large part in the German diet, including the familiar beer and rye bread. In middle and high Germany, agriculture is limited by relief, although the valley floors and basins and the gentler slopes are intensively cultivated and support an efficient and productive system of mixed farming.

Strategy and self-sufficiency

For a highly industrialized country to be 80 percent self-sufficient in food, remembering Germany's handicaps, is commendable. Compared to British food production, it is high. But in Germany's exposed position food has seemed a problem in peacetime as well as during war. It was one more

MAP 41
BOUNDARY CHANGES IN GERMANY

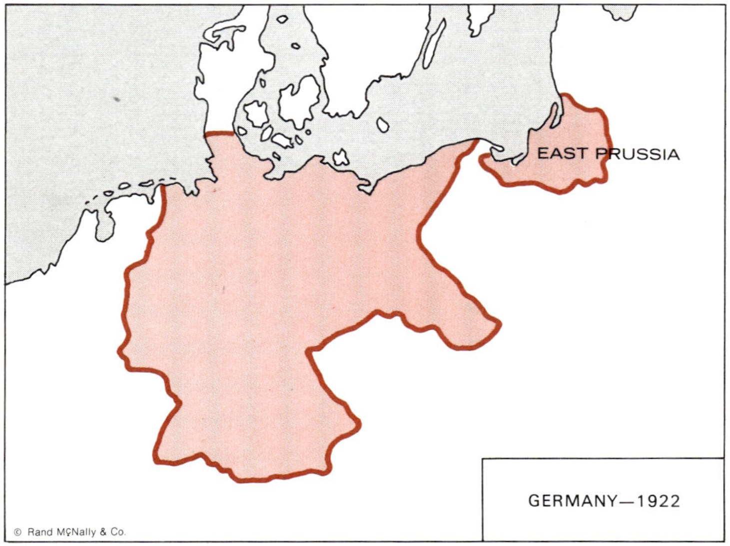

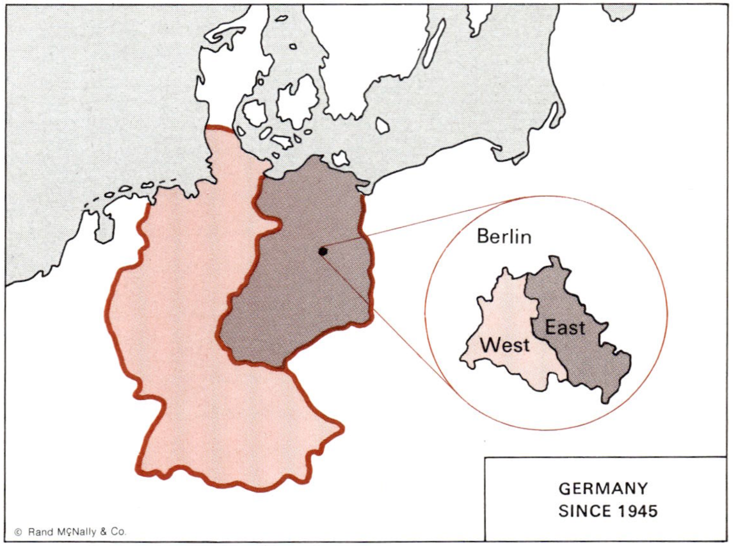

High Germany—Neuschwanstein Castle, near Fussen, Bavaria. This fairy-tale castle was built by mad King Ludwig of Bavaria in the nineteenth century. The valley of the Lech River supports some farming, as this picture shows. *(Photo by Craig Aurness for Woodfin Camp & Associates.)*

essential that Germany had to import from sources over which it had little or no control to maintain economic and military strength against the pressures surrounding the country. Iron ore, most other ores, petroleum, cotton, sulphur, rubber, vegetable oils, and many other basic materials had to be largely or entirely imported. Germany's obsession with the dream of self-sufficiency, or in German doctrine, *autarky,* is understandable. One means to this was the development of synthetics, most of them derived from coal or raw chemicals. The German achievements in synthetics are tremendous and command admiration. But many synthetics are more expensive than the real article, as well as being less desirable in certain cases. Thus, strategic considerations often fostered an artificial and economically unsound solution to economic problems. The Nazi attempt at solution through military conquest brought Germany nothing but catastrophe. To pursue autarky is to pursue a will-o'-the-wisp. No modern industrial economy can ever hope to be self-sufficient. It contradicts the basis of economic development, which is specialization through exchange.

Germany has powerful advantages of cheap coal and chemicals, of location and transportation, of a highly skilled work and managerial force, and of a massive existing industrial structure. These are solid assets that attract a major share of Europe's manufacturing and trade, including the food and raw materials Germany needs. But as a great power with gaping holes in its domestic resource account, it was a consistent temptation for Germany to look eastward into an area rich in many of the things Germany lacked. Eastern Europe was agricultural rather than industrial, and produced food surpluses for export. There was petroleum in Rumania (Romania), ores in Poland and Czechoslovakia. Beyond them was the Ukraine (Ukrainian S.S.R.), a German's dream, since it had rich deposits of iron and coal, nearby petroleum, a productive agricultural base, and a location far from other European or maritime military pressures.

The drive toward the east

German pressures on eastern Europe have gone on for many centuries. But more than autarky was involved in the German drive toward the east, the *Drang nach Osten.* A major factor in German political life from the time of the Teutonic tribes to Hitler was Germany's central position in Europe. Surrounded by different and usually hostile physical and cultural groups, the open plain gave Germany inadequate protection. Expansion was viewed as the alternative to annihilation, and a strong state as indispensable to survival. To the west, earlier-established and larger states formed a firm wall until recently. To the east, for the most part the land was in the hands of Slavic peoples, but seldom controlled by any single large state powerful enough to repel the German drive. Through the Middle Ages and into the modern period, the German *Drang nach Osten* continued, reaching its peak with the final (third) partition of Poland in 1795. This left roughly the western half of

Poland, except for the city of Warsaw (Warszawa), in Prussian hands, where it remained until the Treaty of Versailles in 1919. The Slavs were generally unsuccessful in resisting the German *Drang nach Osten* (at least until Germany became involved in European wars and was defeated by coalitions between Slavic and western European powers). One exception was mountain-protected Bohemia (České) (the western part of what is now Czechoslovakia). Topographic barriers helped to keep Bohemia predominantly Slavic, although the German *Drang nach Osten* rolled on for hundreds of miles eastward on the north and south. Nazi propaganda took advantage of the German settlers in Sudetenland, on the Czech side of the mountains, to make the boundary claims granted at the Munich Conference in 1938. It sealed the fate of Czechoslovakia by placing the mountain frontier in German hands. Nazi maps had shown Czechoslovakia as a dagger pointing at the heart of Germany. It required little imagination, and was more in keeping with the facts, to show Czechoslovakia as a morsel in the jaws of the German wolf, as was done by cartoonists elsewhere.

The Rhine at Bacherach. This is in the heart of the wine district. Notice the concentration of vineyards on the slopes facing south. In addition to water-borne traffic, the river is paralleled by two major rail lines (one on each bank) and by two major highways. The village of Bacherach with its medieval stone castle (left foreground) is typical of the Rhine valley above Coblenz. *(From British Overseas Airways.)*

Encirclement and militarism

Fear of encirclement was understandable, given Germany's geographic position, and it has been a justified fear on repeated occasions. Most Germans remember how Frederick the Great (king of Prussia, 1740–1786) first made German power apparent, and remember also the methods he and his immediate Prussian predecessors used. Briefly, these were the methods of the totalitarian state: universal conscription for military service, a strong central government unhampered by representative or democratic controls and virtually a dictatorship, a national cult of military preparedness or military virtues, an army that ruled itself as nearly as possible or that actually decreed state policy, and an overall authoritarian society, including family relationships and education. These methods were successful in that Frederick was brilliantly effective on the battlefield. They failed in that they contained the seeds of their own destruction, as the defeats of 1918 and 1945 and the dictatorship of Hitler demonstrated.

The Prussian legacy of an army largely independent of civil controls and proud of its power and traditions proved a recurrent obstacle to parliamentary government. The army frequently exercised a dominant influence on foreign policy, and the German General Staff won a notorious, if exaggerated, reputation in modern European history. It is true nevertheless that on many occasions the state was the instrument of the army rather than the reverse, a very dangerous situation. Germany's late political unification and its late start in industrialization and power politics lent urgency to the German drive. It believed it had to catch up with earlier-established rivals and use every means it could find to become a major power. It was a time for authoritarian direction. Everything had to be sacrificed to make the state strong.

Cultural regionalism and political unification

As a cultural entity Germany had begun in the west and moved steadily eastward in the direction most open to expansion. In this process the east tended to gain the upper hand from a military and also from a political point of view. The east led the German push in a contested situation with enemies on all sides. The states of east Germany, Prussia among them, began as military outposts, frontier settlements against the Slavs. These border states were the most military of the German communities. It was largely through their efforts that national unification was ultimately accomplished, specifically by the work of Prussia. The authoritarian, military, totalitarian pattern that had evolved in the east was imposed on the rest of the country. Most Germans of the west regarded this as the necessary price of national unity, one they were generally willing to pay.

Union between the industrialized, commercial, urban, maritime Rhineland area, and the rural, agrarian, conservative, continental east also had its economic logic, since there was mutual interdependence between such different economies. In practice, the east often seemed to make war with the industrial production of the west, but seldom to the west's advantage. While the Ruhr might be devastated by war, its factories wrecked, and its trade connections disrupted, the agrarian east was much less vulnerable to destruction.

The stereotype of a militaristic Germany can easily be exaggerated, however.[3] Germany is by no means the only example of an authoritarian state. Even in western Europe, Louis XIV, king of France (1643–1715), could say, *"L'état, c'est moi"* (I am the state) before the French Revolution began autocracy's downfall in the West. Germany was less directly involved in the changes in Europe, in part because of its geographic position and problems. The revolutionary attempts of the nineteenth century failed to produce a genuine parliamentary democracy even in the Rhineland. But since 1945, Germany has followed a Western-directed path politically, and representative government in the Federal Republic of West Germany has gathered new strength.

Even under authoritarian rule, German contributions to European and world culture have been enviable. No nation has ever equalled the rich, varied outpouring of German music, to which all mankind is indebted. Bach, Beethoven, or Brahms are not good examples of militaristic Germans. German achievements in literature, music, philosophy, and in all fields of learning, especially science, are impressive. They are not the contributions of a narrow, obsessed, or military mind. The scholarly tradition in Germany survived even the Nazis and is still vigorous. The modern study of geography was begun in

[3]For example, see H. C. Meyer, *Mitteleuropa in German Thought and Action, 1815–1945* (The Hague, 1956), for a well-argued and substantiated view of the limited German interest in central Europe.

Germany and continues to owe much to the German school.[4] In return for the problems of centrality, perhaps Germany has received some of the advantages that centrality may bring, especially in easy exchange and a variety of external cultural stimuli.

German unity and dualism

Germans have made their country a national unit, and, however ill-founded it may be in other respects, it is a living entity that cannot be destroyed. The rest of Europe has an important stake in German prosperity. As the largest national economy and the greatest commercial and industrial center on the Continent, Germany is the largest single factor in Europe's economic health. It can function at optimum efficiency only as a unit. Politically speaking, the indefinite denial of German unity may invite a nationalist explosion or a new outbreak of international violence, which few people, including most Germans, want. The bitter experience since 1918 had profound effects, and Germany no longer dreams of conquest and expansion as the solution to its problems. The military tradition and the precedent of an independent army have lost their appeal and have been discredited by the ruin they have brought. The global power role of the U.S.S.R. and the United States further encourages Germany to work with its neighbors rather than to dominate them.

Since 1945 West Germany has made a spectacular economic and industrial recovery without the aid of military subsidies or artificially imposed trade agreements. Membership in the European Common Market has strengthened rather than weakened the leading industrial position of West Germany within Europe, and fed a mounting wave of prosperity. The reduction of artificial trade barriers gives comparative advantage a fuller scope. West Germany has also intensified its agricultural system. Now that food is no longer readily available from eastern Europe, the Federal Republic more nearly feeds itself. In East Germany, Russian-imposed collectivization of agriculture has contributed to food supply problems. However, heavy manufacturing has had an impressive growth, making use of local minerals and sending its production into the Soviet economic bloc.

Germany's basic geographic problems remain, and old solutions to them may be revived. Splits in the German regional fabric may be intensified, as they are at present by the division between the Federal Republic of Germany in the west and the German Democratic Republic in the east. Although a political unification of Germany may reappear, or may be approximated as closer working relations develop between the two German governments, it must contend with the geographic pressures that unity attempts to deny.

Switzerland

It is ironic perhaps that where Germany, despite a common language and culture, has had such difficulty in achieving national unity, Switzerland has had a single national state since the end of the Middle Ages, although there is even now no common Swiss language. Fortunately for the Swiss, their country is not part of the "shatter zone" of central Europe described later. Although Switzerland is a small place surrounded by larger and more powerful states, its mountain base has been enough to protect it, and very early to give it its own separate sense of identity despite its mixture of languages. Roughly the western fourth of the country, including the city of Geneva (Génève), speaks French. Most of the rest speaks German, but there is a small area in the south that speaks Italian. What united these originally separate cultures was their common mountain environment and the way of life it supported. It was different from the patterns developed in the distinctive settings of France, Germany, and Italy. The mountain-based community of interest was strong enough to override the language differences. Indeed, the Swiss have been far more successful in this respect than any other of the many bilingual or multilingual states, such as Belgium or India, where language differences are a chronic source of tension.

Originally a community of relatively poor mountain-forest people, Switzerland has become prosperous as a result of its location between north and south, east and west Europe, hence a meeting place of trade routes. This led also to the development of Swiss prominence in the financing of trade and in banking. The carefully maintained political neutrality of Switzerland has also helped its role as an international banker. Such a tiny country could

---

[4]Unfortunately this record was clouded during the Nazi period by the injurious nonsense and propaganda of the *geopoliticians*, another aspect of the war between politics and geography in Germany. The geopoliticians, too, attempted to twist geography into a self-made political mold.

not afford to antagonize its more powerful neighbors, despite the defensive advantages of the mountains. But it has also been profitable to keep the confidence of all sides and to act as a neutral broker where the major trade routes of Europe cross.

Switzerland's central location is also a strong advantage for the tourist industry, a major part of the Swiss economy. Of course the Alps and the beautiful lakes that decorate the Swiss landscape are attractive. But equally important is Switzerland's easy accessibility from the most densely populated parts of Europe in an age of railway and air travel. Major mountain passes, many with rail lines through them, link this country with those around it. Switzerland, with its unmatched landscape and tourist attractions in all seasons, is only a few hours from anywhere in Europe. Average incomes are also high enough there to permit the relative luxury of a Swiss holiday.

The Swiss still dominate world watchmaking. They are also major producers and exporters of a variety of other kinds of precision machinery. Industrial raw materials are largely lacking and are relatively expensive to import, so the Swiss have specialized in production where skilled craftsmanship and engineering are dominant. Not surprisingly, one of the major specialties is electric-power generators and electrical machinery, reflecting the great importance of hydro power in this mountain country. But Swiss watches still best solve the problems of transport and raw materials. It takes little steel to make them. The finished product is small and light, but extremely valuable, especially in relation to size and weight. Their value comes from the use of skilled labor and experience, qualities the Swiss have long developed.

The capital is at Bern, near the center of the country. But Zürich, the financial capital, is by far the largest city. They, and the other Swiss cities, are not in the Alps, but are on the shores of lakes, along the Swiss course of the Rhine, and elsewhere in the relatively limited valley and plain lying between the Alps on the south and the Jura Mountains on the north. The mountains are relatively empty of settlement except for ski resorts. The traditional Swiss herdsmen of the Alpine pastures and their production of milk and cheese are now greatly outnumbered by urban Swiss of a more modern and more prosperous world. But the Alps nourished and protected the Swiss nation, and still make Switzerland a highly distinctive country.

The Alps near Bern, Switzerland. The huts in the high pastures are occupied only in summer, and used for cheese making as well as shelter. Their roofs are weighted with stones against the strong winds. Crops are raised in the valley, and tourists also contribute to the economy. *(Photo from Swiss National Tourist Office.)*

The shatter zone

The zone between the eastern limits of German and Italian cultures or states and the western limits of the Russian culture and state is both a physical and cultural transition. Sharp or single-line boundaries seldom occur in nature. Even physical differences on each side of a major, well-defined mountain range tend to merge with each other over a zone of a few miles. This is also true of human areal differences. The width of the transition zone depends on many factors, including the presence or absence of sharp physical barriers, the sharp or gradual change from one type of physical environment to another, the surrounding political, economic, or cultural pressures, and the nature of the human occupance. In central Europe all of these factors, with many regional differences, have made the transition zone a wide one, and hampered the development of unity within it. Plains in the north developed differently from those in the mountainous south. But over the area as a whole, there has been a consistent conflict between east and west. It has never been permanently solved in favor of either, but has left permanent aspects of both.

Internally the transitional character of the area, plus serious physical disadvantages especially in the mountainous south, have worked against a common community or a single state strong enough to resist pressures from either side. This area is called the *shatter zone*, a reflection of its political and cultural fragmentation, also partly attributable to pressures from both sides that have had a shattering effect in this intermediate area.

The shatter zone comprises a large chunk of Europe, including the former Baltic states (Lithuania, Latvia, and Estonia), Poland, Czechoslovakia, Hungary, Rumania, Yugoslavia, Albania, and Bulgaria.[5] Austria has been intimately involved as a sovereign power in central Europe in the recent past, as discussed in more detail later, and might logically be included in the shatter zone. Like the other states, Austria has been powerfully influenced from both east and west, like them it has contained a variety of eastern and western cultural and ethnic strands, and like them it has been politically broken apart or shattered by military and political pressures exerted from both sides. The zone, therefore, comprises well over one-fourth, perhaps as much as one-third, of Europe as a whole, an area that has had a tragic history for the most part.

East versus west

Most of the people who live in this zone of central Europe are ethnically Slavic and speak a Slavic language (different in each country and in many subregions). Most of them also have deep and longstanding ties with the West. With some this tie is trade, with others a legacy of Roman influence (as the name *Rumania* reveals, the country is part of the old Roman province of Dacia, and Rumanian is a Latin rather than a Slavic language), or with others adherence to Roman rather than Greek Orthodoxy, or to Protestantism. On the other hand, Slavic nationalism, stimulated by real or imagined threats from non-Slavs, such as Turks, Austro-Hungarians, or Germans, tempted many Slavic communities to look for protection to the greatest of all Slavic communities, Russia. Environmentally western Europe merges in this area with eastern Europe, and the shatter zone cannot be neatly included with either. All of these factors have contributed to the recurrent efforts by both east and west to dominate the shatter zone. Temporarily successful attempts from the west have been followed by temporarily successful attempts from the east, as the military and political fortunes of each waxed and waned. Political alignments, new states, and redivisions have risen and fallen in the shatter zone with each new generation (see Map 43, page 150). The present ascendancy of the east in this troubled area is not likely to be permanent, judging only from the record of the past. It is not surprising that central Europe has often provided the trigger for, if not the basic cause of, so many European wars, or that the shatter zone itself has seldom profited from them.

Physical, cultural, and political disunity

Within the shatter zone so far it has been impossible to create the political unity that could resist external pressures. Ironically, the area as a unit would be larger than Germany and Italy combined, and would be better able to maintain itself successfully against both western and eastern designs on it. But the forces of diversity have been strong, and at most periods the states and cultures of the shatter zone have quar-

[5] Under present circumstances, East Germany may also be regarded as part of the shatter zone.

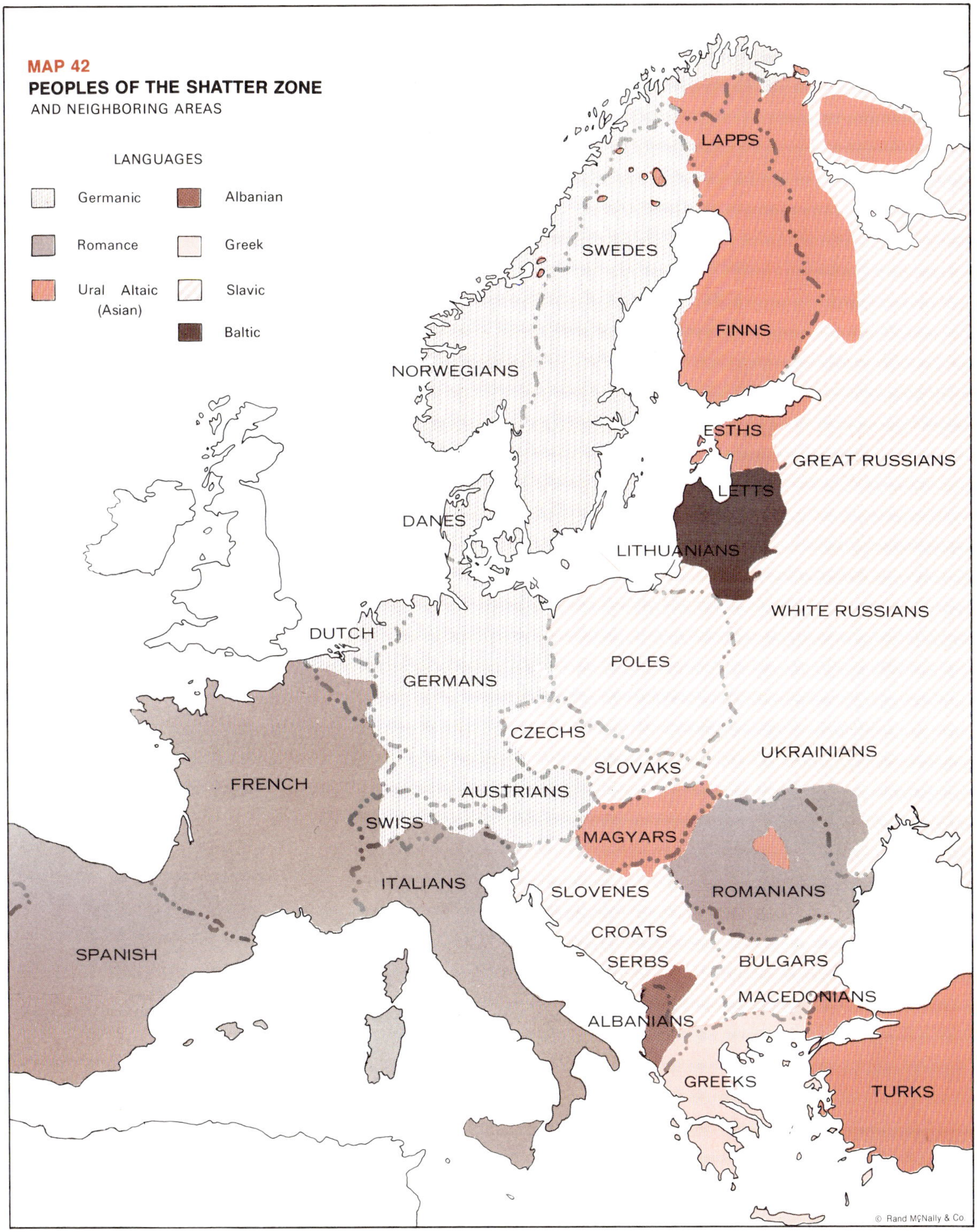

MAP 42

PEOPLES OF THE SHATTER ZONE

AND NEIGHBORING AREAS

MAP 43
BOUNDARY CHANGES IN CENTRAL EUROPE

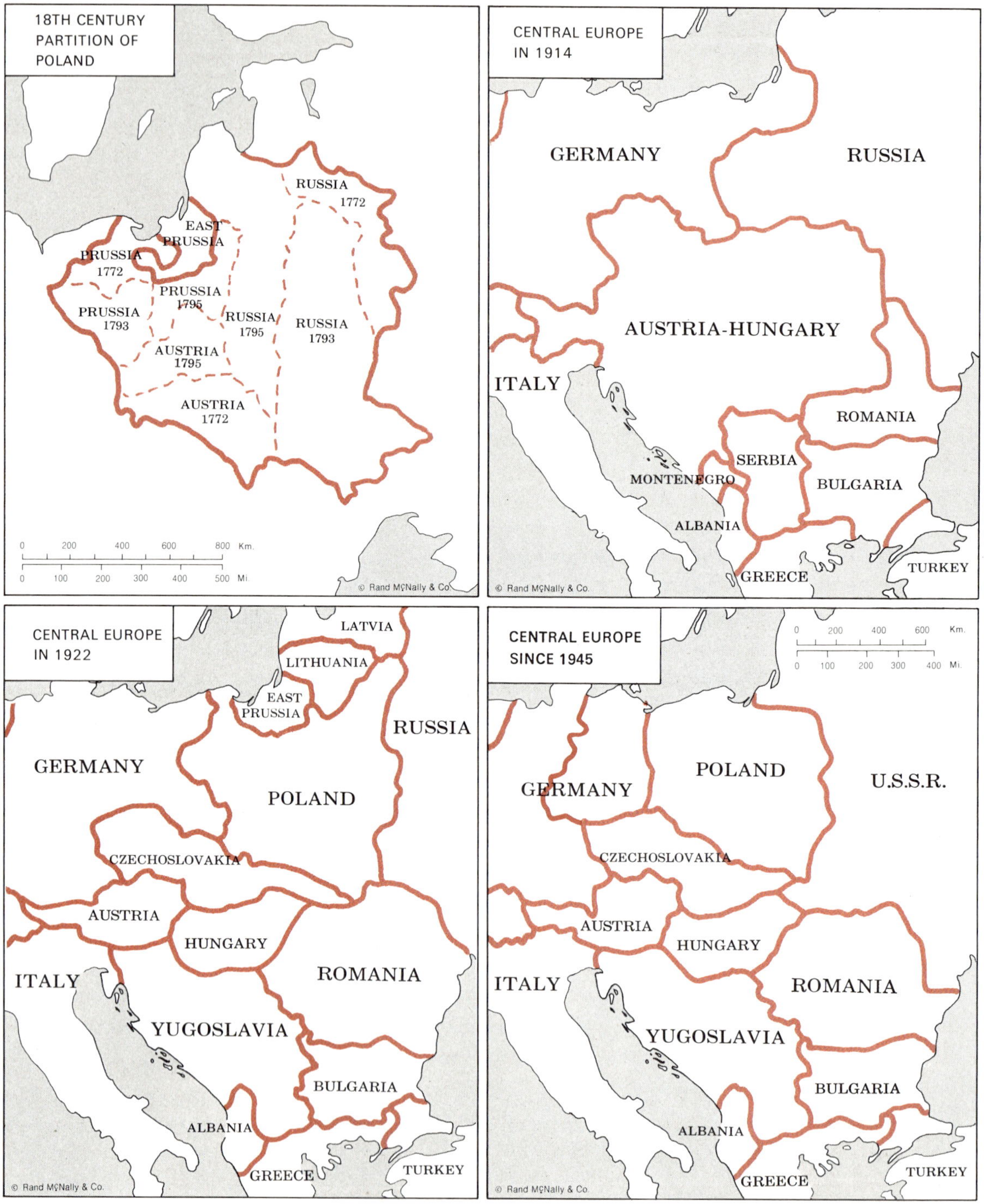

reled among themselves. Just as vigorously, they resented the outside interference they were powerless to prevent as long as they remained disunited. Several central European countries have attempted to enforce unity under their own domination, and all have failed. This is a result not only of strong cultural regionalism, ethnic and linguistic diversity, and more recently, of equally strong nationalism, but of the lack of effective centers for control. It was understandable that the north, mainly part of the great European plain along the Baltic, should foster its own regional and national communities that are distinctly different from those of the south, where mountains are dominant, the human landscape further subdivided, and the sea is the Mediterranean (see Map 14, pages 60–61). There is no single major economic area for central Europe as a whole, and no political base that could maintain its dominance from Estonia to Greece, or even from Poland to Hungary. Poland and the three Baltic states attempt to defy the usual pattern of political entities by existing on a wide, open plain with virtually no physical barriers as protection. As a transition zone, the plain of eastern Europe has had a large share of competing influences, and the physically unprotected north Slavic states have suffered terribly as a result. There is even a lack of well-marked lines across the landscape that could be used to delimit national frontiers. Different cultural and linguistic groups have settled, merged, fought, and regrouped all over the plain. In addition, none of the Baltic states has a domestic economic base adequate to sustain it in the face of these pressing disadvantages. Cultural memory, fervent national traditions and loyalties, and the fluctuations in the balance of power between east and west have kept these states in existence, or, alternatively, annihilated them. Estonia, Latvia, and Lithuania have been reincorporated in the U.S.S.R. since 1945, but Poland's independence is more substantial. In its days of greatness, Poland was the dominant power in eastern Europe and even controlled much of what is now western Russia.

Poland and its resource base

Poland's open position has been strengthened by its mineral wealth (see Map 21, page 71), and enhanced by the addition of Silesia following the Second World War. The most important deposits are concentrated in the south, where the European plain gives way to the mountains of south-central Europe, specifically the Sudetens (Sudetes) and Carpathians. There is a growing complex of Polish industrial centers in the piedmont zone between Krakow, Wroclaw, and Czestochowa. Nowa Huta, a city built with Russian aid around a large steel plant and rolling mill near Krakow, has grown rapidly and is now part of Krakow's population of about 700 thousand. There are large deposits of good-quality coal, some iron ore, potash, and a little petroleum, plus water power from the mountains. Much of the coal is in formerly German Silesia, but even without Silesia there are resource advantages for industrialization in Poland. Its late industrial development is symptomatic of eastern Europe as a whole. But with the help of centralized planning focused on rapid growth, especially in heavy manufacturing, the industrial structure is already large. Heavy manufacturing is done mainly in the south. Consumer goods, notably textiles, are still centered in the older cities: in Lodz, Poznan, and Warsaw, the capital.

Textiles are nearly always the first manufacturing industry to be developed. They require a relatively small capital investment, are not technically complex, do not demand specially skilled labor, and can produce for a wide market from the beginning. The product is relatively low in cost, and the effective demand for it is nearly universal, even before other forms of industrialization are established. Most industrial areas in the world have followed this pattern. The earliest industrial development in textiles has usually occurred in the oldest or largest cities where fuel (or power) and transport costs were lowest and where the market, labor, and capital advantages were concentrated. Even in the later stages of industrialization textiles tend to be market-oriented. They are less sensitive than heavy manufacturing to the location of bulky raw materials or their access. This is characteristic of most manufacturing of consumer goods, which is much more widespread in market centers than the basic elements of a mature industrial structure, the heavy industries.

Warsaw's centrality within Poland and its location near the middle of the most productive agricultural plains make it a good capital. Warsaw also has the best communications site in the country, near the head of navigation on Poland's major river, the Vistula (Wisla). There is good agricultural land, alluvial deposits, in the Vistula valley. The coastal plain of the north is a continuation of the north German plain and shares its generally infertile sandy soils and poor drainage. But the piedmont and hills of southern Poland contain large areas of fertile loess soil, widely used for wheat. On the poorer soils of the north, rye and potatoes are the major crops, and for Poland as a whole, they are the most important (see Maps 24, 25, and 26, pages 75, 76, and 77).

Poland's link with the sea is the Gulf of Danzig, an area with a complex history, since it lies directly in the path of the German *Drang nach Osten* along the Baltic (see Map 40, pages 138–39). Poland now has unbroken control of the Baltic coast from the Oder to the border with the U.S.S.R., with its chief port at Gdańsk (Danzig). But Polish sovereignty generally is under a Soviet cloud. More Polish territory in the east was taken by Russia in 1945 than was received from Germany in the west, although the land gained in the west (including the major coal fields of Silesia) is superior on the whole to what was lost to Russia in the east.

The Baltic states: open and unstable

Lithuania, Latvia, and Estonia have had even more troubled and unhappy histories than Poland, and have fewer domestic resources to sustain them. These are predominantly agricultural economies, but climate and soils are not very good. Commercial and industrial development is concentrated in the Baltic ports: Memel (Klaipéda) for Lithuania, Riga for Latvia, and Tallinn (Reval) for Estonia. All three republics have been plagued by Polish, German, or Russian minorities that have been recurrent sources of domestic dissension and foreign intervention and absorption. The national boundaries have seldom been constant for long. Linguistic, ethnic, cultural, and religious differences among the majority groups explain the presence of three small separate political communities on this physically undifferentiated plain. They are a reminder of the generally fragmented character of the shatter zone, where Eastern and Western influences are so intermixed. Estonia and Latvia are predominantly Lutheran in religion; Lithuania is Roman Catholic. Estonia has close ethnic and linguistic ties with Finland, while the other two use an Indo-European language. Each is proud of its own distinctive history. In the fifteenth century Lithuania was indisputably the greatest power in central and eastern Europe, controlling an area over 500 miles (800 km) wide from the Baltic to the Black Sea, centered on Vilnius (Wilno) and Kiev (Kiyev). Its power collapsed, but the memory of the past remains strong and helps to keep alive the vigor of Lithuanian nationalism in the face of overwhelming difficulties.

The fragile or temporary nature of these conquests and the disastrously shifting fortunes of the states themselves are in keeping with the general record of the shatter zone. Political expansion has often been followed by political obliteration, as both Poland and Lithuania testify. Under Soviet occupation and as integral parts of the U.S.S.R., "Russification" has been stressed and the former German and Polish minorities repatriated.

Geography and the Czech achievement

Czechoslovakia occupies a median position in more than latitude. It strikes a compromise between the disastrous absence of barrier boundaries characteristic of the northern plain and their overwhelming presence in the south. There are three distinct physical units in Czechoslovakia: the mountain-encircled plain of Bohemia in the west, the somewhat interrupted plain of Moravia (Morava) in the center of the country, and the Carpathian Mountains and small basins of Slovakia (Slovensko) in the east. Although Bohemia has been overcome by the crushing assault of both German and Russian power, it consistently retains its cultural identity. Despite the Slavic base of this culture, Czechoslovakia (like its predecessor, the old Kingdom of Bohemia) has solid and long-standing ties with the West. Until its absorption into the Soviet bloc, Czechoslovakia was a vigorous parliamentary democracy, although minorities like the Slovaks in the eastern part of the country did not occupy a fully representative place, in their own view. Czech industrial growth has followed a Western course, and is far in advance of the rest of central Europe. Bohemia's physical penetration into the west and the mountains so neatly outlining the country (see Map 40, pages 138–39) have helped make this distinction possible. There have been other advantages too. The Elbe (Labe) River leads from central Bohemia through Germany to the open Atlantic at Hamburg. Especially before 1939, it helped to orient Czech foreign trade westward. Prague (Praha), the national capital, lies near the Elbe on its tributary, the Vltava (Vltana). In the southeast, Bratislava, the provincial capital of Slovakia on the Danube, is a center of eastern-oriented trade. Brno, now the capital of Moravia, in the past was oriented primarily toward Vienna (Wien) and the Danube basin.

Western and eastern patterns

Thus, Czechoslovakia is integrated with two of the main avenues of European commerce, for the Danube forms a trade highway around which much of the economic life of south-central Europe is gathered. Highlands lie between Prague in the Elbe drainage and Bratislava in the Danube drainage, a fitting line between the western and eastern halves of the country. Water partings such as this are commonly and conveniently used as regional lines. A river basin or watershed often constitutes a consistent region. In this case the division is of greater significance, since it helps orient the two halves of Czechoslovakia in opposite directions, one westward with the Elbe to the Atlantic, the other eastward with the Danube to the Black Sea and Asia. There are also other regional differences. Most of the manufacturing and the larger part of the commercial functions, the cities and towns, are in western Czechoslovakia. The east is more agrarian, more purely Slavic (including its religion and language), and less intimately involved with the institutional and economic structure of maritime Europe.

As it happens, Czech raw materials for manufacturing are also largely concentrated in the western half (see Map 21, page 71). This has helped to heighten the regional distinction. Coal and iron were both originally present in fair quantity and quality on the Czech side of the Erzgebirge, in the lowland around Prague, and to the southwest, where the industrial city of Plzeň (Pilsen) has developed to work them. Czech supplies of coal and iron ore are both approaching exhaustion, but are supplemented by imports. Czech manufacturing still remains competitive in cost terms. There has also been heavy investment in nuclear power production. Per capita industrial output is higher than the average of western Europe. Plzeň is a major heavy manufacturing center even by west European standards. There is also large and varied industrial development in Prague, in Brno, and in Ostrava, near the Polish border, close to what are now Czechoslovakia's major coal fields. A large center of heavy manufacturing developed at Košice in eastern Slovakia. Under Communist control since 1945, a series of five-year plans has greatly increased investment in heavy industry, while maintaining the traditional and highly successful Czech light industrial sector.

Czech culture is highly cosmopolitan, as most

Prague, Czechoslovakia. The Vltava River flows through the heart of the historical part of this beautiful city. The Charles Bridge and some of the Medieval and Baroque buildings are seen in this picture. *(Photo by C. Dittmann for Woodfin Camp & Associates.)*

Balkan mountains—southern Yugoslavia. This is a characteristic Balkan landscape, but roads and railways are only beginning to penetrate the mountains. These developments are more advanced in Yugoslavia than in the rest of the Balkans, where mountains still mean isolation and poverty.

central European cultures are because of their shatter zone location. Czechoslovakian culture is strengthened by Slavic Bohemia's inclination toward the West, the dual trade orientation of the two rivers, and extensive commercial relations with the rest of the world. Multiple external stimulation has played a part in the development of the dynamic, sophisticated culture in Czechoslovakia, as well as its prosperous economy.

But the Czech state lies in central Europe, and it is exposed to the pressures and problems characteristic of the shatter zone. It is a long way from the Atlantic and is surrounded by large, powerful, ambitious states whose rivalries have repeatedly focused on this disputed zone. Thus in recent years Czechoslovakia has been overwhelmed by Germany and then by Russia, as it was included earlier in the old Austrian empire. The independence of modern Czechoslovakia dates only from the Treaty of Versailles in 1919, although Bohemia had been a distinct entity within its shield of mountains for many centuries. Grafting the province of Slovakia onto this old Bohemian-Moravian base was arbitrary and artificial to some extent. But the Slovaks were neither Poles nor Hungarians, and their closest cultural and sentimental ties were with the Czechs, despite the economic differences. The complex cultural and economic mixture of central Europe as a whole makes it an area where there often seems to be no really satisfactory solution to conflicting allegiances and claims.

The Balkans: mountains and competing states

The rest of the shatter zone, aside from Austria, may be lumped together as the Balkans, and includes Hungary, Rumania, Yugoslavia, Albania, and Bulgaria. It has very little unity except that provided by the Danube. Even its generally Slavic culture is only superficially uniform, since language, religion, and most other cultural aspects differ considerably from one small country to another (see Map 42, page 149). In the Balkans there is an oversupply of physical barriers, a jumbled mass of mountains. Seldom do well-defined ranges encircle a whole region or coincide significantly with political lines. It is almost as if the Balkan states had been superimposed on the map in disregard of mountain and plains patterns (see Map 35, pages 124–25). States and cultures have grown up, merged, disappeared, and regrouped around each of several cities, but none of them has been strong enough to dominate the whole of the Balkan region. Few were strong enough to maintain their positions for long, even within their small plains bases. The Danube, which might otherwise form a thread of unity through the Balkans, is used instead as a political boundary along the greater part of its course east of Vienna.

Politics versus geography in the Danube basin

Of all the Balkan states or regions, Hungary is the best situated and has generally been the most successful in using its advantages. It is located on the fertile plain of the Danube and its tributary, the Tisza, with a degree of protection from mountain ranges on the north, east, and west. But even Hungary, with a productive core area and the strategic trade site of its capital at Budapest, has not been able to unify the whole of this lowland region. The great arc of the Carpathians and the Transylvanian Alps (Carpatii Meridionali) on the east crosses the Danube at the famous Iron Gate gorge to meet the mountain chain in what is now Yugoslavia (see Map 35, pages 124–25). The circle is completed in a ring of mountains on the west and north, but they are beyond the borders of Hungary. Modern Yugoslavia, uniting Macedonians, Montenegrins, Serbs, Croats, Slovenes, and a few other smaller groups, has also inherited their possession of the Danube lowland north from Belgrade (Beograd). Rumania, with its capital at Bucharest (Bucuresti) on the other side of the Transylvanian Alps, has been able to retain command of a wide strip of lowland on the Hungarian side of those mountains. Czechoslovakia keeps two large bites out of the northern part of the Hungarian plain, where Danube tributaries come down out of the Slovakian mountains. Productive agricultural areas on these flood plains and trade centers at the navigation heads are thus lost to Hungary. Even Austria retains Vienna as its economic heart, although the plain on which Vienna lies is topographically part of the Danube lowland below Budapest.

German, Russian, and Turkish pressures and invasions have been largely responsible at different periods for this odd political denial of geography. Most major European wars have found the Balkan states on opposite sides. Hungary and Bulgaria unfortunately picked the losing side in the two most recent conflicts, while Rumania, Yugoslavia, and Czechoslovakia fared better in the peace settlements readjusting their boundaries. However, all were overwhelmed by Russian power, which had crushed the previous German domination of the Balkans (see Map 43, page 150).

Balkan mountains, plains, and seacoasts

Political disunity in the Balkans has been a continuing disadvantage to all of the states. Although the mountains gave enough precarious security to help keep the Balkan cultures and states in existence despite their continual problems, mountains have also isolated and fragmented them. Economically they exist on a relatively low level. In comparison with the rest of Europe, this is a backward area where the Industrial Revolution and nineteenth- and twentieth-century change came only slowly. Visitors to the Balkans before 1939 compared the area with China. Its economic level and the dominance of peasant agriculture carried on by ancient traditional methods were very similar to China's. This is least true of Hungary and most true of Bulgaria. Most Balkan states must be content with a small piece of the Danube and a large piece of unproductive but fairly secure mountains. Hungary has large, rich agricultural areas, command of a long stretch of the Danube, and proximity to the major centers of trade and ideas in Austria and the West, brought to it by the river. Bulgaria is mostly mountains and lacks control of the Danube, since the river forms its boundary with Rumania, a stronger and better-situated state. Bulgaria's Black Sea coast is of little use, especially without a major navigable river, and the country is isolated as well as poor. Rumanian Constantsa (Constanta) and Bulgarian Varna, the major non-Russian ports on the Black Sea, are of only local significance.

Yugoslavia is somewhat better off, since it controls a respectable length of the Danube. It is an odd state, composed of at least three cultural and historical groups—Serbs, Croats, and Slovenes—plus several other minorities. Yugoslavia has a seacoast on the Adriatic, but the hinterland is covered with mountains. The mountain area is receiving more attention now in the economic development of its mineral and water-power resources, but the lowland of the Danube is the real heart of the country. Yugoslavia faces toward the Danube rather than toward the mountainous coast of the Adriatic. However, the valuable lowland area east of the mountains has almost no protection against neighboring Hungary and Rumania. Internal physical and cultural disunity are still major problems for Yugoslavia. On the eastern coast of the Adriatic, the city of Trieste, Italian since 1954, provides a significant commercial outlet to the sea. Its good harbor is connected by relatively easy routes with a major pass across the Alps into Austria and central Europe, for which Trieste has been the principal ocean port for many years. Because of Yugoslavia's orientation toward the Danube, Trieste itself has long had an Italian majority in its population. This was a major reason for returning the city to Italy in the 1954 agreement. Some of Yugoslavia's foreign trade still passes through Trieste, but the Yugoslav port of Rijeka (Fiume) some 40 miles (64 km) farther down the coast is increasingly important. Yugoslavia's current partial detachment from Russian influence and its separate version of socialism, however, may be attributed in part to its seacoast facing the West, and in part to the fact that neither the German nor the Russian armies were able to eliminate the Yugoslav guerrillas in the extensive mountains of their country. Yugoslavia was never "liberated" by the Red Army, as were the other countries of the shatter zone.

The Turkish influence

From the fourteenth until the early nineteenth century, the empire of the Ottoman Turks was the major power in this part of the world. Its military conquests, like the earlier advances of the Huns, Avars, and Magyars, reached up the Danube valley to the walls of Vienna. Every Balkan state fell under Ottoman control for longer or shorter periods, and the legacy of Turkish influence is still apparent in all of them. Occasional mosques are still found in many Balkan towns, and the religion of Islam survives in Albania and parts of Yugoslavia. In material endowment Scandinavia is no better off than the Balkans, perhaps less well off. But the Scandinavian people have developed a much higher standard of living, a stable political system, and a creative culture far in advance of the Balkans. To the grinding pressures of east and west Europe on the shatter zone, whose name fits its history so well, was added the even more alien and destructive influence of the Turks in the southern half of the area.

Austria and its former empire

Austria, no longer a great power, once controlled most of what is now Czechoslovakia, Hungary, and Yugoslavia, parts of what is now Poland and

Austrian Alps. The village of Heiligenblut in its magnificent setting is obvious tourist country in western Austria. *(From Pan American.)*

Rumania, and much of northern Italy. The basis and core of this Balkan empire was the Danube valley, to which the Austrian city of Vienna was the commercial and strategic key. Originally an Austrian conquest, the empire became jointly ruled by Austrians and Hungarians (the Austro-Hungarian Empire) in an effort to keep the peace among the variety of nationalities and cultures. The Hungarians, or Magyars, are not Slavs. Their share in the ruling partnership, as the most powerful and prominent of the Balkan groups, was intended to support the Germanic position of the Austrians. But the empire remained a contradiction in every respect except economics. Although it was destroyed by the nationalism of the nineteenth century and by the crushing blows of the First World War in which Austria was disastrously defeated, the empire made far more sense as an economic unit than anything preceding or following it. Austria was a cosmopolitan trading state rather than a military outpost. It originally formed part of the German wedge against the Slavs with an economic base in Danube trade.

Vienna (and Austria) lies midway between agricultural southeast Europe and industrial-commercial central Europe. The greater bulk of trade on the Danube still moves upstream, an unusual situation for a major navigable river. Agricultural goods from the Danube basin were gathered in Vienna and distributed from there by water (including the Danube-Rhine canal by way of Nuremberg) and by rail to Germany and the northwest. Manufactured goods were distributed from Vienna to its mainly agricultural hinterland. The plain of lower (eastern) Austria around Vienna is a crossroads of trade routes. The easiest passes across the Alps, including the Brenner Pass, converge on Vienna. In the past a large share of Italy's trade with north Europe came through Vienna. The Moravian Gate, a mountain pass north of Vienna on the March (Morava) River, provides an important route from the Danube valley into Silesia and Poland. The Danube itself leads northwest into Germany and southeast into the Balkans. Vienna also became a manufacturing center to some extent. It processed agricultural products, produced light consumer goods, and engaged in service industries for its large tributary hinterland, although most raw materials were not plentiful locally.

After the First World War sizable deposits of petroleum and iron ore did become important to Austria, as did development of Austria's hydro potential from the Alps. The modern state is trying with some success to build up a small industrial complex to take the place of its former commercial empire. The Alps cover most of modern Austria. The western two-thirds is commercially isolated and relatively poor in both agricultural and industrial potential. Westernmost Austria is more closely related to Switzerland than to the lowland plain around Vienna. The most important commercial resource of Alpine Austria is tourism. The country as a whole, especially since about 1960, has followed the Swiss model internationally, playing a role of strict political neutrality. This is a change from the past attitude of the Austrian Empire, and one that makes excellent sense, given Austria's present circumstances.

The modern state has made good economic progress, despite its sufferings during and after the two world wars. But adjustment to the drastically reduced size and location of Austria left by the Treaty of Versailles has been difficult. Once a cosmopolitan and prosperous trade center, the imperial capital of a rich empire, Vienna faded when its access to the Danube basin was cut. Johann Strauss could write of the beautiful blue Danube in the nineteenth century and set Vienna dancing to it. After 1919 Vienna's gay and prosperous past became lost in irony and nostalgia. In the heyday of national self-determination, as reaffirmed by Versailles, even a victorious Austria would have had to make large concessions to the many groups in its empire. As it was, they all became independent states, and Austria was left with little economic alternative to an even closer partnership with Germany. The economically logical unit of the old empire was replaced by separate nationalisms and ultimately by the reappearance of a victorious Russia as the arbiter of the Balkans. It remains to be seen whether orientation toward Russia can provide the economic advantages for the Balkans that Austria and Vienna offered.

Questions for further study and discussion

1. What was the main emphasis of the German school of geopoliticians? Why was it unsound?

2. Why is other manufacturing usually attracted to a major iron and steel center? Why specifically to the Ruhr?

3. What is the function of Bremerhaven and Cuxhaven? Why are they not bigger than Bremen and Hamburg?

4. Why are so many German cities on or near the southern margins of the north German plain, where the hills of middle Germany begin?

5. What was the major seaport for the trade of the Danube basin? Why?

6. The Balkans have been called the "Tinder Box of Europe." What does this mean, and how do you account for it? What does *Balkanization* mean? Why?

7. Are there other examples of states or economies organized around a river or a river system? How have they resembled or differed from Austria-Hungary?

8. Yugoslavia has several valuable ores in significant amounts, especially bauxite (aluminum ore) and copper. Why have they not stimulated more industrialization in Yugoslavia?

9 The U.S.S.R.: the gift and the problem of space

St. Basil's Cathedral, Moscow. *(From Bert Hemphill Tours.)*

For centuries Russia has been the largest political unit in the world. The present Union of Soviet Socialist Republics (U.S.S.R.) is only a little larger than the empire of the czars of imperial Russia, which ended with the Russian Revolution of 1917. Russian sovereignty under the empire had extended from eastern Europe to the borders of India and the shores of the Pacific. The Soviet Union, which inherited this territory, covers about one-sixth of the land surface of the entire globe. This vast space is very unevenly settled, and most of it is not settled predominantly by Russians. As Map 45 (page 162) shows, the bulk of the population and most of the large cities are in the west, grouped around the original Russian state, from Leningrad to Moscow (Moskva) to Kiev (Kiyev), all within Europe. Russian people and Russian rule have been expanding eastward into Asia since the fifteenth century, even while contacts with the rest of Europe increased. As early as 1639 Russian explorers and traders reached the Pacific coast of Siberia, claiming all of this huge territory for the czars. Across Asia the Russians found a narrow strip of land they considered suitable for settlement between the arctic north and the desert or mountainous south. Elsewhere most of the land continues to be occupied by a great variety of predominantly non-Russian peoples.

The expansion of Russia eastward across Asia compares with American and Canadian expansion westward across North America. There are certainly many parallels, and the influence of the advancing frontier has been strong in both countries. This may explain some of the common ground between Americans and Russians: their mutual habit of thinking in big terms, their expansiveness in personality, their willingness to tackle outsize problems. Like North Americans, Russians have often seen themselves as tamers of the wilderness, pitted against a hostile environment that must be subdued as the frontier moved forward. There is also much similarity between Siberian landscapes and towns and the dry, grey landscapes of the Great Plains in the United States with their unpainted frontier settlements in the early twentieth century. But Russia, although it had far more space than the United States, had a much less useful territory to expand into. Only a thin wedge of potentially productive land lies between the tundra-taiga (coniferous forest) on the north and the desert on the south, nearly disappearing close to the middle where the Altai Mountains occur. It

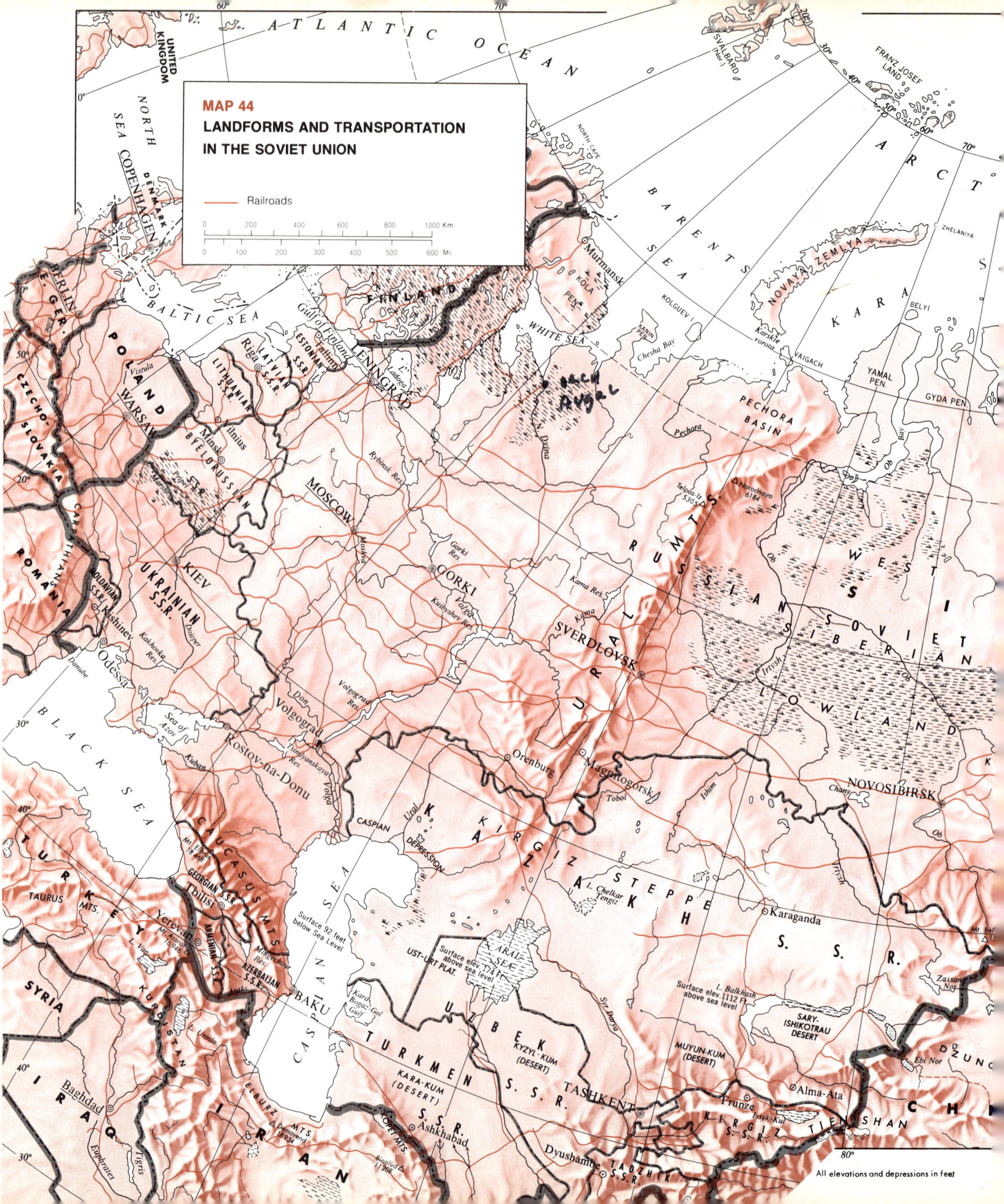

MAP 44
LANDFORMS AND TRANSPORTATION IN THE SOVIET UNION

OCEAN
SEVERNAYA ZEMLYA (NORTHERN LAND)
C. CHELYUSKIN
TAIMYR PEN.
BYRRANGA MTS.
Taimyr
Khatanga Gulf
BEGICHEVA
LAPTEV SEA
NOVO SIBIRSKIE ISLANDS
KOTELNY
NOVAYA SIBIR
FADDEEV
DE LONG
LYAKHOV
NOVAYA SIBIR I.
C. SVYATOI
C. BUOR-KHAYA
EAST SIBERIAN SEA
MEDVEZHY
AION
C. SHELAGSKY
WRANGEL I.
Gulf of Anadyr
ANADYR RANGE
Arctic Circle
KORYAK MTS.
BERING SEA
C. OLYUTORSKY
KARAGIN
KOMANDORSKIE ISLANDS
C. KRONOTSKI
Klyuchevskaya (Vol.) 15912
KAMCHATKA PEN.
C. LOPATKA
PARAMUSHIR
ONEKOTAN
KURIL ISLANDS
ITURUP
KUNASHIR
C. Soya
Kolyma
GYDAN MTS.
C. TAIGONOS
Shelekova Gulf
CHERSKI RANGE
Mt. Chen 10171
VERKHOYANSK MTS.
Lena
Magadan
C. ALEVIN
SEA OF OKHOTSK
Olenek
PUTORANA MTS.
VILYUISK RANGE
Aldan
Vilyui
DZHUGDZHUR MTS.
C. ELIZAVETA
SHANTAR
Udskaya Bay
SAKHALIN (Sov. Union)
C. TERPENIE
Tatar Strait
Nizhnyaya Tunguska
Podkamennaya Tunguska
Mt. Purpula 5377
PATOM PLATEAU
Mt. Polkan 3543
Yenisei
Snezhnyi Pk. 9186
STANOVOI RANGE
BUREYA RANGE
Angara
Amur
Khabarovsk
SIKHOTE ALIN MTS.
L. Baikal
Vitim MTS.
Surface elev. 1535 Ft. above sea level
BAIKAL MTS.
YABLONOI MTS.
NERCHINSK RANGE
LESSER KHINGAN MTS.
Mt. Piramida 10801
SAYAN MTS.
Munku Sardyk 12821
Irkutsk
Lake Khubsugul
Argun
GREATER KHINGAN MTS.
Sungari
Khanka
Vladivostok
HOKKAIDO
TANNU OLA MTS.
Ubsa Nuur
Selenge
Onon
Kerulen
Khara Usa
KHANGAI MTS.
ALTAI MTS.
MONGOLIA
Tsasata Bogda Uula 13,865
CHINA
SEA OF JAPAN
HONSHU
TOKYO
KOREA
Korea Bay
OSAKA
JAPAN
PEKING (Peiping)
Gulf of Chihli
YELLOW SEA
Longitude East of Greenwich
SIBERIA
FEDERATED SOCIALIST REPUBLIC
100°
110°
120°
130°
140°
150°
160°
170°
180°
60°
50°
40°

widens out again toward the east in the valley of the Amur River, but under subarctic climatic conditions. A large part of the Soviet domain comprises what has aptly been called "the dead heart of Asia." The use of this barren land presents a challenge to the Russian state, but over 80 percent of the Soviet population still lives in the less than one-sixth of the country's area within the *fertile triangle* formed by Leningrad, Irkutsk, and Odessa.

Empty distances and excess space

Vast space and a most unevenly distributed population pose problems of communication. Nowhere else in the world are there such distances within a political unit or such gaps between population and resource centers. One graphic measure of the space involved is that there are eleven time zones in the Soviet Union. Not only are the distances great, but they are empty distances to a large extent, without the productive continuity that helps to make transport lines economic. To haul goods a thousand miles through productive country with a chain of cities as trade-generating centers may be very profitable. But to haul them through a thousand miles of desert or forest from one spot of population to another may be seriously uneconomic. The latter is the case with most of the U.S.S.R. outside the Leningrad-Irkutsk-Odessa triangle, and this puts a strain on the economy. Politically as well, this huge area must be held together, administered, defended, and controlled by the major Russian base in Europe, nearly 6,000 miles (9,600 km) from the Pacific. The Soviet em-

MAP 45
POPULATION OF THE SOVIET UNION

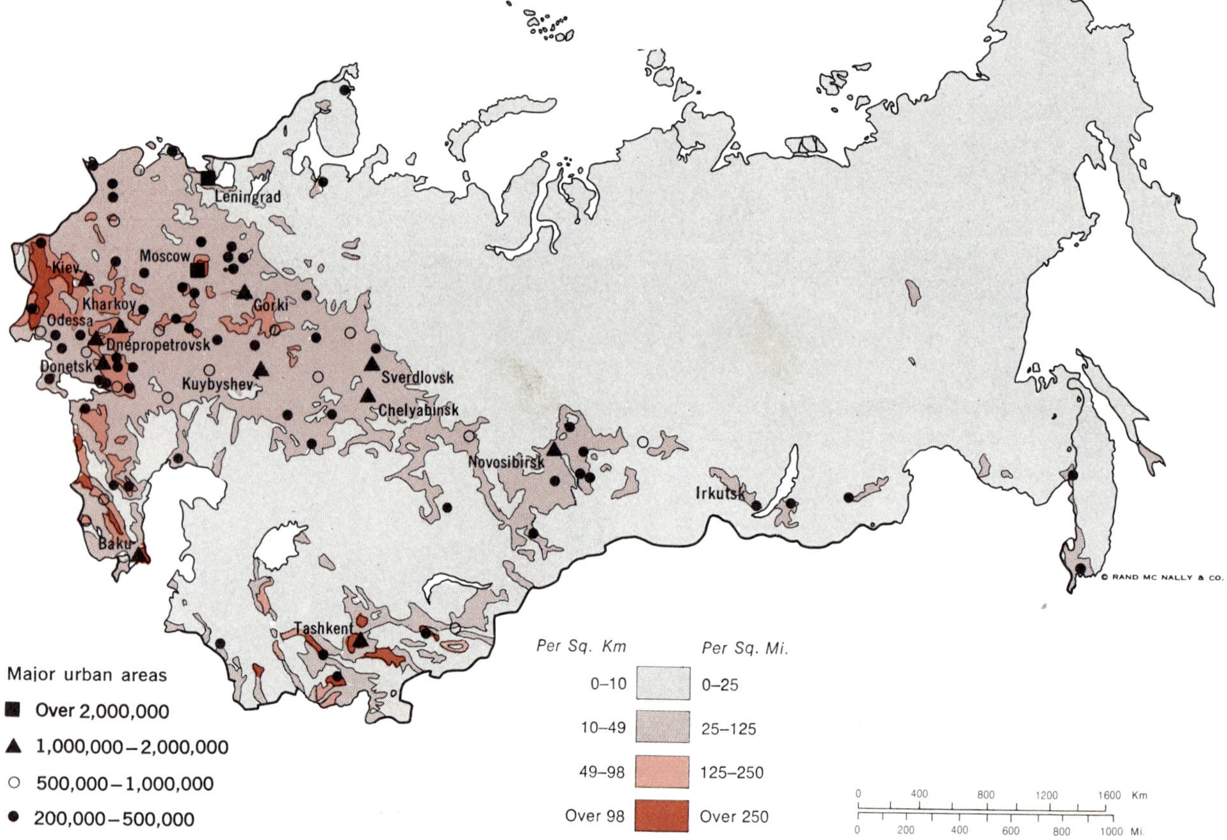

phasis on all-embracing economic planning also poses problems for the administration and development of a huge area from any single center. There have been frequent policy shifts between a system of regional economic administration and central ministerial administration, but no fully satisfactory solution. The U.S.S.R. unquestionably suffers from too much space, and especially from too much empty space. Space may mean power, security, greatness, and a wealth of resources, but there is a point beyond which it becomes increasingly unmanageable. Even in an age of mechanized transport, although the Soviet Union profits in many ways from the gift of space, it suffers also from its disadvantages.

Agricultural land

Good agricultural land in the U.S.S.R. is largely limited to the fertile triangle outlined before (see Map 46, page 164). Much of it has excellent soils, especially in the steppe areas toward the south and west. Under the dry climatic conditions, the soils have retained the mineral nutrients built up by decaying grass and grass roots and formed a deep, dark, rich soil layer. The Russians call this soil *chernozem* ("black earth"). The term is now applied to similar soils elsewhere, since the pioneer Russian classification of soils has been generally accepted. Soils of this type cover most of the Ukraine, the Soviet Union's greatest granary, and reach eastward in a narrowing wedge of patches as far as Irkutsk. Outside this area, with some local exceptions, most soils are of the desert, mountain, forest podzolic (acid), or arctic variety, and relatively unproductive. The pattern of soils reflects the climatic pattern. A little less than half of the Soviet Union is covered by arctic or subarctic climatic types, roughly 30 percent by steppe or desert, and about 20 percent by the sharp-winter, cool-summer climate of northeast Europe, similar to that of Maine or Minnesota.

Where favorable soils occur in the steppe, the climate is not helpful, since rainfall is deficient and precarious (see Map 47, page 166), and the growing season short. Nowhere in the U.S.S.R., even in the Ukraine (where rainfall is often inadequate),[1] is there the combination of favorable soils and climate that

Ukrainian collective. This is Russia's best grain land, a little dry, but level and fertile. Farming methods, including mechanization, are most advanced here. *(From the Soviet Embassy.)*

characterizes other major agricultural regions of the world like the North American corn belt or the Ganges valley of India. Originally coveted as grain land, the Ukraine also proved to have very large deposits of coal and iron, and is still the major industrial area in the Soviet Union. It is one of the greatest grain producers in the world; for many years it provided Russia with a grain surplus for export. Wheat, and timber from the northern taiga forests became

[1]Most of the Ukraine has a modified steppe climate, moister toward the north and west.

the chief sources of foreign exchange, and paid for a large share of Russia's rapid industrialization.

Although the Ukraine is productive, the Soviet Union as a whole has a population of nearly 260 million. As its population has increased, food has become the country's chief economic problem. Most of the good agricultural land is concentrated in one area, separated by thousands of relatively empty miles from the new centers of population in the Asian parts of the Soviet Union. The traditional Russian diet of black bread and borscht (beet and cabbage soup) is a convenient, although now somewhat outdated, symbol of Russian agricultural handicaps. The black bread is made from rye, a crop that will tolerate the poor soils and harsh climate of north Russia, where the modern state and culture first began. Cabbages and beets, especially cabbages, grow almost anywhere, and potatoes have also been an important crop. Wheat is now the leading bread grain in the Soviet Union. However, there has been a large increase in maize production in the Ukraine.

Mineral resources and manufacturing

Soviet mineral resources are very large, and there are few essential minerals with which the U.S.S.R. is not well supplied. But many deposits are located in

MAP 46
MINERALS AND ECONOMIC PATTERNS IN THE SOVIET UNION

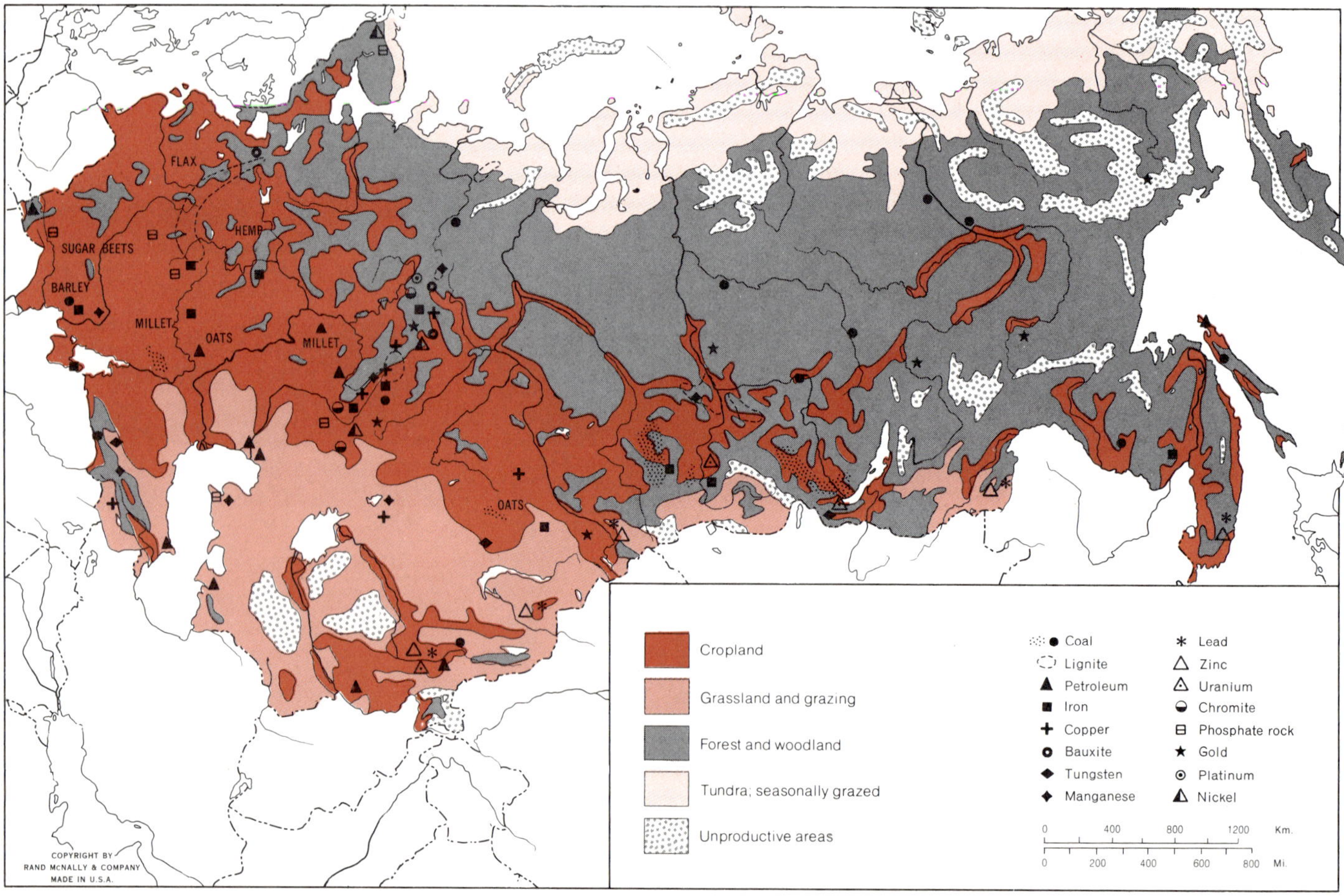

food-poor areas and are far from existing population centers. This puts a heavy strain on the transport system. Coal, iron ore, and petroleum are all relatively abundant and are produced in quantities roughly equivalent to United States production, although good coking coal and high-grade iron ore are less plentiful. Like the United States, Russia is approaching the exhaustion of some of its major high-quality iron ores, but, unlike the United States, it cannot turn so easily to foreign sources with the aid of sea transport. Instead it is increasingly dependent on distant domestic sources of lower-grade ores. Planning decisions have been affected as well by a Soviet bias in favor of autarky (self-sufficiency), and they tend to stress the use of domestic resources even under unfavorable conditions. Nevertheless, the Soviet Union overall has the raw materials for massive industrialization. The major problem is location. Assembly and distribution costs, which form the main part of the total costs of most heavy industrial operations, are high in the Soviet Union because of the large amount of transport necessary due to the scattered or inconvenient locations of many of the mineral deposits.

The Ukraine

There are several major centers of industrial resources and heavy manufacturing. By far the most important is the Ukraine, where coal, iron ore, and petroleum are all plentiful or near at hand (see Map 46, page 164). The Donets Basin, north of Rostov-on-Don (Rostov-na-Donu) between the Dnieper (Dnepr) and the Don, is the principal source of the coal. A large and rich deposit of iron ore lies near Krivoy Rog, less than 200 miles (320 km) west of the Donets Basin, near Dnepropetrovsk on the Dnieper. As the Krivoy Rog reserves approach exhaustion, extensive deposits of lower-grade iron ore in the nearby Kerch peninsula in the Crimea (Krymskiy) are being drawn on, plus other recently developed ores in the Kursk area near Moscow. There is a major oil field east of the Ukraine in the foothills of the Caucasus Mountains and along the shores of the Caspian Sea. An extension of the fields around the Persian Gulf, it is concentrated in the vicinity of Baku, the site of a large refinery center. The cities of Odessa, Dnepropetrovsk, Krivoy Rog, Donetsk (formerly Stalino), Voroshilovgrad, Nikopol, Kharkov, and other smaller centers, plus Volgograd (formerly Stalingrad) and Rostov-on-Don (although they lie just outside the Ukraine), and Kiev, to a lesser extent, owe their industrial expansion to this favorable combination of mineral resources. The Ukraine is unique among Soviet industrial areas in possessing most of the major resources essential for heavy manufacturing within relatively easy distances of each other. It is also a densely populated area, close to markets, transport routes, and the labor force. Equally unique is its food surplus. The booming industrial development that has taken place in the Ukraine has not required expensive long-distance shipments of food to support the population. Indeed the Ukraine is still able to export food to the agriculturally less favored but densely settled north Russian plain where Moscow and Leningrad lie.

Moscow and Leningrad

Although they originally lacked local mineral resources, except for large deposits of low-grade coal or lignite, these two older cities were first to develop a manufacturing complex. It was, of course, manufacturing of consumer goods, primarily textiles. The Moscow area, especially the many industrial towns surrounding it, still leads in textile production, with Leningrad in second place. The Moscow-Leningrad area is also still the country's biggest manufacturing complex in terms of total production and labor force employed in factories. Their production is primarily of consumer goods, and includes, for example, most of the country's book publishing and garment industries.

The Urals

The second most important center of mineral resources and heavy manufacturing, after the Ukraine, is located on the eastern side around the southern edge of the Ural Mountains, especially in the cities of Magnitogorsk, Chelyabinsk, and Sverdlovsk. The Urals area has very little good-quality coal, but there are deposits of high-grade iron ore, chromite, manganese, nickel, copper, and other ores, all within a relatively small area. Although the Urals area faces an acute shortage of high-grade iron ore in the near future, more good-quality ore has been brought into production near Kustanay, about 200 miles (320 km) east of Magnitogorsk. Coal must be brought to Magnitogorsk from great distances at heavy transport costs, mainly from Karaganda and the Kuznetsk Basin (discussed later). There is a large reserve of good coking coal in the Pechora Basin, about 900

MAP 47
CLIMATE IN THE SOVIET UNION

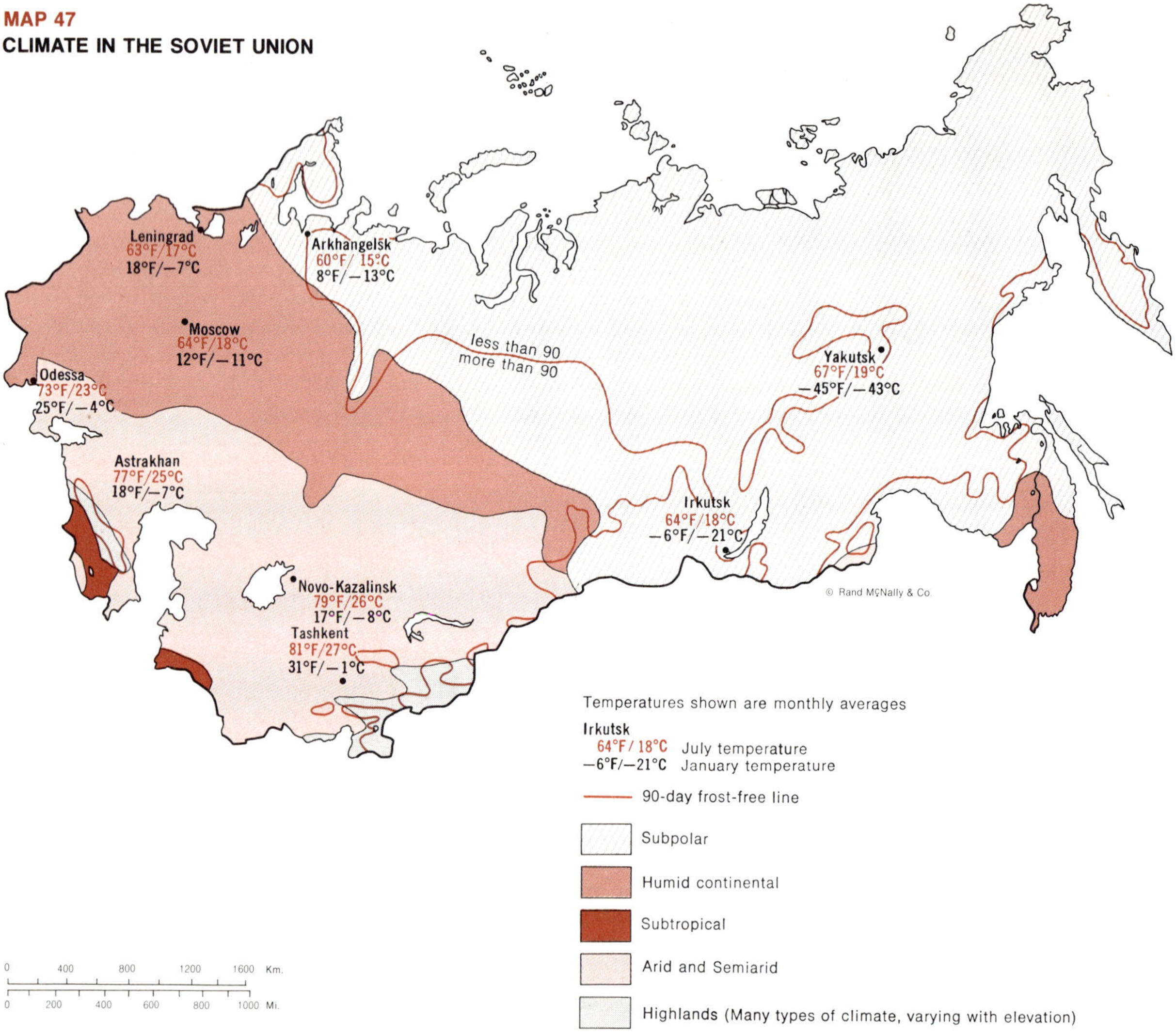

miles (1450 km) north of Magnitogorsk. It will probably be drawn on more heavily, using the new rail line that connects it with the Urals area.

Extensive petroleum deposits have also been developed since the Second World War in a wide zone west of the Urals. It has been labeled the Second Baku, but is perhaps better referred to as the Volga-Urals area. Its reserves and production of petroleum are both much greater than those in the Baku area around the Caspian Sea, and actually account for about 80 percent and 70 percent respectively of the national totals. The presence of oil in the Volga-Urals area helps to lessen the expensive dependence on distant coal sources. Pipelines carry petroleum from the Volga-Urals area westward to the Moscow-Leningrad district; through the Ukraine to Poland, Czechoslovakia, and East Germany; and eastward through the Kuznetsk Basin as far as Irkutsk. More recently, large oil reserves have been found in the middle Ob River valley, not far to the east of the Urals. They have been connected by pipeline to the industrial centers in the northern Urals. But manufacturing centers in the Urals are not close to the main market centers in European Russia.

Heavy industrial centers like those in the Urals are not common in mountain areas. In this case the mountains are so old and worn down that they are more like rolling hills, across which rail lines have been built to tap the ores and support the manufacturing complex. This is especially true of the southern Urals. While there are some more genuine mountains in the northern section, the Urals are a poor line to choose for the boundary between Europe and Asia, as is conventionally done. Physical and human conditions on both sides of the Urals are similar, and a marked change does not come for many hundreds of miles eastward.

Karaganda

The Karaganda area, about 600 miles (960 km) east of Magnitogorsk, contains extensive deposits of coal and some iron ore. Like the Urals area, the Karaganda district must pay heavy transport costs in order to convert its local minerals into manufactured goods. Most of the necessary ores originally had to be brought in, mainly from the Urals and Kustanay. As expected, there is a large reciprocal trade between the two centers. Karaganda coal is exchanged for some Urals and Kustanay ore, providing a return haul for the rail lines. If it were not for this reciprocal relationship, the haulage of such bulky or low-unit-value goods over such great distances would be seriously uneconomic. As it is, the cost of iron and steel made in either place, considering all transport costs in the total, is considerably higher than in the Ukraine or in most other iron and steel centers in the world. Production of iron and steel in Karaganda has risen rapidly, however, and has been able to depend on more recently discovered local sources of ore.

Water is also a critical problem at Karaganda, which lies on the dry margin of the steppe. Water shortages were a severe handicap to the further growth of manufacturing there until completion of a canal from the Irtysh River, an expensive project. Industrial production was originally located at Karaganda partly for strategic reasons. The Soviets want to develop the largely empty areas of central

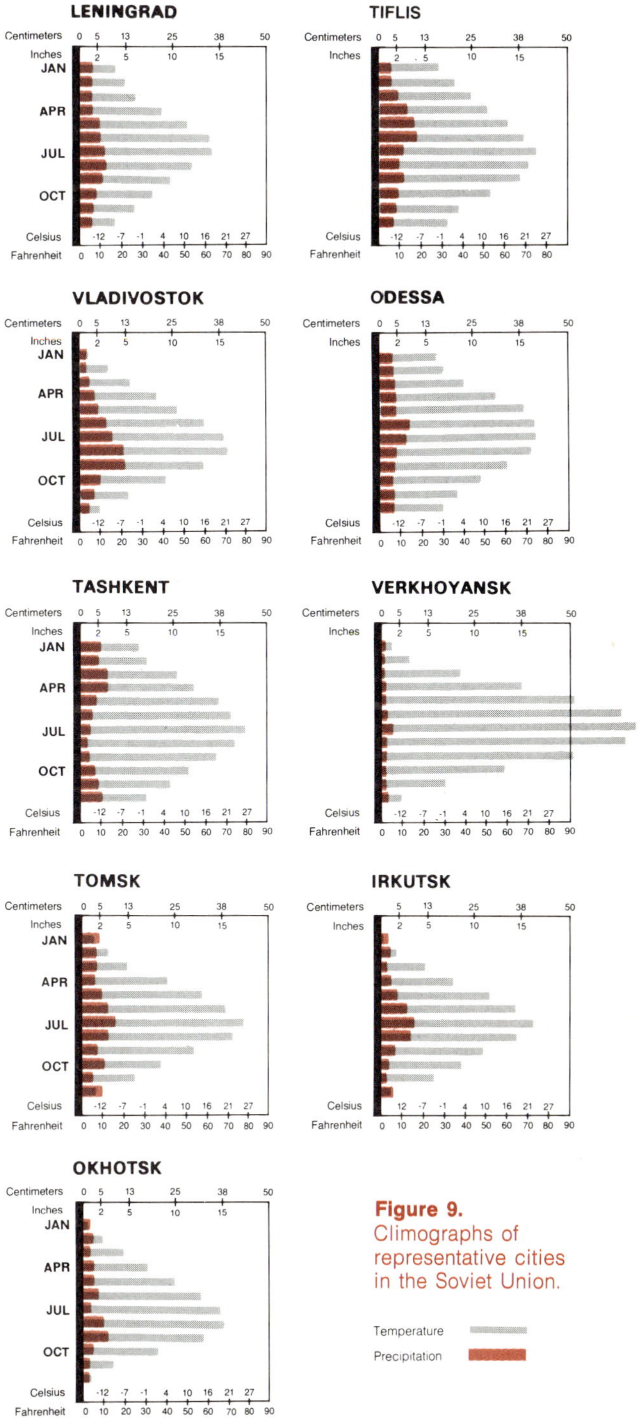

Figure 9.
Climographs of representative cities in the Soviet Union.

Asia and Siberia with their extensive mineral deposits. The Nazi occupation of the Ukraine was the latest and most bitter reminder of its dangerous vulnerability to an enemy from the west. The Second World War greatly accelerated the drive to develop as much manufacturing as possible in the interior of the country where mineral resources existed, as in the Urals and farther east.

The Kuznetsk Basin

The Kuznetsk coal basin, or Kuzbass, as the Russians abbreviate it in a habit they share with Americans, is now the third largest center of heavy manufacturing, after the Ukraine and the Urals area. It lies over 600 miles (960 km) northeast of Karaganda, just east of the city of Novosibirsk on the western edge of the great upland of central Siberia, where the trans-Siberian railway crosses the Ob River. It has the largest coal reserves of any accessible field in the U.S.S.R., and also considerable iron ore. Around the Kuzbass a growing heavy industrial complex has developed since the 1930s in the cities of Novosibirsk, Novokuznetsk (formerly Stalinsk), Kemerovo, Tomsk, and Barnaul. Such boom development of Siberia's resources and the accompanying population shifts are remarkable. But they cost heavily, and took place in the face of economic obstacles that might have prevented development except for the planning decisions of an authoritarian state.

The Kuzbass has huge reserves, good-quality coal including some coking coals, and low extraction costs. But its location might have ruled out its use except for the special circumstances that governed its development. The coal as raw material is far from present markets. Local manufacturers also have a large transport bill to pay, whether they move their products to the main centers of population in European Russia or to new or future markets in Siberia and the Soviet Far East. The new markets in Asian Russia were part of the original rationalization for locating iron and steel production in the Kuznetsk Basin. In time, as the markets of Soviet Asia increase, including the Kuzbass area itself, this investment in heavy manufacturing may pay off. That time is not yet here, even though the development of Kuznetsk will hasten it.

Fortunately, more good-quality iron ore was discovered later in the Kuznetsk area. A little ore is still brought in from the Urals, and a little coal is still hauled from Kuznetsk to the Urals in exchange. On this journey of about 1,400 miles (2260 km) one way, as much as 40 percent of the coal is consumed as fuel for its own transportation, and the railroads have had to be heavily subsidized. Such rail hauls are not found anywhere else in the world, even in the United States where long hauls are common. They take place in Russia not only because distances are great, but because people and resources are widely scattered. Further, a powerful state, if it likes, can disregard for a time particular economic factors in the pursuit of noneconomic objectives in the rapid drive for industrialization, the development of new regions, and military security.

Soviet planners set up the Urals-Kuznetsk Combine in the 1930s according to what they called "the principle of the pendulum." They counted on the two-way haul (ore and coal) and the maintenance of metallurgical plants at both ends of the line to keep transport costs from swamping the venture. If the Kuznetsk ore or the Karaganda coal had been discovered earlier, the Urals-Kuznetsk Combine would probably never have been established. The Urals can now draw coal from Karaganda more cheaply, even though Kuzbass coal is superior in quality. The Donets Basin, or Donbass, in the Ukraine has only about 5 percent of the Soviet Union's coal reserves, but it produces about one-third of the country's coal. The Kuzbass, despite its huge reserves, produces only about one-sixth, a clear indication that strategy is not all powerful.

The Amur valley

The easternmost area of heavy manufacturing is in the Amur valley in the Soviet Far East. Here there is a relatively small amount of coal and larger deposits of poor-quality iron ore, lead, and zinc, plus petroleum on nearby Sakhalin Island. This area is so far from the major Soviet markets and from other Russian sources of raw materials that it must operate as a virtually independent economic unit. Therefore, its total industrial production is small. Even so, incredibly enough, there is some exchange of raw materials by rail with Kuznetsk, at what must be an especially heavy transport cost. The principal manufacturing and urban centers of the Soviet Far East are on the navigable lower reaches of the Amur or on the sea. The main centers are Khabarovsk at the junction of the Ussuri and the Amur, Komsomolsk

The Kremlin. Originally a Medieval fortress on the banks of the Moscow River, the Kremlin has become the governing center of a huge country. The large central building was once the palace of the czars and now houses the Supreme Soviet, the highest government body according to the constitution. *(Photo by Bob and Ira Spring for Alaska Airlines.)*

(Komsomol'sk-na-Amure) (founded in 1932) halfway to the Amur mouth, and Vladivostok on the coast at the extreme southern extension of Russian territory, where its harbor avoids as much ice as possible. A milder and more maritime climate than continental Siberia is a further advantage for settlement and economic activity in the Amur valley. The area is now able to feed itself in most years. For centuries this area has been a pioneer zone for Russian settlement. It developed particularly rapidly in manufacturing under Soviet planning, and new industrial cities have grown up almost overnight. Its disadvantage is the nearly 5,000 miles (8000 km) to Moscow.

Present Soviet plans also include establishment or expansion of additional heavy industrial centers in the middle and lower Volga Basin, in the trans-Caucasus area around Tiflis (Tbilisi), in the Lake Baikal (Ozero Baykal) area around Irkutsk (based especially on hydro power), and in Uzbekistan (Uzbek S.S.R.) around Tashkent. The huge dam at Bratsk on the Angara River (Verkhnyaya Tunguska) about 275 miles (440 km) north of Irkutsk, with a hydro-power capacity much larger than Grand Coulee's, plus iron ore in the same area, is beginning to stimulate rapid industrial growth there. The lower Volga region is also growing fast industrially, including the cities of Kuibyshev (Kuybyshev), Saratov, and Volgograd, from the added stimulus of two big power dams at Kuibyshev and Volgograd, and other dams at intervals along the river's course. In all of these cases local resources are being used in strategically desirable locations, hastening the development of previously sparsely settled areas, an objective of Soviet planning.

Transport problems

People, mineral deposits, and food-producing areas in the Soviet Union are unevenly distributed. Population, agricultural production, and manufacturing are overwhelmingly concentrated in the fertile triangle between Leningrad, Irkutsk, and Odessa, mainly in the western or European half of this area.

Most of the rest of the U.S.S.R. is empty, with only occasional smaller population clusters or centers of commercial activity. Mineral resources are much more widely distributed. In agricultural terms, only the Ukraine feeds itself consistently. The other areas, in differing degrees, must bring in food to support their city and industrial workers, often over very great distances. Siberia is thinly settled because it has a harsh climate. To use mineral resources located there, disproportionate amounts of capital, energy, and time must be expended in exchanging raw materials, food, and finished products over vast and largely empty distances by overland hauls. The government does its best to increase food production in the food-poor industrial or densely settled areas. But it frequently ignores the fact that it may be more expensive to grow food locally in an unsuitable climate or soil than to bring food in by rail from elsewhere. Such decisions are related to the overworked Soviet transport system. The Soviet Union is indeed "an experiment in transportation." Until diesel conversion in the 1960s, Soviet railroads were estimated to consume about one-third of all the coal the country produced. More generally, it is estimated that 20 percent of all the energy and power production of the Soviet Union still goes into railroads (the same figure for the United States is estimated as 12 percent, for the United Kingdom 7 percent). Over twice as much transportation is necessary for each unit of production as in the United States, and three times as much as in the United Kingdom, according to the same estimate. This is one measure of the dead weight of excess space. Transportation is simply a means to an end; it cannot be eaten, worn, or used to make things.

Housing on a collective—Ukraine. These are new houses on a large collective in southern Ukraine. Attached to most houses is a small private plot on which each worker is permitted to grow crops for his own use or sale, after he has fulfilled his work requirements for the collective. *(From the Soviet Embassy.)*

Soviet railroads do a brave job in attempting to overcome this "friction of distance." They run on a broad-gauge track (5 ft. [152.4 cm] instead of the 4 ft. $8^{1}/_{2}$ in. [143.5 cm] standard in western Europe and North America). Consequently, they can use large trains with somewhat greater capacity, suited to bulk hauling over long distances. Total density of railroad freight traffic per mile of line is the highest in the world by far, over three times the average for the United States. Total rail mileage is low relative to the great distances and large population of the country, so existing lines must be very heavily used. Motor roads are not much developed. Trucks are used only for short hauls to collection centers on the rail lines. Roads replace railroads in areas such as eastern Siberia where population density and traffic volume are too low to support a railroad. The per capita income of the population has been relatively low until recently. In the past the economy has cut consumer goods production in order to concentrate on building up heavy manufacturing. By American or even West European standards, there are still few passenger cars. It is not the extremes of climate that account for the lack of roads, but the extremes of distance and the shortage of cars.

The railroads get relatively little help from the rivers in handling the transport problems. Most of the rivers run either into the Arctic Ocean or into more or less enclosed seas like the Aral (Aral'skoye More), Caspian, and Black. Where the rivers are navigable, they are used heavily. Thousands of miles of canals have been built to improve and connect the river systems. But inland water transport accounts for less than one-tenth as much haulage as the railways, and it is heavily concentrated in European Russia. The great bulk of the country is almost entirely dependent on more expensive rail transport for its far greater distances. The Volga alone carries almost half of the total water haulage of the U.S.S.R.

Rivers and the rise of the Russian state

The early growth of the Russian state depended significantly on the rivers of European Russia as highways for communication and expansion. Novgorod, an early center of the modern state, and Moscow, the later center, both owed their leadership to a large degree to their location at river junctions, exploited to the full by the political astuteness of the princes of Moscow. The Volga, Don, Dnieper, and other rivers flowing south to the Black and Caspian seas and north to the Baltic begin in the vicinity of Moscow or in the low uplands northwest of the city, the Valdai Hills. In the originally heavily forested area of north European Russia, where rivers were the easiest means of communication, Moscow and Novgorod could exert their authority and dominate trade over a wide area. Since before Roman times, the Dnieper in the Ukraine was part of an important route between north and south Europe, connecting the Black Sea with the Baltic. Kiev, a much older trade center at a major junction on the Dnieper and near the break between steppe and forest, became a great city and the capital of the early Russian state. Its growth, like Timbuktu's or the cities of northern Italy in the lowland approaches to Alpine passes, owed much to its location where one set of physical conditions gave way to another. But Kiev was destroyed by the Mongol invasion from the steppe. Novgorod and Moscow, protected to a degree from the Mongol horsemen by their shield of forest, river, and swamp, inherited the political leadership of the emerging Russian state. Cultural and economic influence and political control spread over the plain from these two centers with the help of their excellent river connections. By the nineteenth century Russian control again extended over the Ukraine and the shores of the Black and Caspian seas.

The modern Soviet state has invested heavily in improving these natural water routes and in building canals (see Map 44, pages 160–61). One important canal runs between the Volga and the Don, so Volga traffic may reach the Black Sea, and from it, the Mediterranean and the Atlantic. Another goes from the headwaters of the Volga north of Moscow to the Baltic at Leningrad. The network of canals linking Moscow and Leningrad with the navigable rivers of European Russia leading to the Baltic and the Black and Caspian seas is also economically important. The European section is moist enough and flat enough to make heavy investment in waterways worthwhile, especially if they can reach an open sea. Every ton-mile saved from the railway's job here means that much more transport available in the wide areas of the country where railways must do the job largely alone. Unfortunately, of the Soviet Union's major navigable rivers, only the Amur reaches an open sea without a roundabout canal connection. The most important river of them all, the Volga, empties into the Caspian Sea, which has no outlet. None of the rivers emptying into the Baltic are big enough to compare with the major rivers of western Europe in the goods they can carry. The Black Sea, which receives the Dnieper, the Don, and many smaller streams, is still far from the major sea routes, through an awkward exit.

Rivers in Soviet Asia

Although most of Asian Russia, aside from the Amur valley, is arid, most of it is also cold so evaporation rates are low, and several major rivers flow across the area. Unfortunately they drain into the Arctic Ocean, except for the Amur (which perversely bends northward before it enters the Pacific). This makes them much less useful. The Arctic can be kept open for navigation, only with great effort and expense, for barely two months of the year at most of the river mouths. The rivers freeze for an average of eight months of the year, and since they flow northward, they thaw first at their headwaters. The lower river remains frozen much later, so meltwater and blocks of ice pile up every spring for literally hundreds of miles up the river, making the lower areas into one vast marsh. The major Siberian rivers, the Ob, Yenisey, and Lena, despite their disadvantages for navigation, are the principal means of communication for this vast area. Map 45 (page 162) shows clearly how settlement has concentrated in their valleys, and suggests the even more inhospitable and difficult nature of the rest of Siberia. Most of the traffic on these rivers is lumber from the Siberian taiga. But only a little of it can be picked up by ships on the Arctic Ocean during the brief summer of open water. Most of the timber has to be hauled by the trans-Siberian railway, which picks it up from the rivers. The two other major rivers in Soviet Asia, the Amu Darya and Syr Darya, flow into the tiny, landlocked Aral Sea. They are of only limited use for navigation since, like so many desert rivers, their volume fluctuates widely and they are badly silted.

The Ukraine benefits greatly from water power with major dams on the Dnieper and the Don. There is an enormous production of hydro power on the Volga, as mentioned above, and the smaller rivers of north European Russia have been harnessed increasingly to supply power for the large urban markets of the area. The much greater potential power of the rivers of Siberia mainly is wasted, and will be until a local market grows up, or until power can be transported greater distances economically. The Bratsk installation on the Angara, where power-demanding industries are being developed, is an exception. It represents an originally uneconomic investment that counts on stimulating local industrial development to sustain it. Distance and the cost of transport also mean that the greater part of the Soviet Union's timber, concentrated in the taiga, cannot be used commercially. Nature has been overgenerous to the Soviet Union in quantity and stingy in providing good locations for potential resources. The Soviets have attempted to solve this problem by investing huge amounts of capital and labor to force the settlement and development of lands that would be used much less intensively under a less authoritarian socioeconomic system. They have been impressively successful with the use of authoritarian methods. There is no reason to think that the rapid growth of the Soviet economy will not continue until it rivals or surpasses the achievements of Western economies. Since 1945 per capita incomes and total industrial output have both risen more than twice as fast in the Soviet Union as in the United States. But the cost has been and remains high.

The agricultural crisis

Russian industrial accomplishments are far more impressive than their achievements in agriculture. In heavy manufacturing the Soviet Union is now the world's second largest producer, a position it has won mainly since the introduction of rapid industrialization under the so-called Five-Year Plans, beginning in 1928. This represents one of the fastest, long-sustained rates of industrial growth in history, and expansion is still continuing on a large scale.

In agriculture Russia has suffered from a long history of technological backwardness. Efficient industrial techniques have been developed and spread quickly. In many respects the Soviet Union has risen to the forefront of world technological advance. But antiquated methods have continued in agriculture. Agricultural techniques even now are not much above the levels of efficiency and productivity reached in the United States and western Europe several decades ago. Especially since 1950, Russian agricultural methods have improved rapidly. But one major reason agriculture has failed to match the gains of manufacturing and faced recurrent crises over food production has been the difficulty of changing archaic systems of land use in an immense country of nearly 260 million people. It has been much easier to change industrial techniques because there was less resistance to change, and fewer individuals, groups, or traditional regional structures were involved. Further, the Soviets inherited an industrial complex developed under the old regime that had already passed through its infant stages and made significant progress before 1917. Industrialization also had a more dramatic appeal, as it usually has, and the great bulk of new Soviet investment went into manufacturing enterprises, especially heavy industries. Between 1930 and 1970 Soviet industrial output increased more than tenfold. In the same period agricultural output (less easily measured than industrial production) probably increased by no more than 70 or 80 percent.

The physical disadvantages confronting agriculture in the Soviet Union have also been important handicaps. Areas of fertile soil, such as the Ukraine or the Kirgiz (Kirghiz) Steppe in southern Russia, receive inadequate rainfall. Areas with long growing seasons, such as the shores of the Black and Caspian seas or south-central Asia, also have little rainfall. In the northern third of the country, rainfall is more nearly adequate but soils or temperatures are unfavorable (see Maps 46 and 47, pages 164 and 166). The Ukraine is physically similar to western Nebraska in climate, soils, and relief, but the Ukrainian rural agricultural population is much denser, quite apart from the population densities associated with minerals and manufacturing. This is due partly to lower per capita productivity and a lower level of mechanization in Soviet agriculture. But primarily, agricultural production is a matter of comparative advantage. By Soviet standards, the Ukraine is a prime agricultural area; by American standards, western Nebraska is marginal.

Attempts at solution

The Soviet state has tried to increase agricultural production in three main ways: by using new agricultural land on the cold or dry fringes, by improving methods and crops so new and old land will yield more, and by mechanizing and organizing the agricultural system as a whole. Under the Russian agricultural system of collective and state farms, labor, land, and capital equipment are effectively controlled by the state. None of these three methods has been fully successful; food production is only

Soviet power—Lenin Dam on the Dnieper. Hydroelectric power has become one of the U.S.S.R.'s primary resources, with large dams like this one on all of the major rivers in both Europe and Asia. They are appropriate symbols of Soviet industrial growth. *(From the Soviet Embassy.)*

now beginning to keep pace with urbanization and industrialization, although it is still increasing. Most of northern Russia, including the European section, is severely limited in its agricultural potential. *Permafrost,* ground permanently frozen a few inches or feet beneath the surface, is found in much of northern Siberia. It not only presents serious building and drainage problems, but may rule out deep-rooted plants. Some crops can be grown in the surface soil during its brief period of thaw. A few root crops are being grown by the Soviets in the tundra where none grew before. Some drought-resistant crops are also being grown in the arid areas farther south. Irrigation and other improvements are extending cultivated areas substantially, especially in the southern part of Soviet central Asia, sometimes called Russian Turkestan.

But this is poor land for the most part. To invest in it simply makes the crop expensive. Cabbages are grown in northern Siberia but probably at two or three times the cost of growing them in France or in the Ukraine. They are planted in Siberia because the cost of shipping them from the Ukraine would be even higher. Transport bottlenecks and total demand for food also make it desirable for every part of the U.S.S.R. to produce as much food as possible locally. Scarce resources are being used intensively here, but economically the quality of the resources does not justify intensive use. It is done because there is no preferable alternative. Apart from this, where new

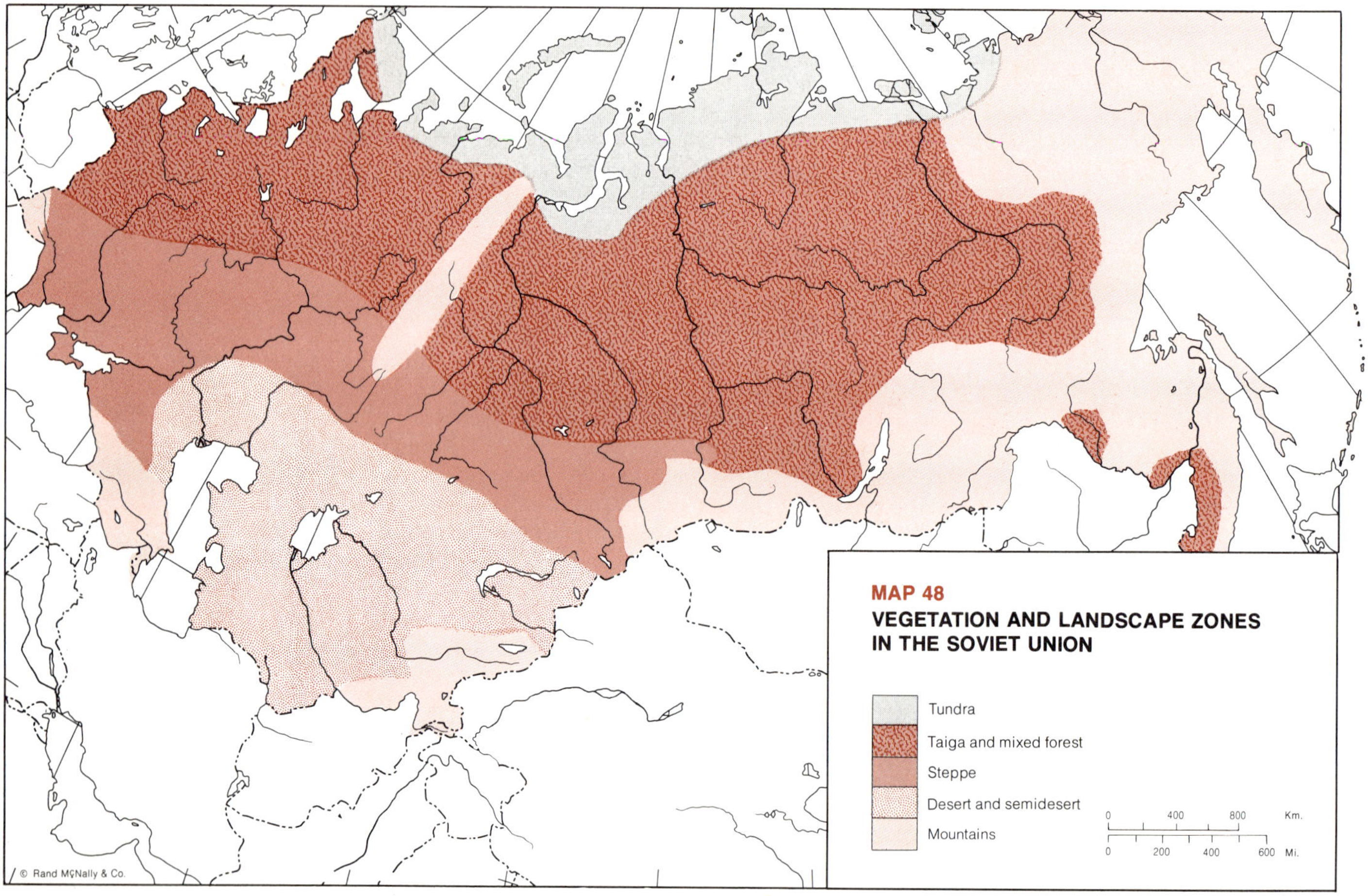

MAP 48
VEGETATION AND LANDSCAPE ZONES IN THE SOVIET UNION

cultivation has been extended into semiarid areas without irrigation (including Siberia), or where original forest and grass cover has been removed by plowing marginal land, future trouble may develop. Soviet dust bowls may become even worse than those in America. Over seventy million acres of virgin land, mainly in Kazakhstan (Kazakh S.S.R.) in central Asia and southwestern Siberia, and in central Siberia, have been brought under cultivation since 1954. Yields are low and crops precarious, but the Russian planners say they will be content with two good crop years out of five in these newly cultivated semiarid areas. They argue that even a small food production from these previously unused lands is better than none or than the low return produced earlier from grazing some of the area.

The extent of marginal land in the Soviet agricultural system, as well as the relative inefficiency of traditional methods, shows in their average grain yields. Grain occupies about one-half of the cultivated land and is predominantly wheat and barley (see Map 46, page 164). Grain is generally grown on the best lands; in Russian Turkestan wheat is widely irrigated. Yet the per capita and per acre yields of wheat are still at least one-third less than those in the United States, where much wheat is grown on what is considered marginal land. There have been frequent crop failures in recent years, resulting from drought and late frosts. But these conditions are to be expected in the difficult environments where most Russian wheat is grown. Large wheat imports, from the United States and elsewhere, have been necessary in many years.

Soviet agriculture is increasingly mechanized. Their wheat and other grain lands, especially the irrigated ones, are being worked more intensively, with human and animal as well as machine labor, than most wheat lands in North America. Although they are rising, wheat yields in Russia remain relatively low by comparison with most other wheat-growing countries.

Apart from the climatic and soil limitations, the closely administered collective farm system tends to support much unproductive labor, especially in administrative supervision, which lowers the per capita yields. Since the workers in the collective system are paid mainly for their time rather than for their output, they have little incentive to increase production. This also works against increases in grain yields. The artificially low prices paid by the state for agricultural products amounts to confiscatory taxation. The proceeds from the sale of these products are often siphoned off for investment in new heavy industry. This may be one of the most important reasons for the lag in agricultural production. Although mechanization is increasing rapidly, many farm jobs are still done less efficiently by human or animal labor. The government has organized numerous reforms and drives to increase agricultural production, but the Soviet Union's food position remains its most pressing problem.

Siberian taiga. Huge areas of taiga (coniferous and mixed forests) cover northern Russia south of the tundra. Most of the area has little population. *(Photo by M. Redkin from the Soviet Embassy.)*

Recent changes

The government has spoken of impressive projects to change the course of Siberian and central Asian rivers, dam the Syr Darya and Amu Darya, and tap other water sources for the irrigation of more and more new land. In much of Turkestan irrigation probably makes good sense, for the soils are good on the whole and the climate is warm. Irrigated cotton, especially, grows well there and is in demand by Soviet manufacturing. New textile centers have grown up in Russian Turkestan. All cotton growing is to be concentrated on irrigated land there, which has already been largely accomplished. The center of wheat production is being shifted as much as possible into semiarid Siberia and Turkestan, which will make room in former wheat areas for more sugar beets. The Ukraine is being used on a large scale to produce maize for cattle feed. Maize will not always ripen because of the short growing season and inadequate rainfall in most of the Ukraine, but it can be cut green and used as ensilage. The Ukraine is to become a grain- and animal-producing area like the North American corn belt, despite the fact that the Ukrainian climate is not well suited to maize. Dairying, poultry, and meat production are being greatly increased in cool, wet, north European Russia with its easy access to large urban markets. Additional fodder from the Ukraine will supplement crops from poorly drained northern soils. But good land remains scarce in most of Russia.

The recent trend toward increased incentives for collective-farm workers providing additional benefits for higher output, and the higher prices being paid by the state for agricultural goods may encourage more intensive cultivation and higher yields. They may even remove the need to focus planning attention on less economic investments in marginal lands. Labor and capital are nearly always more productively invested in good than in poor land or climate.

Industrial growth by 1960 allowed a greater share of new investment in agriculture to begin to make up for the relatively low level of agricultural investment in the past. Agricultural production has been retarded as much by technological backwardness as by the limitations of climates and soils. The revolutionary development achieved by Soviet industrialization has now spilled over into agriculture. The former agricultural deficiency, retarding the growth of the economy and restricting the per capita consumption levels, is beginning to change to a situation of relative plenty. More efficient methods and increased capital investment, even with the physical handicaps of so much of the Soviet Union, have contributed to these increases. This achievement is doubly impressive in light of the disadvantages and the long history of inefficiency and inertia in Russian agriculture in the past. Meanwhile, the earlier rapid population increase has begun to slow down, and the pressure on food supplies is becoming less urgent.

The transformation of nomadism

One of the greatest changes that Russian rule has brought Soviet Asia is the weakening or abolition of nomadism. Until about 1920 most of the area east of the Ob River and the Caspian Sea, south of the Arctic Ocean, and west of the Pacific (about three-fourths of the whole country) was occupied by essentially nomadic people. Russians had settled and established a sedentary commercial economy in the thin strip of more hospitable climate through which the trans-Siberian railroad runs, broadening out in the valley of the Amur. But there were few Russian settlements elsewhere in Asia. Aside from a few ancient Uzbek oases and trade centers in Turkestan (such as Khiva, Samarkand, and Bukhara), the people of this huge area were primarily nomadic hunters and trappers in the taiga and tundra of the north, and pastoral nomads herding sheep and goats in the steppe and semidesert of the south. All were Asian people unrelated to the Russians, but politically absorbed in the nineteenth-century Russian expansion, much as the American Indians were in North America. But these people were far more numerous than the Indians in North America, and they became a political problem. Like most nomads, the native peoples of Russian Asia resisted central control and its other aspects, such as taxation or enforced schooling. Especially in Turkestan they also occupied much potentially productive land, which they were not using very efficiently by the standards of a modern commercial economy.

The Soviets have attempted with some success to transform these people into semisettled commercial producers. They supply meat, hides, wool, furs, and other products to a central processing or distributing plant. From there these products can be sent into the wider Russian market, or exported. The best lands have been taken for agriculture (as happened in North America, too). Mineral resources have been developed, and railways and cities built. The former culture of the nomads is disappearing, yielding to a more settled commercial existence in towns and cities. Now, they often have a definite range with shelters for the animals, houses for the people, and perhaps even an artificial-insemination station attached to the extension worker's veterinary establishment. Nomad resistance to these changes has often been bitter, for understandable reasons, even though better techniques introduced by the Soviets and commercialization of the old economy have brought higher standards of living and wider economic opportunity. Non-Russian peoples are not required to remain in their original areas or to keep their former occupations. They are given an opportunity for a technical or general education. Maintenance of local languages and cultures is permitted as long as this does not conflict with Soviet national interests.

It is not a matter of ethnic or color prejudice. But some of the nomads have felt the conflict keenly and have abandoned their old ways only with reluctance or under force. In some cases they have fled across the border into China, India, Iran, or Afghanistan when resistance was no longer possible. The conflict with the Indians in North America was similar. The major difference is that the Indians have been eliminated, first by slaughter and more recently by absorption as "civilized" Americans or Canadians indistinguishable from everyone else. The Russians are in a weaker numerical position and have recognized the persistence of non-Russian cultures within the Soviet Union. At the same time they are remaking the lives of these people in the Soviet manner and in the Russian interest.

Chinaware menders in Samarkand. These men mending teapots and china in a Samarkand marketplace are obviously not Russians, but their lives are being changed by Soviet policy. *(Photo from Ewing Galloway, N.Y.)*

Russians form less than half of the total population of the Soviet Union. Even if the closely related Belorussians (of westernmost Russia, sometimes called White Russians) and Ukrainians are included, only about two-thirds are Russian peoples. Therefore, some compromise is essential. The Soviet system of federated republics under a strong central government has left the Russians in a dominant position. In time the multinational state, which the Soviets speak of as if Russians and non-Russians were equal partners in a single whole, may weaken or disappear. Already the Russian language is fast crowding out the others.

The U.S.S.R. is composed of fifteen republics, each covering roughly the area occupied by a major ethnic or linguistic group, as shown in Map 44 (pages 160–61). But the Soviet federal structure is not closely comparable with that of the United States, Canada, or India. The Soviet republics have only limited power. To become upwardly mobile in the U.S.S.R., to enter the professions, or to play a role in the national scene, one must master the Russian language, and in effect become a member of Russian culture, regardless of one's cultural origins. Most of the non-Russian peoples are Asian, with fundamentally different cultures and wholly separate histories, until very recently. Many of them—Uzbeks, Kazakhs, Kirgiz, Turkomans—belong to the larger world of Turkish or Turkish-related cultures. But no one group and no coalition can hope to rival or resist modern Russian power or numbers.

Soviet policy does contrast favorably, especially in Asian eyes, with the racist policies of some other national areas containing minority groups, or with South Africa where the dominant whites are a small numerical minority. To most nonwhite people, discrimination or oppression based on race or color is especially hateful. This may influence them to overlook the violence and human misery that accompanied the Russian conquest of their Asian territories and the Soviet transformation of nomadism. The Soviet solution to this problem, a country "nationalist in form and socialist in content," does not necessarily eliminate the separatist national feelings of the Ukrainians, Georgians, Latvians, Uzbeks, or Kazakhs. There is no mutually happy solution when one culture or nation is overwhelmed by another, as has happened repeatedly in many parts of the world. The Soviets claim more for their formula than the nature of human interaction allows.

Russian authoritarianism

It is tempting to speculate on the connection between the persistence of an authoritarian society and state in Russia on the one hand, and the problems of space and environment that have confronted it on the other. Authoritarian control and direction may have been a compelling policy for a state with Russia's problems. Certainly Ivan the Terrible, who ruled Russia in the sixteenth century, or Peter the Great in the seventeenth and eighteenth, were no less forceful rulers than Josef Stalin. The Soviet regime in this respect is a consistent heir of the czarist tradition. The difference is that the Communist party with its efficient centralized control has been able to make authoritarianism newly effective. National pride is also strongly involved. Russian patriotism supports the Soviet state probably more strongly than it ever supported the often corrupt and oppressive rule of the czars. Soviet accomplishments certainly command more support. Soviet citizens may be less free to express complaints, but they are also less inclined to resist or even to resent central authority than most citizens of Western parliamentary democracies. The impressive and exciting development of industrialization under Soviet rule strengthens popular enthusiasm for the Communist government even though investments in heavy manufacturing mean that consumer goods remain scarce. The crusading spirit of national drive toward economic development is widespread. Average living standards are beginning to rise more rapidly as the concrete rewards of industrialization are being realized. There is also wide popular appreciation for the educational opportunities and social mobility the Soviet state offers.

A commercial middle class of private entrepreneurs, as rivals to the economic or political power of the state, never developed in Russia on a west European scale. Opportunities for massive economic development existed in Russia's resources and in the huge continental land mass with no rival states to the east as far as the Pacific. But if these opportunities were to be used, perhaps they had to be forcefully exploited by a powerful state. Of themselves, they did not make development easy in the same way they did in the United Kingdom or the United States, for instance. Russia's late industrialization also adds urgency to the national drive. It encourages the continuation of authoritarian controls as it did in Germany a generation earlier. Soviet emphasis on heavy industry and producer goods and, until very recently, the small output of consumer goods, as compared with Western economies, may accelerate industrial growth. But these policies would hardly be chosen freely by most consumers, who do not like to postpone new shoes or housing for twenty years in exchange for a new steel plant now. The collectivization of agriculture in the early 1930s was designed principally to secure state control over agricultural production so farm surpluses could be used to finance industrial investment.

Since the death of Stalin in 1953, there has been some liberalization of rigid state control within the Soviet Union and some relaxation in the rigidity of Soviet foreign policy. Since 1953, there have been no mass purges or executions as there were under Stalin. Artistic and intellectual expression in Russia have also become freer, although in an inconsistent pattern. They are still far less free than in most Western societies. The brilliant Russian achievements in pure science and technology, where individual research was not tightly controlled, perhaps have underlined the importance of intellectual freedom. Part of the process of industrialization and planned resource use has been the growth of an essentially apolitical managerial and technician group whose goals and methods are not substantially different from those of their counterparts in other industrial societies. Increasing economic abundance in the Soviet Union also tends to bring with it increasing relaxation of tight control and of the earlier sense of urgency. The power of the Soviet state remains absolute to a degree unmatched anywhere except in China, but now it may be somewhat less ruthlessly and widely applied than during Stalin's rule. The tremendous vitality of the Russian people, which has been a major factor in Soviet successes all along, is being given freer rein. The Soviet revolution appears to have passed out of its most violent stage and entered a period of growing moderation.

Since 1917 the U.S.S.R. has gone far toward overcoming many of the obstacles that had retarded its economic development in the past. The human cost has been high, but the accomplishments should not be underestimated, as they often have been in a politically unsympathetic West. Russian accomplishments have been won under an authoritarian regime in a country that was one of the most backward in Europe until forty or fifty years ago. The Soviet Union unquestionably shares with the United States, on more or less equal terms, the political and military leadership of the world. Another decade or two may see the Soviet Union on a par with the United States in economic terms as well. Perhaps Russia will be in a stronger position of influence in the greater part of the world where economic development is lagging behind that of the West. The West has been alternately complacent and fearful about the Soviet Union. These attitudes have hampered Western understanding of the profound developments that have taken place in the U.S.S.R. It is time the West acquired a more thorough understanding of this giant country and of the nature of the world community in which the Soviet Union now plays such an important role.

Questions for further study and discussion

1. Why should new textile factories be established in Turkestan rather than sending all of the cotton to the major markets for manufacturing? Are there any other places in the world where textile manufacturing is located at or near the source of raw material? Why in each case? Why is this arrangement not more common?

2. Which, if any, of the present Russian ports are ice-free? Which are nearly so? How much are they used, or how important are they to the Russian economy (especially in comparison with seaports in other countries), and why?

3. Why are rye, oats, barley, millet, potatoes, and cabbages so important in Russian agriculture? Why is wheat the leading crop? Sugar beets are also important. Why? (Beet sugar is nearly always more expensive than cane sugar, and sugar beets require especially good soil.) Where are most of Russia's sugar beets grown?

4. Why are rivers the chief communication lines and centers for settlement in Siberia, aside from the one railroad line? Why have more rail lines not been built, especially since navigation on the rivers is so limited?

5. Is Moscow a good capital for the Soviet Union? Why, or why not? If not, what alternative capital would be better, and why?

6. How many separate national cultures are recognized in the Soviet Union and given regional political status? Locate all of these republics and autonomous regions on a map.

10 People and land in monsoon Asia

Village girls transplanting rice plants in the wet zone of Sri Lanka. *(Photo from Ceylon [Sri Lanka] Tourist Board.)*

Over half of the world's people live in Asia. Until about two hundred years ago, Asia's share of the world's population was much larger. A century from now it will probably again be close to three-fourths, judging from current population growth rates. Most of these people live in less than a fourth of the total area of Asia. They are clustered around or near the southern and eastern ocean fringes of the continent. Actually, Asia is no more a separate continent than Europe is. The single land mass called Eurasia is also joined to Africa at the Isthmus of Suez, forming what has been called "the world island." The conventional line between Europe and Asia, drawn at the Ural Mountains and through the straits between Turkey and Greece, does not mark strong human or physical differences. But the great steppes and deserts of Eurasia do begin east of Turkey and a few hundred miles (about 320 km) east of the Urals. The large area of steppe and desert and the friction of distance acted as a barrier between western Europe and monsoon Asia, the two poles of world civilization at opposite ends of the Eurasian land mass. The two cultures grew up largely isolated from each other except for the little exchanged through intermediaries.

That isolation has now been broken down. The two peoples have met face to face within the past two or three centuries, after at least three thousand years of largely separate development. The Greek historian Herodotus, in his careful account of the world as known to the West in the fifth century B.C., stressed the lack of knowledge of the East: "Till you reach India, the country is peopled, but farther east it is void of inhabitants, and no one can say what sort of region it is."[1]

The clustering along or near the seacoasts that characterized both Western and Eastern settlement kept the peoples apart when communications were by land. When the West mastered the use of the sea and arrived in force to trade on the coasts of Asia, the cultural exchange increased. Until the time of Marco Polo, and even into the sixteenth century, the West still had only a vague idea of the peoples and cultures of monsoon Asia. And the Chinese, for example, had no accurate conception of the locations and

[1]*The History of Herodotus,* tr. G. Rawlinson, ed. E. H. Blakeney (London: Dent, 1928). Everyman's Edition, Book IV, p. 40.

names of the states of Europe until the nineteenth century. Successive invasions and early trade with Mesopotamia (the valley of the Tigris and Euphrates rivers, now in Syria and Iraq) did bring outside influence and exercise a powerful effect on Indian civilization. Generally, however, Asian civilizations were separated from civilizations in the West by inhospitable distances, developed largely independently of them and even of each other in many cases, and have remained fundamentally distinctive in their nature despite Western influences during the past two or three centuries.

Monsoon Asia

The area where most of Asia's people have always lived, the southern and eastern coastal fringes of Eurasia, is called *monsoon Asia* (see Map 49, page 182). The name is more convenient than accurate, for climate differs markedly over the area as a whole. Broadly speaking it has in common only that rainfall is concentrated in the summer, and that most of the area is warm or at least avoids extreme cold. The word *monsoon* is actually of Arab origin; it originally meant "season," and was probably applied to seasonal winds in the region of the Arabian Sea. Monsoonal winds are much less frequent in some parts of the area than in others, and also occur in the dry areas of southwest Asia. In some cases, Pakistan for example, the culture is closely related to or part of the culture of monsoon Asia although the climate is very different and may not even be mainly monsoonal. But there is a need for some name to apply to the culturally distinctive area where most of Asia's people live. *Monsoon Asia* is the accepted label, whatever its shortcomings, and a better one is not easy to devise. The old terms Far East and Near East are less satisfactory and more ambiguous. Monsoon Asia includes what is still sometimes called *oriental* civilization. It extends from the western borders of Pakistan to the southeastern limits of Indonesia and the northern tip of the Japanese islands, but excludes the massive highlands north of the Himalayas and the great interior desert of central Asia beyond them. The boundaries of monsoon Asia are arbitrary to some extent and follow political lines where they tend to coincide with cultural or physical breaks. Most lines drawn on a map that are not fixed political boundaries or routes are arbitrary, as the original political boundaries may also be. Where arbitrarily drawn lines are used to indicate cultural regions, they may not accurately show the transition zones of varying widths that separate one region from another, as discussed in Chapter 3.

Regional subdivisions

On a broad scale one can most usefully divide Asia as a whole into *Dry Asia,* which includes the Arabian-Turkish-Persian-Mongol world of the Middle East and central Asia; *Cold Asia,* Siberia and the Russian maritime provinces along the Pacific (Tibet and the almost uninhabited mountain areas to the west and north of it are both cold and dry); and *Wet Asia,* which largely coincides with monsoon Asia and also generally has warm temperatures. The characteristics of wetness and warmth fade off gradually on the northern and northwestern margins of monsoon Asia—Manchuria and north and northwest China. There plains rather than high mountain ranges result in a broader transition zone from a maritime to a continental climate. Elsewhere mountains and sea have helped to draw sharper physical lines around monsoon Asia.

Australia and the Pacific island groups of Polynesia, Melanesia (including New Guinea), and Micronesia are obviously distinct places. Not only are they physically separated from monsoon Asia, but the peoples who inhabit them, the cultures they have developed, and their lack of relations with the cultures of monsoon Asia reflect their differences, which are even observable in their native animal and plant life.

Within monsoon Asia as defined above, a further regional subdivision is in order. This is a huge, densely populated area that contains the world's oldest existing cultures. There are four distinct regions, different from one another culturally:

1. India, including Pakistan, Bangladesh, Nepal, and Sri Lanka (Ceylon);

2. China, including most of Manchuria and Inner Mongolia where the Chinese have settled and established their cultural and economic patterns, but excluding Tibet and Sinkiang, which belong to dry central Asia;

3. Japan, including the four main islands and

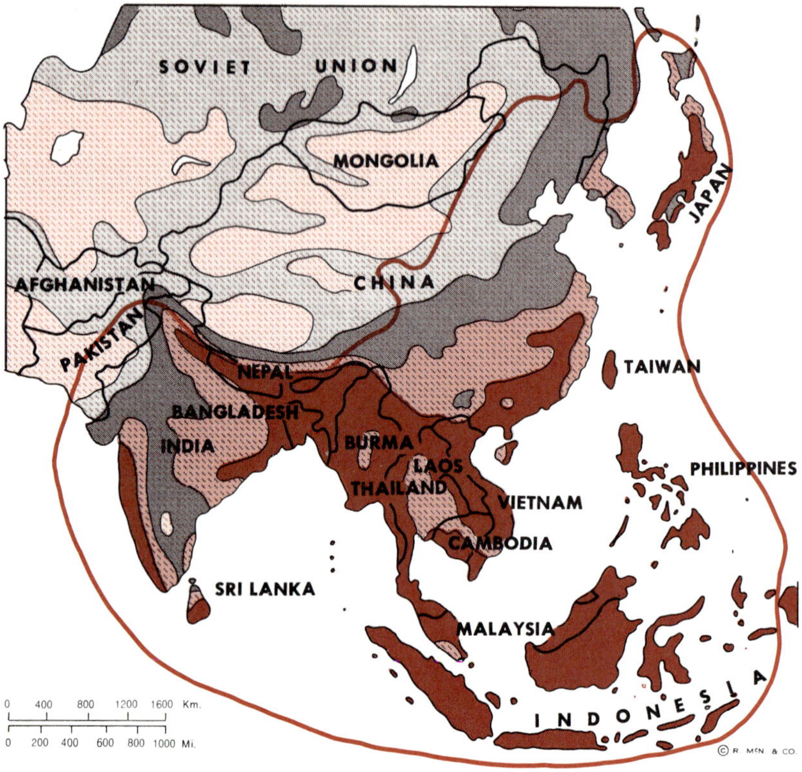

MAP 49
PRECIPITATION IN MONSOON ASIA

Korea, which is closer geographically to Japan than to China; and

4. southeast Asia, a diverse and scattered area with weak regional coherence.

Southeast Asia includes the mainland states of Burma, Thailand (formerly Siam), Laos, Cambodia (Kampuchea), Vietnam, and Malaysia, which share what is called the Indochinese peninsula (lying between India and China), and the island groups of Indonesia and the Philippines.

Physical base

There are several common characteristics that justify the use of a common term for this large and varied piece of the world. The physical base of monsoon Asia is broadly similar in the major areas of dense population (see Maps 15, 49, and 50, pages 62, 182, and 184), similar enough to permit or encourage a broadly consistent agricultural system over most of these areas.[2] The basic characteristic of a monsoonal climate—the concentration of 80 percent or more of the annual rainfall in the four or five warmest months—is common to all of these areas. Most of monsoon Asia is tropical or subtropical. Cold temperatures or steppe conditions appear only on a few of its margins. Generally, rainfall and temperatures are very favorable for agriculture. There are few extensive plains, but they consist almost entirely of highly productive alluvial soils in the basins of some of the world's largest rivers. The monsoonal rainfall north of about 10° N. latitude is notoriously variable from year to year and season to season. Regional

[2]The shifting cultivation on the hillier or more remote lands and the small spots of plantation agriculture are separate systems.

variations in rainfall are also pronounced. The northern and western margins of monsoon Asia have a low average annual rainfall. They lie on the fringes of the rain-bearing monsoonal winds (see Map 49). Rainfall variability or periodic inadequacy is related to the marked concentration of settlement in river valleys, where irrigation water is available. But most of monsoon Asia is well watered, and has a long or unbroken growing season, a further major asset.

This combination of physical circumstances has been accompanied by a highly intensive agricultural system, except for the hill and mountain areas. The agriculture is still largely a subsistence rather than a commercial system. In most areas outside Japan, agriculture represents the bulk of the economy in terms of employment or income. Nowhere else in the world do so many people depend so heavily on agriculture. Nowhere else is agricultural population so dense over such a large area (see Map 50, page 184). Population concentration in river valleys and deltas indicates level land, fertile alluvial soil, easy availability of water, and also the easiest available communication and transport. In most areas in this part of the world, modern mechanized land carriers have not yet been developed on the scale of the West, especially in terms of the number of people to be served.

Agrarian civilizations

On the ocean fringe of each country, commercial-industrial cities have grown up during the past two hundred years, built outright by Western traders or colonial rulers, or stimulated by Western traders. This commercialization and the beginnings of industrialization are increasingly characteristic of monsoon Asia. For the present and the immediate future, however, as in the past, most of monsoon Asia remains primarily agrarian.

The combination of physical advantages has amply repaid intensive agricultural cultivation, and has offered a high return for labor and capital invested in the land. Although farming in monsoon Asia varies widely, its most striking and perhaps most important common characteristic is its intensiveness. Because of the huge population so exclusively dependent on farming, much labor is expended on relatively small plots of land. Capital of all sorts is scarce, since surpluses of all sorts are scarce.

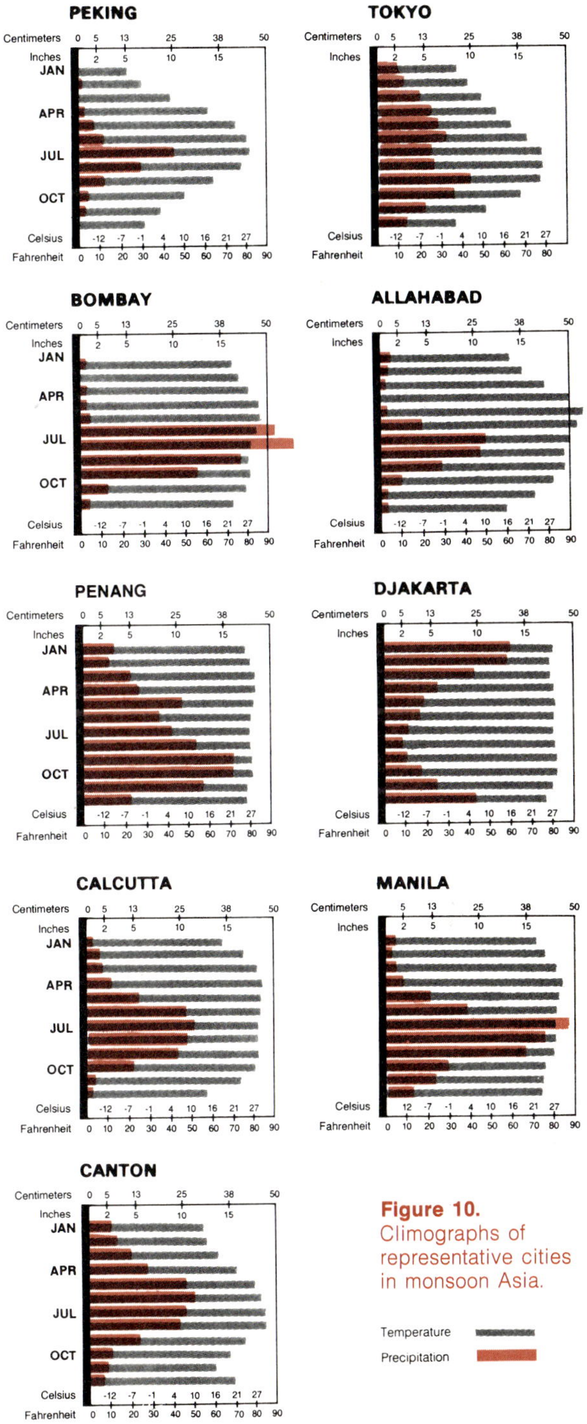

Figure 10.
Climographs of representative cities in monsoon Asia.

MAP 50
POPULATION OF MONSOON ASIA

Per Sq. Km.	Per Sq. Mi.
0–10	0–25
10–49	25–125
49–98	125–250
Over 98	Over 250

Metropolitan Areas

■ Over 2,000,000
▲ 1,000,000–2,000,000
○ 500,000–1,000,000

But what capital is available has traditionally been put into the land, in the form of fertilizers, irrigation equipment, farm equipment, animals, and whatever aids to or means of production the economy can offer and the farmer can afford.

The advantages of climate, soil, and plains that made these investments pay also encouraged the growing of a particular crop, rice. Rice was probably native to southeast Asia, and first domesticated there. It thrives in a hot, wet climate, and if it can be grown in standing water under a strong sun, it is especially productive. Rice probably produces more in terms of calories per acre than any other cereal. It is the dominant crop in monsoon Asia wherever irrigation or simple inundation is possible. Rice is so productive and its traditional place in the agricultural system and the diet is so large that it is grown even in areas that are far from ideal for it physically. Hillsides that must be terraced to be irrigated (see the photograph on page 248), infertile soils that are heavily fertilized to keep the yield high, and areas where the growing season is critically short (in Hokkaido, for example) or where rainfall is excessive and sunshine inadequate (as in the wet zone of Ceylon) are all farmed intensively. In some areas a special variety of rice is also grown on dry fields without irrigation.

Labor-intensive cultivation of rice on the one hand and dense rural population on the other are obviously related. Rice demands much labor and, partly for this reason, is grown in Asia only in areas of dense population. With intensive cultivation, it does yield a great deal of food per acre and can support, or may even encourage, a dense rural population. As population density increases, cultivation becomes more and more intensive. It reaches its peak in the technologically advanced agriculture of Japan, where the ratio of people to cultivated land is especially high. Cultivation is least intensive in southeast

Asia, where population density, except in Java (Djawa), was relatively low until recently.

There are other major regional differences in agriculture and its intensiveness over this very large area, but in general the agricultural system of monsoon Asia is distinctive. Only in parts of Egypt do agrarian rural population densities over any large area approach 2,000 people per square mile (765 per square kilometer), which is common in the major river valley and delta centers of Asian agriculture. There they are almost on a par with urban densities. An agricultural system that can support over 900 million people in China, over 113 million in Japan, 635 million in India, and over 80 million in Java alone must be distinctive, especially for its intensiveness.

Irrigation and its significance

An irrigated crop is an intensively grown crop, if only because irrigation requires a high investment of capital and labor in the land. Irrigation is needed because high production of rice requires ample water. Monsoonal rainfall tends to be concentrated in one season and to vary considerably from year to year. It may come too late or too early to be of best use to the crop, it may be inadequate in amount, or it may consistently not come in a period when warm temperatures would otherwise permit rice growing (see Map 49, page 182). In most parts of monsoon Asia where rice is grown, irrigation has been provided on a large scale. In eastern India and the river valleys of peninsular southeast Asia, simpler inundation or *flood-water irrigation* is often used instead. In many areas the state has taken an important part in providing this important investment. Often only the state had sufficient resources and power to build the large and complex irrigation systems. These irrigation systems were frequently constructed by forced labor or by labor performed as part of the annual tax. Agriculture was the chief source of state wealth and power, thus it was in the state's interest to make the land as productive as possible. Monsoon Asia has traditionally been distinguished for its strong central authority and powerful bureaucracies. In hilly areas or areas of small streams and small pockets of level land, such as in Japan or south China, there were more small-scale local or village projects of water control. In general the physical conditions and the agricultural system did not encourage individualism, even though large estates or large holdings of land were not common, and most of the farmers were small peasant proprietors. Average holdings could not be large when millions of people competed for land as their source of support. Intensive subsistence cultivation is usually characterized by small individual holdings, partly because of the large amount of labor required, and in monsoon Asia partly because of the special conditions of wet rice farming, which do not easily lend themselves to large farm units.

The dominance of agriculture and the generally subsistence nature of the economy left relatively little place for manufacturing or trade. The commercial middle class, which was so important in the development of modern Europe, was relatively small in monsoon Asia, and was often connected with or dependent on the government. Capital was scarce by modern Western standards, and tended to be invested largely in land, in speculation, or in money lending, rather than in productive enterprises in trade or manufacturing. Land ownership brought prestige, security, and a high return in rents (which were frequently exorbitant and often above returns from trade). Land was less liable to confiscation or punitive taxation than commercial or industrial investments, which attracted the attention and the regulation of the all-powerful state. This pattern of investment was crucial in helping to prevent major or rapid increases in productivity or incomes, and in working against the fundamental economic, social, political, and institutional changes that marked the development of modern Europe.

The peasant society

The central figures in traditional monsoon Asia were the peasant farmers, who constituted 70 to 80 percent or more of the population. *Peasants,* as distinct from farmers in Western terms, are essentially self-sufficient producers, born onto the land and into the social, economic, and occupational status inherited from their parents. They seldom move far from their birthplace, and their commercial or other relations with other places or people are limited. They are static traditional figures. In monsoon Asia most of the rest of the population were supported by the peasants. This included officials, merchants, and artisans, but their numbers were small. Even the head of state was viewed as the symbolic head farmer whose role was to represent agriculture to heaven, ensure good harvests, and maintain the efficient operation of the state machinery in the service of agriculture. It was a culture wedded to the land.

Harmony with nature

Agriculturists must be concerned with the rhythm of the seasons and the care and respect of the land that nurtures the people. Although Asian farmers altered nature as people alter it everywhere, particularly cutting or burning off the forests, in general they cared for the land more conscientiously than the commercial farmers of the modern West. In return, the land still supports them after more than three thousand years. Too often agriculture in the West has *mined* the land in its much shorter history, especially in the United States. Farmers in monsoon Asia have transformed the landscape to a greater extent than anywhere else in the world. Almost nothing *natural* is left except the outlines of the mountains. Thirty centuries or more of occupance by a dense agricultural population who used the land hard and shaped it to their needs have left a deep imprint. Dams, dikes, paddies, terraces, denuded hillsides, periodically burned or slashed forests, silted rivers, flood plains, and, recently, plantations largely compose the landscape. It is more human and cultural than natural. But people in Asia were not set against nature, as they often have been in the modern West. They viewed themselves as a part of nature, a small part, who could be happy and successful only if they willingly adjusted themselves to the greater whole.

The acceptance of harmony with nature as the goal of human life and the surest guide to practical success and earthly happiness contributed to the absence of an aggressive attitude toward nature, which may have been part of Western achievement in science and technology. Asian philosophical reverence for nature was often contradicted by peasant actions, such as deforestation and other destructive practices. But such practices were to make nature more fruitful, as stewards, rather than to conquer it as enemies. The traditional East also valued leisure as an end in itself. This attitude is one of the graces of Eastern civilization not yet entirely gone. The long gown and long fingernails of the Chinese gentry were symbols

Warm, wet Asia—intensive rice culture. *Left,* Transplanting young rice plants from the nursery bed to the flooded field (paddy) where they will grow until harvested, a common sight over most of monsoon Asia. This scene is in Taiwan. *Right,* Painstaking weeding of the rice is done by hand several times during each crop season, as shown here in a Taiwan rice paddy. The conical bamboo hats are worn for protection against the hot sun. *(Both pictures from Joint Commission for Rural Reconstruction.)*

of their freedom from manual work. But work of all kinds was regarded as a necessity, to be dispensed with as soon as the greater values of leisure were possible. The modern West has more commonly valued work for itself, and leisure has often been considered merely idleness. R. H. Tawney has aptly labeled the modern West "the acquisitive society." It cast people in the role of doers, the conquerors of nature. It believed in the concept of progress, and tended to measure progress and success in terms of the acquisition of goods and property. These goals were never entirely absent in the East, but they were seldom as important. Happiness and success were more commonly measured in other terms, including the harmonious adjustment to nature rather than its

conquest. The dominant position of agriculture and the respect paid to it minimized the need for and value of change.

Village and field

Most Asian peasants still live in villages, or *nucleated settlements,* rather than in the isolated farmhouses, or *dispersed settlements,* characteristic of farmers in western Europe and North America. A typical village may include five to forty families. The people go out to work every day in the fields adjacent to or radiating from the village. The fields are small and few of them can be very far from the village. Holdings are often split up into scattered pieces of different kinds of land. It is an inefficient system in some ways, but it may represent a useful variety of soils, slope, and water availability suitable for different kinds of crops. Most holdings include some space for vegetables. Sometimes they are grown around the house or near the village so they can be tended easily; sometimes they are grown on the low ridges between rice fields. Near large urban markets vegetables are proportionately more important for sale in the city. But most farmers everywhere grow vegetables for their own use to supplement in a small way a diet heavily dominated by cereals.

Cereal crops

In most areas the total cultivated land worked by a single family averages between three and five acres (1¼ to 2 hectares),[3] except in parts of southeast Asia. This is related, in turn, to the large number of people per square mile of cultivable land, to the high yields per acre the intensive system of cultivation produces, and to the cereal crop (rice, wheat, or other grains) that dominates the system and helps to maximize the food production from each acre. Wheat is grown in large areas of monsoon Asia where the climate is too cold or too dry for rice, or where adequate irrigation is not available, for example, in much of north China and Manchuria, and in central, north, and northwest India and Pakistan. Wheat is also widely grown in rice areas as a dry-field winter crop. Elsewhere, slope or less favorable climate make other, less fussy cereals necessary, especially drought-resistant grains like the sorghums and millets, or crops tolerant of slopes or poor soils. Cereals are stressed because they can produce more food of a more acceptable kind from a given amount of land than any other kind of land use. Potatoes, sweet and white, are highly efficient food producers. They are more tolerant of poorer soils and climates than grains, can yield well without irrigation, and tend to be grown in large amounts in monsoon Asia under these conditions. Potatoes have less prestige, or are valued less as food than cereals, despite their importance in the agricultural system.

The place of animals

Animals are much less efficient food producers per unit of land than field crops. Except for essential draft purposes, they occupy a large place in monsoon Asia's agricultural system only where they can be kept as scavengers and do not represent a drain on the human food supply. It is estimated that, on an average, eight times as much land is needed for grazing or fodder crops to produce a given amount of food from animals than to grow cereals directly for human consumption. Beef cattle, sheep, and goats, therefore, are not kept for meat in significant numbers in monsoon Asia, and almost no land is given over to pasture or fodder crops. This is one of the striking differences in the agricultural landscape between monsoon Asia and Europe or North America. Chickens, ducks (in the rice fields), and pigs are kept in large numbers in many areas, especially in China and Japan.[4] They can forage for themselves on the very lean pickings left by careful farmers who waste nothing and who are living on a thin margin of subsistence. Asian farmers until recently lived economically up to their necks in water and were drowned by the slightest ripple. They must usually keep, or at least be able to borrow from a neighbor, some draft animal for plowing the heavy muck soils of the rice fields, often knee deep in water, or the lighter soils used for other crops. The water buffalo, bullock (ox),

---

[3]All land in China was collectivized by the state by 1957, but average family holdings up to that time were within the range given. The same applies to North Korea and Vietnam, where land is also collectively owned by the state.

[4]Pigs are not numerous in the Indian subcontinent and in most of southeast Asia, partly for religious reasons. Ducks are unevenly distributed. Chickens are more nearly universal.

or bull is too precious to eat until he dies of old age, and he earns his keep not only by plowing but by contributing manure, carefully saved as fertilizer. Where cows are used for draft, as in parts of Japan, they give virtually no milk. Even in India, where they are kept for milk, inadequate diet and breeding mean that most cows and goats give much less than in the West. Horses are not common except on the steppe margins. The horses of monsoon Asia are not large enough for most plowing. Although they serve as pack animals, heavier carts and wagons are more generally pulled by cattle or, in drier areas, by camels.

Night soil and the vegetable civilization

Human manure, or night soil, is also carefully saved and applied to the fields. At least in amount, it is the most important fertilizer the crops receive, although in recent years the Japanese have supplemented it with even larger amounts of chemical fertilizer. In India and much of southeast Asia the use of night soil is still very limited, primarily for essentially religious reasons. In many areas in China and Japan near the towns and cities where night soil is available cheaply in large quantities, the crops are especially rich and green, and they gradually fade farther out into the countryside. Night soil, increasing in amount as population increased, has helped substantially to sustain fertility in fields that have had to be worked more and more intensively as they have had to support more people.

Where there are many people per acre of arable (cultivable) land, even hillsides are plowed, or they may be terraced, as shown in the picture on page 248. The labor required for terracing and for maintaining the terraces and their irrigation is enormous. And the return in crop yields is pitifully small, especially on the steeper slopes where only a few patches of level land can be created. But such land may mean the difference between enough food and starvation. Where slopes are too steep for terracing or annual cultivation, tree crops such as tea or nuts may be raised, or the hillside may be scraped clean of trees, bushes, and even grass to be used as fuel. Little is left that could support big animals like cattle. The destructive grazing habits of sheep and goats restrict their numbers. Sometimes they are kept in small pens and fed on leavings or on grass gathered by hand on the steep slopes. Sheep and goats appear in large numbers in the arid areas where field crops are at a disadvantage, as in central and northwestern India. They have done terrible ecological damage there, as they tend to do in better-watered areas, such as south India, where they are allowed to graze. But monsoon Asia has aptly been labeled the "vegetable civilization," dependent on cereals rather than animals, and on wood rather than metals or mineral fuels like coal.

Limited spatial mobility and exchange

In the traditional setting the nucleated village made it unnecessary for people to move far, except between house and field. On infrequent occasions they went to the nearest market town where there was a somewhat greater variety and specialization of services. Tea houses or their equivalents in the market towns were common and inexpensive centers of social amusement and use of leisure. Even in the village, however, there would often be a blacksmith, a carpenter, or a small shopkeeper, although these might be part-time occupations for some farmer who also worked on the land. Rotating fairs or markets, meeting in turn in each of several villages in a circuit, also provided wider commercial and social exchange. This made it less necessary for most villagers to leave the village. Social opportunities might be as important as trade at these fairs. Lacking the mass amusements of an industrialized or spatially mobile society, peasants made the most of market days, of traditional seasonal festivals like the Chinese New Year and Hindu Diwali, and of village birth ceremonies, weddings, and funerals, the bench marks in the cycle of the agricultural year or of human life. These were the village amusements and the main uses for scarce savings, or often for loans from a moneylender. Such economically unproductive dissipation did not make the social occasions less enjoyable or less necessary.

Most of the villagers' wants were supplied by themselves or by village neighbors: food, housing, farm equipment and tools, and clothing. Spinning and weaving were frequently done in each peasant household. Peasants had considerable leisure when the crops did not need full-time work. Winter or the time between crops might be an especially slack season. This time was needed for rest from hard physical work and for social amusements, but it also gave an opportunity for handicraft production for local sale or barter as well as for personal use. Salt and metals might have to be brought in from greater distances, although small ore deposits were worked locally in a great number of places and supplied many local markets. Most production was consumed locally, as most needs were met locally. Generally only the upper classes depended to any great extent on exchange for the specialties in food, clothing, and other goods they required and could pay for. The state might move goods and people around more widely—troops and their food supplies, grain for taxes, metal for coinage, government officials and their reports—but for the most part it was a spatially static society.

Of course, this describes almost any subsistence economy (as Medieval Europe was also), which minimizes exchange and spatial interaction as commercial economies maximize them. Urban areas are less developed than in commercial economies. Any town or city reflects the complexity, specialization, and variety of economic, political, and social organizations in its tributary area. Tributary areas are also smaller where exchange and spatial interaction are not well developed. There were proportionately fewer cities in monsoon Asia, for example, than in modern Europe (although the total number of cities and towns was large), and much less of the population lived in cities. Some big cities did develop, bigger than any in Europe until the eighteenth or nineteenth century. But they were smaller than the big cities of modern commercialized Europe, and were generally different in their major functions. Mainly they were centers of administration and culture rather than centers of trade or manufacturing, as modern European or classical Mediterranean cities were.

Molecular Asia: village and city

Like most subsistence economies, monsoon Asia still has a molecular spatial organization. There are thousands of little village worlds, all much alike but having relatively little to do with one another. The market towns were commercial centers and were involved in exchange and spatial interaction, but within a limited area and to a limited degree. The few large administrative cities also had commercial functions, but they were generally subordinate. The traditional Asian city was not a center of change or of ideas and values different from those of the countryside, as became typical of Western cities. Even today traditional villages are still the home of most Asians, and traditional village-based culture and its values are still important, especially outside Japan.

Tradition versus change

Weak spatial interaction and molecular organization contributed to the force of tradition and the resistance to change. These attitudes are strong in most agrarian societies, but they were increased in monsoon Asia by the weakness of competing stimuli and the lack of differences and interactions between ideas or institutions. Massive mountain barriers, rain forests, and oceans added to the insulating effects of the desert and kept Japan, China, India, and southeast Asia largely distinct from each other and from the West. Wedded to the land, living in a world where influences from different places were infrequent, people thought it natural and proper to behave and to think as their grandfathers had, to value the experience of the past, and to resist change. The small economic margin on which most peasants lived also made them cautious about any experiments with new methods. This still retards agricultural improvement. The cake of custom is hard to break when it is so old, so sanctified, and so reinforced by an economy that is reluctant to experiment. Of course, there were regional differences within each of the major cultural areas, sometimes very pronounced. But they did not have the effect they would have had in a commercial economy because regional specialization was less possible, interaction between different regions was at a minimum, and most people had to earn their living in much the same way wherever they lived. There was change and economic growth, but it took place more slowly, and never progressed as far as in modern Europe. There economic, social, and political organization was revolutionized, particularly in the agricultural, commercial, and industrial revolutions of the past two or three hundred years.

The Middle Ages in Europe have been described as "the victory of the East over the West." It was in many ways an out-of-character interval during which the West abandoned its changeableness, its concen-

The sophisticated culture of Asia. *Left,* the Golden Pavilion, Kyoto, Japan, a beautiful adaptation of continental Asian architecture and an example of imaginative blending of architecture with landscaping. It is typical of the many beautiful shrines in Japan. *Right,* the Great Stupa, Anuradhapura, Sri Lanka, a Buddhist temple, larger than any of the pyramids of Egypt, except the third at Gizeh. It shows the high technical skill in monumental building found in most of the areas of monsoon Asia. *(Photo on left by Frank Wing for Stock, Boston; on right by R. Murphey.)*

tration on trade, its conquest of nature, and its pursuit of progress. There was a more stable, leisurely, philosophical, intuitive, aesthetic approach to life, which was indeed close to the oriental pattern. But Medieval Europe was totally altered after the fifteenth century. Its change was so radical that one can hardly recognize modern Europe as the same place. Practically all medieval institutions, values, economic bases, art forms, and political systems were changed. That kind of basic change did not come in monsoon Asia until new ideas, institutions, and techniques were imported from the West. Earlier changes were less fundamental, not because Asia lacked the necessary potential resources, or the knowledge of them. Coal was used as fuel and to smelt ores in China centuries before it was used in Europe. Indian steel was famous two thousand years ago. India pioneered such developments in mathematics as the invention of decimals, the concepts of zero and infinity, and the so-called Arabic numerals, which are really Hindu. The traditional societies of monsoon Asia achieved a high technical and economic level very early. Their achievements were superior to those of the West until a century or two ago.

Theoretical science, as contrasted with technology, was less well developed in the East than in the West. The early work of Indian and Chinese scientists was not followed consistently by further developments comparable to the scientific advances in the West since the seventeenth century. The West viewed the physical world objectively, rationally, and theoretically; the East viewed it more immediately and practically, rather than theoretically.

The physical world was explored in light of immediate human values, and the observable cycle of life and death in nature and people was accepted as philosophically right. Having built a culturally sophisticated, technically advanced, and economically productive society, the East was content, and did not attempt radical alterations. If the brilliant theoretical achievements of Greek and modern European science were related to varied external stimuli, then the lack of continuous scientific development in monsoon Asia may be related to the weaker force and variety of such stimuli.

Religious ideas did spread within Asia, particularly Buddhism, Hinduism, and Islam. But these ideas were not accompanied by extensive commercial relations except for short periods between certain areas (south India and southeast Asia, and southeast China and southeast Asia). The conscious borrowing of Chinese civilization by the Japanese was a wholly one-way process and did not bring the two peoples together in any dynamic relationship. Cultural influences from India on southeast Asia were equally strong, but it was also largely a one-way relationship. Just as importantly, there was a very small interchange of trade or ideas with the Western world until the West used its superior technical and military power to force the East to change.

The arable land in Asia was crowded; people lived close together in a tightly structured society that discouraged innovations and preserved the patterns of the past. The principal social organization was the family. Individuals had little freedom and little security apart from it, for it was their provider and their insurance against disaster as well as their judge and arbiter. The traditional family included three generations under one roof, or at least under the supreme authority of the oldest family member. In this system such submission was not questioned and was felt to be naturally right. These societies had a long memory, peasants as well as scholars. The great age of Asian civilizations thus meant that social, political, economic, and cultural patterns centuries old continued to set the standards. Given its solidly agrarian base and its molecular structure, these were appropriate values for the society of monsoon Asia. By Western standards it is hard to realize that India and China as organized cultures were contemporaries of ancient Egypt and Mesopotamia.

The West in Asia—Singapore. This is a typical, originally colonial city, built in Western style and on a Western basis of trade. Part of Singapore's excellent harbor and its dense traffic shows beyond the buildings, constructed with trade profits. *(Photo by P. Gosling.)*

Traditional Asian superiority

However, it is far from accurate to regard Asian societies or economies as backward because their productivity and technical capacities have fallen behind the West in the last two centuries. Despite Europe's achievements in the eighteenth century, until about the beginning of the nineteenth century, Asia probably led Europe in average income levels, in the economic position of its peasants and lower classes, in total productivity, in the quality of its manufactured goods, in the number and size of its cities, in its total volume of trade, and in the material comforts available to its upper classes. This leadership was quite apart from the sophistication and quality of its art, literature, philosophy, and system of government, which are difficult to measure or compare. Asia had occupied this position of leadership and general superiority over the West for at least two thousand years. And Europeans admitted its greatness. Marco Polo recorded an enormous gap between a civilized East and a comparatively barbaric or undeveloped West, and earned the nickname of *millione* because he told such tall tales. Only much later did other Europeans get into China and prove the truth of Marco Polo's supposed dying words that he "had not told the half" of what he had seen.

Even in the seventeenth and eighteenth centuries, when the Europeans had established an extensive sea trade with the coasts of Asia and were beginning to found colonial empires in India and southeast Asia, they met a self-sufficient civilization. In many respects it was still their superior despite its relative inattention to trade. George III of England had a famous exchange of correspondence with the emperor of China, Ch'ien-lung. The English, like all the other non-Chinese "barbarians," were permitted to trade with China only at one port, Canton. In 1793 George III sent a personal ambassador, Lord McCartney, bearing presents to the emperor. The Chinese regarded these samples of British goods as tribute from an inferior barbarian state. They were not impressed by McCartney's request for wider trade relations, and an exchange of diplomatic missions. The Chinese were interested in the trade samples, but rather as a kindly parent would be interested in the work of children, since they could make most of the things far better. They were especially amused by the pieces of Wedgewood pottery Lord McCartney brought. As for diplomatic exchanges, this could have no place in relations between the celestial empire and outer barbarians. Ch'ien-lung's reply was a masterpiece of crushing condescension. He complimented the barbarian ruler on his humility—"I have already noted your respectful submission. . . . I do not forget the lonely remoteness of your island, cut off from the world by intervening wastes of sea." (This viewpoint is interesting when compared with England's locational advantages as discussed in Chapter 5.) But "our Celestial Empire possesses all things in prolific abundance and lacks no product within its borders. We do not need to import the manufactures of outside barbarians in exchange for our own produce." This was the truth in 1793. It ceased to be true in the nineteenth century, although China, in particular, stubbornly clung to its traditional view that it had nothing to learn from or exchange with barbarians. Nevertheless the influence of the West on the East, in which the wheel has come full circle since the European Middle Ages, will destroy much that was good as well as bringing benefits.

Influence of Western traders

The West came to the East at a bad time for Asia, since the Asian states were in one of their recurrent periods of political and economic disorganization and decay. In the traditional pattern, such a period would probably have been followed by a new government, political order, or dynasty in each country that could have restored the vigor and efficiency of the system. Instead the change was provided by the West. Western change had the further advantage of approaching Asia by its less guarded back door on the sea instead of the traditional overland front door. The West came to the East for trade. Much later, territorial colonialism added more complex motives, but trade remained the dominant interest. Trade was Europe's life blood, but the East usually left trade management in European hands, until it realized too late that trade could threaten the entire system and destroy the established order.

Western impact was felt earliest and strongest in India and southeast Asia, because they were nearer to Europe physically and were more directly accessible by sea. Japan, for a variety of reasons examined in Chapter 15, sought knowledge of Western techniques more eagerly. Also, as a small island area, it was more easily influenced than the massive continental area of China. But to all of Asia, the West brought revolutionary ideas and techniques. Trade had moved between East and West for millennia before the Europeans arrived in Asia. Arab traders established trade bases on the coasts of India, China, and southeast Asia as early as the beginning of the Christian era. But the advent of Western traders in the sixteenth century was followed by a great increase in the volume of trade. For the first time goods were hauled all the way from monsoon Asia to European markets by sea without the earlier middlemen.

The European succession in Asia

The Portuguese were the first Westerners in Asia, since they discovered and exploited the sea route around Africa. Vasco Da Gama reached India in 1498, and in the following century the Portuguese built an enormous commercial empire from Goa, on the west coast of India, to the Moluccas (Maluku) in the south Pacific. This did not involve territorial sovereignty, except over small trade bases in strategically located ports. But their commercializing influence was already spreading into Asian economies. Portugal's profits, especially from spices, attracted the Dutch and British, among others. By the seventeenth century the Portuguese no longer dominated Asian trade. Holland took over the East Indies area, Ceylon, and strategic bases on the Strait of Malacca. Spain controlled the Philippines. And the British East India Company was strengthening its position in India.

In the nineteenth century, the heyday of colonialism, Britain controlled India, Ceylon, Malaya, Burma, and Hong Kong (an invaluable trade base). It had the dominant commercial position in Thailand and China, although it faced increasing trade competition from the United States, Germany, and others. American rule replaced the Spanish in the Philippines. The Dutch remained in control of Indonesia, although now on a territorial as well as a commercial basis. The French acquired their Asian empire in Indochina. In the north, Russia dominated Manchuria and might have added Korea to its empire. But the industrializing Japanese, with Western encouragement, defeated the Russians in the war of 1904–1905, and added Manchuria, Formosa (now Taiwan), and Korea to its own colonial holdings. Only Thailand, China, and Japan remained outside the Western colonial empires, and only Japan was fully its own master. Thailand's nominal independence was preserved by the coincidence of power rivalry. China was too big a nut to crack, too resistant to foreign influences even after being defeated by the British several times in small wars, and too much involved with rival European ambitions to become a formal colony. But its foreign trade and finance, its rail and steamship lines, and its small modern manufacturing sector were largely in foreign hands. This grand colonial pattern (see Map 51, page 195) was not destroyed until after the Second World War. Indonesia finally achieved its independence in 1950, and Malaya (now Malaysia) not until 1957.

Under foreign rule, widely different peoples with separate languages, cultures, physical types, and historical traditions were joined as single colonies, as Burma and Indonesia, for example. Partly this was European territorial imperialism, partly it was their construction of logical economic units based on trade. But a demonstration of common interests and destinies among previously separate groups helped to stimulate Asian nationalism, not only as a negative reaction against alien rulers but as a realization of their own wider political communities.

In retrospect, the commercialization of the Asian economies and the influence of Western industrialization proved no more important than the concept of nationalism. In its modern form, nationalism was a European concept that had been largely absent from the traditional East. Except for Japan, huge cultural empires there usually took the place of the European nation-state. In Europe nationalism's development was related to the close connections, rivalry, and interchange among the different, but easily interconnected, regional bases of which Europe is composed. In the East regional distinctions on this scale or of this nature were less evident, and there was much less contact among them. The state may have been powerful as an institution, but it did not evoke a sense of community in the East. The concept of the modern nation-state imported from Europe stimulated Asians to resist and finally to expel colonial domination. Asians now have in their own independent states the same national loyalties, identities, rivalries, and tightly organized political communities. Every aspect of the modern history of Asia bears the strong imprint of Asian nationalism, created or revived because of the example of the national states of the West. The state has become a positive force in Asia and a magnet for the loyalties and energies of vast populations.

Population explosion in Asia

Perhaps the most important economic difference between Asia and the West, apart from the distribution and usability of mineral resources, is the enormous size of Asian populations and their very low average income level for the past century or so, except for Japan. Thus, it is probable that average per capita standards of living may never equal those of the modern West (barring unforeseen technical changes or the self-destruction of war), even though they might reach a level of relative abundance in Asian terms. In the early period of major Western influence, the seventeenth and eighteenth centuries, the introduction of crops from the New World, like potatoes and maize, probably helped to stimulate the growth of Asian populations. Later, as commercialization increased under Western impetus and organization, new commercial opportunities appeared to support Asian workers in agriculture, manufacturing, and trade. The lessons of the European technical revolution in agriculture were also applied, to a degree, in Asian colonial holdings. In addition, European investments to extend irrigation made the land more productive. Although imports of European manufacture, one of the incentives of colonialism, may have destroyed or weakened some traditional handicraft

MAP 51
COLONIAL EMPIRES IN ASIA ABOUT 1900

production and created economic misery for certain groups, the total effect of Western trade was wider economic opportunity.

Finally, colonial rule and colonial investment built hospitals, fought epidemics and floods, introduced new medicines and better public health, and built railroads, steamships, and roads that could alleviate famine by distributing food from surplus to deficiency areas. Famine is usually local, and is essentially a problem of distribution that mechanized transport and effective state administration could solve. Because the death rate fell, and the birth rate remained high, monsoon Asia's population has probably at least quadrupled since the sixteenth century. The critical problem is that production did not quadruple. Europe's and North America's population increased even more during the Industrial Revolution, but it did not increase as fast or as much as production did. Asia's problem is serious; first, there is still a perilously small gap between people and goods (or low income levels), and second, the original number and present size of the population are so immense. Production is increasing, but it is still strained merely to support the annual population increases. Even a growth rate of 2 percent per year, which may be close to the present Asian average,[5] means about forty million births every year.

Western investments and economic stimulation

European investments in agriculture were devoted mainly to increasing the production of commercial crops for export to the West. This was not entirely plantation agriculture, although it was the dominant commercial form in most of southeast Asia. Rubber, vegetable oils, sugar, tobacco, tea, sisal and other fibers, spices, quinine, kapok, and other crops were grown there with Western management and capital for Western markets. Elsewhere plantations were less important than the expansion and commercialization of crops like cotton, jute, wheat, tobacco, tea, silk, vegetable oils, and rice for export. Western capital was used for new irrigation or land development, and Western organization for opening new overseas markets for Asian goods. Previously neglected mineral resources like tin and petroleum were mined and exported. Coal production, with Western methods and organization, was greatly increased. All of this required the development of internal transport and marketing, railroads, steamships, water routes, and port facilities. This expansion was accompanied by growing imports of Western manufactures into the Asian economy. Competing Asian manufacturing was not encouraged in the colonial countries, and machine-powered factories grew slowly. Japan remained outside the colonial pattern and industrialized rapidly and substantially. But it had other advantages that most of the colonial or semicolonial areas lacked. However, Japan was much more heavily influenced by the West, even without the plantation, and more thoroughly adapted its economy as a whole to the new pattern.

Industrialization and social change

Since the Second World War and the coming of national independence or political revolution, Asia has set the goal of industrialization as the most effective means of raising the productivity of the economy to support its huge population more adequately. These new states want to occupy a place in the world commensurate with their size and with their ancient and modern pride. Hard work and even manual labor are now increasingly valued; leisure is viewed as idleness. Manufacturing and commerce are now respectable; the factory worker or engineer has greater prestige than the peasant or landowner. The traditional family system is being broken up into single-generation parts, yielding a greater place to the individual. Children may choose new occupations and new social attitudes different from those of their parents. As arranged marriages become less common, and traditional occupational links are broken, customs and values begin to change. Nepotism and paternalism in business and manufacturing, which have impeded economic growth in the past, are giving way to recognition of individual merit. Old loyalties and patterns are being swept aside. Revolutionary changes are more easily accepted, and may even be sought in contemporary Asia. It is a time of seeking and questioning; the whole set of cultural values has become fluid.

[5] The population growth rate is higher in most of southeast Asia and lower in Japan. But the figure for most of Africa and Latin America is closer to 3 percent.

Agriculture and industrialization

Agriculture dominates all of the economies of monsoon Asia, except in Japan. Despite the plantation and the development of other commercial crops, the widening of domestic and foreign markets, and the development of transport and exchange, it is still a subsistence agriculture for the most part. The success of industrialization will depend to a large degree on the further commercialization of the agricultural system, continuing increases in its productivity, the development of regional specialization based on wider exchange, and a rise in peasant incomes. Only then can a sufficient surplus for industrial investment be achieved, domestic markets created for the products of manufacturing, and the economy as a whole become productive enough to raise average income levels in the face of a still rapidly increasing population. These are still economies of scarcity rather than abundance. Death rates have continued to fall, but birth rates (outside of Japan) remain high. If Western experience (which is not entirely consistent) is any measure, birth rates are not likely to drop or population growth to slow down until scarcity can yield to permanent abundance. Abundance in Asian terms may not have to be as great as in Western or American terms. Meanwhile, continued high birth rates and low death rates will mean continued population growth. But population may be limited again by the supply of goods, especially food, unless productivity can be increased rapidly and substantially.

Not all parts of monsoon Asia have strong advantages for industrialization by comparison with Europe or North America. Many of the new states may have unreasonably large industrial ambitions, at least under present technological conditions. It does not seem likely, apart from textiles or light consumer goods, that most of them will become major centers of manufacturing. The centers already developed or rapidly developing—Japan, northeast China, and the lower Ganges valley region of India—will probably continue to expand. Many of the other states and regions of monsoon Asia may find the plantation or commercial agriculture inherited from colonialism a more productive source of wealth and a more important means for raising income levels. There is nothing degrading about being a prosperous agricultural country. But the lure of industrialization in twentieth-century Asia and its suggestion of wealth, national strength, and prestige are powerful. The supreme position traditionally occupied by the land and the values attached to it have been weakened, even before full-scale industrialization has taken place.

The new Asia

Because Western systems and techniques are being adopted, Asia will not be simply a copy of Europe now or in the future. The traditional Asian societies and their ancient cultures are far from dead. Indeed they are undergoing a vigorous revival under national independence as a reaction against the powerful influences of the West during the colonial period. Western-originated techniques can be applied in more than one cultural or social context. All of the new Asian states are determined that their national distinctiveness shall be preserved. Indigenous art forms, national rather than Western dress, national or indigenous languages rather than the English, Dutch, or French once dominant among the educated groups, and indigenous religions or political ideologies are being consciously revived or revitalized. Monsoon Asia and its separate countries are big, their geographic and economic bases and problems are different from those of Europe or the rest of the world, and their sophisticated traditional cultures cannot be discarded easily. Asians are proud of their history and their achievements. They have taught Europe in the past, and they are now again pursuing their independent paths toward modernization. There is a determination to build a new civilization combining the best of industrial and commercial techniques, the best of the Asian cultural traditions, and a new set of political and social institutions. The cultures of Asia will remain separate and distinctive whether or not Asia once again assumes a position of world leadership.

Questions for further study and discussion

1. How would you account for the "victory of the East over the West" in the European Middle Ages? Why did Europe adopt so many essentially Eastern values and attitudes?

2. Why should tropical monsoon Asia have a

strong competitive position in world trade in agricultural goods?

3. Why did the foreign-founded or foreign-developed colonial ports in Asia become the largest cities in each country?

4. Does *monsoon Asia* seem to you to be a good label? Why or why not? Can you suggest a better one? How would you defend or alter the lines drawn around it on Map 49?

5. Where in Europe are there rural population densities comparable to the highest in monsoon Asia? Why? In what ways does land use differ? Why?

6. Why are the Hindu numerals now used all over the world called Arabic?

11 The Indian subcontinent: between two worlds

Lakshima Narayan Temple, New Delhi, India. *(Photo from Pan American Airways.)*

The Indian subcontinent, as Pakistan, India, Bangladesh, Nepal, and Sri Lanka (Ceylon) together are sometimes called, is dramatically outlined by some of the most formidable physical barriers in the world. The massive mountain and ice wall of the Himalayas rises around it on the north; the sharp arid ridges of the Hindu Kush, the Sulaiman Range, and their southern extensions on the west; the Khasi Hills of Assam and the Arakan Yoma on the east; and the sea occupies the remaining sides. Behind each of the landward barriers is a further forbidding zone: hundreds of miles of cold, arid mountain waste beyond the crest of the Himalayas, hundreds of miles of desert beyond the mountain ranges on the west, and beyond Assam and the Arakan Yoma perhaps the most difficult of all barriers, a tropical rain forest, sharply notched by some of the world's deepest gorges. India seems almost completely sealed off. But this is not true of any part of the earth, and is particularly not true of India. It has had varied and vigorous relations with other places, by land and by sea, relations between India and the west and north, and between India and southeast Asia. The mountain barriers around India have natural openings, and through these passes have come fundamental influences on Indian civilization. Their variety has created an enormously complex civilization, and has largely prevented the regional unity one might have assumed from looking at a physical map. The easiest passes, however, are all through the mountain ranges on the west and northwest. There are not only a number of relatively easy passes (including the Khyber and the Bolan), but also a narrow coastal route between the southern end of the mountains and the Arabian Sea and the Persian Gulf. A long succession of conquerors, invaders, and other cultural influences has come along the coastal route or through the many passes entering India from the northwest. Thus Indian civilization in many respects is related more closely to the West than to its neighboring cultures in the rest of monsoon Asia.

Except for the largely one-way influences India exerted on Indonesia (or the East Indies) and on Burma, Thailand, and Indochina, there was much less interchange between India and the east than between India and the northwest. Indian art forms, Indian trade, some Indian political and social institutions, and Indian Buddhism and Hinduism spread

MAP 52

LANDFORMS AND STATES OF THE INDIAN SUBCONTINENT

into southeast Asia, but there was little influence on India from these areas. China's civilization was much more highly developed than southeast Asia's and at least on a par with India's. But India exchanged very little with China in historical times. It took virtually nothing and sent only the knowledge of cotton and Buddhism (with some of its religious art forms). As useful and important a product as cotton did not come into general use in China until about the thirteenth century A.D. under the influence of the widespread Mongol empire. Cotton had spread west from India perhaps two thousand years earlier. Only because a few intrepid Chinese pilgrims dared the difficult journey across deserts and mountains or by sea through the Strait of Malacca did most of the Buddhist sacred books reach China several centuries after their origin in India, and also several centuries after China first learned of Buddhism.

Rice and the water buffalo may have originated in the Bengal region (now West Bengal and Bangladesh), or possibly farther east in the Indochinese peninsula, and spread from there to China. But this was well before the beginning of recorded history, probably during the Neolithic period. There was some trade by sea between India and China from the second century A.D. on, and occasional tenuous contacts across the Himalayas or through Sinkiang. But interchange between the mature civilizations of India and China was remarkably small. Partly this lack of contact reflected the lack of strong interest by both the Chinese and the Indians in commercial or cultural exchange with other people or other places. But the formidable nature of the mountain, desert, and rain-forest barriers separating the two countries undoubtedly contributed to their virtual isolation from each other. At the same time, communication through or around the mountains on the west allowed a much fuller relationship with Western countries. This has been true for as far back as Indian history can be traced.

The Indus valley civilization and the early invasions

The earliest city civilization discovered so far in India is in the Indus valley, which gave its name to India—it means simply "river." This city civilization appears to have been in existence by at least 2500 B.C. and to have been in contact with Mesopotamia (the Tigris-Euphrates valley). From its well-planned city centers at Harappa and Mohenjo-Daro (see Map 53, page 202), based on a complex irrigation system in this arid area, the Indus culture apparently extended its influence into parts of the Ganges valley and as far south as modern Ahmadabad. The collapse of this highly developed, literate, urban civilization predated slightly the Aryan invasion of India. The invasion was apparently one arm of the great migration of Aryan-speaking peoples, presumably from somewhere in west-central Asia, that went into Europe and the Mediterranean basin at about the same period.

"Race" and the Aryans

Aryan is a linguistic term, although it has been used to indicate race. Race is a much abused word. Technically it means a group who share common physical characteristics based on a common genetic inheritance. The word is often misused to mean a nation or a social or cultural group who may or may not share common physical characteristics. Using "race" in this way is especially erroneous because no race is completely pure. All individuals and all social groups have mixed blood, although the degree of mixture may vary. Therefore, it is highly misleading to refer to the Chinese, or German, or Italian race. In the case of the Aryans, we know that they, too, were very mixed physically. The principal basis for distinguishing them from other groups and for relating them to those who invaded Europe is their language, an early form of Sanskrit. Sanskrit, in turn, is the direct ancestor of most modern north Indian languages. In its pre-Sanskrit form, Aryan must also have preceded Greek, Latin, and most of the languages of modern Europe, since they have clear connections to Sanskrit. This is the source of the name for the language family that includes all these tongues, Indo-European.

Although the Aryans were technically and culturally less advanced than the earlier builders of the Indus valley civilization, they learned rapidly from it and brought new vigor at a time when the old civilization was deteriorating. Ashes and other signs of sudden destruction in the ruins of Indus valley cities suggest that its already ebbing life was finally ended by violence. Possibly the earliest Aryan arrivals were involved.

MAP 53
ANCIENT AND MODERN INDIA

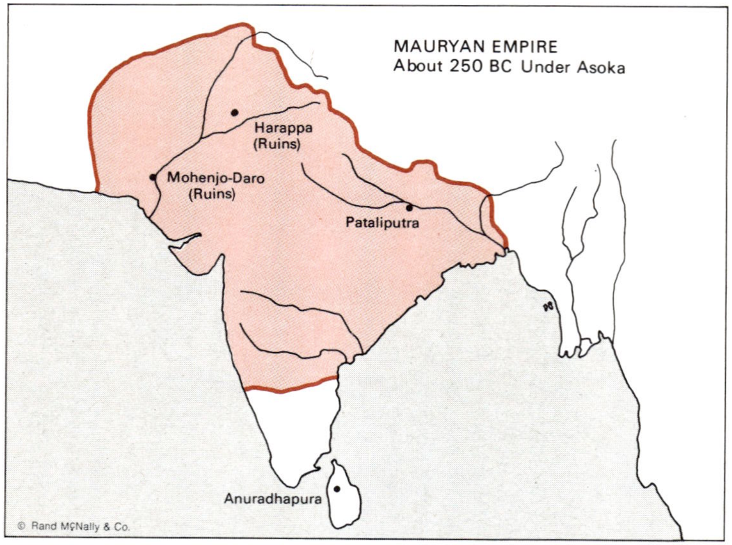

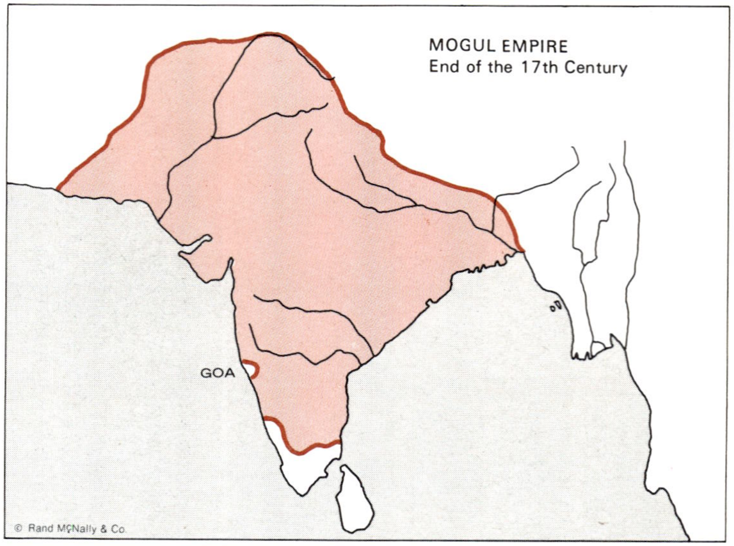

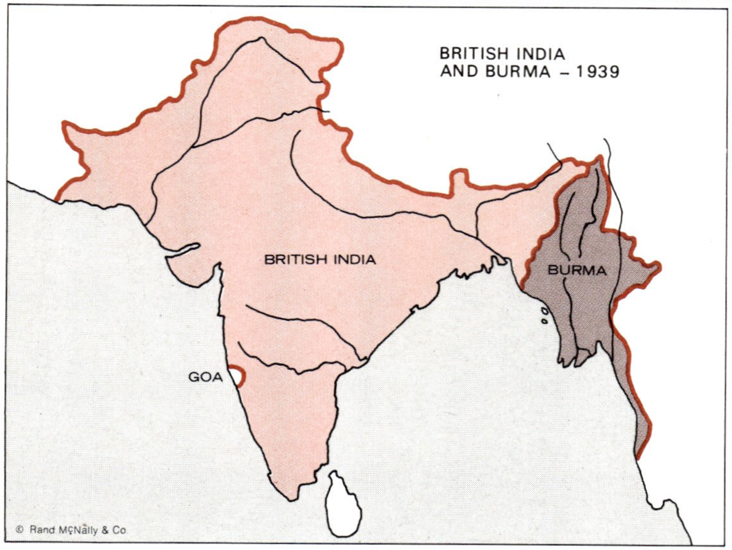

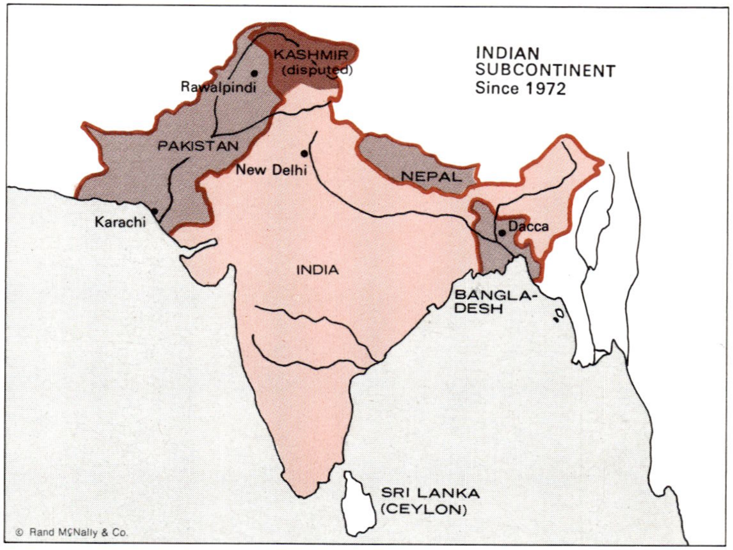

Invaders and empires

The Aryans were part of a long succession of invasions from the West, but the first for which there is precisely datable historical evidence is the invasion of Alexander the Great in 326 B.C. Alexander had a strong sense of mission to spread Greek culture and to widen cultural and commercial interchange over all of the then-known world. His empire did not long survive his death, but the Greek imprint was strong in India. Alexander's invasion and the smaller kingdoms that succeeded him were followed shortly by the first great Indian state, the empire of the Mauryas. The Emperor Asoka (about 274–236 B.C.), by conquest and acculturation, ruled over two-thirds of the subcontinent. The empire's core was the central Ganges valley; the capital was at Pataliputra (modern Patna). It also controlled the Punjab, the Indus valley, most of central India, and large parts of the south (see Map 53, page 202). Asoka's conversion to Buddhism and its doctrine of nonviolence put a stop to his conquests. Otherwise they might have encompassed the whole peninsula and Ceylon. But his conversion was also partly responsible for an even wider spread of Indian culture with Buddhism. Asoka dispatched missions that established Buddhism in Ceylon and Burma, and less successful missions to the eastern Mediterranean. Buddhism is the dominant religion in Sri Lanka and Burma today, although it has almost disappeared or been absorbed into Hinduism in India.

Buddha lived in the sixth century B.C. Born Prince Siddhartha, he relinquished his princely heritage in a small state of the Himalayan foothills to become a wandering seeker of truth. His teachings of reverence for life and the way to enlightenment and release from life's troubles through contemplation and denial of earthly desires were not widely accepted until Asoka aided their spread. Later, Buddhism and Hinduism were carried to most of southeast Asia by Indian traders. Buddhism is still the dominant religion in Burma, Thailand, and Cambodia. In Indonesia, the Indian religions were replaced by Islam, brought in by Arab and Indian Muslim traders. Hinduism still remains on the island of Bali. Buddhism was also taken to China by Chinese pilgrims, and from China to Japan by Japanese pilgrims. After a few centuries, it became a minor religion in China, compared with Confucianism and Taoism. Buddhism is still strong in Japan.

Disunity and regional variety

The Gupta dynasty of the fourth century A.D. revived a smaller measure of the Mauryan order. Its capital was also at Pataliputra, but it was short lived and fell before the invasion of the so-called White Huns (probably Iranians). They were one of several central Asian groups who have invaded India or set up kingdoms within it. India had never had complete cultural unity even under Asoka, and regionalism developed in full force after the end of his empire, except briefly under the Guptas, whose control extended south only to the Narbada (Narmada) River. Regionalism was strong not only because the Indian peninsula is a big place and contains many regional differences in physical terms, but also because of the wide variety of peoples resulting from India's repeated invasions.

The pre-Aryan inhabitants of India were doubtless a highly mixed group, although they may have been predominantly dark skinned. It seems unlikely that the Aryans were the first invaders of India from the west. Dark-skinned people may have come into the south from the Malayan region and even the Australian, Melanesian, and Micronesian areas of the Pacific. South India shows some evidence of this, and Negrito strains remain in the population. But such physical and cultural intermixtures over comparable periods of thousands of years are characteristic of virtually every part of the world. The invasions of lighter-skinned peoples from the west and northwest tended to produce a certain distinction in physical types over the peninsula. The north and northwest became increasingly dominated by lighter-skinned, Western-related people. The darker-skinned and original inhabitants (some of whom were lighter skinned) were increasingly concentrated, or pushed, to the south and onto the Deccan Plateau and its surrounding mountains. These original and darker people are broadly labeled *Dravidian*. Like Aryan, this is properly a linguistic term only, covering a wide variety of related south Indian languages, totally different from the Sanskrit-derived Indo-European languages of the north. However, Dravidian commonly means the varied groups of people who speak these Dravidian languages or share their cultures. The Dravidians are also a highly mixed group physically, as their languages now include many Indo-European elements. They have intermarried widely with the lighter-skinned invaders of India, and their physical types are scattered in mixtures all over the north. Similarly, northern and western physical types are scattered in mixtures all over the south.

Color and caste

Color variation among the people of India does help somewhat to strengthen regional distinctiveness. Perhaps, more importantly, it has fostered social distinctions. The caste system originally may have reflected a color bar between light-skinned invaders and darker-skinned conquered people. Occupational distinctions were an important part of early Indo-Aryan society. Priests and warriors were at the head of the hierarchy, and peasants, merchants, and laborers at varying levels below them. They were not necessarily caste distinctions, but they became involved with caste, and suggest separation between a conquering and conquered people.

Caste, as a social system into which individuals are born and must remain, has been a fundamental feature of Indian life (including Sri Lanka) at least since the Aryan invasions. Each caste usually is limited in the occupations it can choose. There are ritual observances and taboos, including eating together or sharing water facilities, and bans on marrying outside the caste. The caste system is slowly breaking down as a result of the widening and impersonalizing of economic and social relations in modern India, particularly with the increasing urban population. In cities, offices, and factories, people are thrown together closely and anonymously. But the village is the stronghold of caste, and most of the subcontinent's people are still villagers. The disadvantages of caste for the wider interaction and cooperation that must accompany economic and political development are obvious. Its persistence for so many centuries demonstrates its strongly accepted values as a method of social organization and identity in an often disordered Indian social and political setting. It cannot disappear overnight.

Caste is a basic part of Hinduism, but Indian Muslims and other non-Hindus, including Sri Lankan Buddhists, also practice and recognize caste distinctions. Hinduism (which literally means simply Indianism) is hard to define, since it lacks a definite body of dogma. It includes a wide variety of beliefs and practices developed over at least three thousand years, and draws on the epic Sanskrit poems of the Aryan invaders, among other texts. If caste was strengthened or at least more institutionalized and spread by the Aryan invaders, as seems likely, it is a further measure of the influences from the West. India's complex mixture of people seems to have required the superimposed order provided by caste.

Political fragmentation and the role of Hindustan

After the Gupta's collapse, political fragmentation began again. At different periods some of the states controlled sizable areas for relatively long times, particularly in the south. None of these states was able to control the north, or even significant amounts of central India. All of the greatest movements toward political unity have come from a northern base. The northern plain and the basin of the Ganges, sometimes called Hindustan, has been the dominant core of India since the downfall of the early Indus valley civilization. It is still so. This reflects Hindustan's advantages as the premier agricultural area of India, its largest densely populated region, and Hindustan's consistent position as the major line of internal movement and transport. The core area of Hindustan was less exposed to invasion than the northwest, and was more attractive as a center of political power. Its moister climate was also more favorable for agriculture than the climate of the Indus valley, once farming had spread beyond its earliest irrigated origins. Historically, the Ganges valley has been India's main economic nucleus. Successive invaders have established themselves there as the base for their attempts to rule the rest of the peninsula, as domestic unifiers have also done. Culturally, Hindustan is dominantly Aryan, but highly mixed, and has been profoundly influenced from the west on repeated occasions.

The strategic and military key to Hindustan and the rest of India is Delhi, which lies in a narrow lowland passage at a crossing place on the Jumna (Yamuna) River, a Ganges tributary, some 50 miles (80 km) west of the Ganges itself. Desert on the west, mountains to the north, and the Aravalli Range of mountains to the southwest give way through this river valley to the productive river plain and to easy routes south. Delhi was the control point that every invader and conqueror had to command, especially when they came from the northwest. Even the British ultimately had to recognize its importance by moving the colonial capital there from Calcutta in 1911. Delhi is also a conveniently central administrative location. Delhi has risen and fallen at least eight times as forts and capitals were established on this strategic site. Modern New Delhi, built by the British beside the older city, is the most recent in a long succession.

The Muslim contribution

Hindustan, however, has been heavily influenced by the last of the great conquests of India to be accompanied by mass migration, the Muslim advance. All of the previous invaders, although they contributed much to change or enrich Indian life, became thoroughly Indianized in time. This happened partly because Indian civilization was so superior, and in the long run they could only yield to it rather than destroy it. With the Muslims, however, the interaction took a different form. Considerable Indianization has occurred, but after eight centuries, it is not yet complete. Indeed it resulted in the separation of most of the Muslims from most of the Hindus in the newly created states of Pakistan and Bangladesh. Islam is a mature religion that has resisted erosion and dilution nearly everywhere it has spread in the world. It encourages a sharp cultural distinction and a vigorous communal attitude to a greater extent than any other faith. The incomplete Indianization also partly reflected the sophistication of the whole civilization the Muslim invaders brought with them. It was periodically stimulated by new infusions and interactions with Persia and Afghanistan. In effect they superimposed their own Muslim-Indian cultural mixture on the ruins of the Hindu civilization.

The Muslims carried an essentially Persian culture, although it would be more accurate to say that the Persian-Afghan invaders of India between A.D. 1000 and 1530 happened to be Muslims. Artistically, this Persian culture was probably the equal of the older Hindu culture (which may already have been

Village school—central India. India has a variety of physical types partly resulting from frequent invasions from the West. Even in this group of children, some might be mistaken for southern or eastern Europeans. Schools such as this one, attended by villagers of all ages, are common in rural India. Reading, personal hygiene, and new agricultural methods are being taught throughout the country. *(Photo by Homer Page for the Peace Corps.)*

Mogul art—exterior of the Red Fort, Delhi. This monumental building of red sandstone was built in the seventeenth century during the rule of Shah Jahan. This picture shows the Elephant Gate. *(Photo from ROLOC.)*

decaying), although such judgments are difficult. The Muslims destroyed Indian temples and monuments with the conviction that they could build something even finer. And their work at its best was magnificent, as the photographs on pages 206–9 suggest. The Muslim impact, especially in numbers, was concentrated in the north and northwest; only in the northwest did their culture win entirely. Throughout most of Hindustan, Muslim culture coexisted with Indian or Hindu elements without achieving a dominant position, except politically. The south was less disturbed by them, was never fully incorporated, and was held only at scattered intervals.

Splits and fusion

About the twelfth century, after a long period of Muslim raids from Persia and Afghanistan, a major Muslim effort conquered virtually the whole of Hindustan, as well as the northwest. The great Mogul Empire followed four centuries later (1526–1739; see Map 53, page 202). Early in the eighteenth century, the Hindu Rajputs and Marathas were nearly successful in challenging the Mogul rule. But the defeat of the Marathas led to British rule, fixing the existing balance of power and political divisions. India was a hodgepodge of states and cultures, especially in the south. Nowhere did India approach national or geographical unity. British rule was confined to administrative supervision and economic control, and did not attempt to create the full political and cultural union that India had been unable to achieve in the past. This is now taking place for the first time, under independence, in the separate states of India, Pakistan, and Bangladesh.

This brief and generalized sketch of Indian history provides some background for understanding the cultural variety within the subcontinent, the vital importance of India's historical relations with areas to the west, the jumble of regional differences, and the difficulty of generalizing about the present pattern of human occupance.

Major regional divisions

Certain major regional divisions within the subcontinent can be made, however, with reasonable consistency and accuracy (see Map 52, page 200). Forgetting the political divisions into India, Pakistan, Bangladesh, and Sri Lanka, large-scale distinctions may be drawn roughly as follows:

1. *The northwest,* with its core of settlement in the plain of the Indus River, is based on intensive irrigated agriculture, especially wheat and cotton. A wide fringe of arid hills lies on the west and northwest, and desert on the south and southeast. This area was the earliest center of Muslim culture, as well as the site of the ancient pre-Aryan civilization.

2. *The Punjab,* a transitional, mainly steppe region, lies between the northwestern edge of the Ganges valley and the northeastern tributaries of the Indus. It is bounded on the north by the Himalayas and on the south by the Thar (Great Indian) Desert and arid Rajasthan. Punjab means "five rivers," those that drain into the Indus system through the Punjab (the Sutlej, Beas, Ravi, Chenab, and Jhelum, in order from southeast to northwest). Its population has increased rapidly in the last hundred years as irrigation has spread. The leading crops are wheat and drought-tolerant sorghums and millets. The eastern half is moister than the western, which is dependent on irrigation for any major extension of cultivated area. Punjab is Muslim-dominated in the west, but predominantly Hindu and Sikh in the east.

3. *The Ganges plain,* or *Hindustan,* is a distinct lowland corridor and river basin extending from the vicinity of Delhi on the northwest to the joint delta of the Ganges and Brahmaputra in Bangladesh and West Bengal. It is the central core of dense settlement in India, based on wet-rice agriculture in its southeastern half. Jute is a commercial crop in the delta area, but wheat and other drought-tolerant cereals predominate in its drier northwestern half. Rainfall increases progressively toward the sea: Delhi is arid; Calcutta, lush. The low hill area known as Chota Nagpur, south of the Ganges, is richly endowed with coal and iron ore. The lower Ganges plain also contains India's major manufacturing region, especially heavy industry. Iron and coal are combined here with main transport routes and market concentrations. A chain of cities follows the Ganges River from Delhi to Calcutta.

4. *The Himalayan fringe* includes Kashmir (Jammu and Kashmir), the only significantly settled part. Most of the small total population is concentrated in the valley of the Jhelum River, an Indus tributary.

5. *Assam and the valley of the Brahmaputra,* part of the hill and mountain barrier on India's eastern frontier, is largely isolated from the rest of the country. Its population was very small until the development of tea plantations within the last century. Excessive rainfall, periodic flooding, and limited navigability on the Brahmaputra made this valley less productive and less attractive to settlement than the Ganges valley. Lowland Assam is a prominent rice-growing area. Upland Assam, where most of India's tea plantations are, still shelters numerous tribal groups, many of whom resent Indian sover-

Mogul art—interior of the Red Fort, Delhi. An inscription on the wall near this beautiful seventeenth-century screen reads: "If there is a paradise on earth, it is this, it is this, it is this." Notice the channels under the door to carry running water throughout the palace, for beauty and cooling. *(Photo by R. Murphey.)*

eignty. In 1960 the central government granted statehood within the Indian Union to the largest tribal group, the Nagas. This new state, Nagaland, occupies a strategically important area along the border with Burma and China. Upland Assam is a pioneer area for Indian settlement, based on tea and dependent on labor recruited from overpopulated Bengal and other crowded areas. Assam's Khasi Hills run at right angles to the main path of the summer monsoon, making Cherrapunji one of the world's rainiest places, with a yearly average of 428 inches (1090 cm), and 905 inches (2300 cm) in one record year.

6. *Central India,* an upland region between the Thar Desert and the Ganges plain on the north and the higher Deccan Plateau on the south, is a cultural zone of transition between the Aryan-Muslim-dominated north and the Dravidian-Hindu south. Uplands and isolation have also helped to preserve numerous pre-Aryan tribal groups. The area is sparsely settled by Indian standards and produces a variety of crops, wheat and sorghums being the most important.

The Mogul style—the Taj Mahal at Agra. No photograph can do justice to this magnificent seventeenth-century building. Despite its bulk, in reality it appears to be floating. Notice that water is an important part of the plan of the Taj, in the form of fountains and pools that lead to the building itself. *(From ROLOC.)*

7. *The Deccan,* a geologically ancient plateau, is more or less clearly defined by mountain ranges on the east and west (the Eastern[1] and Western Ghats), and by the Tapti (Tapi) and Godavari rivers and the Satpura and Vindhya ranges on the north. On the south the plateau gives way to broken hills and the coastal plain of the peninsular tip. Parts of the northern Deccan are covered with a fertile black volcanic soil known as *regur.* But most of the Deccan is semiarid, or has considerable variation in rainfall, in part because it lies in a rain shadow cast by the Western Ghats. Cotton on the fertile regur and drought-tolerant sorghums and millets, such as jowar and bajra, on the less good soils are leading crops. Mountain barriers and an upland base helped to insulate Dravidian cultures and states on the Deccan against invasions from the north. The Cardamom and Nilgiri hills west and south of Madras near the tip of peninsular India are more rugged than the Deccan. They are more closely related regionally to the southern coastal plains. Tea plantations, resorts, and tribal groups are characteristic of these hills.

8. and 9. *The western and eastern coastal plains* of peninsular India are really two separate regions meeting at the southern tip of the peninsula. Together they support the greatest population concentrations of southern India and two of India's four biggest cities, Bombay and Madras (Calcutta and Delhi are the other two big cities). The warm, moist tropical climate favors agriculture. The major crop of the lowlands is rice, with spices, oil seeds, and tea in the fringing hills. These coasts have been the traditional site of India's trading states, and were the major bases from which traders and missionaries carried Indian civilization to southeast Asia. Trade and cultural relations with the West were carried on

[1]Strictly speaking, *Eastern Ghats* is a misnomer; *ghat* means a narrow pass, which is characteristic of the Western Ghats. The more rounded eastern ranges contain many broad valleys.

through many ports on the west and southeast coasts, especially the southern half of the west coast called Malabar. This area was important from about the third century B.C. through the establishment of the first Portuguese settlement at Goa early in the sixteenth century. Malabar's small and silted harbors are increasingly awkward ports in the modern period.

Bombay and Madras were both developed much more recently, and did not become important ports until well into the British period. They were built to handle the expanding colonial trade of the nineteenth century. Bombay in particular grew rapidly as the main port link with Europe. It is the finest natural harbor in India, almost the only natural harbor adequate for modern shipping. Like New York, Bombay occupies a long narrow island about the size of Manhattan, surrounded by a large and well-protected harbor for ships. It has also expanded onto the adjacent mainland. Cotton from the Deccan area and the northwest also helped to make Bombay India's dominant center of textile manufacturing, and the second largest city. The harbor of Madras is largely artificial, constructed by the British to serve the trade of southeast peninsular India.

10. *Sri Lanka,* although physically similar to south India, has developed a distinctive culture. Under Western control, it added a plantation agriculture of tea, rubber, and coconuts to its subsistence rice cultivation.

Traditional Hindu architecture—Temple of Nataraja in Chidambaram, south India. Traditional Hindu temples in the north were largely destroyed by the Moguls, but many remain in the south. *(Photo by R. Murphey.)*

Larger north-south distinctions

The Indian coastal plain, especially on the west, allowed Aryan influences to penetrate farther southward than on the Deccan Plateau. But, generally speaking, peninsular India as a whole is distinct in important respects from Hindustan and the northwest. The line between them is often drawn along the Vindhya Range in central India, north of the additional barrier of the Narbada River (see Map 52, page 200). But there is a wide zone of progressive change and intermixture rather than a line. A cultural shatter zone extends over south-central India and the northern Deccan, involving a mixture of tribal, Aryan, and Dravidian elements. The Satpura Range, south of the Narbada and north of the Tapti River, presents a further barrier. The Vindhyas, the Satpura, and the two rivers help to draw a regional distinction. Although the mountains are not particularly high, they are wide and very rugged. Their barrier character is revealed in the persistence of aboriginal tribal groups who still inhabit parts of the Vindhyas. Many of these tribes are apparently unrelated to the people of north or south India. The relatively narrow valleys and high banks of the Narbada and Tapti rivers strengthen this isolation. These demarcations coincide reasonably well with the cultural transition

Warm, humid, south India. The coastal plain of Kerala has a tropical climate and is often used for growing rice. In the background of this picture are the Nilgiri Hills. *(Photo by P. Gosling.)*

between Aryan north and Dravidian south. To the south, Dravidian languages are dominant everywhere except in Maharashtra, the skin color is generally darker, and the social structure is slightly different from the north. Most of the south originally had a matrilineal and matriarchal society, for example, which is still preserved along the Malabar Coast. Hinduism is nearly universal, and a distinctly Indian culture is common to both north and south, at least in its broad outlines.

Language and regionalism

Within such broad outlines cultural distinctions are most apparent in the bewildering variety of languages in India. There are at least a dozen separate major languages and literally hundreds of dialects, each associated with a particular area (see Map 54, page 211). No single common language exists, except English. As a result, English is to be more or less indefinitely retained in India as an official *associate* language for certain purposes, particularly higher education and some of the central government business. The official language, however, is Hindi (including its many dialects such as Rajasthani), which is spoken by about 35 percent of the Indian population, and is also the dominant language of the northern area. There are many regional dialects of Hindi, including Punjabi, for example, and a wealth of local variants often not mutually understandable. A standard Hindi is taught in all schools as a matter of national policy. It is thus of increasing importance as a second language in non-Hindi-speaking areas.

Each language group understandably clings to its distinctiveness. The political map of India was redrawn in stages by the central government between 1954 and 1960 to give regional expression to these distinctions in the face of demands that were frequently violent. Map 52 (page 200) shows the new system of linguistically based states. Comparing it with the language map will show that compromises were necessary, not only in the major language areas, but also considering the many dialects and lesser languages too numerous to show on the map. Such division is a blow to Indian cultural and national unity. It was agreed to only with regret by the central government, since it may foster communal and regional rather than national feeling. But the linguistic aspect of regional cultural differences in India was considered so important that concessions had to be made to it. Some local autonomy and recognition of the regional language within a framework of state government was granted. Some regional subdivisions are obviously necessary, in administrative terms. In a country the size of India, the state units, whatever their basis, must be given considerable responsibility for local government. It was hoped that by drawing state lines largely on a linguistic basis, more coherent political regions would be created and a more harmonious national whole constructed. Of course, linguistically based lines do not always coincide with regional distinctions based on economic patterns, physical regions, or efficient administrative units. Language is perhaps the most emotional and most important basis of culture. Its political recognition by the central government, despite the adoption of Hindi as the official language, was probably inevitable, given the strength and variety of language differences in India. In time Hindi may come into general use, although it seems unlikely to displace the regional languages or the Indian version of English as the common language of the educated groups.

MAP 54
MAJOR LANGUAGES OF THE INDIAN SUBCONTINENT

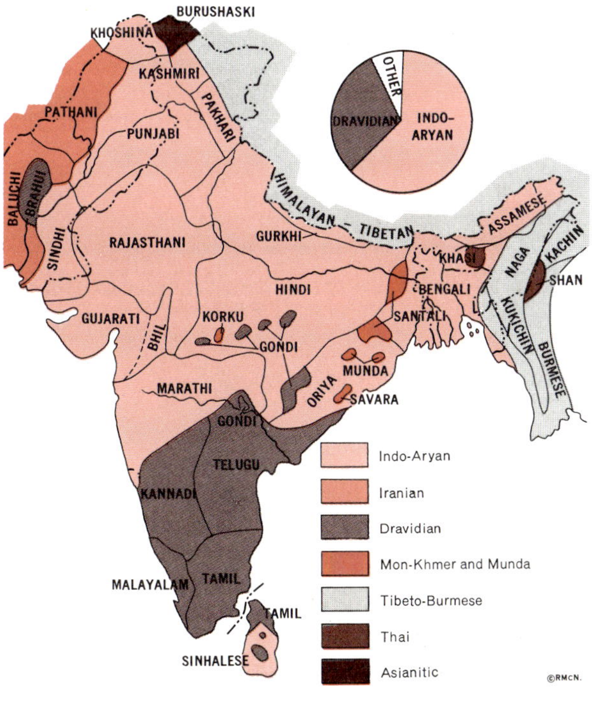

Sri Lanka and Pakistan also have a language problem. As in India, it has been a continual cause of intergroup tensions in both states. Sinhalese, the Aryan tongue of the majority of the Sri Lankan population, has been declared the official language. But Tamil is spoken by about one-fifth of the people, immigrants from south India over the past fifteen centuries and more. The official language of Pakistan is Urdu, almost identical with Hindi in its spoken form, but written with a different Persian script. But Urdu is the native tongue of a minority of Pakistan's population, and language-based rivalry accentuates regionalism. Bangladesh fortunately is free of such linguistic tensions since Bengali is spoken by virtually all its inhabitants. Throughout the subcontinent, although an unofficial language, English remains strong among the elite.

Force of outside influences

These human differences reflect the troubled and varied history of the subcontinent and the powerful influences that have poured in from outside. India might be compared in this respect with the United Kingdom, but there is a crucial difference. In Britain, although outside influences and conquests were periodically overpowering and fundamentally altered British culture many times, actual invasions came to an end in 1066. Britain was able to exercise a degree of choice in the subsequent influences that came in from the Continent and elsewhere. By Elizabethan times it had created a coherent and unified national culture. Even the Celtic outposts, where cultural distinctions still remain, were drawn into the greater English whole. They forsook their own languages for English, and with each century their cultures and economies became less and less distinct. India is much bigger than Britain, and regional distinctions are correspondingly greater and more numerous. But, more important, India has never been left alone long enough to develop a cultural or national unity. In fact, the subcontinent is comparable with Europe in its size and cultural diversity. European unity, cultural or political, has never been achieved, even under the Romans. Its progress in recent years toward more common economic and political institutions has been slow and painful. The lack of Indian unity in the past is understandable and not remarkable in those terms. Actually, in cultural and political terms, for the past three or four centuries India has been more of a unit than Europe has ever been, despite India's greater physical diversity. The achievement of national independence in 1947 began the development of national unity on a deeper and wider basis than ever before. Commercialization, the spread of education, urbanization, and industrialization joined with the new vigor of nationalism to bring India's people closer together in a common effort and a common world of livelihood and interchange.

12 The subcontinent decides and plans

Cap maker in Kashmir. *(Photo by Rupinder Khullar for Van Cleve.)*

The partition of India creating Pakistan, and, later, Bangladesh, has been widely criticized. British India had developed as an economic unit. Although there were cultural differences between Muslims and Hindus, the two groups had coexisted for several centuries. A political split could be expected to heighten tensions and differences, and would impose economic disadvantages for both sides. Muslims had been a minority within India, despite their history of conquest. As independence from Britain approached, there was a growing Muslim movement to establish a separate state on the announced basis of Muslim religion. The movement was too strong to be denied, and the unique state of Pakistan was established at the time of Indian independence in 1947. It was unique because of its religious basis, but also because, until the creation of Bangladesh, it existed in two pieces, separated from each other by a thousand miles (over 1600 km) at the nearest points. These were the two areas in which Muslims were numerically predominant: the northwest and the Indus plain, and the eastern half of Bengal. The spread of Islam in Bengal, separated by Hindu areas from the center of Muslim strength in the northwest, was connected to the early Arab trade relations by sea with the mouth of the Ganges. Bengal was exposed to greater foreign influence, including the Arabs, because of its commercially attractive position at the mouth of India's greatest trade highway, the Ganges.

The Muslim east

The boundaries around the eastern part of Pakistan, which later became Bangladesh, were drawn almost entirely on a religious basis, and excluded cosmopolitan Calcutta, which was Hindu. Calcutta was the dominant market and processing center for eastern Bengal's products, especially jute, and the chief central place for goods and services. The economic waste involved in the partition is typified by the duplication of jute factories and jute-growing areas in both halves of Bengal. This duplication is repeated in many sectors of the subcontinent economy.

Eastern Bengal (*Bangladesh* means simply "Bengal country") occupies fertile deltaic land in a broadly permissive climate with ample rainfall and high temperatures, and is densely populated (see Map 55, page 214). Income levels are low even by Asian standards. The area is seriously overpopulated in terms of present techniques of resource use, as is often the case with highly productive agricultural areas in monsoon Asia. The Ganges and the Brahmaputra and their numerous mouths and tributaries meet in eastern Bengal to form an immense delta. The land has been drained and canals built to aid agriculture, but it is still subject to periodic flooding. Most transport still moves by water.

Rice production in the Bengal area is very large, but is inadequate to feed the population, so there is often a food deficiency. Jute is an important commercial crop in the agricultural system, and occupies land that otherwise might be used for rice. The Bengal area has almost a world monopoly of jute, which thrives half-submerged in river mud in the delta and yields well in a hot, wet climate. Jute's importance here rather than in other similar physical areas, such as the lower Congo or the Amazon delta, is related to the large supply of low-wage hand labor available. Jute has very high labor demands, especially for the separation of the fiber from the stalks. Machine methods have not proved entirely satisfactory for this job. Jute, used for making sacking and carpeting, is now the leading export of Bangladesh. The area lacks mineral resources, and has no significant development of machine manufacturing. Before independence, it had been economically interdependent with the rest of India. Agricultural raw materials had moved to other Indian centers for manufacturing, or to Indian ports for export, with other Indian goods sent into the area.

East Pakistan was only about one-sixth of Pakistan's total area, but had over half of the total Pakistani population. Its people pressed for a larger voice in the national government and a larger share of the budget. In 1971 the voters of East Pakistan elected a regionally based political party to power. The central government in West Pakistan tried to suppress this action, and a civil war followed. With help from the Indian army, the new state of Bangladesh emerged. But the problems of overpopulation and economic underdevelopment remain. The Bangladesh government at Dacca has to manage one of the world's poorest countries, periodically threatened by famine. Bangladesh ironically is both productive and poor. Its rich agricultural base barely supports a population that has outgrown its capacity to feed itself with its existing technology. Further increases in output are still possible with new investment in fertilizers, better seeds, and water control, but such developments have been hampered by political instability.

MAP 55
POPULATION OF THE INDIAN SUBCONTINENT

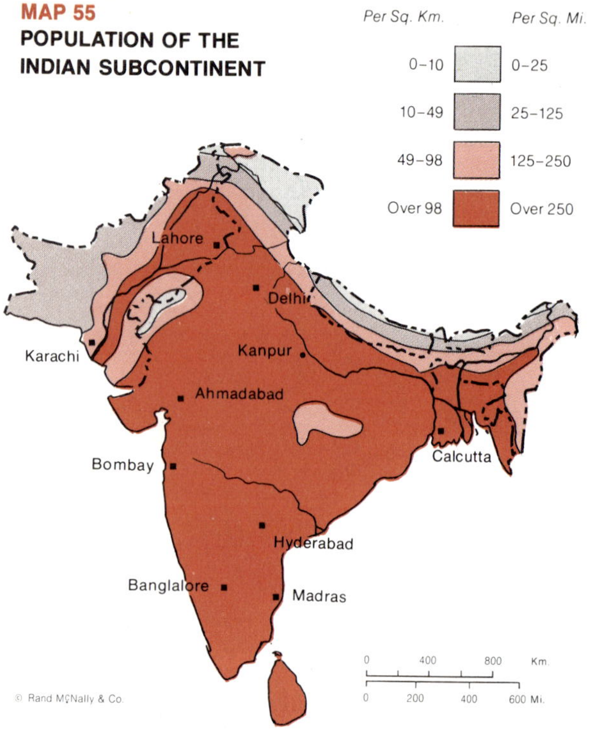

The Muslim northwest

West Pakistan (now Pakistan), the traditional cultural and political center of Muslim India, was the principal base for the movement to create a Muslim state. The Pakistani capital was placed at Karachi, near the mouth of the Indus. Formerly a relatively small city (less than half a million at the time of independence), it more than doubled in size within ten years as it took on new administrative functions. Lahore is a much older and more important center of Muslim culture. But it was inconveniently near the new border with India, and also lacked Karachi's easy communications with the outside world. However,

Karachi became seriously overcrowded. So beginning in 1959 most of the capital functions were moved in stages to Rawalpindi in the foothills of the Himalayas. Rawalpindi's climate is much pleasanter than Karachi's, and it also has the advantage of being close to Kashmir, where sovereignty is still disputed between Pakistan and India. A new capital, Islamabad, was built on an elevated site about 15 miles (24 km) from Rawalpindi.

Most of Pakistan is essentially a desert traversed by the Indus and its tributaries. Large-scale, intensive agriculture based on irrigation, mainly wheat and cotton, is carried on. The southern border with India caused few problems; the Thar Desert creates a population gap between the two states (see Map 55, page 214). But in the north, where Muslim and Hindu have intermixed more widely across the Punjab, it was different. Somewhere through the Punjab, the line between Pakistan and India had to be drawn. It is an area of religious and cultural as well as physical transition. Muslims were in a slight majority in the Punjab as a whole. But there were also some 5 million Sikhs (originally a Hindu sect), and the eastern sections were predominantly Hindu.

Unfortunately, the time of partition was an occasion for widespread riots and mass killings of one religious community by the other, especially as Muslim refugees left the Indian part of the Punjab or Hindu refugees left Pakistan. Violence and bitterness spread, and for one unreal period in 1947, it seemed that much of the subcontinent had gone mad. Apart from the killings, there may have been as many as 15 million refugees moving in both directions. The human misery at this time is difficult to comprehend. After it was all over, 50 million Muslims continued to live safely and happily in India, and some 10 million Hindus in Pakistan and Bangladesh.

Before the partition, what is now Pakistan regularly produced a wheat surplus for export, and was also a major cotton-exporting area. At that time it also sent its agricultural raw materials to the rest of India for manufacturing or export, receiving Indian goods in return. But the area that became Pakistan largely lacked mineral resources, and there was very little manufacturing. A series of good agricultural years helped Pakistan to survive after partition and gave it a favorable balance of trade. This provided some capital for investment in manufacturing and technical improvement of its agricultural system. The program of industrial investment was given new encouragement by the discovery of major reserves of natural gas and some oil at Sui in Baluchistan near the northwest frontier, about 75 miles (120 km) north of Sukkur, site of a large irrigation dam on the Indus. There is very little coal, and mineral ores (except for chromite) are also scarce. A small iron and steel plant at Multan makes use of some domestic as well as imported ores.

Pakistan has nearly 75 million people, but income levels are low. They will probably rise more as the efficiency and amount of agricultural production rises than from increased manufacturing. Investments in extending irrigation (including Punjab), in fertilizers, agricultural equipment and methods, and in market facilities are probably more logical than investment of scarce and expensive capital in heavy manufacturing. Production of textiles and other light consumer goods for the large domestic market may have a more rational basis. But dangerously little progress has been made toward raising the critically low level of per capita incomes, toward significant industrial growth, or toward agricultural improvement.

Pakistan was plagued by dissension between its eastern and western halves, by political disorder, and by a weak development of public responsibility on the part of its citizens and leaders. The severe economic problems caused by partition have continued to retard economic growth and to stimulate dissatisfaction. Heavy military expenditures also have been a crippling economic burden. After more than ten years of political wrangling and ineffective governments, the attempt to establish representative constitutional government was abandoned in 1959, when Pakistan came under the apolitical rule of the commander-in-chief of the army, General Ayub Khan. Under his rather authoritarian administration, some progress was made toward economic growth and the creation of some sense of national community, including the involvement of a larger part of the population in village and regional councils. But Ayub's bypassing of the political processes to which Pakistan, like India and Sri Lanka, had become accustomed under British rule was increasingly resented. He was forced to resign in 1969. The disastrous misadventures leading to the separation of Bangladesh followed his resignation. With the east-west conflict resolved, Pakistan is attempting to

MAP 56
TEA AND WHEAT PRODUCTION IN THE INDIAN SUBCONTINENT

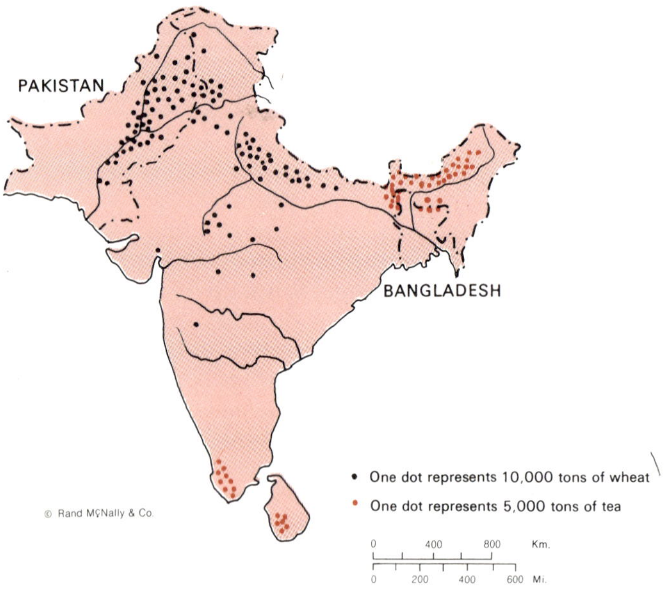

MAP 57
RICE PRODUCTION IN THE INDIAN SUBCONTINENT

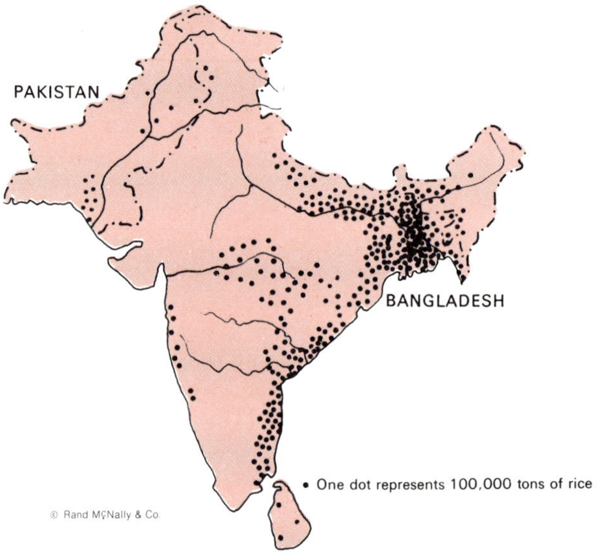

MAP 58
JOWAR AND BAJRA PRODUCTION IN THE INDIAN SUBCONTINENT

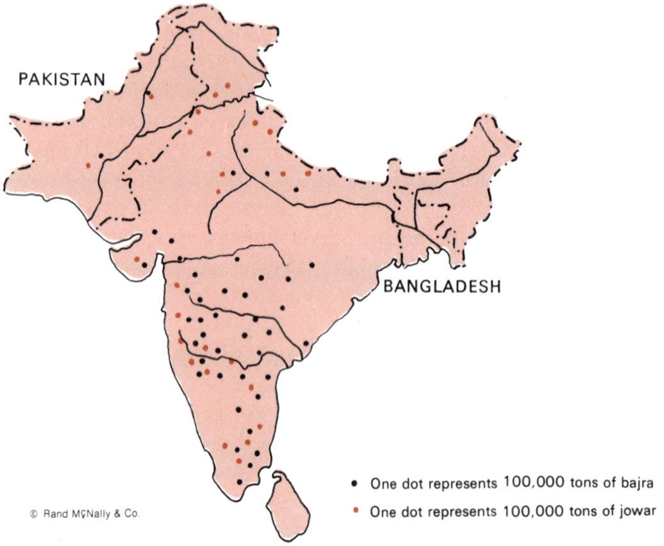

MAP 59
COAL AND IRON ORE IN THE INDIAN SUBCONTINENT

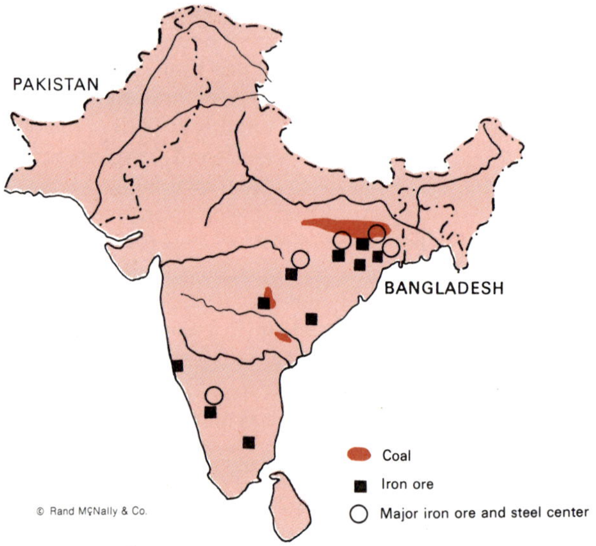

pursue its own development, still under authoritarian direction and with its original parliamentary institutions periodically eclipsed.

Kashmir

Relations between India and Pakistan have been clouded and embittered by the dispute over Kashmir. The comparatively small economic importance of this area has been magnified by such rivalry. Pakistan, as an essentially desert state, has feared Indian control of the major sources of the Indus, which are in Kashmir. Most of Kashmir's population is Muslim, but the ruler at the time of partition in 1947 was a Hindu, who agreed to join India. Military forces from both sides entered Kashmir. India has occupied most of it, including the important Vale of Kashmir in the Jhelum valley, which Pakistan protests. Continued failure to find an acceptable solution to this dispute has often poisoned relations between the two states. There are many aspects to the dispute; a solution to the water problem was agreed to by both governments in 1960. It gave India the use of the three eastern tributaries of the Indus, and Pakistan the use of the three western ones. But the dispute over territorial sovereignty remains. Obviously, their mutual interest requires them to live together closely and harmoniously, which they have not yet achieved. Military buildups by both sides, with American equipment and financial help, have worsened the problem, and re-examination of American policy may be desirable.

Indian industrialization

Iron and steel

In the Republic of India manufacturing has a larger proportional place in the economy, and its role is certain to increase. Heavy manufacturing is concentrated in southeastern Bihar and southwestern West Bengal, less than 150 miles (240 km) west of Calcutta, also a major manufacturing center. The Chota Nagpur district, including the valley of the Damodar River, contains most of India's easily worked, high-grade coal deposits, and virtually the only deposits of good coking coal (see Map 59, page 216). Fortunately, a major reserve of cheaply extractable, high-grade iron ore occurs nearby in the area. The iron and steel plants that have developed at Jamshedpur and Asansol have one of the lowest costs for total assembly of raw materials of any plants in the world. Jamshedpur, site of the largest iron and steel plant in India, is about midway between the iron and the coal, and Asansol is right at the coal. Both are advantageous locations. Large iron and steel plants have been built at Durgapur in West Bengal, Raurkela in northwestern Orissa, Bhilai in Madhya Pradesh (central India), using local coal (see Map 59), and at Bokaro in Bihar, close to the major Damodar coalfield. Chota Nagpur and the Damodar district are becoming an Indian Ruhr. Nearness to Calcutta is an advantage for labor, capital, market for products, and access to more distant markets and supplies by sea. Location on the edge of the Ganges valley also puts these industrial operations near the center of the Indian market as a whole. They are close to India's major transport routes, although problems with transport gaps and shortages still remain. The chain of older cities along the Ganges is becoming increasingly industrialized in metallurgical as well as textile and other production. These cities are major buyers for the iron and steel produced in the coalfields of Bihar and West Bengal. India is unusually rich in high-grade iron ore and may have larger reserves of good-quality ore than any country in the world. Several major ore deposits in central and south India are just beginning to be used on a significant scale.

Fuel and power resources

Coal is much less plentiful than iron ore. Fortunately, most of the best-quality coking coal is in the Chota Nagpur area, where an integrated program is improving navigation, flood control, and hydro-power production in the Damodar valley. This program is part of an overall agricultural, industrial, and commercial development of the river's basin, similar to the TVA in the basin of the Tennessee River in the United States. There is some coal in central India (see Map 59, page 216) and smaller deposits are scattered elsewhere. But the total reserves of coking coal, including the Damodar fields, are critically low. For the present, cheap, good-quality coal for metallurgical coke in the Damodar valley offers a low-cost resource base favorable to heavy manufacturing. But some estimates suggest there is less than twenty-years' supply left at the expected rate of use.

Petroleum production is still relatively small, although more fields may be discovered. The major fields so far are in Assam and in Gujarat on the west coast, but they do not produce enough even for India's present modest consumption. Lack of petroleum is less important to the national economy than presence of coal and iron ore, but it will make industrialization more expensive. The sharply increased costs of oil imports in recent years has created a major problem for India, because oil is an important base of chemical fertilizers. Fortunately, what may be major oil reserves have been outlined off the Gujarat coast. By the 1980s India may become self-sufficient in oil.

Hydro power

India does have impressive potentials for hydro power (or *hydel* as it is called colloquially in India) from its fringing and internal mountains. Seasonal and yearly variations of the monsoonal rains cause problems. But snow meltwater from the Himalayas, and heavy rainfall from the Assam hills and the Western Ghats are more reliable. Development of this potential power is especially logical in view of India's coal position. Bombay, with its large population, major industrial development, and proximity to power sites in the Western Ghats, is the largest single consumer of hydro power by far. The hills of south India are beginning to produce significant amounts of hydro to feed growing industrial centers like Bangalore, Coimbatore, Mysore, and Madras. Hydro in the Damodar basin makes sense because of the local market and the proximity of Calcutta. There is also considerable production of thermal electric power from Damodar coal. Hydro developments in central India and in the Himalayan fringe north of the Ganges valley are planned on the basis of a rapidly growing market from continuing industrialization, commercialization, and agricultural improvement. This is the basis, for instance, of the huge Bhakra Dam about 175 miles (280 km) due north of Delhi on the Sutlej River (an Indus tributary) and similar projects along the southern edge of the Himalayas to the southeast.

High-tension lines are spreading across the plain of Hindustan, making a strange contrast with peasant villages and bullock-drawn plows. But hydro power alone could not justify the dams. Like most large dams anywhere, they are multipurpose, providing water for irrigation and general use, and helping to control floods, as well as producing power.

Urban India—street scene in Bombay. Bombay still reveals its Victorian past. Note the variety of transport methods: human, horse, and motor, and the different languages used in the signs. *(Photo from Trans World Airlines.)*

Development of water power is expensive. Although it may be an inexhaustible resource, an economy still in the early stages of industrialization finds investment capital scarce and many competing uses for it. The shortage of coal and the concentration of its production in one area, with high transport costs to supply the rest of India, make hydro development especially worthwhile, however. Water for irrigation may remain more important than power for some time to come, but cheap power universally available can be a great stimulus to increased agricultural efficiency and production as well as to large- and small-scale manufacturing.

Problems of industrial development

Of the lesser mineral resources, India has much bauxite (aluminum ore), manganese, and chromite (important in steel alloys). Mica has been nearly an Indian monopoly, but is of relatively small importance. Minerals used in nuclear fission (such as beryl, monazite, and uranium), with which India is fairly well supplied, may be of much greater importance in the future.

The first nuclear power station, near Bombay, began to feed power into the commercial grid in 1969. This and the beginnings of an Indian satellite program already in operation suggest some of the contrasts in most developing economies, where agriculture or transport, for example, may include technologically retarded sectors despite a space-age industrial structure. Most haulage in India is still by bullock cart, even while Indian satellites orbit the earth. India has the advantage of size and of a long tradition of scientific and technical accomplishment begun under British colonial rule. Its small but highly educated intellectual elite is on a par with any in the West. Consumer goods, especially textiles, still predominate over heavy manufacturing or producer goods, however. Bombay is the largest single textile center, with important production also in Ahmadabad (north of Bombay), Kanpur (in the Gan-

Smaller towns in India. These pictures show a sample of the smaller towns in India. In the north, crumbling walls surround some towns, and throughout the country there are bullock-drawn carts and wandering cattle. Camels are used occasionally for transport. *(Photo on left by Warren Garst for Van Cleve; on right by P. Gosling.)*

ges valley), Madras, and Calcutta. Calcutta is still the major center for jute weaving, but cotton textiles are also important there. It also has widely diversified manufacturing in many fields, including shipbuilding, engineering, railway equipment, and food processing. Indian planning gives greater emphasis to producer goods than to consumer goods in order to accelerate industrialization, although this involves a clear risk of inflation as well as sacrifices in future consumption to maximize future production.

New heavy industries are being concentrated in the Bihar–Orissa–West Bengal district where iron and steel manufacturing is already established. There is also some new development in central India, based on local coal, and in southern India based on local iron ores and hydro power. Manufacturing costs in India are often higher than favorable resource location alone would suggest, partly because of high transport and labor costs. This is closely related to undercapitalization, that is, to inadequate use of equipment, machines, and power that are initially expensive to buy. Inadequate mechanical equipment makes it necessary to depend on large amounts of poorly trained or unskilled labor. Traditional artisan labor is usually highly skilled, but modern factory labor may have relatively limited skills beyond the knowledge of tending machinery. Many industrial societies have largely forgotten the art of human labor and the standards of craftsmanship. Machines can produce a much greater volume of goods at a lower unit cost than the artisan can produce. In such a situation labor skills become simply mechanical skills, relatively quickly and easily learned. Under any circumstances mass factory production with inadequate mechanization will result in high labor costs. However, rational decisions must be based on the choices that are open. Where capital is scarce, it may be necessary to use more labor.

If industrialization is to succeed on a larger scale, low-wage, high-cost labor must be replaced by mechanically skilled labor, additional machinery, and greater capitalization. This development is only beginning. Indian textiles do compete with growing success on the world market. But the essential heavy manufacturing for an industrial economy (including the capacity of the railways) is still small in relation to the size of the total population, and the domestic market is still limited by very low income levels. Capital is usually expensive when it is scarce, and high interest rates are charged for its use. This retards economic development of all kinds. But capital or savings in themselves do not produce economic development. They can be used productively (to add to the economy's power of production) or unproductively (as simple accumulation). Large amounts of hoarded capital may retard economic development; smaller amounts of capital invested in machinery accelerate development.

Small-scale and handicraft industry

In the early stages of India's industrialization, some small-scale rural industry, in the old handicraft tradition, will be maintained. Some rural industry will also be provided with new machinery and electric power and more efficient organization, similar to the Japanese (see Chapter 15). One of the economic blows of the nineteenth century was the weakening of many handicraft industries all over Asia resulting from factory competition. Originally this competition came from Europe, and later from domestic production. Some small-scale handicraft industry did survive and may even have benefited from the changed conditions, for example, the availability of cheap machine-spun cotton yarn for home weaving. But other traditional hand industries, including the hand spinning of cotton, were weakened or wiped out, bringing economic distress to those groups who had depended on them for part or all of their employment and income. The current plan to preserve some small-scale rural manufacturing has been the source of much controversy in Indian planning circles. The economies of increased size, greater capitalization, and market centrality give a commanding economic advantage to large urban factories in almost every type of production. Local or traditional specialties in art or artisan goods and some local processing of agricultural or other raw materials may be exceptions. The wearing of homespun cloth became a badge of patriotic nationalism during the long struggle for independence, as a protest against the importation of British textiles. Emotional values have become attached to the preservation of handicraft industry, especially cotton spinning, which Gandhi made into a powerful nationalist symbol.

The economic argument for maintaining small-scale rural industry is to help to absorb underemployed labor resulting from the seasonal pattern of agriculture. Rural industry can absorb this labor from the farms more readily than can city-based manufacturing or services, and it can help to raise per capita incomes, which is the central objective of economic development. There are cultural and other, less tangible losses involved in the passing of most traditional artisan production and its replacement by grimy, impersonal industrial cities. In economic terms, only when labor is fully mobile within the economy and there are plenty of jobs for everyone can small-scale rural manufacturing be allowed to decline to the level at which it can compete naturally.

Economic growth and agriculture

India's industrial prospects depend more on economic problems than on mineral resources. Most essentially subsistence economies must raise themselves by their own bootstraps if they are to industrialize or increase production and average income levels. Increases in agricultural efficiency, and production and creation of an adequate low-cost transport network are clearly essential to the creation of large-scale, machine-powered manufacturing. The presence of mineral resources of good quality available at low cost can be a powerful help and may be an absolute necessity in many cases. But it does not guarantee industrialization by itself. India, like Pakistan, Bangladesh, Sri Lanka, and most of the rest

of monsoon Asia, and the rest of the "Third World" of developing societies as a whole, is still predominantly an agricultural country. Most of its labor force is employed in agriculture, and most of its income and goods are produced from agriculture. Agriculture is the principal source of surplus for investment, the chief potential tax base for the government, the major market, and the heart of the economy. Unless it can be made to function vigorously, industrialization is not likely to succeed on any significant scale.

Irrigation

The huge dams near the Himalayas, at Hirakud (near Sambalpur) on the Mahanadi River in Orissa, on all of the major rivers of south India, on the Indus and its tributaries in Pakistan, and at the Gal Oya project in eastern Sri Lanka are designed principally for irrigation. Irrigation allows new land to be cultivated, minimizes drought risks, and maximizes production from existing acreage, which has always suffered from the unpredictable monsoonal rains and the normally dry winter. India's greatest agricultural handicap is drought that occurs nearly everywhere except in Assam and along the west coast south of Bombay. High temperatures, especially the terrible heat that precedes the monsoonal rains in the late spring, parch and brown the landscape. The heat also brings high evaporation rates, and makes a delay in or inadequacy of the monsoon doubly unfortunate. The focus of the Indian village and farm is the well or pond from which people, animals, and fields are sustained. Even the palaces of the emperors were built around water showing its practical and ceremonial importance. One of the oldest figures in the Indian pantheon, probably of pre-Aryan origin, is Yakshi, goddess of the life-giving waters, whose suggestive form is depicted in many fine samples of Indian sculpture. Modern India is following an old tradition and an old and pressing need in building water-control projects. The Bhakra Dam, the biggest, irrigates something like 4.5 million acres (1.8 million hectares), in addition to its power production. It may be the largest irrigation project in the world, although strictly comparable measures are difficult, depending on the degree of irrigation provided. Major dams are supplemented by smaller-scale local projects, particularly by the digging of wells and the restoration and expansion of existing storage tanks (reservoirs) for rainwater and stream water. In Sri Lanka and south India the tank system of water storage between monsoons is extremely old, and is being revived and expanded. In north India and Pakistan, wells are a more important local water source since rainfall is scantier. With government planning and assistance, new and deeper wells are being dug and provided with power-driven pumps or with *Persian wheels,* endless chains of buckets powered by a camel or ox.

Water in rural India. *Left,* carrying water from the village well is one of the most important of the daily household chores. The size of the jar varies with the carrier, and most are borne on the head. Every water jar has a lip at the top so it can be lowered into the well on the end of a rope. *Top,* the village well is the chief social center, especially for women. The well is also used for washing cooking pots. The tree, watered by spills from the well, provides welcome shade on the flat, dry, north Indian plain. *Center,* this counterbalanced log is not a very efficient means of raising water from the shallow well, but it is typical of traditional, peasant irrigation methods. *Bottom,* using a Persian wheel for irrigation is a great improvement over the traditional method shown above, and it is easy to construct. The bullock or camel providing the "power" is often blindfolded. It walks around the circle, turning the wheel that keeps the water jars filling. When they get to the top of the wheel at the left, the water spills out into a channel, and into the fields in a steady stream. *(Photos:* left, *Van Cleve;* top and center, *P. Gosling;* bottom, *David Campbell for Van Cleve.)*

Agricultural techniques and land reform

Community development projects have been set up all over India to teach and demonstrate to the farmers the use of better seeds, better types of crops, better plowing, better equipment, better methods of fertilization, better breeds of animals, and more rational and productive use of the land. Credit facilities, essential to all farmers since their cash returns come in at the end of harvest season and may be long delayed, are being expanded and cheapened. Marketing is being improved so agricultural goods may be sold more widely at lower cost to the consumer and with greater profit to the producer. Sri Lanka, Pakistan, and Bangladesh are carrying out similar programs. In all the countries of South Asia, special attention is being given to the areas where agricultural and social conditions are favorable for development. These are the areas that have benefited most from the so-called Green Revolution, the development of higher-yielding varieties of wheat, rice, and other grains. Unfortunately, these new varieties require larger amounts of water and fertilizer to sustain their yields. Only some areas and some farmers can provide these scarce and expensive new requirements. One result of the Green Revolution has been a further and potentially disruptive widening of the already large gap between rich and poor areas and farmers.

Government efforts to change traditional village cultures and techniques have not always been fully effective and have often met with strong peasant inertia. Change on the scale required for significant economic growth will necessarily take time and a good deal of trial and error in government planning, as well as commitment at the grass-roots level instead of direction from the top. Some reforms are also being made in the pattern of landholding, especially in India, Bangladesh, and Pakistan. The ownership structure inherited from the British period often left landlords, both large and small, in a strong position. The British unwittingly strengthened landlordism and exorbitant rents by creating the *zamindars,* originally tax-collecting (or more accurately, tax-farming) officials who abused their power, pocketed much of the tax revenues, and acquired land from defaulting taxpayers. "Land to the tiller" is now the desired goal, especially since it gives incentives to the peasants to increase their production and to improve their land. The zamindars have been gradually and peacefully abolished by buying their excess land for resale to the peasants. But, unfortunately, little progress has been made on a broader scale toward a greater degree of ownership by the cultivators. Landlessness and tenancy are high and still increasing. This weakens cultivator incentives to raise production as well as creating still more inequality. Only a minority of farmers have done well from the Green Revolution, and it is relative deprivation that hurts.

Traditional Indian agriculture

Emphasis on agricultural improvement is especially necessary in the subcontinent. Population is still increasing, agricultural improvements are necessary for industrialization, and traditional farming methods have been especially inadequate. Agriculture in India is much less intensive than in China or Japan, yields have been very much lower, and much potentially productive land was neglected or underused. In Sri Lanka, Assam, and parts of central India until relatively recently, this may have been due partly to the absence of severe population pressure by Chinese or Japanese standards. For most of the subcontinent, however, the reason is more likely to be found in the area's chronically troubled history and in its overall moisture shortage. Indian peasants have often been accused of fatalism and of disinterest in improving their material circumstances. In fact, in the past they have acted rationally in avoiding investments of labor or capital that might be destroyed during the repeated invasions and internal disorders to which most of the subcontinent has been subjected periodically. Irrigation in particular requires heavy investment in fixed structures. The experience of most of rural India until the last century clearly discouraged all permanent investment in land improvement or in other economic activity. This was especially unfortunate, given the pressing need for irrigation, particularly in the Indian climate. It goes far toward explaining both the relative primitiveness of most traditional Indian agriculture by East Asian standards and its low level of yields. "One for crab and one for crow, one to die and one to grow" runs a traditional Tamil chant at seed-time. Traditional expectations were low, but not irrationally so. Perhaps the most encouraging aspect of this situation is that it is relatively easy to improve such poorly developed agricultural technology, especially with the addition of irrigation now taking place on a very large scale. Water has always been the critical limiting factor. As irrigation plus more nearly adequate fertilization can be supplied, yields can be expected to rise dramatically, as they are now beginning to do in the Green Revolution.

The agricultural record during the last half century of British rule was discouraging, since the British were responsible for large increases in irrigation and new cultivated land, and made attempts to improve peasant farming techniques. Despite these efforts, there was only a very slight increase in yields per acre up to 1947, and only a modest increase in total output (perhaps about 10 percent, including nonfood crops; output of food crops alone may have declined slightly). Consequently there was a definite drop in agricultural production per capita, since population increased by over 25 percent between 1891 and 1947. Since then, efforts at agricultural improvement have been intensified, irrigation facilities have been further extended on a very large scale, and village development projects have probably had a much greater effect on peasant techniques and on yields per acre. But the record of the last fifty years of British control suggests the difficulty and the urgency of the problem in the face of a still rapidly growing population. The traditionalism and resistance to change of the Indian peasants must give way, as they are now beginning to do, to a willingness to adopt new and better methods with the conviction that a rise in living standards is really within the peasants' powers. The problem has been lessened by the success of the new high-yielding strains of wheat, rice, and other grains developed by research and experiment in the 1960s. All four governments have made great efforts to spread the use of these productive varieties. In years of favorable weather, India and Pakistan have begun to produce agricultural surpluses. Their record of overall agricultural growth since 1950 compares favorably with that of most developing countries, including China. The still unsolved problem is that population has continued to grow only slightly less rapidly, and gross inequalities remain in the distribution of goods and income. However, since irrigation and heavy fertilization are necessary to maintain high yields, especially from these new varieties, as well as to ensure against recurrent drought, much more new investment will be necessary before the agricultural problem can be completely and permanently solved. There is still inadequate protection against drought when the monsoonal rains are late or below average. In such years—at least one or two every decade—the entire economy falters and millions of people reach the edge of starvation. Bangladesh suffers not from low yields, but from an excessive and still growing population, so it has chronic food shortages. Sri Lanka's agriculture is dominated by plantation production of tea and rubber, so it must import much of its food.

Making dung cakes for fuel. This is a familiar village sight throughout India. Animal dung is carefully collected, usually by women and children, mixed with dust or straw, and patted into round cakes. They are then stacked to dry in the sun for use as cooking fuel. *(Photo by R. Murphey.)*

Religion and the cattle problem

The frequent unwillingness to use night soil as fertilizer and the preservation of millions of useless cattle have contributed to the relative inefficiency of traditional land use. Pigs are scarce for religious reasons; sheep and goats (kept mainly for wool and milk by Muslims) are most numerous in the areas too dry for farming. Cattle are revered by Hindus, as all life is revered, but especially as symbols of motherhood (connected with their milk production) and in Hindu mythology. Most orthodox high-caste Hindus and Buddhists are vegetarians. Many of them will not eat even eggs or fish, and keeping animals for meat and their slaughter for any purpose is traditionally forbidden. Protein is provided by milk, and especially by dhal (Indian lentils, a leguminous crop), which is very widely grown. But the total diet is still not adequate, although most lower-caste groups do eat meat, fowl, or fish. The traditional prohibitions of other groups appear to be breaking down, together with many other aspects of caste.

These scruples have resulted in a cattle population that is much larger than would be useful. India alone probably supports about one-fourth of the world's cattle population. A large proportion of the animals are unproductive, but they all eat and all represent a drain on the productive resources of the land. Their manure is carefully collected, but it does not wholly right the balance. Too often it is used for fuel or for plastering house walls rather than for fertilizing the starved soils. Manure is essential as fuel, in the absence of adequate supplies of wood or other fuels. Many of the cattle are too weak to serve as draft animals, although most farmers must have such services. In the few areas where good grazing is available or where the cattle can be fed on more nourishing crop residues, some of the draft animals are fine specimens, but most are not. In most areas even work animals have an inadequate diet. Most farmers have too few good animals for their essential needs, and the agricultural system suffers. Most cows give very little milk, and mixed, inferior breeds reproduce themselves freely. Gradually the government is trying to overcome this problem by controlling breeding, by growing more fodder crops, and by elimination or sterilization of excess animals. But it will take a long time to overcome or get around the religious objections. With more intensive cultivation, more rational land use, expanded irrigation, and technological improvements like better seeds and crops, manures, chemical fertilizers, and equipment, agriculture in the subcontinent can certainly increase its total production two or three times and still leave a large labor surplus for employment in manufacturing. Many peasants are already responding vigorously to new opportunities and are becoming innovators themselves, which is the only secure basis for development. The excitement of change and commitment to it are in the air, but mainly only a minority are benefiting so far.

Agriculture is most intensive in the lower Ganges valley and the delta, where population pressure and productive soil and climate are concentrated. Monsoonal rainfall is heavy there and far more regular, especially in the eastern half, than in the rest of the subcontinent. Although the winter is relatively dry, it does receive some rain, enough in Bangladesh to permit a winter crop. The growing

MAP 60

AVERAGE ANNUAL RAINFALL IN SRI LANKA

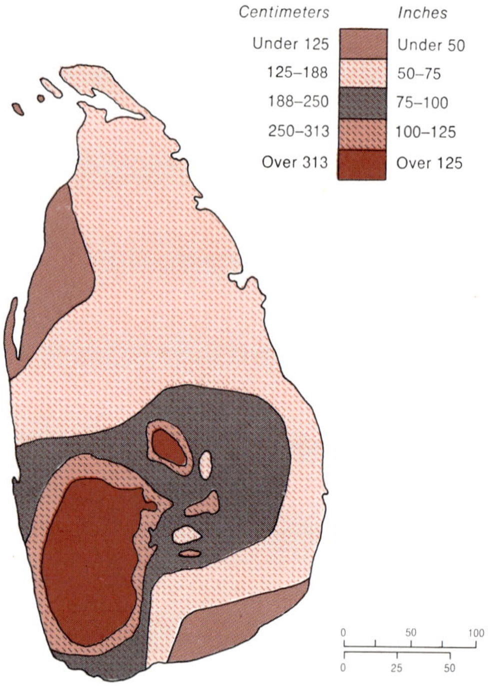

MAP 61

TEA AND RUBBER ACREAGE IN SRI LANKA

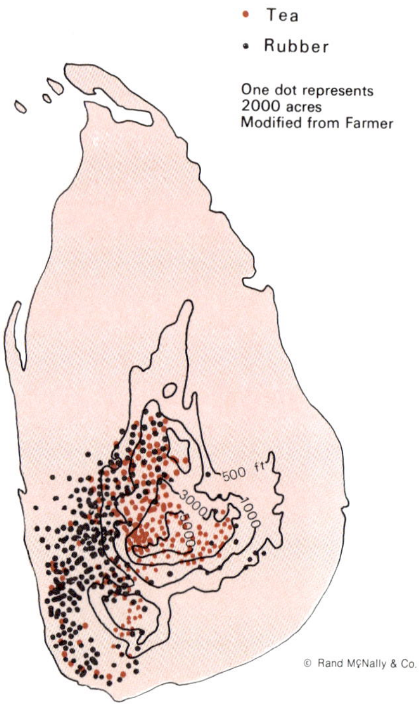

season is unbroken, as it is in nearly all of the subcontinent, but continuously available water is critically important. Rice may have originated in the swamps of Bengal, where some is still grown in water as deep as 10 feet (3 m) during the height of stream flooding. But in the lower Ganges valley as a whole, although the total amount of cultivated land is relatively large, the per capita share is only about half an acre. Per capita incomes are extremely low despite the consistently good yields of rice by Indian standards. The land is badly overcrowded, and there is considerable emigration to Calcutta and Assam. Cattle are plentiful—nearly one for every two people, but they are of low quality and poorly nourished, since pasture and fodder are scarce.

Population growth is both a national problem for each country and a regional one because the most densely settled rural areas tend to be the poorest in per capita incomes (see Map 55, page 214). Where the economic margin is small and constantly threatened, hopelessness may replace innovation and commitment to development. This problem is most acute in Bangladesh, where population densities are greatest and per capita income levels lowest. Since the 1971 separation from Pakistan, chronic governmental instability and repeated coups have increased its problems. Potentially, Bangladesh could easily feed itself and even produce surpluses for export, if it can also slow down the growth of its population. But this requires coherent national planning and organized group action, which have been lacking so far.

Sri Lanka (Ceylon)

Plantations (large cultivated units under single ownership devoted to the growing of one crop for commercial sale outside the local market) are largely absent in both India and Pakistan, except for tea growing. In Sri Lanka[1] they are far more common, and also exhibit the distinguishing feature of the plantation: the presence in each unit of some processing plant that prepares the agricultural product for sale. Although rice is the predominant food crop, plantation crops occupy nearly as many acres and account for about 90 percent of Sri Lanka's exports. Agricultural investment is concentrated in plantations, and Sri Lanka must import about half its food. Three crops dominate the island's plantation system: tea, rubber, and coconuts, in the order of their importance.

Approximately three-fourths of the island, which is called the dry zone (see Map 60, page 226), has rainfall that is inadequate for nearly all field crops without irrigation. Although average annual totals are high (50 to 60 inches [127 to 152 cm] over most of the dry zone), rainfall is highly concentrated in the short period of the monsoonal rains. The remainder of the year is likely to be too dry for crops to survive. Evaporation rates are also very high, since Sri Lanka lies close to the equator. Nevertheless, the dry zone was the site of the first advanced civilization on Ceylon, and population was largely confined there until about the thirteenth century A.D. An invasion of Aryan-speaking people about the fifth century B.C., probably from northwest India, conquered most of the island. They brought the language, culture, and technical achievements of late Aryan India, including the knowledge of irrigation. They became and remain the dominant group on the island, and are still in the majority numerically. Later, mass immigration of Tamil-speaking Dravidians from southern India formed a large minority group.

These Aryan-speaking people, called the Sinhalese, constructed an impressive system of tank and dam irrigation in the dry zone, and made it more productive agriculturally than it is now. Their ruined capital at Anuradhapura is still a striking monument to their artistic and technical achievements. Dry-zone soils are less leached than those of the wet zone, the land is a gently rolling plain rather than a jumble of hills as the wet zone is, and more sunshine also favors the growing of rice. But the whole system of government and society collapsed with dramatic suddenness in the thirteenth century. It may have been caused by the government's inability to provide the large and constant amount of forced labor necessary to build and maintain the irrigation works. Repeated invasions by south Indian kingdoms contributed to the government's weakness, and the dry zone was largely abandoned. Probably somewhat later the spread of virulent malaria in the stagnant pools and intermittent streams of the dry zone was a powerful bar to resettlement.

Originally a neglected and economically back-

[1] Sri Lanka, an ancient native name for the island of Ceylon, was made the official name for the country in 1973.

ward part of the island, since the thirteenth century, the wet zone has been the center of most of the population. In the course of the last hundred years especially, it has become the center of the plantation system on which Sri Lanka's economic well-being came to rest. Hills and mountains lie across the path of the major rain-bearing monsoonal winds blowing off the Indian ocean from the southwest in the late spring and early summer. They help to keep the wet zone wet and the dry zone dry. When the monsoonal flow is reversed in early autumn, and the winds blow out from Eurasia, they pick up some moisture in the Bay of Bengal and deposit it in the dry zone (the northeast monsoon, as it is called in Sri Lanka). The autumnal winds also penetrate into the wet zone to give it rainfall that normally leaves no part of the year seriously dry. Leached soils, shortage of sunshine, and, more importantly, of level land are handicaps for the growing of rice, but do not discourage tree crops. The higher elevations are used for tea, which prefers a cooler climate and constant moisture with good drainage (see Map 61, page 226). The lower elevations grow rubber and coconuts, interspersed with rice fields in the pockets of more fertile or alluvial level land. The British built a rail and road system to serve this plantation sector with port facilities at Colombo, through which virtually all external trade still passes. As the island's commercial economy, dependent on Colombo as its predominant base, has boomed, the city has grown. Sri Lanka's total population increased from less than 1 million at the beginning of British control (1796) to 1 million in 1850, 10 million in 1960, and over 14 million in 1977, mainly as a result of the booming plantation development. Since national independence in 1948, large-scale efforts are being made to resettle the dry zone. The ancient irrigation works are being restored and new ones added to increase food production in the dry zone to or beyond its ancient level. If this can be done, Sri Lanka can better feed a population that is still rising rapidly. Malaria was wiped out between 1945 and 1950, making resettlement of the dry zone possible. But it also led to a population growth rate that was one of the highest in the world, although, fortunately, the rate has decreased since the 1960s. Like other plantation economies, Sri Lanka's fortune depends on demand and prices in overseas markets over which it has no control. This is a further reason for trying to reduce food imports and build a better-balanced domestic economy.

Industrialization has little logical place in Sri Lanka, except for some processing of agricultural goods. Sri Lanka almost entirely lacks the essential mineral resources. It has virtually no coal, iron ore, or petroleum. There is a little graphite and some precious stones, which were once an important element in its foreign trade. Sri Lanka's domestic market is very small. Even rubber tire manufacture would probably not be profitable. Europe and America have economies of scale and the advantage of large markets, so they can make tires and sell them to Sri Lanka more cheaply than they can be manufactured there. Even in textiles, Sri Lanka would do better to buy from India than to attempt much domestic production. Specialization on the basis of comparative advantage is a sound guide for economic development, and Sri Lanka's comparative advantage is overwhelmingly agricultural.

On the whole, Sri Lanka's plantation system is efficient, and represents a highly productive use of the land. There is much room for improvement in traditional Sri Lankan food agriculture, especially rice. Population pressure has been relatively slight until recently; farming has not been intensive, and yields have been low. Until the government-sponsored agricultural intensification began, rice was seldom transplanted, seldom weeded, and inadequately fertilized and irrigated. Sri Lanka has been given a qualified breathing space in which to increase its agricultural production in both the wet zone and the dry zone before the population overwhelms the supply of food. Income levels are still about twice what they are in India, and there is much unused or insufficiently used land. But the time is limited, and the challenge is pressing.

Between two worlds—the British impact

These two chapters have dealt superficially with a huge expanse of varied land and people, about as large as and perhaps more varied than all of western Europe. Attention has been concentrated on the central theme of the subcontinent as lying between two worlds in its spatial interactions, its historical development, and its recent and current revolutionary

Lowland wet zone—Sri Lanka. Rice fields are typical of lowland Sri Lanka as well as south India. The soil is being thoroughly pulverized by driving the heavy buffalo, yoked to a crude plow, through it. This is a wide-spread practice in most of the rice-growing countries, and is the last stage in preparing the fields before the rice is planted. *(Photo from British Overseas Airways Corp.)*

changes. The final invaders from the West, the British, brought in full force the ideas and techniques of modern Europe. From their seventeenth-century trade base in Bengal, the British spread their control over the whole of the subcontinent by the second half of the nineteenth century. They superimposed on it the machinery of a commercial economy, new techniques of production, the example of Western political institutions, and the seeds of nationalism. British commercial activities and investments built the plantations of Ceylon and Assam to serve the markets of Europe. They built much of the irrigation works of the Indus valley and the Punjab, and the jute fields and mills of Bengal. They expanded the cotton acreage, and the network of railways and ports, far in advance of the rest of Asia, except for Japan, to serve their commercial economy.

Before the British period, Bengal and the delta area were not very densely populated and were of relatively minor importance within India. An enormous increase in population and trade resulted from the draining of much delta land, the stimulation of rice and jute growing, and the rise in service functions under the commercial impact of the British occupation. A similar change took place in Assam in the second half of the nineteenth century with the development of tea plantations by the British. Their extensions of irrigation through large dams and canals in the Indus valley and the Punjab to support commercial cotton and wheat growing were also followed by great increases in population and trade. British rule brought an end to chronic warfare and civil disorder. The spread of mechanized transport and unified administration greatly reduced deaths from famine and epidemic. United India had a population of about 150 million in 1800, 250 million in 1871, 433 million in 1951, 533 million in 1961, and approximately 647 million in 1970, with 112 million in East and West Pakistan and 535 million in the Republic of India.

Accompanying commercialization was the growth of a series of large trading cities, mainly created on sites largely neglected earlier. These cities were the service funnels through which the export goods were sent. Each occupied a commanding commercial location with access to the most productive areas. Calcutta, near the Ganges mouth, is the largest city of all, despite its poor harbor some 75 miles (120 km) from the sea on the Hooghly River, a minor delta stream. Bombay, the premier harbor, is the nearest to Europe and the center of the cotton trade. Karachi, near the Indus mouth, is the leading wheat and cotton port for the northwest. Madras has an artificial harbor on the largely harborless east Indian coast, and is the export and service center for the cotton, tobacco, and other products of southern peninsular India. Visakhapatnam (Vishakhapatnam), now a shipbuilding center, also has a mainly artificial harbor. Colombo, the only real city in Sri Lanka, manages the business and exports of the plantations clustered immediately around it in the wet-zone hills. Toward the end of British control, these cities also began to do some manufacturing, especially processing of the raw materials that moved through them. Textiles were, as usual, the first large-scale factory industries. But even heavy manufacturing got a start in India (with Indian capital earned in trade) at Jamshedpur nearly forty years before independence. The port cities became the largest by far, but Delhi became the imperial capital of India in place of Calcutta. It had been the capital often before because of its strategic and more central location, as well as its imperial tradition.

At independence the subcontinent was left with an economic structure fundamentally altered by the British, having the great advantage of a widely spread rail network as an aid in accelerating industrialization. The British legacy of parliamentary democracy, a free press, and impartial law has remained important in all four states, but so far only India has been able to maintain these institutions with full success. These are still critically poor countries, and their populations continue to grow. This adds to the pressures for authoritarian direction. But the basic battle must be against poverty.

Questions for further study and discussion

1. To what extent do language regions in India, new state areas, and broader-based regions agree with each other? How do you account for the discrepancies? The commission appointed to settle the new state areas and boundaries investigated the problem exhaustively for nearly two years, but their decisions were widely criticized. How would you criticize the final result, or what arguments would you offer in its favor?

2. Similarly large and physically even more diverse areas than India have achieved political and, in some cases, cultural unity. What are they? Why has each succeeded in this respect where India has not? Do you think India is likely to do so in the future? Why?

3. Why should productive agricultural areas like the Ganges-Brahmaputra delta become overpopulated?

4. Why should Jamshedpur and Asansol both be advantageous iron and steel centers? Why not just one center, since both use the same coal and ore?

5. Why should the Western Ghats cast a rain shadow over the Deccan?

13 China, the middle kingdom: splendid isolation

The Great Wall of China.

For centuries, and probably for millennia, China has been the most populous country in the world, containing nearly one-fourth of the population of the globe. China also still supports what is probably the world's oldest living civilization. China in 1000 B.C. was enough like China today to say that its civilized tradition has remained largely unbroken, by comparison with other parts of the world. There has been change, of course, in the roughly three thousand years since the beginning of regular written records in China; it could hardly have been absent in such a time span. But probably the greatest changes are those that followed contact with the West. These changes did not gain full force in China until the present century. Institutions were altered to some extent in the past, techniques improved, and population increased and spread areally, but a strong and conscious cultural continuity remained. Chinese settlement spread from an original core in the north into the south. There were also several movements of nomadic groups into China from the border areas nominally controlled by the Chinese state. Otherwise there were no invasions, little mixing of peoples or cultures from outside, and few strong or permanent institutional influences, except those generated within China. Even the Chinese written language, although it has evolved in three thousand years, is still recognizably the same. All writing probably began as pictures or ideographs. But outside of China and the cultures influenced by China, pictographs were discarded relatively early in favor of the much simpler phonetic representation of words by an alphabet. China retained the pictographs, as it retained so many other elements of its ancient civilization, in part because its relations with the rest of the world were extremely limited.

Relations with other cultures

Isolation is never complete, even in Tibet or Tahiti. China is a large country occupying a productive base near the center of monsoon Asia. This has helped to make it less isolated than many other places, certainly less so than Tahiti or Tibet. But China is surrounded by a series of physical barriers and by empty distances comparable to those around India. They have minimized China's interaction with the

70°
75°
80°
85°
90°
95°
100°
105°
50°
55°
45°
40°
35°
30°
25°
20°
KIRGHIZ
STEPPE
KAZAKH
S.S.R.
OZERO BALKHASH
Surface elev. 1112 Ft. above sea level
PESKI MUYUN-KUM
PESKI SARY-ISHIKOTRAU (DESERT)
Semipalatinsk
Ust'-Kamenogorsk
Karkaralinsk
Balkhash
Ayaguz
Zaysan
Urdzhar
Bakhty
Uch-Aral
Alakol'
Kapal
Lepsinsk
Taldy-Kurgan
Panfilov
Alma-Ata
Dzhambul
Chimkent
TASHKENT
Frunze
Tokmak
Ozero Issyk-Kul
Przheval'sk
KIRGHIZ S.S.R.
Kokand
Namangan
Andizhan
Fergana
Osh
Naryn
Pik Kommunizma 24 590
TADZHIK S.S.R.
PAMIRS
AFG.
Leninogorsk
Gorno-Altaysk
GORNO-ALTAY AUT. OBLAST
Marka Kul'
SAYAN KHREBET
Yenisey
Kyzyl
TUVA AUT. OB.
TANNU OLA
Ulaan Goom
Ubsa Nuur
Hirgis Nuur
Tesiin Gol
Cheremkhovo
Angarsk
Irkutsk
Ulan-Ude
Bratskoye Vdkhr.
Zhigalovo
Lena
Oka
BAYKAL'SKIY KHREBET
Munku Sardyk 11453
Khövsgöl Dalai (Koso Lake)
BURYAT
Petrovsk Zabaykal'skiy
Altan Bulag
Kyakhta
Eg'iin Gol
Selenge Gol
Orhon Gol
Lyung
Ulaan Baatar
KENTEI SHAN
Karakorum (Ruins)
HANGAYN NURUU (KHANGAI MTS.)
Jibhglanta
Ideriin Gol
Dzabhan Gol
Hara Usa
Hara Nuur
Jirgalanta
Hobdo Gol
Alot'ai
13 865 Tsast Bogdo Ula
ALTAI MOUNTAINS
MONGOLIA
Baydarag Gol
Ongiin
Ongiin Gol
Sayr Usa
TAILAGEN KHARA
GOBI (DESERT)
T'ach'eng (Chuguchak)
CHUGUCHAK
Ebi Nuur
Telli Nuur
Fuhai
Urungu
Ining
Wusu (Kweitun)
Manassu
Wulumuch'i (Urumchi)
DZUNGARIA
Ch'i'tai
T'ulufan (Turfan)
Shanshan
TURFAN DEPRESSION 505 Ft. below Sea level
Hami (Qomul)
Palik'un
TIEN SHAN
Khan Tengri 22 940
Wensu (Aksu)
Sulo
Wushih (Uch Turfan)
Kuch'e (Kucha)
Yench'i
Bagrash Köl
K'uerhlo
Kuruk Darya
Tarim
K'ashih (Kashgar)
Pach'u
SINKIANG UIGHUR AUTONOMOUS REGION
EASTERN TURKESTAN
TARIM BASIN
TAKLA MAKAN (DESERT)
Khotan
Yingchisha
Soch'e (Yarkand)
24 388 Muztagh Ata
Yehch'eng (Karghalik)
SUGET PASS
Hotien (Khotan)
Keriya
Yütien (Keriya)
Ch'iehmo
Cherchen
Salt incrusted bed of dry Lop Nor
ASTIN TAGH
Sulo Ho
Anhsi
Yümen
Chin-ta
Chiuchüan
NAN SHAN
Ochina Ho
SUIYUAN
NINGSIA
Old Beds of the Huang Ho
ORDOS DESERT
HUI AUT. REG.
ALA SHAN
Yinch'uan
Changyeh
Wuwei
Tsaidam
Swamp
TSINGHAI
Ch'ing Hai (Koko Nor)
Huangyüan
Hsining
Huang Ho
Lanchou
Chungwei
Pingliang
KUN LUN SHAN
(KUN LUN MTS.)
Gilgit
PAKISTAN
KARAKORAM RANGE
Mt. Godwin Austen (K2) 28 250
KARAKORAM PASS
JAMMU AND KASHMIR
Srinagar
Indus
Leh
Chenab
Rudok
Gartok
Amritsar
Jullundur
Sutlej
PUNJAB
Chandigarh
Simla
Ambala
Patiala
Dehra Dun
Saharanpur
HARYANA
Meerut
Almora
Moradabad
Rampur
DELHI
New Delhi
Bareilly
UTTAR PRADESH
Mathura
Aligarh
Shahjahanpur
Farrukhabad
Agra
Chambal
Gwalior
Lucknow
Faizabad
Gorakhpur
KANPUR
Allahabad
Varanasi (Banaras)
Jhansi
Betwa
Ghaghra
Gandak
Ganges
Yamuna
PLATEAU OF TIBET
THANGLHA
TRANSHIMALAYAS
TIBETAN AUTONOMOUS REGION
Seling Tsho
Nam Tsho
NYENCHHEN THANGLHA
Dhaulagiri 26 810
NEPAL
Kathmandu
HIMALAYAS
Brahmaputra
Zhikatse
Gyangtse
Lhasa
Yamdrog Tsho
Mt. Everest 29 028
SIKKIM
Kanchenjunga 28 208
Kula Kangri 24 784
Punakha
Tashi Chho Dzong (Thimbu)
BHUTAN
ARUNACHAL PRADESH
Gangtok
Darjeeling
Ch'angtu
Dre Chu
Yangtze Kiang
Lants'ang Chiang
Mekong
Nu Chiang
Salween
DIPHU PASS
Sadiya
Tinsukia
Sibsagar
Jorhat
ASSAM
NAGALAND
Kohima
Darbhanga
Patna
Monghyr
Bhagalpur
Gaya
BIHAR
Son
Sagar
Katni
Narmada
Jabalpur
Tropic of Cancer
MADHYA PRADESH
INDIA
Cooch Behar
Rangpur
KHASI HILLS
MEGHALAYA
Shillong
Silchar
Imphal
MANIPUR
Rampur Boalia
BANGLADESH
Dacca
Comilla
TRIPURA
Giridih
Asansol
Raniganj
Burdwan
Ranchi
Jamshedpur
WEST BENGAL
Howrah
Chandernagore
Bhatpara
CALCUTTA
Sundarbans
Mouths of the Ganges
Chittagong
Balasore
MIZORAM
Bilaspur
Raipur
Nagpur
Wardha
Mahanadi
Sambalpur
MAHARASHTRA
Chanda
Wardha
ORISSA
Cuttack
Godavari
Indravati
EASTERN GHATS
Puri
Berhampur
Warangal
BAY OF BENGAL
Myitkyina
Mogaung
Hsawnhsup
Chindwin
Bhamo
Mogok
Singu
Lashio
Monywa
Pakokku
Maymyo
Mandalay
Myingyan
Mt. Victoria 10018
Kaladan
Paletwa
Sittwe
Magwe
BURMA
Yamethin
Pyinmana
Kyaukpyu
Salween
Chiang Rai
LAOS
Luang Prabang
Muong Sing
Dien Bien Phu
Lao Kay
Black River
Red River
VIETNAM
Son Tay
Hanoi
Bac Ninh
Haiphong
Nam Dinh
Ninh Binh
CHINA
HSIAN (Sian)
CHIN LING
Fenghsiang
Lunghsi
Ank'ang
TAPA SHAN
Langchung
Tahsien
Chialing Chiang
Nanch'ung
Wanhsien
Fengchieh
Sant'ai
SZECHWAN
CH'ENGTU
RED BASIN
Hoch'uan
Chungshsien
Fengtu
Fouling
CHUNGKING (Ch'ungch'ing)
Yungshan
Yaan
K'angting
Paan
Loshan
Luchou
24 900 Minya Konka
Yalung Chiang
Ipin
Hsich'ang
Tsuni
Ssunan
Chaot'ung
Chihchiang
Huili
Weihsi
Lichiang
Tali
KWEICHOW
Kueiyang
Anshun
Tuyün
Lip'ing
YÜNNAN PLATEAU
K'UNMING (Yünnanfu)
Ch'uhsiung
Paoshan
Shunning
YÜNNAN
Lants'ang Chiang (Mekong)
Chienshui
Mengtzu
Wenshan
Tien Ch'ih
Hung Shui Ho
Ishan
Liuchou
KWANGSI CHUANG AUT. REG.
Szengen
Tienpao
Lungching
Na Cham
Yu Chiang
Nanning
Ningerh
Ssumao
Yenan

MAP 62
LANDFORMS OF CHINA
Longitude East of Greenwich
115°
120°
125°
130°
135°
140°
145°
150°
55°
50°
45°
40°
35°
30°
25°
20°
0 200 400 600 Km.
0 100 200 300 400 Mi.
SEA OF OKHOTSK
PACIFIC OCEAN
SEA OF JAPAN
YELLOW SEA
EAST CHINA SEA
PHILIPPINE SEA
SOUTH CHINA SEA
Tropic of Cancer
Bashi Channel
Korea Str.
Tsushima Str.
Formosa Strait
Tatar Strait
Po Hai
NORTH KOREA
SOUTH KOREA
JAPAN
CHINA
MONGOLIA
UNION
MANCHURIA
HEILUNGKIANG
KIRIN
LIAONING
INNER MONGOLIAN AUT. REG.
CH'AHAERH
HOPEH
SHANSI
SHANTUNG
HONAN
KIANGSU
ANHWEI
HUPEH
HUNAN
KIANGSI
CHEKIANG
FUKIEN
KWANGTUNG
TAIWAN (FORMOSA)
HONG KONG (Br.)
KURIL ISLANDS (Sov. Union)
SAKHALIN (Sov. Un.)
HOKKAIDŌ
HONSHŪ
SHIKOKU
KYŪSHŪ
NANSEI SHOTŌ
YABLONOVYY KHREBET
NERCHINSKIY KHREBET
HSIAOHSINGANLING SHANMO (LESSER KHINGAN RANGE)
TAHSINGANLING (GREATER KHINGAN RANGE)
SIKHOTE ALIN
CH'ANGPAI SHAN
TAPIEH SHAN
MEILING PASS
PEKING (Peiching)
TIENTSIN (T'ienching)
SHANGHAI
SHANGHAI-SHIH
NANKING (Nanching)
WUHAN
CANTON (Kuangchou)
Kowloon
VICTORIA
Macau
T'AIPEI
SEOUL (Sŏul)
P'yŏngyang
PUSAN
TŌKYŌ
YOKOHAMA
NAGOYA
KYOTO
KŌBE
ŌSAKA
KITAKYŪSHŪ
HAERHPIN (Harbin)
MUKDEN (Shenyang)
FUSHUN
CH'ANGCH'UN
Chilin (Kirin)
TSINGTAO (Ch'ingtao)
T'AIYÜAN
Vladivostok
Khabarovsk
Blagoveshchensk
Komsomol'sk-na-Amure
Sapporo
Sendai
Hiroshima
Nagasaki
Kagoshima
Naha
COPYRIGHT BY RAND MCNALLY & COMPANY MADE IN U.S.A.

rest of the world. Mountains and desert on the north and northwest, Tibet and the Himalayas, Pamirs, and Tien Shan (Tyan Shan, or Mountains of Heaven) on the west, and the mountain–rain-forest–gorge country on the southwest combine to make a semicircular wall. It was seldom penetrated by permanent or significant influences from outside. Beyond the mountains is the desert of central Asia. The sea also acted as a barrier until very recently. As important as physical barriers, however, was the friction of distance. China is enormously far away from Europe, and this alone kept interaction between them to a minimum. Even India was a long and difficult way from the eastern lowlands of China, where most of its people have always lived.

The southwest area was settled by primitive tribal people. Other neighboring areas were occupied by satellite cultures, like Korea or Annam (northern Vietnam), whose civilizations were echoes of China's. China felt it had little to learn from these neighbors. Japan borrowed heavily from Chinese culture, but had little to offer in return. In effect, these neighboring cultures held up a mirror in which the Chinese could admire themselves. India, to a large degree, was insulated from China, as discussed in the preceding chapters. Although the steppe corridor on the north and northwest was an avenue of movement into China and periodically an important route for foreign trade, its nomadic inhabitants were dismissed as barbarians. The thousands of miles of travel through steppe and desert before another sophisticated culture was reached minimized exchange of any sort.

The beginnings of China

Influences from the west may have been more important before the beginning of recorded history in China than during the historical period, until the present century. It seems clear that literate city civilization began earlier in Mesopotamia and India than in China. Some indirect knowledge of metal working, wheat, domesticated cattle, horses, sheep, goats, the chariot, and possibly writing and irrigation techniques may have entered China along the steppe and desert route of the northwest around 2500 B.C. or slightly before. Rice, the domesticated water buffalo and pig, and other agricultural innovations probably spread through the south from India or the Indochinese peninsula even earlier. Archeological research has started only recently in China, but the earliest city site with metal and written records found so far is dated roughly 2000 B.C. The evidence shows it to have been a relatively sophisticated culture, already highly distinctive, and not recognizably an offshoot of Mesopotamia. Indeed, there may never have been any significant connection. From about 1800 B.C., beginning with the Chinese dynasty of Shang, one is on more or less firm historical ground. In the approximately thirty-five centuries following, until the nineteenth century, outside influences made little permanent or significant impact on Chinese civilization. During the T'ang and Yuan dynasties (A.D. 618–907 and A.D. 1279–1368) China was unusually open and receptive to cultural influences from India, central Asia, and the West. The T'ang and Yuan capitals at Hsian (Sian) and Peking (Peiching) were cosmopolitan centers where many non-Chinese groups and ideas, including Christianity, were represented. But these influences found no lasting place in Chinese life, and in succeeding dynasties the traditional Chinese pattern was essentially unchallenged. Migrations of people into China were almost wholly absent from the time of the Shang dynasty, and China moved largely in a cultural orbit of its own.

The Middle Kingdom

China was acknowledged as the center of the civilized world by all of the lesser civilizations around it. The very name of the country—*Chung Kuo,* the Middle Kingdom—illustrates the inward orientation of the Chinese and their acceptance of the role as the source of light for less fortunate or less highly developed peoples. The Chinese were only dimly aware of the existence of higher civilizations elsewhere; the intervening distance and physical barriers made this a largely academic matter to them. In their own experience, throughout the world with which they had dealings, they had no peer. Most of the other states of east Asia periodically recognized China's leadership by sending missions of tribute, although the missions were often a respectable cloak for trade. From as far as Java (Djawa) and Malacca (the west coast of Malaysia), and even for a time from Ceylon, tribute and obeisance came to the Chinese emperor, the Son of Heaven and the acknowledged overlord of mankind. This was not the result of overgrown nationalism or military conquests, but of Chinese supremacy in culture and prestige, freely admitted by all who knew the Middle Kingdom.

Such a relationship did not encourage China to learn from others. Trade with the tributary states was relatively small in amount. Its effects were largely concentrated in the few maritime trading centers in southeastern coastal China. In techniques, institutions, ideas, art, or mundane products, China felt it had little to acquire from those who sent missions, or from its nomadic or primitive land neighbors. In time this attitude hardened into a haughty resistance against outside influences. By the nineteenth century, this resistance helped to prolong China's unhappy and hopeless struggle to preserve the illusion of its total superiority over the modern West. Whereas Japan was eager to learn, China refused for too long to admit that these new barbarians were different from the old ones. The high level of traditional Chinese technology and culture, including its productive economy and sophisticated political system, remained ahead of Western achievements by any objective measure until probably little more than a century ago. This long experience of superiority, periodically reinforced by contact with other and inferior cultures, makes it difficult for China even now to accept second-place status especially in relation to the industrial powers of Europe or North America, with their different systems of social and political organization.

China's northern cradle

China's *cultural imperialism,* as it has appropriately been called, began with Chinese civilization in the north. If agricultural and metal-working techniques were derived from indirect contact westward, they would probably have entered China through the northwest corridor. But it seems more likely they resulted from independent innovations by the Chinese. The physical characteristics of the north reproduce in essence the conditions under which civilization began in Mesopotamia, the Nile valley, and the Indus valley—a semiarid river plain of fertile alluvium where irrigation was possible.

The Shang dynasty was centered in the Yellow River (Huang Ho) valley close to the western margin of the great flood plain of north China where it gives way to hills and mountains. This permitted them to safeguard the productive plain against the periodic invasion of enemies from the hills and desert. It was also well placed to take advantage of the agricultural opportunities of north China. Broadly speaking, these opportunities were a highly fertile and easily worked loess[1] and alluvial soil, with a large river system as a potential source of irrigation water. Both the loess and the river's alluvium are very productive if they can be adequately watered. The north was more suited to agriculture in this early period than the moister but heavily forested areas farther south that resisted agricultural settlement for many centuries. Most of the north was too dry for extensive forests; the sparse tree growth that may have covered parts of it originally could be removed easily and did not reproduce itself. It is probable that the trees had been choked out earlier in large areas by the rapid depositing of loess by the wind. The absence of forests also made circulation and exchange easier. The destructive power of water, however, was a major problem in this semiarid and largely treeless area. Rapid runoff and silting of the rivers joined with the marked seasonal concentration of rainfall to produce periodic floods. The need for water control in a marginal climate, for irrigation and as flood prevention, helped to shape the social and economic organization of early China, and emphasize group cooperation, which is still a feature of Chinese society. Once the necessary techniques of water control had been mastered, the north China plain (essentially the lower flood plain of the Yellow River) and the loess valleys of the Yellow River and its tributaries farther west (especially the valley of the Wei River, see Map 63, page 236) were able to support a large agricultural population.

China was able to avoid some of the penalties of easy access that accelerated the downfall of ancient Mesopotamia and northwest India, because it was protected to a greater degree by its relative isolation and the friction of distance. Chinese civilization was also able to shift its center progressively southward, as nomadic pressures increased. Erosion, siltation of irrigation works, and soil exhaustion further decreased the attractiveness of the north. The Yangtze basin was nearby, and south China as a whole of-

[1]*Loess* is fine-grained soil deposited by the wind.

MAP 63

POPULATION OF CHINA

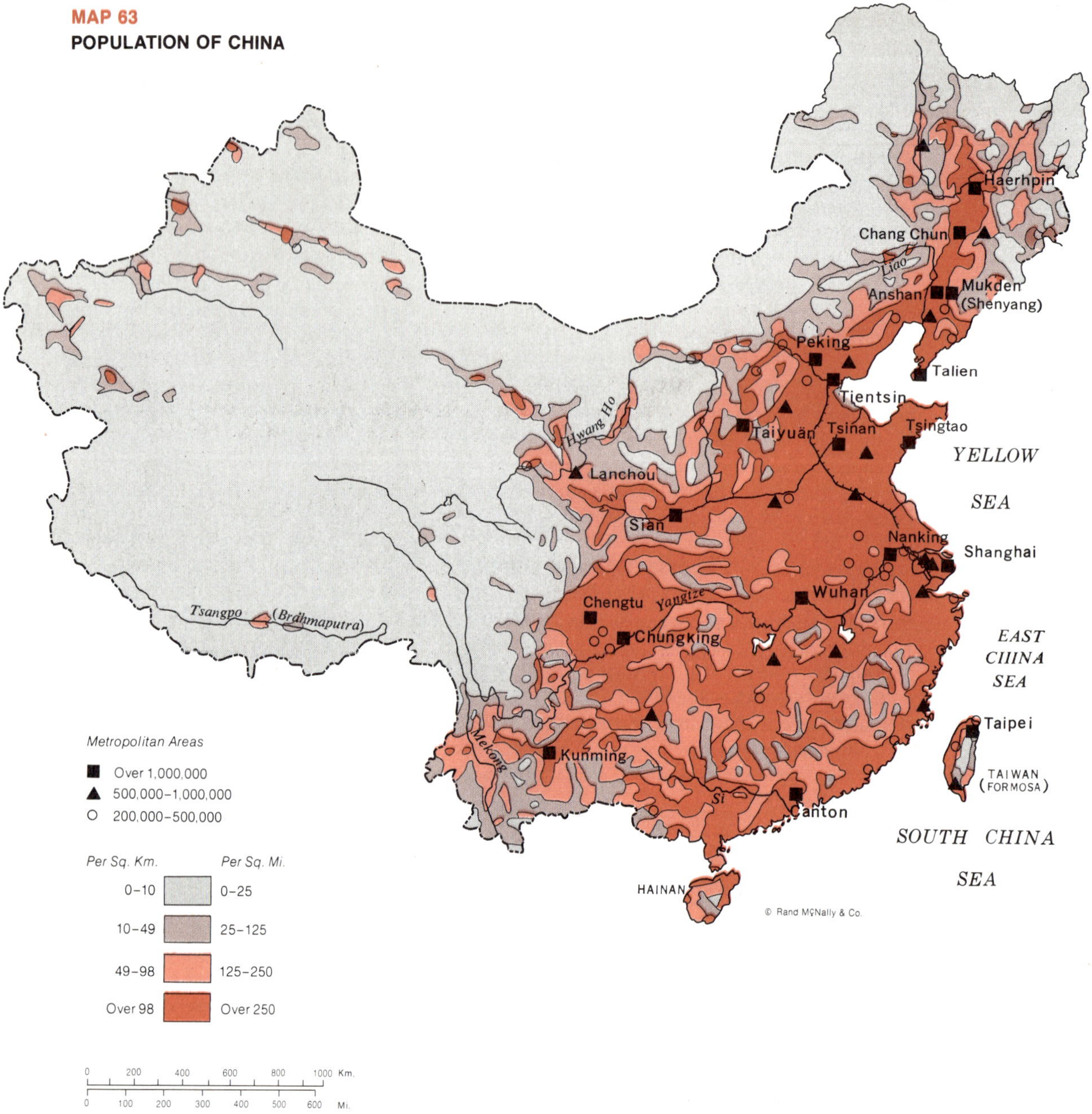

fered a field for expansion, once Chinese civilization and its techniques had matured.

For about a thousand years after 1800 B.C., Chinese settlement was largely confined to the north. The south was occupied by relatively primitive agriculturalists, hunters, and gatherers. Nourished by its productive, developed agricultural base, Chinese population in the north grew, its political organization matured and expanded, and settlement began to spill over into the south. Continuing technological improvements were making agriculture more productive, especially the increasing cheapness and use of iron. By about 500 B.C., iron tools more and more were coming into common use, and iron was no longer reserved for weapons because of high cost or limited production.

Settlement of the south

The first southern area to be occupied was the Yangtze valley, where there was the familiar combination of fertile alluvium and irrigation water. But the moister and warmer climate there supported a dense forest that had to be removed. Once this was done, and the low-lying alluvial lands had been adequately drained, the greater warmth and moisture could support an even more productive agriculture than in the north. In time the Yangtze valley, especially its lower half where there was more level land, became China's chief region of food surplus, and its population grew to rival that of the north China plain (see Map 63, page 236). As commerce increased, navigability on the Yangtze and its tributaries was a further relative advantage, for the silted and fluctuating course of the Yellow River was not usable except by small boats. But level land was not plentiful, especially south of the Yangtze (see Map 62, pages 232–33), and this helped to delay the Chinese occupation. Their civilization rested on agriculture; where farming was difficult, settlement was slow.

The second southern area to be settled in large numbers by the Chinese was the only other extensive river lowland of south China, around Canton (Kuangchou), where the Hsi (Si) River has built a large delta and flood plain. Chinese settlement, or at least the spread of a Chinese culture, may have begun there as early as 500 B.C., and earlier in the central and lower Yangtze valley. But within 100 miles (160 km) of Canton, tribal people still live in the hills. Farther west, in large parts of Kweichow (Kueichou) and Yünnan provinces, non-Chinese people still occupy most of the area. The southwest was the most remote from the original Chinese core in the north. It is also very mountainous and contains relatively little level or usable agricultural land. Some of the original inhabitants of the south were eliminated, some driven up into the hills, and some were forced out of China entirely. Many migrated into the Indochinese peninsula, where their tribes were already important elements in the population and where they are now dominant in many cases. Along the southeast coast of China, broken hill country and mountains also helped to preserve a variety of tribal and regional groups that have contributed to the modern distinctiveness of this area.

Although the north China plain is no longer China's primary economic area, the national capital has remained in the north. From the Shang dynasty's Anyang to the Han dynasty's Loyang on the Yellow River, to the T'ang dynasty's Ch'ang An (Sian) in the Wei valley, to the Yuan, Ming, and Manchu dynasties' Peking, still the Communist capital, the center of military control has been kept in the north. Political tradition is largely responsible for this. The name *Peking* means "northern capital," and the city's large size is still due principally to its administrative functions for a huge country. It is a good example, as was Sian, of a capital located on an exposed frontier to guard the core of the country against nomadic invasions from the steppe, the consistent area of threat.

Expansion of political control

Over half of the area controlled politically by China has never been occupied significantly by Chinese, primarily because aridity, relief, or cold ruled out agriculture (see Map 64, page 238). As the organized power of the Chinese state grew, however, it extended its authority over the nomadic peoples around it, partly for prestige, but mainly for security. The main military threat to the Chinese state came from the nomads. They harried the settled agricultural areas and periodically defeated Chinese armies, or even raided deep into Chinese territory. By the time of the Han dynasty (202 B.C.–A.D. 220), Chinese military power was supreme over nearly all of south China and over most of Korea and Annam as well. Extension of control over the nomadic areas of outer China was a logical accompaniment. Although the political frontier retreated and expanded as China's power weakened and revived in succeeding centuries, Sinkiang, Mongolia, and Manchuria were regarded as politically attached to China from Han times. Tibet was added later, for it, too, was a base for nomadic raids. A further reason for Chinese interest, especially in Sinkiang, and later to an extent in Tibet, was trade routes. Communication with India and with the West was primarily through the steppe and desert corridor of the northwest by the old silk road. It followed the chain of oases in Sinkiang at the foot of the mountains. Military incorporation of Sinkiang by the Chinese state was designed in part to protect its interest in its front door to the outside world. Lesser routes to India led from west China and from Sinkiang through Tibet.

Whenever China was weak, the independence or anarchy of the nomads was reasserted. But only twice, under the Mongol leader Genghis Khan, and under the Manchus, did peoples from the outer areas unite to conquer the whole of China. The victorious Mongols did not maintain their power for long. They held China for about a hundred years in the thirteenth and fourteenth centuries as part of their

MAP 64
PRECIPITATION IN CHINA

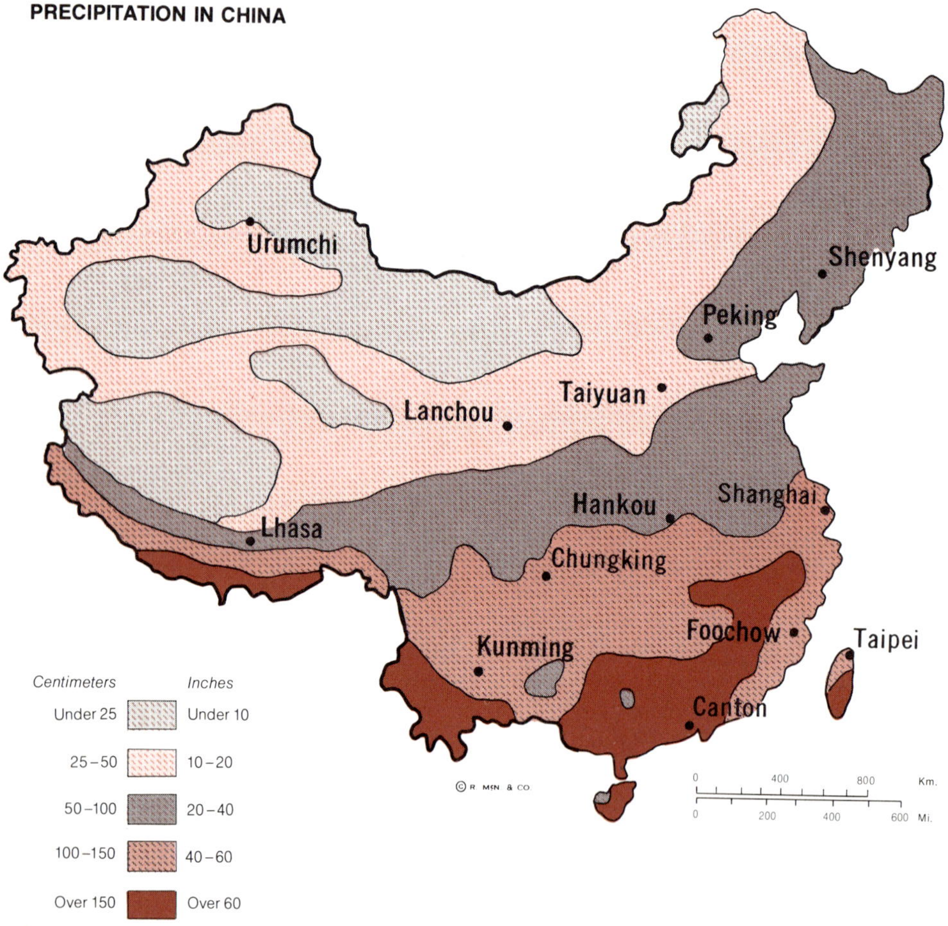

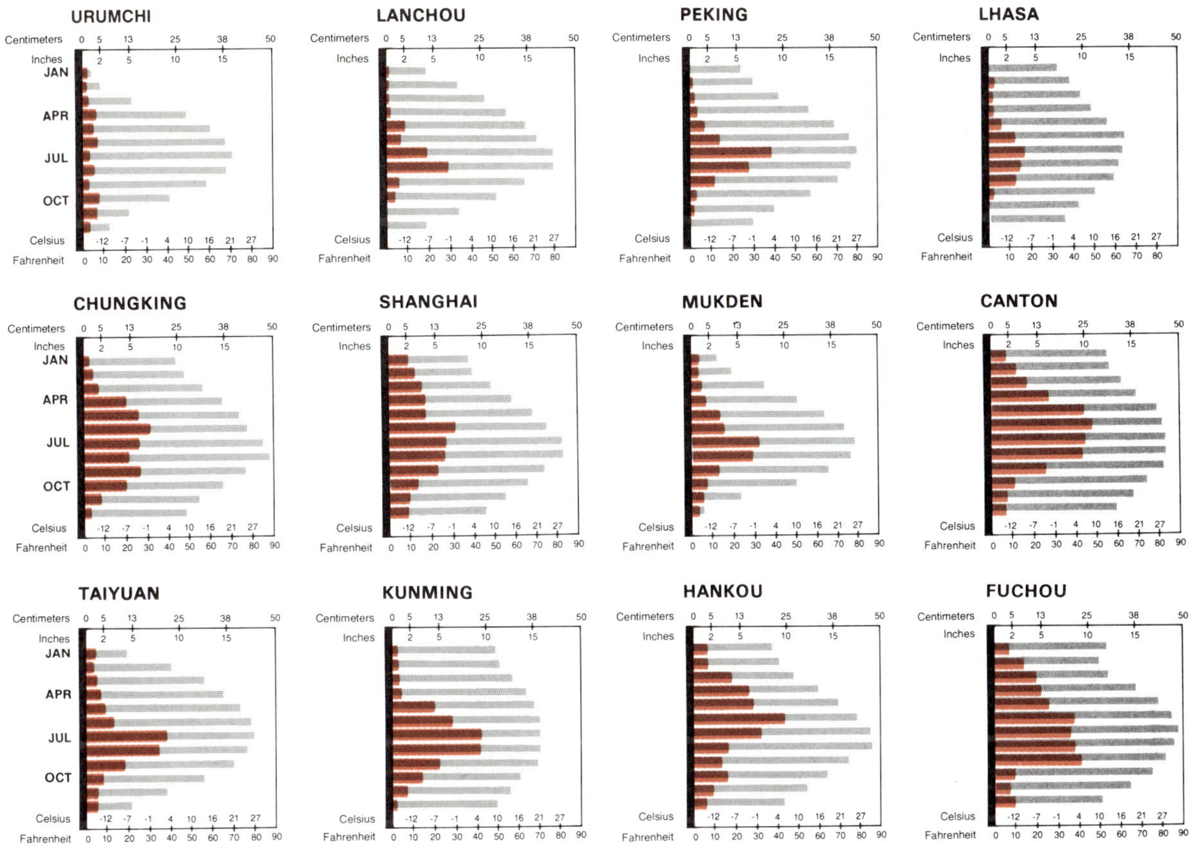

Figure 11.
Climographs of representative cities in China.

Temperature
Precipitation

enormous empire stretching across Eurasia. But they adopted Chinese civilization wholesale, and the rapid crumbling of their power may have been due in part to the loss of their original nomadic vigor. The Manchus, who ruled China from 1644 to 1911, also consciously transformed themselves from Manchurian nomads to cultured Chinese. They were more permanently successful than the Mongols in manipulating political power in the Chinese model, and adopted the Chinese system of administration more completely.

The areal base established under the Han dynasty, the crystallization of Chinese society that took place about the same time as the disappearance of primogeniture and feudal systems of landholding, and the entrenchment of the power of the state and bureaucracy set the pattern for much of China's subsequent development. Even under Communist rule, the Chinese still refer to themselves as "people of Han."

The field and the steppe

The Han people are, more specifically, the Chinese of agricultural China, or China Proper, as it is often called. This is the area of the eighteen provinces inside the Great Wall (see Map 62, pages 232–33), but it includes large hill and mountain areas in the south and west still occupied by non-Han or tribal people. Chinese settlement has also spread into Manchuria (the northeastern provinces) on an enormous scale, especially in the last seventy years, and on a lesser scale into southern (Inner) Mongolia. This immigration was stimulated by the railway, by extension of irrigation, by commercialized agriculture, and especially in Manchuria by large-scale industrialization. Tibet and Sinkiang have offered little for Chinese settlement and are still occupied for the most part by distinctly different peoples. Their population totals are small, since Tibet and Sinkiang are both largely desert. The Great Wall was built in successive stages during the fourth and third centuries B.C. It was not only a military defense against the mounted nomads, but a line to separate the nomad from the farmer, the steppe from the sown, barbarians from Chinese (see Map 62). The wall's course largely follows the line between traditionally agricultural and nonagricultural areas. But it also reflects local differences in rainfall, soil, and availability of irrigation water, or the total usable base for agriculture. It is not surprising that there have been several different walls at differing distances out into the steppe. They were built during slight increases or decreases of annual rainfall, which is especially variable in this marginal area. Climate has periodically encouraged agricultural settlement onto the steppe, and then forced abandonment during a corresponding series of drier than normal years. The remaining Great Wall represents a good average approximation of the limit of agriculture under traditional conditions of technology.

The steppe

With a more effective technology and mounting population pressure, the Chinese have now become the dominant occupants of the area south of the Gobi Desert, which in effect divides Inner and Outer Mongolia. Inner Mongolia has been fully incorporated politically into China, and the area under cultivation is still increasing with the aid of new irrigation. The Mongols of Inner Mongolia are now a minority in their own homeland, outnumbered by Han Chinese more than ten to one. But their place is recognized by the title and structure of the Inner Mongolian Autonomous Region, which includes nearly all of Inner Mongolia, plus some adjacent areas long settled by Han Chinese. The autonomous region is an administrative device that the Communist government has applied to hundreds of non-Han areas of varying size, both within China Proper and in the outer areas. It allows about the same degree of cultural autonomy, especially in language, as in the non-Russian areas of the Soviet Union. While the Han are a strong majority of China's total population (about 94 percent), they are only one of many distinct groups, and their occupance covers only about half of the country. In the interests of greater unity and control, the Communist government has granted fuller recognition to non-Han groups as formal equals, instead of as barbarians, by establishing regions in which their cultures are acknowledged to be important. The largest of these areal units are the Sinkiang Uighur Autonomous Region where the Uighurs, a Turkic people, are the biggest single group, and the Tibetan Autonomous Region, which covers Tibet. Cultural autonomy, however, does not mean political autonomy, and the ruling hand of the central state is strong, as it is in the U.S.S.R.

Outer Mongolia was established as an independent Mongol state in the 1920s, the Mongolian People's Republic. In effect, it is a Soviet satellite. For the most part it is a drier steppe, and nomadic herding in the traditional manner is still important, despite the efforts of the state to substitute collective grazing and increase agriculture and manufacturing. The rail line from north China through Mongolia joining the trans-Siberian railway (see Map 65, page 254) may help accelerate commercialization of the Mongol economy. Another new rail link joins Sinkiang to China Proper through the old northwest corridor.

There is petroleum, coal, and iron ore in Sinkiang, and other local resources may be found. The Chinese state is anxious to develop manufacturing there, particularly in the capital city of Urumchi (Wulumuchi) on the new rail line. Large new in-

The Great Wall at Nankou Pass, northwest of Peking. The arid landscape is typical of the border areas through which the Great Wall passes. The wall was built wide enough to allow troops and cavalry to be moved quickly along the roadway on top. *(Photo by F. G. Clapp.)*

vestments in irrigation are also being made in Dzungaria (the northern steppe areas of Sinkiang) and in the Tarim Basin at the base of the Tien Shan and Altai mountains. This will enable Sinkiang to support a much larger Chinese population. Turkic and other non-Chinese peoples, principally the Uighurs, were almost the sole occupants of Sinkiang until 1956. The Chinese government hopes to strengthen its control by means of Chinese settlement, as well as by increasing mineral, agricultural, and industrial production there. But most of Sinkiang is a desert, and even with a railroad, it is a prohibitive distance from major markets. Bulky raw materials or even finished goods from China face very high transport costs on this line. As in the past, China may feel that its position in Sinkiang is weak without better communication, especially since Russia and its railway system are just across the Pamirs. Russian influence in the past was strong in Sinkiang when Chinese control was only nominal. The increasing hostility between the two governments since 1960, periodically breaking out into small-scale border clashes, provides a further reason for China's efforts to consolidate its hold on Sinkiang.

Tibet had largely broken away from Chinese control in the declining years of the Manchu dynasty. But the Communist government has reasserted China's sovereignty against Tibetan resistance. A revolt by tribesmen in eastern Tibet engulfed the whole of the country in 1959 and led to severe Chinese suppression aimed at the destruction of Tibetan nationalism. Remoteness, mountains, cold, and aridity leave little attractive basis for Chinese settlement or investment. Even for Tibetans, the basis is small in terms of total population and economic development. In the interests of better control, four motor roads have been built into Tibet from China, and one road continues on from Lhasa to the borders of India.

One indirect result of the Tibetan revolt was the flare-up of a border dispute between China and India along the Himalayan front. The only practicable route from China into western Tibet runs from southwestern Sinkiang across a strip of barren mountain and desert waste called the Aksai Chin. The Aksai Chin is located between the Karakoram and Kunlun mountains in northern Jammu and Kashmir, and is an uninhabited area. For many years it was claimed by both China and India. In order to move troops into western Tibet, the Chinese built a road through the Aksai Chin in 1956. Nearly two years later, when Indian patrols discovered what India regarded as an intrusion, fighting ensued. In the autumn of 1962 well-supplied Chinese forces quickly overran Indian positions at several points from Kashmir to Assam. Subsequently they withdrew to the limits of the area claimed by China, suggesting there was never any intention of invading India. In this snow- and ice-covered waste of the high Himalayas no boundary has ever been firmly established, and it is difficult to be certain of exact locations. Even the enfeebled Chinese government of 1914 refused to ratify the boundary agreement proposal made by the British government of India (the McMahon Line). In 1962 the powerful Communist regime welcomed this opportunity to assert both Chinese territorial claims and China's renewed status as a great power.

Manchuria: conflict and development

Manchuria was sparsely settled until early in the twentieth century. The Manchu dynasty had tried to prevent Chinese immigration into its homeland. But with the dynasty's collapse in 1911, a large population shift took place. North China was crowded. A series of famines and floods, aggravated by government ineptitude and general economic decline associated with political disorder, sent streams of immigrants into southern Manchuria. At its height, the migration reached a million a year. It coincided with a period of intense international rivalry over Manchuria, in which Russia and Japan in particular vied for control of Manchuria's impressive mineral and forest resources and potentially productive agricultural land. Manchuria also occupied a strategic location in military terms between Russia, Japan, and China. It was further attractive to the Russians because it could provide a shorter route from Siberia to the ice-free ocean port of Dairen (Lüta) at the tip of the Liaotung peninsula. The Russians built railroads in the northern half of the country, the Japanese built lines in the southern half, and the weakened Chinese government attempted to counter by building a rail line from Peking to Mukden (Shenyang). By its military victory over Russia in 1904–1905, Japan won the upper hand. After the further conquest of Manchuria in 1931 against Chinese opposition, Japanese investments in railways, mining, agriculture, and manufacturing were greatly accelerated. A largely commercial system of agriculture served by railways grew up in Manchuria, producing principally wheat and soybeans for export. Cultivation was concentrated in the Liao valley of southern Manchuria where climate is more favorable than in the dry, cold, and hilly margins on the west, east, and north. Soils in the Liao valley are excellent and very little leached, and rainfall is usually adequate without irrigation, due in part to the low evaporation rates.

Fortunately most of Manchuria's coal and iron ore is also in or near the Liao valley, within about 100 miles (160 km) of Mukden. The mineral resources include a huge deposit of fair-quality coal on or near the surface at Fushun about 30 miles (48 km) east of Mukden, another at Fouhsin about 95 miles (153 km) west of Mukden, and a large deposit of low-grade iron ore at Anshan about 50 miles (80 km) south of Mukden. All are close to the center of the Manchurian market and transport network. Japanese investments and technology exploited these resources and built an impressive heavy and light industrial complex in south Manchuria. The largest center is at Mukden, with iron and steel manufacturing concentrated at Anshan and a variety of manufacturing in other cities. Anshan became the major iron and steel center because the ore there, like most Manchurian ores, is not only of low grade but contains impurities that are expensive to remove. It does not pay to haul the ore to the coal when so much of its weight is waste. Extraction costs are low for the near-surface coal at Fushun, but it must be mixed with better-quality coking coal for making iron and steel. This better coal is brought to Anshan and Mukden from north China and from a rich deposit of coking coal at Pench'i about 40 miles (64 km) southeast of Mukden, exerting a further locational pull away from Fushun.

Manchurian industrial and agricultural production became part of the Japanese economy. Productive Japanese investments in Manchuria were very large. Japan created not only a well-developed rail network, but a significant number of manufacturing plants, including heavy investment in hydro power. Thus, after the Second World War China inherited the largest single industrial complex on the mainland of east Asia. This legacy became the major base of Communist China's ambitious industrialization program. Manchuria's agricultural surpluses were also an important addition to the Chinese economy. China's modern political claim to Manchuria was immeasurably strengthened by the mass immigration from north China. Manchuria's population is practically all Chinese. The Communist government refers to Manchuria simply as *the northeast*, and is anxious to incorporate it as an integral part of China. The checkered history of foreign rivalry and investment of the past half century contributed to the growth of Manchurian regionalism on the part of Chinese settlers who had never been effectively controlled by a Chinese government before 1949.

China Proper

River basins

China Proper, within the Great Wall, still includes the bulk of the agricultural land, the great majority of the Chinese people, and the roots and body of Chinese civilization. Essentially, China Proper is centered on three great river systems, each rising in the mountains of the west and fortunately flowing across the widest area of the country so their usefulness is enhanced. With the addition of the north-south sea route along the east coast, China is well

Shanghai—the waterfront. A misty (and smoggy) morning view of China's largest city, still looking like its semicolonial origins. Shanghai is also still China's largest industrial center, hence the smog. *(Photo by R. Murphey.)*

Loess landscape of the northwest. A town in southwestern Shensi (Shenhsi) on a small stream flood plain in the loess hills, with a small agricultural village in the foreground. Terracing is relatively easy in the soft loess. But the terraced hillsides in the background are not irrigated, since water is present only in the stream valley, and the hillsides grow only drought-tolerant crops. The almost complete absence of tree cover invites dangerous erosion. These problems are increasingly being met through reforestation and new irrigation projects. *(Photo by R. Murphey.)*

supplied with waterways despite its large continental bulk (see Map 66, on page 257). Approximately the northern fourth is drained by the Yellow River, the middle half by the Yangtze, and the southern fourth by the Hsi (West) River.

The Yangtze. The Yangtze is the most important of the three rivers by far. About half of China's people live in its basin, which includes the more productive half of China Proper. The transport usefulness of the Yangtze is increased by the navigable tributaries that enter it from both north and south, each draining its own productive area. The major ones, in ascending order, are the Kan from the south with its mouth at Chiuchiang; the Han from the north with its mouth at Hankow ("Han mouth"), now part of Wuhan; the Siang (Hsiang) from the south entering Tungting Lake and draining into the Yangtze; and the Chialing from the north with its mouth at Chungking (Ch'ungch'ing). As a result of this drainage pattern, much of the trade of north and south China, as well as nearly all the trade of central China, moves interregionally on the Yangtze system. The main stream is navigable for ocean-going ships during most of the year as far as Wuhan, 630 miles (1010 km) from the sea. Steamships, steam launches, or junks can use about 20 thousand miles (32 thousand km) of waterways in the basin. The physical head of ocean navigation is Ichang, at the downstream entrance to the famous Yangtze gorges. Ichang's name may be translated "ought to be prosperous"; its commercially strategic location on the river is not supplemented by a productive hinterland or by lateral access, and it is only a small city. Mountains around Ichang and the course of the Han River have helped to give the advantage to Hankow, which, as a major confluence city, is the dominant break-in-bulk point on the river above Shanghai.

Shanghai itself became China's primate city after the establishment of steamship navigation on the Yangtze. It is still the largest city because the Yangtze links it to one of the largest commercial hinterlands in the world, certainly the most populous. Population and production are particularly concentrated in the level, fertile, irrigable delta, one of the world's most densely settled areas of comparable size. This further contributes to Shanghai's size as a service center. Cheap water routes, by river and sea, mean low transport costs for assembly and for distribution, even for the industrial raw materials lacking in the city's immediate hinterland. Low transport costs, plus the nearness and size of the market, have accounted for Shanghai's predominance in manufacturing, especially before the return of Manchuria. Until then Shanghai contained about half of China's modern factories and industrial labor force, and handled about half of its foreign trade. Railways have been of minor importance in the Yangtze basin, and the two lines that serve Shanghai from north and south carry much less traffic than the water routes.

Yellow and Hsi Rivers. The Yellow is a typical arid-climate river, running through dry northwest China, north in a great loop into the Ordos Desert of Inner Mongolia, and finally down across the semiarid north China plain. Its high silt content, great seasonal fluctuations in volume, and recurring changes of course limit navigability, although small craft ply it along scattered stretches. The river's most dramatic effect on the land has been its frequent floods, hence the name "China's sorrow." Silting of the river bed increased with the cultivation of the hillsides in the watershed. Natural and artificial levees along much of the river's lower course have prevented the regular release of silt. Thus in many places the river bed lies above the level of the surrounding plain, with disastrous results. Silting has also hampered or ruined countless irrigation works in north China. But silt deposits (mainly derived from loess) from the Yellow River and its tributaries over the north China plain have helped to make this the most fertile major agricultural area within the eighteen provinces. The river's course has seldom been constant for more than a century, and major changes have taken it both north and south of the Shantung peninsula, so its delta and flood plain are very widespread.

The Hsi River is a lesser stream than either the Yangtze or the Yellow, and its watershed is restricted by the high relief characteristic of south China. It is navigable by steam launches for about 250 miles (400 km) from its mouth near Canton, but it lacks major navigable tributaries on the scale of the Yangtze, and its basin is far less productive, due to relief. However, as the largest river of south China, its basin also contains south China's largest lowland area, and it is the most obvious core of this extensive region.

Rivers and people

Rivers and their basins are important in China not only as convenient regional units, but for at least two more functional reasons. First, water is the cheapest and often the only practicable means of transport for many bulk goods moving beyond local markets, except in the arid north. Railway development is growing fast under the Communists, but it is still small in comparison with the area and population to be served, and in most cases will remain more expensive than water haulage. Second, aside from the north, China is not well supplied with fertile soils on the whole, and relief further limits agriculture. Fertile alluvium and level land, plus easily available irrigation water, are concentrated in the river valleys. Consequently China's population is also concentrated there (as Map 63, page 236, shows), and reaches its highest and most extensive densities near the river mouths. There the advantages for transport and agriculture are maximized. Shanghai is the outstanding example, but the same factors are responsible in large part for the importance of Canton, Hong Kong, Nanking (Nanching), Wuhan, Chungking, and Tientsin (T'ienching), and for the emphasis on deltaic settlement in both north and south.

Most of the north China plain is in effect the delta of the Yellow River, which is still advancing into the sea at the rate of about 1 mile (1.6 km) every century. The rocky peninsula of Shantung was once an island in the Yellow Sea. Relatively recently, in geologic terms, it has been tied to the mainland by the deposition of silt from the Yellow River. West of the north China plain, settlement is concentrated in the river valleys because of the need for irrigation. Railways were built first in the north because of the lack of navigable water routes, plus the presence of some of China's major coal deposits. But even in Manchuria, railways have not yet been accompanied by major dense settlement outside of the river valleys and deltas, whose agricultural advantages and consequent populations also make them the main market areas. In south China the coincidence of settlement and river basins is even more pronounced, since they make up the level land for the most part. The course of the Yangtze in particular stands out with striking clarity on the population map. The agricultural advantages of its basin are supplemented by advantages for transport and exchange.

Patterns of topography

China has often been divided into the mountainous south and the more level north, but topographic distinctions are also clear between east and west. The north China plain extends westward for the most part less than 400 miles (640 km) from the coast. Beyond are the rugged uplands of the northwest. Local relief in the northwest has been accentuated by the ease with which the loess cover is eroded by the streams. The landscape has a highly carved appearance, well represented in the photograph on page 244. Cultivation has made increasing use of the hillsides, especially by terracing, but irrigation water is most often available only in the deeply entrenched stream valleys. Thus, irrigation cannot be used for crops on the uplands, despite the fertility of their loess soils. On the plain to the east, fertile soil and low relief combine to outweigh the disadvantages of aridity and help to make the north China plain the most extensive area of dense population. Proximity to the sea helps somewhat to make up for the lack of inland water transport, and roads are easily built over this generally dry and level land. A large volume of trade moves overland between the northwest and Tientsin, the commercial and industrial center of north China. More and more trade now moves by rail instead of camel or wagon, and Tientsin is the second largest railway center of China, after neighboring Peking. Peking's leadership as a rail nucleus is political as well as economic. A narrow coastal corridor between the sea and the mountains north of Tientsin links the north China plain with the plain of southern Manchuria. Here the Great Wall meets the sea at Shanhaikuan (literally "mountain sea gate"). This corridor has been a persistent point of entry for invasions, including the most recent one by Japan, and has become a focus of rail lines linking Manchuria with China Proper.

In central and south China, east-west topographic distinctions are less strong but still apparent. In the western half there are only rare and limited breaks in the jumbled sea of mountains. In the eastern half, the lowland basin of the Yangtze extends as far west as Tungting Lake, 700 miles (1130 km) from the sea. In the south the Hsi River valley duplicates this pattern on a much smaller scale. It is true nevertheless that south China as a whole is a land of mountains and that it contrasts clearly with the more uniform and level terrain of much of the north.

There are few well-defined mountain ranges within China Proper. Where they occur, their effects are pronounced. The Chin Ling (Tsin Ling) range, running roughly east-west from southeastern Kansu to central Honan, marks one of the sharpest physical and cultural lines in China, between the dry, brown plain of the north and the green hills of the south. Farther eastward there is no such striking demarcation, although the Tapa and Tapieh Shan (*shan* means "mountain") between Hupeh and Honan help to mark the north-south boundary to a limited degree. Beyond Nanking the open coastal plain begins, where the old deltas of the Yellow merge with those of the present Yangtze and Hwai (Huai) rivers. There north and south meet over a wide zone of transition. Central China, or the Yangtze basin, is not only transitional between north and south in many respects but a distinct region, clustered around the agricultural and commercial core provided by the river. The Yangtze basin lacks both the aridity of the north and the excessive relief of the south. The lowlands of central China include not only the delta and lake plains along the lower course of the river, but also the Red Basin of Szechwan (Ssuchuan), less level but one of China's primary areas of settlement and agriculture. There the Yangtze emerges from the mountain mass of central Asia and is joined by three major tributaries. The agricultural importance of their valleys in the life of the province, China's largest in terms of both area and population, is suggested in the name *Szechwan,* which means "four rivers": the Min, Fou, Chia-ling, and Yangtze.

Southern landscapes

The Fragmented Southeast. South of the Yangtze basin, the only significant gaps in the confused mass of mountains are the delta of the Hsi River and the succession of smaller flood plains and deltas northward from Canton along the coast. Running roughly northeast to southwest, approximately 200 miles (320 km) inland from the coast, are the only major discernible mountain ranges in south China, the Wui Shan (see Map 62, pages 232–33). Difficult broken country extends almost to the coast between Swatow (Shant'ou) on the south and Hangchow (Hangchou) on the north. This succession of barriers has tended to cut off southeast coastal China from the main part of the country. The preservation of pockets of non-Chinese people and culture in this area has already been noted. This drowned coastline includes many fine harbors, and is China's major maritime region. It has had long-standing commercial relations overseas, and a large investment in trade, shipping, and fishing. Piracy has been another traditional specialty of this mountainous coast, based in the multitude of small scattered harbors, and preying on the major trade routes through the China seas. Piracy lasted longer here than in the Mediterranean, and was effectively stamped out only with the advent of the Communist government. The southeast also contains the only remaining forested areas of significant extent in China Proper. The stands are on steep slopes that cover most of the region, are hard to reach, and not very extensive. But this timber helps to support the fishing and trading specialties (as it helped to support piracy), and is a leading regional export. Mountains in this southeast coastal area leave little land for agriculture. Like many mountain areas, the southeast has sent a stream of emigrants overseas, principally to southeast Asia, and in smaller numbers over the whole world. While emigrants from overcrowded and distressed north China moved into Manchuria and Inner Mongolia, Cantonese, Fukienese, and other emigrant groups from the southeast became virtually the only Chinese living outside the political limits of China. In the United States, for example, practically all of the Chinese are from Kwangtung (Kuangtung) or Fukien (Fuchien), speaking their own dialects and offering their own regional food for sale in Chinese restaurants, quite different from what one would find in the rest of China.

The Isolated Southwest. A second discernible topographic region in the south is the misnamed plateau in Yünnan and Kweichow in the southwest. It has such heavy relief that the term *plateau* hardly fits the area. Level land is almost as scarce there as in the southeast, and mostly consists of old lake beds, now largely dry. The area owes its distinct quality to remoteness, to a dry climate, to a high average elevation (over 5,000 feet [1525 m] for most of Yünnan), and to the strong local relief that has discouraged Chinese settlement. These hills are still occupied by non-Chinese tribes. This country, in effect the foothills of mountainous Tibet, continues northward through the western part of Szechwan and into Kansu. The southwest periodically has had close relations with the Indochinese peninsula. Many of its tribal inhabitants migrated or were pushed there by the southward advance of the Han people. The gorges of the Salween, Mekong (Lanisung), and lesser streams, however, plus the tropical rain forest, pose a formidable barrier to the west and southwest, and have helped to limit economic interchange.

Canton, opium, and trade routes

The only easy overland connection between south China and the Yangtze basin, and by far the most important, is the dual route leading from Canton via the Pei River or one of its tributaries across one of two low mountain passes into the headwaters of the Siang or Kan rivers, both leading to the Yangtze (see Map 62, pages 232–33). These passes have helped keep Canton in much closer relationship with the rest of China than has been possible for most of the south. They also furthered Canton's traditional function as the overseas trade center for the whole of China, despite the city's location. A land portage of only about 25 miles (40 km) interrupted the water route from the Yangtze valley to Canton, and the major stream of north-south domestic trade flowed through the two passes. Even before the advent of the Europeans, some trade moved into export markets, mainly to southeast Asia. Until the so-called Opium War of 1840–1842, in which the British forced China to open other ports to foreign trade, including Shanghai, Western traders were restricted to Canton. Chinese exports for some time were paid for in silver, since the Westerners had few acceptable goods to offer China in exchange. This was the origin of the opium trade, designed to give Western traders a commodity to be exchanged for tea and silk. Most of the opium came from India, where it was grown under British management. Before long, the trade resulted in a drain of silver from China since more opium, by value, came to be imported than the tea and silk exports could balance. By the end of the nineteenth century, however, opium had become much less important than imports of manufactured goods from the West. With the opening of other ports to trade, Canton lost its dominance.

In the age of sailing ships Canton derived some advantage from the monsoonal wind pattern. Traders could come through the Strait of Malacca off Malaysia on the inblowing monsoon in late summer, and could leave on the last of the outblowing monsoon of late winter. American ships joined in this trade immediately after North American independence. Its profits made fortunes in Boston, Providence, New York, Philadelphia, and other seaports. Some enterprising captains found an American commodity which the Chinese would buy—ice. Cut from New England ponds in winter and packed in sawdust in the hold, it was carried around Cape Horn. Even after crossing the equator twice, it was still readily salable at an excellent price in Canton, where frost almost never occurs. Despite its loss of trade primacy to Shanghai, Hong Kong, and Tientsin, Canton is still a major port and the largest city by far in mainland south China. Its size is based less on maritime trade than on services for a highly productive and densely populated agricultural hinterland in the delta and basin of the Hsi River.

Taiwan

The island of Formosa, now generally called Taiwan, remains under the control of the defeated Nationalist Chinese government. Like the government in Peking, the Nationalists maintain that Taiwan is an in-

The warm, humid, hilly south. A typical southern landscape of intricately terraced rice paddies in the valleys, interspersed with hills too steep for cultivation. Irrigation is provided mainly from streams or springs farther up the slopes, with the water led to each paddy below in turn. *(Photo by R. Murphey.)*

tegral part of China. Taiwan's inhabitants are overwhelmingly Chinese, but they did not begin to settle on the island on any significant scale until the seventeenth century. Nearly all of them emigrated from nearby coastal Fukien and northern Kwangtung, across the relatively narrow Formosa Strait. Taiwan has become highly productive agriculturally as Chinese intensive techniques have been applied in this subtropical environment of high temperatures and ample rainfall. The western half of the island, roughly speaking, is a broad alluvial plain, rising to high mountains that terminate abruptly in cliffs along the east coast facing the Pacific.

Taiwan became a Japanese colony in 1895, and its agricultural and industrial development was greatly accelerated by Japanese management and investment. The return of Chinese sovereignty in 1945 was followed by massive U.S. aid, and by an influx of refugees after the Communist victory on the mainland. Although the island's political future is uncertain, economic growth has been extremely vigorous. Taiwan has been fully self-supporting since about 1965, and has enjoyed a minor industrial boom. Manufacturing passed agriculture as the major sector of the economy around 1970. Taipei, the capital, at the northern tip of the island, has become a rapidly growing industrial center of over 2 million people, with nearby Keelung (Chilung) as its port.

The coastline

Northern Ports. The geographic distinction between north and south China is re-emphasized by the change in the nature of the coastline that occurs in the area of Hangchow Bay (Hangchou Wan). South of this area is a submerged coast with numerous good natural harbors, especially where the mountains come down to the sea. Northward an emergent coastline provides almost no naturally good harbors, with the exception of the rocky promontory of Shantung. Along this coast the lack of easy transport connections with the hinterland, especially by river, has reduced the usefulness of the major harbors at Tsingtao and Chefoo (Yentai). Tsingtao's easier overland access by rail and its proximity to the Shantung coalfields and the market of the north China plain have made it an important port. Elsewhere along the even, sandy plain of the north China coast, the naturally poor harbors are continually menaced by silt from the arid-climate rivers, and from the Yangtze in the case of Shanghai. Progressive deforestation and expanding cultivation on the hillsides in river watersheds have increased silting. Tientsin, 40 miles (64 km) from the sea on the Yungting River, was forced to remove its port to Taku and Tangku at the river's mouth. Even there growing sand bars made it necessary for deep-draft shipping to load and unload by lighter off the coast. The Communist government has built a new artificial harbor just north of Tangku, but it is still too shallow for the largest shipping. Dairen and Port Arthur (Lüshun), at the rocky tip of the Liaotung peninsula in Manchuria, are now a single urban area, generally called Lüta. On the Dairen side, there is an excellent harbor that avoids the mouth of the Liao River and its silt. It serves the commercial and industrial market of Manchuria, and is China's second or third largest port in the volume of its business. Most of the inland traffic moves by rail, however, since navigability on the Liao is limited.

Shanghai's silting problem is less crucial but still serious. A huge offshore bar at the mouth of the Yangtze has never been dredged deep enough for the largest shipping to enter except at high tide. The river is extending its delta rapidly, and only constant dredging can maintain a navigable channel. The city and most of the port facilities are actually not on the Yangtze itself but on a small tributary, the Huangp'u, which joins the Yangtze at the lower end of the estuary. Although the Huangp'u is narrow and winding, and the city is about 14 miles (22 km) from the sea, silting is less of a problem than on the Yangtze, and the strong currents of the larger stream are also avoided. This site has already been compared with those of Calcutta, Karachi, Venice, and other ports. Shanghai's harbor, even with dredging and other improvements, is physically poor, and, in particular, lacks both adequate depth and space. But the city's commanding commercial location at the mouth of China's greatest trade route, where it can cheaply and conveniently serve the more productive half of China's area and population in the Yangtze basin, more than makes up for the disadvantages of its site.

Southern Harbors. What south China gains in numbers of good harbors it largely loses in ports. Its harbors are inaccessible to a productive hinterland because of mountain barriers, lack of navigable waterways, and the negative effects of mountains on population and production in more general terms. The Hsi River and its basin make Canton an exception, even though Canton's deltaic harbor is poor. A new port has been constructed under the Communists at Chanchiang on Kwangchow (Kuangchou) Bay (opposite Hainan Island, see Map 62, pages 232–33). This is an excellent natural harbor with none of Canton's disadvantages. Access is provided by a new rail line joining it with the recently completed major rail route between southeast China and Hanoi through Liuchow. It is doubtful, however, whether Chanchiang can ever serve as large or as productive a trade hinterland as Canton, despite the differences in their harbors. South China ports north of Canton were relatively more important in the past century than now, partly because the railroad reached Canton and its more productive hinterland relatively early, and came to Amoy (Hsiamen) and Foochow (Fuchou) only in 1956 and 1958, partly because of changes in trade. Tea and silk dominated China's export trade until early in the twentieth century. Both were produced extensively in south-central China, and moved to overseas markets through ports like Hangchow, Ningpo, Wenchow (Wenchou), Foochow, Amoy, and Swatow, as well as through Shanghai and Canton. Tea especially, as a slope crop, was concentrated in the hills and mountains that form the divide between the Yangtze and Hsi river systems, and in the hill and mountain country along and behind the southeast coast. As Chinese tea lost its commanding position on the world market with the increase in plantation production in Assam, Ceylon, Formosa, and Java late in the nineteenth century, as the export market for silk was lost to Japan, and as other commodities became more important in China's total trade, these ports suffered.

Hong Kong is a separate case. When it was conceded to the British in 1842 as a small, rocky island, it contained little more than a fishing village and smuggling base. The growth of China's maritime trade and the breakdown of Canton's traditional monopoly made the port of Hong Kong second only to Shanghai in the volume of its business. Its large and deep island-sheltered harbor just off the estuary of the Hsi River attracted most of the trade of south China, and also made it an excellent entrepôt for the whole of the China coast. Hong Kong is a classic example of an entrepôt, collecting exports from a variety of smaller ports along the coast for transshipment to wider markets, and distributing imports from overseas to them. British sovereignty was a further attraction for trade in the nineteenth century. It provided legal and political security plus the tariff advantages of a free port, which exempts goods in transit from paying customs fees. These advantages, plus the skills and hard work of a population swollen by refugees from China, have helped Hong Kong establish an extensive light-manufacturing industry. This has more than compensated for the decrease in its entrepôt trade with China as a result of the Communist revolution. It has become one of the most crowded cities in the world with a population of nearly 5 million on a confined and hilly site.

Climate

Dry, Cold, North China. Climatically it is difficult to generalize about an area as large as China Proper. It reaches across twenty degrees of latitude and twenty-eight of longitude and is also broken up by mountains (see Map 64, page 238). North China as a whole has a continental climate of hot summers and cold winters, with the extremes being more pronounced farther inland. Continentality is accentuated by the latitudinal position of north China in the zone of the prevailing westerlies. Like northeastern United States, its weather is often imported from the land mass behind it rather than from the sea on which it fronts. The monsoonal wind system, however, while less marked in north China than in southeast Asia or India, helps to produce predominantly inblowing, warm, maritime air in summer when nearly all of the rainfall comes, and predominantly outblowing, cold, dry air from Siberia and inner Asia in winter. The mountains of the northwest help to restrict the penetration of moist maritime air in summer. Close to the sea in the eastern sections of the north China plain winters are relatively mild, roughly comparable to those in Connecticut or Rhode Island. In the northwest winters are rigorous and strongly influenced by outblowing air from inner Asia.

But north China's great climatic problem is rainfall. It is a marginal area agriculturally, more because of the large variation in precipitation than because of the average annual amount. Average annual rainfall variability in north China as a whole is approximately 30 percent; in the northwest it reaches 40 percent. With an annual rainfall average for all north China of between 20 and 25 inches (50.8 to 63.5 cm) (see Map 64, page 238), agriculture is precarious. Annual variability is frequently matched by seasonal variability, with its equally disastrous consequences, including floods as well as drought. Winter precipitation is too slight to support crops even if the temperatures allowed it, although the thin winter snows occasionally help to moisten the soil for spring. The north depends on drought-resistant crops, principally wheat and cotton in the better-watered and level areas, and millet and sorghums on the drier margins. The leading sorghum crop is kaoliang, introduced from central Asia probably about the seventeenth century. Its drought-resistant qualities made possible the agricultural settlement of drier areas previously avoided, and also contributed to significant increases in total food production in the longer-settled areas. New irrigation and reforestation are rapidly changing this traditional pattern. Rice is now widely grown in the north, although wheat remains the major crop.

The Warm, Humid South. The line marking 40 inches (101.6 cm) annual rainfall (see Map 64, page 238) is perhaps the most appropriate dividing line between north and south, although the Hwai River and the Chin Ling Mountains, which closely coincide with it, are more commonly used for this purpose, as they make a clear demarcation on the landscape. South of this line rainfall is generally adequate and reliable; average annual variability is less than 15 percent. Seasonal distribution is also favorable, with a rainfall peak in June, July, and August when it is most needed, and with adequate and usually reliable amounts occurring throughout the year. Cyclonic storms bring gentle showers to the Yangtze basin and south China during the winter and early spring. This rainfall pattern and the milder winter temperatures make winter crops possible, and largely saved the south from recurrent famines in the past. Rice is the dominant crop of China south of the Hwai River, and commands the best land. It is accompanied by winter crops like beans and wheat that may be raised in the dry fields between rice crops. Parts of tropical and subtropical Kwangtung and Kwangsi (Kuangsi) have traditionally grown two crops of rice per year, taking advantage of a virtually unbroken growing season of high year-round temperatures. The present government has promoted the northward spread into the Yangtze valley of double crops of rice, using new quick-maturing varieties. Near the cities vegetables are important. The hill lands may support casually tended fruit, nut, or tea crops, or may be pressed, less appropriately, into use for unirrigated cereals like maize, if they cannot be terraced for rice. During the summer peak of the monsoon, the rain is likely to come as heavy showers. Given the relatively high annual rainfall, the soils of south China are considerably leached. This accentuates the importance of the strips and pockets of alluvial soil in river valleys and deltas. Erosion is less of a problem here than in the north, despite the greater relief. The climate is moist enough to support a more or less dense growth of vegetation that grows again even after repeated cutting.

The relatively mild, moist winters of south China and the long, unbroken growing season that approaches tropical conditions (Canton, for example, lies about 20 miles [32 km] south of the Tropic of Cancer) help to account for the very high population densities developed on the limited pieces of level land. Small areas, like the most favored sections of the Hsi River delta around Canton, may equal or even exceed the Ganges delta in the number of people per square mile supported largely or entirely on agriculture. The human landscape of lowland China is crowded nearly everywhere, but nowhere more so than in these few areas where the physical advantages are so sharply concentrated.

The Southwest. The southwest is a distinct region, climatically as well as in other respects. It is on the fringe of the monsoonal influence and far from the sea. While mountain barriers help to exclude the cold Siberian air in winter, there may be long dry spells. In winter and spring there may be a bare, brown landscape that is reminiscent in some ways of north China. Relief is even more responsible for the fact that the southwest is the most sparsely settled of all the major regions of China Proper. Climatic limitations and remoteness help to explain why the pockets of level land, even in the old lake basins, do not support anything like the human densities characteristic of the great river deltas.

This discussion of China's historical development, of the physical base, cultural outlines, and the variety of human adjustments can be only the briefest survey of a country one-third larger in area than the United States, with nearly four times its population, and with a uniquely long, continuous history. In addition, it is now changing fundamentally in a revolution whose effects are probably more drastic and more rapid than any in the history of the world. Some of these changes, their foundations and implications, are examined in the following chapter.

14 Changes on the good earth of China

Peasants work with hand tools in Yenan. *(Photo by Richard Balzer for Stock, Boston.)*

A country of over 950 million people, whose population is still growing and will probably reach 1 billion before the end of the present century, must make major efforts to increase its production, not only in agriculture but also by accelerating the growth of large-scale manufacturing. As in India or most underdeveloped economies, increases in agricultural production and peasant income levels are essential to industrialization as well as being dependent on it. But a major part must come from a wider use of more varied resources than a solely agricultural economy includes. China's attempts at industrialization until the Second World War were feeble, were carried out largely with foreign capital and direction, and achieved very little in proportion to the scale of the problem, except for the Japanese accomplishments in Manchuria. The Communist government has set as its major economic purpose the industrialization of China. It has made an impressive beginning toward this ambitious goal with the help of the Manchurian industrial structure it inherited. An even more important asset is a fully mobilized population under effective central direction. The rate at which total production has grown since 1953, especially industrial production, compares closely with the growth rate achieved in the Soviet Union, and is significantly greater than the growth rate in the United States in the same period. Increases in agricultural production have been less substantial, as in the Soviet Union, and for essentially the same reason: investment capital until recently was largely concentrated in industry.

There were good reasons for China's past difficulty in building a significant manufacturing complex. With the massive population increases during the mid-seventeenth to the mid-nineteenth centuries, the essentially subsistence agricultural system had lost its ability to keep production increasing faster than population, with existing technology. Few surpluses were generated, and they were commonly invested in landholding, or in speculation and usury, rather than in productive or capital-generating enterprises. The limitations of the transport system made it difficult to move raw materials or distribute finished goods at bearable cost. Bulky goods like coal or grain or ores might double in cost by being transported only 50 to 100 miles (80 to 160 km) overland in the widespread absence of cheap volume movers like railways. Another was the acute shortage of skilled labor, technicians, and management. China's mineral resource base, however, is strongly favorable for industrialization. The problem in the past has been how to put it to work. The Communists believe they can solve this problem by doubling or trebling agricultural production and peasant income levels with an immense program of development in railways, highways, and water transport, paid for primarily from agricultural surpluses as taxes or as exports. At the same time they are also carrying out a mass program of technical and job training for workers, managers, and engineers. It is not wise to be dogmatic about what can and cannot be done to develop the economy, given the revolutionary drive, collective force, mobilization of human energies, and sense of mission that are such prominent features of Communist China. In two particular respects, however, the geographer may comment appropriately about the dimensions of the problem and about what can be done—the resource base for agriculture and the resource base for manufacturing.

Extensions of cultivated area

Although only about 30 percent of China Proper plus Manchuria and Inner Mongolia is under cultivation, due principally to limitations imposed by climate and relief, the proportion is rising, and may ultimately reach close to one-third as a result of technological improvements. Extension of irrigation into areas previously too dry for agriculture is the most important basis for this increase. It is particularly important in areas like Inner Mongolia, the loess highlands of the northwest, and the drier regions of Manchuria west and north of the central valley of the Liao River. Even in more humid China, many arable areas cannot be irrigated adequately by traditional methods, and irrigation can now be made more widespread with consequent increases in yields. Deeper wells, power-driven pumps, longer aqueducts or irrigation canals, and more and larger river dams can make irrigation more effective and more widely available in every area. In north China alone, deeper wells, pumps, and small and large dams doubled the irrigated area between 1950 and 1960. This removed much of the loss from a marginal and highly variable climate. Parts of the arid and semiarid areas that have been limited to grazing in the past will be provided with water. It will come not only from wells, but from large projects like the series of dams being built along the Yellow River and most of the other rivers of China. The dams are intended to prevent or minimize floods, produce power, and enhance navigability as well as providing irrigation. All of these things will be direct or indirect aids to increasing agricultural production within a more fully commercialized economy. Not all of the irrigable areas have the soils, relief, and temperatures favorable enough for agriculture to make irrigation worthwhile. But there are many thousands of square miles that do. In particular there are large areas of the fertile soils commonly associated with semiarid climates where cold-tolerant grains could be highly productive if they had adequate water. Especially in the north and northwest, extensive reforestation is being carried out on slopes as part of soil conservation and river basin management. Other technological improvements will help to increase the amount of land under cultivation by draining swampy land, by deeper plowing and better management of certain soils that it has not been possible to cultivate profitably so far, and by the development of crop types and strains that are more tolerant of special or marginal climates or soils neglected in the past.

Agricultural yields

By far the largest increases may be expected to come from a rise in the yields obtained from the existing cultivated area. Traditional Chinese agriculture was highly intensive. Although the yields per worker were low, yields per acre were relatively high. They have had to be to support such a dense rural population. Yields have been kept high mainly by increasing the intensiveness of the traditional methods: more night soil, more plowings, more crops squeezed out of each field every year, more labor expended per acre. Technical improvements available from the experience of Japan and the West can add to this already highly intensive system to produce very large increases in productivity, as happened in European and American agriculture over the past two or three centuries. The increase in yields per acre essential to economic growth and industrialization, as a source of surplus for investment, can also increase yields per worker, which is equally important if income levels are to rise. China must be able to specialize by diverting growing amounts of its labor force out of agriculture and into manufacturing, trade, and other enterprises.

Technical improvements

There are several immediate and obvious technical improvements that can be made and that are being made on a large scale by the Communist government. Perhaps the most important of these is the development and use of better seeds and better types and strains of crops, each suited to particular climatic and soil conditions, resistant to disease, and having greater yields. Experience in the West, in India, and in Japan suggests that overall yields might be increased by as much as 100 percent by proper seed selection and the distribution of better seeds and plant strains. The Chinese are doing their own research and applying the breakthroughs achieved elsewhere under the label of the Green Revolution, making major progress. Pests and plant disease may have destroyed as much as one-fourth of the average annual crop under the traditional system. They can now be controlled to a much greater extent. While there is no rational basis for an increase in cattle in the agricultural areas, beyond the need for draft animals, large steppe areas in outer China could support a much more productive grazing economy. Areas too dry for field crops or that cannot be irrigated might support healthier and more efficient animals. And extension of transport and marketing facilities can make it possible to distribute their products within a more highly commercialized economy.

Night soil and other organic materials are not enough fertilization to maintain high yields, especially in the many areas of China where the soil has indifferent natural fertility. Many of the new high-yielding strains have heavy fertilizer demands. Chemical and mineral fertilizers can provide another major boost to agricultural production, as they have

MAP 65
MAIN RAILROADS IN CHINA

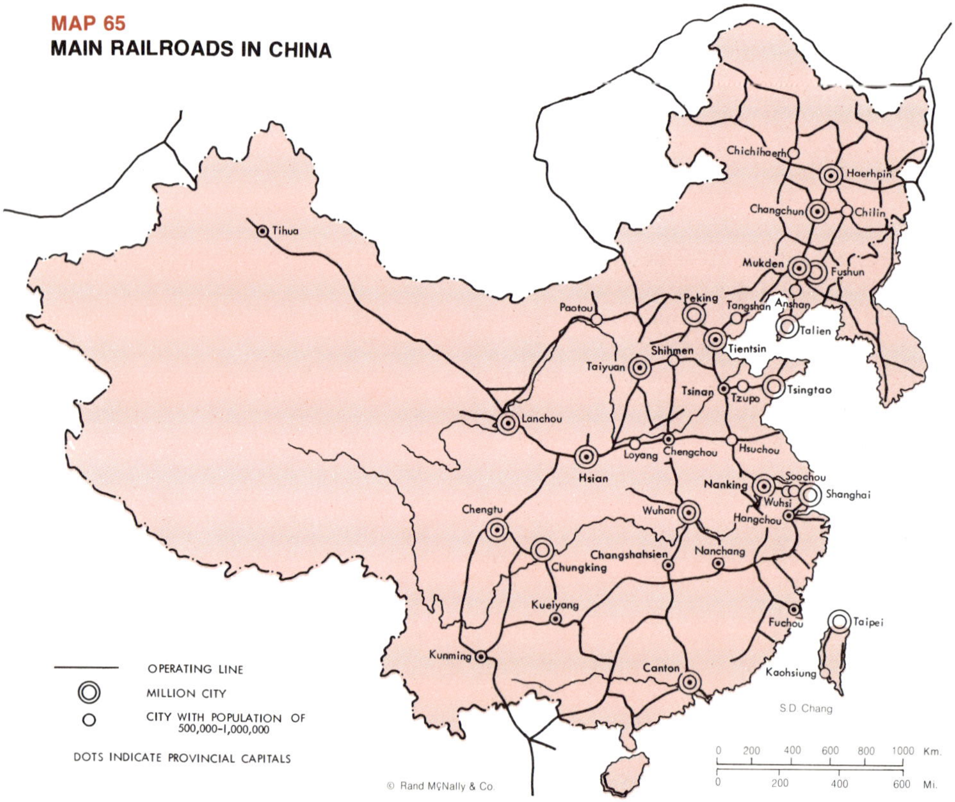

North China village. This village is on a commune near Sian in Shensi (northwest China). As with most villages, its physical appearance (except for cleanliness) has changed little since 1949. *(Photo by R. Murphey.)*

done so strikingly in the West and in Japan. The Communist government has put a high priority on fertilizer production, even if other types of manufacturing must be delayed.

The peasants' hand tools have been few and crude. Significant increases in efficiency and total production can be obtained by providing more and better tools, including deeper plows as well as smaller equipment, especially for irrigation. Considering all possible improvements, including expansion of cultivation into previously neglected areas and increases in the adequacy and extent of irrigation in the existing cultivated areas, China should be able to triple its total agricultural production at least. With over 15 million new mouths to feed every year, the critical question is whether agricultural production can be increased rapidly enough and far enough to maintain and widen the gap between people and goods and still leave substantial surpluses for industrial and transport investment to increase the general productivity of the economy as a whole.

Economic requirements

Agriculture is also dependent on industrialization, not only for the supply of fertilizers, pumps, tools, machines, and other equipment, but for the creation of a national market. Under traditional subsistence conditions, each area, and to some extent each farmer, had to be largely self-sufficient and could not afford to make use of particular local advantages for specialization. Specialization based on cheap exchange is a universal condition of economic growth, in agriculture as in manufacturing. It cannot take place until the economy as a whole is unified, principally through building an adequate network of low-cost volume transport carriers and developing an effective marketing system. Hillsides cannot be used profitably for tree crops or slopes reforested until the grower (state or private) can sell the product profitably and widely, and thus can afford to buy rice or other food from better-suited areas. Peasants cannot cease cutting over their hillsides for fuel until their rising income levels and a widening market combine to make alternative fuel available to them. The rich fish resources of the sea and the rivers cannot become a significant addition to the food supply until they can be cheaply and quickly distributed within a national market.

Mobilization and collectivization

The Communist solution to the land problem is the collectivization of agriculture as part of an overall program of state planning and economic management in order to mobilize human and natural resources to the fullest possible extent. An important part of this program is the rapid creation of an adequate low-cost transport system, which plays such a basic role in agricultural as well as in industrial development. New rail lines and motor roads are being built, waterways are being improved, and the skeleton of a national transport net is beginning to appear (see Maps 65 and 66, pages 254 and 257). But each agricultural unit (commune) and each local area is also designed to be as nearly self-sufficient as possible rather than dependent on long-distance transport or on city-produced goods. Practically all of the rural landscape is now organized into communes, averaging between 10 and 20 thousand people. They are designed to produce their own tools and consumer goods as well as food, with a surplus available to the state for reinvestment or distribution to other communes.

The ancient pattern of individual ownership of small plots, averaging only about 3.5 acres (1.4 hectares) per farm, has yielded to much larger units in which the land and all the equipment are collectively owned by the commune. The technical improvement and reorganization of Chinese agriculture can be carried out far more effectively and quickly through the closely controlled and centrally supervised collectives than through a multitude of small independent farmers. Consolidation of the small scattered individual holdings and the removal of ancient field borders also produce economies. The pooling of labor and equipment (or the provision of more and better equipment) and increasing the size of the unit of production give some of the general economies of scale. In this context of control combined with local self-sufficiency, the Communist government hopes to make the agricultural system productive enough not only to feed a rapidly growing population better than it has been fed in the past, but to raise its income levels, to provide agricultural raw materials for manufacturing and export, and still to leave enough surplus to pay for the costs of industrialization. Basic living standards have risen sharply. Food, health services, clothing, and housing are all more than adequate, and certainly better than in most other developing economies. The transformation of China is little short of miraculous by comparison with the chaotic years preceding 1949. China is appropriately described as a poor country still, but without poor people. This is the result of efficient organization and commitment to equality and mass welfare.

Workers' housing in the new China. Located in the suburbs of Shanghai, this "workers' village," one of many, was designed as a self-contained community. These villages are grouped beside an industrial district, and have their own shopping, educational, medical, and other facilities. The buildings are similar to town houses, and represent an enormous improvement in housing quality for workers. *(Photo by R. Murphey.)*

In the drier areas of China (and elsewhere in monsoon Asia), agriculture is less intensive and more extensive. In dry north China, Manchuria, Inner Mongolia, and parts of the northwest, there are also large expanses of relatively level land. In these areas, fuller mechanization is more likely to be profitable, and to maintain or even increase present yields per acre, while greatly increasing the yields per worker. It will be a long time before China is able to produce the volume of agricultural machinery it needs. Tractors, cultivators, harvesters, and most of the equipment of a mechanized system are already being made in China and are coming into use, especially in the dry north, northeast, and northwest. This is not surprising, for North American agriculture also is more fully mechanized where it is most extensive, especially in the drier areas of the great plains. The areas of more intensive cultivation in the humid east or close to large urban markets are less mechanized. In China the extent to which machines can aid agricultural production also depends, of course, on the success of the industrialization program.

An adequate pool of skilled labor, technicians, and management may be created by intensive education, self-education, and technical assistance as was originally provided by the Soviet Union. But as with India, with China, or with any economy, foreign capital either as loans or gifts cannot create industrialization. It helps to "prime the pump," but the pump must be there or must be built. The economy itself must continuously produce capital in the form of surpluses for industrial investment, and it must have an economic structure that can absorb investment productively. Since the worsening of relations with the U.S.S.R. in the late 1950s and the end of all external aid, Chinese policy has emphasized self-reliance. China has made a virtue of necessity by maintaining a rapid rate of economic and technical development based on the country's own resources and collective effort.

Mineral resources

Coal

The physical resource base is only one of the factors in the industrialization process. Its significance is as much a matter of location, or, more broadly, of economic usability, as of total amounts of raw materials. China is rich in coal, so rich that the precise total of coal deposits is academic even if there were accurate figures. In broad terms, China's total coal reserves are probably the third largest of any country in the world, after the United States and Russia. More importantly, enough of the coal is of sufficient quality and occurs in large, easily worked deposits near enough to major present or potential market centers that industrialization need not be limited by a shortage of coal. Map 67 (page 258) clearly shows China's strong advantage in coal, especially the close relation between several of the major deposits and the principal market centers. Most of the existing and planned industrial developments in China are closely oriented to major coal deposits or to sites where coal is cheaply available by water routes; for example, at Shanghai, which is still the largest single industrial center.

Coal is a major factor in the industrial pattern of Manchuria and in the rapidly growing industrial areas of Shansi (Shanshi), especially in the city of Taiyuan. The similar but smaller industrial develop-

MAP 66
MAIN WATERWAYS IN CHINA

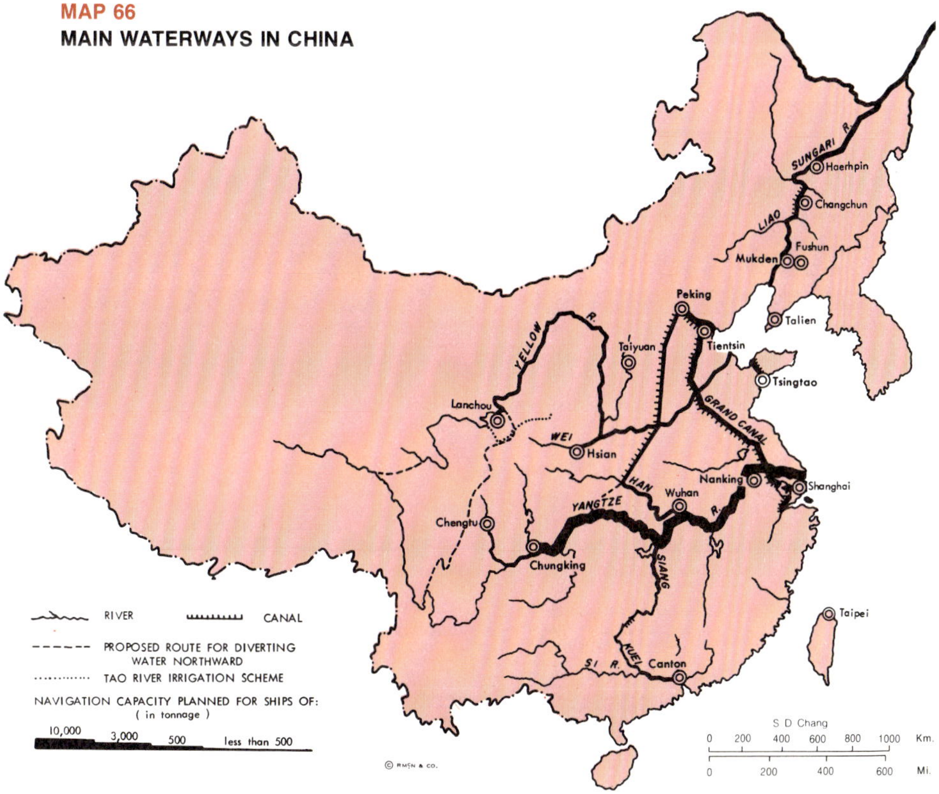

ments at Chungking in Szechwan, at Lanchow (Lanchou) in the northwest, and at Kunming in Yünnan are related to local coal and iron ores. But, more importantly, they are part of the Communist government's plan to spread industrial development more widely and to stimulate the growth of previously neglected regions. The large iron and steel works at Paotow in Inner Mongolia is based on large local deposits of high-grade iron ores and on coal from nearby Shansi, but its development is also part of the plan for dispersal. The older industrial centers in north China, notably at Tientsin and Tsingtao, de-

MAP 67

FUELS AND POWER IN CHINA

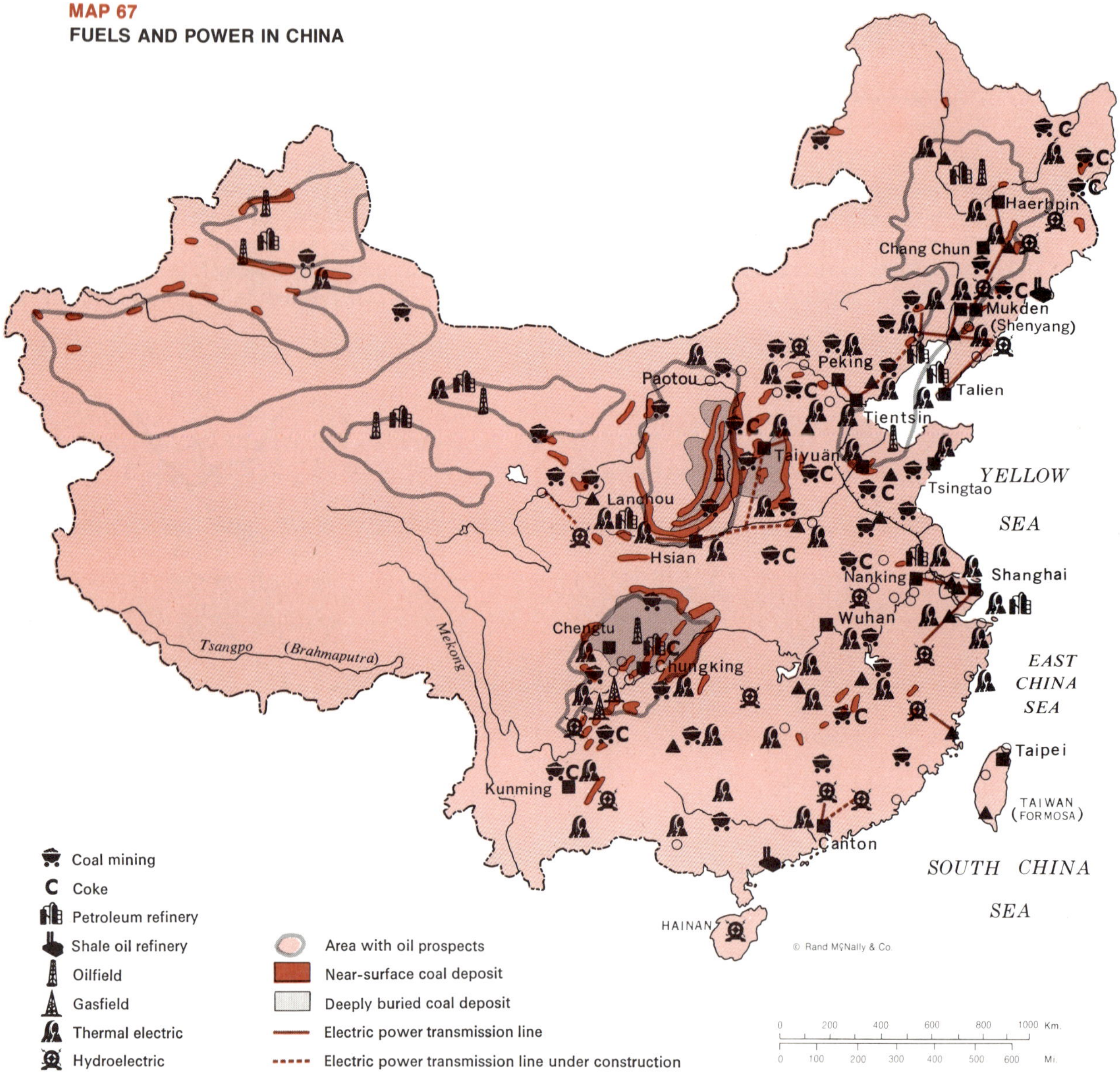

pend on what are still China Proper's major coal mines. They are located in eastern Hopei west of Tientsin, just south of Shanhaikuan at Kaiping, and at the base of the Shantung peninsula, where coal can be distributed by water and rail relatively easily. The remaining important manufacturing center at Wuhan (formed by the three cities of Hankow, Wuchang, and Hanyang) depends on a network of cheap water transport, like Shanghai. It has also become a major iron and steel center, using iron ore from mines at nearby Tayeh close to the Yangtze, and coal from the large deposits at Pinghsiang not far from the Siang River. But coal is important, as fuel or as power, to almost every kind of industrial operation, and its availability plays a large part in shaping the general pattern of industrial location. For example, because delivered coal was cheap in Shanghai and there was a large coal market, the city could make and sell electric power very cheaply and could offer a strong attraction for all kinds of manufacturing. China's coal resources may also be important in the future as exports, especially to coal-poor Japan, India, and southeast Asia. Thus they may contribute further to economic development at home.

Iron and oil

Riches in coal are apparently not matched in iron. China has not yet been surveyed as fully as most Western countries, and it is best to be cautious about what may still be discovered. But present known reserves of usable high-grade iron ore may not be overly plentiful in proportion to the size of the population. Many of the deposits are too small and scattered to be economically usable. Much of the total estimated reserve consists of ores of very low grade or containing impurities that are expensive to remove. This is characteristic especially of the Manchurian ores. Recent discoveries of higher-grade ore in the northwest, in Szechwan, in Sinkiang, in Inner Mongolia, and on Hainan Island have changed the picture considerably. But China is not notably rich in good, usable iron ores in relation to the size of the population. However, this is not a serious disadvantage since no industrialized country can be self-sufficient in all raw materials.

Until recently China was thought to be only modestly supplied with petroleum. Major new discoveries were made in Sinkiang, Manchuria, the northwest, and Szechwan, and a huge pool was

Industry in the countryside—commune power station. This small-scale hydroelectric plant on a commune in south-central China is typical of thousands of similar plants. They use local resources to achieve self-sufficiency and support local industrialization. *(Photo by R. Murphey.)*

found under shallow off-shore waters along the north coast between Shantung and Tientsin. Recent surveys suggest that this off-shore area may contain as much oil as the larger fields around the Persian Gulf. China is already an oil exporter, and its oil will help to accelerate domestic industrialization.

Other minerals, hydro power, and some conclusions

China is abundantly supplied with antimony and tungsten, and has plenty of tin. Over half of the world's known reserves of antimony and tungsten are in China, and exports of these ores can provide welcome foreign exchange. Otherwise they cannot do much to stimulate industrialization. Shortages of other alloys or nonferrous minerals will not greatly impede industrialization, except as domestic shortages necessitate expensive imports. Manufacturing uses small enough amounts of these subsidiary minerals that they are not proportionately important cost items. Shortages of naturally occurring chemicals like sulphur or mineral fertilizers are somewhat more serious since larger quantities of these minerals are needed, and China is not well supplied with them. But vast oil reserves more than make up for this lack since oil is a primary source of chemical fertilizers.

Hydro power is a separate matter, and here China is potentially rich. The Yangtze alone has a potential greater than the total developed hydro power of the United States, and there is a wide variety of other potential sources. The Communist government has already begun to develop this power. But the heavy capital investment is difficult to raise, and the market on which profitable hydro development depends is only beginning to grow. Nevertheless, China's hydro potential is becoming a usable resource on a large scale, and is helping to stimulate the growth of the industrial-commercial economy on which its own success will rest.

Taken in broad view, China's resource base is more than adequate to permit industrialization. The force of concerted state control, national drive, organized group effort, and total mobilization of labor may well be of decisive importance. Chinese technologists have also been quick to adopt and in many cases to improve on Western models. The total record so far is impressive, especially since it has been accomplished almost entirely without external capital or other forms of aid.

The industrial pattern

Until 1949, when the Communists took over, large-scale machine manufacturing was almost entirely confined to a narrow fringe along the east coast, a few spots along the Yangtze River, notably at Wuhan, and the Mukden area of Manchuria. These were the sites where transport costs were minimized and where foreign capital was most safely and profitably invested. To the Communist government, this locational structure was undesirable, both for its association with foreign domination and because it appeared to concentrate the benefits of economic development in a small area at the edge of the country to the detriment of the great bulk of the state and population. A further objection was essentially strategic: the coastal strip was militarily vulnerable. This was emphasized by China's experience in the war with Japan from 1937 to 1945, during which all of the major established industrial areas were quickly lost to the invaders so China's economy and striking power were critically weakened. Only space, mountain barriers, and a determined will to resist kept China from being overwhelmed. The present government is anxious to develop other, better-protected industrial centers. This will not mean the decline of the existing areas of manufacturing, but will concentrate new investment and regional development closer to the center of the country and market, and on its western and northern fringes. Such a pattern of development appeals to China for essentially the same reasons that underlie Russian efforts at economic development of their interior centers in Siberia and central Asia. There is a drive in both countries to exploit previously unused resources in marginal areas and to build up their populations in a more symmetrical and more complete distribution of people and use of total resources. Thus new factories and manufacturing towns, new mines, irrigation projects, power dams, and rail lines are being constructed not only at inland sites in China Proper, such as Taiyuan, Wuhan, Kunming, Lanchow or Chungking, for example, but also in the outer areas, particularly in the northwest, Sinkiang, and Inner Mongolia. The Communist planners are interested in regional development and are willing if necessary to sacrifice some of the speed and total gains of industrialization in order to spread its benefits and strengths more widely.

Industrialization in the New China. Textile manufacturing is very important in modern China. Some is done in large factories such as this one, but smaller plants, often on the rural communes, also make textiles. *(Photo by V. Rastelli for Woodfin Camp & Associates.)*

It should be stressed, however, that although planning emphasizes heavy industry and producers' goods, China's industrial structure is still dominated by consumer goods, specifically by textiles. And about half of the total industrial production still comes from small factories, many of which might better be called workshops, including those on the rural communes. The Chinese say they want to avoid the overconcentration of industrial and urban functions, problems, and benefits in a few giant cities, and distribute them more evenly over the country as a whole. China and India are conscious rivals in their programs of economic development, since they have nearly comparable populations, existing industrial and agricultural bases, and problems. Both are aware that what is being tested in this competition includes the effectiveness of their political systems as agents of economic development. It seems clear that centralized control over a fully mobilized labor force and capital supply can produce economic development at a faster rate than can be achieved by a freer society, other things being equal, and that the benefits of growth can be distributed more equitably. Chinese growth was slowed or even stopped by a series of planning disorders and overextension between about 1959 and 1963. Then growth seemed to resume, although probably at a slower rate and with greater proportional emphasis on agricultural investment. The economy was seriously disrupted again during the disorders and policy disputes of the Cultural Revolution from 1966 through 1968. But rapid growth rates have been achieved in China at a cost in individual freedom, which is considered less important than group welfare and national goals. India, in the absence of a revolution, has allowed far more scope to a free market and to individual choices. Overall economic growth has been similar in both countries in fact, but incomes and living standards are skewed in India, with mass poverty persisting despite development.

China chooses

Any country must make choices in the creation and administration of its political system or its economy, choices that usually involve sacrifices, or at least the elimination of alternatives. In the new countries of Asia, among which one can include Communist

China with good reason, many sacrifices are being made in order to accelerate economic growth and build national strength. China has sacrificed, with few regrets, parliamentary democracy and a large measure of individual freedom. Western-style democracy and the Western concept of freedom were never part of the Chinese experience, in idea or in practice, even under the Nationalist government (Kuo Min Tang). The Communist revolution, in its drive to create a new China, has touched the lives and minds of every individual, and the state is directly involved with the minutest details of daily existence and livelihood. Although such tight control may be resented by some, and although enemies of the state are sometimes loosely defined, the concrete achievements of Communist rule in attacking some of the fundamental problems of the economy at last, in improving the national life and focusing it on the national welfare, in subordinating private or personal gain to the public good, and in regaining for China a place of prestige and influence in the world are enough to win widespread support for the government. The Peoples' Republic is attempting to build not only a new society for China but a system that can challenge all others as a model for human development and cooperation. Many of its goals seem utopian, but its success so far commands attention. As revolutionary passions cool, radicalism may give way to greater moderation. But whatever changes may occur, the new China has unquestionably re-established itself in a new position of leadership in the modern world.

Questions for further study and discussion

1. How did the average size of farms before collectivization differ between north and south China? Why?
2. What are Hong Kong's advantages for processing and manufacturing? Where else in the world have free ports been established? Why?
3. What are Peking's advantages as the national capital? What are its disadvantages?
4. Why should Canton, with its poor deltaic harbor, seem likely to remain a more important port than newly developed Chanchiang with its fine natural harbor?
5. South China and southeastern United States are broadly similar in their physical bases. Compare the human use that has been made of each, and account for the differences.
6. Political China is an enormous and highly varied area. How can one account for its consistently reasserted political unity, especially by comparison with India or Europe?

15 Japanese achievements and the problem of Korea

Mt. Fuji dominates the Japanese landscape. *(Photo from Shostal Associates.)*

Japan has frequently been compared with Britain as an island country that has played a similar role in maritime trade, colonial ambitions, and pioneer commercial-industrial development. Like Britain, Japan is an island (or rather four main islands and many smaller ones) off the coast of a populous and productive mainland, a location with clear advantages for trade. Like Britain, too, Japan has been profoundly influenced in its culture by institutions and ideas from a mainland that developed earlier. Although these similarities are striking, Japan is sharply distinct from Britain in its culture and in many aspects of its economy. Two particular geographic factors help to account for this distinctiveness, despite the locational and insular parallels. First, the Japanese islands are physically very different from Britain in the opportunities they offer for human settlement and use. Second, the continental mainland from which Japan has received so many influences was not the diverse cultural mixture of Europe, but the monolithic civilization of China. Much of Japanese culture has been an echo of China's for most of its history. Written language, art forms, religious and philosophical beliefs, and agricultural systems have been borrowed or adapted to differing degrees from the great neighboring model. Most aspects of Japanese civilization can be traced to a Chinese connection.

These influences were not as varied as those that entered Britain. The influence of Chinese civilization helped to create in Japan a culture that was more like its mainland origins than Britain was like mainland Europe. Relations between Japan and the mainland were never consistently as intimate, however. The Korea and Tsushima straits are 110 miles (177 km) wide, as against 21 miles (33.6 km) for the Strait of Dover, and there is a much larger intervening distance between Japan and China itself. This not only minimized Japanese involvements with the mainland, but helped importantly to keep the Japanese islands free from invasion or immigration. The Japanese people are a far less mixed group than the British, and influences that entered Japan have come largely by Japanese choice. Insularity was accompanied, as in Britain, by an early development of national consciousness and by a constant mental distinction between what was Japanese and what was not. This sense of nationalism and distinctiveness

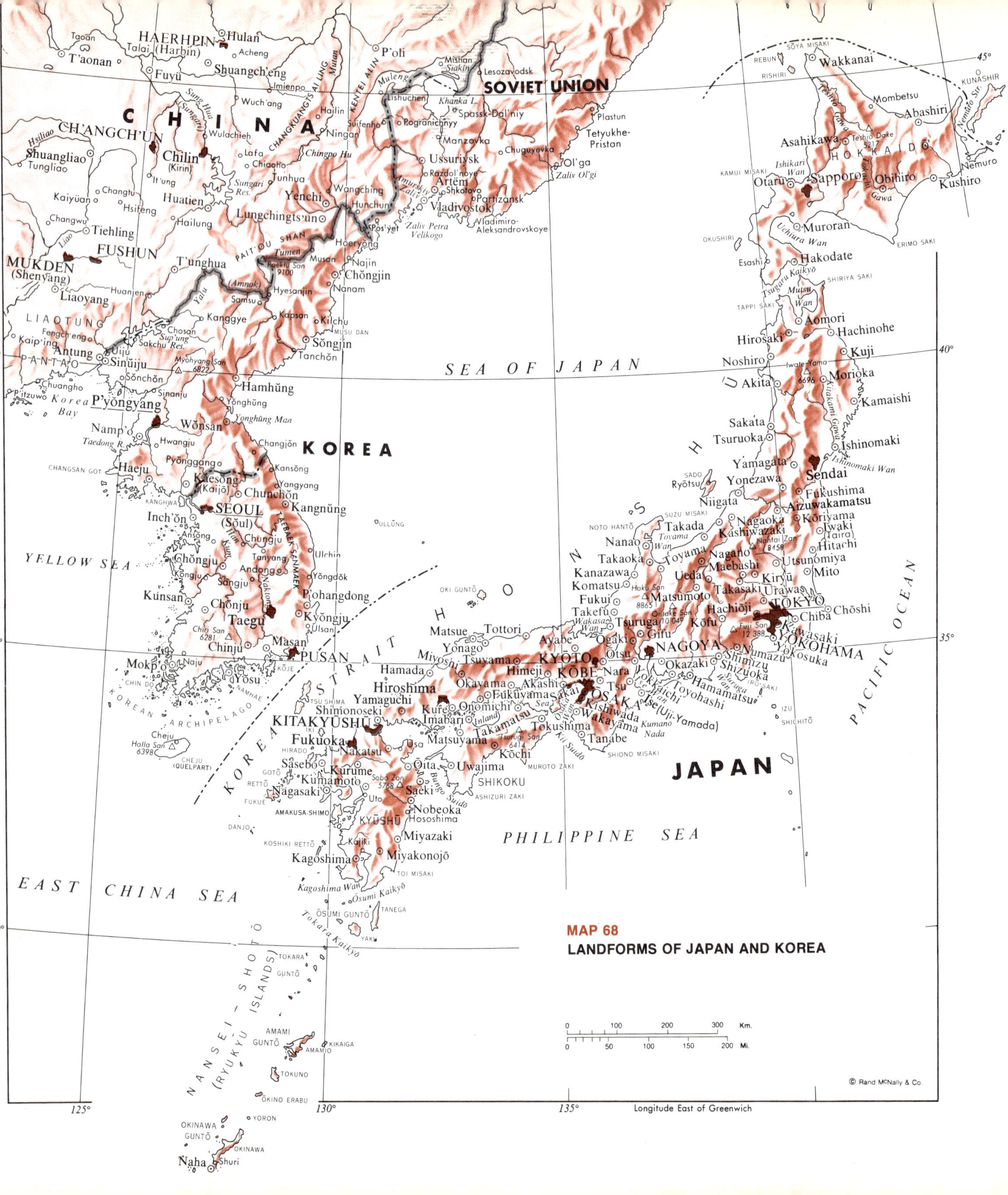

MAP 68

LANDFORMS OF JAPAN AND KOREA

helped Japan shape its version of Chinese civilization into a genuine regional entity, and later helped to ease and accelerate the transformation of Japan as Western techniques were adopted on a massive scale.

Physical framework

The Japanese demonstrate a strong sense of identity with their own physical environment and the dramatically beautiful Japanese landscape, a feeling shown in many aspects of Japanese culture. This may come partly from insularity and the limited contact with other areas by most Japanese until very recently. But whatever the reasons, Japanese attitudes have remained distinctive for their exclusive, absorbing national identity, and yet their openness to outside influences, especially in the modern period.

The overwhelming physical fact about Japan is mountains (see Map 68, page 264). Less than one-fourth of the country is level enough for agriculture of any kind, and only about 16 percent of the total land area is cultivated. Where level land exists, it is worked very intensively. Intensive use of limited resources is perhaps the most striking characteristic of the human geography of Japan. The country is only about half again the size of Britain, yet it supports nearly twice Britain's population, or over 113 million people. Much of this burden must still be carried directly by the limited amount of arable land. And yet the country is nearly self-sufficient in food, unlike Britain. The small amount of cultivated land does not mean that potentially usable areas are being neglected to any important degree. Indeed, much marginal land not well suited to field crops is pressed into use. The omnipresence of mountains has concentrated the large population almost exclusively in a series of small coastal lowlands. National population density per square mile of cultivated land is the highest in the world, a very significant fact. Cities, manufacturing, and trade are also necessarily confined to the limited lowland areas. Map 69 (page 266) shows how settlement in general hugs the sea, and how relatively empty the bulk of Japan is where mountains minimize economic opportunity.

The frequency with which mountains and sea appear in Japanese painting or photographs suggests this fundamental aspect of the landscape. Only on a foggy day or in the center of Tokyo are mountains not prominently visible. For the most part they are fairly "young" mountains, relatively recently uplifted as part of the great mountain and volcanic arc that rims the Pacific, the "circle of fire." Japan still experiences about fifteen hundred earthquakes a year of varying intensity. The mountains generally have steep slopes, rather than the older and more worn-down mountain slopes of much of China Proper. These present a strong bar to agriculture. Most streams have not yet had time to erode very far or to build up extensive flood plains or deltas. Most of the narrow coastal plains in Japan are composed of alluvium, but they are small in size. Most of the rivers are still flowing swiftly when they reach the sea, carrying with them a good deal of fine material that might otherwise be deposited on the land. Even at the mouths of many of the larger rivers there are extensive deposits of gravel and stones rather than the alluvium that helps to make the broader and older river valleys of China and India so productive agriculturally. Nowhere in Japan are there the extensive lowlands, broad river valleys, or large deltaic areas that are so important and so advantageous to the patterns of settlement and the agricultural economies of India and China. Recent volcanic action has covered much of Japan's surface with ash and lava. A good deal of the river alluvium, and hence the soils of many of the small coastal plains as well as the uplands, are derived from this source. Unfortunately, the volcanic materials are mainly acidic in their chemical composition and are not very fertile. Their light, porous structure also makes them vulnerable to serious erosion, makes irrigation more difficult, and heightens the danger of cultivation on the slopes.

Climate

To counteract this unfavorable pattern of landforms and soils, Japan has a climate that is largely free from the disastrous fluctuations characteristic of India or north China, and that is amply moist for the most part. Rainfall must be supplemented by irrigation, especially for wet rice, particularly during the summer growing season. But the annual rainfall total and its seasonal incidence are more favorable and more consistent on the whole than in China or India. No part of Japan is too cold for agriculture. Northern

MAP 69
POPULATION OF JAPAN

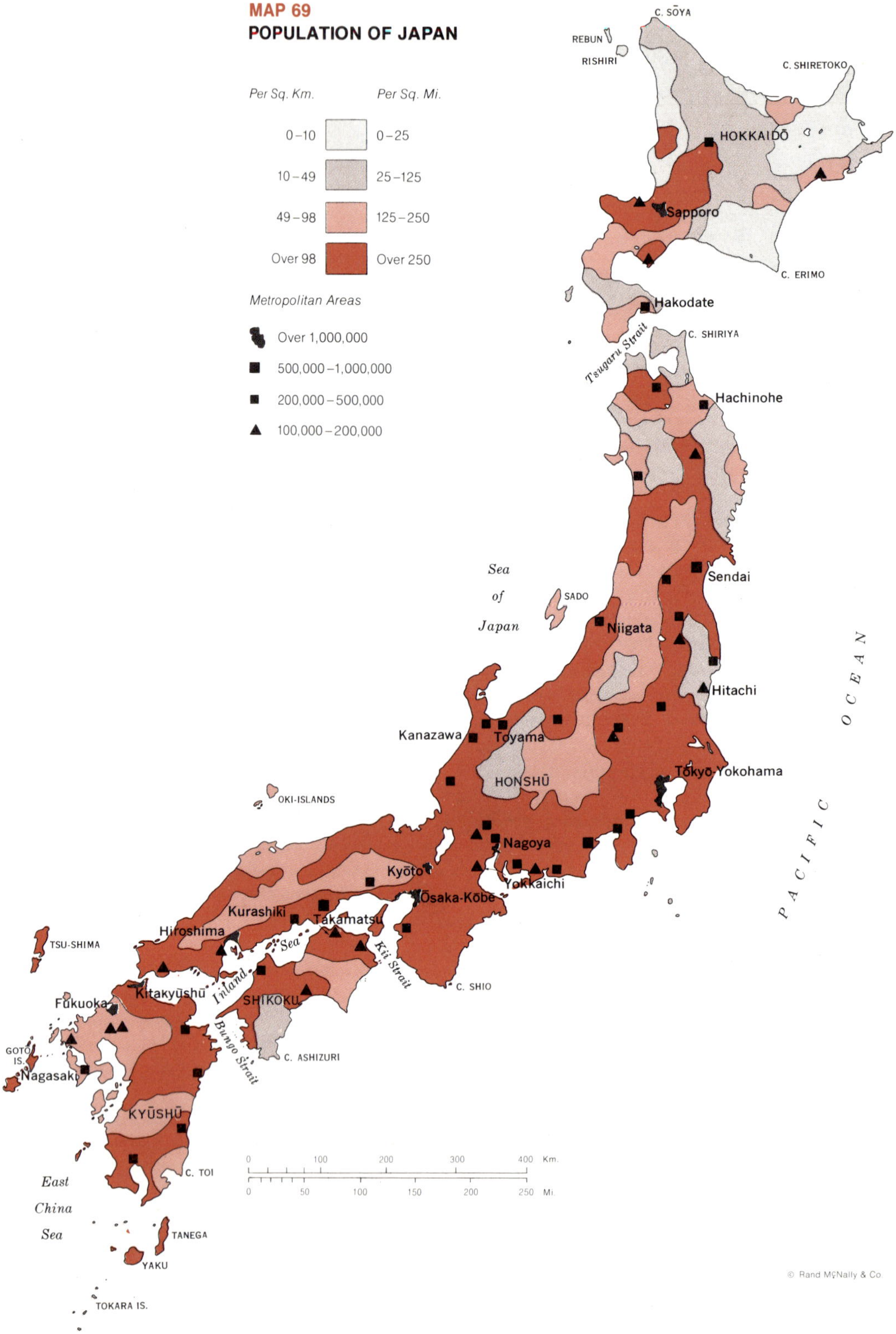

Hokkaido (the northernmost of the four main islands) is climatically marginal for agriculture on grounds of temperature, but this is compensated for to some extent by the subtropical climate and long growing season of the extreme south. As with most islands, the climate is moderated significantly by maritime influences, aided by the pattern of monsoonal winds blowing across Japan to and from the Eurasian mainland. In addition, the islands are bathed by a major warm-water stream, the Japan Current or Kuroshio, flowing north from near the equator past the east coast of Japan and on to the Aleutians and the coast of North America. This parallel to the Gulf Stream helps to create a Japanese climate that is considerably warmer than at comparable latitudes on the Asian mainland. Japan is actually quite far north, opposite Korea and Siberia. The cold, outblowing winter air from the continental landmass does reach Japan, but it is warmed somewhat in its passage across the Sea of Japan. Tokyo, for example, has a winter about like that of North Carolina.

Summers are hot, long, and moist under the influence of the inblowing monsoon, but winter and spring rainfall are also widespread. Rain is derived in large part from the same cyclonic or frontal storms that bring winter and spring rainfall to south China and the Yangtze valley. Summers are much cooler and winters much more severe in Hokkaido. Its climate is also affected by a cold current, the Oyashio, flowing southward down the west coast of northern Japan from the Sea of Okhotsk and the Bering Sea, and meeting the warm Kuroshio off Hokkaido and northern Honshu. Heavy winter snowfall is characteristic of the west coasts of Hokkaido and northern Honshu. There continental air masses, which have been warmed in their passage over the Sea of Japan in winter and have picked up considerable moisture, strike the colder, mountainous land areas as they blow outward from the Siberian high-pressure area. Typhoons are common, especially in central and southeastern Japan where most of Japan's people live, and they do heavy damage every year. But for Japan as a whole, climate is a major asset rather than a problem. China must contend with hills, cold, and aridity. The extent of mountains in Japan perhaps balances out the advantages of its climate, but without those advantages Japanese development would have been very different.

The spread of Japanese settlement

The origin of the Japanese people is not clearly known, but their spoken language suggests that they came from the mainland of north Asia. There may also have been some migrations into Japan from the south. Early Japanese history is cloudy, but it seems virtually certain that the main group of what became the Japanese people entered from Korea and settled first in north Kyushu, the most direct landing point from Korea, about the second or third century B.C. By the fourth century A.D. Japanese settlement had probably spread around all sides of the Inland Sea (Seto-Naikai), and had established its major base in the Kansai region around Kobe and Osaka, near the eastern end of the Inland Sea. This productive lowland area reaching as far inland as Kyoto had easy sea communications, and became the dominant core of Japanese settlement and economic activity (see Map 70, page 268) until the nineteenth century. During the great period of cultural borrowing from T'ang China in the eighth century, an imperial capital on the model of Ch'ang An (now Sian) was built at Nara on the Yamato plain, some 20 miles (32 km) south of Kyoto. In 794 the capital was moved to Kyoto, where it remained until 1869. From the sixteenth century onward, Tokyo (formerly Edo—*Toyko* means "eastern capital") began to grow. It was the seat of the shoguns, the real rulers during much of Japanese history (since the emperor was only a figurehead), as well as a commercial center. It probably reached 1 million in population before the end of the seventeenth century, but its enormous modern growth dates from its selection as the national and imperial capital in 1869.

Tokyo's growth and the development of northern Japan came late in Japanese history. For many centuries the boundary of Japanese settlement lay just north of Kyoto along Lake Biwa (Biwa-ko). Beyond, the land was still occupied by a relatively primitive people, the Ainu. They were unrelated to the Japanese (although there was considerable intermarriage and absorption), and were gradually pushed northward by expanding Japanese settlement. The Ainu were the original inhabitants of the whole of Japan but now are a tiny and dwindling group found only in northern Hokkaido.

The northward and eastward advance of the

MAP 70
LAND USE IN JAPAN AND KOREA

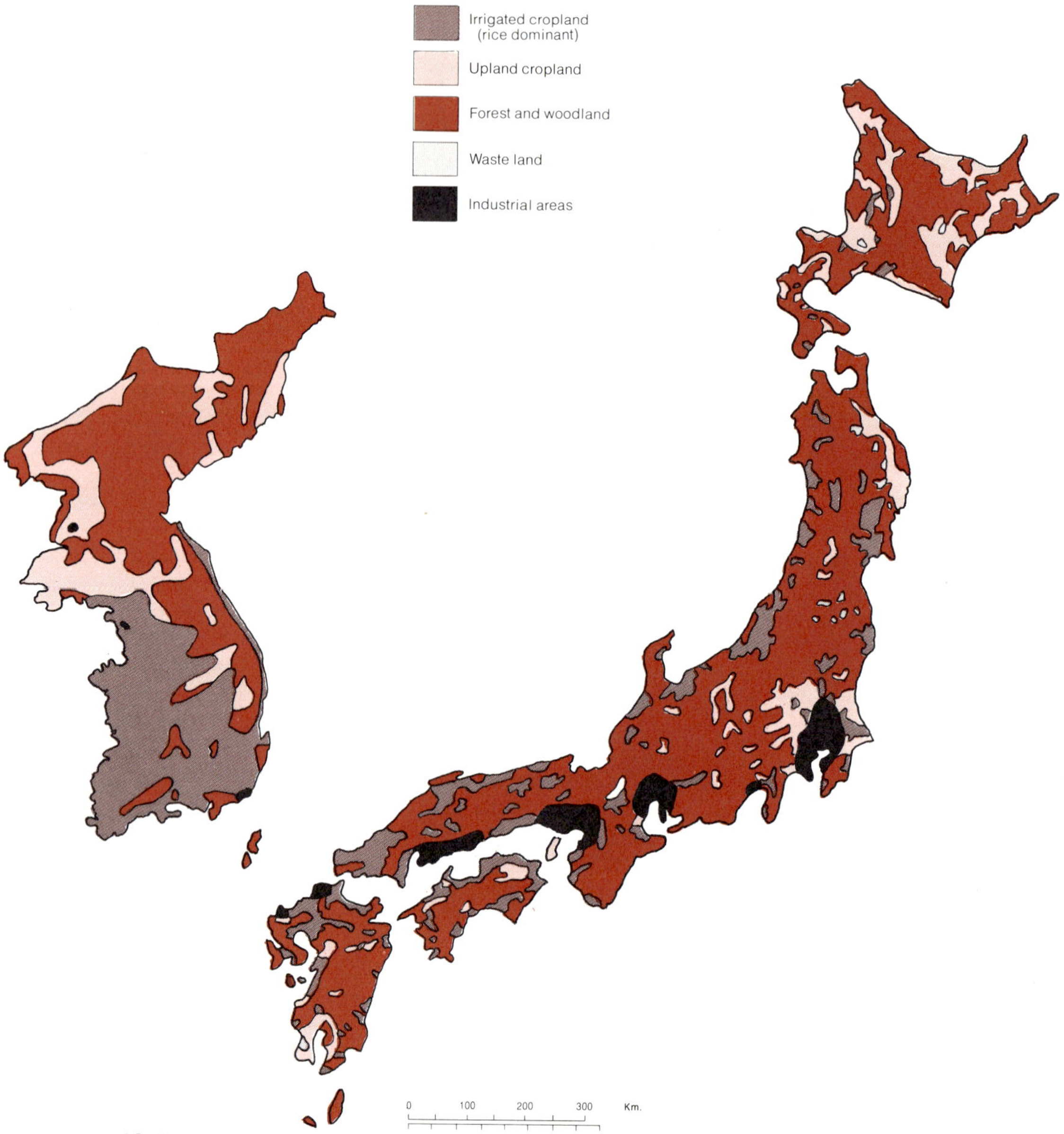

Japanese was delayed partly because of the strong advantages and attractions of the Kansai area and the Inland Sea. It was a productive area and had easy communications with the mainland. There was also much greater difficulty in using the land in northern Honshu and Hokkaido productively. The Kanto plain around Tokyo, although larger than the Yamato plain, has a less permissive climate and is more difficult to irrigate. Most of it is far from hills, which are a great help in distributing irrigation water. Highly porous soils and a low water table in the terraces at the edge of the plain caused problems, especially for rice growing and irrigation. The alluvial soil areas of the plain itself were often poorly drained. Marshes and several small rivers presented a difficult flood problem fully solved by modern techniques only in the nineteenth and twentieth centuries. North of the Kanto lowlands are especially scarce and climate increasingly limiting. Even now there are few large cities (Sapporo and Sendai, the largest, are both only about half a million), and little manufacturing north of Tokyo. The great bulk of the Japanese population is in the southern and western half of the country (see Map 69, page 266).

Northern Honshu beyond Tokyo was not fully occupied by the Japanese until nearly the tenth century A.D., and Hokkaido remains a frontier still. Only since about 1880, with the drive for military preparedness against the Russians (especially from 1890 to 1945), the rapid rise of the national population, and the need for more land and food have Japanese settled in Hokkaido in large numbers. But the island is still sparsely occupied and economically underdeveloped by Japanese standards. There are only about 7 million inhabitants, several million fewer than the city of Tokyo alone. The growing season is critically short. Hokkaido has a net import of food, and usually is unable to pay entirely for its imports by its exports.

Within Japan as a whole, the two major cores of settlement and economic activity, Tokyo and the Kanto plain, and Osaka-Kobe-Kyoto are now nearly one metropolitan area, joined by a string of industrial cities and dense population. Political factors have proved important, and Tokyo has become the giant. But the double core forms Japan's densest knot of population, the most highly urbanized and industrialized area, and the main focus of transport lines. It dominates Japanese commerce and sets the political, social, and cultural patterns.

Economic achievement

The achievement of Japanese economic development, not only in building up manufacturing and trade, but in wringing from the land an amazing amount of food, commands admiration. Despite the rapid increase in Japan's population (it approximately tripled between 1860 and 1957), the four home islands now manage to provide nearly 90 percent of their food needs. Over half of the land surface is still covered with forests. For the most part these areas are so steep or infertile that nothing else can be done profitably. From the small amount of cultivated land and from the sea, the Japanese obtain very high yields by an efficient combination of intensive methods and technological improvements in agriculture. Measured in terms of arable land per capita, Japan is far more overpopulated than any country in the world. In more accurate terms, this is not true, because total production in Japan, from agriculture and fishing and from manufacturing and trade, is great enough to provide relatively high per capita incomes. Japan has long since entered the charmed circle of the developed nations, and is now the third largest industrial power in the world, with a similar rank in gross national product. But Japan is not well endowed with the mineral fuels and ores needed for industrial growth. In all branches of economic life the Japanese have produced a high level of development on a physical framework that is far from favorable. It is also in this context that the commercial and industrial advantages of location, especially for the use of the sea, stand out so strongly as modern Japan's most important nonhuman resource.

Mineral resources

Coal

Fortunately coal is more plentiful in Japan than most other mineral resources. But Japanese coal is relatively small in amount, and much of it is expensive to mine. There is also a marked shortage of good coking coal. The annual consumption of coal in the United States would exhaust Japan's reserves in two or three years, and Japan must be dependent on coal and oil imports to a growing extent. Especially in the early period of Japanese industrialization, the second half of the nineteenth century, coal provided an important advantage. The largest developed deposit was in northwestern Kyushu near Kitakyushu, close to the sea, where it could be distributed cheaply by water transport (see Map 68, page 264). Since most settlement and potential manufacturing sites were located on the coastal lowlands, it was possible to supply nearly every important part of Japan with cheap coal. Population and manufacturing are particularly concentrated around the head of the Inland Sea, where Kyushu coal could be delivered at very low cost: The other important area of coal deposits, in Hokkaido, was developed later. It, too, was near the sea so its coal could be distributed largely by water to supply the northern half of Japan.

As coal mining progressed, the richer and larger seams were exhausted, and extraction costs rose steeply. Japan is no longer an area of cheap coal and must import growing amounts. The occurrence of coal reserves near the sea still helps to keep the total costs of delivered coal down, and has also attracted considerable heavy manufacturing associated with the major coal-mining areas to northern Kyushu. Until about 1956, despite its small reserves and high extraction costs, Japan remained the largest coal producer in monsoon Asia. This emphasizes the critical importance of location, not only for coal production, but also for development of the industrial structure in Japan that created the demand and the means for a wider use of resources.

Iron

Iron ore is scarce and Japan must import over 90 percent of its needs. As in the case of coal, the Japanese make use of low-quality or high-cost domestic deposits that would be ignored as uneconomic in a more richly endowed country. While this may be considered necessary in order to save foreign exchange, it represents added costs and still results in relatively little production. United States' ore consumption would exhaust all of Japan's reserves, regardless of quality, in less than a year. Ore imports make up a large part of Japan's foreign trade. The iron and steel industry also makes extensive use of scrap metal, both from the Japanese market and from foreign supplies. When the Japanese controlled Manchuria and Korea, they had an assured source of ore and pig iron, as well as of coal. The low grade of Manchurian ores made it more profitable to convert them to pig iron and to ship iron to Japan in that form. Japanese concessions in China included mining rights at Tayeh near the Yangtze, and ores were shipped from there as well. With the loss of empire and the rise of an industrializing China, Japan must now look elsewhere for this essential raw material. It is increasingly dependent on more distant supplies from India, southeast Asia, and North and South America.

Oil

Petroleum must be added to the list of raw materials that Japan imports to support its industrial structure. The home islands produce very little, mostly in northwestern Honshu. The annual production of crude oil is less than the daily production of the United States, and the domestic reserves are tiny. The southern half of Sakhalin Island was controlled by Japan from 1905 to 1945, and petroleum developed and produced there by the Japanese was available to the home market. But Sakhalin was returned to Russia at the end of the Second World War. Oil is essential to a variety of modern industrial and transport operations, so Japan must import and pay for it with exports or services. Oil imports and Japanese refining capacity have risen very steeply especially since 1950 to feed a booming industrial development, including the recent rise of large-scale automobile production. Japan is now a major world producer and exporter of cars and trucks, and the automobile age with all its problems has come to the domestic market. Oil has also replaced coal in many manufacturing processes and as the primary railway fuel, as it has done in the United States. It is increasingly used as the fuel in thermal power stations. Japan's relative poverty in mineral resources makes its vital dependence on foreign trade clear. It also emphasizes that minerals are not the controlling element in the industrialization process. Of the other major minerals, Japan is well supplied only with sulphur, and is a net exporter. Domestic reserves of copper, nonferrous

Chinese culture in Japanese hands—the Heian Shrine, Kyoto. A good sample of Japanese adaptation of the Chinese model, and of the Japanese gift for blending architecture and landscaping. Notice how the tree on the right has been trained to echo and emphasize the line of the building. Despite their adoption of Western styles in other respects, most Japanese are still keenly interested in their artistic tradition. *(From Japan Information Service.)*

metals, ferroalloys, mineral fertilizers, and even salt are inadequate in differing degrees to meet Japanese demand, so the balance must be made up with imports.

Hydro power

Even in hydroelectric power, Japan is not rich, although the Japanese have made fuller use of this resource, on a proportional basis, than any country in the world. That is understandable, considering resource poverty in other respects and the high level of technology and market development characteristic of Japan. Nevertheless, it is astounding that Japanese hydro production is about in the same size category as the huge continental producers such as the Soviet Union, the United States, Canada, and China, whose hydro potentials vastly exceed Japan's. A central mountain spine, steep slopes, ample and well-distributed rainfall, and the clustering of the major markets at the foot of the mountains create a total situation conducive to hydro development. Power from this source plays a major role in Japanese manufacturing. Unfortunately, Japan's hydro resources have already been largely developed, and there are periodic shortages of power. Most of the remaining potential hydro sites are high in cost because the volume of water is too small or too irregular, the site is too far from a large enough market, or there are engineering difficulties. The scattered location of Japan's hydro resources on a variety of relatively small rivers and mountain streams is an advantage in some respects. This helped to make power widely available and meant that industrial development was not limited to one or a few sites where coal was cheap. But hydro power may be more expensive than power from coal or oil, especially as the less economic sites are developed. The necessarily small size of most of the hydro plants may also mean higher costs since it limits the economies of scale. Japan is turning to oil-fired thermal plants, placed on the coast where oil imports, refineries, and the market are concentrated, as the chief source of electric power.

Agricultural development

Industrial and agricultural development are closely linked. Japan in 1868 (the beginning of the Meiji period, when the old government was overthrown and a new group of Western-minded reformers came to power) lacked capital for industrialization. Agriculture was an important source of capital. The farmer, in addition to paying for a share of industrialization through taxation, also had to provide food for the rapidly increasing urban and industrial labor force. Therefore, it was critically important to increase agricultural efficiency and total production, despite the physical limits imposed by relief and by soils. Between 1880 and 1940 rice yields per acre doubled, and total food production nearly tripled. Both figures are still rising. Japan's experience demonstrates what can be done, even in an already intensive agricultural system. Food production was increased with the help of technical improvements like better types of seed, chemical fertilizers, and better equipment, and with careful land-use surveys. Meiji Japan felt an urgency about industrialization and the building of economic strength in order to escape the colonial fate of much of the rest of Asia. In the interests of speed, because of the shortage of domestic sources of capital and entrepreneurial skills, the reform government took the greater part of the job upon itself. Although government ownership, management, or subsidy became important in manufacturing and trade, government direction, research, and management were equally important in agriculture. This central planning, combined with a long tradition of direction in Japan's closely ordered society, helped to account for the successful and rapid transformation of Japan from a typically poor and economically undeveloped Asian country into a highly industrialized and efficient economy in about fifty years.

With good reason Japanese agriculture can be called the most efficient in the world, in terms of the use it makes of its available resources. Japanese national average rice yields are exceeded only in a few small, favored areas like the Po valley or parts of Spain, despite Japan's far from ideal climatic and soil conditions for rice. Climate and soil are clearly better, for example, in deltaic south China or southeast Asia. Japanese wheat yields are also high, even though wheat is grown mainly on lower-quality land or as a dry winter crop between rice plantings, and is given much less attention than rice. High yields under less than ideal conditions are typical of nearly every Japanese crop. They result from extremely intensive cultivation plus highly developed techniques. Rising labor costs and the movement of much farm labor into urban and industrial employment, plus the booming technical capacity of Japanese manufacturing, have led to a very rapid growth of agricultural mechanization. In some areas mechanization already exceeds even American levels, as Japanese economic conditions increasingly come to resemble those in the United States. In Japan too, more mechanization has been accompanied by continued sharp rises in yields and total output. These increases have not come primarily as a result of the use of machines themselves, but through technological improvements in crop strains, heavier and more effective use of fertilizers and pesticides, and more efficient irrigation. Three or even four crops a year are made possible, especially close to the market, by protecting plots against cold with huge plastic sheets. In some truck garden areas the landscape has been transformed to appear like an immense greenhouse that makes maximum use of sunshine and warmth. This is intensive use of land close to the market in a situation where all agricultural land is scarce. Other Asian landscapes are crowded, but in none of them do the "many men on little land" make such careful and productive use of what they have.

Wherever possible villages are located on hillsides or on land that cannot be used for agriculture. Houses and gardens are small everywhere, not only to conserve wood but to conserve precious crop land. Less than one-sixth of the population is still employed in agriculture and fishing, although many of these people are also engaged in seasonal or part-time manufacturing. Very high yields help to account for the unusually small size of the farms. The national average, after the land reforms carried out under the American occupation (which did not greatly affect farm sizes), is about 2.8 acres (1.1 hectares). This average includes Hokkaido where, in a less permissive climate, the farms average over 8 acres (3.3 hectares). Small farm sizes—over half are smaller than the 2.8-acre average—further accentuate the intensiveness that is so characteristic of nearly every aspect of Japanese life. Still, it is not surprising that about three-fourths of Japanese farm families have some nonfarm income from factory work, handicraft production, retailing, or other employment, and that there is much commuting to cities.

Crops

Rice is the major crop of Japanese agriculture by far, and is even more dominant than in China, since some special varieties can be grown even in Hokkaido. The virtues of rice as a producer of calories per acre and the traditional cultural value that most Asian societies place on it mean that it is grown in Japan, as elsewhere, even in physically unsuitable places where another crop might represent more rational and more productive land use. But for the most part irrigated rice is the most efficient crop, and it occupies about half of Japan's crop area. Rice is given the best land, particularly the irrigable land, which is also likely to have the most productive soils. It is nearly always lower down on the slope or on the valley floor, where alluvial deposits or soil movement down the slope benefits it at the expense of higher land. It is in rice yields that the major agricultural increases have been won since 1950, as a result of these methods. Two and even three crops of rice per year are now grown over an even much wider area than five years ago. The use of recently developed early-maturing and high-yielding varieties in effect has greatly extended the usable growing season. Japan has become self-sufficient in rice, and has begun to produce small surpluses for export despite its still growing population. However, it imports growing amounts of wheat. Wheat is usually grown as a subsidiary crop or between rice harvests without irrigation during the winter and spring. Barley and other small grains make use of higher, unirrigated, or less fertile land. Vegetables and beans may be rotated with rice or grown on field borders and in tiny kitchen gardens. Near the cities vegetables and fruit, especially oranges, are more important as commercial crops. About half of Japan's population lives in cities of 50 thousand or more. There are about five hundred such cities, heavily concentrated in the coastal lowlands where most of the farm land is. By its productiveness farming must compete for land with streets and buildings. Especially intensive cultivation nourished by urban night soil, particularly vegetables, continues into the suburbs. The Japanese have characteristically made wide use of both sweet and white potatoes, after their introduction from the Western

Hemisphere. Potatoes are an important crop particularly in the poorer-soil areas unfit for cereals, or where the climate is marginal.

Animals

Animals are far less numerous in the agricultural system than in China or India. This is largely due to human pressure on the land. Little can be spared for fodder crops or for pasture. Most of the uncultivated slopes are too steep and sparse for anything but goats. Meat and milk production are extravagant of land. In North America, where meat and milk are important items in the diet, it was estimated that twenty times as much land was needed to feed one person at the dietary standard demanded than was needed in Japan to feed one Japanese at his standard. The same authority estimated that agricultural land in Japan, in terms of total food production, supported between six and seven persons per acre on an average, probably the highest in the world. The figure for India was estimated as barely one person per acre.[1] Traditional Japanese land use, however, is beginning to change, primarily as a result of overall economic growth, and specifically of the steep rise in income levels, manufacturing, and urbanization. Beef and other meat and milk are increasing in the Japanese diet, and dairying areas are growing around most of the large cities. This is a good example of the impact of continued economic development on land use; Japanese agriculture is being transformed by industralization and its many consequences.

Tea and silk

Tree crops have made possible the only significant economic use of most of Japan's agriculturally unrewarding uplands. Apart from timber and fruit, the two most important tree crops in Japan are tea and mulberry (whose leaves are used to feed silkworms). Japanese tea has suffered from competition from cheaper plantation-grown tea, especially in India and Sri Lanka. The silk industry, once the largest single source of foreign exchange and particularly important because most of the exports went to the United States, has been hard hit by the development of rayon and nylon, although it has recovered slightly in recent years. Silk culture is almost a Japanese and Chinese monopoly (Chinese silk exports were largely destroyed by more efficient Japanese competition).[2] It requires prodigious amounts of hand labor, especially in feeding and caring for the voracious worms until they spin their cocoons. This stage of the industry is carried on to a large extent on a household basis, using family labor. It is concentrated in areas of dense population near uplands or other land that can be spared to grow mulberry. In part for this reason, silk production or mulberry growing cannot make use of more than a small part of the uplands. Even setting aside the dwindling silk industry and the loss of tea export markets, it is clear that the best use for most of the uplands is forest.

Forests

Trees are the only possible use for most of the uplands, and, as already pointed out, forests cover over half of Japan's area. Any wholesale deforestation would be disastrous. Trees are essential, not only to conserve soil, water, and hydro potential, but also to supply the economy with wood. These two roles are in conflict. Since the population has grown rapidly since the 1870s, the forests have suffered increasingly. More trees are being cut down than are being replaced by natural growth or by planting, and the cut-to-growth ratio is dangerously close to two to one. This is true even though many of the forest stands are too inaccessible, because of distance or heavy relief, to make their cutting economic. Those that are accessible are extensively cut over. Still Japan must depend more and more on wood imports. It has been called a "wood civilization"; traditionally it has made heavy use of wood, as Medieval Europe and modern Scandinavia have done, because metals and stone were much more expensive, and wood was accessible, plentiful, and cheap. This is no longer true in Japan, but traditional patterns persist. Most houses are built of wood and paper. Many household fires still burn charcoal instead of precious coal or oil. Japanese paper, furniture, and other industries demand huge quantities of wood. Commercial forest cutting is accompanied by a great deal of subsistence cutting by farmers for their own needs, especially

[1] L. D. Stamp, "Measurement of Land Resources," *Geographical Review*, 48 (1958), 1–15.

[2] Over 90 percent of world silk production still comes from Japan (which has about 80 percent), China, and Korea. Production in India, Turkey, Italy, and France has always been minor.

where upland forests and population concentrations meet. In many such areas the original forest cover has been almost entirely removed. It has been replaced by scrubby second growth or by coppice growth, sprouting from the stumps of hardwood trees and cut over periodically as twigs and small branches for cooking fuel or to make charcoal. Forests cut over for fuel or charcoal are no longer usable for timber. After several years of coppice cutting, the exhausted trees die, and the area is taken over by coarse, wild grass and brush, or *genya*. Genya has been spreading for many years, and erosion is a growing problem. Like the Chinese, the Japanese have long understood the importance of forest cover on slopes and the principles of sound forest management. But their situation leaves them little choice. To import more wood in order to avoid overtaxing their domestic forests puts a further burden on foreign exchange, which must pay for imports of the many vital commodities Japan lacks. Wood imports in fact have continued to increase, but so has domestic cutting. Selective cutting and extensive replanting, practiced for many years by the Japanese, must do what they can. But with population still increasing, it is a losing battle, and one on which much besides wood supply depends.

Fishing

The Japanese catch approximately one-sixth of all the fish caught commercially in the world. Fish is second only to rice in the Japanese diet, by weight and by calories. Practically every meal centers on these two mainstays. Fish is eaten in a great variety of forms: salted, smoked, pickled, raw, and cooked. Seafood includes many types seldom eaten elsewhere, such as seaweed, whale, and octopus. Nothing caught is wasted. In addition to being made into animal feed and fish oils, a great deal of the catch, especially the scraps and entrails, is used as fertilizer. The importance of fishing in Japan results from a combination of opportunity and need.

The need is obvious. High population densities in relation to arable land (sometimes called *nutritional density*) result in heavy dependence on starchy cereals and the neglect of protein. The opportunities are varied. Japan is relatively small, the three main islands are somewhat elongated, and no part of the country is far from the sea. All of the largest cities are on the sea or within 30 miles (48 km) of it. There are no towns larger than 50 thousand more than 40 miles (64 km) from the sea without a navigable waterway connecting them to it. Rural population is also concentrated along and near the coast where the arable land occurs and where seafood can be distributed cheaply and quickly to the bulk of the market. The coastline is very long and highly indented, which maximizes contact with the sea and provides numerous good harbors, especially for fishing boats. There is a fairly extensive continental shelf around Japan, where relatively shallow waters and penetration of sunlight encourage the growth of plankton and other organisms on which fish feed. A fortunate convergence of cold and warm currents, especially off Hokkaido where the Kuroshio and Oyashio meet and mix, is highly favorable to plankton and fish growth. Japan extends from 30° to 46° latitude, from near-arctic waters in the north to near-tropical in the south, so a great variety of fish occur within easy range.

As the American corn-belt farmer raises animals, so many Japanese farmers who live close to the seacoast also fish. Villages on the coast are usually engaged in both agriculture and fishing. The fields run at right angles to the beach on which the boats are drawn up. The houses are strung out in a thin line. The husband may fish while the wife and younger children farm, or all may fish in the agricultural slack season and all farm during peak agricultural labor periods. Eels, shellfish, and seaweed are *farmed* in carefully prepared beds close to shore. Fish caught by the small boats from seaside villages are mainly marketed locally. Although there is some drying and preserving for wider sale, fishing by these boats is mostly a subsistence activity. The boats do not range far, can seldom be out overnight, and do not venture into the deep-sea fishing grounds. The bulk of the commercial catch is taken by larger power boats and net trawlers whose range is very wide. They work out of the main ports, and may be at sea for several weeks. Japanese fishermen in fact are active in every ocean in the world, including the Antarctic, where they account for a large part of the world whale catch (restricted by international law). But they get most of their fish from the Pacific within 1,000 miles (1610 km) of Japan.

The northern waters provide salmon, crab, halibut, and other high-priced fish, a great deal of which is exported after canning, much of it to the dollar market. The temperate and southern waters yield a large variety of fish, but tuna is especially important for both domestic consumption and export. The Second World War had serious effects on Japanese fishing, particularly the loss of rich fishing rights off the coasts of Sakhalin, Korea, and the Kuril Islands, plus the use of fishing bases there. Sakhalin and the Kurils were absorbed by the victorious Soviet Union in 1945. The government of South Korea has barred Japanese fishermen from all waters around Korea since 1945 to an extent the Japanese consider unreasonable. But the Japanese record during their control of Korea was not commendable, and is responsible for great Korean bitterness against Japan. Korea is also anxious to increase its own fishing industry and to eliminate Japanese use of nearby fishing grounds. Even with these handicaps, the fish resources available to Japan are rich. But as in the case of forests, they are being depleted rapidly by overuse. More and more effort is necessary to obtain a given catch, and it is not easy to find a solution, for Japan must eat and must export fish to help pay for what it must import.

Food and population

Agriculture is operated at very nearly peak efficiency, and there is little unused land that could be pressed into production. With heavy and increasing use of chemical fertilizers, food production is still rising at about the same rate as the population. Japan, incredibly enough, is now the world's third largest consumer of nitrogenous fertilizers. Naturally poor soils are expensive, especially when they must support so many people. Tenancy and rents were greatly reduced by the land-reform program of the American occupation, most of which remains in force and helps to encourage initiative, land improvement, and higher yields. But the critical problem of supporting too many people on too little arable land remains. Fortunately population growth has begun to slow down. In 1872 the total population was approximately 33 million. It had probably been about the same for the preceding several centuries, with high birth and death rates. The growth of industrialization, commercialization, transportation, health services, and agricultural improvement brought a sharp and rapid drop in the death rate. But this was not accompanied by a significant fall in the birth rate until the 1950s. As a result, the population nearly doubled in the sixty years between 1872 and 1932 and reached 100 million in 1967. With the recent trend toward a declining birth rate, which by 1962 had dropped below that in the United States, it is hoped the population may level off by the end of the present century. This may suggest some measure of the population problem in other parts of Asia now in the early stages of industrialization, and the time lag between falling death rates and falling birth rates. But Japan's rural areas are still heavily populated, migration to the cities is increasing, and the economy is becoming proportionately less dependent on agriculture and more dependent on manufacturing and trade.

Industrialization

Japan's high standard of living is due primarily to its remarkable success in creating an extensive and profitable structure of manufacturing and trade. By 1970 over half of the national income came from manufacturing and mining, and over a third from trade and finance. Japan's annual steel production, now the third largest in the world, overtook West Germany's in 1962. It now exceeds production in Latin America, Africa, and the rest of Asia combined. Japan has also become the world's second largest producer of motor vehicles (and is a major exporter), builds well over half of the world's ships, is second in petroleum refining and the building of television sets, and leads in the production and export of radios. The magnitude of this development and the speed with which it was accomplished contrast especially with the slow and difficult progress of industrialization and commercialization in China before 1950, and with the belated growth in India. Between 1870 and 1900 Japan laid the foundations of a major industrial-commercial structure. By 1930 it was among the world leaders in both respects, competing strongly with European traders and manufacturers in the markets of Asia, and even in Europe and the Americas. Even after the unparalleled devastation of its industrial cities and merchant marine during the Second World War, Japan fully recovered its factory output and trading fleet by about 1951 and was again in a strong position on the world market. This was accomplished despite the loss of the former Japanese holdings in Korea, Manchuria, and China, which had provided most of the necessary coal, iron, and food imports, as well as export markets. Although Japan must now pay higher prices for these essential imports and must compete harder for export markets, its industrial and commercial structure is much greater now than it was at its prewar height. This is partly due to the rapidly growing home market (where income levels, as well as total numbers, are rising). It is also based in part on the same set of advantages that made the rapid transformation of Japan possible after 1870. Japan accomplished in some fifty years the agricultural-commercial-industrial revolution that took three or four times as long in Britain, Western Europe, and America.

Rice threshing in Japan. Although Japanese farms are very small because of the mountainous terrain, farming methods are being constantly mechanized and modernized. Japanese farmers employ modern farm machines designed to fit their needs and requirements. *(Photo from Consulate General of Japan, N.Y.)*

Japan's advantages

An industrial latecomer like Japan had certain advantages in being able to apply immediately the technological and organizational methods developed through trial and error over a longer period in the pioneer industrial areas. But there have been many would-be latecomers with the same opportunity. So far only Germany (which was a pioneer in part) and the Soviet Union have approached the scope and spread of the Japanese accomplishment. What were Japan's advantages?

An early advantage, although a passive one, was Japan's small size and the easy circulation insularity made possible. The contrast with China and India, with Brazil, or with most of the so-called underdeveloped areas is obvious. The parallel with Britain, where the modern Industrial Revolution began, is also significant. National unification was never really an issue in Japan. A national market in commercial terms for many commodities had begun by the seventeenth century. This development helps to explain the large size of both Osaka and Edo (Tokyo) as early as 1700. No part of Japan is far from the sea. Most of the population and most of the production areas are on or near the coast where they can be served easily by water transport. This was a crucial asset in the early stages of industrialization, as it was in Britain. It continues to make an industrial-commercial role possible for Japan even when much of the raw materials must be imported. A highly efficient railway system supplements the encircling sea routes. Since most bulky shipments or longer freight hauls go by water, the railroad system cut construction and operating costs by using a smaller gauge (3 ft. 6 in. [1.07 m] instead of the 4 ft. 8 in. [1.42 m] gauge standard in most of the rest of the world outside the Soviet Union). This gauge is also particularly well suited to the sharp curves and grades required by Japan's mountains. (The ultrahigh-speed Tokkaido line between Tokyo and Kitakyushu, primarily for passenger traffic, is broad gauge.) Until recently motor roads were poorly developed. But rapidly rising incomes and equally rapid increases in automobile production have been accompanied by extensive highway construction. A large number of relatively easily tapped and well-distributed hydro power sites helped Japan's rapid industrialization, especially as the cost of coal rose and its supply dwindled. In addition to being an asset for hydro power, climate was even more important in helping to further Japanese agricultural successes, on which industrialization also depended.

Location

Japan is rich locationally. The four home islands are well situated to assemble, process, and distribute goods to and from most of eastern Asia and the western Pacific. This is an area of varied resources and huge populations for whom Japan is a convenient and skilled supplier of manufactured goods and commercial services. Japan lies directly on the great circle routes across the Pacific between North America and eastern Asia. Cheap sea routes put it within easy reach of the billion or more people clustered near the coasts of Asia and its fringing islands. These locational advantages were undeveloped for many centuries, during which Japan's insularity resulted more in isolation than in exchange. Industrialization and commercialization turned insularity and maritime position into a valuable resource. Location lies at the basis not only of Japan's continued industrial prominence but also of its leading position in maritime trade. Japanese shipping dominates the western Pacific, competes strongly for leadership in the trans-Pacific trade, and carries a large share not only of Japanese imports and exports but of the trade of most of the other countries of Asia. Commercial and financial services accompany the carrying trade, as in Britain, and earn important foreign exchange ("invisible imports") as well as providing about a third of the national income.

Attitudes and directed change

Japanese willingness to use foreign ideas and techniques may also be related to insularity and to the long-established habit of adopting selected aspects of Chinese civilization. Insularity helped to stimulate the early growth of Japanese nationalism and sense of identity. Cultural and technical borrowing may have been easier to practice because Japan did not need to fear for its physical security or for the loss of its regional identity, and because it had always recognized that it could profitably learn from its great neighbor. Borrowing from Western models was not a new experience, and was specifically designed to safeguard Japan and preserve its identity. Modernization and Westernization became a national crusade. Its effectiveness was also greatly increased by the relatively small size of the country and the relative ease with which national life could be organized and mobilized, and innovation diffused. In many ways pre-Meiji Japan was more like feudal Medieval Europe than like monolithic, bureaucratic China. The central state and its bureaucracy, although authoritarian in nature, were much less strongly developed than in China. Japanese social structure was probably less free and more accustomed to both internal and external direction, as well as to strong local and hierarchical authority, commanding organized group effort. The transition from feudal to modern in Japan proved relatively easy. When compared with the prolonged birth pangs of modern China, Japan's expanded autocratic state quickly adjusted itself to a new role of economic direction in close partnership with the emerging class of Western-style businessmen. The role of force in the industrialization process has already been referred to, especially in the discussion of the Soviet Union and China. In Japan until the 1950s, force has clearly been of major importance, especially in preserving a relatively low level of wages and consumption without significant violent protest, and a correspondingly high rate of saving for investment. The maintenance of civil order by a strong central government and of organized, collective cooperation in the drive for modernization in this relatively small and closely integrated country may have been an equally important advantage for economic growth. Finally, Japan had both the luck and the ability to acquire the largest share of the world market for silk. In the second half of the nineteenth century, the devastation of the Taiping Rebellion in China and a silkworm disease in Europe provided the opportunity for world leadership at a time when Japanese industrialization needed capital and foreign exchange. Silk production and mulberry cultivation on slopes could be expanded greatly without taking significant amounts of agricultural land. The First World War, which left Japan free to take over Eastern markets and prevented a threatened financial crisis, was another break at a critical time. Japan was also fortunate in beginning its drive for economic and industrial growth at a time when foreign capital was much more available than it became after about 1905. External loans and investments played a major role in Japan's rapid industrialization.

But it would have been difficult in 1850 to predict Japan's success, especially by comparison with an estimate of China's economic potential as it appeared at that time. The British historian R. H. Tawney has cautioned against interpretation that gives "an appearance of inevitableness by dragging into prominence the forces which have triumphed, and thrusting into the background those which they have swallowed up."[3] Japan was a seclusionist, traditional-minded country even as late as the 1850s. In many respects Japanese attitudes toward change or outside influences were anything but flexible. The critical and distinctive aspect of modern Japanese history is that these attitudes were replaced by an eager searching for the kind of change based on direct Western models that the Japanese felt was necessary for the preservation of their independence and prestige. A German physician, Dr. Erwin Baelz, who visited Japan in the middle 1870s, received a startling reply from a Japanese friend whom he questioned about Japanese history: "We have no history. Our history begins today."[4] This attitude was typical of the extreme adaptability and sense of national mission characteristic of modern Japan. Although the Japanese were ready to reject their own past where they felt it interfered with newly desired changes, they remained strongly nationalistic. Their object was the preservation of Japan, not its Westernization. A great deal of traditional Japanese culture has survived and remained vigorous. Despite China's influence in the past and Westernization in the modern period, both important, there are few national cultures that remain as distinctive as the Japanese. Firmly founded in a sense of national identity, the Japanese have maintained the habit of almost endless inquiry.

Patterns of manufacturing and cities

Tokyo-Yokohama

Manufacturing hugs the sea, and is particularly concentrated around Tokyo Bay, Ise Bay near Nagoya, Osaka Bay between Kobe and Osaka, and the coalfields of northern Kyushu (see Map 70, page 268). The Tokyo-Yokohama area leads, with over one-fifth of Japan's manufacturing, measured by value of production, number of factories, and number of workers. Three factors contribute to its size. A magnificent natural harbor at Yokohama near the entrance to Tokyo Bay provides excellent communications by sea from its central position within Japan. The location of the national capital in Tokyo brings advantages for business and manufacturing (especially with the close connection between government and industry in Japan). And the size of the local market and its specialized services are important. Metropolitan Tokyo, with about 12 million people, is probably the world's biggest city,[5] and certainly is if Yokohama is included. Yokohama is politically separate, but has grown primarily as Tokyo's outport. The two cities have now grown together in one mammoth urban area. Both have highly varied manufacturing. Tokyo is now the largest Japanese steel-producing center, and has deepened its inner harbor to take ten-thousand-ton ships. Tokyo is also the major financial center and the largest rail hub. There are few local resources, aside from the agricultural production of the surrounding Kanto plain, Japan's largest lowland. Even so, food for Tokyo must be brought in to supplement local supplies. As with most large cities, local resources are far less important than wider spatial relations.

Kobe-Osaka-Kyoto

The Kobe-Osaka-Kyoto area is a close second in manufacturing and finance, and is also highly diversified. Osaka, at the head of the Inland Sea where raw cotton imports from China and India could be landed conveniently, was the leading center for cotton textiles until Nagoya overtook it recently. Japan grows no cotton because of climatic and soil limitations plus the need to concentrate on cereals on the limited arable land. But raw cotton is not a large proportion of total textile production costs. Osaka, Nagoya, and other seaports can bring in raw cotton by sea, provide power, skilled labor, and efficiently mechanized plants, and can export the finished goods by sea or distribute them cheaply to the large domestic market. Japanese textiles for many years undersold and largely displaced Western cotton goods, especially in the Asian market. Japan's advantage in textiles is lessening as its costs for power,

[3] *The Agrarian Problem in the Sixteenth Century* (London, 1912), p. 177.

[4] *Awakening Japan* (New York, 1932), p. 17.

[5] This is not the first time that Tokyo has enjoyed such a position. Its population probably exceeded 1 million by 1700, after the Tokugawa Shogunate had made Tokyo (then called Edo) its administrative seat. In the eighteenth century Edo was rivaled in size only by Peking; London and Paris did not reach 1 million until after 1800.

Imperial Palace, Tokyo. Still protected by a moat and closed to the public except on special occasions by invitation, this beautiful sixteenth-century castle and grounds are now surrounded by the bustle of the world's largest city. The center of Tokyo is an appropriate symbol of modern Japan. *(From Japan Information Service.)*

raw materials, and labor increase (especially with the rise in living standards and wages in Japan), and as its Asian and other markets continue to industrialize, beginning in the usual pattern with textiles. Metals and machinery, including ships, have now taken first place among Japanese exports, goods that an industrializing Asia will demand in increasing amounts. Osaka's manufacturing is now more concentrated in chemicals, metals, and machinery.

Steel and machinery are made in all of the major manufacturing centers, but large-scale shipbuilding is concentrated in Kobe and Yokohama. Kobe developed mainly as Osaka's deep-water outport and is still Japan's leading port for foreign trade, measured by the volume of its shipping. Coal from Kyushu, and coal and iron from the continent were assembled at Kobe for making the iron and steel that went into ships and machinery. This specialty has continued although much of the coal, iron ore, and scrap must now come from greater distances. Osaka and Kobe have now merged much as Tokyo and Yokohama have done, although Kobe is still politically separate. Osaka is a leading business and financial center, and was especially important as the main Japanese base for the extensive trade with China and Manchuria.

Kyoto, the old capital, with its magnificent temples and gardens, is carefully preserved as a tourist center. It does a good deal of small-scale manufacturing, especially in high-quality textiles, pottery, and artisan goods, and has a population of well over 1 million. The Japanese are avid tourists, perhaps because so many of them live in crowded cities. They make frequent excursions to view their lovely countryside and mountain landscapes, especially Mount Fuji (Fuji-san), conveniently near Tokyo, which are almost like pilgrimages. Japanese outnumber foreign visitors to Kyoto, Nara, and other tourist centers by at least a hundred to one, including endless busloads of school children.

Nagoya and north Kyushu–southern Honshu

The Nagoya manufacturing complex on the Nobi plain ranks third, with about 10 percent of the national total of production, factories, and workers. Its service area is somewhat overlapped and overshadowed by both Osaka's and Tokyo's. It is also relatively far, by Japanese standards, from coal sources. It is better placed for hydro power and draws heavily on the large number of stations in the mountain mass directly behind it to the north and east, sometimes called the Japanese Alps. Transport equipment (airplanes, cars) and other heavy manufacturing, including oil refining, dominate Nagoya's industrial structure. Pottery is also important, originally dependent (like the British potteries at Stoke-on-Trent or those at Trenton, New Jersey) on fine local clay that is now largely exhausted and supplemented by imports. Lacking most local raw materials, the Nagoya area must specialize in value added by manufacture. Production is widely diversified, including textiles, sewing machines, and textile machinery.

North Kyushu and the adjacent parts of Honshu along the Inland Sea, with main industrial centers at Fukuoka, Kitakyushu, Yamaguchi, and Hiroshima, support the fourth major center of Japanese manufacturing. (Kitakyushu is composed of the former cities of Yawata, Moji, Kokura, Tobata, and Wakamatsu.) Coal and cheap water transport are the attractions. Japan's first big steel plant, and still one of the largest, is at Kitakyushu, right next to the coal. Some of the coal seams actually run under the sea, so it is ideally placed to assemble seaborne imports of iron and coking coal. Kitakyushu and Shimonoseki are important ports for mainland trade. Nagasaki, farther down the Kyushu coast facing the East China Sea and still Japan's westernmost large city, was Japan's "window on the world" during the self-enforced seclusion policy of the seventeenth and eighteenth centuries. Foreigners were excluded from Japan, except for a small Dutch trading factory on an island in Nagasaki harbor, through which a surprising amount of information about the West penetrated into Japan. Nagasaki was later important as a coaling station and terminus for the shipping routes between east Asia and Europe. It is still a major shipping, shipbuilding, and ship-repair center. A big naval base is at nearby Sasebo. Fukuoka produces mainly consumer goods. But Hiroshima, at the eastern edge of the Kyushu–southern Honshu industrial zone, is a center of diversified heavy industry, hence its choice as the target for the first atomic bomb. Hiroshima is now a monument to the horrible folly of war. It still serves as the administrative center for the central and western Inland Sea area, and has rebuilt its heavy manufacturing complex. Generally, the north Kyushu–southern Honshu industrial zone is the only major one in Japan primarily oriented to raw materials. The coal of Hokkaido has attracted some iron and steel production at Muroran, formerly using ores imported from Korea, but it is relatively far from the main markets.

Industrial characteristics and growth

A surprising amount of Japanese industrial production still comes from very small plants. Much is produced by what is really household industry, using power-driven machinery and often family labor or slack-season agricultural labor in one room of a house or in a tiny factory. The widespread availability of electric power has been a great asset in this, and such small-scale production is fairly efficient. Much of it is done on subcontract for large urban firms. Bicycle parts, for example, may be subcontracted to a hundred small producers all over Japan and assembled in Osaka. A good deal of small-scale rural industry also produces traditional foods, clothing, and household equipment for the domestic market. In 1938 over half of Japan's industrial output by value came from factories employing five workers or less, and only 5 percent of all Japanese factories employed over a hundred workers. Even as late as 1965 over half of the manufacturing establishments had five or fewer workers, although over half of the industrial output by value came from plants with three hundred or more workers. Of course, there are some very large plants, especially in the big cities. But the wide distribution of small-scale industrial production is efficient and makes a valuable contribution to the rural economy, where seasonal underemployment is a problem. Small-scale rural-based manufacturing, however, may soon be on the way out, as heavy capitalization becomes increasingly characteristic of Japanese industry. Already nearly three-fourths of the total population is classified as urban.

Heavy manufacturing, especially of ships and munitions, was stressed in the early years of the Meiji era after 1870, for the aim of the new government was the defense of Japan. Machinery, textiles, and consumer goods became more important about the turn of the century. By 1935 Japan had become the world's leading textile exporter, largely at the expense of Britain. After the First World War, and particularly in the 1930s, with Japan's fateful attempt to conquer China and dominate East Asia by force, heavy and military industries were again stressed. Many of them were uneconomic as well as wasteful. The colossal devastation of the war not only destroyed most of the industrial plants, but also convinced most Japanese that force was a self-destroying instrument of international policy. The manufacturing structure that has developed remarkably rapidly since 1945 (by 1956 industrial production was over twice the value for the prewar peak year of 1936, and by 1976 about three times) is widely diversified. Rising demand and growing numbers have made the home market a larger consumer of Japanese industrial goods than overseas markets. Between 1950 and 1960 the Japanese economy grew faster than any national economy has ever grown. Food, textiles, clothing, and other consumer-goods industries lead, including a growing volume of electronic products, light machinery, and domestic appliances for the booming home market. Heavy industries, including chemicals, metals, heavy machinery, and transport equipment, account for about 40 percent of the total value of industrial production and have recovered even more rapidly than consumer-goods industries. Japan can compete better with new Asian producers on the Asian market in heavy, machine, and precision goods than in textiles or light consumer goods. It has unique advantages of location, transport, established plant, and skilled labor, all of which its Asian competitors lack. Textile and consumer-goods production is easier and more widely spread. The wartime destruction of many factories means that most plants are relatively new and highly efficient, also taking advantage of advanced Japanese technology. Japanese steel, for example, undersells American steel on the U.S. west coast despite Japan's need to import most of its ore and despite high U.S. tariffs. Japanese shipbuilding is booming, and ships have become a major export. Japanese yards have contracts from all over the world, including major Western shipbuilding countries, and since 1956 Japan has been the largest producer of ships, including huge oil tankers. Skilled labor, efficient large-scale plants, and cheap assembly costs for raw materials enable the Japanese to undersell their competitors, even though much of the raw materials must be imported. Machinery and high-value specialties like optical goods are also important exports, although textiles and a variety of consumer goods, such as hardware, cameras, toys, and electrical equipment, remain significant. Japan produces and exports in large amounts the world's finest cameras and lenses, sound-reproduction equipment, motorbikes, and cars. Industrialization has come of age technologically in Japan in an amazingly short time, considering the country's circumstances as recently as 1870. In many respects it has shot ahead of the West.

Foreign trade

Exports recovered more slowly. Exports in 1958 were only about 80 percent of those in 1936, and the prewar peak was not reached again until 1959. The health of the Japanese economy can be maintained only if it acts as a factory, importing food and raw materials and exporting finished goods. The slow recovery of exports was due partly to the great increase in the home market, which absorbs more and more of Japan's total production. But large and varied imports are necessary merely to feed, maintain, and employ the Japanese, and these must be paid for by exports. The loss of the former Japanese empire and worldwide inflation, especially in oil prices, mean that Japan must pay a higher price for many of its essential imports. It also lost the former empire markets. Under the empire, China, Manchuria, Korea, and Formosa (now Taiwan) together supplied nearly half of Japan's imports, including 90 percent of its food and sugar, over three-fourths of its coal, and over half of its iron. About two-thirds of Japan's exports went to its empire, where it bought cheap and sold dear to a captive market on its own terms. Now it must buy dear, mainly from much greater distances, and finds many of its old markets closed to it. It must meet growing competition in other Asian and world markets from Western competitors and from industrializing Asian countries. Meanwhile, its population continues to increase. Nearly a million new workers will enter the labor market annually.

Large-scale trade with China is seen by many Japanese as the principal answer to their economic dilemma. But it is doubtful whether Japan can ever again buy and sell in the Chinese market as substantially or as profitably as it did before the Second World War. Other markets and other sources of raw materials must be developed, and Japan must compete by skill, experience, and hard work. These are qualities that the Japanese have raised to a very high level by necessity and long habit. Their success in the past in making much from little augurs well for their future. Japan continues to improve its competitive position as an industrial power that has provided a new model of growth and efficiency.

Korea: the tragic region

The Korean peninsula is physically very similar to Japan, especially the mountainous terrain (see Map 68, page 264). The first Westerners to visit Korea described the jumbled landscape as "a sea lashed by a heavy gale." Korea also has a long, indented coastline that, with its peninsular form, contributes to a strong maritime orientation. Perhaps the most critical difference between the two countries is that the Korean peninsula is attached to the Asian mainland. Thus, it lacks the advantages and, most of all, the security of Japan's insularity. Korea has been overrun, devastated, subjected, and exploited repeatedly throughout its long history. As a peninsula, it is distinct from the mainland, and is further set off by some mountain barriers, including the snowy chain of the Changbaishan (Long White Mountains). The Yalu River, which forms the political boundary with Manchuria, and the Tumen, which marks the boundary with the Soviet Far East, both run in deep gorges along the mountain front and form additional barriers. (They are much more suitable and effective as political boundaries than navigable rivers in broad valleys.) In these terms Korea constitutes an unusually well-defined physical region. Unfortunately Korea's location between the two great power and cultural blocs of east Asia in the past, and between the three great powers and expansionist drives of Russia, China, and Japan more recently has periodically weakened its regional identity and involved it in the power struggles of its neighbors to its disadvantage. With the widening of spatial interaction resulting from the modern transport revolution bringing the spatial widening of political and military structures and policies, Korea has been caught directly in the crossfire of a global conflict.

Perils of centrality

Korea's problems parallel those of the Low Countries. Both are small, they lie between greater cultural

"A sea lashed by a heavy gale"—the Korean landscape near Taegu, South Korea. Mountains dominate Korea as they dominate Japan. Notice the intensive use of the lowlands in the Japanese pattern, and the terracing for rice in the foreground. The mountains have been almost completely deforested, partly as a result of the long Japanese occupation, and the erosion problem has become serious. *(U.S. Army photograph.)*

and power blocs, they have been fought over and caught between opposing pressures, and control of them has been desirable to the great powers for strategic reasons. Korea occupies an obvious invasion route and a key position within the Far East. It is encircled by China, Russia, and Japan, which has won it the nickname of "the cockpit of Asia." But for much of its history, Korea has not even been master in its own house. Where control of the Rhine mouths, a prosperous commercial economy, and successful manipulation of rivalry among the great powers have kept the Low Countries politically independent, Korea has more consistently been a pawn of its great neighbors and a bloody battleground.

The people and their language are of central Asian origin, like the Japanese. Their culture is closely related to China's, to an even greater extent than Japan's. Social systems, art forms, philosophy, political systems, production techniques, and the original written language were all heavily influenced by or slightly modified from China. The beginnings of a sophisticated civilization in Korea date from the time when most of it was a colony of the Han dynasty. Korea was invaded or controlled by Japan in the late sixteenth century and from 1895 to 1945. It has been threatened as well by Russian and Chinese pressures and influences since 1945. During most of its history before 1895, Korea was controlled by or was tributary to China and was able to assert its independence only briefly. The Mongols and the Manchus established their control in particularly devastating campaigns. The Sino-Japanese War of 1895 over the control of Korea ushered in a period of subjugation and exploitation by the Japanese that left the country drained of its economic resources, without political leadership, and poorly prepared for the demanding task of resuming an independent national existence in the troubled postwar world. To these difficulties was added the destructive division of the country into a Communist-dominated north and an anti-Communist south, the direct result of rivalry that completely denied Korean interests and national desire for unity and true independence. To make matters much worse, Korea became a battleground again in the United Nations action of 1950–1953 against an attempt by the government of the north, with Russian and Chinese backing, to take over the south. Thus, to the problems of inherited poverty and dislocation was added a terrible burden of economic devastation and human suffering as the two armies surged back and forth over the peninsula in pursuit of wider objectives having little to do with the immediate welfare of Korea.

The Korean land and Japanese rule

Even without this devastation, whose scope and scale are not widely realized in the United States despite the dominant part played by American forces in the Korean war, the economic level in Korea was low. Agriculture still employs nearly half of the population, and nutritional densities are very high, for there is limited arable land. Agriculture faces the same topographic handicaps as in Japan. Only about 17 percent of the total land area is cultivated, about the same as Japan. Korea is about the same size as Honshu, or a little smaller than New England plus New York. It supports a total population of over 50 million. The Japanese during their occupation of Korea nearly doubled the output of Korean agriculture by applying the same methods they used so successfully at home. But the new output went largely to Japan to feed the growing Japanese population, and the Korean farmer was no better off. The fishing industry was also highly developed by the Japanese, but again, nearly three-fourths of the catch was siphoned off to Japan. Japan's demand for wood, added to the widespread Korean practice of shifting cultivation and burning off forests in the more rugged regions, led to extensive deforestation of Korea's hill and mountain areas. This has greatly worsened erosion and water-shortage problems, and has seriously aggravated the fuel shortage.

Nevertheless, Korea's population more than doubled under the Japanese, from 13 million in 1910 to 28 million in 1945. This did not mean any great increase in economic opportunity or higher income levels, for twice as many people had to share depleted or diminishing resources. While population was doubling, the amount of cultivated land increased by only about 15 percent. The average Korean was probably worse off economically in 1945 than in 1910. The increase in population resulted from the decrease or disappearance of chronic famine with the building of an extensive road and railway network by the Japanese, from the increasing commercialization of Korean economic life with wider exchange, and from the impact of public health measures. Railways, wider exchange, and public health were essential to the Japanese so they might more fully exploit the Korean economy. As in so many colonial areas, the result of such developments was a large increase in population and a net deterioration of the standard of living. Japanese rule in Korea was distinctive because it purposely milked the Korean economy for all it was worth.

Just as important as agricultural surpluses were Korea's mineral resources, which were especially attractive to mineral-poor Japan. Particularly in the north, large deposits of coal, high-quality iron ore, and other minerals were mined by the Japanese for their own use. Most of it went to Japan, but there was also some development of local manufacturing, whose finished products were also exported to a large extent. In 1910 there was no modern manufacturing in Korea. Japanese investments, in both heavy and consumer manufacturing, were large, but the profits and the higher-paid jobs went to their own nationals. Even the local police were largely Japanese, as were the engine drivers and about half of the primary-school teachers. Korean bitterness against Japan is understandable.

MAP 71
POPULATION OF KOREA

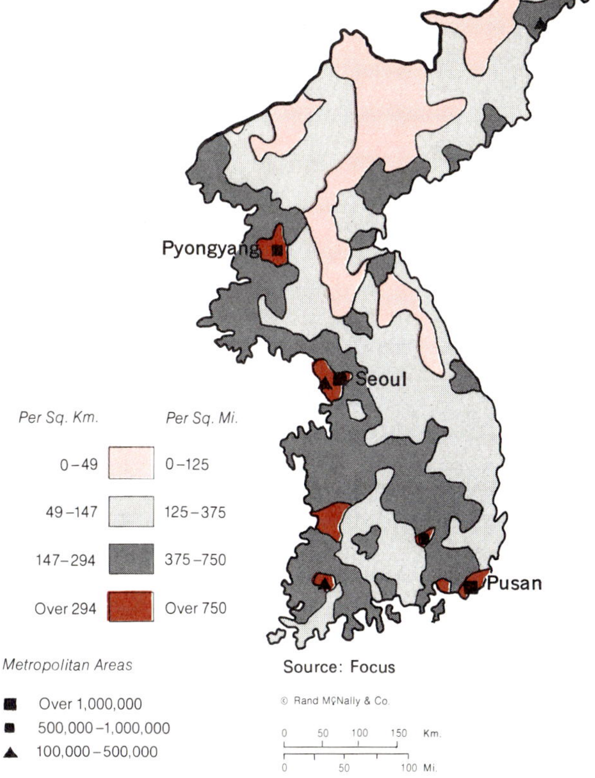

North and south: regional difference and interdependence

It is a tragic irony that Korea may be so neatly divided into two distinct subregions on physical and human grounds. The thirty-eighth parallel, originally intended as a temporary line separating the Russian from the American zone for the purpose of accepting the Japanese surrender, coincides remarkably closely with the line that can be drawn on a geographic basis between the north and the south. As the cold war intensified after 1945, the government of North Korea was strongly influenced by Russia and China, and the repressively anti-Communist government of South Korea received American support. The line between them hardened, and Korea was split in two. The split was against geography, for where there are regional differences, there is usually corresponding regional interdependence. Climatically the north-south distinction is accentuated by the northward widening of the country. The south is peninsular and open to the moderating influences of the sea on three sides. The north is increasingly continental, farther from the sea and more closely a part of the great Asian landmass with its seasonal extremes of temperature. Winters in the north are long and bitter. While the south has a long enough growing season and a mild enough winter to grow two crops a year, the north can grow only one, and must depend on hardier crops like wheat, millet, and barley in many areas.

Relief and population

In landforms, the north is far more mountainous and has much less arable land than the south (see Map 68, page 264). Population and relief are intimately related, as in Japan, and while the north is larger than the south in area (49 thousand square miles as against 38 thousand), it supports less than half as many people (see Map 71, page 285). The major concentration of level land and population in Korea, however, straddles the thirty-eighth parallel on the west coast. The Taebaek range extends along the east coast far south of 38° N. To this extent the regional line differs from the political line, and a sealed frontier running along the thirty-eighth parallel was especially awkward economically. The northern capital, Pyongyang, with about 1.5 million inhabitants, and the southern capital, Seoul, with almost 7 million, both lie on the major west-coast plain. Low hills just north of 38° N. tend to divide the plain into two halves, but the hills present no barrier to the interchange that flowed across them.

Seoul is Korea's largest city because it serves the more populous and productive south and is located on the largest plain. Also its nearly central position within the peninsula and its west-coast orientation toward China and the mainland have made it a good capital and commercial center for the whole of Korea, a role it has played for the past several centuries. Its tributary hinterland normally extended into the north and in many respects encompassed the whole of the country. The series of river valleys that join near Seoul, most importantly the Han River and its tributaries, gave it good access north, south, and east, including a road and railway to the port of Wonsan on the northeast coast. Pusan, with about 2.5 million people, grew especially rapidly after 1910 as the principal port for trade with Japan, and has spurted again as the chief point of entry for American aid and foreign commerce generally. It also serves a relatively large and densely populated plains area, second only to that around Seoul. A line drawn from Seoul to Pusan leaves most of Korea's arable land, most of the plains area, and most of the people south of it (see Map 71, page 285).

Agriculture and food

Food production is heavily concentrated in South Korea. North Korea has a food deficit, formerly made up by shipments of food from the south or from Manchuria. This was matched by an equally vital shipment of mineral and chemical fertilizers from the mines and factories of the north, which were essential to maintain agricultural yields in the south. This important exchange ceased in 1946, and both halves of Korea suffered. Southern agriculture has the advantage not only of more level land but of a more permissive climate. Rice is the dominant crop. Its dominance ceases almost exactly at the thirty-eighth parallel along the west coast, and somewhat farther south in the mountainous area of the east coast. Barley and some wheat, minor cereals, and soybeans are raised in the south as winter crops. In the north millet is even more important than wheat, clear evidence of harsh climate and poor upland soils.

Minerals, power, and manufacturing

An equally striking basis for regional distinction is the concentration of mineral resources in the north. Given the concentration of population and trade centers in the south, this further accentuates the need for regional interchange and makes the division of the country even more awkward. The major area of mineral deposits lies within a 50-mile (80 km) radius of Pyongyang, the chief center of heavy manufacturing for all of Korea. It was built up originally by the Japanese and greatly increased after 1950. A second important mineralized and industrialized area lies close to the Manchurian border in the northwest. Both areas contain large reserves of coal and high-grade iron ore, plus limestone, phosphate rock (for fertilizer and the chemical industry), copper, magnesite, and gold. The coal reserves by now are larger than Japan's total reserves, although production is still less. There is a deficiency of good coking coal so imports from Manchuria are necessary. But North Korea produces over 80 percent of Korea's bituminous coal, nearly all of its iron ore, and a total of about 85 percent of its mineral output. Hydro power is also concentrated in the north, which produces over twenty times as much as the south. Mountains, greater remaining forest cover than in the south due to poorer access and lower population densities, and the presence of the Yalu River help to account for the hydro resources of the north. This regional difference further accentuates regional interdependence. Until about 1946 power was sent directly to the urban centers of the south, especially to Seoul, and indirectly in the form of manufactured goods made in the north. The Japanese built two major hydro plants on the swift-flowing Yalu, and several on its tributaries. The short, swift rivers draining into the Sea of Japan presented the kind of engineering problems and opportunities with which they were familiar at home. Heavy manufacturing—iron and steel, chemicals, pulp, refining—is concentrated in the north, near raw materials and power. But slightly more than half of the total of Korean manufacturing is in the south, especially in Seoul and Pusan. The south is the major market and the dominant center for consumer-goods industries, particularly textiles and food processing. This underlines regional interdependence, for neither half is self-sufficient. Producer goods, machinery, fertilizers, power, and other raw materials need to be exchanged for consumer goods and food.

In some respects South Korea is taking over, on a smaller scale, the former Japanese role as an exporter of light manufactured goods. For the past several years economic growth has been vigorous in both halves of the country, especially notable in the boom industrial development concentrated in the Seoul area. Agricultural output has also been rising steeply, but manufacturing has become the dominant sector.

While the artificial division of Korea lasts, its economy and its people will suffer, as they have suffered so often in the past when they have been involved as unwilling pawns in the rivalries and battles of outside powers. Although the current political separation naturally tends to accentuate and manipulate regional differences and political repression, Korean national consciousness is unified. Unity may again be achieved if the pressures of the cold war lessen. Koreans deserve a chance to work out their own destiny.

Questions for further study and discussion

1. Why do farms in Hokkaido average over three times the size of farms in the rest of Japan?

2. Why should dollar exchange be especially needed by Japan? How is it earned?

3. Why is there a major naval base at Sasebo? What are the locational and site advantages?

4. Why is coal production in Korea less than in Japan, if Korea has larger coal reserves?

5. Why has Korea been less well able than the Low Countries to maintain its independence from foreign domination?

16 Southeast Asia: tropical islands, deltas, and the sea

Kuala Kangsar, Malaysia. *(Photo by Marjorie Jean Jones.)*

Southeast Asia is a regional term of convenience rather than of real meaning. It covers an extremely diverse and scattered physical area containing an equally diverse set of peoples, cultures, and economies. It is different enough and separate enough from the rest of monsoon Asia that it must be treated separately. Yet much of its unity is due to the impact of Indian and Chinese civilizations on it, and later, of the commercializing influence of the West. Most of its religions, art forms, current production techniques, economic and political systems, and many of its languages had their origin elsewhere on the Asian mainland or in Europe. Nearly all of southeast Asia was part of one or another of the European empires of the nineteenth century. For centuries before that, much of it was subject to strong cultural and economic influences from India and China. Although these outside influences have been spread over much of southeast Asia and have given it some degree of cultural uniformity, there are clear distinctions among its various parts.

The most obvious distinction is between the mainland areas of the Indochinese peninsula and the island chains of Indonesia and the Philippines. Further regional distinctions can be drawn as follows (see Map 72, pages 290–91): *Indonesia and Malaysia* (excluding the mountain areas), where the major common ground is the Malay culture and language with its many dialects; *the Philippines,* a clear areal unit on the basis of insularity whose culture is not closely related to the Malay areas except in language, and is heavily influenced by China, Spain, and the United States; *Burma and Thailand,* both heavily Indianized in their culture, peopled to a considerable extent by groups originally from southwest China, and both characterized by a regional core that follows the valley and delta of a master river in each country, the Irrawaddy and the Me Nam Chao Phraya[1] respectively; *Laos and Cambodia* (Kampuchea), closely related in culture and environment to the neighboring areas of Burma and Thailand, relatively sparsely settled by Asian standards, whose earlier Indianized civilizations were largely destroyed by invasions from the north; and *Vietnam,* composed of the valley and delta of the Red River in the north, the narrow coastal plain of

---

[1]*Me Nam* means simply "river." It is used here because it is the name usually applied by Westerners to the Chao Phraya.

Annam, and the delta of the Mekong in the south, which is a densely settled area of intensive rice cultivation, whose culture has been strongly influenced by both China and India.

Soils and population patterns

Regional distinctions based on culture and environment, however, may be less significant than distinctions based on population patterns. Most of southeast Asia is relatively empty, and the few areas of dense population stand out sharply. The people are concentrated where agricultural opportunities are maximized (see Map 73, page 292). There are three types of such areas. By far the most important are the rice-growing alluvial plains, especially the lower valleys and deltas of the rivers. Some of Asia's greatest rivers flow through peninsular southeast Asia: the Irrawaddy, the Me Nam Chao Phraya, and the Mekong. In their valleys and deltas and in those of the Red River in northern Vietnam, the population of the Indochinese peninsula is clustered. There are no large rivers in Malaya, Indonesia, or the Philippines. But the plain of Luzon is traversed by several streams whose alluvial soil is partly responsible for the dense population there.

Next in importance, in terms of the populations they support, are the volcanic soils. Much of the island of Java (Djawa) is covered with deep, fertile volcanic ash and lava soils. Such soils also occur in smaller parts of Sumatra (Sumatera), although unfortunately most of the Sumatran volcanic soils are chemically acidic rather than basic, hence much less fertile. There are also a few areas of basic volcanic soils in the Philippines and on Celebes (Sulawesi), which are much more intensively used and more densely settled than the surrounding areas. Deep soils that are continually renewed and are of recent origin, such as alluvium and volcanic ash or lava, are of particular importance in southeast Asia. The heavy rainfall is spread over the year in much of Malaya, Indonesia, and the Philippines, and concentrated in strong showers during the summer monsoon (see Map 74, page 294). The rain plus the high temperatures throughout the year mean that leaching and oxidation are rapid and extensive and the humus content and fertility of the soil are very low. The plant nutrients in the soil are soluble, as they must be for plants to use them, and tend to be dissolved or leached out in areas of heavy rainfall. Outside the river valleys and the areas of basic volcanic ash or lava, most southeast Asian soils are too poor to support more than a sparse population in this still little-industrialized area.

The major exception is the scattered dots of plantation agriculture. Largely devoted to tree or perennial bush crops, plantation agriculture can be much less particular about soils. Instead it can pick its locations primarily in commercial terms; access to the coast and the sea routes leading to the major markets in Europe and the Americas is especially important. The best example of this locational pattern is the plantation zone of southwest coastal Malaya, which stands out on the population map. The soils of the area are generally too poor for annual field crops, except for scattered pockets of alluvium (which are devoted mainly to subsistence rice growing). However, southwest coastal Malaya supports a prosperous plantation agriculture of tree crops that profits from easy access to sea routes. In Java, where volcanic soils are widespread, lowland, nontree, commercial crops that require fertile soils, such as sugar and tobacco, share the lowland advantages with rice and other food crops. The less demanding tree crops, or those that prefer a cooler climate, make use of the less fertile uplands. Since Java is small and its uplands do not present a major barrier, even interior upland plantations can enjoy relatively easy maritime access. Plantations are also scattered over parts of Sumatra, and to a lesser extent parts of Borneo (Kalimantan), Celebes, Mindanao, Luzon, southern Vietnam, and southern Thailand (see Map 75, page 296). Plantations are of relatively minor importance in the rest of southeast Asia: where rice cultivation is dominant, where population pressures make it necessary to concentrate on cereals, where shifting cultivation makes limited use of the hilly forested areas, or where there are problems of access. Only in Java, with its exceptionally fertile soil, ample rainfall, and unbroken growing season, could large-scale commercial plantations share the land with rice and other food crops without producing a heavy food deficit. Until recently Java has nearly fed itself, despite its dense population. But Malaya, much less densely populated, imports over one-third of its food, since most of its cultivated land is devoted to plantations, and there is relatively little

BURMA
LAOS
THAILAND
CAMBODIA
VIETNAM
CHINA
INDOCHINA
MALAYSIA
BRUNEI
SARAWAK
SABAH
KEDAH
JOHOR
SINGAPORE
INDONESIA
SUMATERA
(SUMATRA)
BORNEO
DJAWA
(JAVA)
INDIAN OCEAN
SOUTH CHINA SEA
SULU SEA
Andaman Sea
Gulf of Thailand
GULF OF TONKIN
LAUT DJAWA
(JAVA SEA)
LAUT FLOR
(FLORES SE
SUNDA ISLANDS
Gulf of Martaban
Mouths of the Irrawaddy
Mouths of the Mekong
Str. of Malacca
Selat Karimata
Selat Makasar
Selat Sunda
Selat Lombok
Balabac Strait
Kyaukpyu
RAMREE
CHEDUBA
Sandoway
ARAKAN YOMA
Yamethin
Pyinmana
Toungoo
Pye
Bassein
Rangoon
Pegu
C. NEGRAIS
Moulmein
Ye
DAWNA RA.
Tavoy
Myinmoletkat 6801
BILAUKTAUNG
Mergui
MERGUI ARCHIPELAGO
Tenasserim
NORTH ANDAMAN
MIDDLE ANDAMAN
SOUTH ANDAMAN
Port Blair
LITTLE ANDAMAN
ANDAMAN IS. (Ind.)
NICOBAR ISLANDS (India)
GREAT NICOBAR
Hanoi
Haiphong
Ninh Binh
Nam Dinh
Luang Prabang
9242 Phu Bia
Chiang Rai
Chiang Mai
Mae Nam Ping
Mae Nam Nan
Vientiane
Udon Thani
Uttaradit
Sawankhalok
Tak
Muang Phitsanulok
Muang Sakon
Khon Kaen
Mekong
Muang Nakhon Sawan
Mae Nam Chao Phraya
Ubon Ratchathani
Surin
Nakhon Ratchasima
Ayutthaya
Prachin Buri
BANGKOK
(Krung Thep)
Rat Buri
Phet Buri
Angkor (Ruins)
Siem Reap
Tonlé Sap
Battambang
Pursat
Kompong Thom
Stung Treng
Kratie
Chanthaburi
Ban Kui Nua
Ban Bangsaphan
Phnom Penh
Kompong Som
(Sihanoukville)
Chau Doc
Kampot
PHU QUOC
Loc Ninh
Bien Hoa
HO CHI MINH CITY
(Saigon)
Long Xuyen
Khanh Hung
CON SON
Thanh-Hoa
Vinh
Dong Hoi
Quang Tri
Hué
Da Nang
(Tourane)
Ky Lam
Quang Ngai
Binh Dinh
Qui Nhon
C. VARELLA
Chu Yang Sin 7890
Nha Trang
Phan Rang
ANNAMITIC CORD.
Chanchiang
(Fort Bayard)
LEICHOU PAN-TAO
Kiungchow
Wu Chin Shan 6165
HAINAN TAO
HSISHA CH'UNTAO (China)
TIZARD BANK AND REEFS
SPRATLY
ISTHMUS OF KRA
Surat Thani
Nakhon Si Thammarat
MUI BAI-BUNG
Phuket
Thale Luang
Ban Kantang
MALAY PENINSULA
Songkhla
Pattani
Alor Setar
Kota Baharu
Pinang
Taiping
Ipoh
7186 G. Tahan
Pahang
Kuala Lumpur
Kelang
Melaka
(Malacca)
Batu Pahat
Johor Baharu
SINGAPORE
Sabang
Banda Atjeh
TG. DJAMBUAIR
Idi
Langsa
G. Leuser 11 178
Medan
Belawan
Tandjungbalai
Danau Toba
SIMEULUE
KEPULAUAN BANJAK
Sibolga
NIAS
PINI
TANAHMASA
KEPULAUAN BATU
TANAHBALA
Dumai
Bengkalis
Pakanbaru
Kampar-kiri
Selat Pandjang
KEPULAUAN RIAU
Bukittinggi
Indragiri
Sawahlunto
Padang
SIBERUT
KEPULAUAN MENTAWAI
SIPURA
PAGAI UTARA
PAGAI SELATAN
G. Kerintji 12 467
Djambi
Muntok
PEGUNUNGAN BARISAN
Palembang
Musi
G. Dempo 10 365
Bengkulu
Tulangbawang
Danau Ranau
ENGGANO
Tandjungkarang-Telukbetung
KEPULAUAN NATUNA BESAR
KEPULAUAN ANAMBAS
KEPULAUAN TAMBELAN
KEPULAUAN LINGGA
Pontianak
KEPULAUAN KARIMATA
Pangkalpinang
BANGKA
Tandjungpandan
BELITUNG
Ketapang
Sukadana
Bandar Seri Begawan
LABUAN
(Br.)
Miri
7799 G. Mulu
Bintulu
TG. DATU
Kuching
B. Rajang
UPPER KAPUAS MTS.
IRAN MTS.
Kapuas
PEGUNUNGAN MULLER
SCHWANER
Bukit Raja 7474
PEG.
Kahajan
Barito
Mahakam
S. Kajan
TG. PUTING
Bandjarmasin
Martapura
TG. SELATAN
KEPULAUAN LAUT KETJIL
Kotabaru
LAUT
Samarinda
Balikpapan
KEPULAUAN BALABALAGAN
TG. MANGKALIHAT
MARATUA
TARAKAN
BUNJU
SEBATIK
Kudat
BANGGI
Kota Kinabalu
Mt. Kinabalu 13 455
Sandakan
Darvel Bay
CAGAYAN SULU
BALABAC
PALAWAN
Puerto Princesa
CAGAYAN IS.
Taytay
Culion
CALAMIAN GROUP
CUYO IS.
MINDORO
LUBANG
Lipa
Batang
MANILA
Quezon City
Subic
Tarlac
Lingayen
Baguio
Bauang
San Fernando
LUZON
Vigan
Laoag
Tuguegarao
7741 Sicapoo
Balintang
BATAN ISLANDS
Zamboanga
JOLO
SIBUTU
TAWITAWI GROUP
SULU ARCHIPELAGO
CELEBES SEA
SULAWESI
(CELEBES)
Danau Poso
Donggala
Bulu Rantekombola 11 286
Madjene
Teluk Bone
Udjung Pandang
(Makasar)
Bonthain
SELAJAR
MASALEMBO-BESAR
KEPULAUAN KANGEAN
KEPULAUAN LUKANG TENGGAJA
KEPULAUAN KARIMUNDJAWA
BAWEAN
MADURA
Bangkalan
Serang
DJAKARTA
Bogor
Sukabumi
BANDUNG
G. Slamet 11 221
Tjirebon
Semarang
Surakarta
Jogjakarta
G. Semeru 12 060 Ft.
10 932
Pasuruan
SURABAJA
G. Raung 10 309
Banjuwangi
BALI
G. Agung 10 308
Denpasar
G. Rindjani 12 225
Mataram
LOMBOK
Sumbawa Besar
Raba
SUMBAWA
FLORES
SUMBA
Waingapu
Equator
15°
10°
5°
0°
95°
100°
105°
110°
115°
120°
COPYRIGHT BY
RAND McNALLY & COMPANY
MADE IN U.S.A.

MAP 72
LANDFORMS OF SOUTHEAST ASIA

fertile area suited to food crops. The Chinese of Malaya, nearly half the total population, also prefer mining, trade, or plantation agriculture to rice growing. On Java, plantations (other than sugar, which is grown mainly on fields leased for a year or two from rice farmers) occupy only about 3 percent of the total area, but they account for over half of Java's agricultural exports and rival the total production of Malayan plantations.

Plural societies and economies

The parts of southeast Asia that have neither fertile soils nor commercial attractions for plantations are almost uninhabited by Asian standards, although population densities seldom fall below two per square mile. Most of this sparsely settled area is still covered with dense rain forest or second growth. Physical isolation and a totally different way of life, in keeping with the demanding nature of the physical environment, help to account for the fact that the people who live in these rain forests are very different ethnically from those who live in the densely settled agricultural areas or the plantation zones. For the most part, the forest dwellers, who include Negrito and Melanesian strains as well as Malays, Thais, and other Mongoloid groups, are not advanced technically. They have little to do with the more advanced peoples living around them who now dominate and govern the new national states inherited from colonial rule.

Cultural and other differences are also marked between the people of the plantation economy and the subsistence growers of food crops. Most of the plantation workers and nearly all of the owners and managers are recently arrived aliens, especially outside Indonesia and the Philippines. The plantation system in southeast Asia is not old and dates only from the second half of the nineteenth century for the most part. It grew very rapidly to supply the booming demands of industrial Europe and America, and is still the major medium of Western commercialization. But plantations are largely confined to small areas of a commercial economy very different from the surrounding sea of traditional subsistence or semicommercial rice growing by small farmers. Except in Indonesia, the Philippines, and Burma, most of the plantation workers are Chinese or Indian; most of the owners, managers, technicians, and financiers are Chinese, Indian, North American, or European.

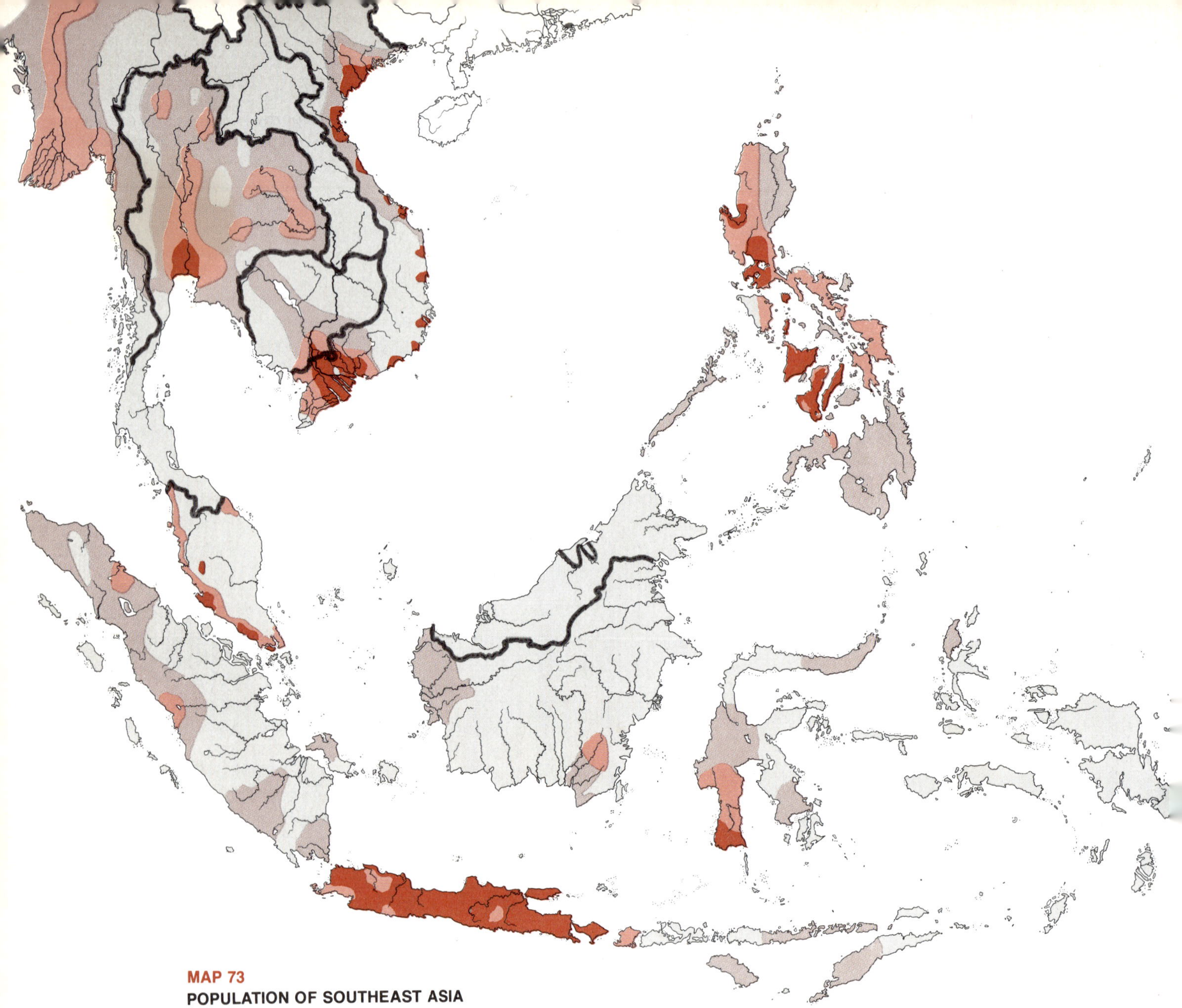

MAP 73

POPULATION OF SOUTHEAST ASIA

Per Sq. Km.	*Per Sq. Mi.*
0–10	0–25
10–50	25–125
50–100	125–250
Over 100	Over 250

0 200 400 600 Km.
0 200 400 Mi.

These differences are also a source of friction in the newly independent national states and are part of the basis for referring to southeast Asia as an area of plural societies and plural economies.

Shifting cultivators

By far the largest number of people in southeast Asia are still essentially subsistence rice growers, living in the alluvial valleys and deltas or on the fertile volcanic soils of Java. The plantation areas support only a small part of the total population and occupy a tiny fraction of the total area. The people of the rain forest are an even smaller part of the total population, but they occupy over half of the area of southeast Asia as a whole (see Map 73, page 292). Most of them are shifting cultivators, supplementing their meager crops by hunting, fishing, and gathering nuts, berries, roots, vines, and other forest products. The poverty of the soil as a result of the climate is increased by the strong relief of the slopes, characteristic of most of the area still in forest, and indeed of most of southeast Asia outside of the limited alluvial plains (see Map 72, pages 290–91). Relief, plus distance from the sea or river transport, is one of the most important reasons why the southeast Asian rain forest has not been cut more extensively for lumber. But relief also contributes to erosion and soil waste, leading to agricultural poverty, especially in this climate of heavy rains.

The shifting cultivators practice an agricultural system that is well adapted to the physical conditions. By slash-and-burn methods, by felling, or by girdling (removing the bark in a thin band around the trunk of a tree so that it loses its leaves and dies), they clear a small patch of an acre or two in the forest. The ash resulting from burning the litter, trees, and branches provides a quick fertilizer. But after one to three or four years (varying with the nature of the local soil), the soil is exhausted, and quick-growing weeds soon invade the clearing. When it will not produce enough to repay the effort of replanting, it is left to return to wild growth, and the cultivators move on to clear another patch. After this brief period of use, each patch is left to rest under its rapidly regrowing cover of jungle for ten to thirty years. The time interval depends on the nature of the local soils, and also on the pressure of population. An area may have to be worked harder or more often than is good for it if it supports too many people. They cannot put long-term maintenance of soil fertility above the more urgent and immediate need for food. Shifting cultivation can support very few people per square mile when one family or small kin group must have perhaps twenty plots of land scattered over many square miles, only one of which is producing a crop at any one time. The crops are relatively small in total amount, both because soils are poor despite the ash fertilizer and because, when the land is to be worked only for a year or two, relatively little effort is put into improving it or into intensive cultivation. Indeed, to an outsider, the fields look haphazard and seem to be given almost no care, especially weeding.

Shifting cultivation has often been called wasteful of land and a crude, inefficient method of agriculture. Actually, under present physical and economic conditions, it is difficult to devise a better method of land use for most of the areas the shifting cultivators occupy. Most of the soils and climate will not support permanent field crops. Physical and economic isolation also rule out commercial tree crops on any significant scale. Heavier fertilization of the small patches in the jungle would certainly raise yields. But under present circumstances, the cultivators do not have the means to afford it nor the animals to provide it, since there is little pasture or surplus food to maintain animals. Plowing and clean weeding of these jungle soils would be ruinous for it would invite destructive erosion. The shifting cultivators stir or disturb the soils as little as possible and usually use only a sharpened stick to dig seed holes and to provide a little aeration. They purposely avoid removing all the vegetation and try to keep the ground permanently covered from the damaging effects of sun, rain, and wind. They weed little, and plant a mixture of crops that will provide effective ground cover. Often the seeds are mixed and sown broadcast, so that a tangled mass of plants results, mixed with the weeds and surviving trees and bushes purposely left in the "clearing." Shifting cultivation has at least two other advantages. It returns relatively high yields per man-hour of labor, since the cultivators do relatively little tending of their crops between planting and harvest. And the diversity of crops grown on any one holding provides insurance against loss, since all of them are unlikely to fail. It is primarily increasing

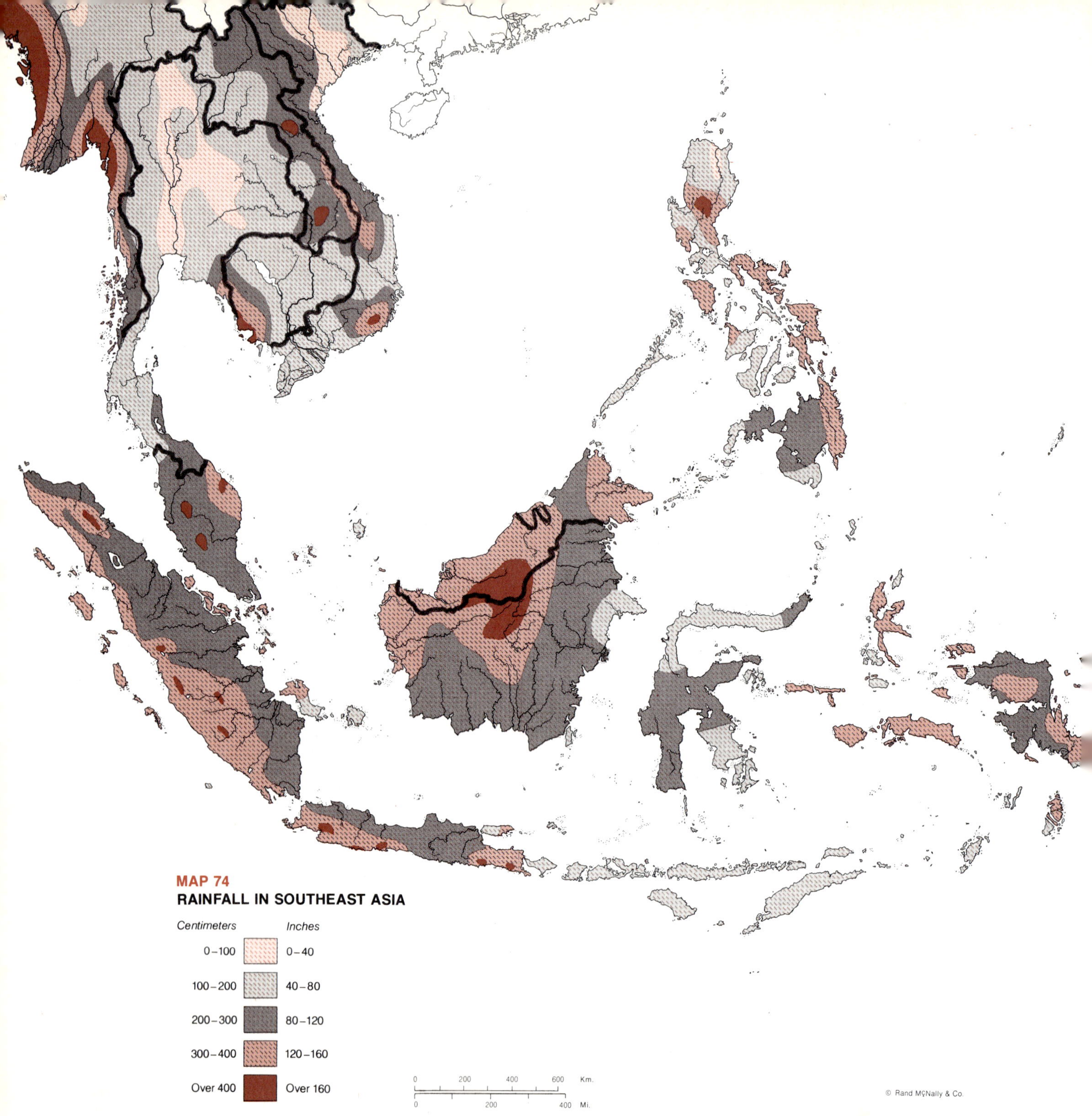

MAP 74

RAINFALL IN SOUTHEAST ASIA

population pressure with its demand for high per-acre yields that displaces shifting cultivation with permanent cultivation, a more laborious system that also requires larger amounts of fertilizer and water.

The most common crops of the southeast Asian shifting cultivators are melons, gourds, eggplants, yams, and squash. Their thick vines and leaves are an excellent ground cover, they are relatively tolerant of poor soils, and they like a hot, wet climate. Maize, upland (unirrigated) rice, and millet yield well in this climate and provide cereal starch and protein. Beans, as legumes, make the most of the soil and also provide good cover. Manioc (cassava) is a remarkable plant, originally native to Brazil. It spread rapidly throughout the tropics soon after the discovery of America, gives very high yields on poor soils under tropical climatic conditions, and can produce more starch per acre than any plant known. It is a tuberous crop, like the potato (which was also native to the New World), but it continues to grow in the ground and to increase in size for as long as four years. Thus, it can be harvested whenever needed and presents few storage problems. Its leaves grow rapidly and form a good ground cover, which also inhibits weeds. The tuber is nearly pure starch, and does not provide an adequate diet alone, despite its high calorie content. But the leaves are a good source of vitamins, minerals, and protein. Manioc is also grown commercially on plantations as the source of tapioca. Yams, sweet potatoes, squash, maize, eggplants, and beans are also grown widely by the subsistence rice cultivators in the more densely populated lowland areas.

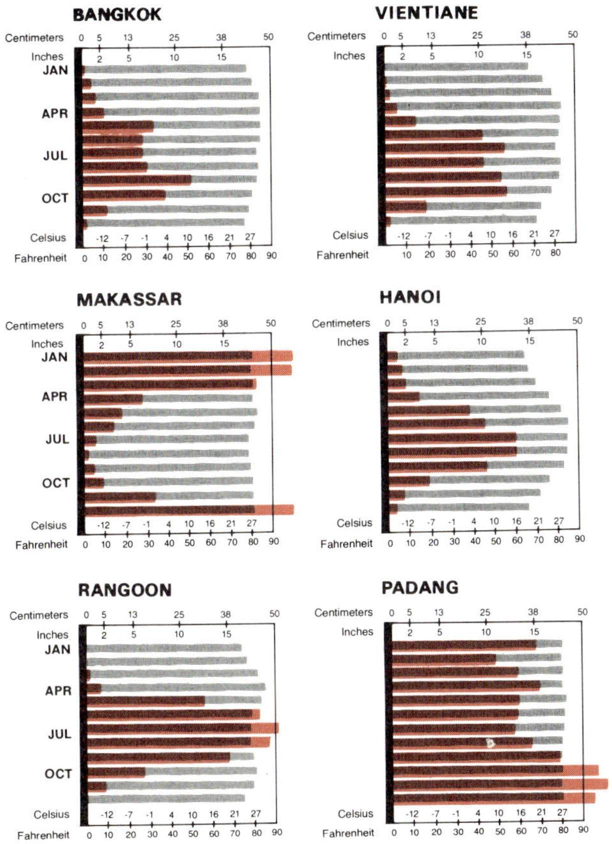

Figure 12.
Climographs of representative cities in southeast Asia.

Temperature
Precipitation

Shifting cultivation in southeast Asia over the millennia it has been practiced has probably almost wholly destroyed the virgin rain forest, except in the most remote or mountainous areas where hunting and gathering displace agriculture in any form. For the most part, the rain forest has been replaced by *jungle*. Strictly speaking, this is not a natural form, but the denser and more mixed second growth that grows when the original forest is cut and the sunlight can penetrate. If left to itself, the jungle would gradually yield to the rain forest again, as the tallest trees spread their thick canopy over everything below them. But the jungle is not left to itself, and is periodically cut or burned in successive patches by the shifting cultivators, unless it was entirely removed to make way for rice fields and the permanent agriculture of the alluvial or volcanic lowlands. By contrast, the plantations, whose soils may be nearly as infertile as those of the shifting cultivators, remove the existing tree cover entirely and replace it with uniform stands of trees or bush crops, which may be grown on slopes as well as on the lowlands.

Use of the uplands

In many tropical areas uplands, and especially upland plains, are favored for agriculture. They may support a greater population than the surrounding

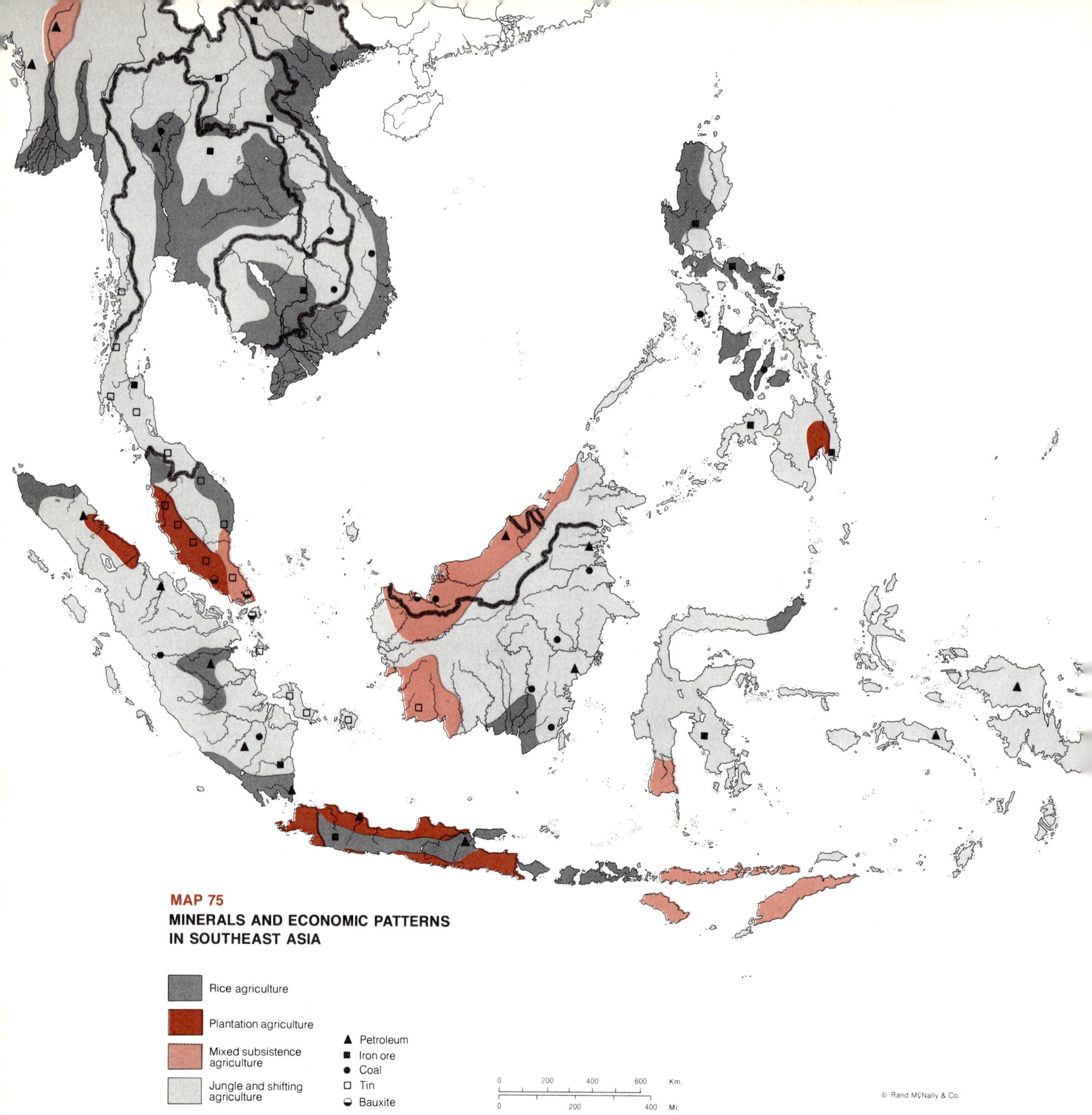

MAP 75

MINERALS AND ECONOMIC PATTERNS IN SOUTHEAST ASIA

lowlands, because the upland climate may be cooler and less troubled either by excess rain or drought. In most of southeast Asia, however, the uplands are very sparsely settled. Partly this is due to the absence of any extensive plateaus usable for agriculture, but also to the powerful attractions and high rewards of farming on the rich alluvial and volcanic soils of the lowlands. Elsewhere in the tropics the lowlands are not characterized by good soils to such an extent, and the uplands are relatively more attractive by contrast. Within southeast Asia only Java has uplands that are densely populated over large areas, for only Java has extensive areas of fertile upland soils. One further factor in the relative avoidance of the uplands, especially in the Indochinese peninsula, is malaria. In this area malaria is most prevalent and virulent, not in the lowlands, but in the hills where the anopheles mosquito that carries the disease thrives in the clear pools of mountain streams. Generally the lowland people shun the hills, and the hill people, culturally quite different, keep as much as possible to themselves.

Pluralism and politics

As is so often the case, political lines do not coincide with cultural and economic lines, nor do they unite the physically, economically, and culturally similar parts of southeast Asia (see Map 72, pages 290–91). Each of the present national states contains hill and lowland cultures, subsistence rice growing, shifting agriculture, and plantation economies. This is partly the result of European colonial rule, which superimposed a single administrative control on the regional mosaic throughout the diverse areas included within each colonial empire. An interest in trade and the prestige, rivalries, and strategic rewards of expanded territorial sovereignty by the colonial rulers resulted in lumping together peoples and economies that had little in common. In part, however, such a political denial of geography also results from the dominant position of the people of the alluvial plains and lowlands. Numerically they have always been the strongest group by far. Technically and culturally they have also been far more highly developed, so it was natural that they should again exercise their political dominance once colonial rule was withdrawn. These lowland peoples were the principal recipients and beneficiaries of the civilizing influence of India and China for at least the last two thousand years. Such influences penetrated relatively little into the uplands and the interior. During the last hundred years a fresh wave of outside influences has swept over southeast Asia, concentrated on the coastal lowlands by its commercial nature. It brought the European and American colonialists with their plantations, commercial cities, railways, and Western techniques to southeast Asia. There was also a massive new immigration of alien peoples, especially Chinese and Indians, attracted by the new, commercial opportunities opened by the Westerners.

The Indian and Chinese impact

Even the most advanced of the southeast Asian peoples were relatively unused to and uninterested in commerce. The trade they had conducted in the past with south India and southeast China had been left mainly in the hands of Indians, Chinese, or Arabs. With the boom of the plantation, the equally impressive boom in mining (especially petroleum and tin), the rapid growth of trade, and the rise of port cities, Indians and Chinese found fresh opportunities in southeast Asia. There they could escape population pressure and poverty at home and find work and trade where there was a need for them, and where there were few local rivals. Indians soon dominated the commercial life of Burma and realized most of the profits from the enormous expansion of its rice and lumber production for export. Chinese equally rapidly dominated the commercial economies of most of the rest of southeast Asia. In Malaya the Chinese became tin miners and prospectors, plantation laborers (work that the Malays, like most of the native southeast Asians, were unwilling to do), and traders and entrepreneurs. Indian Tamil workers also came to Malayan plantations, but in smaller numbers. Tamil merchants never established as strong a position there as in Burma or Sri Lanka (Ceylon). Tamil Nadu, as the Tamil-speaking area of south India is called, has been seriously overpopulated for centuries, and has a long tradition of overseas emigration and trade, especially to southeast Asia. Practically all of the Chinese in southeast Asia came from an area that is demographically and historically similar to Tamil Nadu, the southeast coast of China, especially the vicinity of Canton, Swatow, Amoy, and Foochow (Fuchou). There the tradition of emigration and trade with the eastern part of southeast Asia is fully as old, and population densities are very high in relation to arable land.

With commercial experience behind them and a willingness to work hard, the Chinese built a virtual monopoly of small trade and money lending in all of the southeast Asian countries east of Burma. They also established a strong position even in the export trade, plantation and mining investment, and banking. In Malaya, before the late-nineteenth-century impact of British commercialization, the land was relatively sparsely settled and little developed economically. The Chinese found there an especially large field in which the local Malays, concentrating on subsistence rice growing, were not interested. Chinese became the largest single group in the total population of Malaya (including Singapore). Indian, Middle Eastern, Japanese, British, and other European workers and entrepreneurs also formed significant proportions of the population. In 1957 the newly independent state of Malaysia thus inherited the most evenly and most awkwardly balanced of all of the plural societies of southeast Asia. In Thailand, under a similar set of circumstances, Chinese became about one-fifth of the total population, and about 60 percent of the population of Bangkok (Krung Thep), the capital. In addition to the usual pattern of money lending and small trade, they also dominate the trade in rice, Thailand's chief money earner. In Vietnam, and especially in the cities of Saigon-Cholon and Hanoi-Haiphong, Chinese came to occupy essentially the same position in numbers and in economic matters as in Thailand, and controlled the important rice trade of Vietnam. In the Philippines and Indonesia, Chinese are proportionately fewer in numbers but occupy a similar position economically. They are especially strong in the chief cities and in trade. Any alien group, especially one that specializes in money lending or trade and acquires power over the less fortunate local people, is bound to be resented. The Chinese in southeast Asia, although they number only about 17 million out of a total southeast Asian population of nearly 250 million, cause tensions and pose problems for the newly independent and strongly nationalist states.

Colonial-commercial cities and plural economies

Chinese, Indians, and Westerners concentrated in the cities of southeast Asia through which trade flowed and where investments were financed. Like the plantations whose trade they handled, these cities rapidly became tiny islands of Western-style commerce, alien to the rest of southeast Asia and economically oriented toward Europe and the Americas. In each country these cities were largely new, created by Western-inspired commercial growth or built outright by Westerners. As with Karachi, Bombay, Calcutta, Madras, Tsingtao (Chingtao), and Dairen (Lüta), most of them occupied sites that had either been entirely neglected before by the local people or had supported only small towns. Southeast Asia's relative disinterest in trade in the past led each area to focus its attention not on the sea or on commercial locations but on interior cores, based on agricultural opportunities or strategic protection from seaborne invasions. Beginning with the Maurya period in the third century B.C., extensive maritime empires based on trade flourished for a time on Sumatra and Java under strong Indian influence and control. But they did not last, and were not typical of the rest of southeast Asia. The characteristic southeast Asian pattern was an interior core with the political capital and chief city on an inland site away from the sea and not dependent on trade. In fact there were no real cities except the political capital in each country, which was more an administrative and cultural rather than a commercial center. The largely subsistence economies needed no other cities. Each country was relatively small, by Indian or Chinese standards, so one major center was sufficient.

The port cities that have grown up so rapidly to

One part of a plural culture—traditional long houses in the Padang Highlands, Sumatra. Inhabitants of this village are engaged in some shifting cultivation, some hunting, and some gathering in the forests, as well as maintaining a few small permanent fields. The style of this house is common to much of the Malay cultural area. The traditional long house may be occupied by several families, or by groups segregated by age or sex. With regional variations, this is representative of upland cultures in most of southeast Asia. *(From Indonesian Information Office.)*

serve the new commercial sectors of the economy are all at the best possible commercial location in each country: Rangoon, Bangkok, Saigon-Cholon, and Hanoi-Haiphong. All are near the deltaic mouth of the master river around which the population of each state clusters and which still serves as the major route of internal transport. Singapore is on a small island at the tip of the Malay peninsula where the sea routes from the plantation areas of coastal southeast Asia to and from Europe and the Far East come together in a great knot. Djakarta (formerly Batavia), at the northwestern tip of Java close to the sea lanes through the Sunda Strait (Selat Sunda), communicates by water from this central location with the commercially most productive parts of Indonesia. And Manila, on one of the world's finest natural harbors in Manila Bay, assembles and distributes the export and import goods of Philippine plantations.

Each of these has become a powerful primate city, although Indonesia's area and population are big enough to support some medium-sized cities in addition. Indonesia's many islands also increase the need for separate regional centers, such as Palembang and Medan on Sumatra and Makassar (Udjung Pandang) on Celebes. In the other countries of southeast Asia the primate city is markedly dominant. Both Rangoon and Bangkok had existed as towns before the Western impact, but their enormous growth in the past century was due to their role as colonial ports. They are essentially commercial cities of the Western type. Bangkok became the capital of Thailand at the end of the eighteenth century, but remained a small place until Western and Chinese traders gave it a new economic function. The impressive royal palaces and temples of Bangkok, which attract the modern tourist, are all relatively new, and are out of keeping with the city's major trade status. Saigon[2] and Hanoi were small fishing villages when the French arrived in Vietnam. Cholon grew very rapidly right beside Saigon as a Chinese city, the home and headquarters of the Cantonese and Fukienese who control the rice trade. It is now part of Saigon. Haiphong is the outport of Hanoi, and was the outlet for the colonial export economy of northern Indochina.

Singapore was a largely uninhabited island in 1819 when a farseeing British official, Sir Stamford Raffles, chose it as the site for a trade and naval base. He realized that the shortest sea route between Europe, India, and east and southeast Asia through the Strait of Malacca passed right by Singapore Island with its large and excellent island-shielded harbor. Singapore now has about 2.3 million inhabitants, who live on the trade of the Malay peninsula and southeast Asia as a whole. With the tin rush in Malaya in the 1800s and the rubber boom after 1900, the town of Kuala Lumpur began to grow. Although it is still much smaller than Singapore, it has now become the capital of independent Malaysia, with Port Swettenham as its harbor. Singapore continues to handle most of Malaysia's exports and imports and provides most of its commercial and financial services, but it was not included in the new state of Malaysia. Singapore's population is predominantly Chinese; Malaya without it has a slight majority of Malays. Malay nationalism and a similar Chinese reaction were too strong to permit survival of the short-lived union between Malaya and Singapore within the state of Malaysia. Singapore became independent by mutual agreement in 1965, but the two countries remain closely joined by defense and trade pacts that leave them almost as integrated as they were before the split. Singapore's situation is similar

Hybrid southeast Asia—Kuala Lumpur. This is the main street of the capital of Malaysia, with Islamic-Moorish government buildings on the right, and more familiar British colonial architecture on the left. And the street has a stream of Western-style traffic. *(Photo by P. Gosling.)*

[2]Saigon was renamed Ho Chi Minh City in 1974.

to Hong Kong's in many respects. It has also become a rapidly growing center of diversified light manufacturing. In addition, it functions as a commercial entrepôt and processes raw materials, such as tin and rubber, passing through the free port. The former British colonial territories of Sarawak and North Borneo, now Sabah, which joined Malaysia in 1962, remain in the union at least for the present. They are small producers of rubber and petroleum. Until the period of British control and investment and large-scale Chinese immigration, they were economically undeveloped, sparsely settled, and relatively primitive. Although they are the largest ethnic group in Sarawak and Sabah, the Chinese are outnumbered by a variety of other peoples. Most of the area, especially away from the coast, is still inhabited by hunting, gathering, and fishing groups not closely related to the Malays, such as the Dyaks. The tiny intervening Sultanate of Brunei elected to remain outside Malaysia, but it is similar to Sarawak and Sabah.

Manila had existed as a small town before the Spanish arrived, and owed its development to early Chinese and Arab traders. It grew under the Spanish, but its major growth came during the American period of control when it became the primate service center for the new commercial and export economy. Manila is still the predominant city in its national unit. The site of Djakarta was largely unoccupied when the Dutch established the fort and trading post of Batavia there in the early seventeenth century. These urban commercial sites were little valued or developed until Indian, Chinese, or European entrepreneurs realized their possibilities. Each of the cities they developed has now become the chief focus of the national life of each of the states. They are the economic heart as well as the political capital. It is difficult not to choose as the capital the only real city and the one that enjoys the best access to the populous and productive parts of the country, even if the city was originally an alien creation. The modern cultural life of each state is also focused there. Much of it is Western, even in the very nationalist movements that successfully fought against colonial control and had their roots and power in the colonial port cities. Nationalism and most of the other ideas influencing modern Asia were brought in or stimulated by the West. It is also in these primate cities that the Chinese and other alien groups are strongest, and even many of the buildings are foreign. Thus southeast Asia has been remade and fragmented by colonialists and outsiders. It has inherited from its colonial past a complex pluralism in society and economy that confuses its newly independent existence.

The openness of southeast Asia

It is no accident that southeast Asia has attracted outsiders for many centuries. It probably has the richest economic potential of all tropical areas of comparable size, and has become the most productive by far. Apart from its physical resources, which will be examined later, it has the invaluable asset of access. No part of southeast Asia, except for small and economically unimportant areas of central Borneo and parts of northern Burma, Laos, Thailand, Cambodia, and Vietnam, is more than 200 miles (320 km) from the sea. Practically all of the densely settled and productive areas of southeast Asia fringe the coast or are within 100 miles (160 km) of it. The insular and peninsular character of southeast Asia, which has often accentuated cultural isolation in the past and which still accentuates administrative problems within the insular states, greatly assists the cheap transport of goods. It has helped to make this area easily exploited commercially by outsiders who were accustomed to the use of the sea and had trade ambitions. What the peninsular areas lack in maritime orientation they make up in their river systems. Each peninsular state is drained by a single navigable master river that has become a natural highway for resource transport and the commercial development of the productive areas it serves. In a wider sense, southeast Asia spans one of the most important of the world's sea routes, between Europe, India, east Asia, and the Americas. Sea traffic from China or Japan to India or Europe must pass through southeast Asia. Much Australian traffic does the same, as does the traffic between the Americas and the Asian mainland.

The geographic openness of southeast Asia to outside influences and penetration, especially by sea, has conditioned its history for thousands of years. In many ways southeast Asia still has closer relations with outside areas, including Europe, than it has among its own parts. This weakness of intraregional interaction and the prominence of separate interactions with distant places further weaken the regional integrity of southeast Asia. They may also be partly to blame for the much later development of nationalism and resistance to colonialism than in the rest of monsoon Asia. Cheap, easy access to southeast Asia's resources gave the area a strong advantage over other tropical producers. It was largely responsible for southeast Asia's success in establishing a near monopoly in so many plantation crops. The plantations were developed almost entirely by Westerners, who did not have to contend with strongly organized local states or with local economic competition. The Westerners found a relatively dense local population at a much higher cultural and technical level than was characteristic of tropical areas elsewhere. In large part their advanced abilities were due to the earlier influences of Chinese and Indian traders and cultural missionaries on this eminently accessible region. The labor problems that plagued plantation agriculture in Africa and Latin America were largely avoided in southeast Asia, either by dependence on the local people or by attracting commercially minded and hardworking Chinese and Indians. Above all, tropical or mineral products from southeast Asia could be moved by sea almost from the point of production to the major markets in Europe and North America at low cost.

Modern economic development

By the end of the period of colonial control, or the beginning of the Second World War, southeast Asia produced about 90 percent of the world's rubber, tin, and copra (dried coconut meat as a source of oils), 80 percent of the world's rice exports, and about 40 percent of the world's sugar. It had a near monopoly of other, less important goods such as quinine (from the bark of the cinchona tree), kapok (for insulation, stuffing, and life preservers), and abaca.[3] It also produces large quantities of tea, sisal, palm oil, spices, and tropical hardwoods (see Map 75, page 296). War, postwar disorder, and the recent growth of plantation production elsewhere, especially in Africa, have somewhat reduced southeast Asia's dominance in most of these goods. The development of synthetic rubber, synthetic antimalarial drugs, detergent soaps, and other alternative or cheaper sources of many of the products that southeast Asia produces have also been harmful to its economy. But southeast Asia is still by far the most productive tropical area in the world. Many of its latent resources are still little used. With the exception of Java, central Luzon, and north Vietnam, southeast Asia so far has also been able to avoid the gross population densities characteristic of India or China. Its average living levels are still somewhat higher than those of its great continental neighbors and past tutors. Agriculture outside the plantations is not nearly so intensive, diet in terms of caloric intake is higher, foreign trade is proportionately larger, and most of the economies produce a relatively large surplus for export. The exports are almost entirely agricultural goods or raw minerals. They are sold to and in many cases still financed by industrial economies in the West, where the southeast Asian countries can exert little control over prices or demand. From the southeast Asian point of view, this is an undesirable situation that leaves many of them, despite their political independence, at the economic mercy of the West and without the prestige or power they associate with industrialization. Vietnam, Laos, and Cambodia have undergone a socialist revolution in the Chinese pattern. They are pursuing more self-reliant and self-sufficient goals with little dependence on foreign trade. But most of the other southeast Asian states are still tied to Western markets in varying degrees.

Resource base and resource use

Only Indonesia has a really large domestic market to support large-scale manufacturing. All of the southeast Asian countries lack capital for investment and can earn it most easily by continuing the present pattern of exporting raw materials to the industrial West. All of them, in differing degrees, lack skilled labor, entrepreneurs, technicians, and developed commercial facilities. These lacks are greatest perhaps in Indonesia, so its potentially large market, further divided by insular and subregional conflicts, is not yet a promising basis for industrial development. Equally important is the general lack of coal in southeast Asia. None of the countries is well

---

[3]Abaca, also called Manila hemp, is a fiber obtained from a type of banana plant grown in the Philippines. It is used especially for marine cordage, since it resists rotting in sea water. Sisal is another fiber, of somewhat lower quality and cost, produced from a spiky, pineapple-like plant, used for matting, cordage, and twine, and grown especially in Malaya and Indonesia.

supplied, and only in north Vietnam are there large enough deposits of suitable grade and location to be economically significant (see Map 75, page 296). The coal is mainly anthracite (hard coal), and little is of coking quality for iron and steel production. Hydro power is potentially more important. Most of southeast Asia has heavy rainfall and is close to mountains or hills, so there are plenty of good physical sites. The economic requisites for hydro development, however, are largely absent, and it will be a long time before development on the large scale necessary for major industrialization can pay. Iron ore is present in small quantities in north Vietnam and eastern Malaya, and in somewhat larger quantities in the Philippines (see Map 75). But it is not plentiful or favorable enough to attract manufacturing, and so far has been developed relatively little. More coal and iron ore are exported from southeast Asia than are used within the region, despite the small size of the reserves. Even for export the quantities are relatively small.

Tin and petroleum

The largest mineral reserves of southeast Asia are tin and petroleum. Petroleum is easily and cheaply transported to market centers elsewhere, and tin is relatively high in value and used in proportionately small amounts in most manufacturing. So both minerals characteristically travel long distances to market. Thus neither tin nor petroleum is necessarily a stimulus to industrial development in the area where it occurs, and both minerals lend themselves to the colonial export or drainage economy so important in southeast Asia. About two-thirds of the world's tin is still produced there. The major zone of occurrence runs through southwest China, Thailand, Burma, and Vietnam, down the Malay peninsula, and into the northern islands of Indonesia. The largest single producing area is on the islands of Bangka and Belitung near the southeast coast of Sumatra, although Malaysia as a whole has the largest national total of tin production.

Most of the tin is obtained from alluvial deposits or deposits near the surface. A good part is collected by dredges on rivers or artificial ponds or by hydraulic sluicing, which separates the heavier tin ore

Tin dredge—Malaysia. Most tin "mined" in this area is extracted by these large, floating factories that move through the tin-bearing areas, sucking up alluvial material and using the water to extract the tin. Often the ponds on which they work are artificial. *(Photo by David Harvey for Woodfin Camp & Associates.)*

from the gravel or other material. Extraction costs are relatively low, although the deposits near the surface are being worked out, and more and more mine shafts have to be sunk. But all of the major tin deposits are close to the sea, and transport costs to the refining centers and markets in the West are low. Some smelting is done locally, particularly at the entrepôt of Singapore where tin from both Malaysia and Indonesia can be assembled conveniently for re-export. But southeast Asia as a whole, in the absence of cheap fuel, technical and capital resources, and local markets, has little advantage even for the processing of tin ore when practically all of the tin is consumed eventually by manufacturing in the industrial West. Tin and rubber largely created the commercial economy of modern Malaysia through British and Chinese entrepreneurs and workers. It grew within a largely subsistence, peasant, agrarian economy in which industrialization had little place. The drainage pattern of tin and rubber exports has been superimposed on an essentially unrelated agricultural base. Tin mining and export occupy a similar place in the rest of southeast Asia wherever they occur.

The use of petroleum resources has followed essentially the same pattern, although it began more recently. Southeast Asia's position in world petroleum production is less important than that for tin, but there have been major recent oil discoveries in Indonesia. Production is rising rapidly with Western technological help, and oil is now the country's largest export. Undersea oil fields off the coast of Vietnam may prove important. Southeast Asia is becoming a major oil producer and exporter. Like tin, petroleum is fortunately located within southeast Asia for the most part (see Map 75, page 296). The largest producing areas in northern Sumatra and northern and eastern Borneo are close to the sea. In central Burma, the only other significant producer, the major wells are in the valley of the Irrawaddy from which oil can be transported relatively cheaply to external markets. Like tin, oil is a good example of the extent to which the states of southeast Asia must remain dependent for their economic well-being not only on the markets of the West, but also on capital, technicians, and services borrowed from the West.

Rubber plantation, Malaysia. Latex, or the sap of the rubber tree, drips from incisions made in the trunk and is collected and carried in pails to the receiving station. There it is usually treated and solidified before being shipped to a larger factory for further processing. Many plantations, however, maintain their own processing factories and export rubber sheets or blocks directly overseas. Notice that one of the workers appears to be a Chinese, the other two appear Indian. This suggests Malaya rather than Indonesia, but otherwise the scene might be in either country, or in south Vietnam, Thailand, or Sri Lanka. *(Photo from Uniroyal, Inc.)*

Population, resource use, and colonial economies

In minerals, or practicable bases for industrialization, most of southeast Asia is not well endowed. Its strongest comparative advantages lie in agriculture, by virtue of its tropical climate, alluvial and volcanic soils, and unparalleled locational assets. As populations increase, if per capita incomes rise and if capital, technical, labor, and managerial skills become more abundant, a modest structure of consumer-goods manufacturing for the domestic market may develop. But until these conditions are realized, it is difficult to justify or create more than a small volume of manufacturing in competition with more efficient producers abroad. Agricultural investment also competes for capital allowances and for some time will remain more profitable and more pressing. Although overpopulation on a large scale is still limited to Java and parts of Luzon, and intensification and improvement of agriculture are especially urgent there, population growth rates are as high or higher in most of the rest of southeast Asia. In another generation the whole area may face problems similar to the present crisis on Java and Luzon. Growing food deficits, rising unemployment, and on Java, very little unused land add to the population pressures. The Indonesian and Philippine governments have attempted to encourage emigration from Java and Luzon to the other, relatively sparsely settled islands in each country for many years. But emigration has never been as large as the annual population gain of Java or Luzon, and it is not difficult to understand why. As in Bengal, north Vietnam, or lowland southeast China, these relatively small areas have such strong agricultural advantages that their populations have grown rapidly under Western influences. Particularly by comparison with surrounding areas, characterized by strong relief, isolation, infertile soils, and malaria, the paradoxical situation of human misery where nature seems most favorable has been created. This situation underlines the drastic need for agricultural improvement, and, as in the rest of southeast Asia, for the additional production that can be obtained from land now unused or little used.

Relatively few areas within southeast Asia are well suited to plantation production. Lack of easy commercial access rules out many areas that would be physically suitable. Uncertain future world demand for many plantation products, particularly rubber and copra, makes large-scale additional investment risky. Some of the rubber-producing countries are undecided about the wisdom of replanting their aging rubber plantations,[4] let alone making large new plantings. But it seems almost certain that for many years to come oil and plantations will provide the bulk of the exports of southeast Asia. Mining and lumbering will provide nearly all of the rest. The countries must export in order to finance the import of manufactured goods, raw materials, machinery, transport equipment, and other products they need but do not produce themselves. They must export if they are to make progress toward even modest domestic industrialization or toward agricultural improvement. In this sense, the plantations and mines, the colonial legacies, are at the heart of the southeast Asian economy. Tin, petroleum, rubber, and rice exports, developed almost entirely by foreign entrepreneurs for foreign markets, have revolutionized the southeast Asian economy in less than a century.

Rubber and other plantation crops

The rubber tree, *Hevea brasiliensis,* is native to the Amazon lowlands of Brazil where, like most trees of the wet equatorial lowlands, it occurs scattered through the rain forest rather than in consistent stands. Gathered wild by local tappers, rubber became the basis of a short-lived boom in Brazil at the end of the nineteenth century. But a British explorer smuggled some *Hevea* seeds out of Brazil, and planted them at the Royal Botanical Gardens at Kew near London. Young rubber trees were later shipped to Ceylon, and then spread to Malaya, Thailand, Vietnam, and Indonesia, where climates were similar to the hot, wet Amazon basin. By about 1912 the Brazilian monopoly was broken, and southeast Asia shortly acquired its present near monopoly. Plantation production, with uniform concentrated stands of rubber trees, was much cheaper than gathering the rubber from wild trees by wandering through the rain forest. It also made possible the development of better varieties of trees and better control over the quality of the rubber gathered and processed. Local, Chinese, or Indian labor was much more efficient

[4] The average age limit for profitable, commercial, rubber-tree production is about thirty-five years.

than Brazilian labor, and access by sea from southeast Asian plantations to Western markets was cheaper than access from the Amazon valley. Malaya and Indonesia had their own rubber booms, as did Thailand and south Vietnam on a smaller scale. Overproduction became a major problem, especially after the First World War. Plantation production is less advantageous than wild-rubber gathering because it takes about eight years from the time the trees are first planted for a rubber plantation to begin profitable yields. During that time the market demand or the price for rubber may decline, and the new plantation may prove to be unnecessary or unprofitable. On the other hand, sharp rises in demand or price, such as occurred during the Second World War, may find total plantation production inadequate.

One solution, widely employed in both Malaysia and Indonesia, is to depend on private, small-scale, marginal rubber producers when prices are high or demand is great. Most of these marginal producers, or *small-holders,* gather the rubber from forest trees or from a small, casually tended planting. They can afford to ignore rubber when prices are low, for they have little or no fixed investment. They concentrate on the subsistence food crops or other activities that are their main support. In Indonesia small-holders produce more rubber than the plantations do. They do not contribute to market glut when prices are low, but their activities are not easily controlled by central planning. This makes it difficult to enforce international agreements on the restriction of rubber production when prices are high in order to discourage overproduction and price collapse.

Despite market fluctuations and the development of synthetics, rubber is still the chief export of Malaysia by value, and is an important export of Indonesia and Thailand. South Vietnam's formerly important rubber production was decimated by American military action, especially the chemical defoliation concentrated on many rubber plantations because they provided excellent cover for guerrilla movements. Rubber production in Indonesia has suffered from civil disorders and the disruption of the large plantation estates. Eastern coastal Sumatra and western coastal Malaya are the most important rubber areas, with coastal Borneo, Java, and Thailand following. Most of the rubber plantations are on or near the sea, underlining the chief asset of southeast Asia for competition in Western markets.

Interior Borneo, western coastal Sumatra, the mountainous areas of Thailand and Vietnam, and the islands of eastern Indonesia do not have large plantations of any kind because of distance from the sea, relief and forest barriers, unfavorable soils and climate, or a coastline lacking in good harbors or blocked by reefs. This is true of much of the island of Celebes. The Moluccas (Maluku), also called the Spice Islands, were one of the earliest Western footholds in Asia, and served as a Portuguese trade base from the early sixteenth century. Spices are still produced in southeast Asia, mainly in Java, Sumatra, and Malaysia, where access is favorable. The proportionate economic importance of spices has greatly decreased with the boom in rubber, palm oil, copra, sugar, tea, kapok, and other plantation crops. Spices were more important before the industrial and commercial revolutions, not only to preserve and season food, but also because they were sufficiently high in value to bear the cost of long-distance transport in an age of small sailing ships.

Map 75 (page 296) shows the general outlines of the present pattern of agricultural land use, and gives some indication of the relative importance of the various crops. Rubber is largely absent in the Philippines mainly because it cannot compete with the earlier-established and efficient plantations in the rest of southeast Asia and Sri Lanka. These two areas together can supply more than the world market normally demands. Copra and sugar are the two leading Philippine exports, although sugar growers face difficult competition from other areas. Coconuts are grown throughout the coastal or near-coastal parts of southeast Asia, but small-holders account for a greater production than plantations. Coconut palms require relatively little care, and nearly every rural household maintains a small grove. The nut meat is dried to make copra from which oil is pressed. But the tree also provides almost endless materials for domestic use, including thatch for roofs, nuts and their milk for food, wood for fuel and building, coarse fiber or *coir* from the husks to make rope, sacks, and clothing, and flower stalks as delicacies. Plantation production of coconuts is growing, however. It can often produce copra alone at lower cost. Coffee was once a leading commercial crop in southeast Asia, as the name *Java* still suggests. It was

Another part of a plural culture—rice fields on Bali. Bali is mountainous, but many of the slopes are gentle enough to be terraced for rice, as in this intricately engineered landscape. Palms suggest the unbroken growing season close to the equator. Volcanic soil helps to explain the high fertility of these fields. With regional variations, this picture represents the lowland rice-growing areas of southeast Asia. *(Photo by Deane Dickason for Ewing Galloway.)*

largely wiped out by a blighting plant disease in the 1870s and 1880s, and its place taken by tea and other crops that are more profitable. Brazil recouped some of its rubber losses by building up its coffee production to a near monopoly.

Demand for nearly all plantation crops is subject to great fluctuation. This alone is adequate justification for attempting to diversify the agricultural system, to give more attention to food crops, and to avoid dependence on one or a few dominant exports. The very rapid growth of population in all of southeast Asia makes these developments especially urgent. In the one year of 1953, for example, rubber prices dropped 30 percent as emergency buying and stockpiling resulting from the Korean war came to an end. In the same year the revenue of the government of Malaya decreased by one-third, a decline due also to an equally sharp drop in tin prices. Malaysia is in a special position; it depends on two principal exports and has a large domestic food deficit that makes the import of one-third of the country's food supply necessary. Rubber and other plantation crops make more profitable use of the land than rice or other foods, especially given the shortage of fertile soil in Malaysia where there are no major alluvial basins or volcanic deposits. Such a dangerously unbalanced economy, at the mercy of changes in world demand, becomes a more pressing problem as population increases. Most of the countries of southeast Asia face a similar problem in differing degrees.

Commercial rice growing

Rice is the largest crop in southeast Asia by far, and also occupies the largest share of agricultural land. Most of it is grown by peasant farmers for their own subsistence, but a surplus is produced in some areas that goes to urban markets and may even be exported. Among other things, Burma, Thailand, and Vietnam differ from the rest of southeast Asia in the relatively small importance of plantations until recent years and the major importance of commercial rice growing. Burma's rice production has never reached its pre-1940 levels and its population has continued to increase, so it no longer exports rice. War, civil war, unprecedented devastation by U.S. forces, a tide of refugees, and gross economic disruption have brought the formerly large rice exports of south Vietnam to an end. Northern Vietnam, the Tonkin region, was heavily influenced by Chinese culture from about the time of the Han dynasty. It has been relatively densely populated since that time, and its agrarian rice economy is like China's in that it produces no surplus for export. The Sinicized Annamese, originally inhabitants of Tonkin, gradually spread their control over the whole of Vietnam by migration and conquest. Their conquest and displacement of the earlier Indianized kingdom in the south is still recent enough that population densities there have not yet reached a comparable level of saturation. Until the Second World War, Cochin China (southern Vietnam), Burma, and Thailand were the three leading rice-exporting countries in the world. In these three areas the great expansion of rice acreage on a commercial basis took place at about the same time as the plantation boom in the rest of southeast Asia. It was financed by British, Indian, Chinese, and French capital and drew heavily on Indian and Chinese labor. But production has remained largely in the hands of the local farmers, although tenancy rates are very high in many areas. Farms are still predominantly small, and it is only in financing, marketing, milling, and export that foreign entrepreneurs have been so important. Large areas previously unused in deltaic southern Burma, Thailand, and Cochin China were drained and put into rice production. The new rice surpluses went to feed plantation workers in Malaysia and Sri Lanka, and to make up domestic food deficits in India and, to a lesser extent, in the rest of monsoon Asia.

Agriculture and population densities

Although rice production in peninsular southeast Asia as a whole has continued to increase, exports have declined or disappeared as domestic populations have increased, requiring more rice to be consumed locally. Rice yields per acre in this area are low, despite physical conditions almost ideally suited to rice: alluvial soils in the great river valleys and deltas, high year-round temperatures, and generally adequate rainfall. In the deltaic areas rain occurs to some extent throughout the year, although it is concentrated in the summer monsoon. By Chinese or Japanese standards, relatively little fertilization is practiced. Transplanting, weeding, and cultivating are minimal, and irrigation is predominantly of the crude, flood-water type. Only about 10 percent of the total area of the Indochinese peninsula is cultivated at all. A good deal of the unused land could be used for rice, with proper irrigation or drainage, or for other crops without irrigation. This relatively unintensive use, by Asian standards, of agricultural resources is largely explainable by low population pressure on the land. Intensive land use has not been necessary when less intensive methods not only fed the people adequately but also produced surpluses. It is more difficult to explain the low level of population pressure and the relatively small population totals, especially since the Indochinese peninsula is surrounded by some of the largest areas of overpopulation in the world: Bengal and south India, south China, Java, and Japan. Population is now growing very rapidly in southeast Asia, but the absence of such pressure in the past is something of a mystery.

One reason may be the relatively recent beginning of more than superficial Western contacts with the area. None of the peninsular states became a Western colony, except for small coastal footholds, until the second half of the nineteenth century. Effective British control over Burma was not established until the 1880s. The final French absorption of Vietnam dates from the same period. Thailand never became a formal colony, largely as a result of great power rivalries that the Thais, with British help, were able to use to preserve their nominal independence. Western influence was never more than superficial in most of Laos and Cambodia. They lacked easy commercial access and the obvious exploitable resources that attracted colonialism. Western commercial influence became widespread and fundamental in most of the rest of the Indochinese peninsula, especially with the boom in rice production, the development of rubber plantations in Cochin China, of oil in central Burma, and cutting of tropical hardwoods in Burma and Thailand. But Western influence came relatively late. It seems clear that the Western impact on commercialization, transport, irrigation and agricultural improvements, public health, civil order, introduction of new crops, and other matters was an important factor in the rapid population increases in India, Java, Sri Lanka, China, and Japan. Java, whose population in 1815 was approximately 5 million, had a population in 1950 of about 52 million. Four centuries of Western influence and a century and a half of concerted Dutch efforts to develop a productive commercial economy of plantations, mines, and rice fields to feed the commercial workers led to this population increase. In the outer islands of Indonesia, population has not grown nearly as much. There Dutch investments came later or were smaller, and commercial opportunities seemed less promising because of access, soils, relief, or undiscovered minerals. Malaya, where tin and rubber did not begin to transform the economy until nearly the end of the past century, still has a relatively small population.[5] Although Western impact on peninsular southeast Asia came late, it does not explain why the population densities had not reached the levels of those in most of the rest of monsoon Asia before the Westerners arrived.

India in southeast Asia—detail from the Stupa of Borobudur, central Java. Completed in the ninth century, this large stone temple is an impressive example of the strong influence of India on the culture of Java, in spite of the distance between them. This picture shows a portion of the carvings decorating the many terraces of this temple. *(Photo from ROLOC.)*

Historical disruptions

A further explanation of the low population levels may be found in the earlier history of the Indochinese peninsula. Appropriately named in terms of the powerful cultural influences that have been exerted upon it, this area has tended to occupy a role as a safety valve or an escape hatch during much of Asian history. A succession of migrations and invasions has poured into it, especially by its land approaches from the north. Each invasion disrupted or destroyed the previous civilization. Most of these invading groups came from China or its southwestern border areas. They were displaced by the southward movement of the Chinese, and forced out by the one practicable route into the Indochinese peninsula. Most of the people of peninsular southeast Asia are physically Mongoloid. But the strength of Indian cultural influence is apparent in the dominance of Buddhism and the use of Indian-derived systems of writing, art forms, and literature. Only in Tonkin, or northern Vietnam, did the Chinese cultural influence remain dominant, including the use of Chinese characters for writing and Chinese political, social, and art forms. Later, Chinese influence was carried southward to a lesser degree, as the Sinicized Annamese spread their control by conquest over Cochin China. In addition to invasions from the north, there were smaller movements of Malayan peoples from the south, although there is relatively little historical evidence of them.

Perhaps the greatest of the invasions from the north was that of the Thai people, the inhabitants of modern Thailand. They swept into the Indochinese

[5]Singapore, after its founding in 1819, developed commercially well before the rest of Malaya as an entrepôt for most of southeast Asia and for trade with China and Japan. Malaya as a whole remained relatively sparsely settled by small groups of subsistence rice cultivators on the limited fertile lowlands, and by hunters and shifting cultivators in the hillier forest areas. There was some tin mining before the boom of the 1890s. The east coast of the peninsula has never been as important as the west, since it has only a narrow coastal plain and a largely reef-bound coast off the major shipping lanes.

peninsula in the thirteenth and fourteenth centuries after having been defeated and expelled from southwest China by a Chinese-Mongol army. Their old kingdom of Nan Chao, with its center in modern Yünnan, was destroyed, and several waves of refugee-invaders entered Burma, Thailand, and Laos. Their descendants are still a large proportion of the population of these countries, especially in their hilly northern margins. The Thais, as one major group, not only occupied most of modern Thailand, destroying the pre-existing civilization there, but also destroyed the Indianized Khmer civilization with its principal base in Cambodia and its magnificent capital at Angkor. The destruction was so complete that the city was obliterated by jungle and forgotten until French explorers stumbled on it in the 1880s, hidden in a tangle of forest. But before the Thais, there had been many other invasions from the general area of China, and the peninsula had seldom known any long periods of relative peace.

Intensive, wet rice cultivation is a delicate type of agriculture because it requires careful investment and management, especially of irrigation works. In turn, irrigation depends on civil order, cooperation, and often strong government control. Such conditions existed in Tonkin during the long period of direct Chinese control[6] and the continuation of the Chinese pattern by the Annamese. This area was largely undisturbed by the catastrophic invasions that disrupted most of the rest of the peninsula. This may help to explain why high population density and intensive land use have long been characteristic of Tonkin and absent in the rest of the peninsula. Only within the last century have relatively stable conditions been established outside of north Vietnam. This has also been a period of very rapid population growth and the expansion of wet rice cultivation. The lack of political unity within the peninsula also contributed to its inability to resist invasion or survive invasions with its earlier civilization intact. The outposts of Indian culture established in Cambodia, Cochin China, and lower Burma and Thailand were relatively small. They also tended to be somewhat isolated from each other, and never achieved a common political unity. The Indochinese peninsula is physically divided by the great river basins, which are separated from each other by mountains, hills, and rain forest. Sea communications are interrupted by the southward extension of the Malay peninsula. Population, agriculture, and Indianized culture in Burma, Thailand, Cambodia, and south Vietnam were focused respectively in the valleys of the Irrawaddy, Me Nam Chao Phraya, and Mekong rivers. No one regional center was strong enough to dominate the others. The people of these lowland and deltaic areas were physically more closely related to the Chinese than to the Indians, despite their culture, but they were different enough from the fresh waves of invaders from the north to prevent full assimilation.

Pluralism compounded

This complex pattern of interaction has left each modern state with a plural society that is regionally expressed. In addition to the alien groups and alien economy that resulted from Western colonialism, the succession of migrations and invasions of the past led to a certain regional layering in the Indochinese peninsula. The more recent and less highly developed arrivals occupy the hill and mountain areas that fringe the central river valleys. Cultural (including language) differences between these hill people and the sedentary rice growers of the alluvial lowlands with their Indian heritage are so great that political unity and national harmony are still difficult. Thailand is perhaps least troubled by these problems, since the Thais are strongly dominant numerically within the country. There are also relatively few differences between the peoples of the lowlands and the hills. But for the cousins of the Thais who live in the hills of Burma (people like the Karens, Shans, and Kachins), or in the hills of Cambodia, Laos, and Vietnam, it is not easy to accept the political, economic, and cultural dominance of the people of the lowlands, who are so different from them and who have been their enemies for so long. The presence of Indian or Chinese money lenders, traders, and workers, especially in the cities where they are so numerous, is a further irritant and a further block to political harmony.

[6]This control lasted for a total of approximately seven centuries under the Han, T'ang, Yuan, and Ming dynasties. The Vietnamese subsequently won their independence from Chinese control, as they had for several shorter periods in earlier centuries. Vietnamese nationalism is a powerful force, as the United States discovered.

The future of an independent southeast Asia

For all of the states of southeast Asia, the challenge of independence is urgent and fundamental. Each must build a genuine national entity out of a diverse group of peoples by placing greater stress on nationality and common interest than on racial origin, culture, or regional loyalties. The concept of nationality is new in most of these areas, and it will not be easy to sustain it. Colonial rule bound together areas and peoples who had previously lived apart in many cases. Now that their common welfare exists within the independent national states, they must learn to live and work together fruitfully or face the familiar, destructive conditions of regional conflict or of political separatism. Regional conflict is especially serious in Indonesia, whose scattered islands and widely diversified culture may ultimately result in federation rather than a Java-centered state, opposed by all of the outer islands. The separate states of southeast Asia also have never achieved a workable relationship between themselves. There is little interstate cooperation or common planning. And a legacy of recurrent conflict still exists—between Vietnam and Thailand, also separated now by opposing ideologies; between Indonesia and Malaysia and the Philippines with their conflicting claims to areas in the archipelago; and between power blocs aligned with outside powers, most importantly the United States, China, or the Soviet Union.

But some form of genuine national, international, and intra-regional cooperation is essential for all of the states of southeast Asia, particularly if they are to meet the economic problems bequeathed to them by colonial rule, now rapidly becoming critical. Java, central Luzon, and Tonkin are already grossly overpopulated. It may not be long before the same is true of lower Burma and lower Thailand. Malaysia and the outer islands of Indonesia and the Philippines have a wider margin. But they are still precariously dependent too largely on the export of a few minerals or plantation crops. There seems little rational place for major heavy industrialization in any part of southeast Asia. The most important task would appear to be the improvement, diversification, and intensification of agriculture in every area, both to feed rapidly growing populations and to provide some surplus for export. In agriculture southeast Asia has strong comparative advantages, especially for crops that can be grown only in a tropical climate. No other area is likely to offer stronger competition in these crops, in demand all over the commercial-industrial world. Southeast Asia's advantages of soil, climate, and access place it in a strong position. With fuller and more intensive use of the land, southeast Asia should be well able to feed itself in the future and still leave a considerable margin for agricultural exports of plantation crops, rice, and other food crops. Rice yields, for example, are now only about one-third of rice yields in Japan and could certainly be doubled; much unused land could be brought into profitable production.

There is clear need for greater diversification to avoid the fluctuating fortunes of any one crop or product. With southeast Asia's physical advantages, and with greater attention to food crops for the domestic market, diversification should be possible. Industrialization can come in response to increases in the domestic market. Investment capital can best be earned from mining and agriculture in this situation. As the economy grows, it must also train its own technicians, workers, and entrepreneurs. The future domestic market resulting from a prosperous agricultural system and mineral exports should offer a profitable basis for the production of consumer goods and a limited amount of essential heavy industry. But these are in the future. Few of them have begun to take shape yet. Southeast Asia has both suffered and benefited from outside influences in the past and has largely been shaped by them. Now it has the opportunity of full independence to make what it can of its inherited assets and problems.

Questions for further study and discussion

1. One of the earliest Arab and Western trade bases in Asia was at Malacca on the west coast of Malaya. It was Arab from about the thirteenth century, and its control passed successively from the Arabs to the Portuguese in 1511, to the Dutch in 1641, to the British in 1824; it is now part of the independent state of Malaysia. Why was it so sought after? What is the significance of the succession of political control? Why is Malacca now a relatively unimportant port?

2. What similarities are there between Vietnam and Korea, in terms of artificial political lines im-

posed after the Second World War, and in terms of the geographic implications? How do the two cases differ geographically? (The demarcation line between North and South Vietnam was arbitrarily but temporarily fixed by the Geneva Convention of 1954 along the seventeenth parallel and had no legal validity as an international boundary.)

3. Why was there less cultural resistance on the whole to Westernization in the Philippines than elsewhere in southeast Asia? How has this affected contemporary Philippine culture and political unity? What regional differences in culture remain within the Philippines, and why?

4. Why should Indian cultural influences on southeast Asia have been limited so largely to near-coastal areas, despite the land connection with India?

5. What physical disadvantage of plantation production is suggested by the southeast Asian coffee blight of the last century?

17 At the ends of the earth: Australasia and the Pacific world

Modern Papeete, Tahiti. *(Photo by Milton and Joan Mann.)*

Australia, New Zealand, Papua–New Guinea, and the smaller islands to the east of them were isolated from any continuous contact with the rest of the world for about a million years. Native human, animal, and plant life shows this isolation dramatically, and includes many forms unknown elsewhere, like the kangaroo. New Zealand contained no mammalian species until they were introduced very recently from Europe. There was a land bridge between Asia and Australia–New Guinea across the shallow seas along which fauna and flora originally reached Australasia.[1] But the connection was broken long enough ago that species subsequently evolved independently. Even the native human population, which arrived later after crossing stretches of sea, shows clear differences from the rest of mankind. They are relatively minor physically (although the distribution of blood types is strikingly different), but fundamental in cultural terms.

This isolated world was reached by outside peoples only very recently. Although the coasts were sighted early in the seventeenth century by Spanish and Dutch expeditions, it remained for Captain James Cook's several voyages in the 1770s to reveal the possibilities of European settlement in southeastern Australia and New Zealand. The first permanent settlement by outsiders in Australia was not made until 1788 or the site of modern Sydney. New Zealand did not have a European settlement until 1840. The newly found land seemed so remote and unrewarding to its British colonial claimants that Australian settlement was planned as a secure place to which to send criminals sentenced to banishment. The first settlers were 717 convicted criminals under a large guard. The colonial population grew very slowly. It was not until the present century, well within living memory, that the major geographic features of Australia were known. Some areas still remain partly unexplored. New Guinea, largest island in the world after Greenland, although sighted by European adventurers as early as Australia, even now is not completely known by outsiders.

---

[1]*Australasia* is a term often applied to all the land extending east and southeast into the Pacific from the Philippines and Indonesia, beginning with Melanesia (New Guinea and the island groups lying east and southeast of it as far as New Caledonia and Fiji), and including Australia, New Zealand, and the much smaller Pacific island groups of Micronesia and Polynesia (see Map 76, pages 314–15).

LUZON
PHILIPPINE SEA
1828
SOUTH CHINA SEA
PARACEL ISLAND
Hue
Da-nang
VIETNAM
THANH-PHO HO CHI MINH (SAI-GON)
SCARBOROUGH SHOAL
MANILA
Quezon City
Lingayen Gulf
SIERRA MADRE
POLILLO ISLANDS
CATANDUANES ISLAND
PHILIPPINES
MINDORO
Sibuyan Sea
CALAMIAN GROUP
PANAY
MASBATE ISLAND
SAMAR
Cebu
CEBU
BOHOL
NEGROS
PALAWAN
Mindanao Sea
MINDANAO
Davao
2954 Mount Apo
Moro Gulf
Davao Gulf
CAPE SAN AGUSTIN
TINACA POINT
Sulu Sea
Zamboanga
BASILAN ISLAND
BALABAC ISLAND
Balabac Strait
PULAU BANGGI
JOLO ISLAND
TAWITAWI ISLAND
4424
10497
5576
SAIPAN
TINIAN
ROTA
GUAM (U.S.)
6674
11034
ULITHI
FAIS
YAP
NGULU
8527
SOROL
GAFERUT
OLIMARAO
WOLEA
IFALIK
LAMOTREK
PIKELOT
PULUSUK
PULAP
NAMONUITO
ULUL
HALL ISLANDS
MURILO
TRUK ISLANDS
FALALU
TOL
LOSAP
NAMOLUK
NAMOI ISLANDS
EAURIPIK
OROLUK
SENYAVIN ISLANDS
PAKIN
PONAPE
MOKIL
PINGELAP
NGATIK
UJELANG
PALAU ISLANDS
BABELTHUAP
ANGAUR
4308
CAROLINE ISLANDS
NUKUORO
SONSOROL ISLANDS
PULO ANNA
MERIR
5036
TOBI
HELEN ISLAND
KAPINGAMARANGI
Kota Kinabalu
4101 Mount Kinabalu
Bandar Seri Begawan
BRUNEI (U.K.)
Darvel Bay
Celebes Sea
5520
PULAU SANGIHE
KEPULAUAN SANGIHE
KEPULAUAN TALAUD
MOROTAI
HALMAHERA
Manado
NATUNA BESAR
KEPULAUAN ANAMBAS
TANDJUNG DATU
Kuching
Rajang
MALAYSIA
IRAN MOUNTAINS
Kajan
TANDJUNG MANGKALIHAT
PEGUNUNGAN MULLER
KALIMANTAN
BORNEO
Kapuas
2278 Bukit Raja
PEGUNUNGAN SCHWANER
Balikpapan
Equator
GREATER SUNDA ISLANDS
Bandjarmasin
Barito
Kahajan
Selat Makasar
SINGKEP
Selat Karimata
BELITUNG
TANDJUNG PUTING
TANDJUNG SELATAN
PULAU LAUT
TANDJUNG LUMUT
Laut Djawa
Java Sea
INDONESIA
DJAKARTA
Bandung
Selat Sunda
DJAWA JAVA
3676 Gunung Semeru
Surabaja
MADURA
KEPULAUAN KANGEAN
Laut Bali
BALI
LOMBOK
SUMBAWA
SUMBA
FLORES
Laut Flores
LESSER SUNDA ISLANDS
PULAU LOMBLEN
PULAU ALOR
TIMOR
Dili
Savu Sea
PULAU SAWU
PULAU ROTI
PULAU KALAOTOA
Teluk Tomini
KEPULAUAN TOGIAN
SULAWESI CELEBES
Teluk Tolo
Teluk Bone
Udjung Pandang
PULAU SELAJAR
PULAU BUTUNG
KEPULAUAN TUKANGBESI
KEPULAUAN SULA
5280
Laut Maluku
PULAU TALIABU
PULAU MANGOLE
PULAU BATJAN
PULAU OBI
PULAU SANANA
BURU
SERAM
Laut Seram
MOLUCCAS MALUKU
KEPULAUAN BANDA
KEPULAUAN WATUBELA
Laut Banda
BANDA BASIN
KEPULAUAN BARAT DAJA
PULAU DAMAR
PULAU BABAR
PULAU WETAR
KEPULAUAN LETI
KEPULAUAN KAI
KEPULAUAN ARU
PULAU JAMDENA
KEPULAUAN TANIMBAR
PULAU SELARU
Laut Halmahera
Selat Dampier
PULAU WAIGEO
PULAU MISOOL
DJAZIRAH DOBERAI
Teluk Berau
KEPULAUAN MAPIA
BIAK
PULAU JAPEN
Teluk Sarera
TANDJUNG PERKAM
Djajapura
PEGUNUNGAN VAN REES
PEGUNUNGAN MAOKE
5030 Puntjak Djaja
NEW GUINEA
PULAU ADI
PULAU DOLAK
TANDJUNG VALS
Sepik
4694 Mount Wilhelm
BISMARCK RANGE
Mount Gilewe 4089
Fly
Gulf of Papua
Port Moresby
OWEN STANLEY RANGE
Lae
PAPUA NEW GUINEA
NINIGO ISLANDS
WUVULU
HERMIT ISLANDS
KANIET ISLANDS
MANUS ISLAND
ADMIRALTY ISLANDS
MUSSAU ISLAND
NEW HANOVER
SAINT MATTHIAS GROUP
TABAR ISLANDS
Bismarck Sea
NEW IRELAND
BISMARCK ARCHIPELAGO
LIHIR GROUP
MALUM ISLANDS
TANGA ISLANDS
NUGURIA ISLANDS
FENI ISLANDS
GREEN ISLANDS
Rabaul
2300 The Father
NEW BRITAIN
TAUU ISLANDS
KILINAILAU ISLANDS
BOUGAINVILLE
NUKUMANU ISLANDS
ONTONG JAVA ATOLL
9140
CHOISEUL
SOLOMON
Solomon Sea
VELLA LAVELLA
SANTA ISABEL
NEW GEORGIA
MALAITA
Honiara
GUADALCANAL
WOODLARK ISLAND
D'ENTRECASTEAUX ISLANDS
LOUISIADE ARCHIPELAGO
TAGULA ISLAND
ROSSEL ISLAND
RENNELL ISLAND
Torres Strait
CAPE YORK
Arafura Sea
CAPE WESSEL
66
COBOURG PENINSULA
CAPE ARNHEM
MELVILLE ISLAND
Clarence Strait
Darwin
ARNHEM LAND
GROOTE EYLANDT
Gulf of Carpentaria
Timor Sea
CAPE GRENVILLE
CAPE YORK PENINSULA
Princess Charlotte Bay
4716
Joseph Bonaparte Gulf
CAPE TALBOT
Daly
Roper
Birdum
WELLESLEY ISLANDS
Cairns
1611 Bartle Frere
Townsville
Norman
GREAT DIVIDING RANGE
6459
7450
CHRISTMAS ISLAND (Austl.)
KIMBERLEY PLATEAU
Collier Bay
CAPE LEVEQUE
KING LEOPOLD RANGES
Victoria
BARKLY TABLELAND
Broome
CAPE LATOUCHE TREVILLE
Fitzroy
TANAMI DESERT
Mount Isa
SELWYN RANGE
Rockhampton
GREAT SANDY DESERT
Coral Sea
ILES CHESTERFIELD
NEW CALEDONIA
SANDY CAPE
FRASER ISLAND
MACDONNELL RANGES
Alice Springs
SIMPSON DESERT
Diamantina
Thomson
Brisbane
INDIAN OCEAN
BARROW ISLAND
Fortescue
1227 Mount Bruce
Ashburton
NORTH WEST CAPE
6658
GIBSON DESERT
AUSTRALIA
MUSGRAVE RANGES
1440 Mount Woodroffe
Lake Yamma Yamma
GREY RANGE
Warrego
Culgoa
Lake Eyre
-12
STURT DESERT
Barwon
1615 The Round Mountain
LORD HOWE ISLAND (Austl.)
Lake McLeod
Gascoyne
Carnarvon
Shark Bay
DIRK HARTOG ISLAND
Murchison
GREAT VICTORIA DESERT
Lake Torrens
FLINDERS RANGES
Lake Frome
Broken Hill
Paroo
Darling
Newcastle
SYDNEY
Lachlan
NULLARBOR PLAIN
Lake Everard
Lake Gairdner
EYRE PENINSULA
CAPE ADIEU
Kalgoorlie
HOUTMAN ROCKS
Spencer Gulf
Gulf Saint Vincent
Adelaide
Murray
Canberra
2230 Mount Kosciusko
31
1555
CAPE CATASTROPHE
KANGAROO ISLAND
Encounter Bay
MELBOURNE
CAPE EVERARD
Perth
DARLING RANGE
Geographe Bay
CAPE NATURALISTE
CAPE LEEUWIN
HOOD POINT
TORBAY HEAD
Great Australian Bight
5670
CAPE NELSON
CAPE OTWAY
KING ISLAND
SOUTH EAST POINT
Bass Strait
FLINDERS ISLAND
FURNEAUX GROUP
Banks Strait
86
CAPE GRIM
Launceston
1617 Mount Ossa
Hobart
TASMANIA
5267

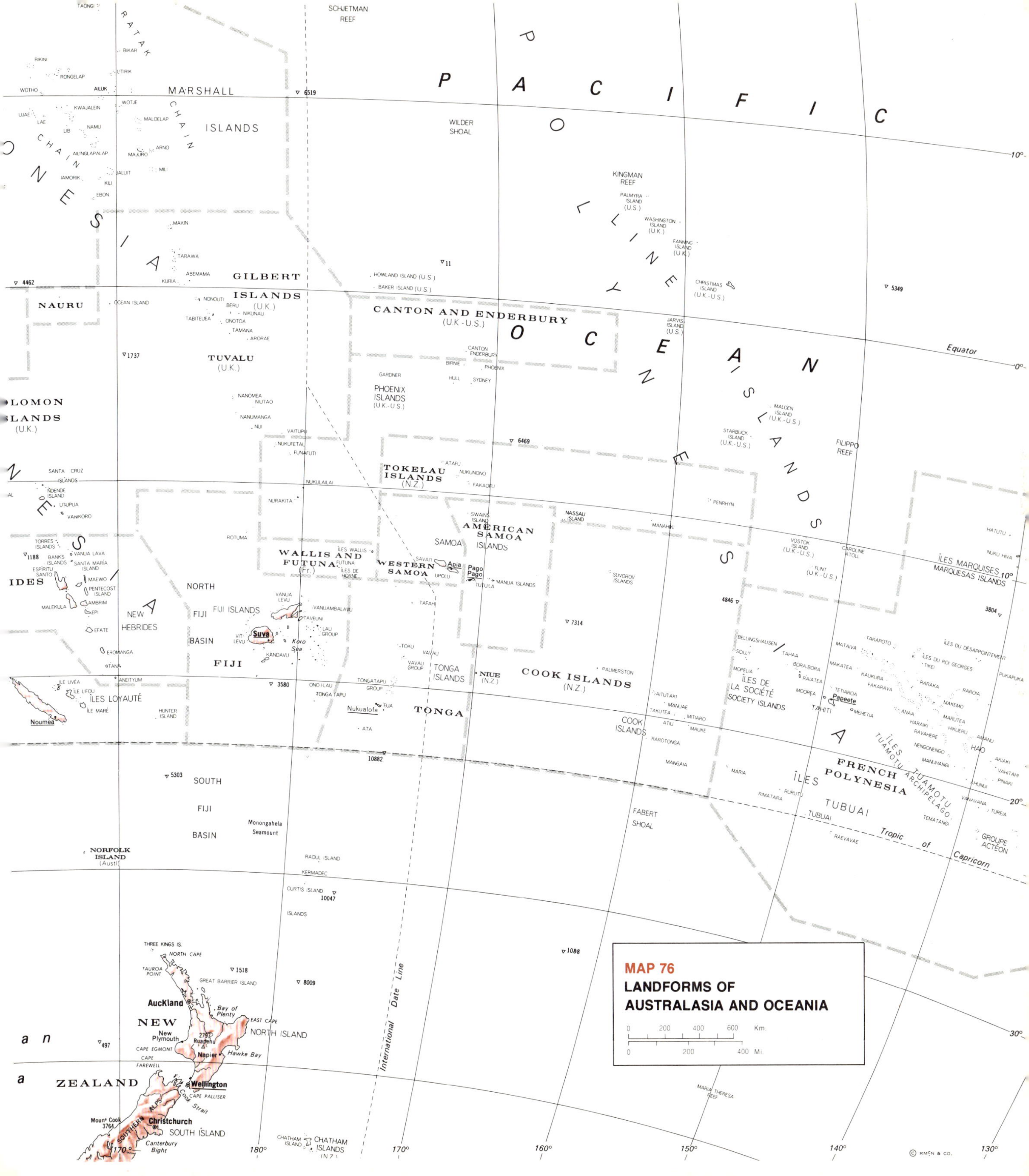

MAP 76
LANDFORMS OF AUSTRALASIA AND OCEANIA
0 200 400 600 Km.
0 200 400 Mi.
PACIFIC
OCEAN
POLYNESIA
LINE ISLANDS
MARSHALL ISLANDS
RATAK CHAIN
RALIK CHAIN
GILBERT ISLANDS (U.K.)
NAURU
TUVALU (U.K.)
CANTON AND ENDERBURY (U.K.-U.S.)
PHOENIX ISLANDS (U.K.-U.S.)
TOKELAU ISLANDS (N.Z.)
AMERICAN SAMOA
WESTERN SAMOA
SAMOA ISLANDS
WALLIS AND FUTUNA (Fr.)
FIJI ISLANDS
FIJI
NORTH FIJI BASIN
SOUTH FIJI BASIN
TONGA
TONGA ISLANDS
NIUE (N.Z.)
COOK ISLANDS (N.Z.)
FRENCH POLYNESIA
ÎLES DE LA SOCIÉTÉ
SOCIETY ISLANDS
ÎLES TUAMOTU
TUAMOTU ARCHIPELAGO
ÎLES TUBUAI
ÎLES MARQUISES
MARQUESAS ISLANDS
NEW HEBRIDES
ÎLES LOYAUTÉ
Noumea
SOLOMON ISLANDS (U.K.)
NORFOLK ISLAND (Austl.)
KERMADEC ISLANDS
NEW ZEALAND
NORTH ISLAND
SOUTH ISLAND
Auckland
Wellington
Christchurch
Napier
Cook Strait
Hawke Bay
Bay of Plenty
SOUTHERN ALPS
Mount Cook 3764
CHATHAM ISLANDS (N.Z.)
Suva
Apia
Pago Pago
Papeete
Nukualofa
Equator
Tropic of Capricorn
International Date Line
10°
0°
20°
30°
180°
170°
160°
150°
140°
130°
© R.McN. & CO.

Isolation and the frontier quality of Australasia are its dominant characteristics, reflected in almost every aspect of its current life and situation. Surrounded by the largest ocean on the globe, in an out-of-the-way corner nearly touching Antarctica, far from any of the routes connecting other parts of the world, this remote area is still undergoing the fundamental change of early, new growth. In some ways, Australia in particular has repeated the earlier experience of North America, also reached late by outsiders and developed as a frontier for a long time. But Australia presents many more obstacles, and fewer rewards. Its modern development will almost certainly remain slower, smaller, and increasingly different in character from that of North America. Australia has lesser resources for human use and is physically remote even in a jet age. As if to symbolize and sharpen Australasia's isolation, it turns its most inhospitable and unapproachable face toward the nearest land neighbor: the rugged mountains, rain forests, mangrove swamps, barrier reefs, and deserts of northern Australia and New Guinea. The Australian continent, almost exactly the same size as the continental United States, still supports less than 14 million people, reflecting both its isolation and its relative resource poverty. New Zealand, with a little over 3 million people, is almost equally empty by comparison with most of the rest of the world, and New Guinea is much more so.

Patterns of resources and barriers

Although remoteness and relative resource poverty are characteristic of all Australasia to differing degrees, they have not produced identical development in each part. New Zealand and Australia were far more open to outside penetration, and to a European type of land use, than any parts of Melanesia or Micronesia. And they also contained some easily exploited resources. Gold and several less valuable minerals were found in Australia, which started a California-style gold rush beginning in 1851. Gold was also found in New Zealand about the same time. In Australia major deposits of lead, zinc, and (more recently) iron and bauxite could be exploited profitably. Australia has become the world's leading exporter of lead, and a major supplier of iron ore. More important, both Australia and New Zealand had the basis for extensive agriculture: wheat and grazing. New Zealand is mountainous, but its mild and moist climate supports excellent pastures. The southeastern fifth of Australia, roughly speaking, is humid enough for dairying, mixed farming, or wheat. The semiarid areas farther out into the Australian steppe have been put to use as immense ranges for cattle and (in the drier parts) sheep. Cattle and sheep are mainstays of both the Australian and New Zealand economies. The much drier Australian base means that the sheep are raised mainly for wool. New Zealand's green pastures produce lamb and mutton that, with wool and butter, are the country's largest exports.

These agricultural and mineral resources were relatively accessible and usable, once the two countries had been opened to European settlement. There were no major physical barriers in the way of expansion of white Australian occupancy from its early base in New South Wales westward, southward, and northward. The Great Dividing Range, running somewhat parallel to the east coast, is not very high anywhere (see Map 17, pages 64–65). In time its difficult terrain was crossed. The narrow strip of relatively well-watered lowland along the coast gives way, beyond these mountains, to increasingly arid steppe, and finally to hot desert. New Zealand's mountains are higher and more rugged, and the two main islands are relatively small. But the coasts are well indented, and sea communication links virtually every important settlement. The northern lowlands around and south of Auckland are the most productive part of the country agriculturally, in part because of their northward reach toward the equator. The lowlands are very accessible and nowhere far from the sea in this relatively narrow extension of the North Island. Understandably, in this agrarian economy, Auckland is New Zealand's largest city, although the capital is at more centrally located Wellington (see Map 17).

By comparison, Melanesia, Micronesia, and Polynesia, with a few minor exceptions, have remained almost completely unattractive to outside settlers, and in addition continue to present almost insuperable barriers to settlement or land use. New Guinea is the clearest as well as the largest such case, dominated by a high and rugged central mountain spine, the Bismarck and Owen Stanley ranges—see Map 76 (pages 314–15). They reach almost to the sea on the north, but are replaced in the south by the even more difficult barrier of a dense equatorial rain forest, which covers all of the New Guinea lowlands in every part of the island. The smaller Pacific island groups are either tiny specks of land (many of them coral atolls), usually lacking in adequate harbors as well as significant amounts of usable land, or somewhat larger entities, such as New Britain or the adjacent Solomon Islands. They reproduce in miniature the repellent rain forest–mountain combination of New Guinea. Many of these islands, such as Guadalcanal in the Solomons, were the scenes of gruesome battles between defending Japanese and attacking American and Allied forces in the later years of the Second World War. They were fought under the impossible conditions imposed by the local climate, relief, vegetation, animals, and related diseases. Large parts of New Guinea remain unexplored by outsiders. It seems virtually certain that at least for the foreseeable future very little of Australasia except New Zealand or Australia is likely to become involved with the rest of the world to any major extent. Fiji is a qualified exception, which will be discussed later. Hawaii, a more obvious exception, is discussed in Chapter 25.

Even this brief comparison may suggest that Australia and New Zealand do not really suffer from resource poverty. But in relation to most areas of comparable size elsewhere in the world (except for the polar regions, the great deserts, and equatorial basins such as the Amazon), neither country is well endowed. Neither has a wide range of usable minerals, actual or potential agricultural resources, or resources of location. Their present population provides the best evidence for this—after a century and a half of growth under heavy European influence, free immigration (for whites), and sustained efforts at resource development. Australia in particular for many years was seen and advertised by many enthusiasts as a boundless frontier of expanding economic opportunity. But its dominantly arid character has placed a sharp limit on economic or population growth. Isolation has compounded these limitations, especially in a world where industrialization and economic growth are dependent on cheap exchange to such an extent. The home markets alone have never offered an adequate base for the support of more than modest industrial development, and seem unlikely to do so for many decades. Small population totals and, especially in Australia, huge areas of little-used or totally empty land may well bring their own rewards. Both factors are reflected in the quality of Australian and New Zealand life. Both countries are equally influenced by their isolation from the main currents of the modern world, centered many thousands of miles away across great stretches of sea and flowing to this remote area only tenuously. For Australians and New Zealanders, this poses a particular problem, since they are almost exclusively of European origin and their major interests and ties are centered not in Asia or Africa but in Europe and/or North America, at the other ends of the earth.

Human interactions

As in North America and parts of Latin America, the white population of Australia has largely dispossessed the native population, an even less difficult feat than it was for the European colonists in the New World. Nineteenth-century European technology, plus the growing weight of numbers, totally outclassed the ineffective resistance of the small groups of Stone Age native Australians. The still more primitive Negroid inhabitants of Tasmania were entirely exterminated by 1876. Even the Polynesians and Melanesians were literally decimated by slaughter, slavery, and European diseases. This record, however, is far from peculiar to Australasia. White behavior has been similar in most of the areas to which European settlement spread. Only when conquest and dominance have become assured has it been possible apparently for whites to recognize that other people, technically or militarily less effective at that point in historical time when they fell prey to white depradations, are nevertheless human beings.

The more highly developed Maoris, whom most Europeans found physically and socially attractive once the rough early years were passed, have now been largely accepted into New Zealand society. In effect they have been absorbed into the European-modeled system with little prejudice. The Maoris, almost 10 percent of New Zealand's population, are a Polynesian group, most of whom had not been in New Zealand long before the whites arrived. They clearly seem to have been the first inhabitants of the country, a dramatic indication of its extreme isolation.

Evidently, they reached it by sea from their earlier island base or bases in the south-central Pacific, some 3,000 miles (4800 km) to the eastward, helped by the amazing seafaring skills for which the Polynesians have long been noted. The first such Maori explorers may have reached New Zealand as early as the third century A.D. But the principal settlers seem unlikely to have come before about the thirteenth century—a moment ago in the history of mankind and even of intercontinental human migration, which is at least 50 thousand years old.

The human imprint on Australia is far older, although until the arrival of the whites it was the faintest of imprints. It is not clear how long humans have been in Australia, but almost certainly for at least 12 thousand years. The Paleolithic culture of the Australian aborigines suggests a long period of isolation from the centers of Neolithic and, later, metal-using, city-building, literate cultures of the Asian mainland and the Malay-Indonesian-Philippine areas. At some stage of their migration from an Asian homeland, however, the Australians did bring with them what may have been a partly domesticated dog. This animal did not evolve in Australia, and the people may have used its help in hunting. The people were not even primitive agriculturists, and were dependent entirely on a hunting and gathering economy. Although they lacked even the bow and arrow, they were superb trackers, expert users of the boomerang (throwing stick), finders of water in arid areas, and hardy survivors of periodic food shortages and cold. They constructed only the simplest temporary shelters and wore hardly any clothing. At the time of the first European settlement, the total native population was about 200 thousand. They were most heavily concentrated along the east and parts of the north coast, where climate and vegetation were most favorable to the animal and plant life on which the economy rested. This population pattern was very similar to that of the present, at least in its location, which is still shaped by the same environmental factors.

Original Australians. Some of the dwindling survivors of white conquest with their throwing spears. They are poised on a desert outcrop, typical of the meager reserves that remain to them. *(Photo by W. Pedersen for Australian Information Service.)*

The white settlers killed or drove off the original inhabitants of any land they wished to occupy or use. Before the end of the nineteenth century the dwindling number of aborigines was almost entirely confined to the remoter areas of the great Australian desert. More recently their numbers have begun to rise rapidly again, but for the most part they are in such close contact with white Australians that their own culture is being eroded. It is a situation somewhat like that of the Indians of North America. However, the native Australians are both physically and culturally much farther removed from their white conquerors, so there has been less intermarriage and relatively slow or ineffective cultural absorption. Well before the end of the present century it seems likely that the traditional culture of the first Australians will have vanished, even if their race survives. Those few who still live the traditional nomadic life, where they have not been entirely emasculated by white pressures, together with the Papuan groups of New Guinea, provide invaluable living examples of the technology, social organization, and ideological system of an old Stone Age culture. In many respects, including even its technological adaptation to an arid environment, their culture bears favorable comparison with contemporary Western culture.

Australia and New Zealand

Extensive land use and urban concentration

The modern growth of both Australia and New Zealand illustrates, among other things, the consequences of an extremely extensive system of resource use. Both countries have a predominantly agrarian economy, in terms of their external trade. Both also have unusually high levels of urbanization. Their populations are heavily concentrated in a few relatively large cities (considering the total size of the national population). The rural areas are comparatively sparsely occupied, especially in Australia. This apparent puzzle is readily explainable when one remembers the nature of the land use. Both countries are engaged in grazing and the growing of wheat and fodder crops. Neither of these activities requires large amounts of labor, although both use relatively large amounts of land. In other words, their agriculture is extensive rather than intensive. The same is true of mining, additionally important in Australia.

MAP 77
PRECIPITATION IN AUSTRALIA

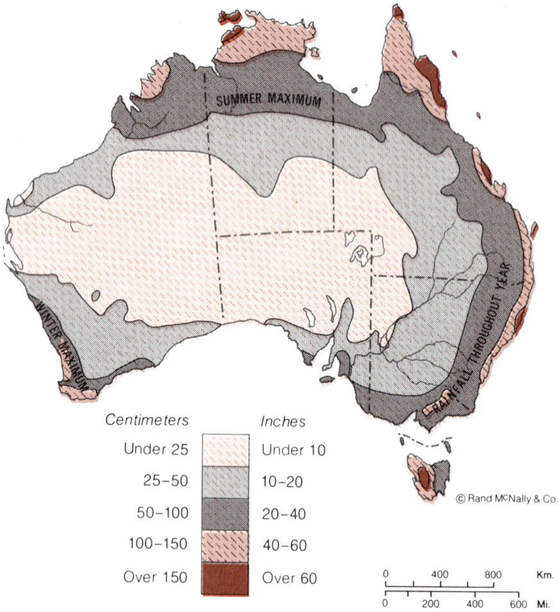

MAP 78
VEGETATION IN AUSTRALIA

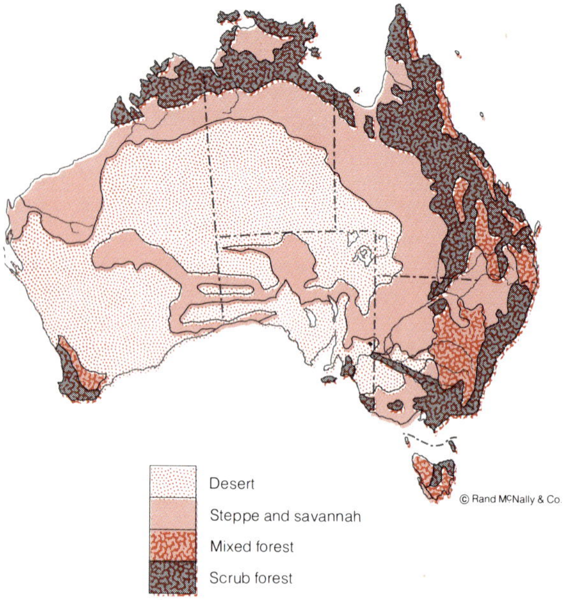

MAP 79
POPULATION OF AUSTRALIA

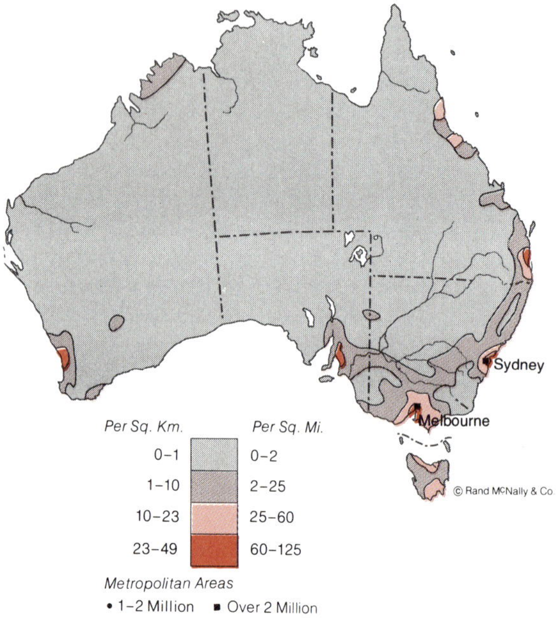

MAP 80
MINERALS AND ECONOMIC PATTERNS IN AUSTRALIA

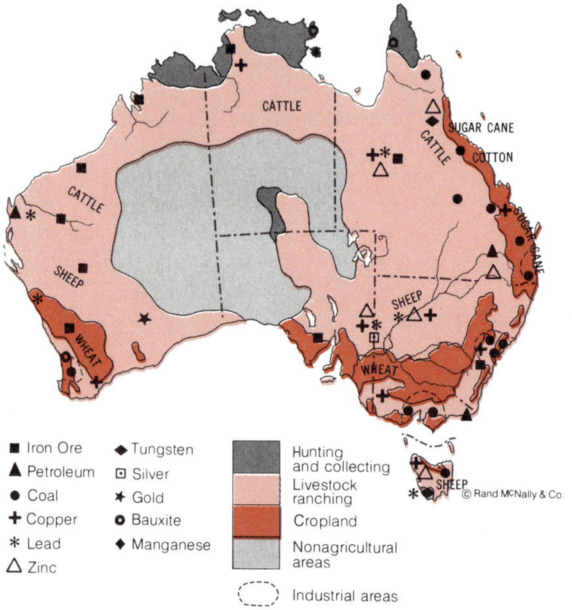

However, both countries are very productive, and they create a large stream of goods requiring processing, something most economically done in a city, where most of the finished product will also be consumed. In addition, a large proportion of this production is destined for export, given the limited size of the home markets. This requires processing, packing, handling, storage, shipping, financing, and other services of an urban character, as well as the presence of a port. Finally, the ranches, farms, and mining centers are dependent on urban bases to provide them with tools, machines, transport links, liquid capital, and other goods and services necessary to their production processes or part of their consumption patterns.

About one-third of Australia's population lives in the two cities of Sydney and Melbourne. If one adds Adelaide, Brisbane, Perth, Newcastle (almost a satellite of Sydney), and Canberra, the only other cities over 100 thousand (see Map 79, page 320), one accounts for more than two-thirds of the country's population. Over 80 percent of Australians live in cities of varying sizes, a proportion equaled only in the most highly industrialized states of western Europe and in Israel. The home market requires some industrial goods of course, and, apart from imports, these are produced, finished, or assembled in Sydney-Newcastle, Melbourne, and the lesser cities. But urban size in Australia is only partly related to manufacturing, most of which is serving consumption in the cities themselves. These urban centers are also concerned, in a more basic sense, with processing and services, direct and indirect, for the vast hinterland of extensive agriculture and mining (see Chapter 1). The New Zealand pattern is similar; Auckland and Wellington together contain about one-third of the total population. However, there are only two other genuine cities in New Zealand, as opposed to towns of less than 100 thousand, Christchurch and Dunedin. The New Zealand environment as a whole is moister and more permissive, especially with respect to rural settlement, and has a somewhat more intensive use of the land. Of course the total population is relatively small, so it cannot support many large cities in any case. In part for this reason, production is oriented toward export even more than in Australia. In both countries, it is notable that every city and most (in New Zealand, nearly all) towns are seaports.

Australian regional patterns

Australia, a continent in itself, has presented modern settlement and political development problems familiar in other large states. Although white occupance is recent and also relatively homogeneous, some degree of regionalism has developed. The country is far too large to be administered as a single unit in all local details, and some regional subdivision of authority is necessary. Rivalries among the states that were created somewhat arbitrarily (see the straight geometric lines on Map 17 marking state boundaries) have tended understandably to center in each state's primate city: Sydney for New South Wales, Melbourne for Victoria, Adelaide for South Australia, Perth for Western Australia, Darwin for the Northern Territory, and Brisbane for Queensland. It was politically awkward to establish the national capital in any of these cities. Of course, since all of them are coastal ports, each is also offside in relation to the others. An inland site for the capital was eventually agreed on at Canberra in New South Wales, about 150 miles (240 km) southwest of Sydney, some 400 miles (640 km) northeast of Melbourne, but still only about 70 miles (112 km) from the Pacific coast. Like Washington and Brasilia, it was planned from the beginning as the capital, in its own surrounding Capital Territory detached from New South Wales. It was built mainly in the 1920s as an almost entirely administrative city, and is also the site of the Australian National University. Canberra has grown especially rapidly since the 1950s and has attracted some light industry in addition to its administrative functions. Although Australia is of continental size, it holds fewer people than metropolitan Tokyo, clustered along the southeastern rim of the country. National administration can be performed by a relatively small city of bureaucrats (just over 150 thousand) located roughly in the middle of this thin crescent of population, and hardly inland in terms of Australia as a whole.

The state-federal pattern superimposed on the Australian landscape, like most such artificial arrangements, has been accompanied by the growth of new regional variety as each state tended to acquire its own identity, pursued its own policies, and attracted the regional loyalty of its inhabitants. The presence in each of a single dominant city has accentuated the strength of regionalism. These cities were the first settlements in each area, and still set the cultural and attitudinal patterns. Railways were built first from each of these cities rather than as part of a national plan, and the rail pattern still reflects such origins. Transport networks and regulations still reinforce state-based regionalism. Systems of land use first established around the early base, slightly different and reflecting local environmental factors in each case, tended also to spread outward.

Sydney and New South Wales, the earliest base, proved also to be the best endowed. The state contains far more usable agricultural land and minerals than any other. It also has the advantage of a magnificent natural harbor at Sydney, which has always been the largest Australian city. Dairying, fruit, and vegetables are dominant in the coastal area, giving way to sheep and wheat as the climate becomes more arid inland. Important deposits of lead and zinc are mined at Broken Hill near the western border of the state well into the desert. There are abundant supplies of good quality coal near the coast, stretching from the south through the Sydney area and up to Newcastle, which is the origin of its name (see Map 80, page 320). The Sydney-Newcastle area forms the largest Australian industrial concentration by far, and in particular dominates iron and steel manufacturing and metal-using industries. Victoria, immediately to the south, although the smallest mainland state, is also well endowed. Virtually all of it is well watered. There is local coal, and the rail network is highly developed. Melbourne's size, consistently a close second to Sydney, reflects these factors.

Queensland, occupying the northern two-thirds of eastern Australia and much bigger than New South Wales and Victoria combined, suffers both from aridity inland and from hot, wet, tropical conditions nearer the coast, for most of it lies within the tropics. This is an advantage for growing sugar, Queensland's leading crop, and dairying in the coastal areas. But it is not well suited to the wheat growing and sheep raising dominant in most of the rest of settled Australia. Cattle ranching has been substituted, although the quality of the beef suffers from tropical conditions and coarse forage. Much of Queensland is still a frontier for white settlement and land use, and the population is strikingly concentrated in and around the capital at Brisbane in the extreme southeastern corner of the state. Cattle ranching spills over into the Northern Territory, which offers the most meager base for settlement of

Australian outback. The Hamersley Range in Western Australia, shown on the left, is true desert, as the landforms suggest, but is typical of over half of Australia. The landscape is understandably empty of human settlement, and is likely to remain so. The area on the right, in the northern part of Western Australia, is only slightly less dry, but carries enough sparse grass to support cattle ranching. *(Photo on left from Caterpillar Tractor Co.; on right, W. Pedersen for Australian Information Service.)*

all the Australian states. Most of the Northern Territory is arid, all of it is hot, and only a negligible amount is cultivated. The total population is tiny, clustered around the small-town capital of Darwin. It is served only by a short rail line that peters out long before it reaches the desert and is not connected with the rest of the Australian system.

The state of South Australia occupies a thin strip of Mediterranean climate along the coast on both sides of the capital at Adelaide, and a large chunk of desert that covers most of the state (see Map 77, page 320). The Mediterranean zone supports the familiar combinations of wheat, grapes, citrus and other fruits, nuts, and vegetables. Some irrigated wheat and sheep grazing extend out into the dry country inland. South Australia's population is even more sharply limited than in the other states to the small coastal area where the environment is permissive, in contrast to the barrier imposed by the dry scrubland desert. There is relatively little climatic transition northward and only a limited intervening band of

steppe-savanna conditions (see Map 78, page 320). A single rail line runs north into the desert to Alice Springs in the Northern Territory. It serves surrounding marginal sheep ranches, but carries little traffic. For good economic reasons it stops about 600 miles (960 km) south of the line leading on to Darwin. South Australia does include one of the country's major deposits of iron ore, near Whyalla on the coast of Spencer Gulf north of Adelaide (see Map 80, page 320). It supplies ore to steel plants in the Sydney-Newcastle area, and also supports pig iron production at Whyalla itself.

Western Australia, the largest state by far, parallels South Australia in that it too includes a coastal area of Mediterranean climate and a much larger area of desert. Most of the state is largely unoccupied. The western and southern margins are used for sheep ranching. Fruit, vegetables, and dairying support a cluster of population around the capital at Perth. There are the beginnings of some mining and industrial activity on the northwest coast, where very large deposits of iron ore have been found in the Hamersley Range and the Yampi Sound areas (see Map 80). Ore is also shipped to Perth to supply a growing iron and steel industry dependent on coal deposits in the southwestern tip of the state. Gold is still mined at Kalgoorlie, and a thin finger of population reaches out to it from Perth along the railway to Adelaide and Sydney.

The standard Australian pattern of a coastal zone of dairying, fruit, and vegetables, grading off into wheat and sheep is evident also in Tasmania, the smallest state and the only one that lacks any really arid area. The rugged mountains that cover most of the western half of the island make up for this to some extent.

The Australian environment

The human use of Australia tempts the climatic determinist, as the foregoing may suggest. Over one-fourth of the country is desert, where even twentieth-century technology finds it difficult to create a productive economic base except for chance spots of mineral deposits. Most of the rest of the continent is steppe or savanna, offering a basis only for extensive sheep or cattle ranching, but not for any more than sparse human settlement. Only about 6 percent of the total land area is cultivated, or seems likely to be in the future. Population accordingly is heavily concentrated in thin coastal strips where climate offers a more permissive base, isolated from each other by intervening gaps of arid country. Most of Australia seems certain to remain largely empty. "The outback," as Australians call it, has rewards of openness and freedom that Australians value highly. It contrasts especially with the growing overcrowding of much of the rest of the world. Its economic value is marginal, which is its protection of course.

Many Australians do not regret that earlier prophets of economic development on an American scale have been proved wrong, and are anxious to retain the qualities of Australian life that vast empty spaces and isolation from the rest of the world help to preserve.

In the urban areas, however, city life and city functions are not notably different from other areas to which the commercial, industrial, and urban models of Europe–North America have spread. Sydney and Melbourne especially, both well over 2 million, are big cities by any standards. Together they constitute, after all, over one-third of the country's total population. Less than one-fifth live outside an urban environment. Nevertheless, the basic aspect of the Australian economy rests on an agricultural foundation, and its exports are dominated by agricultural goods or raw minerals. Wool is about one-sixth of the exports; other animal products, wheat, and fruit, about one-fourth, and ores and coal, another fourth. Manufacturing (including processing) employs far more Australians than agriculture does, including a larger proportion of the total labor force than in the United States. Most of the necessary raw materials are domestically produced, and the products are almost entirely consumed domestically. But this does not mean that, despite the small size of the domestic market and the consequent small total scale of production, it is not economically rational. Although the number of consumers is relatively small, they have very high per capita income levels. The economy is also fueled by the primary products from the outback destined for export through the cities or feeding raw materials into the industrial system. Like New Zealand, Australia has a very high level of per capita foreign trade, made possible by efficient, low-cost sea transport. It has little difficulty in paying for the imports (mainly manufactured goods) that its own economy does not yet produce.

The influence of a still open frontier has often been used to explain the social democracy of Australian life. The society's recent origins, wide individual economic opportunity, and humble beginnings may also help to account for the small importance most Australians attach to status or class distinctions. Such attitudes are reflected in the long tradition of state-supported welfare policies, the strength of labor unions, and the easy and open character of most Australians. Especially since the Second World War, mass immigration has also tended to keep Australian society fluid. In 1947, 90 percent of the population was native-born, but by 1978 this figure had dropped to about 80 percent. Over 300 thousand displaced persons emigrated to Australia in the aftermath of the Second World War. There has also been a large migration from south Italy (see Chapter 7), so the British share of the foreign-born population dropped below half for the first time. It remains official policy to discourage immigration from Asia. The "White Australia policy" is easy to attack, but many Australians fear the possibility that large numbers of industrious, low-wage Asian workers might provide unwanted competition and unwanted crowding. Relatively empty Australia is also hyperconscious of crowded Asia's geographical nearness, its own isolation from Europe and America, and the threat the Japanese forces posed to Australia in the Second World War from bases in Indonesia only a short distance away. There is no clear or simple answer; in the long run, Australia must live with Asia as well as with the rest of the world. If some of the special qualities of Australian life must be sacrificed as a result, that is a small price to pay for the building of a genuine world community.

New Zealand

New Zealand has retained its dominantly British population mix even more strongly in the period since 1945, apart from the continued absorption of the Maoris. The two islands together are roughly equivalent in size to the British Isles, and New Zealanders attempt to reproduce as much as possible a British pattern of society and attitudes. To some extent this is successful, but very different resources, including the vital one of location, have been accompanied by an economic pattern that is almost the reverse of Britain's in some respects. The most important aspect of difference is population density: about 30 people per square mile (12 per square kilometer) as compared with about 600 (230 per square kilometer) for the United Kingdom and nearly 900 (346 per square kilometer) for England alone. There is some New Zealand manufacturing, in addition to the processing of agricultural goods for export, serving the small domestic market almost exclusively. The economy is mainly pastoral, and animal products

Humid, settled Australia. Sheep and wheat, both in New South Wales, are the leading types of Australian land use. Both suggest the semiarid climate, even in New South Wales. *(From Australian Information Service.)*

(wool, meat, butter, cheese) account for about 90 percent of the country's exports; only 4 percent is of nonagricultural origin. The high and rugged mountains that occupy most of the South Island and the central core on the North Island mean that a great deal of the land area is not usable except for forests, which cover about 25 percent of the country, or for recreation. Relatively little of the remainder is cultivated, only about 3 percent of the total land area. Nearly half is classified as pasture. Such a land-use pattern is usually found only in arid areas, but New Zealand is well watered and has a climate not very different from Britain's. The small size of the population helps to explain the concentration on grazing and dairying, for which New Zealand is environmentally well suited. Remoteness is clearly another important factor. The country in effect is almost one big pasture, admittedly an intensively worked one, with a few urban areas to service it.

As in Australia, more people are engaged in

manufacturing and in urban-based services than in agriculture. The pastoral industries, although they employ relatively few people, are very productive. The introduction of refrigerator ships, and later of freezing techniques, made it possible for New Zealand to market the bulk of its agricultural exports in Britain and Europe, and in this respect to overcome the handicaps of its isolation. But isolation is clearly apparent in the still tiny population, and in the extent to which the economy has specialized in high-value agricultural goods for export "home." The heavy concentration of foreign trade with Britain is especially remarkable given the distances involved. It has brought drastic problems of readjustment for New Zealand as Britain has moved closer to the European Common Market and its policies on overseas imports of agricultural goods.

There are limited amounts of coal and iron, enough for the small scale of New Zealand's iron and steel manufacturing. Most manufactured goods are imported, rationally enough; the tiny domestic market imposes high costs. In one predictable respect, New Zealand has an important potential resource: hydro power. It is available through the combination of mountains and rainfall, but its development, although already substantial, will continue to be limited by market considerations. Recent discoveries of extensive natural gas pools may become at least equally important. The North Island also contains many surface or near-surface geysers and hot springs, several of which are used for steam-powered generating stations.

The fact that New Zealand has a high per capita foreign trade in most years symbolizes and reflects most of the aspects of the economy. It also reflects the determination of New Zealanders to maintain the closest possible ties with Britain and Europe with which nearly all its trade is conducted, although there has been a recent, rapid growth of trade with the United States. High incomes, British attitudes, and the influences of a pioneering experience have also contributed to New Zealand's leadership in social legislation. It provides high-quality, public services for the population as a whole, an example that very few of the great powers, especially the United States, can begin to equal.

Polynesia. Coconut palms and a coral beach in a Tahiti lagoon, with two tourist yachts at the dock—a good sample of Polynesia. *(Photo from UTA French Airlines.)*

Melanesia

The inhabitants of Melanesia (*Mela* means "black") and Micronesia are racially varied, but both contain Negro, Negrito, and Mongoloid strains. It is generally accepted that, like the Polynesians and the Australoids, they migrated originally from a homeland somewhere in Asia. But this migration took place over a long period and was largely completed before the beginning of the Christian era, if not much earlier. The Melanesians are mainly Negroid, although there are some pygmy Negritos in interior highland New Guinea. The people of Micronesia are more mixed, while the Polynesians are comparatively homogenous racially, a blend of Caucasian and Mongoloid. Melanesia contains the largest land area by far, including both New Guinea and the clusters of relatively large islands in the Bismarck Archipelago, the Solomons, New Hebrides, Loyalty Islands (Iles Loyauté), and Fiji. Melanesia has been settled longer than any of Oceania (with the possible exception of Australia). Many successive waves of people have passed through it and contributed to its extreme cultural diversity. This is reflected not only in the bewildering profusion of languages but in the wide range of technical, economic, and institutional systems. Some form of agriculture is found in all parts of Melanesia, and yams and other tropical root crops are nearly universal, in addition to coconut palms. Shifting cultivation is the most prominent agricultural form, well adapted to tropical forest conditions. Pig raising is important, especially in New Guinea, and fishing is at least equally important on the smaller islands.

Most communities are small. No large political units existed until the recent and artificial colonialism or trusteeship was imposed by outside powers. There was some trade traditionally in high-value goods, but in many areas, especially interior highland New Guinea, this was minimal. There was often chronic warfare between many groups, accompanied by head hunting and cannibalism. The coastal groups were both more peaceful and more involved with trade. Access to interior New Guinea is dramatically difficult, and even now it is incompletely explored or controlled by any single authority. The eastern half passed from Australian trusteeship to independence as Papua in 1975, while the western half (Irian) has become part of Indonesia. In general, official policy in both areas discourages settlement by outsiders. The total white population is still only about 30 thousand, but there is some pressure for expansion of rubber and coffee plantations and oil prospecting. Native ownership of land and resources is likely to be slowly eaten away. New Guinea is neither an attractive nor a rewarding area for much outside investment, and most of it seems likely to change relatively little even though the outside world has caught up with it. The island is over three times the size of New Zealand or of the United Kingdom. Yet it is without a single rail line and still contains less than 3 million people, many of them living at an early Neolithic technical level. Isolation and the difficulties of the New Guinea environment will continue to retard economic growth even with the help of outsiders.

Fiji is the outstanding exception, not only within Melanesia, but in Oceania as a whole outside of Australia and New Zealand. Fiji has attracted a very large, recent immigration and supports a greatly expanded commercial economy. The native Melanesian population (which included some Polynesian strains) was drastically reduced as a result of European pressures and epidemics. But workers were needed, especially for sugar plantations. A large-scale migration from India began about 1880, and by the First World War Indians exceeded the native population. Most of the Indians came as indentured laborers. But, like the Chinese and Indian migrants to southeast Asia, most have greatly improved their economic status, and many became employers, merchants, or independent cultivators. The immigration largely ended in 1919 as a result of British colonial policy to preserve some Fijian stake in the islands. Indians now substantially outnumber Fijians and occupy the dominant economic position. Independent since 1970, Fiji includes over three hundred islands, only two of which are more than tiny dots on an atlas map of the Pacific. This is the smallest of the island groups of Melanesia, yet it supports over half a million people, far more than any of the others, except New Guinea. Sugar, profiting from fertile volcanic soil as on Java (see Chapter 10), is the mainstay of the commercial sector of the economy. But coconuts are almost equally important, as they are throughout the Pacific world.

Micronesia and Polynesia

Micronesia, a Japanese mandate between the two World Wars, became a United Nations Trust Territory administered by the United States. In 1975, it voted to become a permanent dependency (commonwealth) of the United States. America has pursued a somewhat inconsistent policy of attempting to preserve Micronesian culture from outside pressures, but at the same time pre-empting the island of Guam as a military base and dispossessing the inhabitants of three other islands, Bikini, Kwajalein, and Eniwetok, in order to use these areas for the testing of nuclear weapons. About half of Micronesia's total land area is still held for U.S. government and military purposes. The great distances separating the tiny specks of land from each other, their remoteness, and the lack of significant resources, however, will continue to insulate them from most external contact. In practice, most islands largely manage their own affairs (apart from the exceptions above) and maintain a mainly subsistence economy. The export of copra, by far the leading commercial commodity, is growing fast. There is also a small commercial production of cacao. The employment and material rewards offered by the U.S. military bases and other forms of government subsidy are attractive, and helped to swing the voting for retaining the American tie.

Of the three island-culture groups, Polynesia covers the greatest area, but nearly all of it is water (especially if one excludes New Zealand, originally settled by the Polynesian Maoris). None of the islands except Hawaii are more than the tiniest of widely scattered dots on the vast expanse of the Pacific. Even the largest of them (Savaii in Western Samoa, or Tahiti) are smaller than most American counties. It is understandable that the Polynesians developed early the art of long-distance seafaring in their outrigger canoes. Perhaps only as a result of trial and error, they learned to steer a course and find land. In the course of centuries their voyages were likely to strike most of the habitable islands. Land on most of the islands remains largely in Polynesian hands. Although most of Polynesia is formally administered by Britain, France controls the Society Islands (including Tahiti), the Marquesas, and Tuamotu; New Zealand and the United States share responsibility for most of Samoa; remote Easter Island (Rapa Nui) is controlled by Chile; and Western Samoa and the Cook Islands have become fully independent. Most Polynesians show little interest in politics, and have preserved their own identity in other ways, even while they have been willing to accept European control and many European techniques and culture traits. Copra dominates the commercial sector of the Polynesian economy, but subsistence farming and fishing remain far more important. Tahiti is the most commercialized, including its involvement in the tourist trade. But even there most Tahitians are only partly involved in production or services for sale. Papeete on Tahiti, the largest town in Polynesia, has a population of about 36 thousand. It is the administrative headquarters for all of the Society Islands, as well as a processing and export center for copra and a little sugar and phosphates.

Throughout the Pacific island world as a whole, the pattern of human life is inconceivable without the coconut palm. It reached this area well before man and thrives in poor, sandy, or coral soil close to the sea. Its varied products are even more important here than in southeast Asia (see Chapter 16). It also provides the great bulk of trade commodities with the outside world. Coconut palms can survive even the periodic overwash of the sea on the low coral atolls where little else can endure. From their wood, coir (fibre), leaves, and matting the people build their houses and their boats, often of at least equal importance. On most of the larger, higher islands, coconuts are also the mainstay of the economy.

Although they are not consistently related to culture, the geologic origin and physical nature of the islands are often used to distinguish between them in four categories: (1) the lowest atolls, washed over periodically by the sea so no trees survive; (2) the larger, higher, and tree-covered atolls where soil has

Coral atoll—Marshall Islands. The typical atoll form is a more or less circular reef and low beach nearly surrounding an interior lagoon. The water in the lagoon may be quite deep, although access to it may be hampered by the reef. One of several theories of their origin is that they are the cones of dead volcanoes with the rim of the cone just barely breaking the surface. The reef and beach, on which coconut palms in particular may flourish, represent an accumulation of solid and powdered coral. *(Photo from U.S. Navy Hydrographic Office.)*

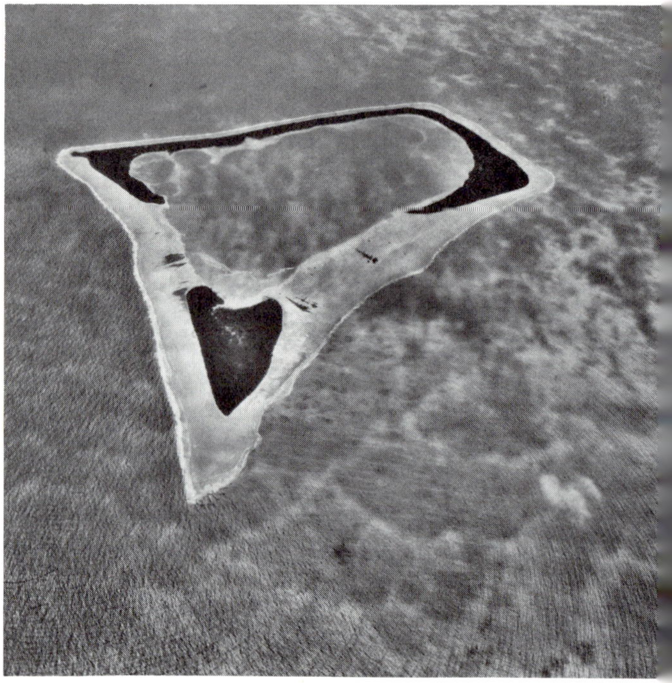

accumulated in pockets as the island has been uplifted by volcanic action; (3) purely volcanic islands with rich soil (where they have been above the sea long enough for adequate weathering); and (4) the larger, continental islands such as most of the larger islands of Melanesia (New Guinea, New Caledonia, Guadalcanal, Bougainville, New Britain) where the sea is no longer overpowering and some culture groups have evolved independent of it. But in almost every sense, the Pacific island world is dominated by the sea, which brought all its inhabitants, late in the history of human migration, and has kept them largely isolated from the rest of the world even in the twentieth century.

Questions for further study and discussion

1. Why has no complete north-south rail line been built in Australia? Why was the east-west line completed? What several factors, negative and positive, were involved?
2. What are the actual possibilities for large-scale Asian settlement in Australia? What effects, on Asia and on Australia, might a large migration of Asians have?
3. Why have both Australia and New Zealand increased their trade, ties, and interest in the United States since 1945?
4. Why is so much of Australia arid, and all of New Zealand well watered?
5. Would any of Melanesia, Micronesia, or Polynesia be better off, economically or in other respects, with independence—or would this be more true of some areas than of others? Why, or why not? Are there any significant feelings or moves toward independence on the part of any of the people of these areas? What does the experience of Western Samoa and the Cook Islands suggest?

18 The Dry World: civilization and conflict

A **land of contrasts.** *(Photo from Israel Office of Information.)*

The world's largest steppe and desert stretches from the north African shores of the Atlantic to India and China, nearly reaching the Pacific in eastern Mongolia and north China (see Map 81, page 332). While there are obviously great physical and cultural differences within this vast area, the limiting conditions of the climate and the proportionately large extent of plain or plateau landforms have been accompanied by a broadly similar set of human adjustments. The great steppe and desert area of Eurasia and north Africa is commonly labeled the *Dry World*. Not only do most of its parts share a physically similar arid environment, but also the people who inhabit it have used their environment in a broadly consistent pattern. Where physical environment offers a small range of choice, as in the Dry World, environmental influences are likely to be clearer and stronger than in better-favored places. This is borne out by the relatively small range of difference in land use and much of the culture associated with it from Morocco to Mongolia.

Perhaps the most striking of the regional differences within the Dry World is the dominance of the Muslim faith and Islamic culture in most of the western three-fourths of the area, and of a separate religion (Lamaistic Buddhism) and its associated culture in Tibet and Mongolia. Iran (Persia) and Turkey maintain considerable separate cultural identity despite their acceptance of Islam. The spread of Islam and much of its associated culture over such a wide area resulted from the Arab conquest of most of the western Dry World in the seventh and eighth centuries A.D. Arab culture also influenced not only Iran and Turkey, where a great deal of non-Arab culture survived, but many of the smaller groups of central Asia who lived on the fringes of the Arab empire. Islam and much of Islamic culture spread by influence and borrowing rather than by conquest into most of what is now Russian central Asia and Sinkiang. It has even left a large group of Chinese Muslims in China's northwestern provinces. The spread of Islam resulted in an even greater cultural unity in the Dry World than had existed earlier, but regional religious and cultural differences remain. One of the important regional variants, discussed in more detail later, is a reflection of climatic differences between those areas where rain-fed agriculture is possible and those where irrigation is essential. Another is the contrast between the maritime fringes of the Dry World, especially the coasts of Arabia and the Levant (the east coast of the Mediterranean) where sea trade is very old, and the much larger areas remote from any maritime contacts. Since about 1920, a further regional distinction has arisen, based on the discovery and use of the oil deposits in the Sahara and around the Persian Gulf (see Map 84, page 356). Algeria, Libya, the Arabian Peninsula, Iran, and Iraq together are estimated to contain about half of the world's proven reserves of petroleum. This has sharpened the distinctions between these areas and the rest of the Dry World, especially because the oil deposits have attracted European and North American investment and influence since the First World War.

Regional lines

Any precise limitation of the Dry World is arbitrary, just as it is arbitrary to draw precise regional lines in any area. If we use the 20-inch (50.8 cm) rainfall line as the boundary, it coincides fairly closely with the limits of steppe climate and steppe or desert forms of land use, and with its associated culture over most of the Dry World area. But to use the rainfall line as the Dry World boundary would exclude large parts of Turkey and the Levant, coastal Algeria and Morocco, and parts of western Iran, all of which are culturally very close to the rest of the Dry World. The rainfall line would also include much of the Ukraine, nearly all of Siberia, and much of Poland, which have very little in common with the Dry World. The boundary cannot be based on the areal distribution of Islam, for this would exclude much of Lebanon and most of Israel and Mongolia and would include Pakistan, Bangladesh, Indonesia, Albania, parts of Yugoslavia, parts of northwest China, and parts of Africa south of the Sahara, most of which could not be considered closely related to the Dry World except for religion. Similar problems and contradictions arise when one attempts to define the Dry World in terms of any single criterion. The problem may be sidestepped by speaking simply of the Arab world, which is often done. But this leaves out Turkey, Iran, and the eastern third of the Dry World; all have much in common with the greater whole. This definition would also omit the people and culture of Lebanon, Egypt, western north Africa, Syria, Iraq, and Israel, all of which are by no means Arab. The labels *Near East* and *Middle East* are even more ambiguous, are seldom defined, and are arbitrary and misleading by any definition. Political lines are simpler, although seldom satisfactory.

The Dry World is defined here to include the countries of Morocco, Algeria, Tunisia, Libya, Egypt, Western Sahara (whose status is currently in question), roughly all the rest of Africa north of about 15° N. latitude (the latitude of Dakar and Asmara-Massaua [Mesewa]) plus a thin strip down the African east coast into Somalia, all of the Arabian Peninsula, Jordan, Israel, Lebanon, Syria, Turkey, Iraq, Iran, Afghanistan, Baluchistan (the western part of Pakistan), Sinkiang (but not the northwestern provinces of China within the Great Wall), Outer Mongolia, and most of Russian Turkestan (but not Siberia)—see Map 81 (page 332). This enormous area, larger than the Soviet Union and nearly one-fifth of the land surface of the globe, has about 7.6 percent of the world's population, or somewhat over 300 million people (see Table 2, page 333).

Bases of unity

What are the similarities that justify considering the Dry World as a single great region and that distinguish it from the surrounding areas? First are the common physical conditions of aridity, varying from nearly rainless desert to long-grass steppe and occasionally better-watered mountain slopes, and the predominance of plains or plateaus. Agriculture is dependent on irrigation or on drought-tolerant crops for the most part, and farming is confined to relatively small areas where surface water or groundwater is available or where local rainfall can support dry farming. Or the land is grass steppe or desert where, in the absence of adequate local water, pastoral grazing has traditionally been the chief or sole support of the population. Most of the *people* of the Dry World have been engaged in agriculture for millennia, but most of the *area* remains thinly populated by groups whose support is derived from flocks and herds. Nomadism is now yielding to a more settled form of livestock ranching or to sedentary agriculture in Turkestan and Mongolia, and may soon follow a similar course in the rest of the steppe areas of the Dry World. But this remains a vast region where man

MAP 81
LANDFORMS OF THE DRY WORLD
DRY WORLD AREA
0 400 800 1200 1600 2000 Km.
0 400 800 1200 Mi.
ATLANTIC OCEAN
A T L A N T I C
O C E A N
I N D I A N
O C E A N
EUROPE
ASIA
AFRICA
SAHARA
SUDAN
UNION OF SOVIET SOCIALIST REPUBLICS
MONGOLIA
CHINA
INDIA
ARABIA
SAUDI
IRAN
TURKEY
EGYPT
LIBYA
ALGERIA
MOROCCO
MAURI-TANIA
MALI
NIGER
CHAD
NIGERIA
ETHIOPIA
SOMALIA
TANZANIA
ZAIRE
ANGOLA
ZAMBIA
RHODESIA
MOZAMBIQUE
BOTSWANA
SOUTH WEST AFRICA
SOUTH AFRICA
MADAGASCAR
PAKISTAN
BURMA
THAILAND
MALAYSIA
INDONESIA
Mediterranean Sea
Black Sea
Caspian Sea
Red Sea
Arabian Sea
Bay of Bengal
South China Sea
Gulf of Guinea
Gulf of Aden
Persian Gulf
Gulf of Oman
Mozambique Channel
Tropic of Cancer
Tropic of Capricorn
Equator
Nile
Niger
Congo
Ganges
Indus
Volga
ATLAS MTS.
URAL MOUNTAINS
HIMALAYAS
TIBESTI
GOBI
THAR DESERT
KALAHARI DESERT
NUBIAN DESERT
AR-RAB AL-KHALI
KARAKUMY
KYZYLKUM
Mount Everest 8848
Kilimanjaro 5895
Mount Kenya 5199
Gora El'brus 5633
Mulhacén 3478
Jbel Toubkal 4165
Ras Dashen 4620
0°
15°
30°
45°
60°
75°
90°
105°
120°
135°
© Rand McNally & Co.

must depend upon animals for the most part, except for the few places where irrigation water is available for farming or where some unirrigated crops can survive. Nomadism or livestock ranching are highly specialized adjustments to a particularly limiting environment that does not permit great variations in the method of land use. Oasis agriculture can be a little more flexible in its form. It depends on local springs or wells, or on water from an *exotic river,* one that has its source in well-watered areas and crosses the desert on its way to the sea. But absence of an extensive market and low-cost volume transport are severely limiting factors.

In all of the steppe and oasis areas of the Dry World, including the exotic river valleys, a similar economy and culture developed. The common physical base aided the wide and rapid spread of Islam from its original nucleus in the Arabian Peninsula and made the Arab conquest easier. Mountains in Turkey, Iran, and Afghanistan, and mountains plus the friction of distance in Mongolia helped to limit the Arab and Islamic conquests. Mountains have also been partly responsible for the differences in land use and culture between these areas and the rest of the Dry World. But the dominance of arid climate has also been accompanied by strong similarities in human adjustment, and the mountains within the Dry World have not prevented continual interaction among most of its parts.

Over most of this vast region, including the plateaus of Turkey, Iran, and Mongolia, the land is rolling or level enough to allow the wide movement essential for nomadism or livestock ranching. Beginning about the tenth century, the Turks advanced gradually from central Asia to control most of what lay between their original home and the Gulf of Suez. And in the thirteenth century, the Mongols conquered most of the Dry World plus China and southeastern Europe (an empire even larger than that of the Arabs). These are only two examples of the movement, interchange, and conquest that have characterized the Dry World for as far back as its history can be traced. Some of the largest of the Dry World empires of the past are shown in Map 83 (page 348). Arabs, Mongols, Turks, Huns, Magyars, Avars, Kassites, and Arameans, among many others, were all originally nomadic peoples whose normal mobility in search of pasture for their animals was closely related to their conquests.

Table 2. Countries of the Dry World and their approximate population*

Algeria	17.6
Djibouti	0.1
Egypt	38.5
Libya	2.6
Morocco	18.1
Somalia	3.3
Tunisia	6.0
Western Sahara	0.1
Dry World areas of	
Chad	3.5
Eritrea	5.5
Mali	0.5
Mauritania	0.4
Niger	2.2
Sudan	9.8
North African Total	108.2
Afghanistan	19.8
Bahrain	0.2
Iran	34.5
Iraq	11.6
Israel	3.5
Jordan	2.8
Kuwait	1.1
Lebanon	3.1
Mongolia	1.5
Oman	0.8
Qatar	0.2
Saudi Arabia	7.5
Syria	7.7
Turkey	41.5
United Arab Emirates	0.7
Yemen Arab Republic	6.8
Yemen, People's Democratic Republic of	1.8
Baluchistan (Pakistan)	1.0
Inner Mongolia (China)	10.2
Kazakh S.S.R. (Soviet Union)	14.2
Kirghiz S.S.R. (Soviet Union)	3.3
Sinkiang Uighur Aut. Reg. (China)	10.0
Tadzhik S.S.R. (Soviet Union)	3.4
Turkmen S.S.R. (Soviet Union)	2.5
Uzbek S.S.R. (Soviet Union)	13.7
Asia Total	203.4
Dry World Total	311.6

*Estimates (in millions) are based on the latest available statistics. Reasonably accurate census counts are available for only a few of the largest countries. For many areas, the estimates are subject to considerable error.

Nomad versus farmer

The Dry World has been a consistent area of conflict. It supported two totally different ways of life, the nomadic or seminomadic cultures of the steppe and desert, and the sedentary irrigation farmers of the oases and exotic river valleys. Urban cultures, especially in the few large trading or administrative cities, might constitute a third category. There are many differences between the nomadic cultures of the steppe and the sedentary communities of the humid lands around it. The outer margins of the steppe and its way of life fluctuated with the climate and the techniques, power, and organization of those who inhabited it. As with all regional boundaries, there is a wide transition zone that might support either marginal farming or pastoral nomadism, or where irrigation might be more or less effectively developed. In some large areas, for example, central Turkey, unirrigated farming *(dry farming)* became dominant. Inner Mongolia in recent years has become more and more closely attached to sedentary agricultural China. As the power of the Chinese state has become greater, irrigation has been increased and extended. Under the Roman occupation north Africa west of the Nile supported extensive areas of irrigated sedentary agriculture. When the area was conquered by the Arabs, it was displaced by nomadism to a large extent. In the past, Mongol power or a temporarily more arid interval periodically reclaimed Inner Mongolia's steppes from Chinese agricultural invasions.

The nomad and the farmer traditionally have represented opposite poles of culture. To the farmers, nomadism represents illiterate, primitive barbarism and lawlessness, beyond the pale of civilization. To the nomads, farming and a sedentary existence in houses and towns under governments and laws represent cramped, stultifying, grubby slavery. They felt that dignity, liberty, and self-respect were to be found only in the wide-roving, unhampered, free life of the steppe or desert.

Nomadic conquerors

Such attitudes have supported the periodic attempts made by farmers and nomads to encroach on or conquer the goods or territory of the other. The sedentary farming cultures have sought to protect themselves against nomadic invasions and to spread what they consider their civilizing influence by enlarging their empires at the expense of the barbarians. The nomads have regarded their sedentary neighbors as fair game, not worthy of consideration as moral equals, but possessed of material riches. The nomads were proud of their material poverty, but at the same time envious of the easy surpluses of the agricultural areas. If their pastures deteriorated, other nomadic groups pressed against them, or the surrounding sedentary areas became politically weak, the nomads were tempted by the farming communities. Necessity, opportunity, and exceptional leadership periodically led the nomads to submerge their individualistic tendencies into a larger and more effective political and military organization of tribes. When this happened, the nomads were virtually irresistible. Their mobility, endurance, and military prowess were welded into an integrated army that nothing could withstand. Often the territorial conquests of the nomads were short-lived because their political organization fell apart or was disrupted by factionalism or rebellion. This was true for example of the Mongol and Seljuk empires in differing degrees. But in many cases originally nomadic groups, such as the Ottoman (Osmanli) Turks, the Manchus and Khitans of Manchuria, or the Kassites of ancient Mesopotamia, transformed their conquests into more permanent empires, adopted sedentary ways, and forsook nomadism for a literate, city-based civilization, irrigated agriculture, trade, and political administration.

The horse was almost certainly domesticated first by sedentary people, like most domestic animals. It was probably used for milk or meat well before it was used for riding. Fully developed pastoral nomadism (as opposed to shifting cultivation, which may have been earlier than sedentary agriculture or contemporary with it), based upon the herding rather than hunting of animals, came later. In many areas, herding is probably an offshoot of earlier farming cultures. By about 2000 B.C. the nomads near the borders of Mesopotamia may have acquired the horse, and soon after, mastered its use for riding. Horses were not only an invaluable aid in herding and movement over the steppe, but were a vital military factor. Speed, range, surprise, and quick maneuverability gave mounted nomads a telling advantage over the much larger armies of the sedentary people. They were effective also for short hit-and-run raids on farming communities. There was never more than an uneasy truce, punctuated by raids, between the nomad and his sedentary neighbors. When larger conflicts erupted, they usually took the form of nomadic waves overpowering the settled communities.

The Mongols, for example, mounted on wiry, little steppe ponies, could cover 100 miles (160 km) a day or more on a forced military march. They could strike hard without warning, and be gone to strike again elsewhere before a counterblow could be prepared. On long marches they could subsist on what dried food could be carried in their small saddlebags. They could travel for days or even weeks without further provisions or without the encumbrances of a supply train. Their cavalry tactics were brilliantly successful in engagements against armies ten times their numbers. They could fire their powerful composite bows accurately while riding at full gallop or could turn in the saddle and fire devastating volleys at their pursuers. The Bedouins and other Arab nomadic groups with their large, fast, and beautiful horses were equally effective against massive opposing armies.

Trade and conflict

A further basis for conflict was the trade routes that ran through the steppe and desert, following the tenuous and scattered chains of oases. The Dry World separates and is nearly surrounded by the two major humid centers of sedentary civilization in the Old World—Europe and monsoon Asia. It borders on both and is the link between them. The nomads of Eurasia have been intermediaries and often transmitters of culture, especially during their own movements and conquests. The communication lines ran through their territory, and they frequently interfered with or blocked the trade routes. Trade was mainly the business of the sedentary people beyond the steppe or in the islands of irrigated agriculture within it. The nomads had little property except their animals, produced little for exchange, and required little they did not produce. Although they were not relatively dependent on trade, the routes that stretched in a thin line through their domain and the small scattered oases that acted as bases and way stations for the trade routes were often tempting to the nomads as a source of easy plunder or regular levies. The warehouses and agricultural surpluses of the oases and exotic river valleys were also temptations, whenever the nomads were strong enough. Caravans traveled across the steppe and desert under guard, but there was seldom an effective defense against a lightning raid by mounted horsemen or camelmen choosing their own time and place. The nomads helped to keep the flow of goods between Europe and monsoon Asia to a thin trickle of very high value commodities. Their price also restricted their markets despite the strong economic complementarity between East and West. Of course, there was some mutually beneficial exchange between nomadic and sedentary groups—animal products for grain and textiles, for example—and the more or less chronic raiding that accompanied or alternated with trading was distinct from the infrequent attempts at temporary or permanent conquest.

Rivalry between the nomads and sedentary farmers may well be one of the oldest conflicts of modern civilization. But the Persian proverb, "Turan conquers Iran," is no longer true. The sedentary states have increased their power with modern weapons and communications. The nomads are increasingly being incorporated into or overshadowed by larger national states dominated by sedentary peoples who now have the further overwhelming advantage of industrial and commercial power behind them. Nomadic or seminomadic peoples are now in a tiny and decreasing minority. Even in the past they were important in terms of the amount of area they occupied rather than in their total numbers. At the height of the Mongol empire, for example, there were probably no more than 1 million Mongols at a time when the population of China alone was over 60 million. In the present century, nomadism is rapidly disappearing. However, it is worth examining because of the important role that proportionately small nomadic groups in the Eurasian Dry World played in the development of early civilization and in the cultural relations between East and West. The nomadic way of life and its attitudes die hard, and still contribute to tensions and maladjustments in nearly all of the contemporary Dry World states and in China and Russia. It is an old struggle, in part because of the great age of city civilization in the oases and exotic river valleys. It seems certain that the world's first cities began in those parts of the Dry World. Literate city civilization, based on irrigated agriculture with regular surpluses, may be nearly as old in some of the oases of Iran and central Asia as in the valleys of the Indus, Tigris-Euphrates, and Nile. Such relatively small islands of surplus sedentary agriculture and commerce were surrounded, and still are, by huge expanses of desert and steppe that are inhabited, if at all, by different and usually hostile people. The earliest recorded history of Mesopotamia is punctuated by attacks and invasions. All of the settled areas of the Dry World and its margins have a similar background although in differing degrees.

The nomadic economy and its base

The nomadic economy revolved around animals. They provided food, consisting chiefly of meat, milk, cheese, and butter; hides and wool for easily movable felt or skin tents, and for clothing and rugs; and horn and gut for bows and bow strings. Animal dung was dried and used for fuel, fat for crude lamps, and leather for thongs, saddles, cords, and reins. These items were also offered in exchange for small amounts of grain, salt, wood or wicker (for tent frames), pots, metal tools and knives, and occasional small luxuries or ornaments obtained in barter with the sedentary people. Dwellings were moved periodically as pastures were shifted according to the season and the quality of the grass. Each group usually regarded a large and often ill-defined area as its territory for grazing and resisted its use by other groups. Arid pasture in particular deteriorates if it is used too often or by too many animals, and such an area might have to rest for long periods between grazings. To this extent the nomad had a sense of property, but, commonly, might made right; pasture, lands, or goods belonged to him who won and held them. Modern attempts to impose rules of property and law and fixed national frontiers onto formerly nomadic areas have met with resistance and many difficulties. There is also a clear risk of overgrazing when nomadic groups are obliged to become sedentary. Population often increases in the absence of continual small-scale wars and with the growth of commercial exchange and public health programs. Extensive and mobile or cyclical use of very large areas to support a few people may be a better technique for preserving the basic pasture resource in this arid region.

The desert

The line between steppe and desert is difficult to draw and is seldom constant as climate fluctuates continually over decades and centuries. Differing peoples with differing techniques may also make different use of the same physical conditions. Most nomadic groups are obliged to cross desert areas from time to time, if not to make direct use of them. And no desert is absolutely rainless. Even the mountain-encircled basins of Sinkiang and central Asia occasionally receive some rain, although it is rare. Yet at Cairo (Al Qahirah), as close as it is to the Red Sea and the Mediterranean, less than half of the years have had any measurable rain at all during the period of modern weather records.

When it does come, desert rainfall may be violent, although it is rarely prolonged or very great in total amount. Nevertheless, it may produce disastrous floods, because of the scarcity or absence of vegetation to retard runoff, the hard-packed nature of the soil, and the speed with which the runoff concentrates in formerly dry stream channels, called *wadis*. The channel may be dry for months or even years, then suddenly fill with water and wash out bridges, fords, or settlements in its path. Even in towns or cities, where most houses in the Dry World are built of mud or sun-dried brick in the absence of wood, sudden accumulation and backing up of water from a rainfall of only an inch or two can result in floods that lead to the collapse of buildings. Except for the major exotic river valleys and the few better-watered regions near the sea, most of the Dry World streams flow into enclosed basins where they evaporate or sink into the ground and do not reach the open sea. This *interior drainage* may also be accompanied by salinity or alkalinity in the basin from evaporation, and may form *playas*, or salt flats. Desert soils also have little humus and may be almost completely infertile. But the absence of leaching plus local accumulation of alluvium may result in high soil fertility in some areas. Plants adjust to aridity by developing characteristics that retard evaporation and transpiration, store water, and repel grazing animals:

Bedouin nomads. These nomads live in black goats-hair tents. Because pasture is scanty over much of the area they inhabit, they often follow regular routes depending on the grazing needs of their flocks. *(Photo by Owen Franken for Stock, Boston.)*

hard, shiny leaves; thick fibers; large, woody stems or trunks; tough outer coverings (sometimes enclosing a fleshy center where water is stored); spikes; thorns; or bitter taste. *Xerophytic* (drought-resistant) and *halophytic* (salt-tolerant) plants dominate most desert vegetation outside the oases.

Desert streams, with their widely fluctuating volume, heavy floods at peak flow, and generally bare watersheds, are characteristically laden with silt, gravel, or even large rocks. They are seldom useful for navigation and are subject to frequent and sometimes drastic changes of course as deposited sediments build up in the old bed and block the channel. Desert air is dry, skies are clear (permitting rapid radiation), and ground surface is usually bare so it reflects heat readily. Thus there is an extreme daily temperature range, with days up to 130° Fahrenheit (54° C). In the same twenty-four-hour period, nights may be freezing. Desert winds are often strong, partly as a result of these rapid temperature changes, and the exchanges of heat and pressure between higher and lower areas. Strong winds and absence of vegetation are largely responsible for the most common type of desert surface, a covering of gravel and rock known as *reg*. The wind removes the finer particles, and only where air movement is less or where there is some resistant barrier does sand accumulate to form an *erg*, or sandy desert.

All of these physical conditions have been adjusted to by the nomads. Their clothing, made of wool from their animals, is good insulation against both cold and heat and can withstand wind. Some sort of hood or face covering, often a shawl, is included for further protection, especially against sand storms. The camel is still better equipped, since it possesses a secondary transparent eyelid permitting it to see even in driving sand. The tents of the no-

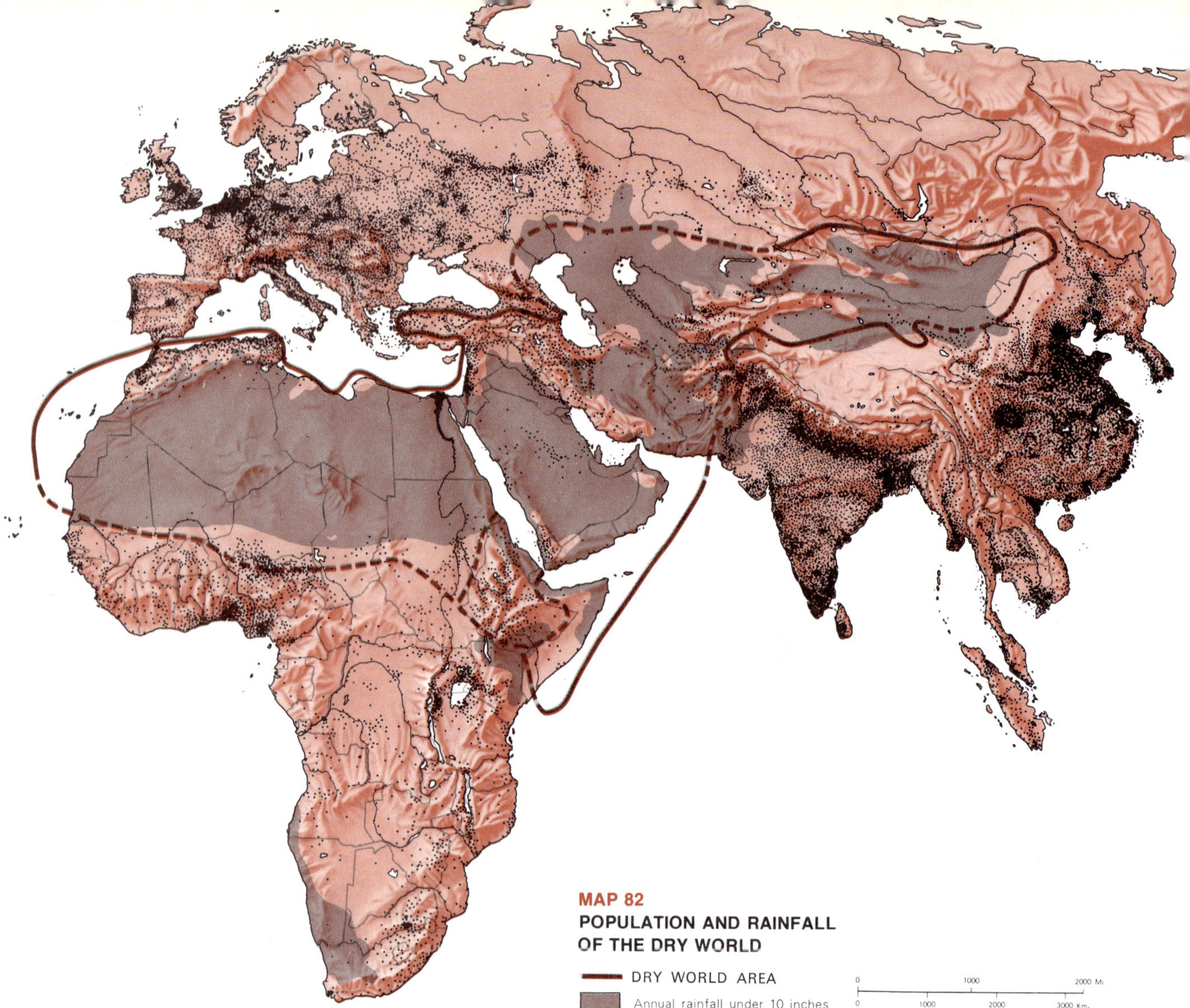

MAP 82
POPULATION AND RAINFALL OF THE DRY WORLD

DRY WORLD AREA

Annual rainfall under 10 inches

One dot represents 100,000 people

0 1000 2000 Mi.
0 1000 2000 3000 Km.
Equatorial Scale 1:90,000,000

mads are made of felt (matted fibers of wool or hair) or of woven goat hair, both of which make excellent insulation, and are provided with a door or overlapping flaps. In Mongolia, well north of the frost limit, with long severe winters, the dominant nomadic dwelling is the circular felt *yurt*. Its shape also fits it to withstand the particularly strong winds characteristic of the area. In the Arab and Islamic parts of the Dry World, the black tent is most common, in a shape more familiar in the West. The animals that support the nomads—sheep, goats, camels, and wiry horses—are as hardy as the people, able to withstand extremes of temperature, sparse pasturage, and prolonged exposure that would be disastrous to bovines or to most domestic animals of the humid or irrigated areas. The nomads are adept at finding water, in distinguishing between potable and nonpotable sources of water, in tracking animals or people across desolate wastes, and in finding their own way from water hole to distant water hole through apparently trackless and featureless areas.

The oasis economy

The economy of the oases has centered primarily on agriculture and secondarily on trade. Wells are still the commonest source of water, especially in the scattered dots of permanent settlement outside the few river valleys. Oases often occur in shallow depressions where erosion or settlement has left the surface of the ground close to water-bearing rock strata that can be tapped with relatively shallow wells or may yield naturally flowing springs. Another device used in many countries of the Dry World, where physically possible, is the *qanat,* or underground channel. It collects water over a large area and carries it underground down a slight slope to a central well or reservoir. Qanats are most common in Iran, where they may have been first built about 500 B.C. But they are found from north Africa to Sinkiang. Where the groundwater supply is promising and the rainfall inadequate, the channels may even be dug out of solid rock, as in parts of north Africa. They are most commonly put along the downslope near the base of an alluvial fan where they can collect water seeping down the slope and convey it to cultivated fields or settlements on the level basin below. Enough water is usually available in the oases to irrigate only a small amount of land, and cultivation is intensive. Irrigation may sometimes bring salts or alkalies to the surface by raising the water table, or may deposit them as a result of evaporation of irrigation water. This may happen especially where rainfall is inadequate to flush the surface layers or where subsurface drainage is poor. In some oases where drifting or blown sand may menace cultivated fields, irrigated windbreaks of trees may be used for protection.

Cereal grains, mainly wheat, barley, rice, and sorghums or millets, are usually the most important oasis crops. They are intensively cultivated, used primarily for local subsistence, and commonly given a share of the best land and the best water supply. Cotton is often the leading nonfood and commercial crop. Much of the weaving of cotton and wool (ob-

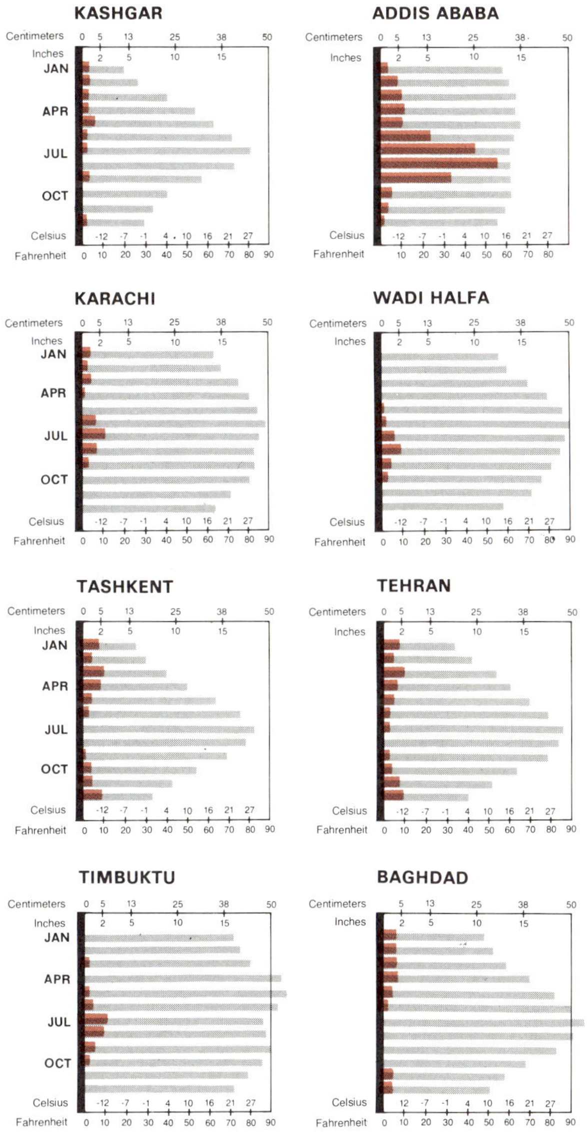

Figure 13. Climographs of representative cities in the Dry World.

Temperature
Precipitation

tained from the nomads) is traditionally done in the oasis settlements. Citrus fruits, olives, grapes, melons, and sugar cane are also common. Date palms are nearly universal in lowland oases south of the limit of frost. All of these crops are sun-loving, most are drought-tolerant, and all profit from a long growing season of high temperatures, especially when the water supply can be controlled by irrigation. Sometimes oasis agriculture attempts to maximize returns from a limited amount of irrigable land by *intercultivation*, growing different crops of differing heights in alternate rows: cereals and melons on the lowest level, fruit and nut trees and grape vines at an intermediate height, and date palms towering above everything else, where they get the most sunshine. The date palm is said to flourish best "with its feet in the water and its head in the fire," a good description of southern oasis conditions. Water is carefully controlled, budgeted, and shared among the users. The Islamic law of the Koran reflects the vital importance of water in the Dry World in its regulations for water use, especially for those on the upper part of a stream or irrigation channel. The Islamic heaven is pictured as a green land flowing with water and with abundant pastures.

Oases are few in the Sahara west of Egypt, in the Syrian desert at the northern end of the Arabian Peninsula, in south-central Saudi Arabia (the Rub' Al Khali [Ar Rub' Al Khali] or "Empty Quarter"), eastern Iran, Baluchistan and southern Afghanistan, central Sinkiang, and the Gobi Desert of Mongolia. These areas are nearly rainless deserts (see Map 82, page 338), and they largely lack other sources of water, such as the permanent exotic rivers provide for the equally arid deserts of Egypt and Pakistan. Mesopotamia (literally, "the land between the rivers"), the lowland in the Tigris-Euphrates valleys, has a slightly greater total of effective rainfall, but it is dry enough to be classified as desert. Its agriculture is dependent on river irrigation and on wells.

Oasis trade centers

Bazaars or markets are also characteristic features of oases. They were and are nearly all centers of exchange along the trade routes through the Dry World. The bazaars also act as central places for the few goods and services the oases provide for their vast, thinly populated hinterlands, and in some degree for trade with other oases. Rugs, the standard and almost only pieces of furniture for the tents of the nomads, often of dazzling beauty and virtually indestructible by wear, are prominently displayed for sale. Dates, raisins, cooking vessels, lamps, knives, fruit, cloth, skins, assorted hardware, and more recently some of the products of the industrial West are sold in oasis bazaars from Sinkiang to Casablanca. All are remarkably similar. Most of these towns or cities are very old, and most have essentially the same dual basis: tiny islands of irrigated agriculture and trade centers. Even a partial list of their names reads like a tale from the *Arabian Nights:* Marrakech, Siwah, Kharga, Damascus (Dimashq) (perhaps the oldest continuously inhabited city in the world), Aleppo, Baghdad (not far from the ancient and now ruined city of Babylon), Qum (Qom), Isfahan (Esfahan), Medina (Al Madinah), Khiva, Bukhara, Samarkand, Herat, Fergana, Tashkent, Kashgar (Kashih), Khotan (Hotien), Turfan (Tulufan), Hami, and Yümen (the Jade Gate near the borders of humid China).

At each end of the route through the Dry World, trading cities developed very early as terminals, distribution centers, and break-in-bulk points where the land routes reached the sea or fanned out into humid and more widely settled country. The old Egyptian, Phoenician, and Philistine trading cities on the Levant coast: Byblos (now Jubayl), Tyre (now Sur), Sidon (now Sayda), and Gaza (now Ghazzah), and the equally old cities of Antioch, Damascus, Palmyra, and Aleppo (see Map 83, page 348), performed such a role. It began not long after the rise of irrigated agriculture, metallurgy, and regular surpluses in Mesopotamia and the Nile valley. Trade flowed between these two primary areas through these trading cities. There were probably similar trade centers, terminals, and way stations along the

Oasis and market. A typical Dry World market in the Saudi Arabian oasis town of Qatif on the Persian Gulf. The architecture is designed to keep out sunlight and heat, with thick walls, few openings, and, in the larger buildings, verandas. The date palms are characteristic. The market, or *suq*, as in many Dry World towns in sparsely settled areas, is held only about once a week, when the town fills with people. For the rest of the week, it may be nearly uninhabited. This is a common spatial adjustment of urban functions to low population density over large areas, especially widespread in the Dry World. *(From Standard Oil Company of New Jersey Collection, University of Louisville.)*

route between Mesopotamia and the Indus, in addition to those listed. Others probably developed along the coast of the Arabian Sea, although little direct evidence of such settlements remains. In Sinkiang, Kashgar and Khotan occupy the lowland approaches to passes across the surrounding mountains into Russian Turkestan, India, and Tibet. Turfan and Hami lie close to the steppe-desert border and serve the northern route into what is now Russian territory. Other ancient oasis trading stations lay along the route through or around the Takla Makan Desert of central Sinkiang. Yümen and Lanchow were terminals similar to Damascus or Byblos.

The camel is another nearly universal feature of the Dry World, associated principally with trade and riding or draft, rather than with herding. Camels were native to Iran or central Asia, but had been spread into north Africa and China by at least the sixth century B.C. The camel's load-carrying capacity is relatively small. But it is well suited to arid-area travel by its ability to store water in the form of fat in its hump and by its wide, padded feet, which do not sink into the sand. Use of the camel greatly increased the range and scope of trade within the Dry World. Camels are still used extensively throughout the region as pack animals, although they are also employed to pull plows and carts, to power Persian wheels or other irrigating devices, and to drive crude mills. The prophet Muhammad (A.D. 560–632) was a camel driver in Arabia in his youth, and later a merchant with his own caravans—a familiar Dry World figure.

The spread of Islam

Islam has spread throughout practically all of the Dry World oases, including those in Sinkiang. It is absent only in the remote steppes like Mongolia, in similarly isolated mountain wastes and dry deserts, and in parts of the Levant where Christianity or Judaism have been maintained from European contacts or settlement (see Maps 9 and 81, pages 54 and 332). The association of Islam with the oases and their bazaars and caravans is understandable because of the Arabs' long-standing involvement in trade. Their home base in the Arabian Peninsula was on or close to the mainstream of commerce in the Old World, and Arabs were pioneers in the trade between the Orient and Europe. Islam spread rapidly along the land and sea trade routes from Arabia, and accompanied the Arab conquest as well. But there was little basis for the advance of Islam into areas with which there was little or no trade. Some of the nomads, like the Mongols or tribal mountain groups, did not wish to borrow any aspect of culture from the Arabs or from their despised agricultural neighbors in the oases, from whom they were so often sharply isolated.

The parts of the Mediterranean coast where Arabic and Islamic culture are weakest, because direct contact with Europe is strongest, are also the parts where rainfall is greatest and land use is most different from the Dry World as a whole. In these areas, agriculture can be carried on without irrigation, and there is an extensive maritime trade with Europe and the rest of the world. Both are a recent legacy of direct European control or colonization, and a system of land use that is closer to that of the Mediterranean basin than to the Dry World. The coasts of Turkey, the Levant north of Israel, the seaward-facing slopes of the Atlas Mountains in Morocco and Algeria, and the narrow coastal plain of northwestern north Africa (mostly in Libya) are such areas. The north African coastal areas receive an annual rainfall averaging from 15 to 40 inches (38 to 101 cm), concentrated in the winter. Most of Turkey, except for the central highlands, receives as much rainfall or more. A similar pattern extends down the coast of the Levant following the mountain ranges that parallel the sea, decreasing from north to south. Beirut in the north has an annual rainfall average of 36 inches (91.4 cm), while Gaza in the south receives only 16 (40.6 cm). Lebanon, which occupies much of the ancient trading state of Phoenicia, has a rain-fed agriculture on the mountain slopes and a thriving foreign trade. The alien nature of Israel within the Dry World needs no emphasis, although most of Israel is too dry for agriculture without irrigation. The Black Sea and Mediterranean coasts of Turkey, most of the highlands of Kurdistan to the east, and parts of northwestern Iran trending southeast from the Black Sea receive well over 20 inches (51 cm) of rainfall on an annual average. In these areas also Arabic influences are relatively weak.

This is not to suggest the simple conclusion that "Islam is the religion of the desert," as is sometimes said, or that it spread only within arid areas.

Bangladesh and Indonesia disprove such an assertion. Islam began in Arabia and was spread by the Arabs as far as their conquests, trade, or cultural influences allowed. These places were attractive to the Arabs or were unable or unwilling to oppose Arab conquest or Islamic conversion for various reasons. Mountain barriers helped to preserve other tribal religions in parts of the Dry World. Close contact with Europe has also been an important factor, plus the determination of some groups, such as the Armenians, to maintain their cultural and religious identities despite the pressures exerted on them.

Neither Islam nor Arab influence is uniformly spread even in the western part of the Dry World. Both are also relatively recent, as history is measured in this ancient home of civilization. For millennia before Muhammad, oases and river valleys supported highly developed economies, dense populations, and sophisticated cultures. Almost certainly the earliest of all literate city civilizations started in the Dry World. The major exotic river valleys stand out from the scattered oases because of their size, the number of people, and the productivity of the economies they were able to maintain. The valley of the Nile remains one of the most densely populated areas in the world. Its small corridor through the desert still contains over an eighth of the total population of the Dry World as a whole. The old land of Mesopotamia, suffering from repeated invasions and more isolated than Egypt from the mainstreams of modern trade, now supports only some 9 million people, probably less than half of its ancient total. Smaller settlements based on irrigated agriculture, but larger than a single oasis, also began very early in Yemen and the Hadhramaut (Hadramawt) in southwest Arabia (where the mountains catch considerable rainfall), and along the Amu Darya and Syr Darya rivers in Turkestan. In southern Arabia and Turkestan, the ancient populations, economic development, and cultural level were probably higher than at present. One explanation for the movement of the Arabs and their conquests in the seventh and eighth centuries A.D. is based on the neglect and collapse of the ancient irrigation systems in Yemen and the Hadhramaut. Or it may be that nomadic groups from drier central Arabia overran the irrigated areas and destroyed or failed to maintain the irrigation works. On the other hand, these areas may have become less productive or may even have been abandoned as a result of cumulative siltation or the building up of salts and alkalines in the soil.

Climatic factors and human changes

Changes in land use, the decline of irrigation, and the abandonment of cultivated land have occurred many times in different parts of the Dry World. During and after the Arab conquest of north Africa, for example, extensive Roman and Persian irrigation systems were destroyed or allowed to deteriorate by a people who had little interest in agriculture and used the oases only where a minimum of effort was necessary to obtain water. In central Asia and Sinkiang there are many ruined irrigation works, cities, and watchtowers along former trade routes, destroyed by invaders or abandoned for other reasons and never reconstructed. The oases in Sinkiang along the edge of the desert at the foot of the mountains have been more permanent. They depend on wells and qanats using streams and groundwater from the mountain slopes. Ruins have often been used as evidence of climatic change. In north Africa, Sinkiang, Mesopotamia, and other Dry World areas, it is argued that the land supporting the ancient agricultural systems, cities, and economies must have been moister than it is now. Egypt and north Africa in particular seem to have declined economically. During the Roman occupation, this area produced large food surpluses for export to Italy and was an important granary for the city of Rome, as well as supporting a large local population. Climatic fluctuations, which occur repeatedly especially in arid areas, may clearly have a disastrous effect on human settlement. But there is no solid evidence to show that rainfall has progressively diminished in any of these areas since approximately 300 B.C., or that the climate of today is markedly different from the climate of Roman or ancient Persian times. Fluctuations of a few inches in annual rainfall may be critical in arid or semiarid areas, and such fluctuations have clearly occurred during the past two thousand years. There is evidence of long periods of a century or more during which rainfall was higher than the present average, and other long periods when it was lower. It is the nature of climate to fluctuate, but sustained progressive change toward dryness seems unlikely during the historic period.

The urban Dry World—the Casbah, Algiers. Whatever its other associations, this street is a typical Dry World bazaar. Bazaars vary surprisingly little from Casablanca to Kabul. *(Trans World Airlines photo.)*

It does seem clear that there was a marked and progressive decrease in rainfall throughout most of the Dry World from about 20,000 B.C. to Roman times, and that in 8000 or even 3500 B.C. rainfall and groundwater were widely available in many areas that have become deserts since then. Neolithic and Paleolithic remains are found in the western Sahara, for example, at sites that would be uninhabitable now on the basis of local rainfall or groundwater. The central Sahara, the Libyan Desert, central Saudi Arabia, the Takla Makan of Sinkiang, and the Gobi Desert of Mongolia, however, seem to have remained as dry as they are now throughout this period. But there was a period of greater rainfall on their margins and elsewhere, especially in the western part of the Dry World. The drying up of these areas may have tended to concentrate the existing population in the oases or exotic river valleys, and may have contributed to the beginnings of irrigated agriculture. Decreases in rainfall presumably resulted from the northward migration of the storm tracks, which had been pushed southward by the strong high-pressure area over ice-covered north Europe during the last glacial period. As the ice slowly retreated, the rain-bearing winds from the Atlantic followed it northward over several millennia and left declining rainfall behind them. These changes had largely taken place before the beginning of recorded history. Since at least that time, the Tarim River in Sinkiang, for example, seems to have become an intermittent rather than permanent stream and to have changed its course radically from time to time, as nearly all arid-climate rivers do. Changes in course and periodic drying up removed the basis for particular settlements or trade routes in many areas. But this does not mean that the climate changed. It seems more likely that the repeated invasions in nearly all of the Dry World may have been influenced in part by temporary dry periods in certain areas. These invasions were far more important than long-range climatic changes (if indeed changes did occur) in accounting for the ruin of settlements, the decline of populations, and the abandonment of irrigation works during the historic period.

Evidence of climatic changes on such a time scale is unsatisfactory. Accurate and continuous weather records begin in northwest Europe only about the beginning of the nineteenth century and are much more recent for the Dry World. Before that time geology, fauna, flora, contemporary accounts, and other sources do not provide precise enough evidence to measure climatic fluctuations with any accuracy on a human time scale of decades. However, they suggest that such fluctuations have occurred and may make approximate dating possible. Major changes, such as accompanied glaciation and its retreat over many thousands of years, are in a quite different category from the minor changes in climate during the past two or three thousand years.

It is likely that settlements in many arid areas have profited from groundwater (or in Sinkiang from meltwater from the surrounding Kunlun, Pamir, and Tien Shan ranges) inherited from an earlier glacial or rainy period, and that these reserves have been progressively depleted. In north Africa, however, it seems reasonably clear that the present rainfall and groundwater resources could support as much irrigated land, as great a production, and as large a population as at the height of the Roman occupation, if the available water were efficiently used. Such a development may well take place by the end of the present century, or the ancient levels may be exceeded.

The Arabs chose to pursue a different type of land use in north Africa, to which the climate was as well suited as it was to irrigated agriculture. Not only did the Arabs generally neglect and underuse existing irrigation works, but they contributed to increased desert by cutting down or failing to maintain the existing vegetation. This vegetation had often been irrigated by earlier occupants and acted as a windbreak against blown sand and a safeguard against erosion and runoff. The unrestricted grazing of the large number of sheep and goats brought in by the Arabs, on which their essentially nomadic economy depended, also had a disastrous effect on the vegetation. This overgrazing clearly contributed to erosion and desert extension as it has done elsewhere in the Mediterranean basin, India, and other dry or semiarid areas. The resulting changes in north Africa and the considerable decrease in population, cultivated area, production, and trade are more readily understandable in terms of human choice and human actions than of climatic change. The same kind of explanation is applicable to the apparent decline of Mesopotamia, southwest Arabia, and the Amu Darya and Syr Darya valleys in Turkestan.[1]

The beginnings of civilization

There is a long-standing disagreement among scholars as to which of the ancient centers of civilization in the Dry World was the first to develop and is the oldest civilized culture on earth. The disagreement focuses on Egypt and Mesopotamia, for it is reasonably clear that they both developed earlier than any of the other oasis or exotic river valley centers, although the Indus valley civilization may be nearly as old. Precise dating is difficult, but by about 3000 B.C., or slightly earlier, written records were being kept in both Mesopotamia and Egypt, and possibly in the Indus valley. By that time there were at least the beginnings of towns or cities in all three areas, based on the surpluses produced by irrigated agriculture and supplemented by trade. The use of metals, starting with copper, which occasionally occurs naturally in nearly pure form, may have begun in Egypt and Mesopotamia about 4000 B.C. The use of bronze, a harder alloy made by adding tin to copper, was common in both areas by about 3000 B.C. It greatly increased the efficiency and usefulness of tools and weapons. Sedentary Neolithic peoples with polished stone tools, plows, domestic animals, some development of trade, and permanent village settlements may have been established in the hilly areas flanking Mesopotamia by about 7000 B.C. Their culture may have been widespread in Syria, Palestine, Iran, and Turkestan even earlier. Wild ancestral forms of wheat and barley occurred in these upland areas with their steppe climates. They seem to have been purposefully cultivated first in permanent fields there. As agricultural techniques and crop yields improved, larger permanent settlements became possible. Some (Jarmo, Jericho) were on the threshold of becoming cities with some division of labor, rather than merely agricultural villages. Once the techniques of agriculture had been mastered, however,

---

[1]For opposing points of view on this fascinating subject, see Ellsworth Huntington, *The Pulse of Asia* (Boston, 1907), and R. Murphey, "The Decline of North Africa since the Roman Occupation: Climatic or Human?" *Annals of the Association of American Geographers*, 41 (1951), 116–32.

the potential productivity of the exotic river valleys with their semiautomatic irrigation and highly fertile, easily worked soils began to attract settlement. In the lower Tigris-Euphrates and Nile valleys, techniques and crops that may originally have been Iranian or Syrian found favorable circumstances for their further development. Population multiplied and development became more rapid, especially city building, metallurgy, and writing. They accompanied the growth of agriculture and trade. In Europe Neolithic culture was not well established before about 2500 B.C., and copper and bronze did not appear there until about 2000 B.C. The Australian culture was still at the Paleolithic (Old Stone Age) level when Captain Cook reached Australia in A.D. 1770.

Mesopotamia and the Nile

Physically Egypt, Mesopotamia, and the Indus region are similar. In Mesopotamia two permanent exotic streams, the Euphrates and the Tigris, originate close together in the rainy mountains of what is now Kurdistan. They follow a parallel course across a joint flood plain some 150 miles (240 km) wide, and meet about 80 miles (128 km) from the Persian Gulf to empty through a common mouth. The lower Nile valley of Egypt is drier and is almost rainless. But rainfall has little agricultural significance in either area. Both profited immeasurably not only from the more or less automatic irrigation provided by seasonal river floods, but from the continual deposit of fertile and easily worked alluvium. The Nile, beginning in the wet, tropical, or equatorial highlands of Ethiopia and central Africa where it receives heavy monsoonlike rains, was far more regular than the Tigris-Euphrates. Its annual flood could be precisely anticipated, not only in its time and duration, but in the amount of land the flood covered. The spring floods of the Tigris-Euphrates were less easily predicted or managed, varied considerably in time and extent, and were often destructive.

The beginnings of civilization cannot be dated with any accuracy in either area, nor is it easy to draw a valid or significant line between *civilization* and *barbarism* or Neolithic culture. Clearly there was a long transitional period everywhere during which people learned the techniques of settled agriculture, irrigation, building, and bulk transport. They became able to produce consistent surpluses as a basis for cities and the further refinement of techniques and arts. Our word *civilization* is derived from a Latin word meaning "city." It is both convenient and appropriate to date the beginnings of civilization from the rise of the first real cities, whenever that may have occurred.

Writing probably followed the development of large consistent surpluses, and the means of transporting, storing, and exchanging goods. The increase in trade led to problems of management, distribution, and administration of goods not consumed immediately by the direct producer. It seems likely that economic and cultural development in each area was independent to some degree, at least in its early stages. Before the beginning of recorded history, however, there was much interchange in trade and ideas between the Nile valley and Mesopotamia, and between Mesopotamia and the Indus valley civilization. The weight of evidence at present seems to favor Mesopotamia as the most important single center of trade and probably the earliest center of literate city civilization. The Mesopotamian impact on early Egyptian civilization and the Indus valley is clear. But it is not entirely one-sided and does not suggest that early Egypt was an offshoot of Mesopotamia. Both civilizations were highly distinctive and approximately contemporary in their origins.

Communication between China and the early civilizations in the Dry World was made difficult by distance and mountains. Irrigation was less vital in China, and there was a long, cold winter. For these and other reasons, it seems clear that city civilization with metals and writing was not firmly established in China much before 2000 B.C. How much of the Chinese development began there and how much resulted from influences or indirect transmissions from Mesopotamia, northwest India, or southeast Asia is still uncertain, although the evidence so far favors its independent development.

Agricultural origins

It is not known and it is unlikely ever to be discovered where settled agriculture first began, perhaps ten thousand years ago. It seems probable that it did not develop first in the exotic river valleys or oases, but that it spread there from other areas. One likely area would have been on or near a warm, humid seacoast, perhaps in southeast Asia, where there was a profusion of wild plants including tropical tubers. At the same time there was some means of obtaining food from the sea and from local game regularly enough to make a permanent settlement possible. The inhabitants might gradually have learned to gather wild plant products as a supplementary source of food, and then to plant and tend crops in permanent plots or patches. Another likely area might be the hilly or upland steppe regions in the Middle East, where wild ancestors of wheat, barley, sheep, and goats were present, and where hunting and gathering could supplement the food supply. Animals were probably first domesticated in such early agricultural settlements rather than by nomadic hunters. The pastoral nomad appeared relatively late and was an offshoot of the agriculturist rather than a separate development.[2] Although the desert was not a likely place for the origins of agriculture, the exotic river valleys offered physical conditions that were highly favorable to agriculture. So it is not surprising to find the densest and technically most highly developed populations of the pre-Graeco-Roman world there. The techniques of irrigation were almost certainly first developed in these valleys. Probably the first large and consistent agricultural surpluses were produced there, making possible the first cities, the beginnings of economic specialization, and the development of regional exchange. Continually renewed soil fertility, easily workable and, in effect, preplowed land, an ample and, more important, a controllable water supply, and the strong sun and high temperatures of an unbroken growing season composed agricultural advantages that were peculiar to the exotic river valleys in the ancient world. The early importance of Egypt, Mesopotamia, and northwest India in the development of civilization is understandable in the light of these advantages.

Siltation of irrigation works and salinization of the soils were factors in the decline of ancient Mesopotamia, as they have been in the decline of many semiarid or desert areas elsewhere. In Mesopotamia siltation was increased by continued cutting of the forests in the Tigris and Euphrates watersheds, and by the violent and irregular flooding pattern of both rivers. Many irrigation channels became choked with silt, and the difficulty of keeping them cleaned out was greatly increased by invasions and political disorder. There is good evidence that large, formerly productive agricultural areas in lower Mesopotamia were also abandoned because their soils became too salty. The salt was probably brought to the surface by the irrigation raising the local water table, accumulating salts on the surface as a result of high evaporation and inadequate rainfall. By the middle of the twelfth century A.D., "most of the Nahrwan region [east of Baghdad] already was abandoned. Only a trickle of water passed down the upper section of the main canal to supply a few dying towns in the now hostile desert. Invading Mongol horsemen under Hulagu Khan, who must first have surveyed this devastated scene a century later, have been unjustly blamed for causing it ever since."[3] But for at least three thousand years before the Mongol conquest, siltation and salinization had been building up gradually in Mesopotamia. It contributed to the fading of Mesopotamian civilization well before the Christian era.

The succession of empires

The empire of the Persians succeeded in controlling most of the ancient Dry World outside of Sinkiang and Mongolia, from Libya to the Indus (see Map 83, page 348). Modern Western sympathies are with the Greeks whom the Persians failed to conquer. But the Persian empire, with its capital at Persepolis, was probably an improvement on earlier empires, certainly on the ruthless and tyrannical rule of the Assyrians. On the whole Persia governed its huge territory fairly and well, with a large degree of local autonomy, and the benefits of a wide dispersal of Persian culture.

The Persian empire was largely destroyed by Alexander's conquests in the fourth century B.C. His empire was also short-lived and fell to pieces soon after his death, although several local Alexandrine kingdoms or satrapies survived. The Romans reintroduced a high degree of order, unity, and economic

[2]For an analysis of this problem by a geographer, see Carl Sauer, *Agricultural Origins and Dispersals* (New York, 1952).

[3]T. Jacobsen and R. M. Adams, "Salt and Silt in Ancient Mesopotamian Agriculture," *Science*, 128 (1958), 1251–58.

MAP 83
DRY WORLD EMPIRES

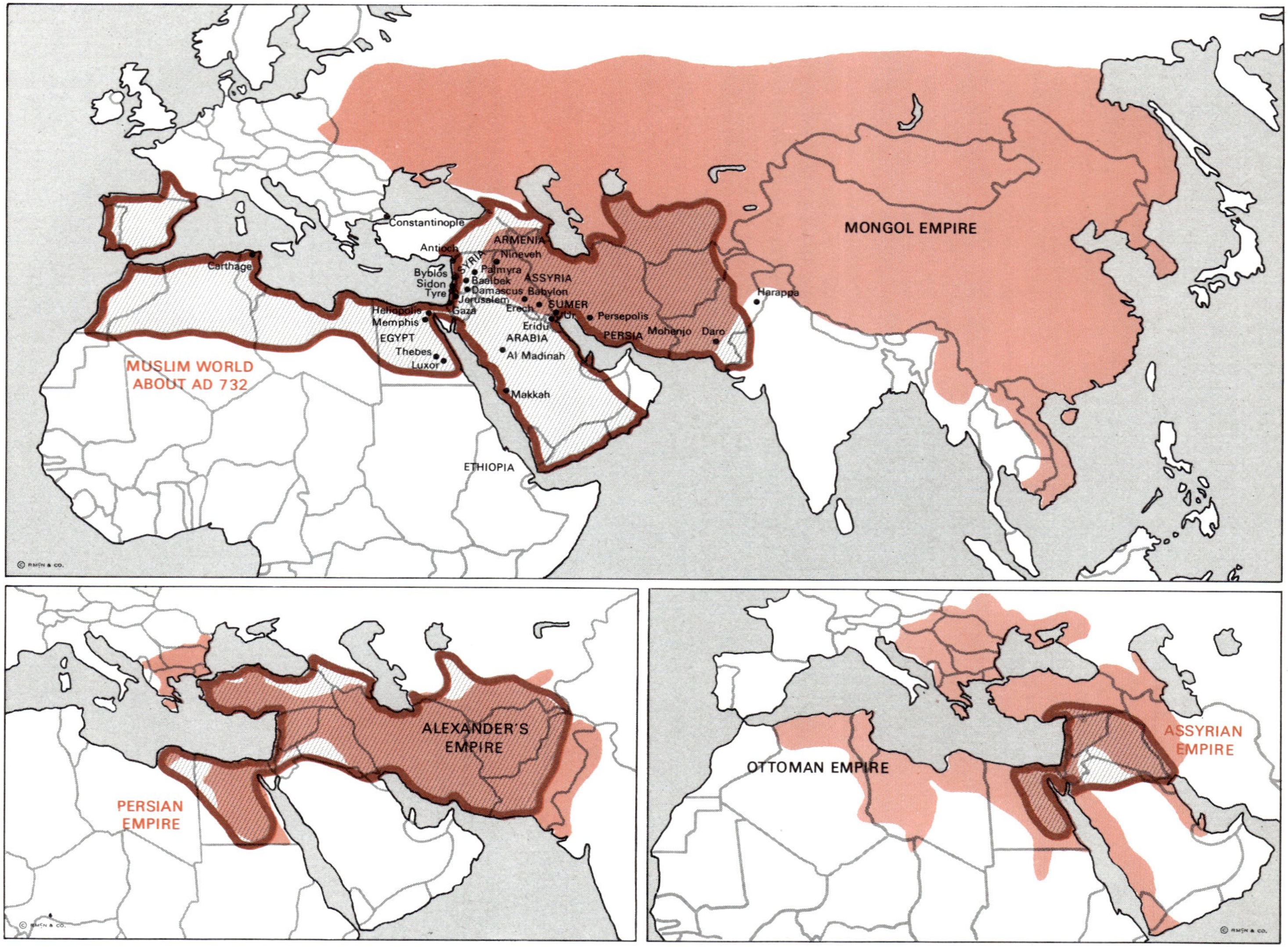

and technical development with their conquest of north Africa and most of the Levant and Mesopotamia, although they held the lower Tigris-Euphrates valley only briefly. These Asian holdings of the Romans contributed heavily to the support of Rome and its empire. They outlived the fall of the city of Rome and Roman Europe, and remained an island of sophisticated civilization, especially the city of Constantinople (now Istanbul), until the Arab and Turkish conquests.

The Arab empire at its height in the mid-eighth century reached from southern France to the borders of Rajasthan in India (see Map 83, page 348). The Islamic Moors, partners with the Arabs in their conquests, were not finally driven out of Spain until 1492. The Turks, expanding both from their original home in central Asia and from their newly won base in the Anatolian peninsula, rebuilt an Islamic empire under the Ottomans (lasting from the fourteenth century to the First World War). By the sixteenth century, their empire included virtually all of the Dry World west of Persia and east of Tunisia, plus considerable territory in southeast Europe (see Map 83). Earlier conquests by the Seljuk Turks in the eleventh and twelfth centuries, encompassing most of modern Turkey, the northern Levant, and most of Mesopotamia, were destroyed by local revolts, Mongol attacks, and the rise of the Ottomans. Out of the wreck of the Ottoman Empire the contemporary states of Turkey, Syria, Lebanon, Jordan, Iraq, and Saudi Arabia were constructed. Greece, the Balkans, and Egypt established their independence from Turkish rule in the nineteenth century with Russian, British, and other European help. The state of Israel was created out of the British Mandate of Palestine, a former part of the Ottoman Empire, inhabited largely by Arabs until 1948. Morocco, Algeria, Tunisia, northern Sudan, and the steppe areas along the southern edge of the Sahara acquired increasing local autonomy under their own sultans from the time when the political force of the old Arab empire faded after about the tenth century until the expansion of French and Spanish colonial control in the nineteenth century. The small and ill-defined states, protectorates, or sultanates of Yemen, People's Democratic Republic of Yemen, Oman, the United Arab Emirates, Bahrain, Qatar, and Kuwait around the coastal fringes of the Arabian Peninsula, remain as political anachronisms in the modern world, or as territorial hostages to the strategic interests and oil companies of the industrial West.

With such a history, it is not surprising that the people of the Dry World, except for remote Mongolia, insulated Egypt, and, to a lesser degree, mountain-protected Iran, are enormously mixed, despite the wide spread of Islam and Arabic culture and language. The blood and culture of countless earlier rulers, groups, invaders, and conquerors are mingled in the contemporary inhabitants. They are probably no more closely related to the ancient Babylonians, Sumerians, or Assyrians than are the modern inhabitants of Europe to the Romans, Greeks, or Goths. The different physical types have intermarried widely; there is no uniform composite type. There are important physical characteristics that originated in distant places—Mongoloid, European, Negroid, and other strains. Western north Africa was invaded and settled by Germanic Vandals and Goths in the fifth century A.D. in the course of the great barbarian wanderings after the collapse of Rome. Negroid and tribal groups are scattered throughout north Africa, and the succession of invasions of southwest Asia by Mongoloid and Turkish peoples needs no further emphasis.

Turkish identity and the persistence of Iran

Apart from Mongolia, two major areas within the Dry World, for differing reasons, have successfully resisted many of the waves of invasion and conquering empires: Turkey and Persia (now Iran). Both have profited from a plateau base and the protection of surrounding mountains. Both have had the advantage of a seacoast, although this was of much greater importance in Turkey than in Iran. Once the invading Turks established themselves in Anatolia (after several earlier Arab raids through it), they were never successfully invaded. In the modern period, with the loss of the Ottoman Empire in Europe and Asia, the Turks have strengthened their cultural contacts with the West, and reaffirmed their identity separate from the Arab world.

Iran lies close to the center of the Dry World. It has important lowlands on two sides, rather than occupying a separate peninsula. Its coastlines on the Caspian Sea and the Persian Gulf are of relatively minor commercial value. The main route of land trade between the East and the West ran through Iran or along its margins for thousands of years. It was difficult for the Persians to detach themselves from the Arab world as successfully as the Turks have done. Despite the protection of the Plateau of Iran, Persia was conquered by the great wave of Arab expansion in the eighth century A.D., although Arab control was relatively short-lived. The Plateau of Iran nevertheless has helped to preserve the distinctiveness of Persian culture through the long centuries of invasion and disruption. Modern Iran is the direct cultural heir of ancient Persia. Its continuity and persistence cannot be found over such a long period or to such a degree anywhere in the Old World except perhaps in China. Iran, in fact, has been called the "China of the Dry World," a comparison that is valid both in terms of cultural and political continuity, and also in terms of its cultural influence on surrounding areas. Persian influence was especially noticeable in northern and northwestern India and Afghanistan, and to a lesser extent in Turkestan and the western half of the Dry World as far as Libya. The magnificent art and architectural styles of Persia, Persian literature (known in the West mainly through the *Rubaiyat* of Omar Khayyam in its beautiful English translation by Fitzgerald), the technique of the qanat, Persian achievements in mathematics, and the imperial Persian traditions of administration and road building are perhaps the most striking of these influences. In the prehistoric period ancient Iran may have played a basic role in the original development of civilization. The plateau may have been part of the homeland of the Aryans and some of the ancient peoples of Mesopotamia, and even of Egypt. Persian culture, it is true, has been heavily influenced from outside at several periods. Greek and Turkish influences have been particularly important, especially in architecture and literature. The Arabs, by contrast, have been essentially borrowers, carrying techniques and culture to the West from China, India, and Persia. The West has often given the Arabs credit for inventing these. Most of "Arab" mathematics, numerals, and steel-making techniques were Hindu or Persian. The compass was Chinese, and most of "Arab" art, architecture, and literature was Persian. The only significant innovation of the Arabs was the Islamic religion.

Ancient and post-Roman Persia had its periods of larger empire, and periodically controlled parts of the western Dry World after the collapse of the great Persian empire of the fourth and fifth centuries B.C. But, except for the limited period of Arab control, the Persian core on the Plateau of Iran remained largely undisturbed. The Parthian kingdom, which occupied most of the plateau and periodically controlled surrounding areas (including Mesopotamia), defied all attempts at conquest by the Romans. Alexander's control of Persia in the fourth century B.C. lasted for only a few years. Persia embraced Islam after the Arab conquest and has retained it. In most other respects its cultural identity has remained. Even in religion, elements survive of the old Persian Zoroastrianism, now represented also by the Parsees (a corruption of *Persians*) of India. Most of the Muslims of Iran belong to the Shiah sect, which rejects many of the orthodox tenets of Islam. It is considered heretical by the orthodox Sunni Muslims who are dominant in most of the other Islamic countries. Iran is thus no more Arab than is Turkey. The political character and orientation of both states in the modern period bears this out, and is examined in more detail in the following chapter.

Afghanistan, Iran's neighbor on the east, also occupies a highland base, but it is far more rugged in its topography. Despite its area, it still supports only about 20 million people, and its economy and culture bear the marks of extreme isolation as well as physical poverty. Afghanistan and the adjacent Baluchistan region, which are physically and culturally similar, form the largest of the mountain areas in the Dry World remaining largely outside Arab conquest or the spread of Arab culture. Even Islam is not universal in Afghanistan; some of its people still follow an essentially tribal religion and culture. What influences have penetrated into the country have been mainly Persian, Greek, or Indian rather than Arab, except for Islam. Indeed, western Afghanistan occupies part of the Iranian plateau. The modern borders of Iran on the west and of Afghanistan and Iran on the north still follow very closely the outer limits of the plateau whose protection and isolation have contributed so much to the history and culture of both countries.

The paradox of Dry World history

Modern civilization originated in the Dry World. Until nearly the beginning of the Christian era, for some three out of the approximately five thousand years of recorded human history, it was probably the most advanced area in the world in technical, cultural, economic, and political terms. For the past several centuries, especially since the beginning of the Industrial Revolution, it has probably been the poorest area in the world economically, and one of the most backward politically, technically, and culturally among lowland areas of comparable size. How is such a reversal explained? No simple interpretation is possible. But it should not be assumed that there has been a progressive or absolute decline in the Dry World in any sphere except political power, which is itself a relative concept. There has probably been an average improvement in per capita living standards since the ancient days of greatness. But this improvement has been completely dwarfed by the much greater increases in production, trade, technology, cultural change, and political power in areas outside the Dry World. The impressive monuments, conquests, cities, and works of art produced in ancient Mesopotamia and Egypt came from a society whose low per capita incomes were disguised by the magnificent ruins and records that have been preserved. Most of the surpluses and their benefits were enjoyed by a small ruling elite, especially after about 2000 B.C. This in itself may have helped to retard further technological and economic development as the line between the rulers and the ruled hardened. The artisans and actual producers had been the main innovators. Reduced to a level of servile labor under the control of a strong central authority, they may have lost their capacity and opportunity for innovation. The small ruling group, on the other hand, could depend on the mass labor they controlled. Since it was adequate even for building on the monumental scale of the pyramids, they did not need to seek labor-saving techniques. As the new civilization matured, respect for the past or for the established ways of doing things retarded change of any kind.

It seems significant that the great bulk of technical innovations that led to city civilization, such as the development of irrigation, the use of metals, the wheeled vehicle, writing, the plow, the arch, and a calendar, were produced immediately before or in the early stages of city development. In the two or three thousand years that followed, very few developments of comparable importance took place. The major new spurt in pure science and techniques took place, not in the Dry World, but in the very early history of Greece, before the growth of the privileged ruling group and its slave base during the later periods of Greek history. There is also no doubt that continual invasions and conquests by the agrarian empires and by the nomads slowed or destroyed material progress. It is ironic that before the beginning of city civilization, there is relatively little evidence of weapons of war or of organized human conflict. War may be one of the products of civilization, and in the Dry World may have contributed importantly to its downfall.

Economic, social, and political democracy have still not been achieved in most of the Dry World. The grinding poverty of the mass of the people contrasts sharply with the steep rise in average living standards that has accompanied substantial industrialization and commercialization. Commercial development was greatly hampered in the Dry World by the discovery and use of cheaper sea routes between the East and the West; overland trade soon shrank to a trickle. The Dry World did not participate to any significant extent in the commercial revolution that followed these discoveries. The Industrial Revolution has also scarcely begun to affect it, partly because of its relative commercial isolation and lack of capital and markets. Also, with the exception of oil, there are no really important mineral resources in the area in amounts needed by large-scale industrialization, particularly very little coal or iron ore. Oil alone is not necessarily a stimulus to industrial development. Instead, in the Dry World, oil resources have attracted a colonial type of foreign investment with a history of Western political and strategic manipulation of Dry World affairs.

With the exception of oil, most of the Dry World still has a subsistence economy. Dates, cotton, and a few other goods are produced for commercial sale and export in a few areas. Israel is a tiny island of Western commercial-industrial economy. Since the collapse of the Ottoman Empire at the end of the First World War, Turkey has made rapid strides toward a modest industrial-commercial structure. Lebanon's ancient trade base still includes a vigorous commercial sector. Otherwise, Dry World economies are largely devoted to subsistence. Oasis agriculture, farming in the larger river-basin areas and dry farming, urban-based barter, and nomadic and seminomadic herding or sedentary grazing are still the most important occupations (see Map 7, pages 52–53). The modern national state is only beginning to emerge in the Dry World, and it still ranks close to the bottom of the list of underdeveloped areas by most economic criteria.

Greatness in any field or by any measure is not necessarily permanent, nor is poverty or underdevelopment. It seems highly unlikely, however, that in the foreseeable future, the Dry World will again occupy the position of leadership it held for so long. For the most part it does not offer a promising areal base for the economic developments that have led to leadership in the modern world for other areas. But as change has occurred in the past, so it will occur in the future. The new states that emerged after the downfall of the Turkish empire and the older states that survived from an earlier period are already changing rapidly. The impact of commercialization, industrialization, and nationalism is beginning to be felt. Some of these developments in the states of the Dry World will be examined in the next chapter.

Questions for further study and discussion

1. Cairo is a relatively recent city. Why? Where were the capitals of ancient Egypt? Why?
2. Why is dung used so nearly universally over the Dry World as a fuel?
3. What common pattern of city location is illustrated by Damascus, Antioch, Yümen, Omaha (Nebraska), and Timbuktu among others?
4. Why has a location on the Suez Canal not promoted more commercial development in Egypt?
5. How does the shortage of low-cost volume transport in most of the Dry World affect the nature of oasis agriculture?

19 Modern states and issues in the Dry World

Drilling for oil in the Rub' al Khali, Saudi Arabia. *(Photo from ARAMCO and Saudi Arabian Information Service.)*

The modern history of the Dry World, especially since the downfall of the Ottomans, has been dominated by two issues whose roots are in outside areas: the scramble for oil, and the great power rivalry over control of or influence in Dry World states. As pointed out in the last chapter, the Dry World contains about half of the known world reserves of petroleum, a resource in great demand in industrial Europe, North America, and Japan. But there are relatively few uses for it in the Dry World itself. The Dry World also occupies a median location between Europe and monsoon Asia, along the southern flank of the Soviet Union. Maritime powers in Europe have been anxious to maintain control over the trade lanes, sea and land, that run through or beside the Dry World. Russia and the Western powers have struggled for influence in the Dry World areas that lay between their own territorial holdings. They are perceived as safeguards against invasion, as footholds for further expansion, or simply as bulwarks to prestige and power. At the eastern end of the Dry World, Russia and China have been rivals for influence over the large arid areas that separate them, inhabited mainly by nomads or by oasis agriculturists. Since the collapse of its older empires, the Dry World has been largely unable to resist these outside pressures. None of its modern states is politically or militarily strong enough to withstand outside pressure with full success. Most of the states are relatively new as organized, political communities in the modern or Western sense.

Colonialism

In the late nineteenth century this power rivalry led to widespread colonial control by Western powers. Morocco, Algeria, Tunisia, and most of the northern parts of western and equatorial Africa became part of the French colonial empire. Spain controlled a piece of northern Morocco, plus the loosely defined and very loosely administered tribal territory of Rio de Oro (Western Sahara), sparsely inhabited mainly by nomadic groups and essentially a desert. French influence was also important in Syria and Lebanon, even while they were nominally controlled by the Ottomans. Both became French mandates after the First World War. Egypt came within the French sphere of influence to a large extent until the British established a stronger position there, in part by financing and assisting the decadent Egyptian government and in part by purchasing a controlling share of the stock in the Suez Canal (built by the French and completed in 1869). The canal was operated by an international company until the Egyptian government nationalized it in 1956. Libya, largely a desert, was finally controlled by the Italians, latecomers in colonial expansion. Turkey retained its full independence through a domestic revolution and the establishment of a reform government after the First World War.

In the nineteenth century, Persia (Iran) was divided into British and Russian spheres of influence. The British influence was paramount until the Second World War, although Persia's nominal independence was retained. The Sudan in effect became British territory in the late nineteenth century, although rule was formally shared with British-dominated Egypt, and the area was known as the Anglo-Egyptian Sudan. Somaliland and Eritrea, on the east coast of Africa, were divided among Italy, France, and Britain. Although the Italian attempt to conquer Ethiopia in 1896 failed, Ethiopia lost its independence under the brief Italian control from 1935 to 1941.

With the breakup of the Ottoman Empire in 1918, Saudi Arabia and Yemen became independent. The areas known as Mesopotamia, Transjordan, and Palestine became British mandates. Iraq and Transjordan became independent in the 1930s, but Britain still provided heavy subsidies to Transjordan. The name *Jordan* was generally adopted in 1949. Iraq was strongly influenced by the British until after the Second World War.

Afghanistan kept its independence, with British and Russian agreement, as a barrier and buffer between India and Russia, but British influence was heavy and Afghan political control weak. Baluchistan was incorporated in the Indian empire of Great Britain, and Turkestan was conquered by the expanding Russians in the nineteenth century. Outer Mongolia was attached very loosely to dynastic China and took its opportunity to declare independence when the Chinese dynasty fell before the republican revolution of 1911. It is now effectively a satellite of the Soviet Union. Traditional Chinese control over Sinkiang was weak, so Russian influence became strong after about 1850. But Sinkiang became more firmly attached to China after the Communist revolution there. Germany made strenuous efforts after 1870 to establish a position of influence in the crumbling Ottoman Empire in Morocco and Persia. But plans for a Berlin-to-Baghdad railway were never carried out. The Germans had visualized the route as a corridor of German-dominated territory to rival the British-dominated sea lanes to monsoon Asia, and as a possible spearhead for a German advance farther east.

Independence

Politically most of the Dry World has now become at least nominally independent. The tiny, neutral, and internationally administered territory of Tangier, guarding the African side of the Strait of Gilbraltar, has been incorporated into independent Morocco. Bahrain Island and the nearby peninsula of Qatar are independent states. South Arabia has absorbed Aden and is now the People's Democratic Republic of Yemen, sometimes called South Yemen to distinguish it from the Yemen Arab Republic that borders on the Red Sea. Oman now also includes the sultanate of Muscat, and the United Arab Emirates is a republic composed of seven sheikdoms formerly under British protection. Russian Turkestan is firmly in the Soviet Union, Sinkiang in the Chinese People's Republic, and Outer Mongolia remains a Russian satellite. The rest of the Dry World now consists of more or less sovereign states, most of them newly independent or newly organized as states within the last thirty or forty years. Most of them are still relatively weak, and outside powers still tend to regard the area as a whole as a *power vacuum*. This is an overused, ambiguous, and often inappropriate phrase that, in most cases, means simply that outside powers are stronger than those in the region itself and its territory and resources are considered fair game. This has certainly been the attitude of most of the Western powers toward the Dry World since the nineteenth century at least, and it is resented understandably by the sensitive new nationalisms in the Dry World states.

The strategic significance of Dry World territory has often been exaggerated because outside influences could be established there relatively easily. Bitter rivalries thus developed between contending outside powers. A secure route from western Europe to monsoon Asia need not depend on territorial control or influence along the route, except when there is a threat that rival powers may attempt to block the route by establishing their own influence in the area. The best solution to such a problem is clearly the establishment of strong, independent, and well-governed states in the Dry World itself. Until this happens, outside pressures and contentions are likely to continue.

Petroleum and politics

The discovery of oil deposits around the Persian Gulf and in northern Iraq provided another powerful incentive for outside influences on that part of the Dry World and the routes leading from it to the industrial West. Saudi Arabia, the sultanates and sheikdoms along the east coast of Arabia, Iraq, and Iran all became targets of outside influence and granted large concessions for petroleum production. Bahrain and the small neutral zone adjoining Kuwait are British or Anglo-American territories in effect, as a result of oil concessions or political influence and the consequent flood of British and American technicians and administrators. Americans are now responsible for most of the oil exploitation. Saudi Arabia derives most of its revenue from oil royalties from the Arabian-American Oil Company, or Aramco (a holding company of several American oil corporations), paid on the amount of oil produced from the areas where Aramco has concessions. The peoples of the Dry World were technologically unable to produce the oil themselves and had no industrial or commercial need for it, so its exploitation was left to Westerners. With the oil concessions, legal rights granted to oil companies for the production and sale of petroleum, came political influence. Areas known to contain oil or where there was a likelihood of oil discoveries were eagerly sought after by Western oil companies, often with the implied, but unspoken, backing of their national governments. Political sympathies or compliance of the Dry World states were sought to facilitate oil concessions and to ensure the flow of oil on favorable terms. Large-scale Western investments were made in the oil producing areas, including refineries to ease the problem of bulk transport to distant markets.

Western attentions were also concentrated on the routes from the oil fields to Europe and the United States, where most of the production still goes. Pipelines were built across Syria, Iraq, Jordan, Lebanon, Saudi Arabia, and Palestine to the Mediterranean, and attempts were made to control the routes or to protect them (see Maps 12 and 84, pages 56–57 and 356). The Suez Canal became newly important as the principal route of oil flow to Europe. The British built up their military and naval bases in Egypt and Cyprus to guard the canal and protect the land routes and pipelines through the Levant. British military power declined after the Second World War, and strong nationalism grew in Egypt. In 1956, Egypt took complete control of the Suez Canal. The United States has strengthened its forces in the Mediterranean area and has given strong support to a military alliance that joins Turkey, Iran, Pakistan, and Britain in a treaty of mutual defense and assistance, primarily against a possible threat from the Soviet Union. The sheikdom of Kuwait, which had been a rich, British oil protectorate, became politically independent in 1961, despite continuing attempts by the revolutionary government of Iraq to incorporate it.

In western north Africa, Morocco, Algeria, and Tunisia won independence from France, and Libya from Italy with the Second World War, after a short period as United Nations trusteeships. Western political influence in north Africa has been greatly reduced, and the former French and British military bases there have been lost. But recent discoveries of very large oil deposits in the Algerian Sahara and the Libyan Desert (see Map 84, page 356) and the building of several pipelines to the coast have promoted a renewal of outside investments and pressures in north Africa.

The Dry World's oil resources, however, also provide a weapon against outside pressures or efforts at exploitation. Rising nationalism in all of the oil-rich Dry World states led to the formation in 1974 of the multinational Organization of Petroleum Exporting Countries (OPEC). It joins Algeria, Libya, Saudi Arabia, Kuwait, Iraq, the United Arab Emirates, the small neighboring sheikdom of Qatar, and Iran[1] in an agreement that tries to set a uniform price for all exported oil. The price tends to be more than double the previous average of free market prices,

---

[1]Nigeria, Gabon, Indonesia, Ecuador, and Venezuela, as major oil exporters, are also members of OPEC.

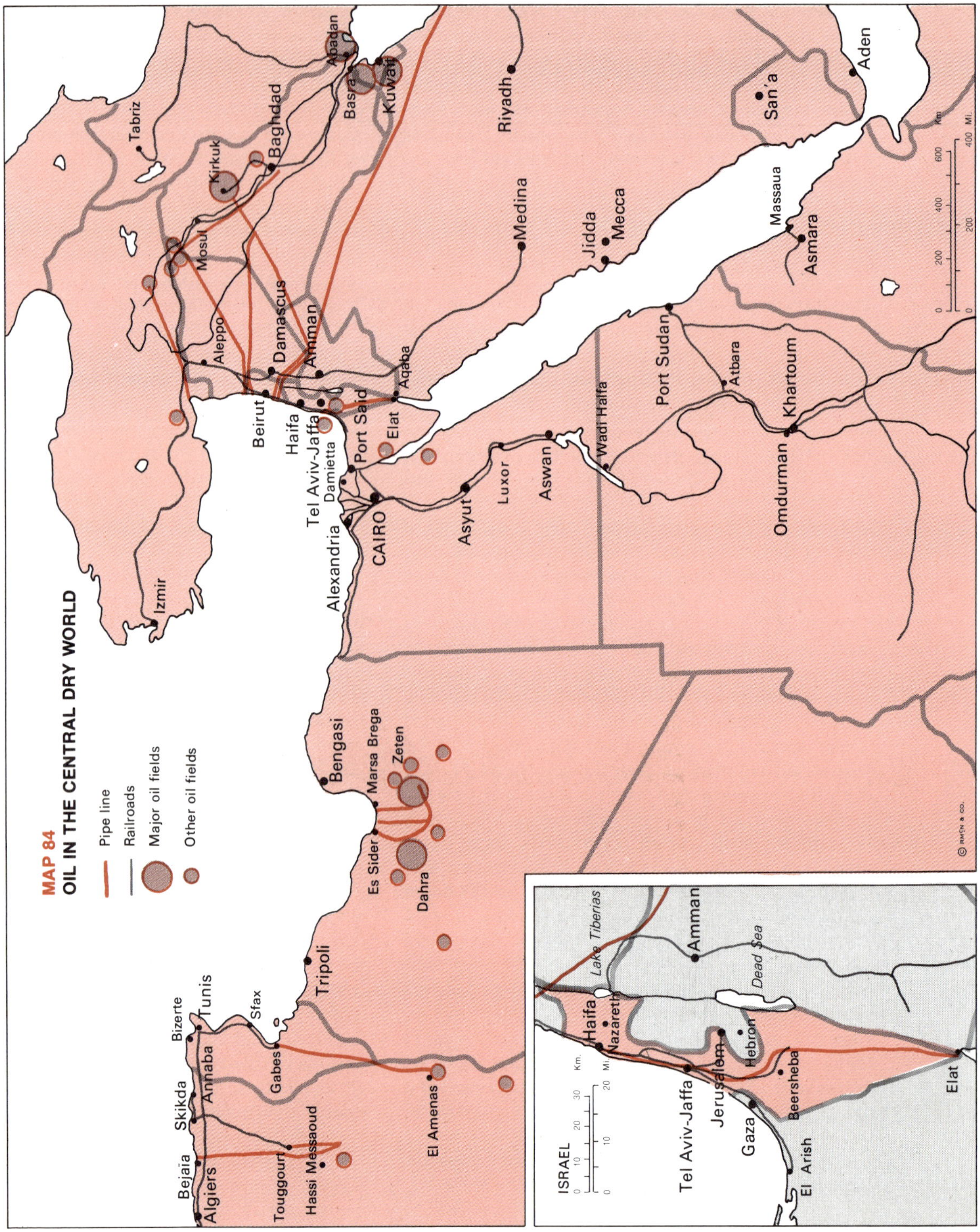
MAP 84
OIL IN THE CENTRAL DRY WORLD
Pipe line
Railroads
Major oil fields
Other oil fields
Tabriz
Kirkuk
Baghdad
Abadan
Basra
Kuwait
Riyadh
Mosul
Aleppo
Damascus
Amman
Aqaba
Beirut
Haifa
Tel Aviv-Jaffa
Port Said
Elat
Damietta
Alexandria
CAIRO
Izmir
Medina
Jidda
Mecca
Asyut
Luxor
Aswan
Wadi Halfa
Port Sudan
Atbara
Khartoum
Omdurman
Massaua
Asmara
San'a
Aden
Bengasi
Marsa Brega
Zeten
Es Sider
Dahra
Tripoli
Tunis
Bizerte
Sfax
Gabes
Algiers
Bejaia
Skikda
Annaba
Touggourt
Hassi Messaoud
El Amenas
0 200 400 600 Km.
0 200 400 Mi.
© RMcN & CO.
ISRAEL
0 10 20 30 Km.
0 10 20 Mi.
Lake Tiberias
Dead Sea
Haifa
Nazareth
Amman
Tel Aviv-Jaffa
Jerusalem
Hebron
Gaza
Beersheba
El Arish
Elat

reflecting both the growing world shortage of petroleum as demand continues to rise and the determination of the oil exporting states to use their new bargaining strength to ensure that they reap the benefit. As the first really successful effort of these new states to work together, if only on the immediate, practical issue of oil prices, OPEC has brought new political strength to this area that has so recently emerged from colonialism and outside domination. The success of OPEC may lead to additional efforts at multinational cooperation in their common interest. It is already giving this bloc of Dry World states a new power and validity in relations with the industrial countries. Outside oil buyers, corporations, and governments are obliged to deal with OPEC as an administrative body; thus, it is acquiring an important political as well as an economic function.

What are the geographic bases of the Dry World states, and what are their prospects for economic and political development that could further strengthen their independence?

North Africa and its new states

North Africa west of Egypt shares a largely common physical environment: a narrow coastal lowland strip, or *tell*, of winter rainfall and Mediterranean climate, and an arid desert stretching south to the savanna or steppe landscape on the borders of Black Africa. The rainier coastal zone of the tell widens in Tunisia, Algeria, and Morocco, partly because it is somewhat farther north, hence closer to the winter storm tracks through the Mediterranean, and partly because the Atlas Mountains force the air masses to rise. The orographic rainfall the mountains cause is fairly heavy on these mountain slopes, reaching 40 inches (102 cm) annually in small areas. But most of the coastal plains of Algeria and Morocco receive 20 inches (51 cm) or more. The Atlas do not extend as far east as Libya, but they constitute in effect a separate region between the tell and the desert to the west. Libya and Egypt largely lack a tell region and are almost entirely desert. A few isolated highlands in the southern Sahara, such as the Tibesti Massif, the Aïr and the Ahaggar mountains and the Tassili-n-Ajjer plateau, catch a little rainfall, but it is generally not enough for agriculture. The few inhabitants of these areas are relatively primitive Negroid or mixed groups who are mainly nomadic. They have often supplemented their meager production by raiding trade routes and oases. There is almost no trans-Saharan trade now, except for the recent development of pipelines from the oil and natural gas deposits. West African trade goes by sea, and the Saharan areas themselves produce or need almost nothing for exchange. Motor routes are relatively easy across reg terrain, but, in the absence of need for bulk haulage except for oil, most of the slight traffic now goes by air. There is obviously no basis for a railroad, except for the short spurs from the coast to serve the tell and irrigated areas and to connect with the pipelines. Irrigation from the permanent streams beginning in the Atlas Mountains and from wells and qanats is rapidly increasing in the tell. It may soon equal or exceed the irrigation of Roman times. Irrigation is used to ensure and increase yields on the coastal plain, and to extend the cultivated area inland. Rainfall is scarce or absent in summer on the coastal plain, and some areas, especially toward the east, have inadequate annual rainfall for field agriculture.

Agriculture is of the Mediterranean type, with barley, wheat, citrus fruits, olives, grapes, nuts, cork, and some cotton, tobacco, and dates as the main crops. Wine was the leading export of Algeria until it was overtaken by oil. Morocco, Algeria, and Tunisia also export olive oil, citrus fruits, some tobacco, and wool and hides. The sheep and goats graze on the mountain slopes or on the steppe borders of the cultivated land. High-quality dates are produced in the Algerian Sahara, especially along the short railway spur leading to Touggourt, south of Skikda (formerly Philippeville). Dates depend both on groundwater and on irrigation; there is some commercial date production in the small Algerian oases to the southwest. In Libya the humid or irrigated coastal strip shrinks to a few miles in width and disappears altogether east of Tripoli. There is just enough precarious rainfall on the Cyrenaica (Barqah) plateau for hardy cereals and olives. Around this plateau are the arid areas where the chief battles of the desert campaign in the Second World War were fought. The terrain is ideal for tanks but of little or no use economically. Most of the wadis, a feature of the landscape of much of arid north Africa, do not reach the sea or a permanent stream, and are largely useless as sources of water.

Large deposits of phosphate rock, crushed and used as fertilizer and in the chemical industry, occur in Morocco, Algeria, Tunisia, and Western Sahara. Phosphates are an important part of the economies of all of these countries; they dominate world trade in phosphates. Mined mainly along the seaward side of the Atlas Mountains, phosphates do not present a great transport problem. Morocco has extensive deposits of manganese, iron, and lead; the iron-ore deposits of Algeria, mainly near the coast not far from the Moroccan border, are even more substantial. Iron ore and lead occur in smaller amounts in Tunisia. Unfortunately very little coal has been discovered. Petroleum alone cannot make up the deficiency as far as domestic industrialization is concerned, although it yields large revenues for investment and gives the whole economy a boost. For the present, both the ores and the oil are important exports. Hydro potential is limited by the sharply seasonal character of the rainfall, and economically by the small local market. There is a modest basis for hydro development in Morocco and Algeria.

People and cities

Native Berber peoples are still numerically in a slight majority in most of north Africa west of Egypt. But the population is highly mixed and includes Arab, Negroid, and Germanic strains, among others. Vandals, Goths, and other Germanic groups invaded and settled in north Africa during the great barbarian wanderings after the fall of Rome. Traditionally this area was called Barbary, from the Berbers, and the Barbary Coast meant the tell from western Morocco into Libya. The Carthaginians, although originally Phoenician colonists, intermarried with the Berbers. After the fall of both Carthage and Rome, a Berber empire ruled most of the area west of Egypt, including the Sahara and its oases, until the invasions by the Arabs. Arabs and Arab strains are included in the present population. Although most of Barbary was also included in the Ottoman Empire, Turkish control and influence were nominal. Arabic has become the dominant language of most of Barbary. But French is strong in the cities, and the Berber language survives as well in many areas. During the modern period of French, Spanish, and Italian control, considerable permanent European settlement took place, and the population of the area as a whole is now nearly 10 percent European, heavily concentrated in the cities. Permanent French settlers in Algeria (*colons*, or colonists), numbering slightly over a million, made the granting of independence to Algeria especially prolonged and difficult. Algeria had long been regarded as an integral part of France, sending its own elected representatives to the Chamber of Deputies in Paris. The Algerian oil discoveries in the middle and late 1950s further stimulated French desire to retain control of the country. Bitter fighting and terrorist activities between Algerian nationalists and French forces drained French strength, poisoned French democracy, and delayed the eventual political settlement.

Europeanization, of a basically Mediterranean sort, is very strong in all of the cities of north Africa and less strong in the rural parts of the tell. But it has

The facade of north Africa—Tripoli, Libya. Tripoli, with its white buildings gleaming in the sun, is typical of the coastal cities west of Egypt. The Mediterranean plain was once the granary of Rome, with a larger irrigated area than now. The city shows the influences of the many different cultures that have occupied it during its existence. *(Photo by Jim Holland for Stock, Boston.)*

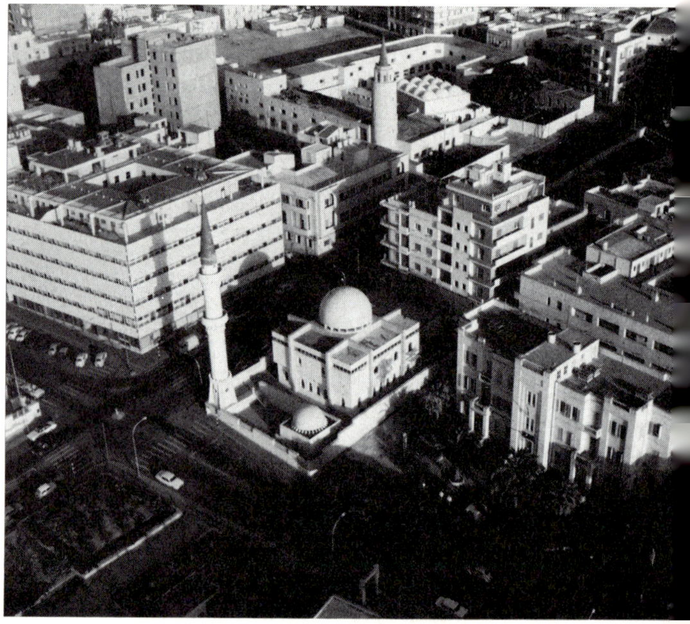

made relatively little impact on the arid, unirrigated areas where settlement is very sparse and the few inhabitants are still largely nomadic. The tell belongs culturally, physically, and economically as much with the Mediterranean basin as with the Dry World. But it covers only a thin strip around the African part of the Dry World, as the Levant is a narrow margin of the Asian part. The tell was the base for the famous Barbary pirates. Their ships anchored in the harbors of Algiers (Alger), Tunis, Tripoli (Tarabulus), and other coastal centers, preying on Mediterranean trade from the days of Rome whenever the major maritime powers were weak. Final suppression came only with the French conquest of Algeria after 1830 with subjugation of the inland bases of the pirates and raiders in the Atlas Mountains as well as on the coast.

Cities grew rapidly under colonial rule, as centers of trade and European settlement, in somewhat the same way as the colonial ports in southeast Asia. By north European standards the cities are exotic, but their facades of white walls, flat roofs, olive groves, and narrow streets have much in common with other Mediterranean and Levantine cities. There is a strong flavor of Arab culture, especially in the presence of many mosques, and the characteristic Dry World bazaars are also found there. Algiers and Tunis, although the capitals of their respective states, until recently had a majority of European inhabitants; there are still large European groups in most of the cities. Many Italian as well as French settlers came to Tunisia, but Italians are almost the only European settlers in Libya. Their small numbers decreased still more after the Second World War. The largest city of Morocco is Casablanca on the Atlantic coast, the major gateway for European trade. The Moroccan capital is at Rabat, the seat of the sultan. Algiers, with its excellent bay harbor and central position along the Algerian coast, may soon overtake Casablanca in size as Algerian economic development continues to boom. Tunis, a few miles from the ruins of Carthage, which stand near the entrance to the inner harbor of Tunis, is the third city of Barbary in size. It is growing rapidly, both as a national capital and as the major Tunisian center for foreign trade. There are no real cities in north Africa west of Egypt except on the coast, aside from Fès and Marrakech in more humid Morocco on the Atlantic side of the Atlas Mountains. Marrakech was once the largest Moroccan settlement, but both it and Fès lost importance with the development of maritime trade. Otherwise there are no large inland cities to rival the Europeanized coastal cities, now also the national capitals.

Egypt

Egypt has probably changed less during its long history than any of the other settled areas of the Dry World. The ancient empire of the pharaohs fell over two thousand years ago, but Egypt has remained the largest area of dense population in the Dry World. It still depends almost entirely on agriculture fed by the Nile waters. Over half of the population is still engaged directly in farming, and all of the cultivated land is irrigated. The greatest changes have taken place since the latter part of the nineteenth century: the completion of the Suez Canal in 1869, the period of strong British influence and investment, and the development of cotton growing for export. After 1860, when the American Civil War reduced American cotton production and export, Egypt became third after the United States and India as a cotton producer (although cotton production in China and Russia is now larger). Almost their entire crop is exported. Cotton accounts for nearly 60 percent of Egypt's exports, although it occupies only about one-fifth of the cultivated land. Yields are high, favored by fertile alluvium, high temperatures, strong sun, and controlled water supply through irrigation, an ideal set of conditions for cotton. Long-staple cotton for the weaving of high-quality cloth is especially suited to these conditions, and is an Egyptian specialty. There is only a small domestic development of cotton textile manufacturing, since power, labor, capital, and market advantages are concentrated in Europe, Japan, and North America, all of which purchase raw cotton from Egypt. Egypt's population and that of neighboring areas is relatively small, and its demand for manufactured goods of any kind is limited by low incomes. Wheat, corn, rice, and sugar cane are all produced in larger quantities than cotton, but all are essentially subsistence crops. Egypt does not feed itself, however, and must import growing amounts of rice and other foodstuffs, as well as wood, coal, and manufactured goods, to support a rapidly increasing population.

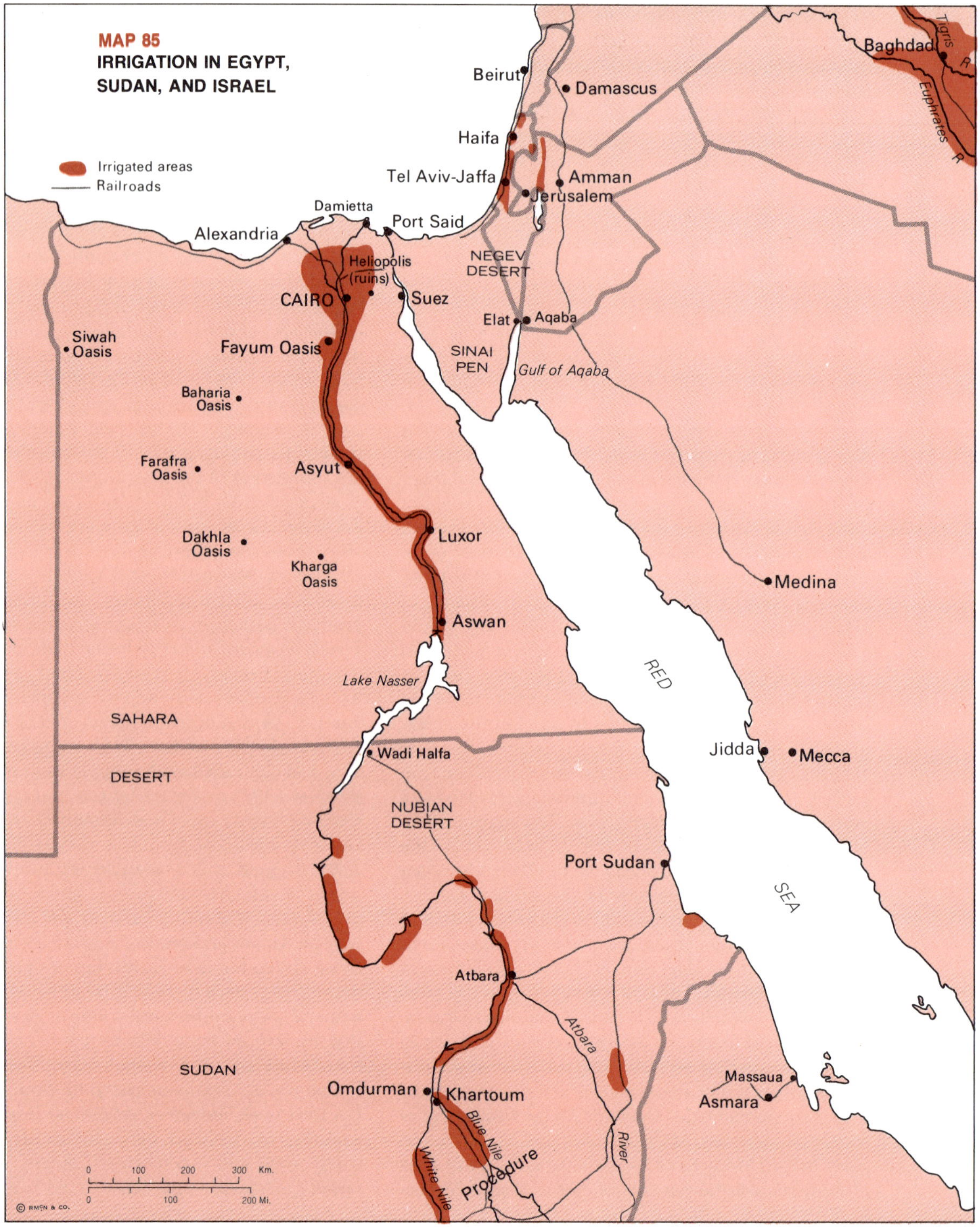
MAP 85
IRRIGATION IN EGYPT, SUDAN, AND ISRAEL
Irrigated areas
Railroads
Beirut
Damascus
Baghdad
Tigris R.
Euphrates R.
Haifa
Tel Aviv-Jaffa
Amman
Jerusalem
Damietta
Port Said
Alexandria
Heliopolis (ruins)
NEGEV DESERT
CAIRO
Suez
Elat
Aqaba
Siwah Oasis
Fayum Oasis
SINAI PEN
Gulf of Aqaba
Baharia Oasis
Farafra Oasis
Asyut
Luxor
Medina
Dakhla Oasis
Kharga Oasis
Aswan
RED
Lake Nasser
SAHARA
Jidda
Mecca
Wadi Halfa
DESERT
NUBIAN DESERT
Port Sudan
SEA
Atbara
Atbara
SUDAN
Massaua
Omdurman
Khartoum
Asmara
Blue Nile
River
White Nile
Procedure
0 100 200 300 Km.
0 100 200 Mi.
© RMcN & CO.

Agriculture is largely limited to the narrow valley of the Nile, where irrigation is possible. The sharply confined pattern of settlement is apparent in Map 82 (page 338). The major irrigated area, a strip paralleling the Nile only about 30 miles (48 km) wide, extends downstream from the first cataract (rapids and falls) at Aswan, spreading out into the fan-shaped areas of the delta (see Map 85, page 360). The mouths of the Nile and the seaward advance of silt deposition in the quiet coastal waters of the Mediterranean are responsible for our word *delta,* given to this area by the Greeks because of its triangular shape like the Greek delta, Δ. Smaller irrigated areas, with land use and crops characteristic of Dry World oases, are maintained in the oases of the Egyptian desert west of the Nile, especially Kharga (Al-Wahat al-Kharijah), Dahkla (Al-Wahat ad-Dakhilah), Farafra (Al-Wahat al-Farafirah), Baharia (Al-Wahat al-Bahriyah), and Siwah. Altogether only about 3 percent of the total area of Egypt is cultivated, and all but a tiny fraction of the cultivated land is in the Nile valley and delta. The Nile has no perennial tributaries for the last 1,700 miles (2700 km) of its course. Its volume decreases markedly downstream, both from evaporation and from withdrawal of irrigation water. Thus, the river is navigable only for small boats, except during the limited high-water period, as far as Luxor (Al Ugsur), and Cairo is not an ocean port. But there is dense traffic of small boats and shallow-draft motor launches. The last perennial tributary is the Blue Nile, originating in the wet highlands of Ethiopia and joining the main stream at Khartoum (Al Khurtum), the capital of Sudan. Monsoon rains in Ethiopia create a much greater flow during the summer. The main stream of the Nile above Khartoum, usually called the White Nile, begins south of the equator and is nourished by heavy year-round rains. In Egypt the river rises sharply in late June, after the Blue Nile floods occur, and the high-water season is over by November.

The traditional agricultural system of Egypt was organized around the seasonal flooding of the Nile. Fields were flooded during the summer, and crops, planted in the autumn after the water receded, were harvested in the spring. This ancient system of *basin irrigation* was gradually replaced by *perennial irrigation* during the nineteenth century. Storage dams were built along the Nile, and in 1902 the British completed a large dam at Aswan, which was subsequently raised in 1912 and again in 1924. Water drawn from the reservoirs behind the dams became available for irrigation throughout the year in most of the Nile valley, and two or three crops are now raised each year in the same field. Mechanical pumps have made Nile water available for larger areas, flooding of fields is controllable, and silt deposition is much reduced. A decline in crop yields resulted, and soil exhaustion was hastened by raising more than one crop each year. Increased amounts of fertilizer have become necessary, but total agricultural production has risen. The government of Egypt constructed a higher dam 4 miles (6 km) south of Aswan in order to increase the area under irrigation still further. Hydro production is also being increased, but the effective market for it is still limited. Population has risen more rapidly than agricultural production, and the Egyptian *fellah* (peasant) is still wretchedly poor with few economic opportunities outside agriculture. Most farms are uneconomically small, and equipment and methods are generally limited and crude. Agriculture is highly labor-intensive, but not capital-intensive, except for irrigation. The loss of the former annual silt deposits from the flooding, now prevented even more by the new high dam, creates pressing need for scarce and expensive fertilizer. The Aswan high dam, in fact, is a classic case of technological impact on a traditional agricultural system. As in so many other cases, it suggests that change cannot be effectively introduced piecemeal, but must be carefully planned as a single whole.

As in Bengal and elsewhere, favorable environment and high productivity per acre have been accompanied by mass economic misery. The population of Egypt increased about six times between 1860 and 1960 and is still growing. There has not been a commensurate rise in total production or economic opportunity. The high dam at Aswan, a mammoth and expensive project financed largely by the U.S.S.R., was originally expected to raise living standards substantially. But it now appears that the population of Egypt will continue to increase so rapidly that crops from the newly irrigated land will barely be able to feed the new consumers, with little rise in the perilously low level of consumption. The Nile valley is now, as it has been for thousands of years, probably the most densely populated rural area of comparable size in the world. Egypt has been described as "a vast rural slum." The country has few

Fellah agriculture in Egypt. Threshing is done with this crude sledge that has spikes and sharp stones on the bottom. It also cuts up the straw. This threshing technique is virtually unchanged from the time of the early pharaohs. Primitive methods still restrict agricultural production in Egypt, which cannot keep pace with the population increases. This scene is near Cairo. *(Photo from United Nations.)*

nonagricultural resources and few important mineral deposits. There are oil fields east of the Nile, and a little phosphate rock in the same area along the Red Sea. There is some iron ore near Aswan, and a small iron and steel plant has been built near Cairo, although the coal must be imported. The oil reserves are sizable, but foreign investors have been hesitant to become involved in what they regard as a risky political situation, and Egypt itself has inadequate capital and technical means to develop the oil resources.

Modern Egypt attempts to play the role of a political leader for the Arab world, but the United Arab Republic, joining Syria to Egypt in 1958, ended with Syrian withdrawal in 1961, and that name was dropped in 1971. Domestic poverty and the absence of full economic or social democracy and political maturity in a largely illiterate population cannot be overcome or disguised simply by expressions of Arab[2] nationalism on the part of the political rulers. The Egyptian government is now attempting to even out landholding and stimulate growth in agricultural and industrial production.

Cities

Most Egyptian settlements are peasant villages or small market towns strung along the Nile in a garden landscape of irrigation ditches, fields, and palm groves. Only Cairo, Alexandria, and Port Said have become large cities in the modern period, and all are essentially products of the Western commercial impact. They include relatively small textile and food-processing industries, but are mainly centers of trade and administration. All three are considerably Europeanized in physical appearance, although camels, mosques, and the mud hovels of the poor contrast with the large modern buildings, hotels, cars, and buses. Alexandria, with a population of about 2 million, was an important British naval base until taken over by the Egyptians in 1954. It is still Egypt's major commercial port and handles the bulk of imports and exports. Location at the western edge of the delta helps Alexandria to minimize silting problems. There is no important settlement or major port at any of the shifting, shallow, and silting mouths of the Nile. Port Said, administrative headquarters for the Suez Canal, was built by the French to a large extent, and is now a cosmopolitan city handling some foreign trade. French was widely spoken and written in Port Said. Under the revolutionary government, and especially after the ill-advised and abortive Anglo-French attack on Port Said in 1956 (an attempt to prevent sole Egyptian control of the canal), anti-Western feeling increased greatly, and efforts were made to Egyptianize the culture of Port Said, Alexandria, and Cairo.

Suez, at the southern end of the canal, is a smaller-scale version of Port Said. Oil is by far the largest single item of cargo moving through the canal. Total cargo traffic is more than double that of the Panama Canal. No locks are necessary, since the Isthmus of Suez is flat and the Red Sea and Mediterranean are at the same level. But drifting sand, wave wash, and silting of the channel make constant maintenance and dredging essential. To the surprise of Britain and France, Egypt managed and maintained the canal successfully, and the flow of traffic continued to increase until it was blocked by the Arab-Israeli war of 1967. The canal was reopened in 1975 but may have trouble regaining all of its former oil traffic. The present supertankers are too large to use the canal, and follow the route around the Cape of Good Hope. Loose desert sands, with occasional irrigated strips or palm groves, cover the land on both sides of the Suez Canal. From the deck of a ship in transit or from the banks, the ships appear to be sailing across the flat desert.

Cairo, with over 6.5 million population, is the largest city in Africa and in the whole of the Dry World. Located at the head of the Nile delta, it is well situated to serve as the major central place for the populous area of Egypt. Its administrative, financial, and commercial services are far more important than

[2]The Egyptians, strictly speaking, are not Arabs, although their present language is Arabic, and they tend to consider themselves part of the Arab world. *Arab* is admittedly an ambiguous term, but it may be least misleading to apply it only to natives of the Arabian Peninsula and their direct blood descendants elsewhere.

its small manufacturing. Easy overland routes (now served by rail) to Port Said, Alexandria, Suez, and the Nile valley give Cairo connections in every important direction. The ruins of Memphis, traditional capital of lower Egypt, are a few miles south of Cairo on the opposite bank of the Nile. The landscape between is dotted with ancient pyramids, all of which help to attract the tourist trade to Cairo and its Western-style hotels. Throughout Egyptian history this area has been a focus of routes and a convenient capital site. Size alone gives Cairo some basis for its claim as unofficial capital of the Arab world as a whole. It does exercise some supranational functions as a cultural center and as the seat of an old and famous Muslim university, the al-Azhar, which continues to draw students from many Muslim countries. Cairo is also the headquarters of the League of Arab States. The city is on the right bank of the Nile, reversing the more usual pattern of left bank sites in Egypt. The right bank has somewhat higher ground that is normally free from floods except in the low sections of the city. But access to the Red Sea, the Sinai overland route, and the traditional centers of the Arab world and its empire is also easier from that location. Cairo was founded about A.D. 968 by the Arab conquerors. Most of its growth has come since the beginnings of modern commercialization after about 1860: the opening of the Suez Canal, and the investments made by the British in irrigation, transport, and commercial-financial services.

Together, these essentially alien factors have reshaped modern Egypt. But they have made very little impression on the average rural fellah, whose living conditions are still much the same as under the pharaohs. Cairo exemplifies modern, Westernized, commercialized Egypt, but also exhibits, even within the city, the frightening contrasts between wealth for a few and abject poverty for the rest often characteristic of the Dry World. It is nevertheless worth remembering that, although this is an old tradition in the Dry World, it was far from absent in the history of the West, even though in the West the contrasts between the few rich and the many poor may never have been so pronounced. The West has spread economic and other benefits more evenly among its population only within the last two or three centuries. Such changes may well come in Egypt and in the rest of the Dry World in the coming century, as their independent governments become more effective, and as commercialization and resource development proceed.

Desert east Africa

Desert continues down the African coast of the Gulf of Suez and the Red Sea and around the horn of east Africa nearly to the borders of Kenya. The relatively narrow coastal strips of lowland in Eritrea and Somalia belong physically and culturally to the Dry World. The people are dominantly Muslim and have long been heavily influenced by Arab control, trade, and settlement, although the dominant racial strain is Negroid. It is a sparsely settled, arid land, much of it almost entirely empty, and inhabited mainly by nomads. There are a few small ports, some of them of older Arab origin, some more recent colonial outposts or strategic bases, such as Djibouti, the largest. Ports on the Red Sea, some founded by the ancient Egyptians, have decayed since the opening of the Suez Canal. But Port Sudan (Bur Sudan) has revived considerably as the major maritime outlet for the independent state of Sudan. It has a railway connection with Atbara (Atbarah) and Khartoum. Massawa (Mesewa), with port facilities largely built by the Italians, serves as the outlet for part of Ethiopia and highland Eritrea. But most of Ethiopia is more directly served by the rail route from the capital at Addis Ababa to Djibouti. Arab influence is old all along the coast of east Africa north of Dar es Salaam. Trade with Arabia and the Persian Gulf coast was relatively easy from the beginning of Arab commercial ventures. It was in the trade with east Africa that the Arabs first learned to recognize and make use of the monsoon winds. The same sailing dhows are still used along the coasts of most of the Dry World. On the east African coast there was little local economic or political development, and the Arabs easily established coastal footholds, trading factories, and small local emirates or kingdoms.

Ethiopia

Mountainous Ethiopia is different enough to be excluded from the Dry World. Much of it is humid, and some parts are excessively rainy where the summer monsoon strikes the mountains. Islam has never penetrated significantly into the Ethiopian highlands. Most of the inhabitants have preserved their local form of Coptic Christianity since it was first introduced by missionaries from Egypt in the fourth century A.D. It is true, however, that the Ethiopians generally prefer livestock grazing to agriculture, and in this respect the country has affinities with the Dry World. Cultural and economic ties are also predominantly with the arid north rather than with the humid south. There is relatively little nomadism and little irrigation, although some irrigated cotton, dates, and sugar cane are grown in the limited areas of relatively level land below 5,000 feet (1525 m). Much of the country as a whole is too rough for agriculture, and this is largely responsible for the emphasis on animals. But cultural preference is also a factor, and has led to the neglect of agricultural opportunities that a farming culture might have pursued. Mineral deposits and their exploitation are of little consequence, and modern commercialization is just beginning. Addis Ababa has a population of a little over 900 thousand, and contains a few large, modern buildings erected by the Italians. But much of the rest of Ethiopia is still essentially tribal, preliterate, and isolated.

Sudan

Sudan is more closely involved with the rest of the world, although subsistence livestock raising is still the principal occupation, and many of the people of Sudan are nomads. Most of the country is a vast, semiarid plain, physically and culturally transitional between the Dry World and Black Africa. The southern third of Sudan, close to the equator, is moist enough for tall grass and low trees, and is dominated by Negroid people. The arid north is inhabited by Arabic-speaking Muslims of mixed origin. British control after the 1880s and British investment in transport and irrigation have been accompanied by rapidly growing commercial production, especially of cotton for export. Egyptian hopes and attempts to control the Sudan were frustrated when Sudan voted for independence after the withdrawal of British control in 1953. There have been border disputes and conflicts over water rights since. The Nile receives all of its major tributaries within Sudan. Sudan has agreed not to construct dams that would restrict water for Egypt without Egypt's consent. But most agricultural expansion in Sudan depends on increased irrigation. The largest irrigated area, with cotton as the leading crop, is the Gezira, a triangular-shaped region between the Blue Nile and the White Nile. Just above Khartoum, it is supplied with water from the Sennar Dam built by the British (see Map 85, page 360). Some cotton is grown without irrigation farther south. Cotton provides almost half of the country's export revenue.

Sudan is an enormous area, nearly a third as large as the United States, but with a population of about 18.5 million. Economic opportunity is limited by aridity and also by the shortage or absence of mineral resources and locational advantages for trade. But there are wide possibilities for extension of agriculture, both irrigated and unirrigated, and for commercial livestock production, which are being energetically pursued by the government to the limit of its meager financial abilities. In some respects, economic development from an essentially tribal and primitive base presents fewer social and political problems than development in areas of older agrarian culture, large landowners, large populations, and traditional ways of doing things. Sudan has few attractions for foreign influence or interference, except for the issue with Egypt over Nile water. About 13 percent of the population lives in towns, and a large part of that percentage is in the one real city, or rather three associated cities of Khartoum, Omdurman (Umm Durman), and Khartoum North (Al Khurtum Bahri), which make a single metropolitan area. Khartoum is the capital, conveniently central, with rail links north and south and to the only other sizable town, Port Sudan.

Turkey

Turkey has been the most successful of the modern Dry World states in developing an effective national government, spreading literacy, maintaining its independence from outside influences, and applying Western techniques of commercialization and industrialization toward raising the economic level of its people. The collapse of the decadent Ottoman Empire and the loss of its territories were followed by a political revolution in Turkey in the early 1920s led by a group of Western-minded reformers under Mustafa Kemal, who became known as Ataturk ("father of the Turks"). In addition to building a strong, responsible administration and attacking economic problems, the new government reduced the dominant cultural influence of Islam where it retarded economic and political progress. The custom of *purdah,* where women are cloistered at home, forbidden to show their faces in public, allowed to appear outside the home only heavily veiled or wearing a cloth over the face, and denied education or normal social intercourse, was outlawed. Men were required to shave off their beards, and prohibited from wearing a fez. Emphasis was placed on a secular state and society. A modified Latin alphabet was adopted, and the old Muslim law was replaced by a Western legal code.

Turkey still has far to go before it achieves a complete political, economic, or social democracy, and before its resource use has developed to a Western level. But the change from the conditions and attitudes of the corrupt and reactionary Ottoman Empire is revolutionary indeed. Turkey has minimized its formerly close ties with the Arab world, emphasized its own distinctive identity apart from the Arab states and cultures, and strengthened its cultural, political, and economic ties with the West. Turkey is a member of NATO, although this is also partly the result of a common boundary with the Soviet Union and of Turkey's position on the straits leading to the Black Sea. International agreement recognizes the right of free passage through the straits for nonmilitary vessels in peacetime. But they are closed to warships, and Russia for at least a century has attempted to negotiate and apply pressure to remove this bar. The several Russian wars against Turkey in the nineteenth century (satirically referred to as "wars of the one-eyed against the blind") and the repeated attempts of Western powers to preserve international control of the straits against Russian pressure form part of the background of the present situation.

Agriculture is still the dominant Turkish occupation and is likely to remain so. About two-thirds of the population is engaged in agriculture or livestock production. About one-third of the total land area is cultivated. Turkey is a mountainous country, and even on the central Plateau of Anatolia there are relatively few large expanses of level land. The most useful lowlands are on the somewhat broken coastal plain facing the Aegean Sea, which helps to turn Turkey westward. The Aegean coastal plain was predominantly Greek in population for many centuries. Most of the Greeks were returned to Greece after the revolution in Turkey, in exchange for smaller Turkish minorities transferred from Greece to Turkey. But there are still strong evidences in the coastal towns that this area was part of Greece for more than two thousand years. The loss of the skilled Greek artisans and merchants was an economic blow, but the Turks valued the benefits of a more nearly homogeneous national population. There is a small coastal plain along the Black Sea, which has quite a humid climate (see Map 82, page 338). But the Black and Mediterranean coasts are hilly or mountainous except for a few short stretches. The Plateau of Anatolia, the heart of Turkey, has a severe winter and a somewhat precarious rainfall. Agriculture there is limited by climate as well as by relief, and wheat growing is the main form of land use. Nomadism or seminomadism is still relatively common, especially toward the east, where climate and relief grow more severe. There the proportion of Kurdish people among the population increases until they dominate the mountain area of Kurdistan. The Kurds are an essentially nomadic group, distinct from the Turks. The population of Turkey in general is racially mixed, even after the expulsion of most of the Greeks. The Turkish language, however, is now dominant, except among the Kurds, and the government has made progress toward building a more unified nation with an integrated Turkish culture. The Armenians, inhabiting mountainous northeastern Turkey and adjacent parts of the Soviet Union, retained their own Indo-European language and Christian religion, but they suffered greatly from bitter Turkish persecution and killing. Most of the surviving Armenians in Turkey emigrated to Europe and the United States; there are now fewer than 65 thousand left in Turkey, almost all of whom live in Istanbul.

The coastal lowlands have an essentially Mediterranean agriculture in keeping with the climate and terrain. Wheat and barley are the main subsistence crops, and tobacco, cotton, grapes, figs, olives, nuts, and citrus fruits are the leading commercial ones. Turkey is the world's leading source of hazel nuts (mainly from the Black Sea coast). Tobacco, also a Turkish specialty, is an important export, grown along all of the coasts. Raisins, wine, and dried figs are also important in the country's foreign trade, particularly through the old Greek port of Smyrna, renamed Izmir by the Turks, which still does some entrepôt business for the Aegean Islands. The exports suggest the predominance of agriculture in the Turkish economy and the relatively small progress of economic development. Since the 1920s there has been a great expansion of wheat acreage with some increase in yields. In recent years with Western technical assistance, cotton has become the leading export crop. There is a small development of domestic textile manufacturing. The Plateau produces considerable wheat and barley, plus hides, skins, and wool, including mohair from the Angora goat, which is woven into high-quality rugs and fabrics. Fortunately there are few landless farmers, and the Turkish peasant is much better off materially than the Egyptian fellah, although his methods of cultivation are probably little less primitive, especially on the Plateau.

There are some significant mineral deposits, including a relatively small amount of fair-quality coal, some petroleum, a little high-grade iron ore, and some extensive deposits of chrome ore. Turkey is a large producer of chrome. With the exception of chrome and some coal mining, mineral exploration is still very limited. There are only the beginnings of an industrial structure, almost entirely limited to textiles and consumer goods, although there is a small and growing production of iron and steel.

About one-third of the people live in cities. By far the largest city is Istanbul (Constantinople), a major trading center since at least the seventh century B.C. and throughout the time of ancient Greece. It was selected as the capital of the eastern Roman Empire by the Emperor Constantine in A.D. 330 and bore his name until the Turkish conquest in 1453. It was changed officially in 1930 to Istanbul. After the fall of the Ottoman Empire, Turkey became a republic. The revolutionary government moved the Turkish capital to Ankara, in the center of the Anatolian plateau, in 1923. Istanbul, on the European side of the straits, was considered too cosmopolitan and too much on the fringes of the new Turkey to be an appropriate capital, especially after the loss of its former large European holdings and its determined drive to build a new and genuinely Turkish nation. Ataturk personally stressed the virtue of Ankara as the national capital on the grounds that the Turks are Anatolians. Ankara had been a small interior town without particular advantages except for its central position and the Turkish character of its hinterland. It is now a rapidly growing city of over 1.5 million, based mainly on administrative functions. Road and railroad building to link it with the rest of the country have progressed fairly rapidly. Istanbul still does a large entrepôt trade with Black Sea and Aegean ports. It has a few light-industrial establishments, but is also Turkey's principal cultural center. The Dardanelles (Canakkale Bogazi), leading from the Aegean into the Sea of Marmara (Marmara Denizi) and the Bosporus (Istanbul Bogazi), connecting the latter with the Black Sea, plus hills to the north and an easily defended peninsula, gave Constantinople outstanding advantages for pro-

Istanbul and the Bosporus. The Dolmabahçe Mosque on the right now houses a marine museum, appropriate to Istanbul's long-standing trade function where Europe and Asia meet. This view is from the European side. *(From Turkish Tourism and Information Office.)*

Baalbek, Lebanon. This small town is located on the rich plain of Beqaa, east of the Lebanon Mountains. This productive plain produces fruit and dairy products, and is extensively irrigated. Roman ruins show the town's history. *(Photo by Owen Franken for Stock, Boston.)*

tection as well as for trade. Land routes also converged there to cross from Europe to Asia, and the shores of the Golden Horn (named for the rich profits in trade and fishing) at Constantinople supported one of the largest and wealthiest trading cities and bazaars in the ancient and medieval world. Despite its loss of capital functions to Ankara in 1923, Istanbul is growing rapidly as Turkey's primate city.

Arab Asia

Syria and Lebanon

The new states formed from the wreck of the Ottoman Empire in Asia have little unity despite their common heritage of Arab occupancy and Turkish rule. Lebanon, with a small Christian majority, has remained distinct from its neighbors. Since at least the time of the Phoenicians, it has had an interest in trade, perhaps the oldest such continuous tradition in the world. The state of Israel is entirely different from its neighbors, an extension of essentially Western-style commercial-industrial development. Saudi Arabia, Yemen, and the countries along the east coast of Arabia are nearly solidly Arab in culture and ethnic composition. To that extent they are distinct from Mesopotamian Iraq and Syria, which have their own ancient, pre-Arab lineages. The state of Jordan, a strip of desert under an Arab king, dependent on heavy British financial subsidy until 1955, is a recent artificial creation that is only now developing a sense of national integration.

Syria and Lebanon speak different dialects of Arabic, and are physically and economically dissimilar. About half of Syria is desert, but there are some areas of forest and pasture in the southern uplands, considerable unirrigated Mediterranean agriculture in the humid coastal zones, and much dry farming in an arc along the east side of the mountains and the northern border. Irrigation is concentrated along the Euphrates in the eastern desert and along a few small streams in the west. A large new dam on the Euphrates, completed in 1975, has greatly extended the irrigated area, and also produces electric power. Most of the present population is agricultural, and most of the agriculture is basically Mediterranean. There are also large areas in the arid interior occupied by nomadic Bedouins.

Traditionally, *Syria* referred to all the territory between the Euphrates and the Levant coast south of the Taurus Mountains (Toros Daglari) of Turkey and north of the Sinai Peninsula. The present boundaries of Syria reflect the greater power of surrounding states and the influence of Western powers. Lebanon was periodically independent of Syria in the past, and is so now. Mountains have served as a refuge for dissident religious groups and buttressed the political independence of Lebanon. The mountains of Lebanon, paralleling the coast, catch considerable rainfall, and are covered with snow for about six months in winter (*Lebanon* originally meant "white"). They are responsible not only for the greater rainfall of the coastal area and its distinction from the interior, but also for the aridity of the interior to some degree. Most of the original tree cover and the famous cedars of Lebanon were cut off long ago. But the slopes of Lebanon are carefully terraced for rain-fed agriculture in many places. These terraces supplement the limited humid area of the coastal plain, where irrigation water from the hills is still extending the cultivated area. The coast itself is still an important trade base, as it was for the Phoenicians. But the old ports of Tyre and Sidon, although still in existence, have been overshadowed by the modern development of Beirut, the Lebanese capital, and Tripoli as important ports. Beirut has recently become additionally important as an air center on the main route between Europe and the Far East. Its principal business is still commercial, port, and financial functions for much of the Middle East as well as for Lebanon. Per capita incomes and literacy are relatively high in Lebanon as compared with other Dry World states, and Western influence is relatively strong. More recently Lebanon has been torn by chronic violence, especially in Beirut. Islamic and Christian groups, each identified with rival political units and policies, have fought for power. Lebanon's traditional cosmopolitanism seems likely to yield to a more communal and doctrinal Islamic order.

The present territory of Syria reaches the coast north of Lebanon but is largely devoid of important ports or good harbors; the old trade center of Antioch (Antakya) is now included within Turkey. The Syrian coast has a Mediterranean climate and supports a fairly dense agricultural population. A high proportion of both Lebanese and Syrians are urban traders in a very old tradition. Damascus, the Syrian capital, occupies an oasis site on the arid, landward side of the mountains of Lebanon. The old caravan routes that led to Damascus have been supplemented by railroads and highways, with a road connection to Baghdad. Aleppo (Halab), somewhat smaller than Damascus, is an even more important route junction. Syria has no oil and no other significant mineral deposits. But the pipelines passing through it from Iraq to the sea, with major terminals and refineries at Tripoli, are in a sense a continuation of Syria's ancient function of transit trade.

Jordan and Iraq

Jordan has existed as a separate state too short a time for its inhabitants to develop a real sense of national identity. Jordan and Iraq have a common frontier, a similar desert base, and a long history of close association under Arab rulers and earlier empires. Jordan has nearly 3 million inhabitants, many of whom left Palestine with the establishment of Israel in 1948. It is a seriously overpopulated desert, a bare land of nomads with a few small oases or sparse steppe regions. Less than 15 percent of its total area is cultivated.

Iraq is underpopulated, or could support many more people. With constructive investment of its large oil revenues and extension of irrigation, it could be made much more productive. The irrigated area of Iraq could be increased three or four times relatively easily, as the existing water in the Tigris and Euphrates rivers is now most inadequately used. Floods could also be controlled better, and the swampy areas covering many square miles, especially near the rivers' common mouth, could be drained and cultivated. Jordan has begun to make some use of the Jordan River by constructing irrigation canals with financial aid from the United States. But the possibilities of agricultural expansion there are limited. Significant mineral deposits, including oil, are almost entirely lacking, except for some mineral salts in the Dead Sea.

Petroleum is the most important product of Iraq, composes 98 percent of its exports, and is found mainly along the Iranian border. The largest fields are in the north around Mosul (Al Mawsil) (near the old Assyrian capital of Nineveh) and Kirkuk (see Map 84, page 356). There are no other minerals of consequence. But the oil deposits are rich, especially in financial terms, for an economically backward country of only some 11.5 million people. The oil is produced and marketed almost entirely by Western interests: British and some American, French, and Dutch companies. They pay royalties to the government of Iraq on what is produced. Iraq is extending and improving irrigation, its agricultural techniques are becoming more efficient and better equipped, and welfare and educational services are being increased. There are still gross inequalities in landholding, and the political system is frequently oppressive and inefficient, as in Saudi Arabia, Iran, and lesser neighboring states. Most of the population remains illiterate, wretchedly poor, and burdened with disease. Most irrigation is still relatively crude, and much of it is merely primitive floodwater irrigation, as first used some eight thousand years ago. There is a long road ahead before Iraq can measure substantial progress by modern standards. The revolution of 1958 and periodic shifts in power groups and policies since then have slowed economic development, which requires a political stability not yet achieved in Iraq.

Many of the people of Iraq are sedentary farmers, working irrigated land in the valley of the two rivers and their joint mouth, the Shatt-al-Arab. Dates, grown along the lower course of the Shatt-al-Arab where the water table is high, are a commercial specialty, and Iraq provides about 80 percent of the world's date exports. Wheat and barley are the most important cultivated crops, and some cotton is also grown. The populated parts of Iraq are limited to the valley of the two rivers (see Map 82, page 338). Most of the rest of the country is desert, very sparsely inhabited by nomadic Bedouins. The extreme northeast receives somewhat more rain than the rest of the country, but there is little field agriculture. Most of the unirrigated land is used for grazing. For the country as a whole, about 12 percent of the land area is cultivated, and it is concentrated in the valley of the two rivers, as in the Nile valley of Egypt. In the Dry World, areas that have water stand out sharply from the rest. Greater Baghdad, the capital of Iraq, has nearly 3 million people, a large city for an underdeveloped country with only some 11.5 million inhabitants. It owes its size to its traditional trade function on the old routes through the Dry World and to the commercial activity created by oil. The Shatt-al-Arab is not navigable for large ocean shipping. River steamers reach Baghdad, and smaller ocean vessels can use Basra (Al-Basrah), Iraq's only seaport, which handles nearly all the exports of dates, wool, and barley, as well as some of the petroleum.

Arabia

Arabia Deserta

The Arabian Peninsula is often divided into *Arabia Deserta,* desert Arabia, and *Arabia Felix,* fortunate Arabia. Arabia Deserta is still very sparsely inhabited, most of it by predominantly nomadic groups of Arabs, loosely governed on an almost tribal basis. Outsiders are virtually unknown over large areas, especially in the erg desert of the Rub' al Khali (Ar Rub' Al Khali) in the south, "the Empty Quarter." Politically, most of Arabia Deserta became the domain of the Arab King Saud and his family, an absolute monarchy in the old style. But the territorial limits of Saudi Arabia are hazy, and in many areas there are no definite boundaries. This has been a source of trouble, especially since the discovery of oil in the 1920s. Nomads commonly disregard political boundaries, but oil concessions must be more specific. The boundaries between Saudi Arabia and Yemen, the People's Democratic Republic of Yemen, Oman, and the United Arab Emirates have never been clearly fixed, and there have been frequent conflicts arising out of the scramble for oil concessions. Oil rights might be granted in territory claimed by someone else or where there was no real sovereignty, and local sheiks frequently defied outside control. Oil concessions in Saudi Arabia are dominantly American; those in Oman are mainly British. Power rivalries have been in the background of many of the boundary and territorial disputes.

The principal seat of Saudi Arabia's government is at the oasis of Riyadh located on an old caravan route across Arabia Deserta between Mecca on the west and the United Arab Emirates on the east. A few years ago Riyadh was little more than a glorified tribal encampment, but it is now a fast-growing town whose modern buildings reflect the oil boom. Mecca is also recognized as the capital, but is far more important as the birthplace of Muhammad, the Prophet. Mecca is the premier holy city of Islam to which every devout Muslim must make a pilgrimage. The city supports itself largely on the pilgrims from all of the Muslim world. The seaport of Jidda (Juddah) nearby serves this function also. Foreign diplomats are stationed at Jidda since they are not permitted to enter Mecca. Despite its name, Arabia Deserta supports more sedentary farmers than nomads. The amount of cultivated land is very small and is concentrated around the coasts (where rainfall is greater) and in a few scattered oases. Sorghum, wheat, and millet are the major crops, but there is very little export trade and most production is subsistence in nature. The government of Saudi Arabia, however, has constructed a small steel plant at Jidda in conjunction with a small oil refinery (to be supplied with crude oil by tanker from Dhahran [Az Zahran]), and a seawater desalinization plant.

Oil has become the heart of the Arabian economy. Indeed it has been said that even the government of Saudi Arabia "floats upon a sea of oil." The main fields are along the shore of the Persian Gulf (see Map 12, pages 56–57). Oil has also been discovered in the political twilight zone between Oman and Saudi Arabia on the south and elsewhere in the interior. The boom town of Dhahran, with facilities largely built by the Americans, is the principal base of operations. Some of the oil is refined there before shipment by tanker to Europe through the Suez Canal or around Africa, or by pipeline through Jordan to Sidon on the Mediterranean coast of Lebanon. Oil produced and refined in the British concession on nearby Bahrain Island or shipped in crude form uses the same routes. But the oil fields in Kuwait (mainly British) and the small adjacent neutral territory depend on tankers.

Arabia Felix

Arabia Felix includes the coastal fringes of the southwest and southeast where mountains catch enough rainfall to support field agriculture or to make irrigation possible from mountain streams and in small oases. The highlands of Yemen, the People's Democratic Republic of Yemen, and Oman receive enough reliable rain, mainly of monsoonal origin, to support some field agriculture without irrigation and to nourish numerous irrigation works. Sedentary agriculture and irrigation are both ancient in these areas. To this extent their culture differs from that of Saudi Arabia, although their populations and language are predominantly Arab. Mocha (Al-Mukha) in southwestern coastal Yemen gave its name to a world-famous coffee, once a regional specialty, but now of minor importance in the face of Brazil's near monopoly. Sorghum, millet, barley, cotton, and coffee are now the principal crops. There is also still much nomadic or seminomadic herding. The British-developed coaling station and entrepôt of Aden still dominates the trade of Arabia Felix, and is slightly larger than Mecca in size. Its excellent natural harbor and its central location along the Arabian Sea coast and opposite east Africa are important advantages. Aden and the area formerly claimed by the British as the Aden Protectorate (largely the Hadhramaut) became independent in 1967 as the People's Republic of Southern Yemen (later the People's Democratic Republic of Yemen). Most of it is desert north of the mountains, and it is almost uninhabited. Its volcanic mountains and desert have been graphically described as "Hell with the fires put out."

Oman has an old tradition of maritime trade between the East and the West, and was probably the earliest base of Arab traders and navigators. Later it was the major base for Arab trade with east Africa. The present inhabitants are highly mixed, and are somewhat distinct from the Arabs of Arabia Deserta, probably because of their long history of interaction with distant peoples and places. Trade is now largely with India, but it is quite small and is conducted largely out of the tiny capital at Muscat. In 1971 the seven small sheikdoms of the Trucial States along the Persian Gulf joined to form the United Arab Emirates with its capital at Abu Dhabi (Abu Zaby). It has become the richest country in the world per capita, based on extensive offshore and inland oil fields.

Israel

The state of Israel was established by a United Nations commission in 1948 after many years of Jewish attempts to achieve a national homeland. It was promptly set upon by its five Arab neighbors (Lebanon, Syria, Jordan, Iraq, and Egypt) who understandably resented the creation of a Jewish state from territory inhabited for centuries by Arabs. Israel was completely successful in repelling this joint invasion and, in fact, was able to secure some additional territory in the south, in the Negev Desert. The Arab states have continued to resent the existence of Israel, so it has grown up in an atmosphere of unbroken tension, repeated violence, and hostility to which there seems no easy or equitable solution.

Israel is a tiny territory. Well over half of it is nearly rainless desert, yet it supports over 3.5 million people (plus about 1 million Arabs, largely in areas occupied by Israel during the four-day war of June, 1967, whose status and future are unclear). Its population is growing rapidly, especially by immigration. Perhaps the most remarkable aspect of its development is the speed and efficiency with which the Israelis have been able to construct a prosperous economy and a strong state out of desert and seacoast, in dramatic contrast to the other Dry World nations. Heavy subsidization has come from Jewish communities in the West, but the Israelis have built their economy out of almost nothing by concerted hard work and careful planning. Cooperative farming communities (kibbutzim) have been organized to cultivate the dry, stony soils and the somewhat better, more humid strip of land along the northern and central coastal plain. Irrigation facilities have been increased many times, and water has been supplied to farms even in the Negev Desert by means of long, surface aqueducts. Millions of trees have been planted to conserve soil and moisture. Israel would like to develop a much larger hydroelectric and irrigation project to use water from the Jordan River. But the river begins in Lebanon and Syria, and flows mainly through Jordan. So far, Syria, Lebanon, and Jordan have refused to agree even to a joint project.

Israel's major commercial crop is oranges and other citrus fruits grown mostly in the coastal strip of Mediterranean climate, and largely exported. The familiar combinations of barley, wheat, olives, and grapes are also important, but food production is not keeping pace with population increases. Physically Israel is a poor land as well as a tiny one. Industrial development has been pursued vigorously with the help of a population that includes a high proportion of skilled workers. Precision industries and crafts (diamond cutting and optical goods, for example) are dominant, and manufacturing as a whole employs many more people than agriculture. But manufacturing that depends on a relatively high proportion of raw materials is at a crippling disadvantage. Except for some oil deposits in the Negev, Israel is devoid of mineral resources for the most part. Diamonds can be imported for cutting, and goods that require mainly skilled labor can be made competitively. The only significant domestic raw materials, apart from oil, are the salt, magnesium, and potash of the Dead Sea where lack of an outlet and high evaporation have concentrated them. Production of other minerals from the Dead Sea is being developed, and there is a growing cement industry. Despite the raw-material problems, a small steel plant, dependent on imported coal and ore, was built at Tel Aviv, mainly for strategic reasons.

Political tension is increased by the plight of the

The challenge of Israel. A kibbutz, or cooperative farm, occupies the lowland on Lake Kinneret. The irrigated land is intensively cultivated growing vegetables, grapes, fruit, and other high-value Mediterranean crops. The bare desert hills rise in the background. *(Photo from Israel Information Service.)*

A kibbutz farm—Israel. Three intensively cultivated crops—vegetables, grapes, and dates—show against the desert from which this now-productive land has been reclaimed. The kibbutz system is run like a work cooperative, and young people make up much of the work force. *(Photo from The Israel Ministry of Tourism.)*

Arab refugees. Over a million people who had lived in Palestine for centuries were uprooted to make way for Israeli immigration. Most of them went to Jordan, which is particularly unable to support them. Others went to the Gaza Strip, a finger of territory along the coast northeast of the Sinai Peninsula, then controlled by Egypt. The rest scattered among Egypt, Lebanon, and Syria. Many of the refugees have not established permanent new homes or are still living in refugee camps. Most hope to return to their former homes, and are supported in this hope by the governments of the Arab states, although a United Nations commission has most of the responsibility for maintaining the refugees. Another problem is the status of Jerusalem, a holy city to Muslims, the ancient capital of Jewry, and a city of strong Christian interest. Jerusalem was inconveniently partitioned between Israel and Jordan until the Israelis occupied it during the four-day Arab-Israeli war of June, 1967. Like the six-day campaign in 1956, the 1967 war ended in a total Israeli victory. The so-called Yom Kippur war of September, 1974, when Egypt and Syria attempted to reclaim their territory lost in 1967, in effect ended in a quick stalemate. Israel's boundaries with its Arab neighbors, including territories occupied for centuries by Syrians or Arabs and seized by Israel as necessary for its own defense, remain in dispute. Instead of trying to settle their own people in these disputed territories, as they did along the west bank of the Jordan River, seized in 1967 from Jordan, it may be wiser in the long run for Israel to return these captured territories in exchange for genuine peace and an end to continued insecurity.

The Israelis have made Jerusalem into their national capital, but the largest Israeli city and the leading industrial center is the combined city and port of Tel Aviv–Jaffa near the coast. Haifa, to the north, is the leading port, and formerly the terminus of one of the pipelines from Iraq. The large oil refinery at Haifa was partly dependent on the flow of Iraqi oil, which was stopped for political reasons in 1948. In 1957 an oil pipeline was completed from the Israeli port of Elat on the Gulf of Aqaba (the eastern inlet of the Red Sea) to take Iranian oil from tankers and pipe it to the Haifa refineries. This pipeline is useful not only to Israel, but as a possible alternative to the Suez Canal. Oil now flows through this pipeline from the smaller oil pools in the Negev.

Israel is a continuing threat to the peace of the traditionally troubled areas around it, and hence to the peace of the world. Great power rivalries have become involved, and the issue has been complicated and worsened by oil politics and strategy. Israel's fight for existence and for economic development commands admiration. But a permanently peaceful solution depends, like so many other things, on the development of strong states, economic health, and political maturity among Israel's neighbors, which could promote a genuinely cooperative and mutually tolerant relationship. As the strongest and most highly developed state, Israel must give the most if such a solution is to be achieved. Israeli military success and power have preserved the state in the short run, but the bitterness of the Arabs continues to increase, threatening peace and even Israel's existence.

Physical distinctions within Israel—the better-watered north, including the Plain of Esdraelon and Samaria between the coast and the Sea of Galilee, and the drier and more rugged south, including the Plateau of Judea—were reflected in the ancient Jewish communities. The Biblical state of Israel centered in the north, in productive Samaria. The southern Kingdom of Judah, with its traditional frontier close to the desert boundary, occupied the more arid plateau of the south. Jerusalem, which may rival Damascus in length of time as a continuously inhabited city, was midway between these two ancient kingdoms, although it is on the Plateau. It was the capital of Judah both before and after it became the capital of a united Jewish state under David and Solomon in the tenth century B.C. Jerusalem was protected by its upland site on a local hill, and was strongly walled. Nazareth (Nazerat) is in the north, above Samaria on the plain of Galilee, but Bethlehem (Bayt Lahm), some 6 miles (9.7 km) south of Jerusalem, is now in the part of Jordan occupied by Israel in 1967. Much of the area of ancient Hebrew settlement is now Jordanian territory occupied by Israel. Much of Israel's territory, including most of the coastal areas, was not in Hebrew hands in Biblical times.

Iran

Iran has remained largely aloof from Arab-Jewish conflicts, not only because it is physically remote but because it is not an Arab state. It coincides largely with the Plateau of Iran but includes narrow strips of lowland along the Shatt-al-Arab, the Persian Gulf, and the southern shore of the Caspian Sea. The Plateau continues eastward into Afghanistan and Baluchistan. Most of the country is desert or semiarid. There is enough rainfall for unirrigated agriculture on a significant scale only on the narrow lowlands of the northwest around the Caspian Sea and in Iranian Azerbaydzhan (see Map 82, page 338). Parts of the southwestern edge of the Plateau near the Iraq border receive 20 to 30 inches (50 to 76 cm) of orographic rainfall, but relief permits little field agriculture. Most of Iran is very sparsely settled by nomadic herders. Not quite half of the population is engaged in subsistence agriculture, especially in the small area of more humid lowlands around the Caspian Sea. The climate and agricultural system there are essentially Mediterranean, with wheat, barley, rice, cotton, olives, tobacco, citrus fruits, and grapes as the main crops. Rainfall decreases eastward; well over half of the country receives less than 5 inches (12.7 cm) on an annual average. Southeastern Iran and Baluchistan are nearly rainless, searingly hot in summer, and almost uninhabited.

A range of high mountains, the Elburz, with peaks above 18 thousand feet (5490 m), runs parallel to the south shore of the Caspian, between 5 and 30 miles (8 and 48 km) back from the sea. Precipitation in the Elburz reaches 30 to 40 inches (76 to 102 cm). The mountains also help to make the narrow and interrupted coastal strip relatively well watered. On the south or inner side of the Elburz, mountain streams make considerable irrigation possible on the fertile plain and support a fairly extensive oasis agriculture. Tehran, the modern capital, lies close to the south side of the Elburz. Although cut off from the Caspian lowlands, it is protected by the mountains against invasions from the north, and has good access to the bulk of the country in the south. It occupies an old crossroads site at a meeting of caravan routes leading to the east and the oasis of Meshed (Mashhad), to the south and the oases of the Isfahan area and the Persian Gulf, and to the west through Baghdad and Anatolia. There is now a rail connection around the eastern end of the Elburz to the Caspian, another northwest to Tabriz and Soviet Azerbaydzhan, and one south to the delta of the Shatt-al-Arab and the Iranian port of Abadan, built largely by the British.

Isfahan on the Plateau of Iran. Water for this ancient oasis town comes through qanats drawing water from the mountains near the western edge of the plateau, and also from wells. The mosque is unmistakably Persian. There are several other beautiful reminders of a rich past in Isfahan. *(Photo from Pan American World Airways.)*

Ancient Persia was centered principally in the south at oases like Isfahan (on a small river) and Persepolis, the capital of the old empire. They were nourished by streams and groundwater derived from the mountains at the western edge of the plateau, and had easier access to the Persian Gulf and to Mesopotamia. Qanats may have been first developed in ancient Persia and are still widely used, especially on or near gentle slopes leading from the rain-fed mountains. Ancient Persepolis was supplied with water mainly by qanats. Modern Tehran, with a population of nearly 3 million, still depends largely on them for its water. Some of them are as long as 45 miles (72 km). They suit the topographic conditions of Iran, as they suit the topographic conditions of Sinkiang. Qanats also restrict evaporation in these hot, dry climates, and for that reason as well they are preferable to canals.

Modern Iran, as compared with the ancient empire, has retreated onto the Plateau to a great extent. Some of the old oases in the south have been abandoned or allowed to deteriorate. The Elburz on the north and the mountains to the west of the Plateau have helped to create the very arid desert of central Iran, southwest of Tehran, which lies in a rain shadow to some degree. Settlement has been confined for the most part to the sides of the mountains on the north and west of the country, including the small strip of Caspian lowland, which benefits from some orographic rainfall.

Iran's modern economic development is dominated by oil. This has resulted in a renewal of Iranian interest in the south. Petroleum now accounts for practically all of Iran's exports by value. The main Persian oil fields follow the western edge of the Zagros Mountains and the Shatt-al-Arab, at the head of the Persian Gulf. The wells and the world's largest refinery at Abadan, plus port facilities, were developed by the British Anglo-Iranian Oil Company. By 1950 Iran was the leading oil producer in the Dry World. Anti-Western agitation and the nationalization of the oil facilities in 1951 led to a prolonged dispute with the erstwhile British owners, and production virtually ceased until a compromise solution was reached in 1954. Kuwait and Saudi Arabia have now overtaken both Iraq and Iran in oil production. Iran suffered from the decrease in oil royalties, but the loss was soon made up. However, income levels are low, as in neighboring Dry World states, and there are still gross inequalities in landholding. Although the government of the Shah has made some efforts at land reform and improvements in agriculture, education, welfare, and industrialization, Iran is still backward in every sphere, including its political system. Hunger, disease, illiteracy, and grinding poverty make it more difficult to institute changes or lift traditional farming methods out of their primitive and underequipped condition. There is very little industrialization and only a small development of textile manufacturing, despite domestic production of raw cotton. Magnificent rugs are still made in Iran, the traditional handicraft production characteristic of the Dry World, using wool from the nomads and artisans from the oases. But the chasm between a few rich and the masses of the poor is perhaps even more pronounced in Iran.

Except for oil, the country is poor in minerals and produces only a small amount of coal from the Elburz. Considerable hydro development is possible, but relatively little has been accomplished. Probably the most obvious and immediate method for economic improvement lies in the extensive possibilities of increasing irrigation, as in Iraq. But capital, direction, planning, techniques, and a favorable social and political atmosphere are all lacking. The Plateau of Iran has kept its cultural continuity and a reasonably homogeneous population, largely avoiding the devastating invasions and conquests that contributed so much to the decline of Mesopotamia. Still modern Iran is a pale reflection of the power and prosperity of ancient Persia. The ruins of Persepolis symbolize the enormous, relative changes that have come about in the Dry World in the last two thousand years.

Afghanistan

Russian Turkestan, Sinkiang, and Mongolia were discussed in previous chapters. Western Afghanistan and Baluchistan (western Pakistan) are physical continuations of the Iranian plateau and desert. Eastern and northern Afghanistan is more mountainous and more humid, including the main ranges of the Hindu Kush and their extensions into the central part of the country. Large areas in the eastern mountain valleys are suitable for agriculture, and there is a good deal of essentially subsistence production of wheat, barley, other cereals, tobacco, and fruits. Many of the Afghans, however, prefer grazing to agriculture. Land that could support crops is often used instead for animals, especially the fat-tailed sheep for which Afghanistan is famous. Kabul, the capital, is a city of about half a million, located in the most important of the eastern mountain valleys. This area is important not only for its agricultural resources but because it commands the route to the Khyber Pass. Despite the mountain passes that connect Afghanistan with the Indus valley and Turkestan, which are mainly responsible for the state's political significance, Afghanistan and its capital are remote and primitive. Politically Afghanistan is a traditional reactionary state, like Saudi Arabia. There are no railroads in the country, and in the more rugged mountain areas the population is still essentially tribal, including the warlike Pathans and Tadjiks. Both groups are mainly nomadic and both spill over the frontiers of Afghanistan, the Tadjiks into the Soviet Union and the Pathans into Pakistan, where they periodically demand independence or local autonomy. Afghanistan was maintained as a buffer state between British India and Russia, a function that would have been hindered, not helped, by the building of railroads. The passes are primarily of political or strategic importance, since there is little trade through or with this largely unproductive and isolated area. Russian and American counterpressures have now largely displaced the earlier British influence; Afghanistan has received technical and economic aid from both powers. There are some little-used deposits of copper, lead, and iron, and a little coal and petroleum. But the major occupation is livestock grazing, and the country exports many skins, hides, and wool, including the famous *Persian lamb* (or karakul) pelts with their tightly curled fleeces.

Afghanistan is another Dry World area where irrigation could be greatly increased and agricultural and other economic opportunities more fully pursued. Outside influences and economic aid are beginning to make some impression there, as in the rest of the Dry World. Although it seems highly unlikely that any of the Dry World will ever regain its former relative position of power and prosperity, the achievement of greater economic health, equality, and political maturity seems certain to be retarded as long as outside pressures continue to take the place of genuine cooperative efforts by the peoples of the Dry World themselves.

Questions for further study and discussion

1. Why should favorable environment and high productivity be accompanied by mass economic misery in Egypt or Bangladesh? Why is population increasing so rapidly in both areas despite poverty and low health standards?

2. Why has Tunisia, occupying the old Cartha-

ginian base potentially commanding the Mediterranean sea lanes, not developed into an important maritime power? What is, or was, the significance of the modern city of Bizerte (Binzert), also near ancient Carthage?

3. Why was Cairo selected as the site for an Egyptian iron and steel plant? Can it be economically competitive with imports of iron and steel?

4. Why did the Suez Canal have over twice the traffic volume of the Panama Canal? Why did the blockage of the canal in 1956 produce an economic crisis in western Europe? What alternate routes are available? What other effects elsewhere resulted from the blockages in 1956 and from 1967 to 1975?

5. What is the significance of the Gulf of Aqaba to Israel and to Jordan? What has been the nature of the disputes over it? What are the functions and nature of the port cities of Elat and Aqaba (Al Aqabah)?

6. Why should there be a relatively high degree of urbanization in Lebanon and Syria when there is such a small amount of industrialization?

20 Sub-Saharan Africa and its modern development

Masai warriors get a new chief. *(Photo from British Information Services.)*

Africa south of the Sahara—often called Black Africa from the darker skin color of its inhabitants—has had a fascinating but paradoxical history. It now seems clear that the earliest evolutionary developments toward the emergence of humans took place there, probably in eastern Africa around what is now Rhodesia (Zimbabwe), Zambia, and Tanzania,[1] beginning perhaps six million years ago. Much later, when mankind was beginning to evolve out of the Neolithic stage, some of the earliest developments of metal working (especially in iron), and animal and perhaps plant domestication seem to have taken place in Africa in some of the same areas. We know unfortunately little of this and subsequent periods of African development, primarily because there was no written language or written records, although there were some beginnings in a few places. With the development of city-based civilizations in Asia and the Mediterranean basin, the chief focus of human development and the growth of more advanced technology shifted away from Africa. Trade links remained between Africa and the Arab area of Asia, but Arab contacts were largely limited to a narrow strip along the east coast. There was also trade across the Sahara to north Africa, and a wide spread of Islam southward into the northern Black African countries. But the rest of the huge continent south of the Sahara remained largely isolated from the outside world until little more than a century ago. The expanding Europeans bypassed it for the most part in their effort to reach Asia. What they saw of the coastal areas did not suggest that Africa offered the trade profits or the rewards of conquest that they saw in India, Indonesia, China, or Japan.

The most highly advanced African cultures and states, in fact, were heavily concentrated in inland areas. The continent south of the Sahara Desert is dominated by a great inland plateau, interspersed with broad basins, that ends relatively abruptly in escarpments (steep cliffs) only a few miles from the seacoast. There is only a narrow coastal plain in most areas, and the cliffs guarding the inland areas are both high and steep. There is also a marked deficiency of adequate natural harbors along most of the coast, except in the extreme south. There are several

[1]The political situation in Sub-Saharan Africa is changing almost daily. Insofar as possible in this book, the geographical names used will be the most recent ones available.

long stretches of hundreds of miles without an acceptable natural shelter for ocean-going ships. Only one of the great African rivers, the Congo (Zaire), has an estuary of deep water, but falls and rapids a few miles upstream greatly decrease its usefulness as a harbor, or as a transport route. Rivers usually provide the easiest links between inland and coastal areas, but in Black Africa most of them have falls or rapids close to the coast where they tumble off the plateau and over the escarpments. Elsewhere, long, offshore reefs block access from the sea to and from the coasts. In many areas the tidal margins are covered with mangrove swamps. Mangrove trees thrive in brackish water and form a dense network of roots above ground that lift the trees above the tidal level, forming an almost impenetrable barrier.

It is hardly surprising that African attention was concentrated on the far more rewarding inland areas. The most impressive developments were centered there, and contact with the coast or with overseas areas was minimal. Nor is it surprising that when the expanding Europeans did try to establish themselves on the coast, they learned little of what the more advanced part of Africa was like, and developed a prejudiced opinion of Africans as "primitive." This was further emphasized by the Europeans' chief business on the coasts—slave raiding and the slave trade. These origins of modern interest in Black Africa have continued to warp European and American attitudes, in addition to the terrible damage inflicted on African societies. Coastal west Africa's population especially was decimated. By the time the slave trade was finally suppressed, about 1840, perhaps as many as 90 million Africans had been enslaved and transported, or killed in the course of this brutal enterprise. There was also a high death rate of the human cargo on the slave ships, where people were chained and packed into miserable holds. The brunt of the slave trade fell on the coastal and near-coastal areas. For the most part they were the least highly developed technologically and the least able to defend themselves against slave raiders armed with guns and a wholly ruthless drive for profits.

At the high tide of the European effort overseas, beginning at the end of the nineteenth century when the West rode the wave of its own industrial success, the European impact was especially powerful in Africa. Indigenous technological and economic development there, once ahead of the rest of the world, had fallen behind. In any case it had not produced the kind of power that could resist European aggression in the age of industrial-based imperialism. European domination of Black Africa lasted less than a century in most areas. But it created a radically new set of circumstances, restored Africa's links with the outside world, and introduced the commercial and industrial techniques that are now being pursued by the independent African states.

In the fifteenth century the Portuguese established a few small bases on the coasts of southern Africa to provision their ships trading with the Orient. British and other European traders did not appear in any numbers until the end of the sixteenth century. Until the latter part of the nineteenth century, they had minimal relations with or knowledge of inland areas, and were confined to a few coastal footholds. European attention focused on west Africa. It was closest to them, and also most productive of the high-value goods they sought. Four commodities dominated the trade they managed during these three centuries of marginal contact: gold, ivory, slaves, and spices (the most important spice was grains of paradise—*Aframomum melegueta*—a drug as well as a spice). Europeans named sections of the west African coast for each: Gold Coast (now Ghana), Ivory Coast, Slave Coast (coastal Nigeria and what is now Benin), and Grain Coast (meaning grains of paradise, coastal Liberia).

All these goods were produced for sale by the local people and had been traded overland by the interior Sudan states with the Moors and Arabs for centuries before the Europeans arrived. This trade was diverted to the coast by the opening of sea routes to Europe (see Chapter 1). The goods were all of high value in proportion to their weight, and could bear the high cost of transport to the coast by human porter and to Europe in the small carriers available. European commercial management or investment was not necessary except at the tiny coastal ports, and the trade was highly profitable. Trade volume, however, was relatively small, and except for the slave trade did not directly affect the economies or cultures of the major inland African areas.

The slave trade did have a profound influence. Slaves were in particular demand in the sugar and tobacco plantations of the West Indies and the southern American states, forming part of a "trade triangle." British, American, and other European

MAP 86
PRECIPITATION IN AFRICA

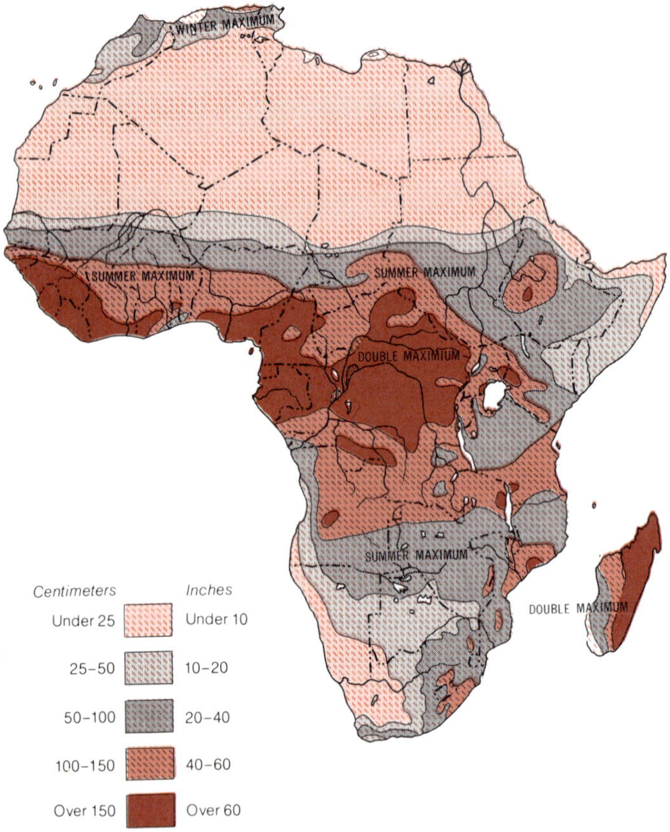

MAP 87
VEGETATION IN AFRICA

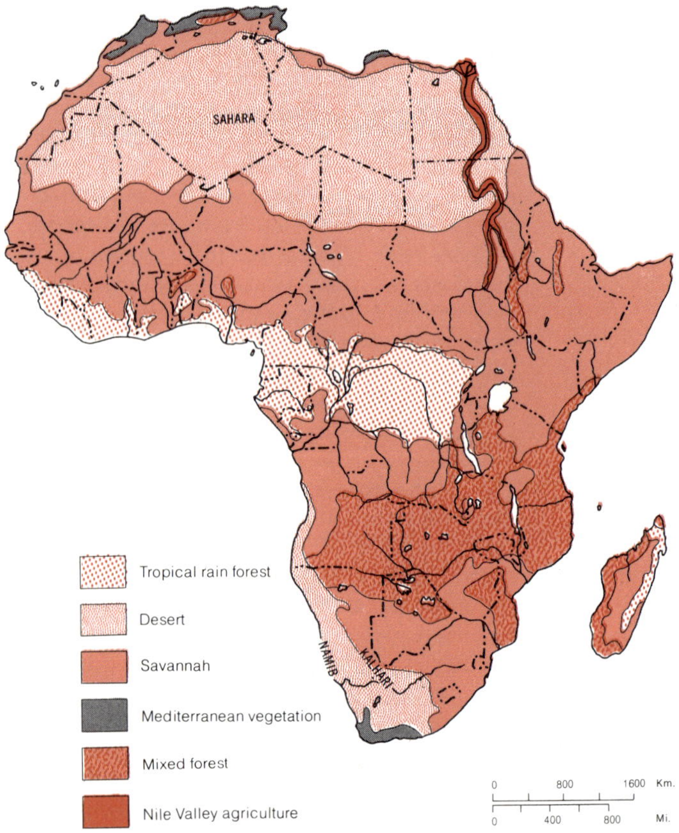

ships sailing to west Africa exchanged trade goods: trinkets, guns, liquor, etc., for slaves. They carried the slaves to the Caribbean or to North American ports, such as Charleston, where they were exchanged for sugar, molasses, rum, tobacco, and cotton. Then these plantation products were sold in England or Europe, in return for money and for more trade goods. New England became involved also, because it provided fish as food for slaves in the West Indies, and many New England ships were engaged in the triangular slave-rum-fish trade. By the end of the eighteenth century, at least a hundred thousand slaves a year were being exported from west Africa, although not that many survived.

In the nineteenth century, slavery and slave trading were generally outlawed. But there was a rise of territorial imperialism, and a great expansion of Western markets with their demand for raw materials and food. Economic motives and desire for prestige and conquest were both involved in the relatively sudden expansion of the original European footholds on the coast about the end of the century and in their efforts to explore, claim, and develop the interior. Sea transport at about the same time became much cheaper, faster, and more efficient, which had the effect of widening the market for African goods. The steamship and railway also increased the possibilities and profitability of trade with the interior

MAP 88
POPULATION OF AFRICA

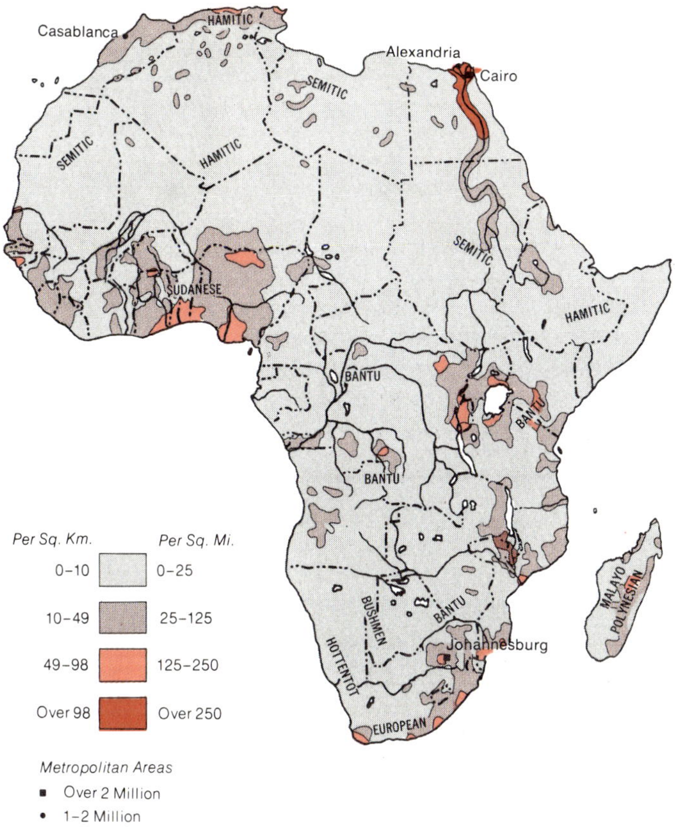

MAP 89
MINERALS AND ECONOMIC PATTERNS IN AFRICA

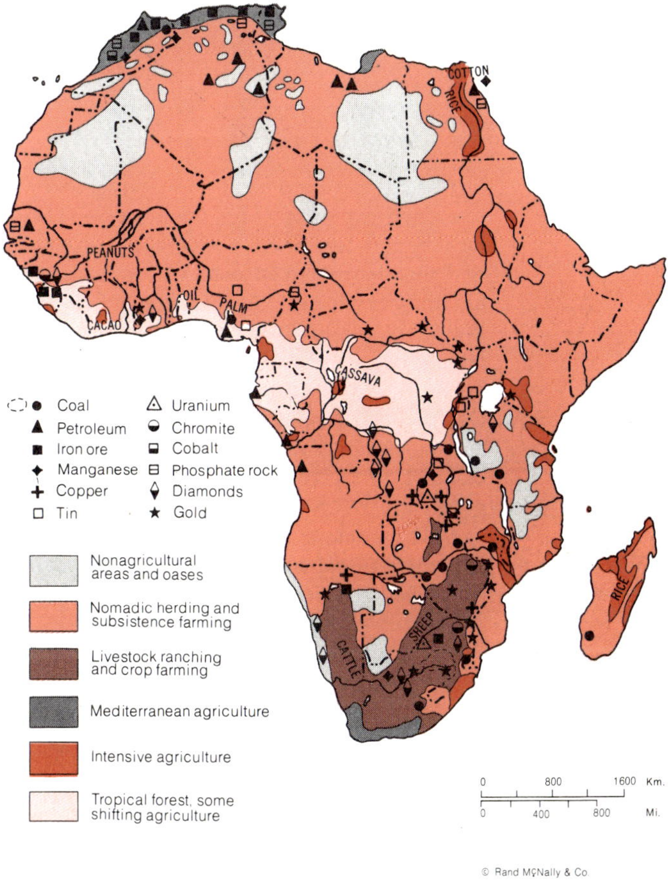

in goods other than the limited high-value commodities of the earlier period. Gold still retains an important place among African exports, but new items have taken the place of the other old staples: palm oil,[2] peanuts, cacao (chocolate), diamonds, copper, tin, cobalt, manganese, uranium, and small but increasing amounts of cotton, sugar, rubber, hides and skins, coffee, and wheat. Production of these goods encouraged capital investment by Westerners, Western techniques, and often Western commercial management. Most represented resources that the Africans had not developed on a commercial basis. West Africa, with its tropical coast and richly varied vegetation, was probably one of several early centers of agricultural development in the Neolithic period, or even before. Subsistence food supplies were greatly increased by the spread of manioc (see Chapter 16), maize, potatoes, and other crops from the Western Hemisphere and the Asian tropics as a result of early European contacts with Africa. But for the spread and efficient production of

[2]Palm oil is pressed from the pith covering (pericarp) of the fruit of the oil palm, a tree of the equatorial rain forest native to west Africa. The oil is used principally in the manufacture of soap, margarine, and tin-plate, and is consumed in large quantities in industrial economies. Palm kernel oil, pressed from the inner kernel of the fruit, is of higher quality.

commercial crops for export, like palm oil or cacao, or for the exploitation of mineral deposits, Western techniques or large-scale commercial organizations were important.

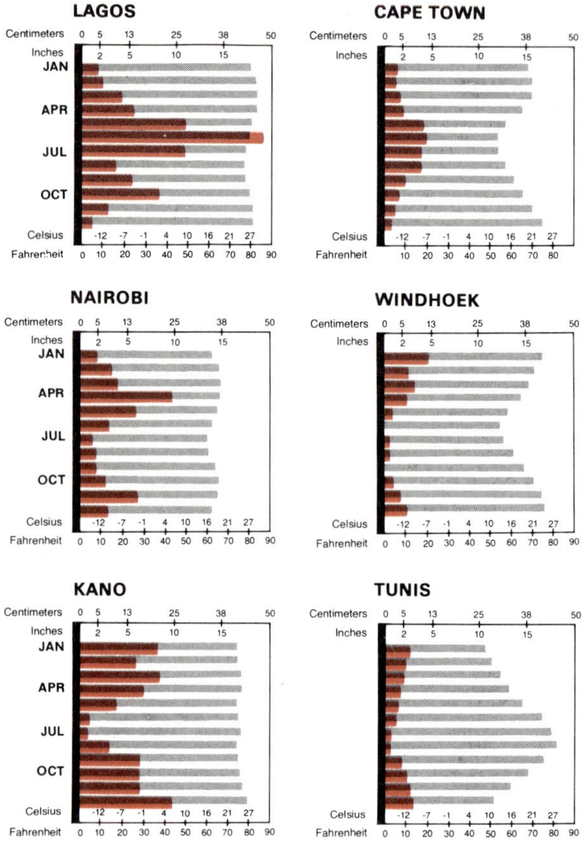

Figure 14. Climographs of representative cities in Africa.

Temperature
Precipitation

Ownership and use of resources

European-managed trade and development in Africa remained exploitative for the most part. Most of the local people live on an essentially subsistence basis, and most commercial production, both agricultural and mineral, is designed for export. Since the Second World War and the independence of the African states, domestic capital and domestic management are replacing foreign capital and management, as well as foreign political control. But domestic capital and technicians will remain scarce for many years; in much of the continent south of the Sahara, Europeans are often still present as technicians, managers, financiers, and exporters. With the rise of African nationalism, such a situation understandably promotes tension and questions about the ultimate ownership of Africa's resources.

African development

Minerals

Black Africa's mineral resources particularly lend themselves to a drainage type of export economy, because the great demand for them is in the industrial West. The states of central and southern Africa derive most of their income from minerals. Mining is an important part of the economy in Ghana and Nigeria (including tin and oil in the Niger delta). Coastal and offshore oil production is also important in Angola and Cabinda. Such dependence on mineral exports is rare outside Africa, except for the oil-rich areas of the Dry World or Venezuela. Very little coal has been found in Black Africa, except in the Republic of South Africa and Rhodesia (called Zimbabwe by its African inhabitants who are in the vast majority). There are modest deposits of iron ore in Sierra Leone, Liberia, Guinea, and South Africa, and few important chemical deposits. Among other things, this means that there may be few strong prospects for the development of substantial competitive heavy industry, and that the export of raw materials may continue.

There are already the beginnings of domestic industrialization in many African countries, and ambitious plans in virtually all of them. But potential hydro power, of which Africa is estimated to have about 40 percent of the world total, cannot make up for the deficiencies in coal and petroleum, and the absence of economic factors favorable to the development of water power. The great bulk of the potential power sites, and also of the uranium deposits, are concentrated in central Africa where the market is least developed so far. Uranium ore is even more easily and cheaply transported than oil, considering the amount required for a given output of energy. The presence of uranium deposits is even less likely than hydro sites or petroleum pools to attract manufacturing in the absence of a large local market or other necessary economic factors. The uranium ore is being and will be mined for export. Most of the hydro sites are still undeveloped, since power in that form cannot be exported to distant markets.

Black Africa leads the world by a large margin in the production of several precious or relatively rare minerals, most of them exposed along the escarpments of the plateau. Gold, diamonds, and platinum are among them; all have industrial as well as ornamental uses. Africa produces about 98 percent of the world total of diamonds by weight. Most of them (over four-fifths) are used for cutting tools and abrasives, and only a small number for gems. For a long time African gold production exceeded that of all the rest of the world. Now it is a little less than 40 percent of the world total. Africa also accounts for large proportions of the world production of cobalt, chromite, manganese, uranium, copper, tin, and asbestos. Most of these minerals are useful or in demand on a large scale only in industrial economies. The first important uranium mine was in central Africa, and for over twenty years it remained the dominant source. Production of uranium has now greatly increased in Europe, Russia, North America, and elsewhere. Although details are withheld for security reasons, Africa probably still accounts for about a third of world uranium output.

Heavy dependence on mineral exports is not usually a sound basis for an economy. Mineral deposits become exhausted (several of the African copper deposits are already worked out, and the cost of gold extraction is rising sharply). Also, general local economic development aside from minerals may be limited, and there may be too few economic alternatives when the export market for minerals declines. Many other mining areas elsewhere have been abandoned or have deteriorated economically because no other adequate local means of support had been built up. Ideally, profits from mineral exports should be invested in the development of other, more lasting resources: water power, agriculture, transport, and manufacturing. This is now beginning to happen in several of the African states.

Agriculture

Commercial agricultural production is still much smaller than total mineral production by value. Most agriculture and livestock raising are still subsistence in nature. Crop farming is hampered by tropical soils of low fertility or by rainfall that is often yearly or seasonally inadequate. There are large areas still unused, much of the Sudan, for example. They could be made productive as population continues to increase, but many of them are marginal for profitable or prosperous agriculture. There are also large areas where overuse or misuse of the land has led to serious erosion or soil exhaustion. As in most of the rest of the world, increases in production must come primarily from increased output from the present cultivated land, through irrigation, fertilizers, and improved crop strains, rather than from new land. Tropical Africa is not an unlimited storehouse of potential agricultural resources; in fact those resources are severely limited. Black Africa is not yet overpopulated on the scale of monsoon Asia or Egypt, but its population is growing very rapidly. It has probably doubled in the past century to reach the present figure of over 300 million (see Table 3, page 397).

As in most areas of the world, the total cultivated land is greatly exceeded by land that is suitable only for grazing animals because of aridity, relief, or poor soils. Large areas in the hot, wet, tropical parts of Black Africa, however, are infested with the tsetse fly (see Map 90, page 385). It still rules out most grazing animals, and further contributes to economic poverty, although some progress is now being made in eradicating this pest. Much of the grazing area free from the tsetse supports only poor pasturage. Grass is scanty and of low quality in the semiarid regions north and south of the equator. The savanna grasslands that border the wet equatorial belt, where the rainfall is concentrated in the high-sun period, are often characterized by tall, rank, harsh growth unfavorable for cattle. In the dry season of the year savanna grasses dry up, and through most of the year, high temperatures present a further obstacle to raising good-quality cattle. Wherever possible, cat-

African savanna near Mount Kilimanjaro. Much of the savanna grassland has drought-tolerant low shrubs and trees, since there is a definite dry season. Grass-eating animals, such as the zebra in this picture, live in this area. *(Photo from ROLOC.)*

tle are kept in large numbers, but often as a mark of wealth and prestige rather than for economic reasons. There is some commercial grazing of the European-settled highlands in southern Africa, and traditional, productive use of cattle by a few east African groups. Elsewhere cattle are kept almost as if they were money. This has tended to restrict production and led to overgrazing of inadequate ranges by thin, bony animals that contribute little economic support to the human population. Like the rest of the agricultural system, animal husbandry is now being improved and developed as a more important source of food.

European investment and connection with overseas markets have been responsible for large-scale development of plantations and commercial crops for export, especially cacao and palm oil, and some rubber, cotton, sugar, wheat, coffee, and tobacco. But such development has been in relatively small islands of commercial production in the sea of subsistence agriculture. The highlands of southern Africa, occupied in part by Europeans and devoted to an essentially European system of commercial agriculture and grazing, support a much larger population, but are a relatively small proportion of the total land area.

In summary, African resources under European exploitation have not been developed by or for Africans. Although some minerals are plentiful, they have not greatly benefited the African economy. Average income levels are very low, industrial-commercial techniques are not yet highly developed in African hands, and the economy is still predominantly of a subsistence nature. There is much potential for future economic development, spurred by the enthusiasm of new nationalism. But it will require careful planning and wise, stable leadership.

Environment, location, and African cultures

Long before the arrival of the Europeans, African social and political organizations were advanced in many inland areas. Especially in the Sudan and the savanna country of the north bordering the Sahara Desert and in adjacent parts of western Africa, there were several advanced African tribes that had an extensive metal technology, such as the Hausa of northern Nigeria. These groups had many opportunities for local and more distant cultural and economic exchange, as in the case of Timbuktu (see Chapter 1). They also occupied a physical base with relatively fertile soils, a more permissive climate, and local mineral deposits. Their location helped them develop a more productive economy and more advanced technology than was possible in the coastal areas or the equatorial forests farther south. As suggested before, some eastern and southern African areas that were similarly open zones of movement and high productivity, continued their pioneering development of iron technology, and also evolved well-organized political states.

Elsewhere in Black Africa (excluding Ethiopia), there was no fully developed system of writing, and even the wheel was not known until it was introduced by the Europeans. The low productivity of wet, tropical soils and the absence of grazing and draft animals in the tsetse country were serious disadvantages. The shortage of native animals suitable for domestication, by comparison with Eurasia or southeast Asia, was also a serious handicap, as it was in the Western Hemisphere. Agriculture was commonly supplemented by or even subsidiary to the less productive occupations of hunting and gathering.

In Black Africa the large remaining stands of natural vegetation provide a clue to climate, and, indirectly, to soil fertility. But natural vegetation is also an important factor in itself, as it may restrict movement and exchange (see Map 87, page 380). The barrier character of the rain forest needs no emphasis. Within the large areas of heavy year-round rains and constant high temperatures, soils are badly leached. Only on the savanna and steppe margins, where six months or more may be relatively dry, are there reasonably fertile soils. But agriculture there is limited by water deficiencies. Some of the Africans themselves traditionally divided their country into two kinds of landscapes: the big bush (rain forest) and the small bush (savanna). This division recognized not only the obvious difference in vegetation but the complex of environmental elements and human adjustments. Tribes of the big bush were fundamentally different in their culture and economy from the tribes of the small bush. In the small bush, particularly toward the north in the wide zone of transitional steppe between the moister savanna and the desert, large political kingdoms developed. They were able to organize considerable areas, to build cities, and to maintain both local and extraregional trade, as well as to support relatively advanced technologies and productive economies. Toward the desert and on the drier margins of the steppe, pastoral nomadism replaced agriculture. In this part of the Dry World, small nomadic empires flourished periodically.

MAP 90
TSETSE FLY INFESTED AREAS

The transitional steppe and its moister southern margins in Africa are called the *Sudan*, an area stretching in a wide band south of the desert from the Atlantic to the highlands of Ethiopia, not limited to the modern political state of Sudan. *Sudan* means literally "land of the blacks," and its inhabitants have always been dominantly Negroid. Since at least ancient Egyptian times, the Sudan has been repeatedly influenced by non-Negroid peoples from the north, and is much more mixed, in physical population types as well as culture, than the lands of the big bush farther south. Little is known of outside influences that may have affected the big bush or the savanna country south of the equator. Outsiders were barely aware of their existence. The Africans said that "the big bush is stronger than the cannon of the white man," as indeed it was until very recently. The rain-forest barrier did preserve a refuge area for some African groups, but the effects of isolation were equally important in technical and economic retardation.

European settlers

Permanent European settlement has influenced Black Africa in only a few areas. Before the nineteenth century the only such settlement was a small Dutch colony south of the tropics at Cape Town where the Dutch East India Company had established a port and way station on the route to the East. Dutch agricultural settlers, the Boers (*Boer* means "farmer"), gradually made their way inland and by the nineteenth century had largely abandoned their original maritime commercial functions. Britain acquired the Cape Colony as a result of the Napoleonic wars, and British settlement grew. It was not until the Boer War of 1899–1902 that British control was established over the whole of what is now the Republic of South Africa. People of Dutch descent are still in the majority among the European population, but Africans outnumber Europeans in South Africa by more than four to one. There are also nearly 1 million Indians there who occupy much the same position in trade and money lending as the Chinese in southeast Asia, and who are similarly resented, both by the Europeans and by the Africans. The *Cape Colored,* descendents of African-European (and sometimes Afro-Asian) intermarriages, total about 2 million in South Africa.

Nearly all of the Europeans who are permanent residents of Africa south of the Sahara live in South Africa. The very small European settlements in other

Tropical rain forest. The high canopy of leaves from the tallest trees restricts the sunlight that reaches the forest floor. Rain forests grow throughout the year because there are no marked seasons, so their barrier character changes little. *(From Photo Congopresse.)*

African areas have a highly uncertain future, such as those in the former British colonies of Kenya and Rhodesia, and the ex-Portuguese colonies of Angola and Mozambique. Europeans in Kenya and Rhodesia (Zimbabwe) are a tiny, fast-shrinking minority as African control is rapidly replacing white political and economic domination. The Portuguese traditionally intermarried widely with the local people in all of their colonial territories, so it is not easy to distinguish European settlers in Angola and Mozambique, but permanent residents of dominantly European descent are in a tiny minority. In the rest of Black Africa Europeans are present only in very small numbers, predominantly as temporary residents for administration, trade, and professional services.

Most of the areas of European settlement are concentrated in the highlands. The lowland part of Kenya was left largely to the Africans, and there is almost no lowland in south Africa outside of Mozambique. The central plateau in tropical Kenya and the more extensive, relatively level uplands of southern Africa known as the *veld*[3] offered opportunities for developing the kind of economy that was attractive to European settlement. That the upland climate was physically more pleasant for the Europeans than the climate of the lowlands was less important than the economic opportunities of the uplands. Commercial agriculture and stock raising were economically possible, requiring or profitably employing large numbers of Europeans. An exclusively plantation type of labor is not suited to the operation of large, commercial farms. Although a great deal of African labor is used, there is also a profitable place for a great many Europeans, both as managers and workers. In South Africa rich mineral deposits in the veld also attracted European settlement and helped to support it. Large cities of the European type have developed to serve a commercial, agricultural, mining, and manufacturing economy. All are numerically dominated by Africans, although not as much as in the lowlands. For the Europeans, these upland areas are their home where nearly all of them were born.

Racial conflict

The conflict between black and white, which has become such a tragic issue in the areas where European settlers live, is not easy to resolve. These parts of Africa are in a dangerously unfortunate and unique position. In most of the Western Hemisphere and in Australia–New Zealand, Europeans were able to eliminate or absorb the native inhabitants. Even the relatively highly developed American Indian groups, such as the Mayas or the Incas, were largely overcome and their traditional cultures destroyed by the Europeans. The cultures of North and South America and of Australia and New Zealand are now dominantly European in character. This resulted from conquest, intermarriage, and imposition of commercial-industrial techniques on a relatively small and technically less developed group of local people (if they were not wiped out). In Asia, the only other area where Europeans have gone in significant numbers, the local cultures and states were at least as highly developed as in Europe, and the inhabitants were far more numerous. Despite the powerful influences exerted by the West, Asia was never overwhelmed. It has retained its cultural identity, regained its political independence, and has never supported any significant permanent settlement by Europeans.

In Black Africa the highland settlements of Europeans have not been and will not be able to eliminate the local people or their culture. They are increasingly outnumbered by a more rapidly growing African population. The Europeans in the Republic of South Africa, for example, are determined to remain there. Because of their originally superior technology and economic and political strength, they have been able to dominate the local people even though they could not eliminate them. The Africans, in the end, were militarily unable to resist European conquest, but not so weak that they could be wiped out or absorbed as the American Indians were. Except for the Portuguese and the special situation of the Cape Colored, the Europeans have generally not intermarried with the Africans, so the cultural and physical line between black and white is sharply drawn. Thus, while the whites prevailed in the Americas, and in Asia were largely kept out, in the highlands of South Africa they are attempting to

---

[3]*Veld,* strictly speaking, describes vegetation. But in common usage it has come to refer to the level or rolling uplands at different elevations in South Africa specifically. The word is of Boer origin, and may be compared, for example, with *prairie, savanna,* or *steppe,* also vegetational (or climatic) terms that have acquired a landforms sense as well, especially in specific regional settings.

maintain themselves in control of, or at least in the midst of, the African people. But the African people grow more numerous and (in a European sense) more competent every day, and they increasingly resent white domination. Technology, skills, ambitions, and the attraction of a higher standard of living cannot be arbitrarily confined to the Europeans, however hard the Europeans may try to do so.

Africans everywhere now demand either the control of their own countries, or at least a representative share of the control and the rewards of the economy. In the Republic of South Africa the response has been the establishment of a number of *Bantustans* or *homelands*[4] to which Africans now living in white areas will theoretically be repatriated, and where they will have some autonomy. Opportunities for education and economic advancement for Africans are extremely limited. They are denied the vote, and are excluded from the white universities and from jobs with better pay and prestige. If the present policies of the white minority continue, it is hard to see how ultimate mass violence can be avoided, since improvement in African welfare can come most effectively through free access to the education, privileges, and modernization now denied the Africans. Even the small amount of opportunities now open to them may lead in time to some modification of present policy, if only to avoid violent and disastrous confrontation. Any realistic look into the future poses questions that are not easy for the white South Africans to answer.

In the Dry World and monsoon Asia, colonialism is largely gone. Control by outside powers in both areas had the effect of stimulating economic development, spreading cultural and technical influences, and promoting the growth of more effective local states. But, especially in South Africa, the cultural transfer and interaction taking place, despite white efforts to restrict them, can lead only to an intensified conflict. The large African majority rejects the alien control of a white minority who are not stewards or traders in the traditional colonial pattern, but would-be absolute rulers.

[4]Many Africans and many observers see the Bantustans as merely labor camps where Bantus are separated from white residence areas but used to provide most of the manual labor the economy requires (see the further discussion in Chapter 21).

Colonialism in Black Africa

The policies of the Belgian, French, and Portuguese colonial governments in Black Africa have differed from those of the British. British policy encouraged the political development of the Africans aimed toward granting self-government and independence. It frequently attempted to avoid changing those elements of African culture that did not retard responsible political development and also tried as much as possible to leave local administration in the hands of the local people (*indirect rule*). British control was regarded as a check against abuses like slavery or intertribal wars and against the domination or exploitation of minority groups. It was also the most effective channel toward self-government and self-determination by the Africans. French policy, by contrast, long tended toward the assimilation of colonial areas into a wider French Union. They hoped local cultures would give way to French culture, and ultimately local people would become full French citizens. This objective was sometimes epitomized as transforming other peoples into "little brown Frenchmen." But such a policy could not be permanently successful, especially when Africans in neighboring British areas were achieving full national independence and cultural self-determination. After the Second World War, French policy moved rapidly toward independence for its African territories. Its goal of assimilation became instead one of a looser association with France and the rest of *France Outremer* (France Overseas).

In the Congo Free State (now Zaire), Belgium pursued an essentially paternalistic policy, similar in some ways to that of France, but much less effective and often repressive. Political development was not encouraged. Instead, the emphasis was placed on encouraging Africans to participate in the commercial enterprises of the Europeans, partly by offering them a limited amount of technical training, education, medical care, and acquaintance with European methods. These French and Belgian policies worked reasonably well until the late 1950s. But, as African nationalism spread, France and Belgium were forced to grant independence to their colonies. The Portuguese in their colonies gave little attention to political, economic, or social development of their African subjects, and concentrated on a policy of divide and

Africans on the move—bus stop in northern Nigeria. This truck is bringing yams to a market town, but it carries more passengers than yams, at least by weight. This is a characteristic sight on African roads. *(From Information Service, Department of Nigeria.)*

rule. Their policy has also been characterized as one of neglect, but African nationalism grew rapidly. Mozambique, Angola, and Guinea-Bissau became fully independent in 1975.

Disruption of African cultures

Whatever the policies pursued by the colonial powers, the presence of Europeans in Africa and the spread of commercial activity have produced revolutionary effects on African society. Perhaps the most important is the progressive modification and reform of the traditional tribal culture. These changes did not occur by European design but by the more effective impact commercialization makes on any society. It frequently takes time and some fundamental readjustment of values to persuade the Africans to work for money wages and to seek the essentially Western rewards that money can bring. But there are relatively few who have been able to hold completely aloof from involvement in the European's commercial economy. In west Africa this involvement has been more nearly one of free African choice, and Africans are themselves responsible for the bulk of commercial production. But all over Black Africa the local people have been drawn on for labor in the plantations, mines, farms, and cities originally dominated by Europeans. The commercial economy could not function without African labor, and the Africans are becoming dependent on the employment provided by the European-style commercial system. The present system requires adult males to leave their families in the tribal areas, sometimes for a few months, sometimes for two or three years, sometimes permanently. They work in some commercial or industrial enterprise that may be a hundred or more miles away. Many of the companies whose activities are located in originally sparsely peopled areas are obliged to do very long-distance recruiting, and to provide the African workers with housing where they work. Wages, conditions of work, housing, and other amenities vary greatly from region to region. Some are good, some are bad, but all are part of a system that is basically disrupting African family life and traditional African cultures. Sometimes whole families or tribes move more or less permanently to *locations* (areas of housing for African workers), or to shanty towns in the slums of cities or mining centers. Under these circumstances, too, traditional African culture is rapidly changing.

Africans have become highly mobile, traveling by bus, train, boat, and on foot back and forth between tribal reserves or local villages and the plantations, mines, farms, cities, and towns where they find work. Ties are maintained with the tribal homeland wherever possible, and there is a great deal of labor turnover and absenteeism as a result.[5] Sometimes it seems that all of Africa is continually on the move. Cultural stability is difficult to achieve under these circumstances, and new or revolutionary ideas circulate and grow. African cultural change, including the development of new leadership and political parties, is almost entirely concentrated in the cities. There interaction with the outside world is strongest and commercialization farthest advanced. The role of the city in Europe, Asia, and the Americas has been essentially the same.

To what extent and in what ways African culture will be changed is still to be seen. What is clear is the flexibility and resilience of African cultures under the impact of new pressures and circumstances. The Africans have proved able to adapt dynamically and selectively to the powerful forces at work in their land during the past century, and especially in the years since independence. Some of the independent states and some sections of African society have responded more effectively than others. In many there are still incompletely solved problems of political stability, rational planning for development, and adequate guarantees of civil liberties. In other African states these problems seem to be under better control. They are making encouraging strides in improving agriculture, extending industrialization, providing health and educational facilities, and in lessening dependence on mineral exploitation or export trade in favor of more self-sufficient and communal development. Nearly all of Black Africa is in the midst of an exciting ferment of change, experimentation, and the challenges of independent growth. After millennia of relatively isolated development, this original cradle of mankind is now involved once more with the rest of the world, but increasingly on its own terms. Africans have reclaimed control of their own destinies and are planning their own patterns of modernization. Their solutions to the problems that confront them will be African solutions, not merely adaptations of European models.

[5]In South Africa this pattern is reinforced by the Bantustan system.

Questions for further study and discussion

1. Does the presence of rich uranium deposits in central Africa offer any advantage for the ultimate local development of manufacturing? Why or why not?

2. What are the reasons for the banding of climate and vegetation zones in Africa north and south of the equator?

3. Why is ivory no longer an important African export? What accounts for the decline in the importance of African spice exports?

4. What factors have been responsible for the increase in Black Africa's population during the past century? Is the population likely to grow equally rapidly in the future? Why or why not?

5. Lenin argued that "imperialism is the last gasp of capitalism." How does this assertion fit the historical and present situations, and the prospects of colonialism in Africa?

21 African regions

Tanzanian farmers harvesting millet. *(Photo from the Peace Corps.)*

Black Africa is commonly divided into four primary regions, each distinctive on a large scale, and each covering an area nearly as great as the United States (see Map 91, page 392). The boundaries between them are arbitrary to some extent, but should be thought of as relatively wide transition zones, as with boundaries between most regions anywhere. Despite the broad generalizations of the last chapter, Africa is not alike everywhere, as should be expected in a continental area of this size. This chapter outlines some of the more important aspects of regional difference.

West Africa

West Africa was the base of the most highly developed traditional societies of Black Africa. Rain forest occurs only in the coastal zone (see Map 87, page 380). There is a larger area of coastal plain and relatively level interior than in any of the other major regions of Black Africa. These physical advantages were further strengthened by west Africa's relative nearness and accessibility to the cultures of north Africa and Europe. Moors, Arabs, and others could move more easily across the Sahara and into the Sudan of west Africa than into any other area south of the desert. Elsewhere Nile swamps, rain forest, or mountains barred the way. The Sudan itself was an easy highway of east-west as well as north-south movement. Relatively fertile savanna and steppe soils, and easy movement and exchange characterize much of the interior. Large political kingdoms developed in the savanna and steppe, and there was considerable internal trade. Agriculture was relatively productive, and technology was far more advanced than in most of the rest of Black Africa. Some of the premodern cities of west Africa were very large. In addition to Timbuktu (see Chapter 1), Kano in northern Nigeria (now over 350 thousand population) may have had nearly half a million inhabitants in the eighteenth century. It and other west African cities were walled and were both trade and administrative centers. But much of their population may have been transient or may have been engaged in agriculture in surrounding areas to which people commuted from the cities, as they still do to a large extent.

Internal regional contrasts

Urbanization and the economic development, exchange, and political organization on which it de-

MAP 91
REGIONS OF BLACK AFRICA

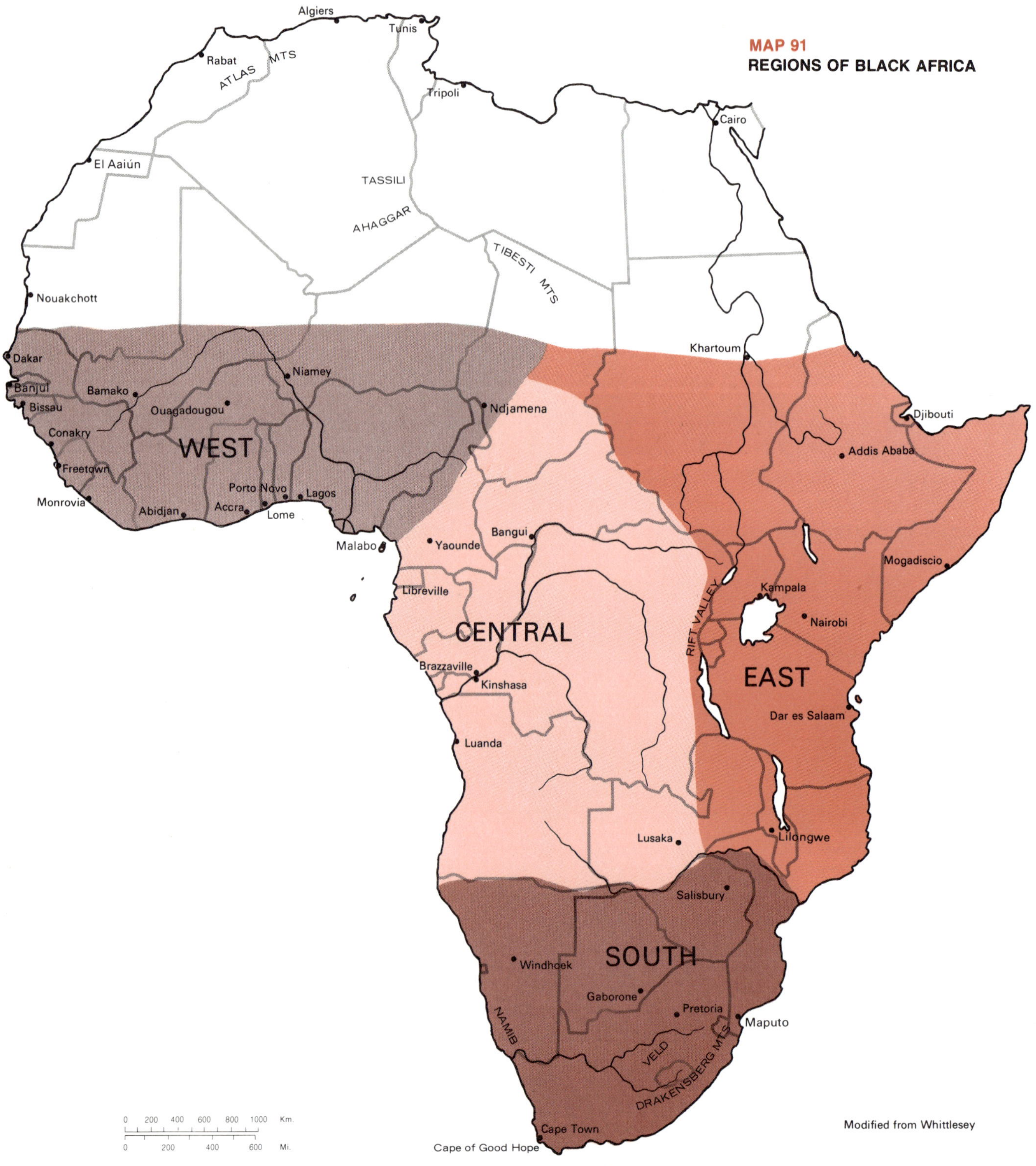

Modified from Whittlesey

pended were concentrated in the open savanna country of the north. The Yoruba and other relatively highly developed groups in southern and central Nigeria also built cities, some of which (Ibadan, Ife, Benin City) were also large. The maritime trade has now brought an even greater development in the coastal area, despite its original rain-forest cover. In Nigeria, for example, Lagos, the seaport capital, and Ibadan, a short distance inland from the coast, are now both much larger than Kano, and are better served by railways. As in much of monsoon Asia, an original landward or internal orientation has been replaced in west Africa by an orientation to the sea. This resulted from the new stimulus to trade and commercialization, especially of the areas most accessible to the sea. The same pattern is observable in Ghana, Liberia, Sierra Leone, and the scattered, but vast, formerly French colonial areas in west Africa. The longest coast of west Africa trends east and west. Thus a line drawn inland from the coast passes successively through rain forest, savanna, steppe, and desert at increasing distances from the equator. This regular banding of climates and attendant vegetation and soil types is still accompanied by pronounced regional differences in culture. Among other things, it helps to account for the bewildering variety of groups and languages within most of the west African states. Nigeria alone, with a population of about 65.6 million, has over two hundred different languages or distinct dialects. Even the small state of Ghana (with a population of about 10 million) has over thirty-five languages in an area a little smaller than Oregon. In many west African areas, people speak several different languages as a matter of necessity, now that originally separate groups interact more closely in the new national states. As elsewhere in Africa, there is also wide use of a *lingua franca*, a "common" commercial language, often English (as in India) or French.

It was in west Africa that political independence from European control came first. Several centuries of outside influence and increasing cultural diffusion contributed further to the political and technical development of the local people. The British colony and protectorate of Gold Coast became the independent state of Ghana in 1957, taking the name of an old Sudan empire that flourished in the northwestern part of the area from about A.D. 300 to 1076. Nigeria, largely self-governing since the early 1950s, became independent in 1960 as a federal republic. French West Africa and Equatorial Africa by 1960 became a group of autonomous republics.[1] Liberia, a rather special case, was founded as a homeland for freed American Negro slaves in 1822 and was declared independent in 1847. But Sierra Leone, founded by the British as a settlement for freed slaves in 1799, remained a British colony until 1961, largely because it was considered necessary to safeguard the rights of the original inhabitants. The Liberian record suggests that some such outside supervision was appropriate.

Especially in Ghana and Nigeria, which extend far inland from the coast, the regional differences between the northern savanna-steppe and the coastal rain forest with their attendant cultural distinctions are now sources of internal conflict. The now relatively backward northern territories feel dominated or exploited by the coastal areas. Nigeria recognized these differences by adopting a federal form of administration, granting considerable local autonomy, and making federal funds available for the assistance and development of its economically more backward northern areas. Under colonial rule the northern territories of Ghana and Nigeria were administered as protectorates, separate from the small, commercially important coastal areas which formed the *colony*. The northern areas were better organized politically and militarily, and were able to resist European attempts at control until about the end of the last century.

After a century or more of European colonialism, the coastal or southern areas had become more highly developed economically and more sophisticated politically. Cities in the south were cosmopolitan centers of trade with the outside world, as opposed to the local market centers of the north. The proportion of the total population living in cities (the degree of urbanization) was much higher in the south; literacy rates and incomes were also well above those in the north. Large areas in the north are still occupied by seminomadic groups or subsistence farmer-pastoralists, and Arab and Muslim cultural influences are very strong. These contrasts between

[1]Former French West Africa is now the autonomous republics of Guinea, Ivory Coast, Togo, Benin, Mauritania, Niger, Upper Volta, Senegal, and Mali. Former French Equatorial Africa has become the autonomous republics of Congo, Central African Empire, Chad, Cameroon, and Gabon.

Kano in northern Nigeria. Kano and some similar cities of the open Sudan may have had a semipermanent population as large as half a million in the eighteenth century. The city still has a distinctively African appearance, and shows a great deal of Dry World influence. *(From British Information Services.)*

north and south delayed the coming of independence. Indeed many northern groups opposed it in the fear that a national state would be dominated by the more advanced south. Where independent Nigeria adjusted to the problem by federalism, Ghana imposed strong central control. The federal system in Nigeria has since been questioned by sectional and tribal political struggles and power shifts. In an atmosphere of contention and reprisals, the Ibo group felt increasingly threatened, and in 1967 broke away from the federation as the independent state of Biafra, incorporating the Ibo homeland along the coast east of the Niger delta. The secession was contested by the federal government, and a long war with great loss of life was finally ended by an agreement that provided for increased autonomy for the Eastern Region within the Nigerian federation. But south and north remain different worlds.

Nation-states in the modern sense cannot be created overnight. They imply not only a unified central government with effective administration of a precisely defined area, but also the development of a truly national and relatively uniform culture. Almost every state grows as one dominant group or core area imposes its control and its culture over a wider and originally culturally diverse area. This has happened in practically all of the European states, in China, and in Japan. It is progressing rapidly in the rest of Asia and in the Dry World. More recently it has been completed or is in progress in the Western Hemisphere. It has only begun to happen in most of Africa. Nigeria, for example, might be compared with Great Britain or France at the end of the Roman occupation. They were beginning the long process by which the people and culture of the English lowlands or the Paris basin spread their political control, language, and economic organization over Scotland and Wales, or over the Midi and the whole of France. It will take time before Nigeria, Senegal, Ghana, Guinea, Cameroon, or other independent African states become nations in this sense.

Commercial development

The economic development stimulated by the Europeans was concentrated in the coastal areas not only because of easy access but also because the goods the European traders sought were mainly produced in or near the rain forest. Actually, the Niger River and its

Cacao grove—Sierra Leone. Cacao is grown under the shade and protection of larger trees, duplicating natural rain-forest conditions. This also reduces the risk of wind damage to the heavy pods attached directly to the trunk and limbs of the tree. These men are cutting ripe pods. *(From British Information Services.)*

major tributary, the Benue, and the Volta River in Ghana to a smaller extent, cut through the rain forest and the Sudan. The Niger reaches north in a great arc into the desert. But its navigability is severely limited by silting and shifting shallow channels even after the serious navigational problems of the delta are passed. The Europeans did not make extensive use of these natural avenues of inland penetration, or even the railroad lines that now carry most of the traffic, until relatively recently. Commercialization is still heavily concentrated in the rain-forest region of the coastal plain. Gold, ivory, and spices occurred there originally, and the commodities that began to replace them toward the end of the nineteenth century were also produced for the most part in or on the margins of the rain forest: cacao, palm oil, rubber, industrial diamonds, manganese (Ghana), and iron ore (Sierra Leone and Liberia). Cotton and peanuts, two other recently developed and important commercial products, come largely from the savanna and steppe country, but move through the rain forest to the coast for export.

Nigeria and Ghana together produce over half of the world's cacao (the source of chocolate). For Ghana alone cacao dominates commercial agriculture and the country's exports, accounting for about a third of world production. Nigeria alone accounts for about one-fourth of the world's production of palm kernels and palm oil, and also grows some peanuts (used mainly for their oil). Cacao, rubber, and oil palms are plants of the tropical rain forest and cannot be grown profitably elsewhere. The heavy pods of the cacao beans, growing directly out from the trunks and limbs of the trees, are easily damaged or broken off by wind before they are ripe. The crop is grown only in sheltered spots (often within a forest of other, larger trees) in the zone of equatorial calms or doldrums (see picture on page 395). Most of the iron ore, gold, and manganese deposits of west Africa happen to be located within the rain forest or on its margins. Tin, coming mainly from northern Nigeria, is beyond the rain forest, but not very far away. Significant petroleum resources have been found along the Nigerian coast near the Niger delta, which will further accentuate the commercial importance of this area.

Plantation production of tree crops in west Africa is less important than small-scale production by local farmers, who sell their crop to trading companies or to processing plants. European capital and management are largely limited to the export trade and to mining. Since the beginning of the modern commercial era in west Africa, the Africans have managed most of the actual agricultural production and have continued to perform most of the middleman functions. Domestic market systems were well developed in west Africa before the arrival of the Europeans. Commercial and subsistence production is frequently mixed. Individual farmers may have a small grove of cacao or oil palm trees or a field of peanuts or cotton and also a patch of maize (in the more humid areas), millet (in the drier parts), yams, manioc, beans, or other food crops for their own consumption. There is a small and growing development of cacao, oil palm, and rubber plantations, often under European ownership, and a very large American investment in rubber plantations in Liberia by the Firestone Company. But generally west Africa is responsible for most of its own economy, as it is now responsible for its own government. The difficult highland areas of Cameroon, southeast of Nigeria, and Equatorial Guinea are culturally more isolated and economically much less developed.

Port and city development

With the rapid increase in overseas trade, especially in the past twenty or thirty years, efforts have been made to improve the harbors along the west African coast. There are now several mainly artificial harbors between Dakar and Lagos where, before the Second World War, there were none equipped to handle ocean-going freighters on a stretch of coast almost as long as the whole eastern seaboard of the United States. Dakar, capital of Senegal, has an excellent natural harbor, continually being improved, which still supports the biggest west African port in terms of trade volume. Dakar's hinterland is enormous in size, even if not very productive per square mile. Banjul, capital of the tiny river state of Gambia, completely surrounded by Senegal, has the advantage of an estuary harbor at the mouth of the Gambia River. It has been deepened and provided with better port facilities. Freetown, the Sierra Leone capital, was originally located on a river-lagoon harbor that

The harbor at Lagos, Nigeria. The traditional fish traps in the foreground contrast with the many modern buildings and ocean-going ships. The buildings show the long and strong British influence. *(Photo by Marc and Evelyne Bernheim for Woodfin Camp & Associates.)*

has been dredged to take bigger ships and is better equipped for cargo handling. An almost wholly artificial harbor has been built at Monrovia, capital of Liberia, which was formerly dependent on surf boats.

Accra, the Ghana capital, has improved its originally small and shallow harbor to some extent, but much of Ghana's port business is being transferred to a large, artificial harbor at Tema. A direct rail line has been built linking the two ports with inland Kumasi, and with the twin port cities of Sekondi-Takoradi. Accra has the advantage of occupying a tiny piece of arid climate, almost desert, on the coast, related to a cold, offshore current. The local absence of rain forest and its attendant problems was a strong attraction for early European settlement at Accra. It still gives advantages to the city and its environs. Like Lagos and all of the other west African capital cities or centers of colonial administration, Accra was largely European in origin. It remained the chief focus of the European influences that have shaped the modern commercial and political development of west Africa. The chief seat of local power in all of the west African countries was traditionally in the interior. Kumasi in Ghana, former capital of the Ashanti Federation, is now much smaller than Accra, although it does lie close to the center of the major cacao-producing region. It has declined relatively, as Kano has. In both countries, as in most of southeast Asia, the modern capital has been placed in the European-founded city that has now become the chief focus of the national life. These modern capitals not only dominate in economic terms but also in terms of the national orientation toward the markets and cultural patterns of the West. In Nigeria the lagoon harbor of Lagos, second only to Dakar in the volume of its trade, has been greatly improved. Extensive work has also been done at Port Harcourt in the Niger delta, so it is now accessible to much ocean shipping.

Resource development

As harbor improvement proceeds and this primary obstacle to trade is removed, the volume and variety of trade increase. In turn these increases promote further improvements in transportation, stimulate economic development in west Africa, and widen the range of commercial production. West Africa is rela-

Table 3. Countries of Black Africa and their approximate population*

Angola	6.2
Benin	3.2
Botswana	0.7
Burundi	3.9
Cameroon	6.6
Central African Empire	1.8
Congo	1.4
Equatorial Guinea	0.3
Gabon	0.6
Gambia	0.5
Ghana	10.3
Guinea	4.6
Guinea-Bissau	0.5
Ivory Coast	7.0
Kenya	14.1
Lesotho	1.1
Liberia	1.6
Malawi	5.2
Mozambique	9.3
Nigeria	65.6
Rhodesia (Zimbabwe)	6.6
Rwanda	4.4
Senegal	5.1
Sierra Leone	2.8
South Africa, Republic of	26.5
South West Africa (Namibia)	0.7
Swaziland	0.5
Tanzania	15.8
Togo	2.3
Uganda	12.1
Upper Volta	6.3
Zaire	26.0
Zambia	5.1
Black African areas of	
Chad	0.7
Ethiopia	23.5
Mali	5.3
Mauritania	1.0
Niger	2.6
Sudan	8.6
Total	300.4

*Estimates (in millions) are based on the latest available statistics. Reasonably accurate census counts are available for only a few of the largest countries. For many areas, the estimates are subject to considerable error.

MAP 92
CENTRAL AND SOUTH AFRICA

tively densely populated, more so than any of the other primary regions of Black Africa. The threat of overpopulation in terms of resource use is becoming urgent. Various schemes have been proposed for increasing the effective use of resources, and several are being energetically pursued by the governments of Ghana, Guinea, Nigeria, and Senegal. Probably the most impressive is the project for producing hydro power from the Volta River near Tema in Ghana. Foreign loans have been obtained for a large-scale development designed in part to produce power for the reduction of aluminum ore for export. Ghana has rich deposits of bauxite that have not been used significantly so far. The Volta project is seen as the beginning of an ambitious industrialization program. Economic progress in west Africa, however, will depend just as importantly on increases in agricultural production and improvements in agricultural efficiency, both for export and for domestic consumption. In several crops, as well as in raw minerals, west Africa has a strong commercial advantage. Increased production for the domestic market can also be obtained by extending irrigation in the north, by clearing new land, and especially by the improvement of farming methods. This may be a more practical solution than the glamorous but less solid prospects of heavy industrialization.

Central Africa

The People's Republic of Congo, Zaire, Angola, and Zambia occupy the center of Black Africa. The basin of the Congo River is the principal focus of the region. It straddles the equator and includes tributaries that drain all of Zaire and Congo and the moister, northern parts of Angola. The central Congo Basin, close to the equator, has a hot, wet climate throughout the year (see Map 86, page 380), and is covered with dense rain forest. But north and south of the equator total rainfall decreases, and becomes seasonal and more concentrated in the high-sun period. Big bush gives way to small bush, and finally to steppe and desert. The savanna north of the equator has its maximum rainfall while the southern savanna is having its dry season, and vice versa. So the northern and southern tributaries of the Congo complement each other in their high-water periods and help to keep the flow of the main stream relatively constant. Most other African rivers, such as the Niger or the Zambezi, fluctuate widely from season to season. The deep Rift Valley on the east, occupied by a broken chain of great lakes, and the high mountains that parallel it, mark a clear transition between this and the separate region of east Africa. The political boundary between Zaire and Zambia follows the divide between Congo River drainage and Zambezi drainage. But in terms of rainfall, vegetation, and African occupance, Zambia is more similar regionally to central Africa than to south Africa.

The Congo and its basin

The Congo itself remains the principal means of access for most of central Africa. The river is deep and broad from Kinshasa, the capital of Zaire, to Kisangani. But below Kinshasa and above Kisangani, navigation is interrupted by a series of falls and rapids that make the river unusable for shipping. Matadi, the chief port for ocean traffic, just above the head of the estuary and below the falls, is hampered by a shortage of level ground for port facilities and by strong currents and whirlpools in the river downstream. Even below the falls the river flows very rapidly down a relatively steep gradient to the sea. A railway through difficult hill country around the falls links the port of Matadi with Kinshasa. Another railway takes traffic around the falls and rapids above Kisangani. Two more rail portages are necessary around river obstructions to reach the mineral-rich upland area of Shaba (Katanga) near the headwaters of the Congo in the southeastern part of the country (see Map 92, page 398). The Ubangi River, flowing into the Congo from the north, and the Kasai, coming from the south, are also both major navigable streams that greatly increase Congo drainage and the area accessible by river transport.

The Congo River system is partly responsible for the development of commercial plantations in the rain-forest heart of the basin, since river transport provides access to distant markets for the otherwise nearly impenetrable equatorial region. The largest plantation crop of the Congo Basin is palm oil, with a growing development in rubber. Plantations have been established particularly along the main river and along the Kasai. Each plantation has its own dock, with the oil palms or rubber trees planted close to the river. Zaire has never been more than thinly settled, and recruitment of an African labor force is often difficult. This is one reason for the wide range of inducements offered, including company housing of a comparatively high standard, health and medical services, and some degree of education and technical training.

However, as in so many other areas where Europeans attempted to establish commercial enterprises requiring local labor, it has not always been easy to convince the African people that better housing, higher incomes, or other amenities valued by Westerners are desirable. The difference in values between two totally different cultures may be symbolized by the following interchange: The African asks why he should work as a paid laborer, and the company representative replies, "Why, so you may earn money and buy things." "But I do not want money and there is nothing I want to buy." "Ah, but when you have money, then you discover all kinds of things you want to buy." The answer, of course, is true, if depressing. Many societies with a subsistence economy, especially where population pressure is relatively low (the south Pacific islands or parts of Latin America, for example), exhibit attitudes similar to the African's. A common device to recruit labor is imposition of a money tax on the population as a whole. This may be a reasonable return for the various kinds of government services provided, but it also has the effect of requiring the local people to earn money, which frequently is not possible under subsistence conditions. To earn money it is necessary either to work as a paid employee of a commercial enterprise or the government, or to produce commercial goods for sale. Both solutions serve the purposes of development, and also hasten the process of commercial economic growth that all African nations want.

Away from the river in the zone of dense rain forest, population is even sparser. The more remote sections are occupied by aboriginal tribes, including the pygmies, who live mainly by hunting and gathering. But most of the rain forest is thinly settled by shifting cultivators who practice the same kind of land use as the shifting cultivators of southeast Asia described in Chapter 16. About half of the Africans in the Congo Basin still live in tribal villages. Some of the shifting cultivators do produce small amounts of cash crops, including a little cotton and palm oil or palm kernels gathered wild. There are very few roads. The few cash crops are commonly carried by porter along the narrow forest trails to the nearest stream, or to a crude road where a motor truck operated for a trading company may pick up the goods. Heavy rainfall makes road building and maintenance difficult, and large investment in roads is not reasonable in the absence of extensive commercial opportunities. There is only a small commercial production of tropical hardwoods, despite the rich forest stands, since access to market is extremely expensive. The rain forest itself and the characteristics of tropical hardwoods are in part responsible for the high costs of transport. The interruptions to navigation near the mouth of the Congo further increase shipment costs.

Outside penetration of the Congo Basin was made later than in any other part of Black Africa, and dates only from about the First World War on any significant scale. Falls and rapids on the lower Congo, the barrier character of the big bush, and the apparently limited commercial opportunities of the area as a whole retarded outside interest and exploitation. A Belgian trading firm did become interested in the Zaire region in the latter part of the nineteenth century, strongly influenced by the Belgian king, who personally owned the area for several years. A small but profitable trade in the few products of the rain forest, especially ivory and rubber, was built up. But ruthless methods of labor recruitment stained the Belgian record and attracted an international investigation. As Africa was successively divided up among the colonial powers, the Belgian state annexed the Zaire region in 1908. They held it until 1960, when it became independent.

Development of Shaba

The mineral deposits of highland Shaba, incompletely known before the First World War, transformed the economy of the Congo Basin and made it one of the most valuable colonial territories in the modern world. The Congo-Zambezi watershed contains some of the richest copper deposits on earth. The same general area also yields large amounts of cobalt, uranium, zinc, lead, and vanadium. Most of these deposits continue under the border into Zambia and support an extensive mining industry there as well. Gold is also mined in the northeastern part of Zaire, tin in the east, and industrial diamonds in the southwest. Copper is the most important mineral product of Zaire, both in volume and in value. (It may have been exceeded by uranium for which no accu-

Roan Antelope copper mine—Zambia. The "location," or area of housing for African workers, occupies the foreground. This is the usual pattern at large mines or factories in Africa. The "copper belt" of central Africa runs under the boundary between Zambia and Zaire, with the largest share of mines in Shaba on the Zaire side. *(From British Information Services.)*

rate figures are available.) In boom towns like Lubumbashi and Kambove in Zaire, or Ndola and Kabwe in Zambia, giant ore smelters dominate the urban landscape. Now there is housing in modern quarters for African workers where even in 1920 there was almost no permanent settlement (see the photograph on page 401). Shaba was originally very thinly settled, and labor has had to be recruited, often from great distances. Another problem is fuel for the reduction of copper and other ores before shipment. Copper especially loses weight in the manufacturing process, and even the rich ores of Shaba are only about 6 percent pure copper, so some concentration is essential at or near the mines. The nearest usable coal deposit is 400 miles (640 km) away at Wankie in Rhodesia (just south of Victoria Falls). The cost and delays of rail shipment have promoted a large development of hydro power in Shaba, making use of streams that drop down from this highland area into the Congo Basin.

The greatest problem for the economic development of Shaba, however, is the difficulty of access. No other part of Black Africa is so difficult to reach from the sea, and the markets for its minerals are almost entirely overseas. The upper tributaries of the Congo drain Shaba, but they are not navigable above Bukama, well over 150 miles (240 km) short of the main mineralized area. The main stream below Bukama is interrupted by falls and rapids in two stretches before it finally tumbles off the plateau to the sea below Kinshasa. Short rail portages were built around the falls on the upper and middle Congo just before the First World War. They give Shaba a transport connection by rail from the Zambian border to Bukama. Goods then move by river, with two rail portages from Bukama to Kinshasa, and again by rail to Matadi, Zaire's seaport. The great distances, slow schedules, and several expensive transshipments of this route prompted the building of a rail line from Bukama direct to Ilebo on the Kasai, from which shipments move by river to Kinshasa (see Map 92, page 398). But the shortest and quickest route to the sea from Shaba is the rail line through Angola to the port of Lobito, where there is an excellent harbor that contrasts with the confined and inadequate facilities at Matadi. This route also has the advantage of leading to the west coast. Since nearly all of the mineral exports are destined for Europe and the Americas, shipment from west coast ports is much more efficient. The Lobito rail line has carried a large share of Shaba's export trade since the early 1940s, especially as port congestion at Matadi has grown. An earlier rail route leads south from Shaba through Rhodesia to the ports of Beira and Maputo (Lourenço Marques) in Mozambique. Despite the longer route and the terminals on the east coast, this line connects the ore region with its source of coal at Wankie and with the mining and agricultural economy of Rhodesia. In 1975 Chinese engineers and workers completed a new line, the Zamtam Railway, from Dar es Salaam to the Zambian copper belt and connecting with Shaba, as an aid project to the governments of Zambia and Tanzania. It provides Zambia with its own rail outlet, but also serves Shaba as an additional rail link.

Thus, the magnet of Shaba minerals and the adjacent Zambian copper belt have drawn railroads from several ports on both coasts, like the spokes of a wheel (see Map 92, page 398). International rivalries have also contributed to the building of these rail lines. Paradoxically Shaba, barely known to outsiders fifty years ago, has become one of the most highly developed areas in Black Africa. Its cool, upland climate has also made it attractive as a resort center. Increased access by rail lines, and more recently by airlines, has fostered this development, and the area supports the somewhat strange combination of mining and smelting and resort establishments, including numerous Western-style hotels with swimming pools and sports facilities.

Angola and north-central Africa

The upland southern and extreme eastern parts of the Congo Basin and most of Zambia have a climate that is fundamentally affected by altitude. Much of Angola and north-central Africa is seasonally or continuously hot, lowland steppe or desert. The coastal zone of central and southern Angola is particularly dry as a result of the cold Benguela Current, which flows northward just off its shores, and causes frequent fogs. But it is difficult for a marine air mass that has already been cooled by the Benguela Current to be cooled further when it strikes the land, where temperatures are usually a good deal higher. Angola's population is slightly over 6 million, but less than 1 percent of the country as a whole is cultivated. Corn and cotton are grown mainly in the coastal areas. There is a growing development of commercial coffee production in the north, and coffee is now an important export. Angola also exports diamonds, which made it profitable to the Portuguese. Major oil resources have been found along the northern coast and in the Cabinda area. Oil now dominates Angola's exports, attracting outside capital as well as less welcome political pressures from the West.

The People's Republic of Congo, the Central African Empire, and Gabon occupy part of the Congo Basin rain-forest country, with a narrow strip of Zaire territory through this area to reach the sea. These sections of the rain forest are physically like the Zaire sections, but have not been developed economically and are still thinly settled. Brazzaville, just across the Congo River from Kinshasa, is the capital of Congo, and much smaller than Kinshasa. Its administrative hinterland is enormous in area, but sparsely populated, and it has little commercial production or trade. The state of Gabon has extensive forests of commercially valuable timber, an important resource. The Central African Empire includes large tsetse-free upland areas where cattle can be raised. But the whole of north-central Africa has less than 12 million people. Brazzaville does have the advantage of a rail link with the port of Pointe-Noire, and a location that can serve trade on the Congo and the Ubangi rivers. But there are no other rail lines and few roads in the huge remaining area. Kinshasa's situation and hinterland are much more favorable.

The port of Pointe-Noire handles most of the little external trade from north-central Africa, including small exports of cotton, rubber, peanuts and peanut oil, tropical hardwoods, palm oil, and other minor forest products. There is a small and increasing development of cotton and peanut growing in the savanna and steppe areas and of palm oil and rubber plantations in the rain-forest areas. The African rain forest, on the whole, is physically as well suited to rubber as southeast Asia or the Amazon basin, but about 90 percent of the world's rubber still comes from southeast Asia. Southeast Asia's critical advantages are access from its rubber-producing areas to overseas markets and a large, competent labor force. They contrast sharply with the problems of plantation production and marketing in central Africa. North of the Congo Basin and southward in Angola, the climate becomes progressively drier until the margins of the Sahara and the Kalahari Desert, in southern Africa, are reached. Neither of these steppe and desert areas ever supported societies that were technically or politically as advanced as those of the west African savanna and the Sudan, for which isolation is to blame in large part. The relatively low level of local development, the aridity of the climate, the absence of large or important mineral deposits (except for some diamonds), and the exploitative attitude of the Portuguese all contributed to the unimpressive political and economic growth. By comparison, Zaire has almost 26 million inhabitants and a large foreign trade with a surplus of exports.

Most of the commercial economy of the Congo Basin has been dominated by large corporations licensed by the government. Under colonial control, technical training was offered to the Africans. Although this furthered the European commercial interest and supplied better labor, it also was a means whereby Africans could increase their standard of living and find employment well above the level of manual work. Europeans still fill many technical posts, but their presence is regarded as a temporary and mutually convenient expedient.

South Africa

The southern tip of Africa reaches into the middle latitudes and includes climates that are not only physically pleasant for Europeans, but offer them attractive economic opportunities. The climate of the tropical parts of south Africa is moderated by the generally high plateau elevation. Roughly speaking, mild climate covers Africa south of the Zambezi River and east of the desert areas of southwest Africa, except for the lowland plain of Mozambique. The southwest desert includes the mandated but disputed territories of South West Africa (Namibia) and Botswana, the former protectorate of Bechuanaland, an area reserved for Africans and independent of the British Colonial Office since 1966. The former British colony of Rhodesia borders and is physically similar to the Republic of South Africa. This country is called Zimbabwe by its African inhabitants, who are demanding control over the government since they form 96 percent of the population.

By every standard of measurement, the south is the most highly developed part of Black Africa. It includes the largest cities and busiest ports, the only real network of railways, the only large-scale development of manufacturing, and the largest sector of commercial agriculture. The coast has an excellent natural harbor in spacious Table Bay at Cape Town, and there are fewer barriers hampering access to the interior than anywhere else in Black Africa. The stiff climb up the plateau escarpment has not greatly handicapped the building of railways. Several port and inland cities have grown impressively large and very similar to European cities. Johannesburg, the metropolis of South Africa, has many modern skyscrapers that give it an American look (see the picture on page 404). In all of these cities Africans outnumber Europeans; Africans compose over three-fourths of the total population in South Africa. About a third of the Africans still occupy the tribal lands left to them as reserves by the Europeans, generally in the lower, drier, hotter, or more rugged areas. About one-third live and work on European farms as permanent employees or migratory laborers. The remaining third are employed in European mines and in a variety of manual jobs in the cities, including domestic service. In the tribal areas a small number of Africans are engaged in skilled crafts, but most are subsistence agriculturists and pastoralists. A very few Africans have made places for themselves in business and the professions in African communities. Otherwise Africans are barred both by pressure and by explicit law from most high-status jobs. The general refusal by the Europeans in South Africa to do manual work is in part responsible for this policy, but it also rests on the determination of the whites to "keep the African in his place." Thus economic development of Europeans and artificially restricted development for the Africans coexist uneasily, at least for the present.

Minerals and mining

Mining dominates the commercial economy of South Africa and Rhodesia. Commercial agriculture is not far behind, and is much greater than in any of the other three regions of Black Africa. As with central Africa, most of the valuable minerals are exported, although coal and iron are largely consumed domestically. Gold is the leading mineral produced, measured both by value and by labor employed. It occurs in a number of areas in South Africa and Rhodesia, but much of the output still comes from the province of Transvaal where an underground reef more than 70 miles (112 km) long occurs, known as the Witwatersrand. Johannesburg has grown up near the center of the reef, and a string of satellite mining towns stretches to the east and west. Johannesburg is the leading rail center of the whole of Africa. The Witwatersrand district contains about a sixth of the European population of South Africa. Very deep mining is now necessary to reach the gold ores, and the workable grades of ore available at depths that do not impose prohibitive costs are approaching exhaustion in the foreseeable future. Proportionately more gold now comes from mines elsewhere in the Republic,

Johannesburg. Skyscrapers like those in North America break the veld horizon, symbols of the revolutionary changes and concurrent problems associated with European settlement in southern Africa. In the foreground are tailings from the Rand gold mines, which were the principal factor in the growth of the city. *(Photo by Marc and Evelyne Bernheim for Woodfin Camp & Associates.)*

especially in the Orange Free State to the south. On the Witwatersrand itself old gold mine tailings have proved to be workable sources of uranium. South Africa is also the world's leading producer of platinum.

Johannesburg and its satellite towns have not grown entirely on the basis of gold, although gold first prompted the railway building and urban development. The city is conveniently central to what has become one of the most productive agricultural areas of the high veld, the district known as the "maize triangle." There is also a large coalfield about 80 miles (128 km) northeast of Johannesburg at Witbank, and the coal deposits of Natal province to the southeast are not far away. A deposit of good-quality iron ore lies about 100 miles (160 km) north of Johannesburg. Fortunately these industrial minerals are located close to the center of the largest market in South Africa, an area already well served by the rail lines that focus on Johannesburg (see Map 92, page 398). Pretoria, 40 miles (64 km) north of Johannesburg, is South Africa's administrative capital (the legislative capital is at Cape Town). It is also the site of a major iron and steel works, since it is closer than Johannesburg to both the iron and the coal sources. But steel is also made at Johannesburg and in some of the smaller centers of the Witwatersrand. Including its suburbs, Johannesburg has a population of about 2.5 million, and its rapid growth seems unlikely to be slowed much if and when the local gold ores are exhausted. For the economy of South Africa as a whole, however, gold has remained a basic source of wealth. Agriculture, grazing, manufacturing, transport, and other mining all profit from public funds derived from gold mining and export.

Diamonds were also partly responsible for the rush of European claims and development toward the end of the nineteenth century in what is now South Africa. They were mined at first largely from a huge open pit at Kimberley, roughly midway between Cape Town and Johannesburg. Surface or near surface deposits there are long since exhausted, and mining has had to go deep underground. Production of diamonds from alluvial deposits in the Transvaal (the northernmost province; the name means "across the Vaal" [river]) and the Cape Province has become increasingly important, but both central and west Africa have now outproduced the south for some years.

More significant in the long run than gold or diamonds are South Africa's sizable reserves of coal and iron ore. Coal reserves are especially plentiful; the Republic ranks sixth among the countries of the world, although its rank in terms of coal production is lower. In the almost complete absence of petroleum and natural gas deposits, and with the relatively small hydro potential in this generally dry country, coal wealth is particularly valuable. The largest deposits are near Newcastle (hence its name) in Natal province, at Witbank in the Transvaal, and in the Orange Free State. Iron-ore reserves are smaller, but are adequate for the relatively modest industrial structure of South Africa and for its immediately forseeable development. Industrial growth will remain limited by the small size of the domestic market. South Africa is too remote from other markets elsewhere to compete significantly in the export of manufactured goods. There is a fair-sized deposit of high-grade iron ore about 75 miles (120 km) northwest of Pretoria, and much larger deposits of medium quality in the Transvaal that have not been exploited to any significant extent. The coal of Natal is very cheaply extracted and is of good quality as steam coal. Since the First World War it has been exported in large volume through the port of Durban, which was used as a bunkering port for coal-burning ships until the almost universal conversion to oil about the time of the Second World War.

The Boers

The development of mining is very recent. It was of no consequence at all until the discovery of diamonds in the Orange Free State in 1867, and of gold in the Transvaal in 1886. Mining of coal, iron, and lesser minerals came even later. Agriculture and stock raising by Europeans have a much longer history. The station to supply food to ships on the route to the East Indies, established by the Dutch East India Company at Cape Town in 1652, was responsible for the first European agricultural venture in Black Africa. By the end of the seventeenth century Dutch agricultural settlers had spread nearly 100 miles (160 km) from Cape Town, and were providing a surplus of wheat and wine for export to Java. The Boers killed or drove away most of the sparse local population of Hottentots and Bushmen, and enslaved the survivors. More slaves were brought in from the East Indies, India, and from west and east Africa to supplement the small numbers available locally. Thus the groundwork was laid for the regrettable tradition in south Africa that the white man is exempt from manual labor and is free to oblige dark-skinned men to work for him. The Boers had a religious conviction that men are not created equal, and this belief has been tragically perpetuated in modern South Africa. Permanent settlement by British groups began during the Napoleonic wars, when Britain acquired the Cape Colony. British-Boer tension was greatly increased by the British decision in 1833 to abolish slavery throughout their empire. The Boer response was a mass migration out of the Cape Colony (although it involved only about one-fourth of the total Boer population). They conquered and settled lands in the interior and on the east coast, where they could be free to practice their own way of life. The "Great Trek," as the Boers called it, began in 1836 and led to the founding of the independent republics of Natal, the Orange Free State, and the Transvaal. Natal was absorbed by the British in 1845, but the other two republics remained independent of British control until the Boer War (1899–1902), which was started by the rush of British prospectors and investors into the diamond and gold areas.

There is a narrow zone of Mediterranean climate with winter rainfall around Cape Town and eastward along the coast. Inland the climate is generally too dry for agriculture until the higher land of the veld is reached. Most of Natal lies in the zone of the trade winds blowing off the Indian Ocean and bringing adequate and reliable rainfall, averaging over 40 inches (102 cm) annually. Southward, westward, and northward from Natal rainfall decreases. But it is adequate for cereal crops over most of the veld. In the drier sections there is commercial grazing (see Maps 86 and 89, pages 380 and 381). During the eighteenth century Boer pioneers had explored some of the veld and Natal. They saw an opportunity there to establish a prosperous agricultural commonwealth. Their northward migration from the Cape Colony in the Great Trek met with vigorous resistance from Bantu groups, who had themselves moved into the veld from the north only recently. But the guns of the Boers ultimately prevailed against Zulu and Matabele spears, and these Africans, too, were wiped out, driven off, or enslaved. They and other relatively well-organized tribes had developed small territorial states, although these were largely destroyed by the Boers, and survive now only in the weakened form of African reserve areas.

The Boers did not enjoy their hard-won promised land in peace for long. But the Union of South Africa (the Republic of South Africa since 1961), established in 1910, granted Boers full equality with the British settlers. Later *Afrikaans,* the Boer dialect of Dutch, was recognized as an official language on a par with English, and Pretoria, former capital of the Boer Transvaal, was made the administrative capital of South Africa as a concession to Boer sentiments. Afrikaaners (those of Boer descent, speaking Afrikaans) are a rapidly growing majority of the white population of the Republic. They are beginning to outnumber British settlers even in the cities. In the rural districts Afrikaaners form nearly 90 percent of the white total. In keeping with the history of their settlement, the British are concentrated in the port cities: Cape Town, Port Elizabeth, East London, and particularly Durban. The Cape Colored, descendants of mixed marriages between the Dutch and Africans and imported slaves, are also prominent members of the population of the Cape Province, and they comprise over half the population of the city of Cape Town.

The veld

The veld is commonly divided into high, middle, and low. The high veld, between 4,000 and 6,000 feet (1220 to 1830 m) in altitude, generally receives the greatest effective rainfall, 20 to 30 inches (51 to 76 cm), concentrated in summer. Most of the veld is suited to commercial agriculture or grazing. High veld covers most of the Orange Free State and the southern half of the Transvaal, and is drained by the Orange, Vaal, and Limpopo rivers. Toward the west, north, and south the plateau elevation falls more or less gradually into the middle and low veld, and northern Transvaal merges into a tropical savanna. The veld as a whole is mainly a long-grass steppe, but with a relatively small annual temperature range (see Map 86, page 380, and Figure 14, page 382). It is by far the leading agricultural and grazing area in South Africa; indeed it leads the whole of Black Africa as a nonplantation, commercial system. The major crop is maize, much of which is fed to animals, although some is exported to Europe. Maize ("mealies") is also the staple diet of the Africans. Livestock grazing is extensive, and some veld areas specialize in wheat (for export as well as for domestic consumption), tobacco, fruit, and vegetables for the local markets. The veld also supports South Africa's major center of mining and manufacturing and its largest urban complex. The low veld is significantly hotter, and was plagued by malaria in many areas until recently. High evaporation rates decrease the effectiveness of the rainfall, and parts of the low veld also lie in a rain shadow cast by surrounding higher areas. Even with the elimination of malaria, the low veld is not as well suited to commercial, mixed grain–animal agriculture. It has more generally been left to the Africans, although European settlement there is increasing. The veld as a whole, and especially the high veld, is crisscrossed by roads and rail lines, befitting a highly commercialized area. They serve both the agricultural and the manufacturing functions concentrated in the Orange Free State and the Transvaal.

Natal

Natal is better watered and has a largely frost-free climate, at least on the narrow coastal plain, so it specializes particularly in tropical crops. The most important commercially is sugar cane, but vegetables and citrus fruits are also grown. Most of Natal, however, is hilly, and the relatively narrow coastal plain is sharply defined by the steep, high front of the Drakensberg Mountains, cresting about 10 thousand feet (3050 m). The mountains have hampered access to the interior and made railway building difficult. But, once across them, local relief is smaller and the landscape is hilly or steeply rolling plateau rather than mountain. In inland Natal maize and cattle replace the sugar, fruits, and vegetables of the warmer, moister, and more fertile coastal plain. The development of sugar plantations in Natal in the second half of the nineteenth century brought a large immigration of Indians as indentured laborers. Most of them chose to remain after their indentures were up, and their numbers have multiplied despite the ban on Indian immigration since 1913. Most of them are in Natal (especially in the Durban area), but there are also sizable numbers in the cities of the Witwatersrand and in Cape Town. They occupy a strong position in local trade and small market gardening. Their alien roots further contribute to the racial tensions of South Africa. Durban, the leading port for cargo (Cape Town handles more passengers), serves the Witwatersrand and the veld as well as Natal, and includes a small and growing manufacturing complex, mainly consumer goods.

Cape Province and South West Africa (Namibia)

The Cape Province (Cape of Good Hope) is the largest of the four provinces of South Africa, but contains the smallest agriculturally usable area. It is mainly arid steppe or desert, except for the coastal belt of Mediterranean climate. The climate becomes more humid toward Natal on the east, where it is more subtropical than Mediterranean. The city of Cape Town is second in size to Johannesburg. Its hinterland is too arid to offer much economic support, and the city of a little over 1 million clusters around the magnificent harbor of Table Bay. The area of Mediterranean climate is small and largely limited to the southwest Cape. It is agriculturally productive in the familiar Mediterranean crop combinations of winter wheat (dependent on winter rainfall), grapes and citrus fruits, with widespread irrigation. A good deal of wheat, citrus fruits, grapes and wine, and vegetables are exported to Europe, together with fruits and vegetables from Natal. South Africa has the advantage of being in summer while the northern hemisphere is in winter, so it can market seasonal produce in Europe at a time of peak prices, which help to defray the transport costs.

Northward and northwestward from the Cape the climate is increasingly arid. South West Africa, long a protectorate of South Africa against the protests of the United Nations, is almost entirely desert or arid steppe. Large parts of it are almost uninhabited except for small aboriginal groups of nomads and a few spots of European population engaged in mining or administration. South West Africa became at least nominally independent as Namibia in January, 1978. A ruthless German colonial policy in South West Africa largely eliminated the local Herero, Hottentot, and Bushmen tribes. The few still left are pastoralists and isolated hunters. (The African Bushmen are similar in some ways to the Bushmen of Australia.) The coastal Namib Desert broadens inland. Most of South Africa, however, except for coastal Natal, also suffers from a water shortage, and a great deal of the land is too dry for unirrigated agriculture. Veld rainfall is close to the workable margin, and serious drought occurs in many years. West of the veld is drier and drier desert.

New states and African reserves

In various areas within South Africa, reserves of land were set aside for the Africans. Three similar but larger adjoining areas, formerly protectorates of the British crown, have now become fully independent. Bechuanaland, larger than Texas, changed its status in 1966 from that of a British protectorate to the African state of Botswana. In 1968 a degree of British responsibility yielded to complete independence for two smaller areas of African land: Basutoland, now Lesotho, and Swaziland, on the western and northern borders of Natal respectively. Both are dependent on South Africa as a market for cattle products, as the major source of imports, and for jobs for male migrants. Over half of the adult males work in the Republic. Swaziland is relatively rich in exploitable minerals, and has a climate similar to Natal's that permits some of the same commercial agricultural developments, especially in sugar cane. Botswana and Lesotho are much less well endowed, largely desert or arid scrub and mountains, although there is hydro-power potential in Lesotho. The Kalahari Desert, which covers nearly half of Botswana, is barely moist enough for a little thin grass and scrub in some areas.

South African reserves (often referred to as "black spots") are generally much smaller in size. They vary greatly and include many tiny tracts, totaling about one-eighth of the area of the Republic. Europeans took the best land; the reserves have less productive soil and more erosion. Many of them are also in dry or hilly areas. The reserves, in effect, are labor camps for African workers, and are overpopulated. The *apartheid* (separateness) policy of South Africa attempts to segregate the Africans on these reserves. Total population has increased rapidly under the European rule that brought a marked decline in the death rate and an end to intertribal wars. The thin soil of the reserves is overworked, and far too many animals are grazed. Soil erosion has become critical. European investments in irrigation, along the Orange River, for example, provide water largely for European-owned land. Some efforts at resource development in the reserves (including Transkei) have been made with European help. But they are far from adequate, especially given the rapid growth of population.

Since the Second World War the restrictionist policies of the white minority in South Africa have become increasingly rigid. The long-standing and wide-ranging color bar, the denial of the franchise to Africans, Colored, and Indians, and the prohibition against the owning of land by Africans outside the reserves have been accompanied by increased emphasis on *apartheid*. As mentioned in Chapter 20, areas are to be established for Africans where they will have some autonomy. But a rigid set of boundaries will be imposed to segregate blacks from whites still further, with little or no concern for African welfare. The first of the "Bantustans," the Transkei, was established in 1976, and nine others are planned. They will comprise well over a hundred scattered areas surrounded by higher-quality land reserved for whites. South Africa will continue to control foreign policy and defense, so the "independence" spoken of officially hardly exists actually. African and European areas are to be politically separate from one another, but economically interdependent. Within South Africa, Africans may not move about freely, must carry a pass at all times, and their living areas are already being further set apart from those of the Europeans. In some cases whole villages or towns of Africans, Colored, and Indians have been forcibly moved away from what are considered to be European areas. There is courageous resistance among a few of the Europeans to these immoral and dangerous policies.[4] But there is no really effective political opposition, and South Africa becomes more and more like a police state. African resentment continues to grow. The entire structure of the European economy is dependent on African labor and could not exist without it. This gives the Africans a power that they are now beginning to realize, and also reveals the impracticability of *apartheid*. African incomes in South Africa, already the highest on the continent, are rising rapidly. The existing policies do give Africans some economic opportunity, but in doing so Europeans may be preparing for their own replacement. The income gap between Africans and whites continues to widen, however. Equally significant is the fact that by the year 2000, at current growth rates, South Africa will contain about 5 million Europeans (mainly of Dutch descent), about 21 million Africans and Colored, and over 1 million Indians. Somehow the European must realize that genuine reconciliation with the African is the only safe guarantee of his own future.

[4]For a well-known sample of the liberal European view in the Republic, see the magnificent novel by the South African Alan Paton, *Cry the Beloved Country* (New York, 1948). It is a realistic story of life in the African slums of Johannesburg and on the reserves, and of the tragic problems that arise when the old values of tribal life are destroyed and Africans living a rootless and miserable life in the new cities meet with Europeans. A more recent effort at objective appraisal is Douglas Brown, *Against the World: Attitudes of White South Africa* (London, 1968), which is both sympathetic and critical.

Rhodesia (Zimbabwe) and Malawi

Most of Rhodesia is covered by an extension of the high veld, a rolling, high plateau with a nearly unbroken growing season and a generally adequate rainfall (see Maps 86 and 92, pages 380 and 398). As on the veld in South Africa, the leading crop is maize. There is considerable commercial stock raising as well as some production of wheat, tobacco, and citrus fruits. There are rich deposits of chrome ore, mined almost entirely for export, and one of the best-quality coal deposits in all of Africa at Wankie (see Map 89, page 381). A moderately well-developed rail net focuses on highland Salisbury, the capital, which has a strikingly European look. Rhodesia is still only thinly populated in terms of its economic resources, and its great population and economic growth lies in the future. Two hundred fifty thousand Europeans attempt to dominate over 6 million Africans. Only a tiny fraction of the Africans are allowed to vote, and almost no education is provided beyond the elementary level. The apparatus of a police state has not been enough to maintain this oppressive structure indefinitely, for all its profitability to white settlers, and Rhodesia seems certain to revert to African control in the near future. The border with Zambia follows the Zambezi River, whose deep gorge makes an appropriate regional line. Zambia is more like central than south Africa, and is more strongly tropical in climate.

Malawi (formerly Nyasaland) is transitional between south and east Africa. It has a narrow belt of hot, wet lowlands on the west shore of Lake Nyasa, and a somewhat wider belt of hills and upland plateau, where the climate is moderate. A small commercial production of tobacco, tea, cotton, and sugar has developed. The small size and isolation of the country have fostered strong regional feelings, but on broader regional grounds Malawi belongs with east Africa.

East Africa

Africa's highest mountains, including the famous snow-capped peaks of the Ruwenzori Mountains and Kilimanjaro, help to distinguish the region of east Africa. A discontinuous group of mountains and highlands paralleling the deep rift occupied by the chain of great lakes (Albert, Edward, Tanganyika, and Nyasa) forms the regional boundary with central Africa. Mozambique, also transitional between south and east Africa, has hot, wet, tropical lowlands in the southern half and a progressively arid plain and low hills toward the north. It is thinly settled, and economic development is minimal. Its most important commercial product is cotton, plus smaller amounts of cashews, sugar, and tea, but tropical, subsistence agriculture is dominant. The ports of Beira and Maputo (Lourenço Marques) do some entrepôt business, and also serve as outlets for central Africa. Beira handles some of the maritime trade of Zambia and Shaba. The hot, humid climate of the southern half of Mozambique, in the path of the trade winds off the Indian Ocean, is not typical of the rest of east Africa, most of which lies beyond the trade winds in an arid belt straddling the equator, except where highlands help to produce orographic rainfall.

The economy of east Africa is predominantly agricultural, in the absence of major mineral resources. Commercial agriculture is the chief basis for economic growth.

The states of Kenya, Uganda, and Tanzania dominate east Africa. Uganda and Kenya became British colonies around 1890, but colonial control was relatively brief, and both states became independent in 1962. Tanganyika was seized as a German colony in 1885, and became a British mandate in 1919. In 1961 it became an independent state, and joined with Zanzibar in 1964 to form Tanzania. These three large political units combine a great variety of originally separate cultures, smaller traditional states, and languages. Some of the earlier states were highly developed, carrying on considerable interregional trade as well as trade with the Arab-dominated areas along the east coast. Swahili developed as the language for trade over much of east Africa, and is still used. English is also widespread as a second language, especially in national politics and in the cities.

Kenya

The relatively small area of well-watered highlands in the southwestern quarter of Kenya supports commercial mixed farming. Nairobi, the capital, is a pleasant small city of European appearance in many respects, connected by rail with Kenya's port at Mombasa. It also does some entrepôt business along the east African coast. Well before independence the Europeans attempted to cooperate with the Africans rather than to pursue a policy of "white supremacy." The vote was extended to Africans, education and skilled employment were open to them, and Africans began to participate in governmental administration. With Kenya and its development in African hands, the small and dwindling group of Europeans are able to continue as commercial and technical consultants by mutual agreement.

Progress toward cooperation between black and white was interrupted, and then hastened by the bitter and unhappy Mau Mau terrorism. Mau Mau was a secret society sworn to the elimination of whites by murder and violence. Between about 1952 and 1957, often by force and threats, it increased its membership among the Kikuyu tribe, which had originally occupied most of the highlands taken over by the Europeans. Kenya lived virtually in a state of siege for several years. Finally security forces and troops were able to wipe out the Mau Mau strongholds in the more rugged highlands and break the Mau Mau hold on the many thousands of Africans who lived and worked on European farms or in the towns. The Europeans, with some justice, were accused of ruthlessness and illegal procedures in stamping out Mau Mau, but the terrorists themselves must accept some blame for the violence and misery they inflicted on their fellow Africans. Both sides learned from this bitter experience that genuine cooperation requires a great deal of mutual tolerance, sensitiveness, and concession, and that the future of Kenya would be served by avoiding bitterness and violence through common efforts. Clearly the Europeans had to give the most, and in the end it was the Kikuyu who furnished most of the political leadership for independent Kenya. The few Europeans remaining are gradually and peacefully leaving the area.

As so often happens when two different cultures come together, much of the bitterness over land in Kenya resulted from misunderstanding. The Europeans believed that they had purchased or occupied the land in the highlands in good faith. The Kikuyu believed in equally good faith that they had merely leased or loaned the land, which they were willing to do since their system of shifting agriculture and herding left large areas unused for long periods. When the Europeans arrived, much of the highland area was not being used. But the Kikuyu still regarded it as belonging to them and bitterly resented the permanent European occupation. The Kikuyu attitude is typical of a great many non-Western cultures, including other Africans, the American Indians, and several Asian groups. To these people, permanent transfer of land is impossible and unthinkable. Western legal forms and Western concepts of private property have not been understandable to them.

The Kenya highlands have a cool, savanna climate, and agriculture has specialized in maize, wheat, coffee, and sisal. Pyrethrum, a flowering plant grown mainly as a source of insecticides, is also a specialty. There is a good deal of commercial grazing and dairying, aided by the generally cool climate and rich pastures. The agricultural system and commercial foundation established by Europeans has now largely been taken over by Africans, and their commercial farming sector is the backbone of the Kenyan economy. On the coastal plain there are sugar plantations and some production of sisal and fruits. Indentured Indians were also brought to Kenya to work the sugar plantations, and many came to work on construction enterprises, including railway building. In Kenya the Indians are concentrated on the coastal lowland and in Mombasa. In Tanzania and Uganda they are more widely spread and, as in South Africa, have won a strong position in trade.

Large areas of the middle uplands in Kenya were originally occupied by the Masai, a warlike pastoral and hunting people. They invaded the area from the drier north, and killed, drove off, or enslaved the previous inhabitants. Instead of using the land for subsistence agriculture, they chose to use it for grazing cattle, sheep, and goats instead. As elsewhere in Black Africa where the tsetse fly does not rule out grazing animals, herds were kept largely as a mark of prestige and a store of value. The adult males were hunters and left the care of domesticated animals largely to the women. Some of the Masai still retain their traditional culture. They probably have changed less than any other major east African group, but most have begun to practice a little agriculture, despised by their forebears. Population in east Africa as a whole may have more than tripled under European control. But agricultural production and improvement of agricultural techniques have not kept pace, and there is urgent need to raise the level of the African economy and increase the efficiency of its resource use.

Uganda and Tanzania

Uganda is more densely populated than Kenya, and sedentary agriculture and cattle raising dominate the economy. There are some small tribes that are still seminomadic pastoralists. The traditional native

East African landscape—Tanzania. The dry scrub character of this area is apparent here. It is typical of most of Tanzania and east Africa. *(From the Peace Corps.)*

kingdoms of Uganda, including Buganda and Busoga, were developed on a par with some of the advanced west African states on a similar base of relatively open, sudanlike country. The modern movement toward self-government and independence in Uganda has followed these traditions, as did the modern peoples of west Africa. There is also a vigorous commercial development in Uganda. Under the British protectorate the politically sophisticated people of Buganda in southern Uganda were the principal force behind indirect rule for the country as a whole, including the much less developed northern parts. With independence and a national electorate, the new political power of northern traditionalists confronted the more progressive nationalists of the south, a situation similar to west Africa's. Like Nigeria, Uganda has become a complex federation joining the former southern kingdoms with the less-developed parts of the country in the north. Most of the country is much lower than the Kenya highlands, but it has a fairly heavy rainfall and an unbroken growing season. Commercial agricultural production has been increasing rapidly. Uganda now produces large amounts of coffee for export through Mombasa, as well as some cotton.

Tanzania includes some highland areas, especially near the border with Kenya in the north. But most of the country is semiarid lowland or middle-altitude plateau covered with brush, scrub, and occasional small bush. Three-fourths of it is still plagued by the tsetse fly, and there are few areas where soils and rainfall are strongly favorable to field agriculture. If some means of simple protective inoculation of cattle against tsetse infection or of elimination of the flies can be found, there will be an enormous increase in production potential, not only for Tanzania but for most of central and parts of west Africa as well. Most of Tanzania's commercial agricultural production comes now from plantations on the coastal plain, where the leading crop is sisal. But for the country as a whole, most agriculture is still subsistence in nature, and only about 13 percent of the total area is cultivated.

Just after the Second World War Tanzania was the site of a large-scale British project to grow peanuts, principally for their oil, during a world shortage of vegetable oils. Extensive areas were planted in peanuts, but the scheme was not successful because the soils selected were not really suited to field agriculture and the rainfall was inadequate. The partial and short-term climatic records and soil surveys, not surprisingly, were found to be unrepresentative. This experience pointed out, in Africa and elsewhere, the vital need for a thorough survey of any area before using it permanently for any purpose.

The port city of Dar es Salaam, terminal for the Zamtam railway from Zambia, is the capital and largest city. Tanzania is known within Black Africa for its progressive policies of agricultural development, its greatest potential resource. There has been encouraging success in particular with a new form of communal agricultural villages called *Ujamaa*. They emphasize local and regional self-sufficiency and promote improved methods of production, introducing the techniques of the Green Revolution. Tanzania's ties with China have been close, as the Zamtam railway project suggests, and the *Ujamaa* idea owes something to the experience of the Chinese communes. It may serve, in turn, as a model for other African states as they seek to promote their own development and lessen dependence both on foreign markets and on Western-style commercial-industrial concentration.

The nearby islands of Zanzibar and Pemba, an ancient Arab trading base, were British protectorates, although ruled by a Muslim sultan. Their short-lived independence came to an abrupt end in 1964 with a massive uprising of Africans against Arab domination. Its eventual outcome was the formation of Tanzania. Zanzibar has declined since the suppression of the slave trade, which had been its commercial mainstay for centuries. Clove production has taken the place of slave trading, and Zanzibar still supplies about four-fifths of world consumption. Its warm air is fragrant with the spice of cloves, but the city and the island are slowly declining in commercial importance. Most of their formerly vigorous entrepôt trade has been lost to Mombasa and Dar es Salaam with their better harbors, port facilities, and landward rail connections.

Ethiopia

Ethiopia, also discussed in Chapter 19, is distinct both from Black Africa and the Dry World in many respects. It escaped colonial control except for the brief Italian interlude from 1935 to 1942, and retained its Coptic Christianity (as well as Islam and paganism). It has stood largely aloof both from the

Africans to the south and the transitional Sudanese to the north and west. Ethiopia's highland and mountain base has also been distinctively used by its inhabitants. A form of contour ploughing on the slopes and a generally well-balanced system of grazing and field agriculture were developed there long ago. There are large areas of fertile volcanic soils in Ethiopia that have supported permanent field agriculture for centuries without exhaustion. The highlands receive much more rain than any other part of east Africa. Although agricultural equipment is relatively primitive, the crop complex of teff (a cereal grass grown for grain and hay), barley, sorghum, wheat, and coffee is reasonably productive on the whole. The coffee plant was probably native to the Ethiopian highlands and is still grown commercially, accounting for a little over one-fourth of its exports. Ethiopia shares with the Dry World and east Africa a heavy dependence on grazing animals, even in many areas where field agriculture would be possible. But well over half of the country is too rugged or too dry to permit any other use of the land.

East Africa's prospects

East Africa is awkwardly located for foreign trade. The local population is not large enough, nor does most of it have high enough incomes, to constitute a promising market. Overseas markets are remote and expensive to reach. Suez Canal dues are high enough to discourage shipments of any but high-value or strongly competitive goods, and the route around the Cape of Good Hope is nearly as expensive because of greater distance. Most shipments from Mombasa to Europe cost nearly the same through the Suez Canal or around the Cape of Good Hope. Commercial isolation need not retard east Africa's economic growth if attention is concentrated on the development of domestic resources. Barring the always present possibility that significant and valuable mineral deposits may be discovered, east Africa is likely to remain dependent very largely on agriculture. Tanzania and Kenya have been fortunate so far in developing stable, progressive, and effective governments. They are pursuing national integration and economic growth more successfully than most of the rest of Black Africa. Africa's future will depend on the growth of this kind of responsible local management in all of its regions. With the fading of the brief period of European domination, these east African states are charting a course of independent planning that has encouraging implications for all of Black Africa as it pursues its own development.

Questions for further study and discussion

1. Why should political and military organization have been more effectively developed in the savanna and steppe country of west Africa than in the rain-forest country near the coast, or in the savanna and steppe areas south of the equator?

2. Why is Cape Town a smaller port for freight than Durban, despite its superior natural harbor? Why does Cape Town have a larger passenger traffic?

3. Why does the political boundary between Zaire and Zambia cut right through the area of rich copper deposits? What is this kind of boundary called?

4. Why did Britain retain control of Botswana, Lesotho, and Swaziland against protests from the Republic of South Africa?

5. Is Africa likely to rival southeast Asia as a producer of rubber? Why or why not?

6. Why has west Africa been able to achieve such a dominant position among world producers and exporters of cacao? What are west Africa's advantages in competition with other cacao-producing areas elsewhere?

22 Latin America: a blend of Indian and European

Ruins of Machu Picchu, Peru. *(Photo from Braniff International.)*

The continent of South America and the areas north of Panama and south of the United States, known as Middle America, are still more Indian than European in the composition or mix of their populations even after nearly five centuries of European settlement. The first permanent settlements of Europeans outside their homeland were made in Latin America immediately following the first voyage of Columbus. This was over a hundred years before the first European colonies in North America (except for a few small Spanish settlements in Florida and California), and several centuries before significant European settlement in Africa, Asia, or Australia. In North America people of European descent dominated the continent relatively quickly. They have virtually eliminated the original inhabitants, although for the first two centuries, their settlements were largely limited to a narrow fringe along the east coast. In Latin America, Spanish and Portuguese control spread very rapidly in the first century after Columbus and well before the landings at Jamestown and Plymouth Rock. But people of purely European descent are still in a minority, and the modern culture of Latin America is a blend of Indian and European.

The most obvious reason for this cultural difference is that in Latin America Indians were originally far more numerous, and European immigration far less than in North America. Middle and South America together are still called Latin America, despite the minority position of the Europeans.[1] The area was conquered and has been most profoundly influenced by Latin peoples, the Spanish and the Portuguese. Canada and the United States are often referred to as Anglo-America because of the English origin of the dominant influences responsible for the area's development in the past three centuries. But the number of original Indians and the origins and totals of European immigration are not the only factors involved in the cultural differences between Latin and Anglo-America. Indians were more numerous in Latin America since they were able to make better use of varying environmental conditions that differed substantially from those to the north. From the European point of view, and espe-

[1]Europeans are a minority only in racial terms. European culture is probably as strong as Indian culture for most Latin Americans.

cially in terms of European methods of land and resource use, Latin America did not seem attractive or promise wide economic opportunity for Europeans. North America proved to have the most lavish and easily exploited resource base in the entire world. The North American combination of superabundant and varied mineral resources and superabundant and productive agricultural land contributed enormously to heavy immigration from Europe and to rapid population growth internally. Latin America could not match these resources, and European population has grown much less.

Climate has played a part as well in contributing to the productivity of North American agriculture as compared with the relatively lower productivity of most of Latin America. The parts of Latin America that were originally relatively densely settled by culturally advanced Indian groups were nearly all in tropical latitudes: central and southern Mexico, Yucatan, and what is now Guatemala (the empires of the Aztecs and the Mayas), and the Inca empire covering parts of what is now southwestern Colombia, Ecuador, Peru, western Bolivia, and northern Chile. Agriculture in these areas (several of which are at temperate altitudes) benefited from the absence of a cold winter. But it depended also on large-scale irrigation, which was developed to a high technical level. In some ways, therefore, Latin America compares with the early centers of civilization in the Old World. North America before the sixteenth century compares with ancient north Europe, both of them culturally and technically backward, sparsely populated, and economically unproductive. The development of Western civilization was concentrated in the subtropical centers until the Age of the Discoveries. Then the technological revolution and the growing importance of mineral resources gave an advantage to the northern lands. Europeans arrived in the Americas in the sixteenth century, bringing these developments to an area where technological change of that sort had not begun and where the most highly developed regions were still in the tropics. It is not surprising that the Europeans assessed the relative usability of the Western Hemisphere differently from the Indians. They were able to develop the resources of North America and to accelerate its growth far more rapidly than in Latin America.

The Spanish and Portuguese rulers

Latin cultural predominance was also a factor in the relatively smaller economic development of Latin America. Spain and Portugal, the major sovereign states involved, were not notable centers of the ferment characteristic of northwest Europe after the sixteenth century. Nor were they effective carriers or practitioners of the revolutionary changes in technology, resource appraisal, resource use, and the rights of the common man that helped the civilization of northwest Europe conquer and develop most of the world. Spain especially, by the seventeenth century, was increasingly a cultural, technical, and economic backwater (as discussed in Chapters 4 and 7). It was both unwilling and unable to make of its colonial domains what north Europeans made of theirs. Both Spain and Portugal were also relatively small bases, especially in population and in domestic economic development. They could not sustain a large flow of emigrants to the New World or support a growing trade pattern that might have stimulated colonial development. From the beginning their interest was almost entirely exploitative; they were not primarily concerned with permanent settlement or the development of local resources. They concentrated nearly all of their efforts on the extraction of precious metals and the conversion of Indians to Roman Catholicism. Portuguese energies were also diverted to their more lucrative trade enterprises in Asia. Spanish garrisons in the New World in the early period were frequently faced with starvation because they refused to make any effort to raise their own food in areas where agriculture was almost ridiculously easy. Enslavement of the Indians saved the Spanish from the degrading tasks of farming and digging for the gold and silver that were their chief objectives. But it also helped to perpetuate an exploitative and cruel system that contributed little to the economic development of Latin America.

It is significant that the beginnings of substantial economic growth and widening resource use in Latin America have come only since Spanish and Portuguese colonial control was thrown off (between 1810 and 1825) in a series of revolutions. In the latter part of the nineteenth century, north European capital and immigrants, and more recently North American capital, have flowed into Latin America.

MAP 93
PRECIPITATION IN CENTRAL AMERICA

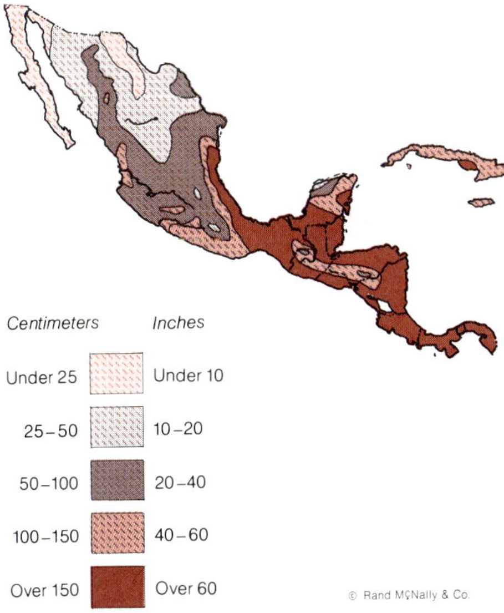

MAP 94
VEGETATION IN CENTRAL AMERICA

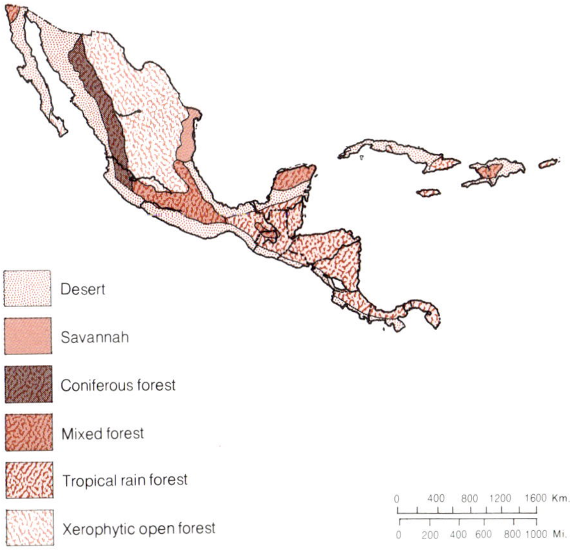

MAP 95
POPULATION OF CENTRAL AMERICA

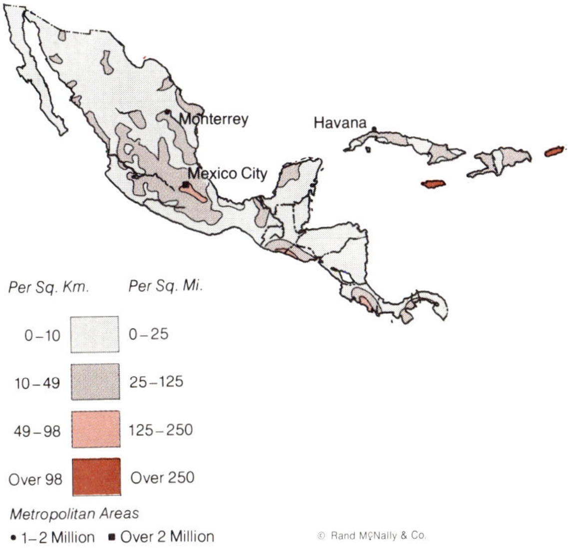

MAP 96
MINERALS AND ECONOMIC PATTERNS IN CENTRAL AMERICA

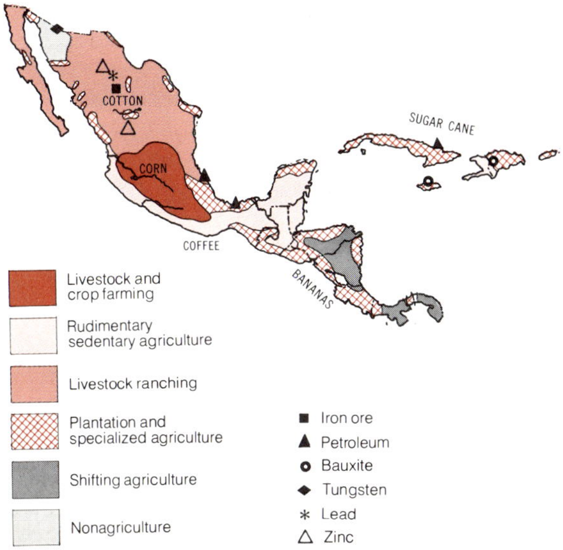

Under the old colonial system both Spain and Portugal had excluded the nationals of other countries from their domains and kept a tight governmental monopoly on trade. Under their control Latin America remained very largely a neglected area, whose total resources and economic opportunities were not effectively used.

Of course, there were good reasons for the Latin predominance. The most obvious reason is that Spain and Portugal are closer than north Europe to the southern part of the New World. And the Portuguese and Spanish were leaders in the great discoveries because of their location where the Mediterranean joins the Atlantic. The coast of South America may first have been sighted by a Portuguese ship blown off course while attempting to round Africa in the long Portuguese search for a sea route to the Orient. Columbus sailed for the New World under the Spanish flag and, despite four voyages across the Atlantic, believed until his death that he had reached an outpost of Cathay, not a new continent. The trade winds and accompanying currents also played an important part in the fact that Latin America was found, explored, and settled earlier than North America, and that the Spanish and Portuguese discovered and continued to dominate it. The first land in the New World seen by Columbus was an island he named San Salvador (the modern Caicos?) in the Bahamas just northeast of Cuba. His ships used the trade winds off the west coast of Africa and sailed with them due west. He attempted to return to Spain by a more northerly route and made the trip very rapidly with the help of the prevailing westerlies. But he nearly came to grief in a terrible storm like those common in that part of the Atlantic in winter. This experience encouraged him and others to stick to the southern route, despite contrary winds on the homeward voyage.

Farther north, in the zone of the westerlies, contrary winds on the outward voyage toward America presented a major problem for all of the early explorers and emigrants. The southern route using the trade winds carried Spanish and Portuguese fleets to the West Indies and the mainland with relative ease and security. But they never came in the numbers that flocked to North America after the eighteenth century as the rich resources of the north were rapidly developed.

The focus of Latin American urban culture—church and plaza. The Cathedral of Lima, in Peru, seen from across its plaza, is a sample of a universal aspect of Latin American cities. Urban life tends to revolve around the plaza, with its seventeenth- or eighteenth-century monumental architecture, in the tradition of Latin Europe. *(Photo from Braniff International.)*

Racial and cultural mixtures

As a result of these varied conditions and influences, Latin America is still mainly Indian by blood. Subsistence agriculture in the traditional Indian pattern is still the most important sector of the economy, in terms of area covered and numbers employed. The present inhabitants of Ecuador, Peru, Bolivia, Paraguay, and Guatemala are predominantly Indian. European blood is predominant only in Costa Rica, Uruguay, and Argentina. In the other countries of Latin America, people of European descent are in a minority. Most of the populations are of mixed Indian and European blood, or *mestizo* (in Brazil, *mameluco*), with the Indian inheritance predominating. The *Ladinos,* those who claim European descent or have adopted a European way of life, frequently dominate the cities, but are rare in rural areas. Culturally the Latin influence has been greater, especially in language and religion. Except for relatively small groups of primitive forest dwellers and isolated tribes, most people in Latin America speak Spanish or Portuguese, and most are Roman Catholic in religion. In both respects some original Indian culture survives: native languages replace Spanish or Portuguese in the home, especially in rural areas; and animistic or tribal religious beliefs persist, as in so many peasant or rural societies elsewhere, despite formal acceptance of a more modern institutional religion. Traditional Indian art forms have also been increasingly revived, especially in Mexico, but the physical appearance of almost all Latin American cities is dominantly European.

There were few cities before the Spanish arrived, and most were destroyed by the *Conquistadores.* As in monsoon Asia and Africa, European influences were concentrated from the beginning in cities and are still strongest there. Most of the cities were founded and built by Europeans, and all but the most recent tend to repeat an essentially Spanish or Portuguese pattern. From small market towns to regional and national capitals, most have a central square or plaza dominated by a church or cathedral, and most were built during the high period of Spanish power and wealth. Their consequent scale, magnificence, and impressive Spanish baroque architectural style have helped to maintain church and square as a nucleus, even though most of the cities and towns have grown appreciably, and in some cases, tremendously. The culture of the rural areas, most of their economic structure, and most of their systems of land and resource use, as well as the composition of their populations, remain more Indian than European, except for Costa Rica, Argentina, Uruguay, and parts of Chile and Brazil. The relatively limited areas of plantation agriculture in Middle America and large-scale mining of industrial raw materials in Venezuela, Chile, Brazil, and elsewhere constitute separate cases that are discussed in Chapter 23.

Lack of accurate geographical knowledge at the beginning of the European expansion into the New World lay behind the artificial division of Latin America into Spanish and Portuguese spheres. As a result of arbitration by the pope, the Treaty of Tordesillas between Spain and Portugal in 1494, only two years after Columbus first saw the new land, gave Portugal sovereignty over all land east of the fiftieth meridian of longitude. The assumption was that this would include as much as would fall to Spain west of the same line, the Line of Papal Demarcation. The papal line actually left only a small part of the Western Hemisphere to Portugal, but this did not become apparent for some time afterward. Meanwhile Portuguese interest was much more heavily concentrated on trade and conquest in the comparatively more productive Orient. Brazil, to which Portugal had fallen heir, received even less attention on the whole than the Spanish colonies, but it remains a dominantly Portuguese-speaking area.

Population patterns and political fragmentation

The economic development of Latin America has unquestionably been retarded by the multiplicity of political divisions. When Spanish control was thrown off, regionalisms became nationalisms, and political fragmentation was multiplied. This was not simply perversity, for this enormous area is extremely diverse physically. It is further divided by a series of difficult barriers: tropical rain forests, the towering Andes mountain chain, and extensive deserts (see Map 19, page 67). The only large lowland plains are in the equatorial basin of the Amazon (Rio Amazonas). Economic opportunity and cultural development there are even more limited than in the similar Congo basin. The northern part of the smaller Parana basin in Argentina and Paraguay is also of limited usefulness because of excessive heat, seasonally uneven rainfall, and isolation, similar to the African Sudan. Except for the *pampa* of Uruguay and Argentina (originally a long-grass steppe), most of this limited plains area is characterized by the usual shallow tropical soils of low fertility outside the areas of alluvial deposits. Close to the rivers, there are severe drainage problems and extensive swampy areas unfit for agriculture or settlement. It is not surprising that South America as a whole is still characterized by settlement on or near the coasts. This pattern lent itself to the development of numerous separate political units, each associated with one or more clusters of population, and each isolated from other states by intervening physical barriers or sparsely inhabited areas (see Map 99, page 419).

The Spanish, in their search for precious metals, converts, and slaves to work the fields and mines, avoided sparsely settled areas and those thought to contain no gold. This further fragmented the settlement pattern and also led to the neglect of areas that

MAP 97
PRECIPITATION IN SOUTH AMERICA

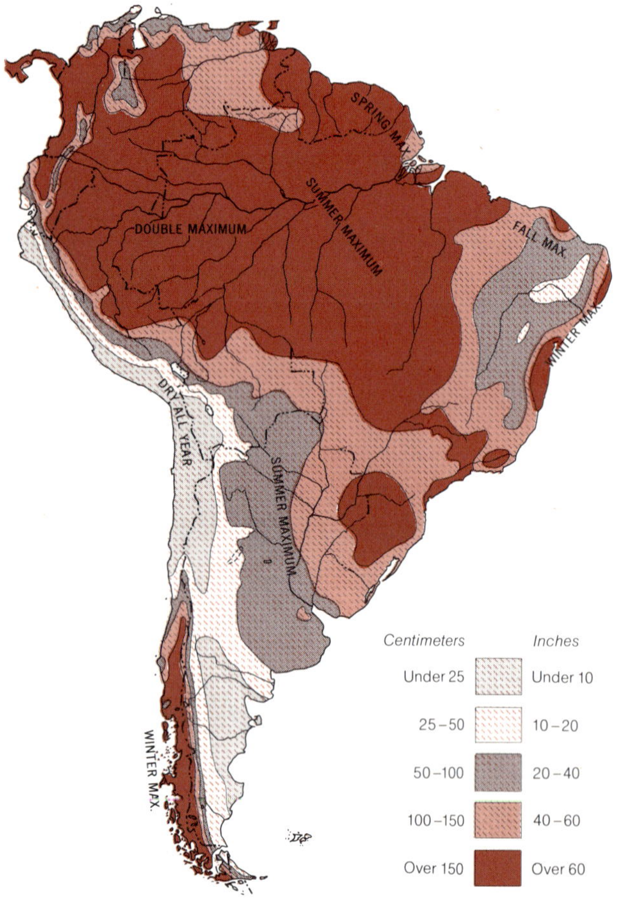

MAP 98
VEGETATION IN SOUTH AMERICA

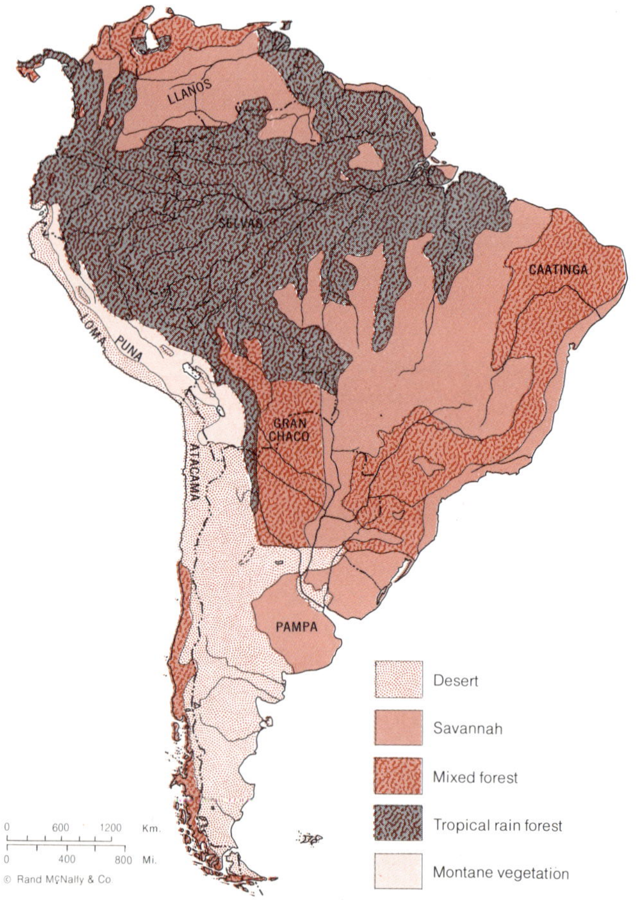

later proved to be extremely productive, including the pampa. It had a very sparse Indian population and its potential agricultural advantages were of no interest to the Spanish Conquistadores. But many of the patterns of settlement and regional structures in Latin America are still essentially as they were under the Spanish, and in turn reflect what they were when the Spanish found them. The huge area of the Amazon basin, covering well over a third of South America, is still essentially empty except for a few small clusters of population near the river's mouth and at some of the major confluences. The Amazon basin contains the most extensive single area of rain forest remaining in the world, and it is questionable whether it will ever support more than sparse settlement. The southern fifth of the continent is about equally empty, although here it is aridity, cold, and relief that are the principal obstacles to settlement. Aridity is also largely responsible for the belt of sparse settlement running roughly between 20° S. and 30° S. latitude down the middle of the continent to the east of the Andes. These barrier areas occupy the heart of South America, leaving a large basis for settlement only on the fringes.

MAP 99

POPULATION OF SOUTH AMERICA

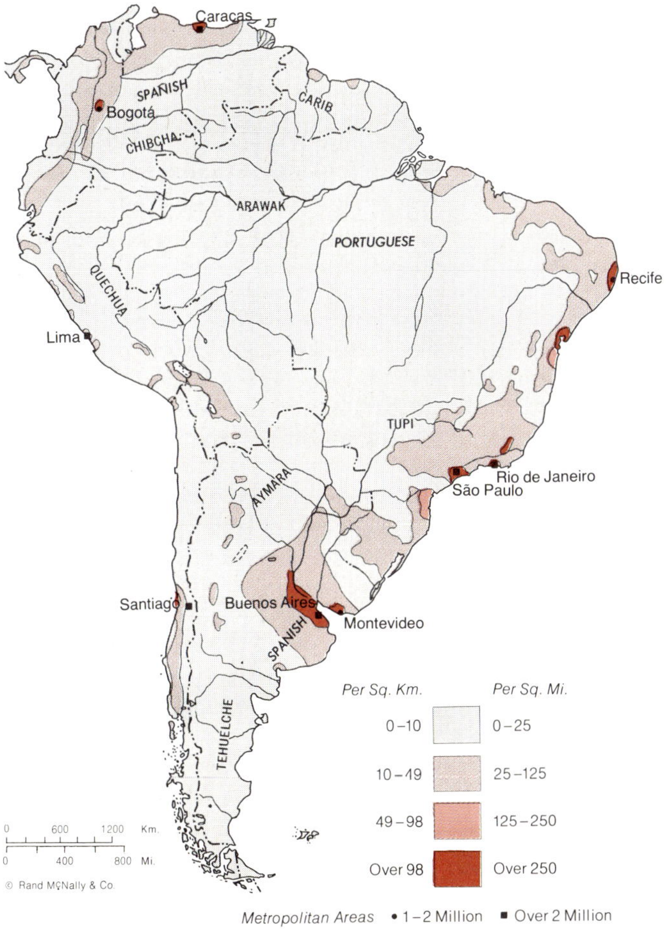

MAP 100

MINERALS AND ECONOMIC PATTERNS IN SOUTH AMERICA

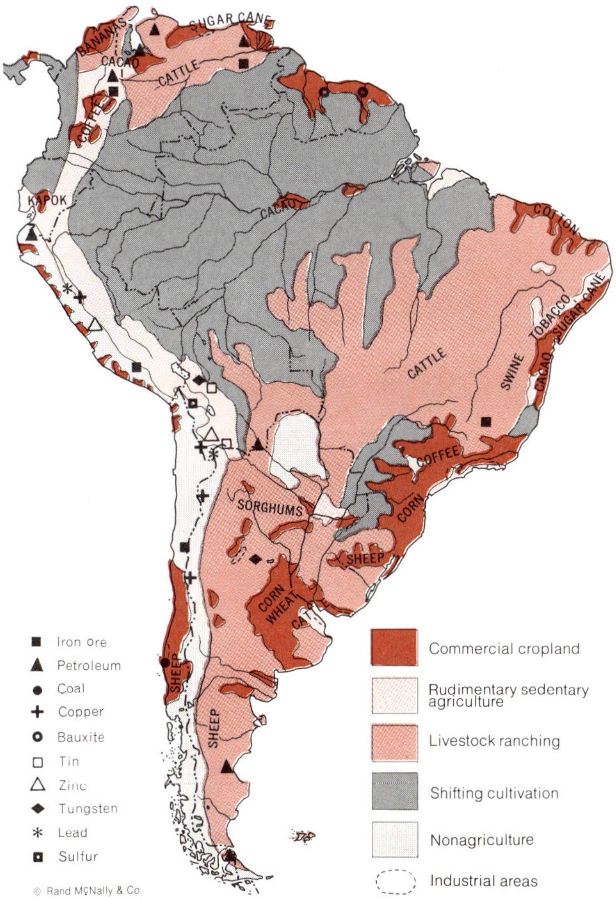

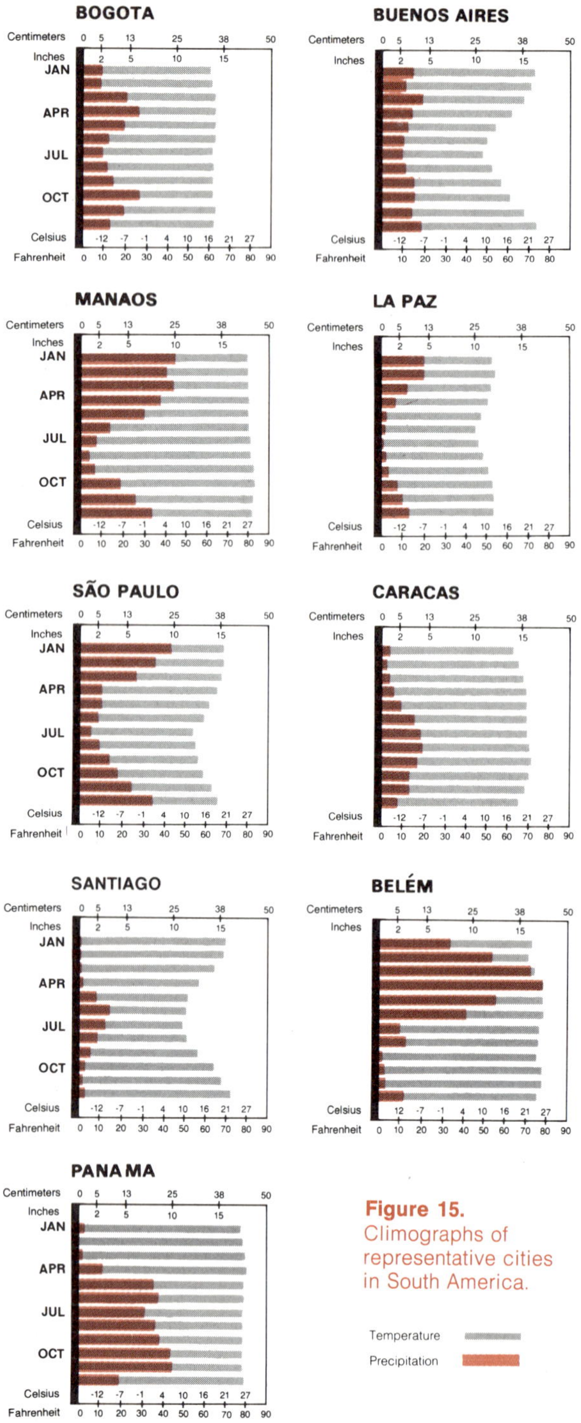

Figure 15. Climographs of representative cities in South America.

Ancient Indian civilizations

The three major areas of dense population before the Spanish arrived are still important on the population map: the central plateau of Mexico (center of the Aztec empire), the lowlands and highlands of Guatemala and some adjoining territory (center of the Mayan empire), and the long chain of settlement from Colombia south to Chile following the Andes (center of the empire of the Incas and Chibchas). Except for lowland Guatemala (and lowland Yucatan, once densely settled but now relatively empty), these are upland areas within the tropics for the most part. Sea-level temperatures and, frequently, rainfall conditions are moderated and improved at these higher elevations. Where level land is available, as it is in many places, there are strong agricultural advantages, especially by comparison with the lowlands or with the heart of the South American continent where rain forests or aridity discourage settlement (see Map 97, page 418). In these highland bases the Indian tribes created impressive civilizations, completely different from the relatively primitive cultures of the South American forest-dwelling or steppe Indians, and from the equally primitive Indian groups of North America. To some extent it is puzzling that the civilizations of the Aztecs, Mayas, and Incas did not spread more widely over the Americas by conquest or by cultural diffusion. One assumes that they developed from the particular set of conditions in their respective home bases and could not be transferred easily to the very different areas elsewhere, especially considering the formidable mountain barriers. Although these cultures were advanced according to the standards of their time and hemisphere, they lacked the techniques that made Europe's conquest or influence over so much of the world possible: firearms, machines for production, transport, and communication, and a fully developed, written language. They also lacked the large numbers of people and, more importantly, the highly developed system of trade that contributed so much to the expansion of Europe. The Inca, Maya, and Aztec economies were almost exclusively agricultural and were largely subsistence in nature. There was only a little interregional trade (the primary transmitter of culture). Although the Inca and Aztec civilizations were contemporary, they were not really aware of each other's existence. They were separated by rain forests and unable to use the sea to the extent that Europeans had done since the time of the ancient Greeks.

Probably the most striking accomplishment of these native American civilizations was their calendars, remarkably accurate and presuming a knowledge of the solar system considerably ahead of Europe's at that time. Development of a calendar is a logical accomplishment of an agricultural society because precise knowledge of seasonal changes is of vital importance. Ancient Mesopotamia, Egypt, and China had almost equally accurate calendars for the same reason. Another achievement, especially of the Incas but also of the Aztecs, was the development of a complex and efficient system of irrigation. It was on a par with the ancient Chinese system and accompanied by a similar growth of highly organized, bureaucratic government. With the help of irrigation and the use of maize, which probably originated somewhere in Middle America and can produce high food yields per acre, these civilizations supported relatively large populations, although they probably never approached the densities of ancient Mesopotamia, Egypt, India, and China. Their agricultural economy, however, did produce substantial surpluses, enough to build impressively large cities and to feed the nonagricultural urban dwellers and administrators.

It is remarkable that this culture developed largely without the use of metals, wheeled vehicles, or a complete system of writing. The Aztecs and Mayas were essentially a Stone Age people. Metals entered their culture quite late, and then they used only precious metals as ornaments. Equally striking, neither had the enormous help provided by domesticated animals, except for a half-wild turkey, the bee, and a voiceless dog. The absence or minimal importance of animals in the Aztec and Maya civilizations may have been a major factor in retarding development, as it was in Black Africa.

The Incas were somewhat better off because they had domesticated the native llama and alpaca, wool-bearing, cud-chewing animals resembling a cross between a large goat and a small camel. Meat and wool were important animal products, but the llama and alpaca seem to have been used only as pack animals rather than for draft. The Western Hemisphere as a whole contained relatively few native animals suitable for domestication. The absence of cattle, horses, pigs, sheep or goats, and chickens was a serious disadvantage. There was undoubtedly a great deal of useful plant domestication in pre-Columbian America, especially maize and potatoes. The Incas also made somewhat greater use of metals than the Aztecs or Mayas, and had learned to make bronze. But the Incas were still in the beginning stages of a metal culture when the Spanish arrived, aside from their well-established smelting of gold and silver. Metal tools were not widespread, and the Incas were not far removed from the late Neolithic era characteristic of the Aztecs and Mayas.

The absence of the wheel from all of these cultures is perhaps even more remarkable, especially in view of what they accomplished: huge stone buildings, aqueducts, canals, a system of roads for local transport and relay runners who carried messages with amazing speed, and the cities themselves. The Aztec capital at Mexico City, the Mayan city of Tikal in lowland Guatemala, and the Inca capital at Cuzco in the high Andes, 350 miles (560 km) southeast of modern Lima, were all 2 or 3 square miles (5.2 to 7.8 sq. km) in area, with stone-paved streets, sewers, and monumental stone buildings. This record

The Maya civilization—temple at Cichen Itza in Yucatan. These huge stone structures are comparable in size and technological level with the pyramids of Egypt. The human figures at the base of the steps give some idea of the size. *(From Mexican National Tourist Council.)*

suggests caution about the too-easy assumption of "requisites" or conventional sequences in the development of civilizations. It also clearly indicates that native American civilizations were critically hampered by the absence of some basically useful techniques and aids. It seems likely that the Mayas were shifting agriculturists, an almost incredible fact. Yet they clearly produced large surpluses and were able to combine them with organization to build cities and monuments as impressive as those of the Incas and Aztecs, indeed as impressive as any in the premodern world. Shifting cultivation was well suited to the rain-forest base in lowland Guatemala. Mayan culture appears to have originated there in an area that is now much less densely settled than the adjoining uplands, where coffee has become the major commercial crop. If Mayan culture did have a rain-forest origin and spread from lowland to highland Guatemala and to lowland Yucatan (also a rain forest), it was a remarkable and probably unique development. This brief discussion can mention only a few of the major characteristics and puzzles of the ancient American civilizations. It is a reminder of the extent and variety of areal differences and of the dangers of applying the standards, conventions, and patterns of Eurasian development to all areas.

The conquest

The Mayan civilization was decaying when the Spanish arrived, and its sites are only now beginning to be rediscovered and excavated in the heavy jungles of lowland Guatemala and Yucatan. The full reasons for its decay or collapse are still unclear. But the Inca and Aztec empires were flourishing at the beginning of the sixteenth century and were conquered and largely destroyed by proportionately tiny Spanish forces with unbelievable swiftness. The Aztec and Inca emperors, Montezuma and Atahualpa, who were considered to be gods by their subjects, were murdered by the Spanish. This tended to demoralize or destroy resistance to the conquest. Horses were unknown in the New World, and the Spanish mounted warriors struck fear and terror into the stoutest Indian hearts. The Spanish adventurer Cortes conquered Mexico with some six hundred men, seventeen horses, and ten cannon (plus some Indian allies). His fellow adventurer Pizarro did the same in the even larger and stronger Inca empire with a force less than half as great. Within fifty years after Columbus first sighted the new land, the Spanish and Portuguese had conquered or explored practically all of Latin America outside the rain-forest areas, as well as large parts of what is now southwestern United States and Florida. Probably millions of Indians were killed, although more died as a result of diseases imported from the Old World and the atrocious conditions of the new slavery in mines and fields than in the brief military campaigns. It seems certain that there are even now fewer people in highland Peru than there were before the Spanish conquest. However, this decline reflects greatly changed economic conditions and resource use as well as actual losses caused by the Spanish.

There was a large population originally, and it was not eliminated. The total Indian population of North America in 1492 is estimated at about 1 million. But Latin America at that time is estimated to have had between 30 and 35 million. How long these people had inhabited the Americas is far from clear. It is generally accepted that they came originally from Asia, probably by way of the Bering Strait and perhaps by other routes as well, as long ago as 30,000 or 40,000 B.C. The short crossing at the Bering Strait would have been relatively easy even for primitive people, especially at the height of the continental glaciation. At that time, the average sea level was low enough or sea ice widespread enough that people could move from Asia to the Americas without crossing water. There may also have been some contact across the central or southern Pacific, perhaps at a later time. There is a great variety in languages and physical types, as well as in cultural levels, among the American Indians. This suggests that they either entered the Western Hemisphere over a long period of time from different areas, or that they lived in scattered areas for long enough to have developed differently, or perhaps both. In Latin America total population was large enough, and in some respects advanced enough, to survive the impact of the Spanish and Portuguese even in most of the areas where the European impact was concentrated at first, the old empires of the Aztecs and Incas. In the areas that the Europeans ignored or turned to later, and in the barrier country of the rain forest, original Indian population was much smaller, but fewer inroads were made on it.

Negro mixtures

The death of so many Indians, especially in the early stages of European conquest, and their unsuitability for forced labor, under which they quickly died, led very rapidly to the introduction of African slaves, especially to work the plantations. The first cargo of Negro slaves arrived in the West Indies in 1501, nine years after Columbus's first voyage. Negro slaves came almost as early to Brazil, where sugar was the first important product. The Latins refused to do manual labor, like the Europeans in many parts of Africa. It was not until the independence movements of the nineteenth century in Latin America that slavery was abolished. Meanwhile, slavery added another important element to the Latin American population. Negroid racial mixtures are present in large parts of the area now, and are particularly prominent wherever plantation agriculture is or has been important. In Brazil especially, as in all Portuguese colonial territories, there was no color bar and intermarriage was common. Within the past fifty years, particularly with increased immigration from Europe and elsewhere, the Brazilian population has become even more mixed. Elsewhere in Latin America, color and racial lines are more sharply drawn, despite considerable intermixture. The areas of dominantly Indian, European, mestizo, Negro, and mulatto (mixed Negro and white ancestry) populations can be more or less easily distinguished. The Negro elements are concentrated in the plantation zones of the West Indies and northern, tropical South America. In these latter areas the original Indian population was small and primitive, but they received a heavy share of the early European impact because of location. On most of the West Indian islands, and in parts of South America's north coast, the Indian population was largely annihilated within a few years. Their places as laborers were taken by Negroes, and as landowners by the Spanish. Today, for example, the people of Haiti are almost entirely Negro, and neighboring Puerto Rico is largely Spanish, with some Negro mixtures. By contrast, Peru, which did not support an extensive plantation development until recently, still has a dominantly Indian population. Argentina, very sparsely populated before the modern period and without plantation agriculture, is dominantly European.

Resource use and the Latin American system

The Inca, Maya, and Aztec areas fitted the Spanish prescription: obvious sources of gold and silver supporting dense Indian populations available as labor and potential converts. Attention to other areas in Latin America came later. Gradually the Spanish emphasis began to shift from its exclusive concentration on precious metals to include some attention to agriculture. The name *Argentina,* which means "silver," is a reminder of the early objectives pursued in this area even in the absence of precious metals and in the face of a resource base whose advantages were overwhelmingly agricultural. The first important agricultural development in Latin America under Spanish and Portuguese control was in sugar. It was a crop that Europe demanded but could not grow. Tropical Latin America had strong advantages of climate and relative nearness to Europe. The West Indies became the first boom agricultural area in the modern world, and was the original home of the plantation system as it has developed in the past three centuries. Even earlier, sugar came from plantations along the tropical coasts of Brazil, Venezuela, and what is now Guyana, Surinam, and French Guiana.

Other crops, for their own use and for export, were grown only reluctantly by the Spanish, and wider agricultural development came slowly. The pampa, neglected at first, was a potentially rich agricultural and grazing area. The nearly pure Spanish population found few nonagricultural alternatives there. An essentially aristocratic, stratified system of social organization and landholding, imported from Spain, became established, especially in Argentina. It has tended to serve as a model for most of Latin America still dominated by people of Spanish extraction. Most of the land is owned in large estates or *haciendas* by a small group of self-styled aristocrats, and most of the labor is performed by employees or by tenants, known as *peons*. Social distinctions are sharply drawn between these two groups, and there is relatively little farming done by small owner-cultivators. The hacienda differs from the plantation in growing a variety of cereal and other crops and in raising animals. In Latin America the hacienda system is largely concentrated south of latitude 30°, a distribution consistent with similar land use in

North America, Eurasia, and Africa. Plantations are concentrated in the tropical areas in Latin America, as elsewhere. In the absence of extreme population pressures in Latin America, except for a few areas, land is often held and used as much for social prestige as for economic gain. It is generally used extensively, or even wastefully, by Asian or European standards. Land that could grow grain is used instead as rich pasture for purebred cattle and horses whose value lies in social prestige rather than in production. The great majority of the population of Latin America, however, are peons. Most are poor, subsistence tenant-farmers, growing crops for their own immediate support and very little involved with commercial exchange.

This grossly unbalanced pattern of landholding and incomes and the concentration of political power in the hands of the small landowning group contribute to the familiar Latin American cycle of unstable governments and frequent revolutions. Too often they simply reshuffle political control among the same small group. The Indians, like many non-Western peoples, found it hard to understand or accept European notions of land ownership and transfer. To them the land was free and universal, like the air, and could not be owned, bought, or sold. This attitude still persists to some extent, but it has not prevented the growth of a landholding system based on prestige and power, leaving the Indian elements in a peon role. In Brazil development in the Portuguese tradition was somewhat different. Prestige or unprofitable landholding was much less attractive than the speculative profits that could be won by buying and selling land. This also has obvious evils and has contributed to less than constructive land and resource use. There have been boom-and-bust cycles of speculative activity: first in sugar, then in gold and diamonds, then in coffee, and most recently in present and potential urban and industrial real estate. Each phase has been accompanied by soil exhaustion and rapid shifts of capital and labor from one enterprise to another.

The Spanish and Portuguese heritage and the European-Indian-Negro population that has developed in Latin America under these circumstances have contributed to the present political picture. Portuguese and Spanish influences have been important especially in the role played by the Roman Catholic Church in social and political matters, and in the degree to which politics are personalized or dominated by the idea of the "boss." Political patronage, the spoils system, and one-party or one-group rule have become characteristic of Latin America in part because of the Spanish-Portuguese tradition. A more genuinely democratic society and political system in Latin America has developed in areas where small independent land owners are relatively common, or where there is a large, urban middle class. These factors were also associated with the rise of modern parliamentary democracy and socialism in Europe. Great discrepancies in wealth including landholding, as in the Dry World and elsewhere, have been accompanied in Latin America by political authoritarianism and oligarchy or boss rule. The rapid growth of an urban middle class in industrializing Brazil and the presence in southern Brazil of many European immigrants as small independent landowners may ultimately lead to a more healthy growth of Brazilian democracy. But it would probably be fair to say that in the whole of independent Latin America[2] genuine parliamentary democracy is well established only in Costa Rica. Elsewhere, small groups of Spanish or Portuguese landowners or mining interests exercise an essentially undemocratic, monopolistic control. Armies play an inappropriate and important role in politics, and dictators or oligarchies continually succeed one another as small upper-class factions jockey for power more or less unscrupulously.

Yet economic development in Latin America has come a long way since the days of the Conquistadores and their search for gold and converts. The pampa has become one of the world's major surplus producers of wheat, maize, and cattle. Brazil, Middle America, and Colombia now dominate world production of coffee. Middle America has become the world's leading area of commercial banana plantations. Chile rose to prominence as a major supplier of copper ore and nitrates. Since the Second World War Venezuela has increased its industrialization on the basis of its extremely rich deposits of iron ore and petroleum. Brazil has the beginnings of an industrial revolution. Rubber had its brief boom in the Amazon basin in the middle of the nineteenth century, al-

[2]That is, excluding Puerto Rico and the few tiny remaining British, Dutch, and French colonial areas. In the former British West Indies, especially Jamaica and Trinidad, parliamentary democracy is relatively healthy.

though most rubber production is now being done on the plantations of southeast Asia. Sugar, the first plantation crop of all and the first major agricultural venture of Latin America, is still a mainstay of the West Indian economy. Mexico also has the beginnings of an industrial revolution, and produces oil, iron, sulphur, and other minerals, as well as the silver that first attracted the Conquistadores. Thus the limited pattern of resource use followed by the original Indians and the early Spanish conquerors has expanded. Latin America is still in the early stages of a major period of economic growth. Latin American mineral resources, although not as great as those of Europe or North America or even Asia, are beginning to be used more widely. The agricultural resources, already drawn on commercially by the plantation system and the development of the pampa, are being exploited more fully. They will have to be, even without commercialization, as Latin America's population continues to increase at its present very rapid rate.

Population growth and economic development

Latin America as a whole, with the exception of the West Indes, may still be regarded as one of the world's largest underpopulated areas, in relation to potentially available resources. It will probably not have this distinction much longer. Latin America's average population growth rate is now the highest in the world for an area of comparable size, and is over twice the world average. Average Latin America standards of living, in terms of per capita incomes or per capita consumption, are very low, and are threatened still further by rapid population increases. As is often the case elsewhere, population appears to be growing rapidly in some already relatively crowded and poor areas. Their birth rates are traditionally high and their death rates have fallen rapidly in the past half-century as a result of outside influences that have not yet had time to make a significant impact on the birth rates. By the end of the present century, Latin America's total population will probably have reached or exceeded 500 million. Unlike Asia, or at least to a greater degree, Latin America has a relatively large margin of presently unused but easily usable resources that can be and are increasingly being employed to support its rising population. In general and per capita terms, Latin America is the only so-called economically underdeveloped area where per capita economic expansion is now rapid and where there are excellent prospects for sustained and accelerated growth. In this sense it is fortunate that Spanish colonial policy and other factors have tended to retard population growth in the past. Now substantial and rapid economic growth can occur before the population has reached the unwieldy numbers that are such an obstacle to raising per capita incomes in monsoon Asia. There seems a good chance that increases in total production in Latin America can be kept permanently and increasingly ahead of population expansion. Already it is estimated that per capita production in

Undeveloped South American wilderness—Colombia. Much of Latin America is very thinly settled and the people have been affected little by the Europeans. This farmer's home in a jungle clearing near an exploratory oil well may soon change. *(Photo from Standard Oil Company of New Jersey.)*

New settlers in Colombia. In most of the pioneer areas of Latin America, the land has been neglected in the past. Although greater efforts are being made to increase the cultivated land, the improved techniques to make it profitable cannot be provided quickly. *(Photo by Nicholas Saphieha for Stock, Boston.)*

Latin America is rising more rapidly than in the United States, especially since 1945. The relatively small total size of Latin America's population is an important factor in this achievement.

Resource base

Population density in relation to total land area and known usable resources is still low. Only Africa and Australia have lower total population densities per square mile. It seems likely that Latin America has at least as promising a usable resource base as Africa. Present population densities and growth rates are highest in the West Indies and Middle America, the first areas to receive the full impact of European and American influence, including the plantation system. The rest of Latin America is still relatively thinly settled, and available resources have been drawn on only superficially for the most part. There is a very large margin for increased agricultural production. More efficient and intensive use of existing cultivated lands (especially the pampa) and extension of cultivation into areas not now cultivated, including the sudan- and savanna-like regions that cover vast areas of interior Latin America north and south of the Amazon, can be accomplished. Certain problems remain to be solved, such as the development of pasture grasses better suited for commercial grazing,[3] the extension of irrigation (generally a neglected technique in modern Latin America), and the provi-

[3]Success with African grasses introduced into the savanna areas may suggest a large part of the solution.

sion of adequate transport networks and marketing facilities. But total agricultural production could clearly be at least tripled, and with much greater speed and ease than in most other parts of the world.

For industrial development the prospects seem less certain and less impressive. Unfortunately, Latin America largely lacks significant deposits of good-quality coal (see Map 12, pages 56–57). How long and how seriously this will remain a major disadvantage depends on future technological development, especially of substitute fuels. But it will unquestionably continue to retard the growth of manufacturing. Total potential hydro power is also relatively small in proportion to the large area involved. It has been very little developed, and some economic gains can be realized here. But the problems of hydro development are much involved with the nature of the market and with location. In the vast area of Brazil where almost half of Latin America's total hydro potential is situated, figures of potential horsepower or kilowatts mean relatively little. It will be a long time before even a small part of this potential is translated into electric power. Venezuela and some of the small islands off its coast are abundantly rich in petroleum. Petroleum also occurs in significant amounts in Mexico, Colombia, Peru, and Argentina (see Map 12, pages 56–57). But oil is not an economic substitute for coal for many industrial uses. Except for Venezuela and Mexico, which export the great bulk of their oil production (as the Dry World oil states do and for the same reasons), the quantities found so far are not large enough to earn much foreign exchange.

Brazil, Venezuela, and Chile are all rich in iron ore; manganese is plentiful in Brazil. Bolivia produces considerable tin, and Chilean copper is still of world importance. But none of these minerals constitutes a striking advantage for industrialization, in the absence of an adequate mix of the other factors necessary for major economic growth, although they may be profitably exported. In Brazil vast iron-ore reserves are stimulating rapid industrial growth. Elsewhere in South America the scattered presence of mineral ores cannot lead to significant industrial development unless and until the complex economic factors become more favorable to industrialization. Such a change is already beginning. As population increases, markets broaden correspondingly, per capita incomes rise, and more capital is generated and accumulated. A skilled labor force begins to take shape, transport lines are improved, and the whole process of economic growth begins to gather momentum. Although it seems doubtful that Latin America will become a major industrial center on a par with Europe, Russia, Japan, or North America, or even with India and China, substantial growth is likely to occur from the present very small base.

Continental averages of almost anything can be seriously misleading. Of course, there are gross regional differences in Latin America, in economic matters as in others. Thus, while average per capita incomes are very low, Argentina and Venezuela are relatively well off despite a high cost of living. This is in part because of rich and easily exploited agricultural and mineral resources, and in part because of relatively small total populations. But nearly all of the Latin American states share the economic problem of heavy dependence on a few export commodities, a situation like that of most of the countries of southeast Asia described in Chapter 16. In both regions a superstructure of commercial production, agricultural and mineral, has been added relatively recently to the still-dominant basis of subsistence agriculture. In both regions, too, this development has been due primarily to foreign capital investments, and has been designed primarily for supplying overseas markets with food and industrial raw materials. Foreign capital has largely built the commercial sectors of Latin American economies, as it has in southeast Asia and Africa. All three regions are commercially dependent to a disturbing degree on a small list of commodities. Although this trade is profitable, it is psychologically unsettling, given the frequent fluctuations in market demand and price. Its lopsided nature may be a symptom of relatively immature economic development. There are increasing attempts in every Latin American country to diversify commercial production and export. However, it is not always reasonable or profitable to do so at the expense of the one or two highly profitable specialties on which the export trade now depends.

Latin America as yet has few advantages for other types of production, in competition with industrial and agricultural areas elsewhere. There is no doubt that diversification can and should be pursued, especially where land is left idle or is underused because it is not suited to the production of a particular commodity, or because cattle raising and horses have more social prestige than cereal farming. In addition to the relative scarcity of its mineral resources and the economic disadvantages common to all underdeveloped areas, Latin American production faces high transport costs in many cases. Lands south of Brazil are remote in distance from the major markets concentrated in the Northern Hemisphere. The west coast, although it is actually closer to the north Atlantic, faces an economic barrier in trade with the north Atlantic because of canal dues at Panama, a situation similar to that of east Africa. Internally, physical barriers are reflected in high transport costs. For example, despite Panama Canal dues, it is sometimes cheaper to ship goods from coastal Peru to eastern Peru through Panama and up the Amazon than to carry them across the Andes. Vast rain forests and deserts in Latin America, as well as mountains, constitute both economic and physical barriers to exchange. The generally coastal pattern of settlement tends to increase the gaps between trade-generating centers and to widen the economically empty distances that separate them.

Growth prospects

As in many parts of the world where governments are anxious to stimulate rapid industrial or economic growth, agricultural development is frequently neglected. Manufacturing has great prestige in the modern world, and is a source of political power. But for Latin America, as for southeast Asia, the greatest comparative advantages are in agriculture. Neglect of the land problem, for any reason, is neglect of the heart of the economy as it almost certainly will remain for the foreseeable future. Agriculture will continue to be the dominant occupation and the primary source of domestic income, foreign exchange, and capital surpluses for investment in economic growth. Except for the limited plantation areas of Middle America and most of the coffee and commercial grain- and animal-producing areas in Brazil and the pampa, Latin American agriculture is predominantly subsistence in nature, and is carried out with primitive techniques by poor peons. It often makes inefficient or even destructive use of the land, and produces low per capita incomes. Its productivity could be very greatly increased with planning, assistance, and investment from government. If the methods used in Japan during the past seventy or eighty years could be applied (including not only the introduction of better techniques and research stations, but provision for an effective agricultural extension service to demonstrate improved techniques to the farmer), the entire face and economic situation of Latin America could be changed in a generation. In a sense the problem of Latin America is a social and political one. Effective organization and planning have been largely lacking, with some notable exceptions. There are signs that many of the Latin American countries are beginning to realize the nature of their problem, its urgency in the face of rapid population growth, and some of the solutions that can be applied. The coming century will be a critical one. Latin America has been given an economic margin that was largely denied to Asia. It remains to be seen whether this opportunity will be seized effectively.

Questions for further study and discussion

1. Why should it be undesirable for a country to have one or two commodities dominate its export trade? How does Latin America differ in this respect from the southeast Asian countries?
2. Why should it seem reasonable not to expect major industrial growth in Latin America on a par with Europe, Russia, North America, India, China, or Japan?
3. How can one account for the fact that Latin American population growth rates are among the highest, if not the highest, in the world at present?
4. Why is central coastal Chile a desert? Why do arid conditions also appear on the east coast in southern Argentina, while southern Chile is humid?
5. How has it happened that Africa, although close to Europe, was conquered or explored so much later than Latin America, that Europeans are still in a tiny minority there, and that African culture and political control seem certain to prevail? What has made the modern history of Latin America so different?

23 Latin American regions

A Peruvian couple goes to market. *(Photo by Barbara Klutinis for Van Cleve.)*

Regional distinctions are probably less sharp and less profound in the New World than in the Old World of Eurasia, if only because the Western Hemisphere has been settled only recently by the people who now dominate it. There has been less time for regional distinctions to develop, at least in cultural terms. The tremendously increased facility of communication in the modern world has tended to aid the spread of a single, dominant culture more or less uniformly over very wide areas, especially where there were few sophisticated local cultures to resist it. The isolation characteristic of the Western Hemisphere before Columbus, which was responsible for a good deal of backwardness or primitiveness by the standards of the Old World, suddenly yielded to a large-scale flow of powerful influences from Europe. Despite marked regional differences in physical environment, population density, and more recently in economic specialization, there is considerable cultural unity over most of Latin America, as compared with Europe or with monsoon Asia. Both of them are far more diverse culturally, in part because their regional cultures grew up many centuries before the modern revolution in communications. But these are, of course, matters of degree; North America is culturally far more uniform than Latin America. The variety of Indian, Negro, and differing European influences on Latin America, plus physical differences, distance, and barriers to communication, have resulted in marked regional variations. There are important cultural differences within Latin America, and important differences in political forms and sovereignties. This chapter will attempt to establish half a dozen major regional entities and discuss their characteristics.

The West Indies

The islands that neatly ring the Caribbean Sea are still called Indies, perpetuating their early confusion with the Orient on the part of Renaissance explorers looking for a sea route to India and Cathay. "West Indies" distinguishes them from insular southeast Asia, which is also called the "East Indies." The name is suggestive of wealth, and "the riches of the Indies" is a phrase still used. In terms of human well-being, it is tragically inappropriate, probably more so for the West Indies than for the East. At the

PACIFIC OCEAN
GULF OF M
MEXICO
Tropic of Cancer
ARIZONA
NEW MEXICO
OKLAHOMA
ARKANSAS
MISSISSIPPI
LOUISIANA
TEXAS
BAJA CALIFORNIA NORTE
BAJA CALIFORNIA SUR
SONORA
CHIHUAHUA
COAHUILA
NUEVO LEON
TAMAULIPAS
SINALOA
DURANGO
ZACATECAS
SAN LUIS POTOSÍ
NAYARIT
JALISCO
GUA
HIDALGO
MEX
MICHOACAN
GUERRERO
OAXACA
VERACRUZ
TABASCO
CHIAPAS
CAMPECHE
YUCATAN
QUINTANA
GUATEMALA
BELIZE
EL SALVADOR
HON
BOLSÓN DE MAPIMÍ
SIERRA MADRE DEL SUR
Golfo de California
Bahía de Campeche
Golfo de Tehuantepec
Golfo de Fonseca
ISLAS REVILLAGIGEDO (Mex.)
ISLAS MARIAS
CABO CORRIENTES
C. SAN LUCAS
C. SAN LAZARO
Bahía Magdalena
Bahía Sebastián Vizcaíno
Laguna de Términos
Phoenix
Tucson
Tijuana
Mexicali
Ensenada
Nogales
Bisbee
Douglas
Cananea
Deming
Las Cruces
Socorro
El Paso
Ciudad Juárez
Pecos
Hermosillo
Guaymas
Ciudad Obregón
Navojoa
Santa Rosalía
La Paz
Culiacán
Mazatlán
Chihuahua
Ojinaga
Hidalgo del Parral
Gómez Palacio
Torreón
Durango
Monclova
Piedras Negras
Eagle Pass
Del Rio
Saltillo
Monterrey
Nuevo Laredo
Laredo
Matamoros
Brownsville
Corpus Christi
San Antonio
Austin
Houston
Galveston
Beaumont
Dallas
Fort Worth
Waco
Corsicana
San Angelo
Lubbock
Wichita Falls
Childress
Lawton
Fort Smith
Memphis
Little Rock
Hot Springs
Pine Bluff
Texarkana
Shreveport
Vicksburg
Jackson
Natchez
Baton Rouge
New Orleans
Mobile
Pensacol
Ciudad Victoria
Tampico
Ciudad Mante
Zacatecas
San Luis Potosí
Aguascalientes
Tepic
Guadalajara
León
Guanajuato
Querétaro
Celaya
Morelia
Pachuca
MEXICO CITY
Toluca
Cuernavaca
Puebla
Tlaxcala
Jalapa Enríquez
Veracruz
Orizaba
Tehuacán
Chilpancingo
Acapulco
Colima
Manzanillo
Ciudad Guzmán
Oaxaca
Juchitán
Tehuantepec
Salina Cruz
Coatzacoalcos (Puerto México)
Minatitlán
Villahermosa
Tuxtla Gutiérrez
Ciudad de las Casas
Comitán
Campeche
Ciudad del Carmen
Mérida
Progreso
Valladolid
Ciudad Chetumal (Payo Obispo)
Belmopan
Quezaltenango
Mazatenango
Guatemala
Antigua
Cobán
Santa Ana
San Salvador
San Miguel
Comayagua
Citlaltépetl 18 701
Popocatépetl 17 887
Nev. de Toluca 14 409
Tacaná 13 417
30°
25°
20°
15°
10°
5°
110°
105°
100°
95°
90°
© Rand McNally & Co.

MAP 101
LANDFORMS OF CENTRAL AMERICA
Railroads
0 200 400 600 Km.
0 200 400 Mi.
ATLANTIC OCEAN
NORTH CAROLINA
SOUTH CAROLINA
GEORGIA
FLORIDA
Knoxville
Raleigh
Mt. Mitchell 6684
Charlotte
CAPE HATTERAS
Wilmington
CAPE FEAR
ATLANTA
Columbia
Augusta
Charleston
Savannah
Jacksonville
Tallahassee
St. Augustine
Apalachee Bay
Ocala
CAPE CANAVERAL
Tampa
Tampa Bay
W. Palm Beach
Lake Okeechobee
MIAMI
CAPE SABLE
Key West
FLORIDA KEYS
Straits of Florida
BAHAMAS
GRAND BAHAMA
GREAT ABACO
ELEUTHERA
Nassau
ANDROS
CAT
SAN SALVADOR (WATLING)
LONG
ACKLINS
CAICOS (Br.)
TURKS (Br.)
GT. INAGUA
HAVANA
Guanabacoa
Matanzas
Marianao
Cárdenas
Pinar del Río
Santa Clara
Sancti Spíritus
Cienfuegos
Trinidad
Ciego de Avila
Nuevitas
Camagüey
Holguín
CUBA
C. SAN ANTONIO
ISLA DE PINOS
Manzanillo
Guantánamo
SIERRA MAESTRA
Santiago de Cuba
PUNTA MAISI
C. CRUZ
GRAND CAYMAN (Br.)
GREATER ANTILLES
WEST INDIES
Windward Passage
Cap-Haïtien
Gonaïves
ÎLE DE LA GONÂVE
HAITI
Port-au-Prince
Pico Duarte 10,417
DOMINICAN REPUBLIC
Puerto Plata
Santiago de los Caballeros
C. SAMANA
Sánchez
Santo Domingo
C. ENGAÑO
HISPANIOLA
Mona Passage
Mayagüez
San Juan
Ponce
PUERTO RICO (U.S.A.)
VIRGIN IS.
ST. THOMAS
Charlotte Amalie
SAINT CROIX (U.S.A.)
ANGUILLA (Br.)
BARBUDA (Br.)
ST. KITTS (Br.)
NEVIS (Br.)
ANTIGUA (Br.)
MONTSERRAT (Br.)
V. Soufrière 4869
Basse Terre
Pointe-a-Pitre
GUADELOUPE (Fr.)
DOMINICA (Br.)
LEEWARD IS.
MARTINIQUE (Fr.)
Fort-de-France
ST. LUCIA (Br.)
BARBADOS
ST. VINCENT (Br.)
Kingstown
Bridgetown
GRENADA
WINDWARD IS.
LESSER ANTILLES
Montego Bay
Mt. Denham 3236
Port Antonio
Spanish Town
Kingston
JAMAICA
CARIBBEAN SEA
MEXICO
Trujillo
Coco
Managua
Bluefields
Lago de Nicaragua
San Juan del Sur
CENTRAL AMERICA
COSTA RICA
San José
Limón
Puntarenas
Cartago
CANAL ZONE (U.S.A.)
Colón
Portobello
ISTHMUS OF PANAMA
Golfo de los Mosquitos
Panamá
PANAMA
David
Antón
Santiago
Golfo Dulce
COIBA
PEN. DE AZUERO
Gulf of Panama
Golfo del Darién
ARUBA (Neth.)
CURAÇAO (Neth.)
BONAIRE (Neth.)
C. SAN ROMAN
Willemstad
PUNTA DE GALLINAS
PENÍNSULA DE GUAJIRA
PEN. DE PARAGUANA
Golfo de Venezuela
Coro
ISLA LA TORTUGA
ISLA DE MARGARITA
TRINIDAD AND TOBAGO
TOBAGO
TRINIDAD
Port of Spain
Carúpano
Santa Marta
Barranquilla
Ciénaga
Soledad
Cartagena
Maracaibo
Cabimas
Lago de Maracaibo
San Felipe
Puerto Cabello
La Guaira
CARACAS
Cumana
Puerto la Cruz
Maracay
Valencia
Barquisimeto
Maturin
Lorica
Sincelejo
Mompós
Magangué
Montería
Trujillo
Valera
Mérida
CORDILLERA DE MERIDA
Guanare
Calabozo
El Tigre
Orinoco
Ciudad Guayana
Ciudad Bolívar
Cerro Bolívar
Puerto de Nutrias
San Fernando de Apure
Apure
Ocaña
Cúcuta
San Cristóbal
Pamplona
Arauca
Meta
VENEZUELA
Cerro Icutu 7800
Salto Angel
Barrancabermeja
Bucaramanga
Medellín
Sonsón
Manizales
Pereira
Tunja
BOGOTÁ
COLOMBIA
Río Magdalena
Cauca
Atrato
Río Orinoco
Caura
Caroní
Ventuari
GUYANA
Kaieteur Fall
L. COCO (Costa Rica)

beginning of Western colonial empires, the Caribbean islands did yield an impressive financial harvest. For two or three centuries they supported a major economic boom that brought fortunes to colonial proprietors and investors, even though much of the local population was legally in slavery and economically in misery. The basis of the boom was sugar, a crop in which the West Indies had compelling advantages for the supply of the rapidly growing European market. Relative proximity to Europe, cheap sea transport, local tropical climate, and numerous areas of productive soils made an effective combination. Sugar had a very high value in Europe, where it could not be grown. Until the West Indies entered the market, it was a luxury far beyond the reach of all but the very wealthy. Even many of them used honey, the only sweetener available to most of the population. The West Indies did not produce the gold the Conquistadores hoped for, but sugar created even more wealth than was looted from the Aztecs and the Incas. The boom was not permanent, however. Since about the beginning of the nineteenth century, the Caribbean islands have increasingly become a depressed area. In the eighteenth century Londoners spoke of being "as rich as a West Indian planter"; in the 1950s the West Indian Federation of former British colonies was described as "a pooling of economic misery."

Two factors were primarily responsible for this devastating change. Probably most important was the growth of competition from other tropical areas, most of them also part of European colonial empires. Increasingly their sugar flooded the market and depressed prices, a problem still far from solution. The tropics cover an immense area, and large parts of it are suited to growing sugar cane. There is much more sugar land than is needed to supply world needs, and so far it has proved impossible to regulate cultivation and production effectively. All producers hope to win an advantage in the market so they may continue to employ their existing sugar labor force or sugar investments, and each growing area is reluctant to curtail its own sugar production so others may benefit. Many tropical producers enjoy direct or indirect subsidies as a result of preferential agreements with former colonial masters: for example, the Philippines, Hawaii, and Puerto Rico. All three, as colonial territories, were allowed to sell sugar to the continental United States without import duty or with special tariff concessions or quota provisions (advantages they still retain). All profited from heavy American investment in sugar production. But few sugar producing areas anywhere in the tropics have enjoyed much prosperity during the past century.

The second factor was the development of a beet-sugar industry in Europe and, more recently, in the United States. Stimulated by the blockades of the Napoleonic wars, and alarmed by the prospect of being cut off periodically from tropical sugar sources, Europe encouraged and subsidized the cultivation of sugar beets. Before the end of the century they came to supply about half of the continent's sugar. In the United States sugar beets are only slightly less important. United States sugar-beet production is subsidized, as it is in Europe. This understandably irritates Latin American sugar producers, who complain that the United States' professions of concern for their economic welfare are not sincere, and that trade is preferable to aid. Beet sugar is still more expensive to produce than cane sugar, but artificial subsidies in both cases obscure real costs and tend to penalize the consumers.

Political patterns

Sugar and its changing fortunes were involved both in the early prosperity of the West Indian economy (at least for the owners and planters) and in the economic doldrums of the past century and a half. Sugar also contributed to the present composition of the population by providing the need for the import of Negro slaves, whose descendants now dominate most of the West Indies. Several of the most productive areas came under British and American rather than Spanish control, in part because sugar made the West Indies much more desirable than South America to outside powers. French, Dutch, and Danish colonial holdings were also established in the Caribbean area, but were never extensive. Only a few remnants now remain (consult a political map of the area—the Virgin Islands passed from Danish to American control by purchase in 1916). The Caribbean, however, was attractive to many European powers, not only for its potential riches in sugar, but also for strategic reasons. Long before a canal was cut through the Isthmus of Panama, the Caribbean was a major avenue of communication between the New World and the Old. Gold and silver from the west coast of South America and goods from China by way of Manila were carried by pack animals across Panama on a specially constructed road through the rain forest and reloaded on ships for transport across the Atlantic. Forts in the West Indies could dominate or threaten these important sea lanes, since the island ring almost encloses the Caribbean and leaves only narrow sea passages. The Greater Antilles (Cuba, Jamaica, Haiti and the Dominican Republic, and Puerto Rico) encircle the north, and the graceful curve of the Lesser Antilles south and east of Puerto Rico, including the Leeward and Windward islands, blocks the east.

This situation was also a standing invitation to piracy. Early Spanish forts at Havana (la Habana) and at Cartagena, on the Colombian coast, were designed to guard the movements of gold and silver by galleon fleets to Spain from all of the Spanish territories in the New World. But they could not keep the whole of the Caribbean free of raiders and rivals, despite its name, "the Spanish Main." Pirates, based in the innumerable tiny islands of the Caribbean, plagued the rich fleets. The English, in particular, occupied and retained a large number of Caribbean islands, attracted by sugar, by strategy, and by the chance to embarrass the old and bitter Spanish enemy.

Britain won control of nearly all of the Lesser Antilles (except for the islands of Martinique and Guadeloupe, which remain French), plus Trinidad, Guyana (formerly British Guiana) on the South American mainland, Belize (formerly British Honduras) on the east coast of Central America, the sugar-rich island of Jamaica, and the island chain of the Bahamas. In 1956 the West Indies Federation united all of these British territories except the Bahamas, the British Virgin Islands, Guyana, and Belize. After a long dispute the federation capital was located at Port of Spain in Trinidad. English is the language of Jamaica and the Lesser Antilles, except for French Guadaloupe and Martinique, although the population of all of the British and French areas is dominantly Negro. But progress toward unity within the new federation was hampered, and destroyed in the end by the isolation of its numerous small islands from each other, by the very strong regionalism characteristic of most of them, and by the fact that similarity of production makes them competitors for trade rather than partners. Their strongest ties traditionally have been with London rather than with each other, and the sea has tended to divide rather than to join them. Local political party structures especially in Jamaica and Trinidad-Tobago, by far the largest two units in the Federation, also worked against the union. Jamaica and Trinidad felt they had little to gain from federation as opposed to the separate independence they chose instead in 1961. The smaller island units would have benefited more from a wider unity, but the Federation could not be kept alive.

Haiti, the western part of the island of Hispaniola, was a French colony until the end of the eighteenth century, and was long the leading sugar producer in the New World. The French were driven out by a series of slave revolts culminating in national independence in 1804, although French is still the language of Haiti. The eastern part of Hispaniola, where sugar was less important, Negroes less numerous, and Spanish blood and culture proportionally more important, established its independence from Haiti in 1844 as the Dominican Republic. It still regards more densely populated, Negro Haiti as a threat. The Spanish kept control of Puerto Rico and Cuba long after they had lost their colonial possessions on the mainland. Puerto Rico fell to the United States after the Spanish-American War in 1898, and in 1917 it was made a United States territory. It is an integral, although largely self-governing, part of the union rather than a colony. Cuba never became United States territory, but following the Spanish-American War, its commercial economy was heavily influenced by United States' investments in sugar production. Most of Cuba's culture is characteristically Latin American. It is distinguished from most of the rest of the West Indies by Spanish language and customs (including bull fighting), and the population is mainly mestizo. The succession of corrupt and oppressive oligarchical rulers was replaced in 1959 by a Socialist government under Fidel Castro, which carried out land reform and nationalized most large commercial and industrial enterprises.

People and resource use

The total area of the West Indies is small, but population totals are large. Cuba, by far the largest island of the West Indies, is about the same size as Pennsylvania; Jamaica is approximately as big as Delaware. Practically all of the West Indian islands are densely populated, more so than any other part of Latin

America. The region as a whole contains nearly 30 million people in a total area smaller than the state of Oregon. The original Carib Indian inhabitants (hence the name *Caribbean*) were largely exterminated by the Spanish, so the present population is alien. The largest single group are descendants of Negro slaves. In Puerto Rico and Cuba, where Spanish control lasted longer and where sugar plantations were not developed as early or as extensively, mestizo elements are dominant. East Indians are also present in significant numbers, descendants of indentured laborers brought in to work the sugar plantations in the British-controlled islands. They form nearly two-fifths of the population of the island of Trinidad. Chinese, Europeans, and North Americans are important minorities, and there is a large element of mixed Negro-white and Negro-mestizo population.

The region as a whole is one of the most overcrowded and poorest in the modern world, in terms of per capita incomes. The duration and effectiveness of European influences, as in the former colonial areas of monsoon Asia, have resulted in a population explosion by lowering the death rate relatively rapidly and failing to affect the still-high birth rate significantly. There is a large continuous stream of emigration from the West Indies—primarily to Britain in the case of the British areas, and primarily to the United States from Puerto Rico and Hispaniola. But emigration does not keep pace with population increases. With the sugar industry in poor economic health for the past century or more, and with few prospects for its adequate recovery in the future, the West Indies faces discouraging economic prospects. Sugar is still the region's dominant export and the chief source of commercial domestic capital and income.

Attempts at economic diversification are hampered by limited usable resources. Tropical agriculture is highly competitive, and there are relatively few enterprises in which the West Indies has solid advantages. One of these is sugar by-products, espe-

Bait for tourists—Barbados beach. Such scenes have obvious appeal for North Americans, especially in winter. Fortunately, they are typical of much of the West Indes, and tourism is now the area's largest single industry. *(Photo from American Airlines.)*

cially molasses and rum, which are produced almost everywhere in the region and exported in relatively large amounts. Another is tobacco, although even Cuba must be considered a marginal producer of tobacco to a large extent in the face of North American, Mediterranean, and Near Eastern production. Coffee production has grown considerably but faces stifling competition from Brazil and Middle America. Bananas have also become an increasingly important commercial crop, especially in Jamaica, but Middle American competition is vigorous. Cacao, coconuts, citrus fruits, and other tropical crops are grown, but most of them do not seem to be good prospects to support the mushrooming population on an adequate scale.

Fortunately there are some mineral deposits whose importance can be expected to increase, although they do not constitute a favorable basis for local manufacturing. Bauxite (aluminum ore) in Jamaica, petroleum in Trinidad, and nickel in Cuba will continue to be exported rather than supporting major local industrialization in the absence of other minerals, of an adequate domestic market, and of other necessary factors. Most of the farmers of the West Indies, like those of Latin America as a whole, are still essentially subsistence producers, barely supporting themselves on what they can produce by primitive techniques from an overcrowded and often poor land. A great deal of the land, especially outside of Cuba, is in slope. A large area is semiarid, despite the sea and the reliable trade winds blowing off the Atlantic. Aridity results from rain-shadow effects in the lee of hills or mountains and from the absence of topographic barriers or opposing air masses to cool and obtain rain from the trade winds. They are often colder than the air over the land, especially in summer.

One relatively promising economic enterprise is tourism. The northern parts of the West Indies are especially well located to attract profitable business from the United States. Climatic attractions are particularly strong in winter, and the Caribbean is an increasingly popular winter tourist area. Urbanization is still slight, except for Cuba. Havana has grown to over 1.7 million, attributable in part to American tourists until the advent of Castro and in part to heavy past American investment in sugar production and urban-based processing. Over 60 percent of Cuba's population is classified as urban, but elsewhere in the West Indies, cities are few and small. San Juan, the Puerto Rican capital, with about 1.3 million people, is second in size to Havana. It also owes much of its size to American tourists, American investments, and recent small-scale industrialization. The small degree of urbanization is one measure of the still-dominant subsistence nature of the West Indian economy and the economic depression that has been characteristic of the area for so long. Puerto Rico, whose rapid economic growth and rising income levels since 1945 result mainly from heavy United States investments, is a qualified exception that probably will not be duplicated elsewhere in the West Indies in the near future.

Mexico and Middle America

The narrow isthmus between the United States and South America is an isolated area, cut off on the north by desert and mountains and on the south by the rain forests adjoining Panama. But Middle America was the first part of the American mainland to be penetrated by Europeans, and of all of Latin America outside of the West Indies, it is now the most closely involved with "the colossus of the north," the name many Latin Americans use to refer to the United States. The commercial economies of the countries of Middle America, especially those south of Mexico, are dominated to differing degrees by United States investments in mining and plantation agriculture. And United States political influence has often been applied with a heavy hand. Middle America's connections with South America are tenuous economically and politically as well as in transport terms. Cultural similarities are closer and more substantial, but Middle America may legitimately be regarded as a separate region with Latin America. European as well as North American influences are stronger in Middle America than in South America as a whole, especially in investment and trade, and to some extent in immigration. European powers made several attempts to establish control over Middle American areas in the nineteenth century. France fought a losing battle to dominate Mexico, and both British and German territorial ambitions or spheres of influence in southern Middle America were also thwarted. The United States' position has been explicitly stated in the Monroe

Doctrine, which claims in effect the whole of the Western Hemisphere as an area of United States' concern from which all outside powers are to be excluded. The rising power of the United States, especially after the American Civil War, has tended to produce such a result. But it is resented in many respects, understandably, by most Latin Americans, despite the United States' role in preventing European colonialism.

Mexico dominates Middle America in size and population. For the most part it has been better able than the smaller states to resist United States' pressures despite the common border along the Rio Grande. The border marks a strikingly sharp cultural line. Although the physical landscape on both sides of it is essentially uniform and does not begin to change substantially for at least 100 miles (160 km) on either side, Latin America begins abruptly at the Rio Grande, and Anglo-America ends as abruptly. Spanish-Indian and British-European cultures, subsistence and commercial economies, poverty and wealth, agriculture and manufacturing, rural and urban civilizations, peasant and businessman face each other across the border.

Population and economic patterns

The northern third of Mexico is arid and mountainous, and relatively sparsely populated by poor subsistence peons. It provides a tremendous contrast with Texas and California across the border, or even with New Mexico and Arizona. Large parts of these four states were part of Mexico until 1848. They have developed extremely differently since their inclusion within the rich, rapidly growing, powerful, and English-influenced United States (by conquest in the United States–Mexico war of 1846). Mexico's major core is in the mountain basins and valleys on the high Mexican plateau in the southern part of the country around the capital at Mexico City, and between it and the city of Guadalajara about 300 miles (480 km) to the northwest. About half of the total population is concentrated in this core area. A tongue of the core reaches southeastward from Mexico City to the coastal port of Veracruz, which serves the capital and handles most of Mexico's foreign trade. Outside the core area most of Mexico is arid or semiarid, plagued by strong relief, and agriculturally marginal. Most Mexicans are subsistence producers. About a third of the population is pure Indian. Mexico City is a cosmopolitan Ladino metropolis with over 13 percent of the total national population. Some of Mexico City's size is attributable to American and other tourists. They are attracted by a bracing climate of sunny skies and mild temperatures on the central Mexican plateau at an elevation of about 7,500 feet (2288 m). But the city's growth is due more importantly to its share of urban functions within Mexico, to widespread mining activities financed and channeled through Mexico City, and to the beginnings of manufacturing.

The size of the capital is a political issue in Mexico, related to the dispute between the supporters of rapid Western-type industrialization and those who favor the retention of a more distinctively Mexican culture. Mexico City, as the principal example of Westernization and industrialization, is regarded by some of the nationalists as an alien drain on the Mexican economy and an undesirable cultural intrusion, especially because of its swollen size. Since the anticlerical and Socialist-inclined Mexican Revolution, which began in 1910, the Indian element in Mexico and its version of Mexican nationalism have played an increasing role in political, social, and cultural life. The bold mural paintings of the Mexican artists Diego Rivera and José Orozco and their "proletarian frescoes" adorn the walls of many public buildings in Mexico. They are perhaps the sample of Indian and revolutionary influences on modern Mexican culture best known to North Americans through examples of their art in museums and public buildings in the United States.

Subsistence agriculture still dominates the Mexican economy. But there are significant mineral resources that are now being more fully worked, and manufacturing is growing rapidly. Mexico still produces considerable silver, plus important amounts of lead, zinc, sulphur, and manganese. More recently, petroleum production has become important, mainly in the Tampico area along the central east coast. Several other new pools of oil have been discovered, and Mexico now has more than enough to meet its domestic needs. Relatively small reserves of good-quality coal and iron ore are mined in increasing volume in the vicinity of Monterrey (northeast Mexico). The main iron and steel plants are at Monterrey and Monclova, although most of the iron ore is mined near Durango, some 300 miles (480 km) west of Monterrey. The bulk of industrial production is still consumer goods, mainly textiles and food processing. It is still concentrated primarily in the core area, around Mexico City. A German firm has built a steel plant on the central west coast, using rich local ore, which produces half a million tons a year.

The sharp, high mountain chain of the Sierra Madre and its southern extensions, marking the edge of the Mexican plateau, tends to hamper access from the core area to the west coast. There is no major port on the Pacific and only a narrow and semiarid or desert coastal plain, which supports few people. The somewhat broader eastern coastal plain, widening into the rain forest of the Yucatan peninsula, is also relatively thinly settled. Tampico and Veracruz, with their special functions, are the only real cities. Roughly the northern half of the coast is arid; the south is excessively wet and hot. In most of Yucatan, settlement is virtually absent, and the rain forest poses an additional barrier. The origins and flourishing of the Maya civilization in Yucatan and adjacent Guatemala remain a puzzle.

Middle America as a whole illustrates the advantages that may accrue to highlands in the tropics: moderate temperature and rainfall patterns, and often soil fertility superior to the lowlands. The Spanish recognized three altitudinal zones in tropical America and gave them names that are still used: (1) *tierra caliente*—sea level to 2,000 or 3,000 feet (610 to 915 m): tropical, heat-loving, and moisture-loving crops (*caliente* means "hot"); (2) *tierra templada* ("temperate")—up to about 6,000 feet (1830 m): coffee, maize, cotton, and subtropical crops; (3) *tierra fria* ("cool")—up to about 10 thousand feet (3050 m) (higher near the equator), with occasional frosts: wheat, apples, barley, potatoes. Such zones and gradations in crop patterns are clearly observable in most of tropical America. With a few exceptions outside the West Indies, the largest and densest population clusters are in the highlands, or generally in the zone of *tierra templada*.

The dominant crop of Mexican agriculture is maize, originally native to Middle America and still strongly favored by cultural preference. It is often grown in areas not really suited to it, and occupies about one-third of Mexico's cultivated land. Most of the Mexican plateau is really too dry for maize and is better suited to wheat or to grazing. Only about 12 percent of the total area of Mexico is arable, and some of that is marginal. Few areas are free from aridity or from too much rainfall. Excessive slope is a nearly universal problem, so erosion is widespread, especially in fields planted in a poor cover crop like maize, which leaves the soil exposed between rows. But without maize the Mexican peons would be bereft indeed. Their traditional diet revolves around it in a limited variety including tortillas and tamales (various forms of simple cakes and dumplings made from corn meal), supplemented by beans, squashes, and small amounts of green vegetables, hot peppers, eggs, poultry, and a little scrawny beef. Mexico produces some cotton, a good deal of it irrigated, to supply the domestic textile industry. Tobacco, sugar cane, sisal, coffee, and bananas are important in the zone of *tierra caliente*. Wheat acreage is increasing in the dry areas of the plateau, especially with the help of high-yielding dwarf varieties originally discovered and developed in Mexico. They have spread more recently to many areas throughout the world as the original basis of the "Green Revolution." But the agricultural system as a whole is still not far from what it was when the Spanish arrived, and improvements are threatened by population growth. It will be many years before industrialization by the Mexican government and its program of land redistribution and improvement of agriculture can gather momentum and make a large impact on the economy in terms of major increases in per capita income, especially for peasants and peons.

Middle America

Guatemala, Belize, Honduras, El Salvador, Nicaragua, Costa Rica, and Panama lie between Mexico and South America, occupying an area a little smaller than Arizona and Nevada combined. They support nearly 20 million people whose numbers are increasing even more rapidly than those of the West Indies. Each of these independent republics is characterized by a small core clustered around its capital city, leaving the rest of the area relatively sparsely settled. The tiny state of El Salvador, relatively thickly populated over most of its territory, is the exception. The plateau that runs down the center of Middle America and contains most of the capitals and cores lies in the zone of *tierra templada* for the most part. It generally receives an adequate rainfall, concentrated in summer. The low coastal plains are small, except in Nicaragua, and are characterized by excessively heavy rains and high temperatures, especially on the Caribbean side. The plateau contains large areas of volcanic soils, and the present settlement cores on the plateau coincide fairly closely with the original Indian cores that first attracted Spanish attention. Soils on the coastal plains are much more heavily leached and do not support nearly as many people. In Nicaragua this pattern is interrupted to some extent. The original center of Indian settlement and the present population core spill over onto the Pacific coastal plain. Panama is an artificial state, created by revolution from Colombia with American support, at the time of the American demand for a territorial concession to include the Panama Canal Zone. Colombia refused, and the state of Panama agreed. Its artificial nature is shown by the concentration of population almost entirely in the Canal Zone, which includes the old Spanish-founded cities of Panama and Colon. The rest of the country is nearly empty, and most of it is covered by dense rain forest or bush growth.

Middle America south of Mexico, like the rest of Latin America, is still dominated economically by subsistence peon agriculture. But, during the past century, two commercial crops have become important. After about 1850 coffee growing in the *tierra templada* of the plateau became significant and is still the area's chief export. The volume of production is relatively small, compared with Brazil, for example, but the coffee is of high quality and brings a good price. By 1900 banana plantations had begun to make use of some of the coastal lowlands and have now become of great importance, second only to coffee. Forests were cleared, harbors improved, rail lines built, and Negro workers imported from the West Indies. Until recently banana plantations were largely confined to the Caribbean coast for easier access to the main United States market. About 1940, however, the spread of plant disease, aggravated by the concentration of a single plant species on plantations, forced the abandonment of large areas of banana production. The west coast, where the disease was restricted by control measures, now has more banana plantations, although the location increases transport costs to market.

The development of commercial production of coffee and bananas has been financed and managed almost entirely by foreigners: mainly Europeans in the case of coffee, and almost entirely by Americans in the case of bananas. The United Fruit Company and Standard Fruit have a virtual monopoly of banana plantations and marketing facilities in Middle America, including their own fleets of refrigerated ships. A large part of the area's rail lines and port facilities have been built by United Fruit, and the company's commercial and even political power in these small countries is significant. The political instability characteristic of most of the Middle American republics made the United Fruit Company's political influence even more effective in the past, as it could tip the continually changing balance of power between rival groups.

Coffee has been developed to a greater extent by small individual planters, most of them Europeans. The population as a whole, however, remains more Indian than European. Guatemala is still mainly pure Indian. The other republics are dominantly mestizo, except for Costa Rica, where Spanish and more recent European immigrants are in a majority. Costa Rica is also remarkable within Latin America as a genuine political democracy. It is characterized by independent smallholder farmers rather than large estate owners and peons, a high literacy rate, relatively high and relatively equal incomes, and the complete absence of an army. Unfortunately Costa Rica is not typical of Middle America, although, like its neighbors, coffee and bananas constitute most of its exports.

Northern South America and the west coast

Colombia, Venezuela, Ecuador, Peru, and Bolivia are all Andean in part. Except for landlocked Bolivia, each also has a strip of tropical coastal plain and an interior tropical lowland behind the mountain chain. The Guianas, split into the three political units of Guyana, Surinam, and French Guiana, are considered here because they have even less in common with Brazil or with the West Indies. Venezuela and the Guianas may also be regarded as parts of a single physical region, sharing the same hot coastal plain and the same interior highlands. In the Guianas and eastern Venezuela, population is largely confined to the coastal plain. But in western Venezuela and in the other countries that form this tropical region, the principal settlement clusters are in the highlands (see

Llanos landscape—Venezuela. This is tropical savanna country, still largely unused, although agricultural settlements such as this pioneer village are beginning to increase with government support. *(From Standard Oil Company of New Jersey Collection, University of Louisville.)*

Map 99, page 419) for the same reasons as in tropical Middle America. The equator runs through Ecuador and Colombia, and in these low latitudes large parts of the lowlands are both uncomfortable and economically unproductive. The region as a whole is characterized by isolated population clusters in separate highland areas where relief is moderate, and a lesser population on the coastal plain where there are local agricultural or mineral resources. As elsewhere in Latin America, these separate nuclei of population are associated with separate political states, set off from each other by relatively empty intervening areas.

Colombia

Colombia's population is concentrated in the Andean highlands, where Bogotá, the capital, is located. A second center is near the mouth of the Magdalena River on the Caribbean coast. Petroleum and banana plantations in the Magdalena valley have contributed significantly to the growth of this lowland population cluster. But the chief commercial crop and export of the country is coffee. Coffee is a crop of the *templada* and in Colombia is grown mainly in the upper highland valleys of the Magdalena and its tributaries. There is some commercial grazing on the generally poor tropical pastures, but coffee supplies about half of Colombia's exports. The bulk of the population, however, remains on a subsistence level of agriculture. Bogotá, with nearly 3 million inhabitants, is a surprisingly big city for such a country, whose total population is only a little over 25 million. Bogotá has been a cultural center for northern South America for centuries despite its relatively isolated location in a high Andean basin. Part of its importance resulted from its location along the route used by the Spanish and the Indians between the Caribbean and southern South America. The forests of the Amazon basin remain a far more formidable barrier than the Andes. Until the modern development of sea transport, traffic from eastern Argentina, for example, moved northward and to Spain by the highland road along the main chain of the Andes through La Paz, Quito, and Bogotá to the old port of Cartagena on the Colombian coast of the Caribbean. Barranquilla on the coast northeast of Cartagena and the interior highland cities of Medellin and Cali in southwestern

Colombia are now as important commercially as Bogotá or Cartagena, however.

Venezuela

The northern extensions of the Andes chain reach the Caribbean coast of Venezuela, leaving the swampy basin of Lake Maracaibo (really a gulf of the sea) open to the Caribbean, but largely enclosing an interior plain, the *llanos*. Beyond the llanos, in the southern half of Venezuela, are the rugged highlands that also extend into the Guianas (see Map 19, page 67). Thus, Venezuela may readily be divided into two highland and two lowland areas, with population principally concentrated in the northern highlands of the Andes chain. Caracas, the capital, is near the center of the coastal zone where the highlands reach the sea and begin to degenerate into a low plateau of slight relief. The rich petroleum pools of the Maracaibo basin are contributing to a rapid increase in the population of the lowland area around the lake, which was extremely sparsely settled until about 1925. Swamps are being drained and new towns are growing up. Petroleum now provides over 90 percent of Venezuela's exports (much of it obtained from wells sunk into the shallow lake). With about 6 percent of the world oil production, Venezuela has one of the highest per capita incomes in Latin America. This has contributed to the impressive growth of Caracas, whose metropolitan population is about 2.4 million. Growth is being further accelerated by the discovery of very large and rich deposits of high-grade iron ore on the edge of the southern highlands, especially at Cerro Bolivar ("the iron mountain") in a previously neglected area, fortunately close to the navigable Orinoco River (see Map 19). The development of these iron-ore resources is being financed with United States capital, and a small Venezuelan steel plant has been completed. But most of the rapidly growing ore production moves by water to steel plants along the northeast coast of the United States. Until recently most of the petroleum has been exported in crude form, to be refined mainly on the two islands of Curacao and Aruba off the Venezuelan coast by British, Dutch, and American companies. But there is now a growing number of refineries in Venezuela.

The exploitation of the iron-ore deposits will also accelerate the transformation of the llanos, which is, in effect, the basin of the Orinoco River. Some of the capital earned will be invested in agriculture and grazing. Venezuela, with millions of acres of unused potential agricultural land, imports food. Agriculture has been seriously neglected. Investment and attention have been concentrated on mineral resources, and to a smaller extent on the growing of a few commercial crops for export: mainly sugar, tobacco, and cotton on the limited coastal lowlands, and coffee in the highlands. The llanos especially has been neglected (except for booming oil production), although scrawny cattle have been grazed there since they were introduced by the Spanish in the sixteenth century. Much could be done to improve the llanos pastures. Productive African grasses could make the area a major surplus producer of meat and hides of reasonable quality. The highlands of the south, frequently called the Guiana Highlands, are little

The altiplano. Llamas are primarily used as pack animals in this area. The man's felt hat is worn partly as protection against the cold at these high altitudes, and is typically Peruvian. The people of the altiplano are noted for their longevity. *(From Standard Oil Company.)*

known, very sparsely populated by primitive tribes, and remarkably isolated. But they contain large *templada* areas where expansion of commercial agriculture is possible.

The Guianas

The Guianas—Guyana, Surinam, and French Guiana—are still very sparsely settled. They lack the Andean highlands and petroleum-rich basins of Venezuela, but share its isolated southern highlands and steamy coastal plain. Only a tiny fraction of the land surface is under cultivation, although agriculture supports most of the small population. The three colonial owners were attracted to this area for its potential as a producer of tropical plantation crops. There has been some development in sugar, bananas, and coconuts, but it has remained small. Rice is the chief crop in fact, and most of the agriculture is subsistence in nature, although the presence of plantations has resulted in a dominantly Negro population. The former British and Dutch areas, now independent as Guyana and Surinam, have one major mineral resource—bauxite (aluminum ore). They produce about 20 percent of the world total, most of which moves to the United States and Canada, where it is refined electrically. French Guiana lacks this asset, is isolated from transport routes, and has been developed very little economically.

Ecuador

Ecuador (Spanish for "equator") is dominated by the Andes, which cover over half of the country and tower 20 thousand feet (6100 m) above the limited coastal and isolated interior plains. Settlement is highly concentrated in small mountain basins and valleys rather than on the equatorial lowlands. The people are still predominantly pure Indian. With few mineral resources to attract the Spanish and little productive agricultural land, the Indians were largely left to continue their traditional subsistence farming, still the major sector of the economy. In the present century there has been a significant growth of lowland settlement with the discovery of petroleum there. The increase in cacao plantations, favored by equatorial climate and the absence of strong winds, also encouraged lowland settlement. Cacao may have been native to Ecuador. Bananas and sugar are also grown on the coastal lowlands for export. Coffee is grown in the highlands and has long been an important export, although it is now greatly exceeded by bananas. Lowland Guayaquil, the chief port, is now a larger city than Quito, the capital. It handles the bulk of Ecuador's foreign trade, which is surprisingly large for so small and commercially underdeveloped a country. Quito is remarkable because, although it is only a few minutes from the equator, it has a cool climate at an altitude of over 9,000 feet (2745 m), well within the *tierra fria*. Quito is principally an administrative rather than a commercial or industrial center, and its modern growth has been hampered by isolation and transport costs. Quito's population is only about half a million. The mountain valley in which it is located, and the other tributary valleys of the Andean highland, are not commmercially productive enough to support a large city. But administrative functions have centered in Quito on the ancient Andean highway since the latter days of the Incas, when Quito was made the capital of the northern half of their empire, shortly before Pizarro landed.

Peru and Bolivia

Peru and Bolivia have an Indian majority, and both were part of the Inca empire. Despite the discovery of gold and silver in Peru and the hoards of the Incas, the area was not attractive to large-scale Spanish settlement. Precious metals proved to be limited, and there was little land suitable for plantations. Other agricultural opportunities were of small interest to the Spanish. Although the Conquistadores were probably responsible for a permanent decrease in the population of the highlands through slaughter and forced labor, they never formed a large part of the population themselves. Traditionally the Indian population clustered in the highlands, including the extensive plateau that covers a large part of Peru and Bolivia (the *altiplano*). The Inca capital of Cuzco and most of the Inca agricultural areas were on the altiplano or in smaller mountain basins and valleys. The modern border between Bolivia and Peru actually bisects a single population cluster of more or less homogeneous Indian settlement. This is a rare phenomenon in Latin America where political states are generally associated with separate settlement cores. The Spanish established their colonial capital at coastal Lima rather than in Cuzco, in part for easier communication with Spain and Middle America.

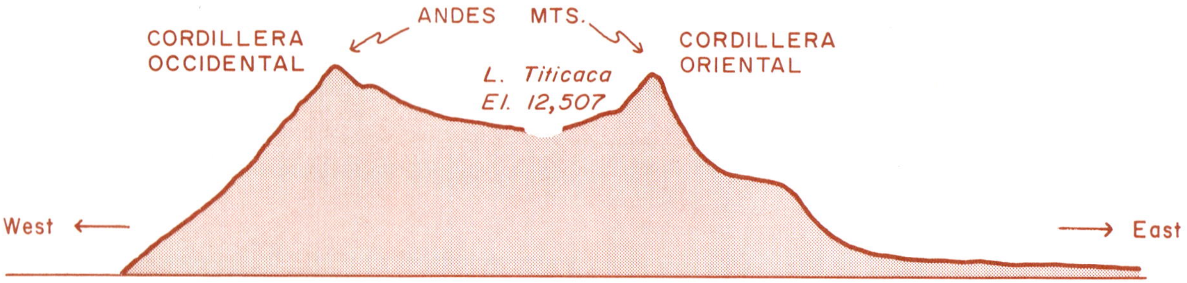

Figure 16. Basic Andean landforms of Peru and Bolivia.

Bolivia later broke off from Peru from which it felt awkwardly isolated.

Both countries are dominated by the Andes, not only because most settlement is in the highlands, but in purely physical terms. The Andes are distinctive as the longest mountain range of comparable height in the world, and in Peru and Bolivia they reach their most impressive scale. The large, fresh-water Lake Titicaca, on the border between the two countries and over 3,000 square miles (7770 sq. km) in area, is at an altitude of 12,507 feet (3815 m). Peaks around it reach 21,000 feet (6405 m), yet the altiplano in this district supports a large population cluster, as it did even before the days of the Incas. The basic landform pattern of Peru and Bolivia—a coastal plain, two Andean ridges enclosing the altiplano, and an interior lowland—are represented diagrammatically in Figure 16 (page 442).

Rivers tributary to the Amazon have cut into the east face of the plateau, and many become navigable on the interior lowland. But there is little lowland settlement in this hot, steamy, and heavily forested area that also suffers from excessive isolation. Primitive tribes, including head-hunters, are thinly scattered over thousands of square miles, as over most of the Amazon basin. About three-fourths of Bolivia's people live in the highlands, especially on the altiplano, although highlands cover less than half of the country's total area. Peru has a port on the Amazon, Iquitos, through which most of the small trade of eastern Peru passes. The Amazon is navigable for large river steamers or small ocean-going ships of thirty-five hundred tons as far as Iquitos, 2,000 miles (3200 km) from the Atlantic. This river route is still used for some shipments between eastern and western Peru by way of Panama. The Andes barrier has now been pierced by a road, but only shipments of high unit-value goods can use it economically. Bolivia has a somewhat less difficult route by rail from the capital at La Paz on the altiplano to what are now the Chilean seaports of Arica and Antofagasta, in territory owned by Bolivia until it lost a war with Chile in 1884. Bolivia became a landlocked state, although Chile administers Arica as a free port for Bolivia. La Paz, with about 700 thousand people, is the largest city in the world at such an altitude.

Peru's coastal region has become increasingly important for commercial irrigated crops. There is a little petroleum on the north coast, growing urbanization, and some consumer-goods manufacturing in Lima and lesser towns. The narrow Peruvian coastal plain is a desert, cooled by the Humboldt Current (or Peru Current) that flows parallel to the coast from south to north and produces frequent fogs. But it discourages rainfall by cooling air masses before they reach the land. The Incas and their predecessors had developed widespread irrigated agriculture along the coast in scattered oases fed by streams dropping from the Andes. But the Indians generally preferred the altiplano and the highland basins. Rainfall on the altiplano is adequate for agriculture for the most part, although irrigation is helpful and fairly widespread. Highland agriculture also makes use of extensive terracing, faced with stone walls and irrigated by mountain springs farther up the slope. These stone-walled terraces, dating probably from pre-Inca times, are called *andenes* and gave their name to the great mountain range that extends so far beyond the altiplano.

All agriculture on the coastal desert plain depends on irrigation. There is wide use of qanats identical with those in the Dry World of Eurasia, drawing on groundwater at or near the base of the mountains. It is not certain whether the qanat system was introduced by the Spanish, who doubtless learned of it through the Moors, or whether it was developed independently by the Incas or their predecessors. At one time, perhaps in the pre-Inca period, irrigation on the coastal plain was probably even more highly developed than at present. A number of ruined cities have been found in what is now, and probably was then, complete desert. As in north Africa, climatic change does not seem a likely explanation. But it does seem clear that the pre-Inca civilization was at least as highly developed as that of the Incas, who spoke of themselves as the heirs and trustees of an earlier culture that they regarded as superior to their own. Unfortunately almost nothing is known about it. In the later period of Spanish control and since independence, irrigation of the coastal lowlands has received growing attention. There is significant commercial production of irrigated cotton, sugar cane, rice, and citrus fruits, some of which is exported. The highlanders grow potatoes (which were originally native to the central Andes and have spread all over the world since the fifteenth century), maize (which may also have been native here in a wild form as well as in Middle America), and barley. They also graze sheep and llamas, but primarily in a subsistence pattern.

As in Venezuela and the Guianas, the export trade of Bolivia is largely in minerals. Over half is tin (accounting for about 12 percent of total world production); tungsten, silver, lead, and zinc are also important. Practically all of the mining was developed with foreign capital,[1] and most of the production is exported. Most of the rest of the economy is little affected and remains largely on a subsistence agricultural basis. Tin, lead, and zinc commonly occur together, and in the Andean mountain area are obtained from lodes rather than from alluvial deposits as in southeast Asia. This tends to make Bolivian tin somewhat more expensive than Malayan or Indonesian. Its importance to the United States is due to its relative proximity and more secure access, especially in wartime. Peru exports greater actual amounts of minerals than Bolivia; over one-fourth of its foreign trade is in copper, zinc, and iron ore. Agricultural exports have risen sharply since the Second World War with sugar accounting for a substantial share of them. Peru is the leading Latin American country in fishing and catches over 15 percent of the world total. Some of it is processed into fishmeal, another important export. Tuna is plentiful off the Peruvian coast, and adds a valuable item to Peru's foreign trade. This is responsible for Peru's early insistence on a 200-mile (320 km) limit to its sovereignty from the coast. This rule has since been followed by many other countries, including the United States. Nearly all of the Peruvian export trade passes through Lima and its port of Callao. Callao is a largely artificial harbor behind breakwaters, an offshore island, and a promontory on the smooth Peruvian coast. Lima became a prominent administrative and cultural center for Spanish America, as the Spanish successor to Inca Cuzco. It now has over 3 million inhabitants (including Callao), surprisingly large for a country of about 16 million whose economy is still primarily of a subsistence nature. Recent and rapid growth of consumer-goods manufacturing in Lima and increases in foreign trade passing through the city are responsible in large part for its present size.

[1]The tin mines were nationalized in 1953.

Brazil

This immense country, exceeded in size only by Russia, China, Canada, and the United States, covers about half of South America and contains about half of its population. Most of Brazil's area, however, is nearly empty, and only about 4 percent of it is cultivated. Nearly three-fourths of the country is very sparsely occupied by primitive Indian tribes over whom the Brazilian government has little or no control, and who have almost no contact, commercially or otherwise, with the outside world. The Amazon basin, which occupies nearly half of Brazil, contains the largest area of equatorial rain forest in the world. In an area nearly as large as the continental United States, the basin supports only some 4 million people. Communication is extremely difficult except along the Amazon and its navigable tributaries. Most of the basin has not been explored thoroughly, let alone surveyed. Its emptiness is likely to be permanent, at least for the areas removed from the river system. Its potential agricultural resources cannot be very great, given the general nature of tropical soils.

Minerals that may occur there face prohibitive transport costs despite the new program of limited road building.

The only large-scale commercial venture in the Amazon basin failed after a brief boom.[2] The commercial rubber tree is native to this area, and when a demand for rubber appeared about the middle of the nineteenth century, the Amazon basin was the only significant producer. The rubber was gathered from *hevea* trees growing naturally, scattered through the rain forest. It was crudely refined over wood fires, and shipped to Belém (Pará) near the mouth of the Amazon, or to Manaos (Manaus), the principal port for ocean navigation 700 miles (1130 km) from the sea, for transshipment to North America and Europe. The volume of production increased rapidly but collapsed almost overnight after 1900 with successful development of plantation production in Malaya and elsewhere in southeast Asia (see Chapter 16). Local Indian labor, inadequate in amount and quality, plagued the Brazilian wild-rubber industry during its brief heyday, and has also plagued the few attempts to establish rubber plantations in the Amazon basin. The most important of these was the Ford Motor Company's enterprise at Fordlândia (Brasilia Legal) in which millions of dollars were invested, but which was ultimately abandoned as a failure. Lack of adequate labor, plant diseases, transport problems, and the trying effects of the local climate all contributed to this disaster. The Amazon basin's advantages do not compare with those of southeast Asia or of coastal west Africa.

The Amazon, an enormous stream well over a mile (1.6 km) wide hundreds of miles from its mouth, remains navigable nearly to the Andean escarpment in the west. But it is used commercially relatively little by comparison with the much smaller rivers of Europe or those of eastern United States or monsoon Asia, because its basin produces and demands so few exchange goods. Even the valuable tropical hardwoods that are so plentiful in the basin cannot be brought out at a bearable cost in competition with southeast Asia. There is only a trickle of exports, including some hardwoods, nuts, resins and gums, orchids, and other products gathered wild. What little nonprimitive settlement exists is almost entirely confined to small, precarious clearings in the dense vegetation wall along the main stream and its tributaries.

Roughly the southern half of Brazil is covered by highlands, nowhere at great elevations, but including large areas of strong relief as well as plateau landforms. For the most part the highlands are an irregular plateau between 1,000 and 3,000 feet (305 to 915 m) above sea level (see Map 19, page 67). The bulk of Brazil's population is on this broken plateau, which extends practically to the coast south of the Amazon basin. A relatively abrupt escarpment on the east rises from a narrow coastal plain where most of Brazil's large cities are located, although São Paulo is actually on the edge of the plateau rather than on the coast. The highlands peter out on the edge of the Amazon basin. There is a considerable stretch of equatorial coastal plain in northeastern Brazil, the site of the first sugar plantations in the Western Hemisphere in the very early days of Portuguese control. Sugar is still grown there, although its boom days ended abruptly in the eighteenth century with the rise of West Indian production. There is also some production of other tropical and subtropical crops, including cacao, rice, cotton, and fruits, and a small grazing industry. Grazing is more important in the western part of the Brazilian plateau, especially the province of Mato Grosso.

The boom in gold and diamonds, which followed the sugar boom, was principally in the central highlands, where a little of both minerals is still produced. But the great bonanza of the Brazilian economy in the past century has been coffee, mainly a crop of the southern highlands in the area tributary to São Paulo. The moderate, frost-free highland climate is favorable for coffee, and the soils of this area are also especially suitable, including the famous *terra roxa* soils. A combination of these advantages has led to a virtual Brazilian monopoly in world coffee production and to the phenomenal growth of São Paulo (over 7 million) and its port of Santos. There has also been a rapid increase in cotton production in the same area, some of which is exported. Many European immigrants have been attracted to the vigorously growing economy of São Paulo and its coffee-cotton hinterland. But recent European settlement has been even more important farther south, beyond the highlands, in the province of Rio Grande do Sul. A German colony was established there in the first

[2]The development of manganese mines north of the Amazon mouth might constitute an exception.

Interior Brazil—Mato Grosso. This formerly forested land on the savanna margins in the southern part of the Amazon basin is now being used as low-quality, unimproved pasture. Most of the trees have been felled or girdled. The agricultural potential, or good-pasture potential, of much of this land is considerable, but it has been developed very little. *(From W. R. Grace and Company.)*

quarter of the nineteenth century, and there has also been heavy Italian immigration. An essentially European pattern of land use has developed, and relatively recent European arrivals form a majority of the population. The area is far enough south of the equator so temperatures are moderate. The lowland agriculture can profitably follow a European pattern: the cultivation of cereals (mainly corn and rye) and the raising of animals (mainly hogs), the familiar mixed farming of Europe. The settlement and commercial development of Brazil has thus progressed generally southward, from the early Portuguese sugar plantations on the north tropical coast, to the gold and diamonds of the central highlands, to the coffee and cotton estates and smaller farms of the São Paulo highland region, to the predominantly small farms of the far south with individual proprietors raising cereals and animals.

Brazil is making rapid strides toward industrialization, however. Already the total value of industrial production is about twice the value of agricultural production, and increasing amounts of industrial goods are being exported. Mining of gold and diamonds has long since been dwarfed by mining of iron ore, the relatively limited coal of the extreme south, and manganese, with which Brazil is well supplied. The steel plant at Volta Redonda, on the Paraiba River between Rio de Janeiro and São Paulo, is by far the largest in Latin America and is growing fast. Iron ore is brought from the Belo Horizonte district of Minas Gerais, mainly from the chief ore deposits near Itabira about 300 miles (480 km) north of Rio (see Map 13, pages 58–59). Coal is brought from the deposits in the southern provinces of Santa Catarina and Rio Grande do Sul mainly by sea to Rio de Janeiro. Unfortunately most of the coal is of low grade, and large imports will probably be

Rio de Janeiro—former capital of a booming country. Sugarloaf Mountain stands at the entrance to a magnificent natural harbor, around which an enormous modern city has grown. Level land is severely limited and the hills have restricted the expansion of Rio. In some ways the city has a situation similar to San Francisco's. *(From W. R. Grace and Company.)*

necessary to sustain significant industrial growth. Brazil has an impressive total of potential hydro power, but most of the sites are remote from market areas in the Parana basin in the extreme south or in the Amazon basin. There is increasing development of hydro along the eastern edge of the central highlands. The steel plant at Volta Redonda is near the center of the rapidly growing Brazilian market, in terms of urban population, and at a median-cost location between the sources of iron and coal. Increasing steel production will contribute to the growth of Rio and São Paulo, and will further strengthen the dominance of this market core for the whole of Brazil. There is a secondary population cluster on the north coast between the cities of Recife and Salvador, commercially dependent mainly on cotton, sugar, maize, and tobacco. The Rio de Janeiro–São Paulo area is growing much more rapidly and includes the largest industrial concentration south of Texas. Production is still dominated by consumer goods, but it will not be many years before Brazil is making most of its own capital goods and exporting finished products in large quantity.

Rio owes its size (nearly 5 million) to its former administrative functions as Brazil's capital, the development of consumer-goods manufacturing, and its function as the port, trade, and service center for the central Brazilian highlands, plus the narrow coastal plain and its commercial agriculture. Rio's magnificent natural harbor, at the mouth of a large bay, has contributed to the growth of its port functions. But Santos does a larger volume of freight business, since it serves a commercially more productive hinterland. Rio leads in passenger traffic (much of which is tourist trade), a situation similar to that in South Africa where Cape Town and Durban parallel Rio and Santos in this respect.

Rio was proportionately more important in the boom days in gold and diamonds when the central highlands were the chief focus of commercial activity. It is profiting now from the new development of Minas Gerais iron ore and the expansion of Volta Redonda, but its growth will undoubtedly lose some momentum with the shift of capital functions to the new city of Brasilia. An interior location for the capital was discussed for many years. Brazil is a huge country, and settlement is concentrated along the coast even more than in the rest of Latin America. Probably most of the now unused or little used interior outside the Amazon basin could be made productive. It is certainly the largest empty area of potentially productive land in the world. The moving of the capital to Brasilia, in the central highlands about 700 miles (1130 km) northwest of Rio in a direct line, was in many respects a gesture of faith in the future development of Brazil's empty areas. Brasilia was planned entirely as a new city on a grand scale in a former wilderness. Nearly all other Brazilian cities are water-oriented: Belem and Manaos on the Amazon, and all the others on the coast around the margins of the country. At the present rate of economic and population growth in Brazil, and with the increasing development of the interior highlands, Brasilia may not be isolated for long. Brazil's population, already about 112 million, could, and, within the next century, probably will triple in size without resulting in overall overpopulation, as resource use is extended. The problem will be distribution, and rational use of Brazil's vast total resources.

The pampa-Chaco region

Northern Argentina, Uruguay, and Paraguay share the great grassland plains of southern South America, known as the *pampa* and the Chaco. To the south, the pampa fades off into the steppe, desert, and mountains of southern Argentina (see Map 102, page 448). On the west, the pampa is limited by the Andes. On the north it gives way to the essentially savanna country of the Gran Chaco, where tropical temperatures and rainfall reproduce conditions similar to the Venezuelan llanos north of the equator. Paraguay occupies most of the Chaco; the drier northwestern part was won from Bolivia after the war of 1932–1935. A large part of the Chaco is also included in northern Argentina. The Brazilian highlands and the beginnings of the Amazon rain forest limit the Chaco on the north.

The pampa is Latin America's major area of commercial grain and animal production, and one of the most important areas in the world in terms of exports. It is only lightly populated on the whole, and has large surpluses available for foreign trade. The actual totals of cereal and animal production are much smaller than for most of the other major commercial grain and animal regions of the world. The pampa is physically excellent for maize, alfalfa, wheat, and commercial stock raising. Soils are generally of high natural fertility in this climate of moderate or slight rainfall and moderate temperatures. Chernozem and chestnut soils like those of the grasslands in North America, Eurasia, and Australia are widespread. The pampa has never been intensively used. Agricultural production could be greatly increased and the region could support many times its present population. Corn and fodder crops dominate the moister, northeastern sections of the pampa, roughly the state of Uruguay and the plain within a 300-mile (480 km) radius of Buenos Aires. This area, known as the humid pampa, contains over three-fourths of the commercially productive capacity of Argentina and virtually all of that of Uruguay. Westward and southward from the humid pampa, rainfall decreases, and wheat and grazing replace fodder crops and corn. Northward into the Chaco savanna, conditions are increasingly unfavorable for cereals, and the grazing of low-quality cattle on generally poor, unimproved pastures is the major form of land use.

Corn and alfalfa are grown on the pampa, as in the American corn belt, as feed for animals (cattle and hogs) raised and fattened locally, and for export as grain, principally to food-deficient Europe. Europe also buys large quantities of meat and animal products from the pampa. The region's two major cities, Buenos Aires (capital of Argentina) and Montevideo (capital of Uruguay), are both located on

BOLIVIA
PARAGUAY
URUGUAY
BRAZIL
ARGENTINA
CHILE
PACIFIC OCEAN
ATLANTIC OCEAN
MATO GROSSO
MINAS GERAIS
SÃO PAULO
PARANÁ
SANTA CATARINA
RIO GRANDE DO SUL
GRAN CHACO
JUJUY
SALTA
TUCUMÁN
CATAMARCA
SANTIAGO DEL ESTERO
CHACO
FORMOSA
MISIONES
CORRIENTES
SANTA FE
ENTRE RÍOS
LA RIOJA
SAN JUAN
CÓRDOBA
SAN LUIS
MENDOZA
LA PAMPA
BUENOS AIRES
NEUQUÉN
RÍO NEGRO
CHUBUT
PATAGONIA
PUNA DE ATACAMA
ATACAMA TRENCH
Antofagasta
Tocopilla
Chuquicamata
Calama
Taltal
Chañaral
Caldera
Copiapó
Vallenar
Huasco
Freirina
Coquimbo
La Serena
Tongoy
Ovalle
Illapel
Los Vilos
Quillota
Viña del Mar
Valparaíso
SANTIAGO
San Antonio
Melipilla
San Bernardo
Rancagua
San Fernando
Constitución
Cauquenes
Talca
Linares
San Carlos
Parral
Talcahuano
Concepción
Coronel
Chillán
Lota
Los Angeles
Lebú
Angol
Victoria
Lautaro
Temuco
Valdivia
Corral
La Unión
Osorno
Puerto Varas
Puerto Montt
Ancud
Castro
ISLA DE CHILOÉ
Golfo Corcovado
ARCHIPIÉLAGO DE LOS CHONOS
Puerto Aisén
PENÍNSULA DE TAITAO
Golfo de Penas
Tupiza
Tarija
Villazón
Yacuiba
Tartagal
Orán
La Quiaca
Jujuy
San Pedro
Salta
Metán
Tucumán
Monteros
Andalgalá
Tinogasta
Catamarca
Santiago del Estero
La Rioja
Chilecito
Córdoba
Villa Dolores
Río Cuarto
Villa María
San Juan
Mendoza
San Rafael
San Luis
Villa Mercedes
Asunción
Concepción
Villarrica
Encarnación
Resistencia
Corrientes
Formosa
Posadas
Santa Fe
Paraná
Rosario
Concordia
Salto
Paysandú
Rivera
Tacuarembó
Melo
Mercedes
Fray Bentos
San Nicolás
Pergamino
Junín
Zárate
BUENOS AIRES
Avellaneda
La Plata
Colonia
MONTEVIDEO
Maldonado
Rocha
Minas
Río de la Plata
Mar del Plata
Necochea
Tandil
Azul
Olavarría
Bahía Blanca
Santa Rosa
General Pico
Neuquén
Zapala
General Roca
Choele Choel
Carmen de Patagones
Viedma
San Antonio Oeste
Golfo San Matías
PENÍNSULA VALDÉS
Puerto Madryn
Trelew
Rawson
Esquel
San Carlos de Bariloche
Nahuel Huapi
Comodoro Rivadavia
Golfo San Jorge
Puerto Deseado
Pelotas
Rio Grande
L. dos Patos
Porto Alegre
Caxias do Sul
Passo Fundo
Santa Maria
Uruguaiana
Florianópolis
Curitiba
Paranaguá
Ponta Grossa
Londrina
Bauru
Campinas
Sorocaba
SÃO PAULO
Santos
São Vicente
Jundiaí
Ribeirão Prêto
Franca
Juiz de Fora
Petrópolis
Nova Iguaçu
RIO DE JANEIRO
Tropic of Capricorn
MAP 102
PENINSULAR SOUTH AMERICA
Railroads
© Rand McNally & Co.

the deep broad estuary of the Parana River, which is somewhat confusingly called the Rio de la Plata (River Plate). They owe their growth to the productivity of the pampa and the ocean trade with the Northern Hemisphere. Although population densities in their hinterlands do not compare with those for most other large cities, the pampa is highly productive, and a large part of what it produces is processed or exported through the chief urban centers. Despite the fact that this is predominantly an agricultural region, the metropolitan areas of Cordoba, Santa Fe, Rosario, La Plata, Buenos Aires, and Montevideo contain well over a third of the total population. Extensive commercial agriculture for export has thus been accompanied by a high degree of urbanization, as in Australia (see Chapter 17).

Mineral resources in the pampa are very few. The original Indian population was sparse and extremely hostile. As a result, the Conquistadores attached little importance to this region, and its economic development was delayed until the second half of the nineteenth century. The agricultural resources the Spanish disdained became of major importance with the rapidly rising demand for food in industrializing Europe. The development of railways and cheap sea transport (especially refrigerator ships) reduced the former economic remoteness of the pampa. The growth of mechanized agriculture, whose techniques were suited to the pampa, as to the great plains grasslands of North America, Australia, southern Russia, and south Africa, was equally important. Buenos Aires, which had been a small town until about 1850, grew rapidly to become the largest city in the Western Hemisphere south of New York. It now has a population of about 9 million. Ironically, the original base the Spanish established on the site of Buenos Aires in 1535 was abandoned. The Spanish moved up the valley of the Parana-Paraguay River toward a denser Indian population, and founded their first permanent settlement at Asunción in 1581. Navigation on the river is limited, especially above Rosario, and water transport has never been an important commercial asset. It was not until the end of the eighteenth century that Buenos Aires achieved any importance. It became the seat of a viceroyalty, and began to have direct communications with Europe by sea, instead of going through Peru and Panama. The post-1850 growth of this city is a vivid illustration of how changes in resource evaluation and use, as well as changes in spatial interaction, may affect human settlement patterns. Nearly three-fourths of Argentina's population is now concentrated in Buenos Aires and its immediate hinterland, the area of the humid pampa.

Most of the stimulus to commercial development of the pampa came from Europe. Heavy British investment in particular played a major role in building the dense net of railways that covers the humid pampa and indicates the commercial nature of the agricultural system, sending its products to Buenos Aires. Port facilities, shipping lines, and commercial services were also built with British capital, and there was large German and Italian immigration. Since the Second World War, the United States has taken over much of the formerly large British investment in southern South America and in much of the rest of Latin America. It has also displaced Britain as the area's chief trade partner. A large part of the manufacturing that developed in Buenos Aires, Montevideo, and the few lesser cities is the processing of agricultural goods, especially flour, canned meat, hides, wool, and cotton.

Consumer-goods manufacturing is growing. The region's market is not large numerically, but is significant in terms of income levels. There is also a substantial production of cotton from the Chaco, plus wool from sheep and goats. They are grazed in the far western and southern areas that are too dry even for wheat. Flax (for linseed oil) is also an important crop of the humid pampa. At one time most of the cotton, flax, wool, and hides were exported raw, but they are now increasingly supplying domestic textile and leather-goods manufacturing. Heavy manufacturing is little developed, in the virtual absence of significant coal or iron ore, although Buenos Aires has a small steel plant. A larger one on the Parana River above the city depends almost entirely on imported ores and fuel. Argentina has some petroleum deposits on its southeast coast near the Gulf of San Jorge (see Map 12, pages 56–57). There may be some very large petroleum pools in the northern and northwestern Chaco. Development of these resources has hardly begun, but they are more likely to be exported than to stimulate local industrialization. Unrealistic heavy industrialization plans under the Argentinian dictator Peron and his equally illogical drive for economic self-sufficiency in Argentina seriously damaged the economy. As much as any in the world, it is vitally dependent on foreign trade and the export of agricultural goods to industrial economies from which it buys capital goods and finished products. Argentina's comparative advantages, like those of southeast Asia and most of Latin America, are solidly agricultural.

Uruguay

Uruguay has a similar economy and physical base, except that a hillier and moister version of the pampa covers nearly all of it. Commercial fodder growing and grazing are dominant. Most of Uruguay is more intensively used than most of Argentina. In fact, Uruguay is the most densely populated state in South America, although it is also the smallest. The population is even more European than in Argentina, and is mainly of more recent origin. Uruguay was originally established as a buffer area between Portuguese Brazil and Spanish Argentina, but Spanish attention was minimal until the eighteenth century. Montevideo was founded in 1726 to counter Portuguese pressure from the north and to establish a Spanish settlement on the left bank of the Plate estuary. Previously Uruguay was very thinly settled. Its population grew primarily as a result of conflicting Spanish and Portuguese efforts to claim an area whose advantages for commercial grain and animal production did not become apparent until the latter half of the nineteenth century. Heavy European immigration accompanied Uruguay's modern commercial development, and Montevideo is now a city of over 1 million, a smaller-scale Buenos Aires. In contrast to neighboring Argentina, Uruguay for a time became a parliamentary democracy with social legislation as advanced as any country's in the world. This was attributable at least in part to the numerical majority of non-Spanish European settlers, to the generally adequate income levels, and to the absence of large-scale poverty or depressed Indian groups in Uruguay. In recent years, unfortunately, its political system has lapsed into the more familiar Latin American pattern of coups, countercoups, and repressive rule by power cliques.

Paraguay

Paraguay is in effect a backward province of Argentina, with a heavily Indian and mestizo population. It was long contested by the Portuguese and the Spanish, and later by Brazil, Argentina, and Bolivia. This insecure status led to neglect of economic development, and Paraguay's comparative economic value is still limited. Principally, it is remote, landlocked, and hampered by slow and expensive transport connections with other areas. Paraguay's economy is primarily subsistence in nature, and it has one of the lowest income levels in Latin America, despite a climate and soils that could support a much more productive agricultural system. Grazing is the most important activity, but, as in the Venezuelan llanos, the quality of the cattle is low. The west is arid, but the humid subtropical climate of the east and south could be much more intensively used on a commercial basis, if location were not so unfavorable. A trickle of high-value goods is exported through Buenos Aires: dried beef, quebracho (a wood extract used for tanning leather), canned meat, hides, cotton, and tobacco. The Paraguayan Chaco includes petroleum deposits that may be of significant size, especially in the northwestern area won from Bolivia, but little has been done to develop them. The ill-equipped but well-led Paraguayan army defeated a numerically superior and better-equipped Bolivian invasion force in the war of 1932–1935. Paraguay's political position has been secure ever since, but its economic backwardness remains.

Chile and the tip of South America

Chile is a peculiar state in spatial form. It also includes large areas of hot desert, Mediterranean climate, humid temperate forest, taiga, and cold desert in its thin strip of territory stretching along the west coast of southern South America from 18° S. to 56° S. latitude. At its widest, Chile is only a little over 200 miles (320 km) wide from east to west. Its average width is only about 100 miles (160 km), and it is over 2,600 miles (4160 km) long from north to south. Such a spatial arrangement is awkward administratively and economically. Sea communication has helped to hold Chile together, and the barrier of the Andes, which largely coincides with Chile's eastern border, also contributes to the country's legitimacy as a separate political region. Fortunately the Mediterranean climate occurs about in the middle of the country, and is productive enough to give the state political and economic stability despite the fric-

Santiago de Chile and the towering front of the Andes. The picture shows the somewhat Mediterranean character of Chile's capital city. It is close enough to the Andes that the mountains completely dominate the eastern horizon. *(Photo from Braniff International.)*

tion of distance at the extremes of north and south. A railway runs north into the desert, but there is neither railway nor through road in the southern third of the country. Chile's neighbors on the north are not strong enough to represent an effective threat, and the Argentinian core in the humid pampa is far enough away so Argentina is not an active contender for sovereignty over the remote Chilean territory in the south. Disputes between the two countries over the southern tip of the continent have been settled largely in Chile's favor. The famous Christ of the Andes stands at the summit of Uspallata Pass across the Andes between Mendoza and Valparaiso, marking the agreed boundary and the willingness of both countries to forget their boundary disputes.

Remoteness and limited economic value are probably the most important guardians of Chilean sovereignty over the southern third of the country. (The southern third or fourth of Argentina is treated here as part of the same region.) In this tip of South America the Andes gradually decline so they no longer present a strong barrier. Patagonia, the Argentinian share, is semiarid or desert, colder toward the south, and very thinly settled by grazers of sheep and goats. Some wool and mutton from this area are exported. On the Chilean side there is more rainfall, especially along the coast. But the inland areas are arid and the local economy is much the same as in Patagonia. Tierra del Fuego, shared by the two countries, is a cold, rocky desert, the southernmost land next to Antarctica, only a little over 600 miles (960 km) farther south. The southern third of Chile is mainly a highly fragmented fjord coast, like that of British Columbia, which lies in a similar latitude. Communication is almost exclusively by sea. The distances to any significant market areas are great, and the local landscape presents few opportunities. Level land is virtually absent, there is extensive glaciation, workable soils are scarce, and the only important resource of the area is the forests along the fjorded coast. They have been used relatively little as a result of the poor quality of the wood, distance from market, shortage of low-cost volume transport, low density of local settlement, and Chilean preoccupation with more promising resources in the central and northern parts of the country. Despite its southern forest stands, Chile has a net import of wood. Settlement has gradually been moving southward from the central core, and there is a small development of dairying and mixed farming (wheat, potatoes) in the cool, humid coastal zone just north of the fjord country.

Nine-tenths of Chile's population is concentrated in the central core of the country, sometimes known as the Vale of Chile, an area of Mediterranean climate running approximately from 30° S. to 36° S. latitude. To the north is the great Atacama Desert (Desierto de Atacama), and to the south the heavy forest of the humid coast. The vale itself, properly speaking the valley of the Aconcagua, lies between the low coastal mountain ranges and the main Andes chain along the eastern border of the country. The Inca empire included this part of Chile, but did not reach beyond it. The south was thinly occupied by primitive and hostile Indian tribes, some of whom still remain. The early Spanish settlements in Chile were severely isolated and were of minor importance. The Spanish were little interested in agriculture, even in a physical setting in the vale so similar to that of their own homeland. In the nineteenth century there was considerable German immigration and a great increase in the cultivated land in the cool, coastal zone north of the fjords and in the vale. There agriculture follows an essentially Mediterranean pattern of grazing, wheat growing, vineyards, and irrigated citrus groves. Wine, citrus fruits, and wool are exported, but the volume is relatively slight as a result of economic remoteness from major foreign markets. Chile does have the same seasonal advantage in Mediterranean agriculture enjoyed by south Africa. But the distance to most Northern Hemisphere markets is greater, and canal dues at Panama present an additional obstacle. By far the most important agricultural export is wine, for which Chile is developing a reputation. It is sufficiently high in value to bear the costs of transport. Santiago, the Chilean capital, is understandably near the center of the vale; Valparaiso is its port. Hydro power is developed from the Andean slopes, and there is a small development of consumer-goods manufacturing in Santiago and Valparaiso. But Chile must import most of its manufactured goods and also imports considerable food. The vale is Chile's only substantially productive agricultural area. But it covers only a small part of the country as a whole and suffers from the Mediterranean climatic pattern of rainy winters and dry summers.

Mineral resources and climate

Chile's northern third, the Atacama Desert, is the least settled but most productive part of the country in terms of exports. The Atacama contains the world's largest deposits of commercially exploitable nitrates. Their mining after 1850 led to an economic boom, which collapsed after the discovery in 1912 of the Haber process for producing nitrates electrically by fixation from the air. Chilean nitrate production is less than one-fourth of what it was before the First World War, although still significant. Copper, mined mainly on the eastern edge of the Atacama in the Chuquicamata district, succeeded nitrates as Chile's leading export. Like nitrates, it is almost entirely exported to industrial economies in the Northern Hemisphere. The copper and silver deposits in the same area were discovered and exploited only about 1850, but since then minerals have dominated Chile's export trade. Iron ore is also mined in the southern Atacama and is of relatively high grade. Most of it is exported, but some ore also goes to supply a small iron and steel plant near Concepción, on the south-central coast. There it can draw on low-grade coal deposits south of the city. The controlling role played for many years by American investment in Chilean mining, the backbone of Chile's commercial economy, had obvious politically awkward implications, as in Venezuela, where American capital is equally heavily invested in iron ore and petroleum. In Chile, U.S. pressures have often had a major influence on politics, as in the overthrow of the Allende government in 1973.

The Atacama is exceedingly dry and contains very few oases. In several places weather stations have never recorded any rainfall. This arid area is a continuation of the coastal desert of Peru, and occurs in the same latitudinal position as the coastal deserts of southwest Africa, northwest Africa, western Australia, and lower California. All are characterized by strong and relatively stagnant high-pressure areas off the coast with very little air movement. In Chile the topographic barrier of the Andes contributes further to the aridity of the coastal lowlands. The cold Humboldt Current running along the coast also discourages rainfall on the land, as in coastal Peru. The Mediterranean climate of the vale is duplicated at similar latitudes in south Africa, southern Europe, southern Australia, and southern California. It yields poleward, in the zone of the prevailing westerlies, to the same mixed forest as succeeds the Mediterranean climate poleward in Europe, California, Australia, and New Zealand. In south Africa there is no land far enough south to be in the zone of the westerlies. In southernmost Chile the forest gives way to the tundra that also succeeds the forest poleward in Europe and North America.

Few parts of the world, including Antarctica, are as isolated culturally or have received so little human attention as the southern tip of South America. This can hardly be attributed to climate, since similar climates elsewhere have been accompanied by much greater human development. Chile as a whole suffers from the same isolation, although to a lesser degree. This isolation is apparent in most aspects of its national life. Location in the Southern Hemisphere is generally a handicap, given the preponderance of land masses, people, cultural development, and economic activity in the Northern Hemisphere, and for this fringe of South America, the handicap is magnified.

Latin American cultural unity

The degree of cultural unity within Latin America as a whole is surprising, considering the great range of regional differences in other respects and the enormous area involved. Even in Chile, Spanish language and Spanish culture remain dominant, and there is the same mixture of European and Indian elements that is characteristic of Latin America as a whole, including a large mestizo population. For the most part, Latin America has not been nearly as attractive to human settlement, nor as productive economically, as the earlier-developed and culturally more diverse parts of the world. In Latin America, with the exception of the Caribbean area, the Spanish and Portuguese had a free field in effect, with little competition from other foreign groups. They had little difficulty in supplanting indigenous Indian control or in transplanting their own culture as the dominant one except in the Amazon basin rain forests. They could not eliminate the Indians, but were able to produce a cultural and racial blend that varies surprisingly little from Mexico to Chile. This similarity prevails over a distance of some 7,000 miles (11,200 km), an area that encompasses high mountains and great plains, deserts, rain forests and savannas, plantation agriculture, commercial grazing, commercial grain growing, mixed farming, rich mineral deposits, and the widespread subsistence farming that is still the most important economic activity. Cultural diversity is greater in Latin America than in North America. But the area is considerably larger, regional differences are more marked in physical terms, and Latin America contains twenty-five sovereignties instead of the two in North America. A more appropriate comparison would be with Africa, where the much greater degree of cultural diversity in an area no larger and no more physically diverse suggests that relative isolation from the chief centers of world civilization has been an important factor in distinguishing the pattern of Latin American development.

The challenge of the future for Latin America centers on whether population increases can be curbed and potential resources more fully used so living standards can rise. Part of the solution will have to include some redistribution of population and the settlement and exploitation of the large inland areas still mainly neglected. Settlement is still primarily along the coasts and waterways, and is characterized by isolated clusters, many of them already overcrowded. Vast inland resources go unused. As economic growth proceeds, efforts must also be made to correct the present grossly uneven distribution of wealth. The record of Latin American development so far has not been very encouraging in any of these respects, but the challenges of the future may stimulate a more effective set of responses.

Questions for further study and discussion

1. Belém is the largest city of the Amazon basin. Why? Why is it not located on the Amazon? Why is Manaos the head of most ocean navigation on the Amazon?

2. Why has the city of La Plata developed so close to Buenos Aires, really in the same metropolitan area? What is La Plata's function?

3. Why is Uruguay the most densely populated country in South America?

4. Why did the cities of Panama and Colon develop centuries before the Panama Canal was built?

5. What are the long-term possibilities for the development of the Amazon basin?

24 North America: a modern prodigy

Agricultural wealth—cornfields in Iowa. *(Photo © by Dom Franco from Van Cleve.)*

It is difficult to write about one's own country; perspective is hard to achieve. And in the case of North America,[1] historical perspective is more limited than for most other parts of the world. One of the chief purposes and benefits of geographical and historical study is to provide perspective and, incidentally, to increase understanding of one's own area and one's own time. Thus, it should be possible to assess North America in the light of some knowledge of other areas in their spatial and temporal frame. This chapter concentrates on examining certain selected regional characteristics and distinctive qualities that set North America off from the rest of the world.

Europe and America

This land, which lay hidden from the knowledge of the Old World for so many thousands of years, was settled from Europe at about the same time the rapidly developing technology and expanding population of Europe were becoming ready to make effective use of the new continent. If North America had been reached by the Phoenicians, the Romans, the Arabs, or the expeditions of the Ming dynasty, it is doubtful whether any very significant settlement or economic development would have followed. The experience of the Vikings would tend to confirm this. It seems certain that Norse explorers based in Greenland landed on the coasts of what is now New England at least as early as the eleventh century A.D. It is probable that they established a few temporary settlements. There is strong evidence to suggest that they traveled up the St. Lawrence River and through the Great Lakes as far west as Minnesota. Norse settlements in the New World were all abandoned (as was Greenland) by the fourteenth century, probably because the already marginal climate temporarily became harsher. The knowledge of the new continent appears to have been forgotten or kept secret in Europe, although it is possible that Columbus and other early explorers may have heard tales of the Vikings'

[1]Technically the continent of North America includes everything north of the Isthmus of Panama. But, as indicated in Chapter 22, the area between Panama and the United States is commonly labeled *Central* or *Middle America*, leaving *North America* to refer only to Canada and the United States.

Vineland. To Medieval Europe, northeast America would probably not have been attractive enough to warrant trying to reach it, even if its existence had been known. Indeed, the early permanent European settlements in Virginia and Massachusetts barely survived the first several years, especially during the unfamiliar, harsh winters. An earlier attempt at permanent settlement in Virginia had failed. For a time the settlers were largely fugitive criminals, black-sheep adventurers, convicts, and those who had suffered religious persecution at home. Few others could be persuaded to accept the hazards, difficulties, and apparently minimal rewards of life in the strange, bleak, hostile, meager, new land. But before the end of the seventeenth century rapid changes were beginning. They would make North America a land of opportunity par excellence, as judged by its present occupants and most of the rest of the world.

Before Columbus, the population of Amerindians[2] was sparse, scattered, highly fragmented in language and culture, and mainly still at a preagricultural, Stone Age level of development. They were much less advanced than the Aztec, Maya, or Inca groups of Central and South America. A few North American Indian groups were beginning to grow crops in permanent fields, mostly to supplement hunting, gathering, and fishing. The dog had been half-domesticated as an aid in hunting and a guardian from attack, but there was no animal husbandry, and few permanent settlements. Parts of the southwest were an exception, with adobe houses (sun-baked clay) forming small towns whose inhabitants tilled permanent fields. East of the Mississippi or the Appalachians, which was all the early European settlers saw, the Amerindians were predominantly forest hunters and shore-based fishermen and clam diggers. They had made little impact on the thick, virgin forest that still covered the landscape. Most of their encampments were temporary or were shifted periodically. Their culture and technology were hardly impressive to a seventeenth-century European. However, they were effective for some time as defenders of their forest strongholds, even raiding outlying European settlements. Their military abilities, together with the relatively small numbers of white settlers, kept the Europeans confined to the coastal or near-coastal areas until well into the nineteenth century. By the first and second decades of the 1800s, however, European people and technology began to sweep westward across the Appalachians and up the St. Lawrence River and the Great Lakes. Many cleared the forest and killed or drove the Amerindians before them as they went. The changes that European occupation brought about in North American resource use, as compared with what the Amerindians had done, are little short of incredible, especially remembering how brief a period was involved.

Improvements in navigation and ship building soon made the voyage across the north Atlantic much faster and less hazardous. The ships were better able to carry supplies and trade goods. This great age of sail, necessarily preceded by the Age of the Discoveries, probably could not have happened much earlier on such a scale. It was accompanied and followed by the beginnings of the agricultural, commercial, and industrial revolutions in Europe, and helped to transport these techniques effectively to the new land. As colonists in North America began to learn more about the physical properties of the land, especially about its climate, they developed more suitable methods of agriculture. Many of them were modified from those of the physically different Old World and made large use of productive native American crops like maize and pumpkins. The colonists' numbers increased and the settled area began to expand. By 1800 it included almost all of the Atlantic coastal plain from Nova Scotia to Florida, and the lower valleys of the St. Lawrence, Connecticut, Hudson, Delaware, Susquehanna, Potomac, James, Savannah, and lesser rivers. The great fish resources along the east coast were of vital importance from the beginning of permanent European settlement. They saved the tiny colony in Massachusetts from extinction more than once, as well as contributing to its prosperity after the first hard years were past.

Until the nineteenth century, however, the prodigal resources of North America were still largely undiscovered and unused. In 1750 the total population was about 1 million; European immigrants by that time totaled little more than the decrease in original Indian population through slaughter and disease. The population was about 6 million by 1800.

[2]Columbus was convinced that he had reached some outlying area of the East Indies, so he called the people *Indians,* a label that has stuck. In professional literature it is often altered to *Amerinds* or *Amerindians.*

MAP 103
RAILWAYS IN NORTH AMERICA

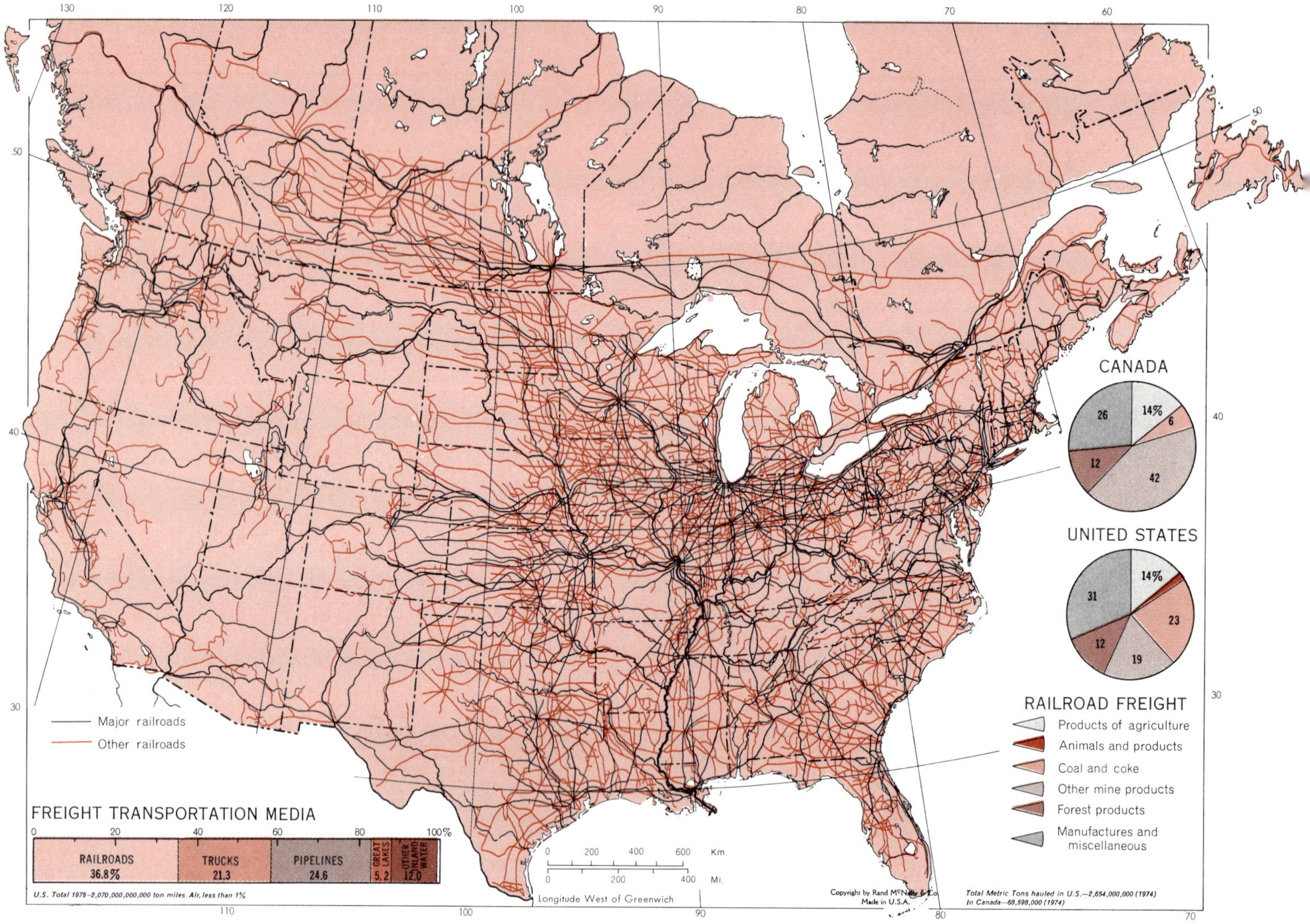

Even after nearly two centuries of permanent European settlement, it was still confined almost entirely to the eastern coastal plain and the major river valleys east of the Appalachians. The economy was still almost exclusively agricultural, with fishing as an important adjunct. There was some commercial hunting and trapping, mainly by French *voyageurs* in the vast wilderness of the Great Lakes area and the Ohio-Mississippi-Missouri basin. And there were a few scattered British outposts in central Canada and on Hudson Bay. The surviving Spanish settlements in Florida, California, Arizona–New Mexico–Texas, and the New Orleans area were numerically very small, separated by much empty space, and integrated with the Spanish holdings in Mexico. It was not an impressive record for two centuries, especially by the standards of Europeans. North America seemed to them a wilderness with a precarious fringe of settlement along the east coast. American culture was commonly dismissed as a pale and crude colonial echo of Britain's, then approaching the height of its power and wealth. Boston, New York, and Philadelphia were substantial cities serving their agricultural hinterlands east of the mountains and conducting a growing trade with Europe. But even in their own estimation, they were small and culturally backward by comparison with London or Paris. Few people in Europe, or even in North America, guessed a fraction of the potential riches of the new continent or the prodigious human development that was to take place in the succeeding years.

Resources

North America proved to be almost ideally suited physically to the industrial-commercial pattern of resource use developing in northwest Europe from the seventeenth century onward. No continent and no comparable area of land in the world possessed the variety and richness of resources of North America. At the beginning of the nineteenth century, although still mainly unknown and unused (in the absence of effective transport among other factors), North America had the following assets (see Maps 7, 12, and 13, pages 52–53, 56–57, and 58–59):

About one-third of the world's known reserves of commercially workable coal, most of it in large deposits of thick, easily worked seams, and most of it conveniently concentrated in the Missouri-Mississippi-Ohio basin close to what became the chief market area and population cluster.

About one-fifth of the world's reserves of iron ore, conveniently concentrated in the area of the Great Lakes, cheaply extractable, and of high grade.

About one-sixth of the world's reserves of economically recoverable petroleum.

About one-tenth of the world's potential hydro power, based on normal minimum flow (firm power base).

Superabundant deposits of sulphur, phosphate, potash, nickel, copper, lead, zinc, gold, silver, helium, natural gas, molybdenum, titanium, salt, magnesium, lower-grade aluminum ores, asbestos, fissionable materials, and a long list of other, less important minerals. Potentially serious mineral shortages were to become apparent only in tin, tungsten, chromite, manganese, antimony, high-grade aluminum ores, and some lesser items.

By far the largest and potentially most productive area of unused agricultural land in the world, including a wide range of climates and soils, most of which was well suited to commercial agriculture as developed in the nineteenth century.

The world's largest commercially usable forest reserves.

Fish reserves off both coasts that probably included at least half of the world's commercially usable marine products.

A well-placed and extremely useful network of internal waterways, including the Great Lakes–St. Lawrence, the Missouri-Mississippi-Ohio system, the Columbia, and the large number of navigable east coast rivers and estuaries.

A location between the Atlantic and the Pacific that afforded protection against outside pressures without seriously hampering migration and influences from Europe. North America was relatively free to develop in peace without the need of providing for its own defense until very recently, an economic, political, and social burden that most other areas of the world have had to carry to their own loss.

Patterns of resource use

What use has been made of these resources? A glance at any statistical atlas or statistical survey of world production reveals that contemporary North America accounts for a disproportionate share of the industrial-commercial economy of the globe. With a population just short of 240 million, or about 5 percent of the world total, North American industrial production outranks its closest competitor, non-Soviet Europe, with well over twice North America's population. Nearly one-fourth of world steel production alone, perhaps the most important single manufacturing industry, takes place in North America. It also produces nearly 40 percent of the world's electricity (thermal and hydro combined), nearly half of the motor vehicles, about one-third of total cement production, and nearly half of the total production of machinery and electrical goods. North American

MAP 104

GENERALIZED FARMING TYPES IN NORTH AMERICA

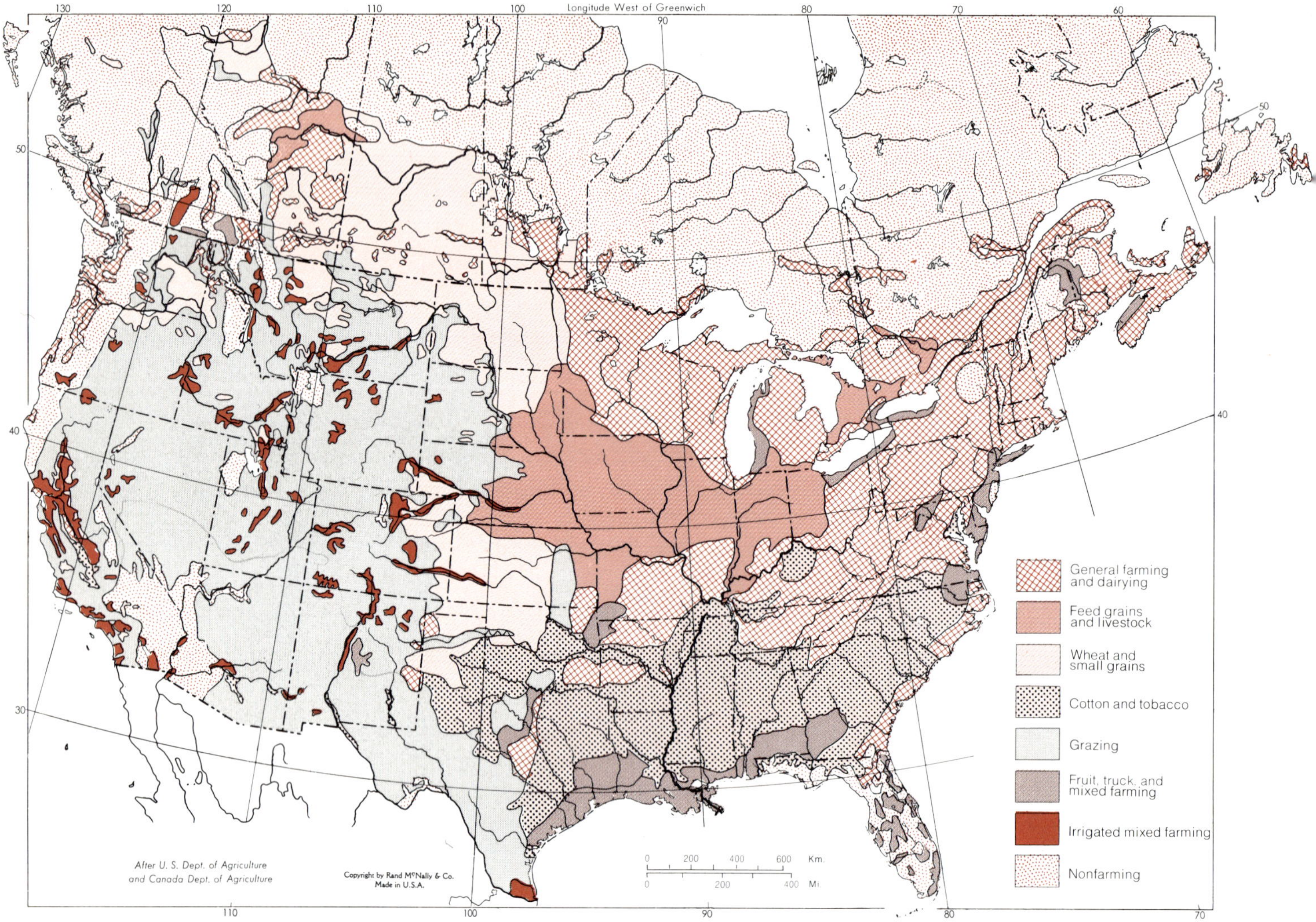

railway freight traffic accounts for nearly half of the world total in ton-miles, and its internal air traffic (passengers and freight) for almost two-thirds. It accounts for about one-fifth of the world's petroleum production, one-fourth of the coal, and about one-fifth of the iron ore. It produces about 40 percent of the world's timber and wood pulp, nearly one-third of the world's citrus fruits, nearly one-fifth of the world's wheat, a third of its oats, nearly half of its corn, one-fifth of its cotton, two-thirds of its synthetic fibers (nylon, rayon, etc.), and one-fifth of its tobacco. Equally striking, by comparison with most of the rest of the world, is the very high per capita share of this prodigal production, since the total population density does not yet approach that of other commercially productive areas or of the subsistence economies of monsoon Asia.

North Americans often say or assume that this

most impressive economic achievement comes from their own efforts, technological advances, hard work, form of government, economic system, or other cultural characteristics that have been proved superior by the record of their economic progress. Many Europeans, Asians, Africans, and Latin Americans will agree that credit can be given to these characteristics and demonstrate their agreement by applying methods they frankly recognize and label as *American*, especially in industrial production. There are two important points, however, that Americans in particular need to remember. First, most of the basic techniques and organizational structure of the modern commercial-industrial world originated not in North America but in Europe. (In earlier forms, to a minor extent, they were found in India and China.) These techniques were transplanted to North America when they were already well developed. North Americans have merely applied them on a newly large scale, and developed certain refinements of technique and organization better suited to this scale. Second, North America possessed usable resources that had and have no equal. At the very least they have played as great a part in the area's economic development as any cultural abilities that may be called American or Canadian, rather than European. These resources and their fruits have been made available to a population that, despite its very rapid growth, is still relatively small in proportion to the economic support it can command. The people-resource ratio is uniquely favorable. When the cards were dealt, North America held the strongest hand by all odds. Although that hand has been played with some skill, it is misleading to give sole credit to the player. It may also be said that, given such a strong hand, it is not surprising that North Americans have become commercial-industrial specialists and have contributed so much to the refinement of commercial-industrial techniques.

A further advantage that should be considered is the size of the domestic market. No area in the world could match it until the recent Common Market in western Europe. Of course, numbers are greater in India and China, but the extent of the market there is limited by low per capita incomes. Comparison with European states is economically more appropriate. The economies of scale the domestic market of the United States makes possible have been the envy of every European country for years. Although North America, like any other area, is not self-sufficient and must depend on the import of a number of essential raw materials and other commodities, the degree of that dependence is smaller than for any other commercial-industrial economy except the Soviet Union. Within this superbly endowed resource base, production takes place for a market of the magnitude of the whole of non-Soviet Europe. It is the size of the domestic market, together with abundant and varied industrial resources, that is principally responsible for the North American development of techniques of large-scale, low-cost production, high capitalization or mechanization of industrial processes, and large-scale, low-cost techniques of distribution and marketing. Heavy capitalization of industrial processes is helped by a large volume of production, rapid turnover, wide distribution, and very large numbers of consumers. North American production has become noted for its ability to produce in large volume at low cost. Its techniques of mass production are made possible by heavy capitalization or mechanization, and rest fundamentally on the economic conditions offered by the size of the domestic market. Successful mass production, using what is widely called "the American method" and dependent on heavy mechanization, is perhaps the most important single development of the continuing industrial revolution. It further increases the economy's ability to produce more goods at a lower unit cost with increasing dependence on power-driven machinery.

The role of transport

The size of the domestic market, which contributes so much to North American economic development, could not have become an asset, however, nor could much of the fabulous resource base have been built, without the accompanying development of low-cost, volume transport. Like the Soviet Union, although on a smaller scale, the United States may be called "an experiment in transportation." The same is equally true of Canada. It was only after the railroad and the steam engine were developed that these large areas with their abundant potential resources could be integrated into a market whose size was an advantage and whose potential productivity could be realized. This development is still only a little over a century old; its major beginnings in North America date only from the latter part of the nineteenth cen-

tury. The first transcontinental rail link was completed in 1869, but it was nearly the end of the century before the main outlines of the present rail network of North America were laid down. The second half of the nineteenth century was the great age of railway building, especially in North America and European Russia. In both places vast spaces were progressively knit together and widespread rich resources economically exploited in a way and to a degree never before possible. Had it not been for the British creation of the steam engine and railway, the United States, Canada, and the Soviet Union could probably not have developed as single political units. They could not have built their large domestic markets, made significant use of their extensive commercial-industrial resources, or achieved the position of commercial-industrial leadership they now enjoy.

The railways still carry an enormous and essential volume of traffic. But it is more apparent every year that North America, and especially the United States, overinvested in railways in the nineteenth century. At that time, railways were felt to be not only essential to economic growth, but almost its guarantees and the soundest possible investment.

Mechanized American agriculture. On the left, combines harvest hybrid grain sorghum in Texas. Stubble remaining after harvest will protect against wind erosion and increase organic matter in the soil. On the right, mechanical pickers harvest cotton in Louisiana. Each of these harvesters picks as much in an hour as a man could pick in seventy-two hours. *(Photo on left by Bob Kral, and on right by Larry Rana, both for U.S. Department of Agriculture.)*

Fortunes were made in that highly competitive and somewhat anarchistic age, but there was much wasteful duplication of lines. The meteoric rise of motor vehicles and the development of air transport, unforeseen as late as 1900, have basically altered transport flows. More and more rail lines have been abandoned every year, since about the time of the First World War, even while the traffic volume on most of the remaining lines has continued to rise. Over 80 percent of intercity passenger travel in North America is now by private car or bus, and well over half of the remainder is by air. Railways still haul more freight than all other internal carriers combined, but their share is still decreasing, especially as pipelines are extended, the cargo capacities of motor trucks continue to increase, and relative haulage costs diminish. The railways' province will probably be increasingly restricted to dry cargo of bulk commodities in large volume: coal, ores, bulk grain, construction materials, heavy machinery, and heavy equipment. Rival land carriers for this kind of traffic do not seem likely. Thus the railways will continue to perform their original and still basic function: bulk transport in large volume at low cost, the essential prerequisite for industrialization.

The automobile has played a large role in North American industrialization. It has deprived the railways of passenger traffic, and changed the growth, spacing, and spatial form of cities and their mushrooming suburbs. It has led to the blight of older, inner areas in the cities and the growth of suburban shopping centers. The farmer's life has been transformed; the family farm is disappearing and the rural hamlet is yielding to larger, more widely spaced service towns. New resort areas are growing; many formerly wild or natural landscapes are disappearing under the onslaught of mechanized vacationers or commuters. North Americans are increasingly mobile in their employment, residence, shopping, amusement, and recreation. Road congestion in and around cities threatens to choke them, and the increase in atmospheric pollution from exhaust fumes may have even more dangerous consequences than we now recognize. These results of the automobile have spread in Europe and Japan, and in time may engulf more of the world. The use of the private car in North America and its effects on the economy, society, and the pattern of settlement are directly related to the high level of North American per capita

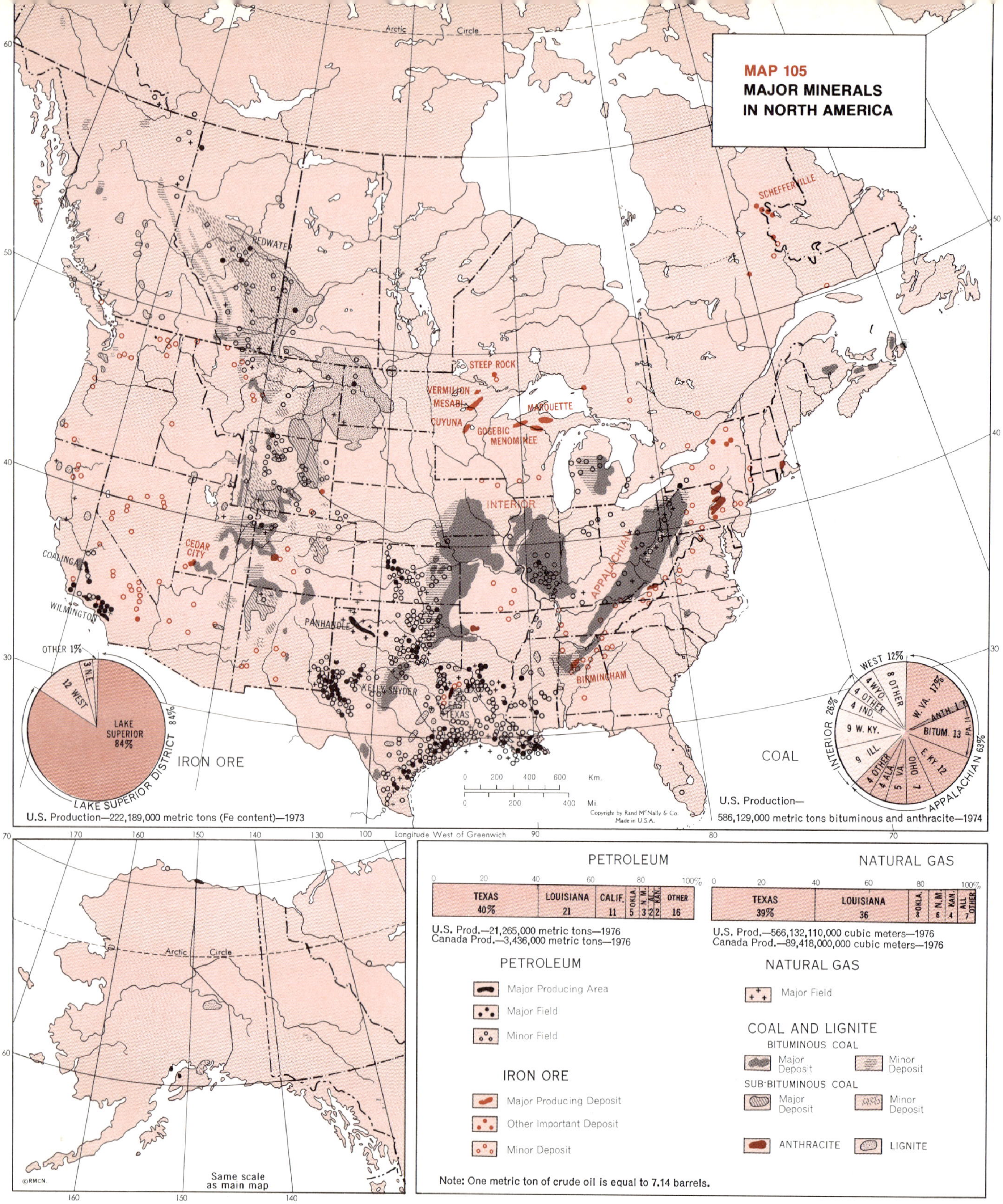
MAP 105
MAJOR MINERALS IN NORTH AMERICA
Arctic Circle
SCHEFFERVILLE
REDWATER
STEEP ROCK
VERMILION
MESABI
CUYUNA
MARQUETTE
GOGEBIC
MENOMINEE
INTERIOR
APPALACHIAN
CEDAR CITY
COALINGA
WILMINGTON
PANHANDLE
KELLY-SNYDER
EAST TEXAS
BIRMINGHAM
OTHER 1%
3 N.E.
12 WEST
LAKE SUPERIOR 84%
LAKE SUPERIOR DISTRICT 84%
IRON ORE
U.S. Production—222,189,000 metric tons (Fe content)—1973
0 200 400 600 Km.
0 200 400 Mi.
Copyright by Rand McNally & Co. Made in U.S.A.
COAL
WEST 12%
8 OTHER
4 WYO.
4 OTHER
4 IND.
9 W. KY.
9 ILL.
4 OTHER
4 ALA.
5 VA.
7 OHIO
12 E. KY.
BITUM. 13
ANTH. 1
PA. 14
W. VA. 17%
INTERIOR 26%
APPALACHIAN 63%
U.S. Production—586,129,000 metric tons bituminous and anthracite—1974
Longitude West of Greenwich
Same scale as main map
©RMCN.
PETROLEUM
TEXAS 40%
LOUISIANA 21
CALIF. 11
OKLA. 5
N.M. 3
AL. 2
KAN. 2
OTHER 16
U.S. Prod.—21,265,000 metric tons—1976
Canada Prod.—3,436,000 metric tons—1976
NATURAL GAS
TEXAS 39%
LOUISIANA 36
OKLA. 8
N.M. 6
KAN. 4
ALL OTHER 7
U.S. Prod.—566,132,110,000 cubic meters—1976
Canada Prod.—89,418,000,000 cubic meters—1976
PETROLEUM
Major Producing Area
Major Field
Minor Field
IRON ORE
Major Producing Deposit
Other Important Deposit
Minor Deposit
NATURAL GAS
Major Field
COAL AND LIGNITE
BITUMINOUS COAL
Major Deposit
Minor Deposit
SUB-BITUMINOUS COAL
Major Deposit
Minor Deposit
ANTHRACITE
LIGNITE
Note: One metric ton of crude oil is equal to 7.14 barrels.

incomes. There is well over one private car for every three Americans and Canadians, primarily because individual North Americans can afford to buy and maintain this expensive luxury. The effects of the automobile on public transportation and the spatial form of North American cities and rural service centers, however, have increasingly made it a necessity. This is not yet true anywhere in the world outside North America, with the possible exception of Australia–New Zealand, where per capita incomes are also very high. North America's share of the world's goods is disproportionately large, due not so much to their own efforts as to circumstances and the bounty of nature. They need to keep this constantly in mind, especially as they choose or are obliged to live increasingly closely with the rest of the world.

The agricultural dilemma

Another outstanding characteristic of North America is its bountiful productiveness, especially in agriculture. Starting with superabundant agricultural resources and a relatively small, mainly agricultural population, by about 1920 Americans were producing far more than they could consume, and even more than they could sell abroad at normal market prices. Since that time the number of North Americans employed in agriculture has fallen steadily, and there has been a slight but steady decline in total harvested acreage. But production has continued to mount. It more than doubled between 1900 and 1950, and since 1950 it has continued to rise. About a million North Americans are now leaving agricultural employment every year to seek other work. This may not be enough, and the agricultural work force still appears to be too large. Although it comprises less than 2 percent of the population and grows smaller every year, it continues to produce large surpluses for export.

A number of factors have contributed to this situation. North America has much more good agricultural land than it needs to support its present population or to meet present commercial export demands. A great deal of marginal land is left uncultivated or has been abandoned (large parts of New England, for example). This land would be intensively used if real population density were greater, as it is in Japan or in the Soviet Union. In spite of this, North America still has a much higher per capita share of good-quality cultivated land than any other area of comparable size in the world. As older and less productive land has passed out of cultivation, new and much more productive land has taken its place, including a great deal of irrigated land in the approximately one-third of North America that is too dry for unirrigated farming. Even though the net total of cultivated acreage has declined slightly, its average productivity has continued to rise.

Then, there is mechanization. Machine cultivation does not necessarily increase yields per acre, but it does free the land from the necessity of producing feed for draft animals. It is estimated that between 1900 and 1950 the equivalent of seventy million acres of cultivated land (exclusive of pasture) formerly used to sustain farm-animal power was released for food production as a result of mechanization. But by far the major factor is the increase in yields per acre, irrespective of new land. In the United States only, the yield per acre of four leading crops, corn, wheat, cotton, and oats, showed an average increase of nearly 100 percent in the period between 1900 and 1970. This record is generally typical of all North American agriculture. It can be attributed to the development and use of better varieties of crops, animals, and seeds (such as hybrid corn), increased use of fertilizers, better control of plant and animal diseases, pests, and weeds, and increases in irrigation, not only in the arid areas but over the country as a whole. All present evidence strongly suggests that increases in yields per acre and per worker will go on rising in the foreseeable future, and that they will continue to increase faster than the population. Investments of labor and capital in agriculture will have to be proportionately decreased, or additional outlets for agricultural surpluses must be found abroad (especially in food-poor Asia and Africa), or both.

Adjustments will also have to be made in the excessive use of chemicals—pesticides and fertilizers—if the natural environmental base is to be preserved (see Chapter 2), and better efforts made to prevent or control erosion. North American agriculture also makes excessive use of energy from fossil fuels to achieve its high yields. By global or perhaps by current American criteria, this is both wasteful and dangerous, and will need to be modified. The North American emphasis on raising animals for meat is also wasteful by comparison with food crops. It reflects the same combination of abundant land and other resources with a relatively small and high-income population that has shaped all of North American resource use. But it is increasingly out of tune with changing circumstances and changing perceptions.

Labor and resource use

In a land of bountiful nonhuman resources, large area, and proportionately small population, it is understandable that labor has tended to be scarce and expensive by comparison with most of the rest of the world. Until the Great Depression of the 1930s, with few and minor temporary exceptions, labor was chronically short in North America, characteristically better paid than elsewhere, and necessarily used as sparingly and efficiently as possible. Relatively high pay and wide economic opportunities acted as a powerful magnet for mass immigration, especially from overcrowded southern and eastern Europe. There were not only urban jobs in servicing, manufacturing, and transport, but also millions of acres of good agricultural land available at very low cost or free to would-be settlers. They have helped give North America the mixed population and culture that is one of its greatest strengths and most admirable characteristics. But the immigration, which reached a million a year at its height about the beginning of the present century, was never enough to meet the insatiable labor demands of a booming economy. Not only was it expanding westward spatially, but it was widening its pattern of resource use. Mechanization wherever possible became the major solution and it is still strongly characteristic of North America in virtually every aspect of its life. The degree to which daily life as well as production processes have been mechanized is one North American characteristic that seldom fails to impress foreign visitors. It is clearly related to the scarcity of labor, for most North American inventiveness has been devoted to techniques that save labor. In other countries, Japan, for example, or even Europe, the reverse has often been true. Their attention has been concentrated on saving or making more efficient use of nonhuman resources with less regard for labor investment. North America's profusion of resources waiting to be exploited encouraged techniques that made maximum use of limited labor. It also tended to encourage wasteful use of apparently unlimited nonhuman resources.

Influence of the frontier

Bountiful resources, especially agricultural land, were available from almost the beginning of European settlement until the present century to anyone in North America who was willing to pull up stakes and move westward. A whole continent, largely unoccupied, most of it potentially very productive, was open to the advancing waves of settlement that progressed in stages from the Atlantic coast to the Pacific. The influence of this expanding, moving frontier on the structure of North American life has been profound and perhaps has done more than anything else to bring about the social equality that has been characteristic of white North America since its founding. Equal opportunity and democracy in social, economic, and political life, ideals of the eighteenth-century philosophers of Europe, were brought closer to reality by the expanding horizons of the frontier. There status and fortune were won by one's own efforts. Such circumstances inevitably had their impact also on the longer-settled areas in the east. They found it difficult to maintain social stratifications as the moving frontier opened wider and wider opportunities for any who wished to strike out on their own. In political terms, too, it made its voice heard. The election of Andrew Jackson in 1828 was rightly regarded as a victory for the frontier democracy of the west, although staid easterners despaired of the country's fate with a band of rough frontiersmen and "levelers" in control of the national government. Jackson's election, however, was only one stage in a process of leveling that has continued ever since. It helped make North American democracy distinctive as a society in which a degree of social equality has been realized and accepted among most whites. Now, at long last, it is beginning to be extended painfully and slowly to blacks and other minorities. The informality that characterizes life in the United States may be related to the influence of the frontier, too, for it is characteristic of Canada as well. Modern North American society is a far cry from its seventeenth- and eighteenth-century European origins, or from its pattern in the New World of the eighteenth century. Before the influences of the

Mechanized America—cattle roundup by helicopter. This ranch in northwestern Texas is so large that sophisticated equipment must be used to cover it. Note the limited vegetation in this area. *(Photo from Standard Oil Company of New Jersey.)*

moving frontier became strong, formality in dress and manners and tendencies toward social stratification were important issues, as in the period during and following the American Revolution. The Revolution not only severed political ties with Britain, it marked the beginning of a new kind of society.

The North American experience with the moving frontier has been compared with the somewhat similar experience of Australia, where a similar society also developed, and with Russia, which was different, despite some important similarities. In Russia the expansion of settlement was frequently under the control of the state. Many of the pioneer communities were organized for essentially political or military purposes, such as the Cossack communities in the Ukraine and farther east. As pointed out in Chapter 9, the land available for Russian settlement eastward was much less productive than that in North America. Russian settlement did not appeal to a pioneer spirit of rugged individualism because of its difficulties and the need for outside assistance or support as well as direction. Nevertheless, the frontier has exerted some of the same leveling and equalizing influences on Russian society, although to a smaller degree.

In North America, rugged individualism as a way of life had profound consequences in economic and political terms, and was closely related to the freedom and opportunity of the moving frontier. But individualism was not good preparation for the world of the twentieth century. The United States in particular has had political power thrust upon it as one result of the two world wars. Its people must live increasingly closely with the rest of the world. Unlimited economic opportunity and unbridled individual freedom are replaced in most other countries by the need for common effort, the fight against poverty, and the demands of public responsibility rather than individual license. In economic terms, the frontier also contributed to wasteful resource use. Land, minerals, forests, or animals could be ruthlessly depleted because there was "always more farther west." In this respect also the experience of the frontier was harmful, and North America is still paying for it.

But perhaps the greatest adjustment that North America must make in the present century, and certainly the most important, is discarding unrestrained individualism and accepting the responsibilities, humility, and cooperative effort that wealth and

MAP 106

VALUE ADDED BY MANUFACTURE IN NORTH AMERICA

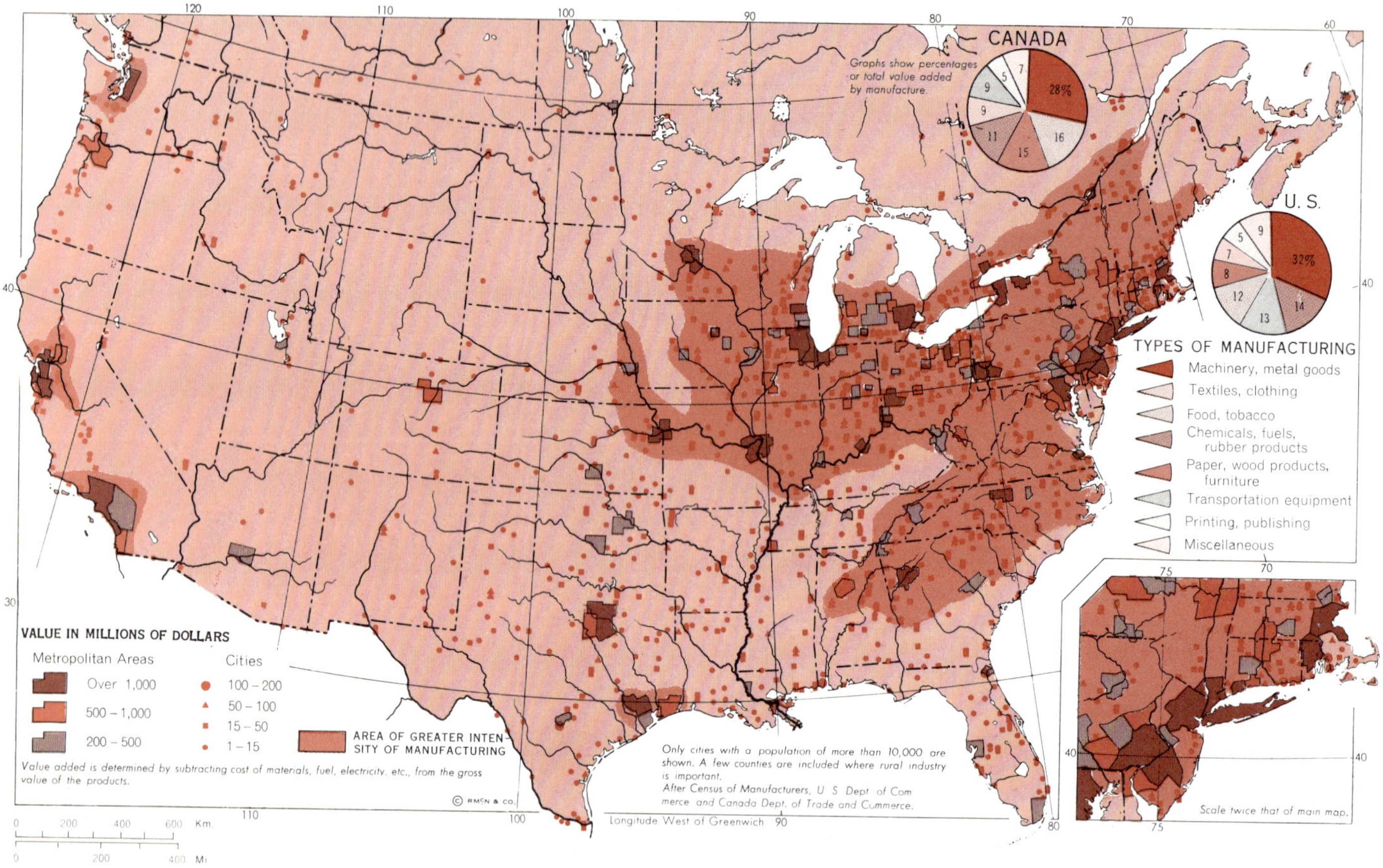

power must bring, if they are not to be destructive in a world that now lives as a single community. The attitudes and actions of the rich and the powerful are no longer only their own concern, as they may have been in the days of the frontier. They have become of fundamental importance to everyone in the world. In the long run the welfare of North America itself will depend on the wisdom and responsibility with which it manages its wealth and power within the world community. The role played by the United States abroad since the Second World War has not been a very encouraging beginning. Nor have Americans yet learned how to manage many of their urgent domestic problems: racial tensions, urban and rural poverty, educational deficiencies, and environmental pollution. A sense of social responsibility and humility, essential preconditions for successful solutions to all of these problems, has yet to develop adequately. Self-satisfied and assertive rugged individualism, inherited from America's past and its special conditions, exacts a high price.

Urbanization

One of the striking phenomena that accompanied the transformation of the agricultural North America of 1800 with its 6 million inhabitants into the present, highly commercial-industrial area of nearly 240 million was urbanization. Cities existed before the beginning of recorded history. But they changed in certain important respects with the advent of the Industrial Revolution and the improvements in transportation that accompanied it. Cities are more necessary in a commercial-industrial economy than in a subsistence agricultural economy. They are larger and more numerous, and they contain a much greater proportion of the total population. Almost all of the inhabitants of a commercial economy demand services and goods that are most conveniently provided by cities. The rapid growth of cities in North America paralleled the somewhat earlier experience of northwest Europe in this respect, but with some important differences. Urbanization, the proportion of the population living in cities, has not yet reached the high levels of England, Belgium, or Germany, but it probably will do so before the end of the present century. Nearly three-fourths of the population of the United States lives in what are called Standard Metropolitan Areas. They are defined as cities of 50 thousand or more, together with the counties in which they are situated, and adjacent, relatively densely settled areas, most of whose inhabitants commute to or have direct dealings with the city. About a fifth of the United States population lives in Standard Metropolitan Areas of 3 million or more; the proportions in Canada are similar. The most striking differences between the cities of North America and those of most of the rest of the world resulted from the fact that most American cities began or had their major growth in the age of the railway and motor vehicle.

Any city must depend fundamentally on access, hence, on transportation. It cannot perform services or provide goods for a wider area without transport at bearable cost. Nor can it feed and maintain its own population without transport, except on the very small scale that could be supported by farming within half-a-day's walk of the city. Ancient cities grew to significant size primarily at locations that had good access by water transport. Thus, ancient Rome and Peking were fed, and medieval Venice, Antwerp, Copenhagen, and Calicut were sustained as major centers of trade. Until the development of railways, the even larger cities in northwest Europe that grew to giant size in the early period of the Industrial Revolution, such as London and Paris, were closely tied to water routes, both sea and river, as most ancient cities had been. In North America until about 1850, the same was true. The only sizable cities

Urban America—midtown Manhattan at night. Although New York owes much of its growth to an extensive railway network joining it to the rest of the country, port functions are still of major importance. Most of the goods passing through New York or consumed in the city enter and leave Manhattan by water. Tunnels and bridges across the Hudson River and Long Island Sound carry mainly passenger traffic. *(From New York Convention and Visitors Bureau.)*

MAP 107
PRECIPITATION IN NORTH AMERICA

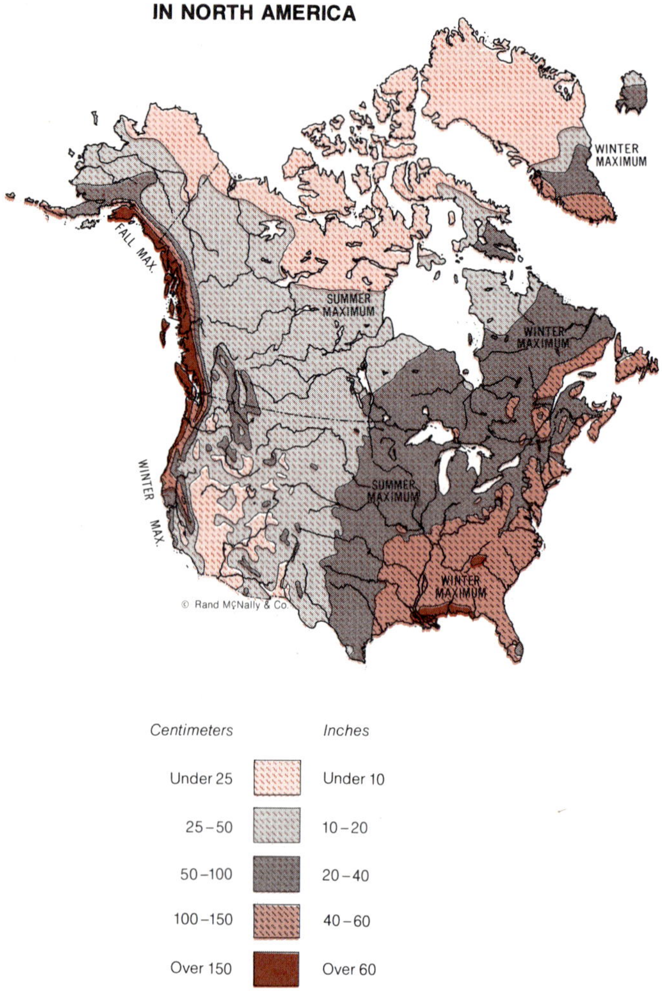

MAP 108
VEGETATION IN NORTH AMERICA

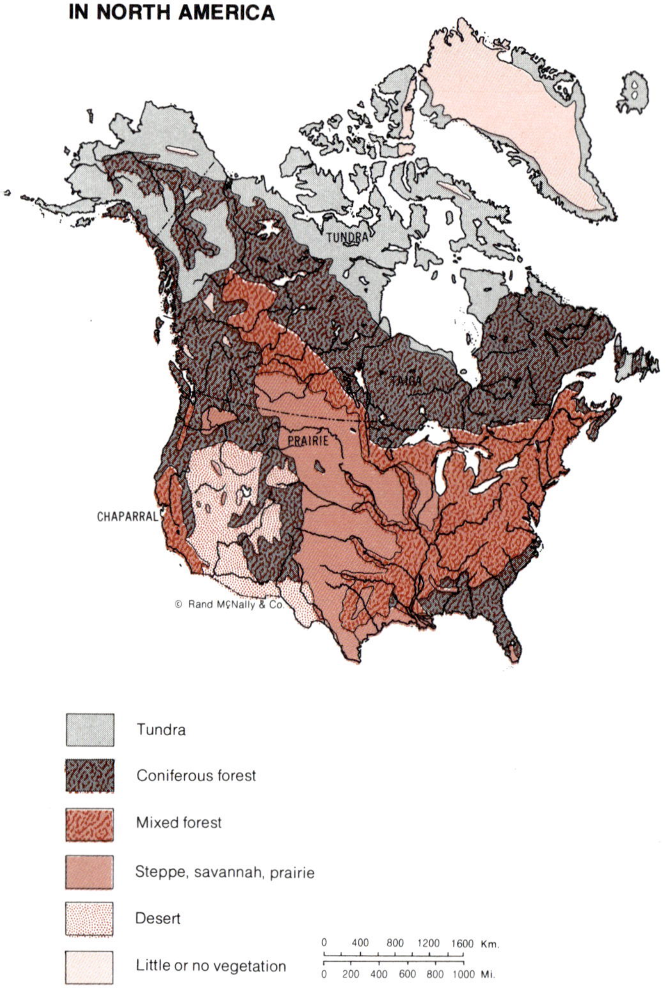

were on the east coast or on major rivers and their estuaries. The original core of all these cities was directly on the water. They have grown away from it only with the advent of railways and motor vehicles. This pattern is clearly observable in New York, Philadelphia, Boston, Baltimore, Montreal, Quebec, and many other lesser cities. Most North American cities either did not exist or had not grown to large size before the transport revolution. Therefore, their present form reflects that revolution much more fundamentally than the older cities of Europe or Asia. Many of their cities were approaching or had exceeded 1 million in size, and were relatively firmly established before 1850. Railways have come to them as a later adjunct. But in North America most cities have been shaped by the railways, and, in the case of newer cities, by the motor vehicle to a much greater extent.

The extremely rapid growth of many North American cities is directly related to the fact that

MAP 109
POPULATION OF NORTH AMERICA

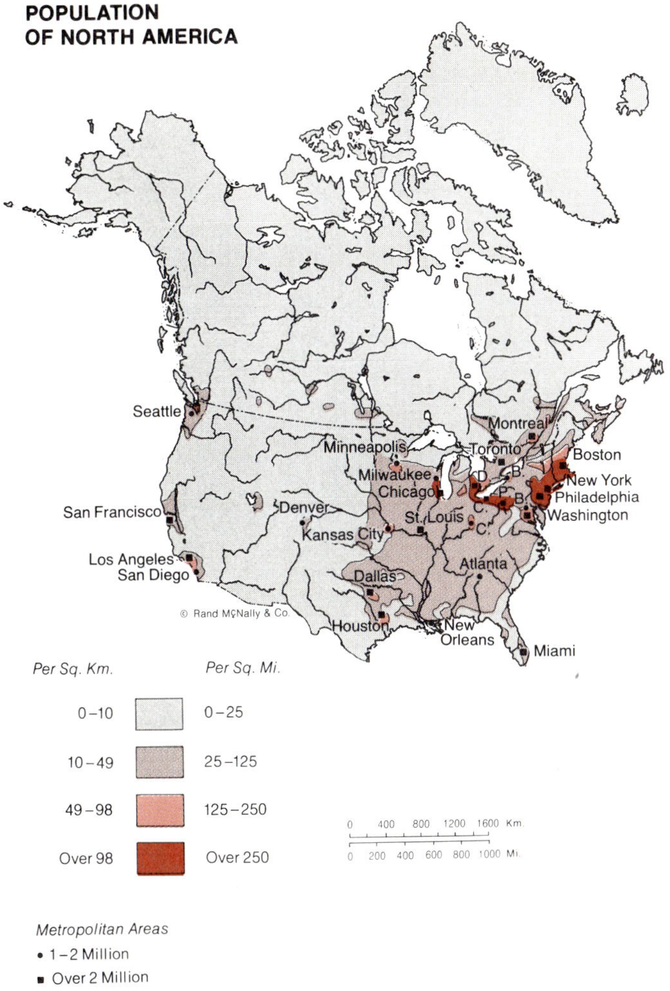

most of them began in a previously almost uninhabited, or at least nonurban, area at the beginning of the transport revolution. Transport has made enormous cities possible for the first time. Thus Chicago grew from nothing in 1804 to a tiny village in 1831, to a city of over half a million in 1880 (about thirty years after the first rail lines reached it), to over 2 million in 1910, and, including its vast contiguous suburbs, to about 7 million now. Los Angeles, its form more a product of the motor vehicle than the railway, grew

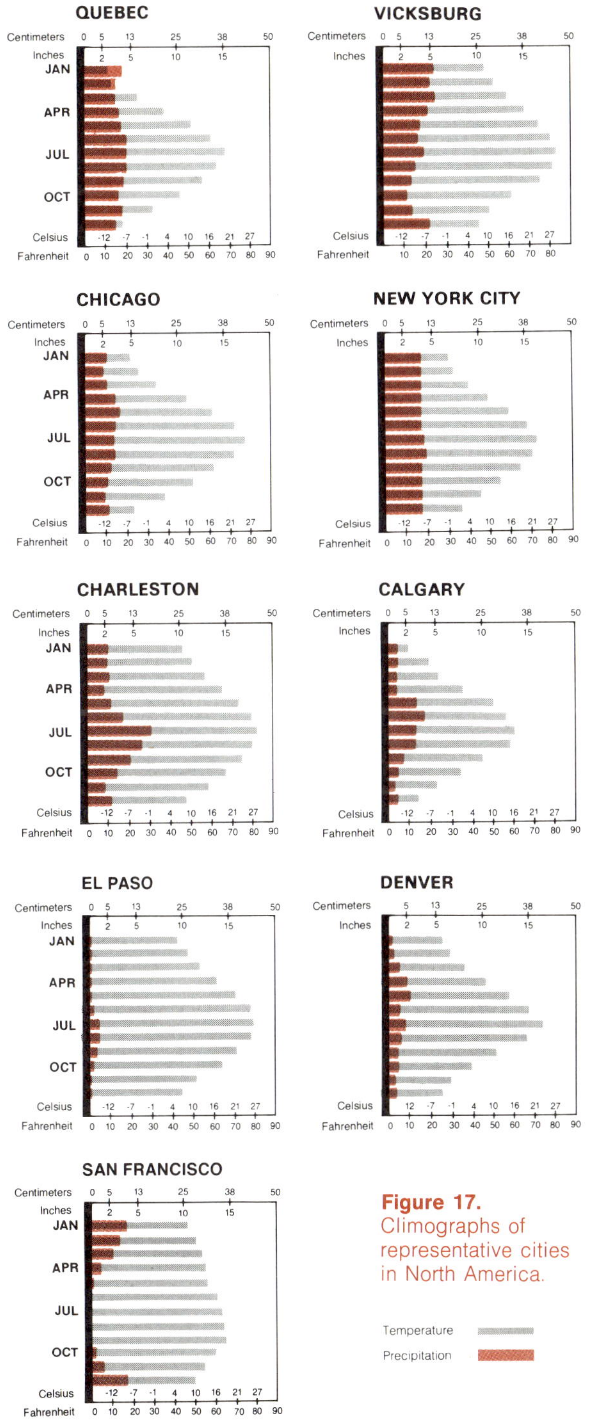

Figure 17. Climographs of representative cities in North America.

from a tiny Spanish village in 1850 to a town of about 100 thousand in 1900 (at the dawn of the motor vehicle). It passed the million mark about 1925, and is now, including its sprawling urban area, even larger than Chicago. The spatial form of Los Angeles—great distances within the urban area, a multitude of semi-independent communities and shopping centers, and weak dominance of the central city area—suggests the strong influence of the private car. Although all North American cities have experienced this trend, most of them had grown to some size before the advent of the automobile. The urban sprawl and exploding suburban development that now characterize cities have overlaid the earlier water-oriented or railway-oriented base with a central city core that still retains considerable dominance. Detroit is probably the closest parallel to Los Angeles among large cities. Chicago, the outstanding railway city (see Map 103, page 456), and the older cities of the east coast still preserve a good deal of their original waterfront cores. The essential role played by transport improvements in accounting for the rapid growth of large cities, however, should not obscure the fact that the transport lines were built to serve the city and its tributary area. It was the city and its commercially productive hinterland that built the transport lines, not vice versa. The Chicago skyline includes the tall tower of the Chicago Board of Trade, topped by a statue of Ceres (goddess of the harvest). It is a reminder that this enormous city exists primarily because the corn belt is a highly productive commercial agricultural area that required a city to serve it and railways to link it to the city. Carl Sandburg presented the reasons for Chicago's growth accurately when he called it "hog butcher to the world," and characterized it as "playing with railroads."

The characteristic of North American cities most closely related to the use of the automobile—the spread of suburbs—is like the automobile in that it, too, reflects high per capita incomes. No longer is there a large or readily definable group of poor workers, as in the nineteenth century (although perhaps one-fourth of the total population still lives in at least relative poverty). The spatial form of most North American cities mirrors this change. Some of the older slum areas near the centers of the cities have been taken over by commercial functions, demolished, or replaced by more desirable multiple dwellings. Massive urban slums remain where both black and white live in substandard housing with grossly inadequate public services. But the major spatial trend has been outward to more expensive and more spacious housing. This trend developed not only because of the automobile, but also because average incomes have risen. This development still has growing pains, and it has presented new problems as well as solving others. It seems certain that by the end of the present century North American cities will change at least as much as they have since 1900, in spatial form as well as in other characteristics. They will probably continue to be distinctive, although many cities elsewhere may reproduce North American urban forms to differing degrees and at differing rates as local circumstances warrant.

However, it seems reasonable to regard suburban sprawl as related fundamentally to high per capita incomes. Suburban sprawl is accompanied by the private automobile and the search for additional luxuries or amenities, and thus it becomes a peculiarly North American phenomenon.

Broader spatial patterns

Two broader spatial characteristics of North America seem fundamental:

1. The sharp three-part distinction between the humid east, the arid west, and the arctic and subarctic northern fringe (actually almost half of North America in area, but with less than 1 percent of its population).

2. The pronounced and dominant extended core area, an approximate rectangle bounded roughly by St. Louis, Washington, Portland (Maine), Montreal, Ottawa, and Madison (Wisconsin) (see Map 110, page 474).

The cold north

The cold north is a true pioneer area, playing much the same role as the Soviet arctic and subarctic. But its significant human development lies almost entirely in the rather distant future, if it will ever be of more than strategic or minor economic significance by comparison with the rest of North America. It covers most of Alaska and Canada, although the population of both areas is almost entirely confined to the coastal or southern edges where a more moderate climate prevails. About 95 percent of Canada's population is crowded into a relatively thin, irregular ribbon along the southern border of the country. Nearly 85 percent of Alaska's much smaller total hugs the southern coast from Seward and Anchorage to Ketchikan (see Map 109, page 469). The enormous timber stands of the northern taiga and the minerals that may lie beneath them are not usable resources under present conditions, for the most part. The area's agricultural possibilities are severely limited by soil as well as by climate, although, as in Russia, new crop strains and farming methods have been developed that continually push the agricultural frontier northward on a small scale. In the absence of severe population pressure, and with the abundance of good agricultural land in North America and its high agricultural production, it is even less reasonable on economic grounds to invest in the cultivation of marginal land, such as all of northern Canada and Alaska. Except for the large oil resources of northern Alaska (which will continue to be almost entirely exported), it does not seem likely that most of the area will ever be more than an unproductive wilderness from the human point of view.

Humid east and arid west

Distinction between the humid east and the arid west rests primarily on climate, but has very widespread ramifications. Broadly speaking, North America east of the hundredth meridian and south of the cold northern area—the humid east—is green and urban, intensively farmed, and highly industrialized. It is thickly interlaced with rail lines and motor roads, densely populated, and highly productive. North America west of 100° W.—the arid west—is brown and rural, extensively farmed or uncultivated, used for stock ranching, open range, or no production at all. It is industrialized very little, thinly populated except in a few spots, and accounts for less than a sixth of the total commercial production of North America, although it occupies about half of the area south of the taiga (see Maps 18, 103, 104, 107,

The green, humid east. This is Pennsylvania farmland on the edge of the Appalachians. It is typical of wooded, rolling, eastern North America, with a white church steeple marking the village in the valley. Both corn and wheat are being grown; the corn is primarily used for animal feed, and largely for dairy cows. *(Photo from U.S.D.A.–Soil Conservation Service.)*

109, and 110, pages 66, 456, 458, 468, 469, and 474). The distinction between the humid east and the arid west is much more significant and far-reaching than the more conventional distinctions, often based on the line of the Rocky Mountains or the Mississippi River. The Mississippi has an immense basin, and, in broad terms, its parts are more like each other, and like the humid east, than like the arid west.

Homogeneity within the humid east and the arid west, of course, is far from complete. In the east most of the characteristics outlined above are interrupted by the thick wedge of the Appalachian Mountains, running from northern Georgia through northern New York to northern Maine. Even in these mountains, however, the east remains green. Many coal-rich sections of the Appalachians support large-scale manufacturing, urbanism, and dense transport networks, for example, the Pittsburgh area and northern West Virginia. The Canadian Shield, an area of nearly bare rock where glaciation has removed most of the soil, extends south to the northern shores of Lakes Superior and Huron. It supports little or no agriculture, no cities except for Thunder Bay (formed from Port Arthur and Fort William) and Sault Ste. Marie (which lie directly on the lakes and are engaged in the lake trade), almost no manufacturing, and one of the sparsest population patterns in North America south of the northern area. Northern Maine, Vermont, New Hampshire, and New York, in a heavily glaciated, cold, poor-soil area, exhibit many of the same characteristics as the Canadian Shield in their human patterns. So do the northern thirds of Michigan, Wisconsin, and Minnesota. In the southeastern United States, urbanism declines sharply south of Washington and St. Louis, and manufacturing is generally secondary to agriculture, although it is growing rapidly, especially near the base of the mountains and in Atlanta and Birmingham. In the southeast, transport lines are less numerous, population density is lower, and total productivity is less than in the rest of the east as a whole. In all of these respects the southeast is nevertheless distinct from

The brown, arid west. The trees in the residential areas of Amarillo, Texas are artificially watered. The real nature of the climate shows in the dry wheat fields beyond the grain elevators—a typical Great Plains landscape. This landscape is essentially the same from Texas to Saskatchewan and from Kansas to the Rockies. *(Photo by Fred S. Witte for U.S. Department of Agriculture.)*

the arid west, and is less like it than like the humid east.

The unbroken greenness of the east, from Hudson Bay to Key West and from central Nebraska and Kansas to the Atlantic (except where it is interrupted by grey cities), is its most obvious characteristic to a traveler from the arid west. The east is appropriately labeled humid, and its humidity is related to all of its other characteristics. It is productive in part because it is humid; and, because it is productive, it is also urban, industrial, agriculturally intensive, and well served by transport lines. To take only one example, it is not surprising that all the larger cities of Texas are in the eastern half of the state and that population and production are largely concentrated there. Texan spatial patterns are also closely related to the rich petroleum and sulphur deposits in the eastern half of the state. But the outlines of the settlement pattern were essentially the same before the discovery of the minerals, and there is almost as much petroleum in west Texas as in east. The direct relevance of rainfall is also suggested by the same eastern orientation of settlement, cities, and production in Oklahoma, Kansas, Nebraska, the Dakotas, Manitoba, and Ontario (see Map 109, page 469).

In North America west of 100° W. and south of the cold northern area, the generally arid, rural pattern is interrupted by two large and three medium-sized urban-industrial-trading complexes (Los Angeles, San Francisco, Portland, Seattle-Tacoma, and Vancouver), and by the generally humid strip of narrow coastal plain they occupy. In actual fact the arid west ends near the crests of the Sierras and the Cascades, although aridity extends to the sea in the Mediterranean climate of southern California. Los Angeles maintains itself on water imported from great distances. Imported water supports the city itself and the extremely productive agricultural hinterland that helps to maintain the city and that largely depends on irrigation. The humid strip of coastal plain from central California north into southern British Columbia, and the heavily irrigated interior valleys and coastal plain of southern California, in some respects support a smaller-scale replica of the humid east. In these areas productivity per acre, population density, value added by manufacturing, and urbanism are relatively high, agriculture is generally intensive, and the transport net is relatively well developed.

This coastal pattern, however, is more interrupted as well as smaller in scale than that of the humid east. There are large gaps, especially in mountainous northern California and southern Oregon, where, despite adequate rainfall and green landscape, most of the other eastern humid characteristics of agriculture, manufacturing, urbanism, population density, and transport are absent (see Maps 104, 106, and 109, pages 458, 466, and 469). In the cool, humid northwest, the familiar dairying pattern of the cool, humid, glaciated northeast reappears, although on a smaller scale. Its local market is much smaller, and it cannot compete significantly with the northeast to supply the markets in the middle west or the south. Mineral resources play a relatively minor role in the industrial structure of the Pacific coast. Coal and iron ore are limited and mainly of low grade (see Map 105, page 462). Even the petroleum deposits of southern California, although important, are not the dominant factor in the industrial complex of the Los Angeles area. The coastal cities are primarily service centers for their productive agricultural hinterlands. They ship and process wheat, fruit, timber, fish, vegetables, and cotton (from California). They handle trade with the Orient. They are seats of state and federal branch governments. And they manufacture products not closely tied to local raw materials or transport lines—aircraft, films, machinery, and diversified light consumer goods. There was no integrated iron and steel works on or near the Pacific coast until late in the Second World War, when a plant was established at Fontana near Los Angeles, drawing coal and ore mainly from Utah. It was built because the west coast market had grown big enough to support such a plant. But its production is still dwarfed by that in Gary (Indiana), Baltimore, Pittsburgh, Bethlehem, Morrisville (Pennsylvania), and other large eastern establishments.

The rest of the west is truly arid, but the sparse human pattern is speckled, primarily as a result of irrigation (see Maps 104 and 109, pages 458 and 469). Virtually all of the scattered clusters of relatively dense population, including Salt Lake City, Denver, Boise (Idaho), Yakima-Wenatchee (Washington), and Phoenix (Arizona), coincide with irrigated areas that are still increasing but that occupy only a tiny fraction of the arid west. Deserts with little or no vegetation cover far more area than does irrigation, especially in the southern half of the land between the Pacific coast ranges and the Rockies. In this area, desert and scrub desert combined occupy about half

MAP 110
CORE AREAS OF NORTH AMERICA

(All figures approximate)

AREA 1	(% U.S.) 7.7% area, 50% income, 70% persons in "Who's Who"
AREA 2	(% Canada) 0.4% area, 20% population, 33% industrial employment
AREA 3	(% U.S.) 7.0% area, 7% income
AREAS 1 and 2	(% U.S. & Canada) 3.7% area, 41% population, 66% industrial employment
AREAS 1 and 3	(% U.S.) 14.7% area, 50% population, 57% income, 73% industrial employment
AREAS 1, 2, and 3	(% U.S. & Canada) 7.0% area, 48% population, 71% industrial employment

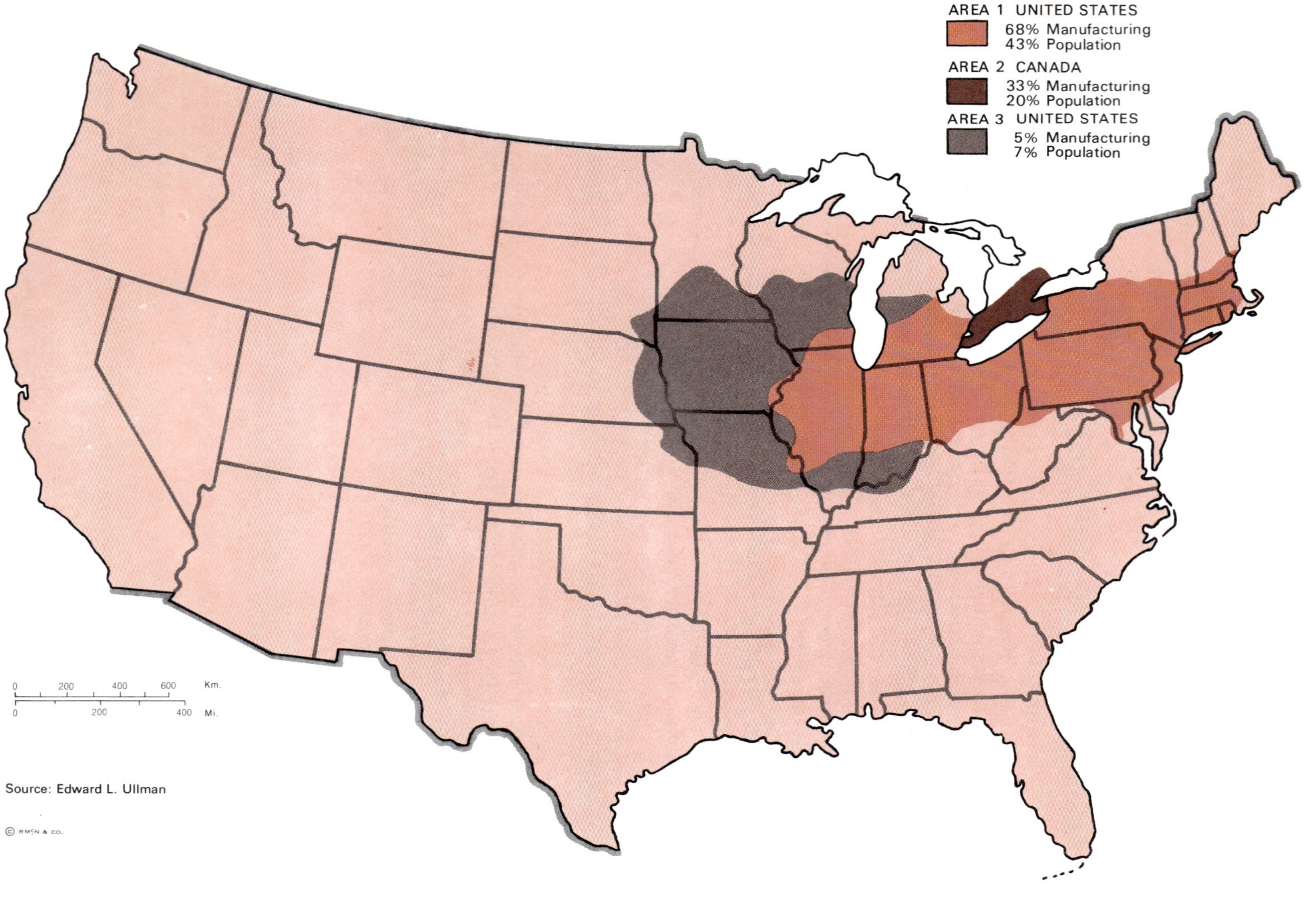

Source: Edward L. Ullman

of the total area; most of the rest is steppe. Growing iron and steel production at Provo, Utah and Pueblo, Colorado, based on nearby coal and iron ore and serving the rapidly growing western market, supports significant population clusters. But they are still relatively small and are not associated with large, local urban growth. Population density and the density of the rail net actually increase west of 100° W. on the Canadian side of the border. This is due to a combination of other factors, including lower evaporation rates resulting from generally lower temperatures than those south of the border, the shortage of good agricultural land elsewhere in Canada, and the beginnings of development of rich petroleum, coal, and natural gas deposits, although the minerals are still largely exported. Nevertheless, Canada west of 100° W. and east of the Pacific coastal ranges is predominantly brown rather than green, rural rather than urban, extensive in its land use, agricultural and pastoral rather than industrial, and much more like the arid west of the United States than like the humid east of Canada.

Travelers crossing North America from east to west can hardly fail to appreciate the significance of the transition from humid to arid. As they approach and pass the hundredth meridian, or the twenty-inch annual isohyet that closely coincides with it (see Map 107, page 468), the towns become smaller and farther apart. The farms become larger; railroads and rail crossings are less frequent and trains fewer and longer. The motor roads become narrower, less used by traffic, and much less numerous. The price of gasoline rises; spare parts are more difficult to obtain. In a great variety of ways it becomes more and more apparent that this land has limited economic opportunity. Its flatness is relieved only by the grain elevators that rise at long intervals like skyscrapers above the plains, and by the cattle-crossing signs that warn the wary driver. The greenness of the Pacific coastal strip, when it is finally reached, is especially striking after the hundreds of miles of brown and grey, as are its cities, towns, truck and dairy farms, factories, traffic, trees, and sea beaches.

The dominant core

The only other regional distinction or spatial pattern in North America that rivals in significance the threefold division into cold north, humid east, and arid west is that between the dominant extended core region and the rest of the United States and Canada, which acts as its hinterland to a large extent. Map 110 (page 474) shows this core area, and although different lines could be drawn around it—for example, Minneapolis–St. Paul might be included or Portland, Maine excluded—the major outlines are reasonably clear. The North American core region plays much the same role and contrasts with the rest of the area in much the same way as the similar extended core areas of Japan, Italy, the Soviet Union, the United Kingdom, France, Argentina, and other cases discussed in previous chapters. Nearly every country and nearly every region in the world exhibits this pattern of concentration. It is not surprising to find it in a pronounced form in North America where communication, commercialization, and industrialization are highly developed, since all of these favor concentration. The core region shown in Map 110 contains about 7 percent of the area of North America, but approximately 48 percent of its population. It has about 71 percent of the industrial employment, well over half of the total income and retail sales, and three-fourths of the rail traffic. It includes both national capitals. Practically all of the book publishers and the headquarters of large corporations are located in this core area, as are over 70 percent of those listed in *Who's Who*, over three-fourths of the universities and colleges, and all of the cities with a population of substantially over 1 million, except Los Angeles and San Francisco.

The marked dominance of this core is partly of historical origin. But its spatial limits have been moving rapidly westward for at least the past century, and are much more closely related to current economic forces than to historical precedent. Washington, D.C., as the capital of the United States, may not seem well located within the whole area of the country, but considering the present distribution of population and production, Washington is still not far from its center. The physical center of population in the United States is still moving westward but has not yet passed Illinois. The east coast ports, the cities of New York, Philadelphia, Washington, and Chicago, the Great Lakes waterway, the huge coal deposits of the Appalachian region and Illinois-Indiana, and the productive commercial agriculture of the eastern half of the corn belt and the middle-Atlantic coastal plain have all played a part in the dominance of this core area. But it is most strongly a region of industrial cities. Its spatial limits, in fact, coincide extremely closely with those of the manufacturing belt, measured in terms of industrial employment or value added by manufacturing (see Map 106, page 466).

Like most large cities, and especially like most primate cities, core areas seem to have a self-perpetuating quality. Simple bigness or quasi-monopoly brings with it economies of scale, advantages of specialization, strong market attractions, and a pyramiding complementarity of services and production processes. The biggest market continues to attract the greatest amount of new commercial-industrial enterprises. The biggest commercial-industrial complex continues to attract most of the new complementary trade and industrial services. These, in turn, support denser populations, more extensive noncommercial and nonindustrial services, higher income, more capital accumulation, and more new investment. Areas outside the core suffer in this competition. In North America, however, distances are relatively great. This tends to give some assistance to commercial-industrial development and the growth of a small subsidiary core in coastal California, serving all of the west. As the California core grows in size, in turn it will begin to acquire some of the same self-perpetuating qualities. But it seems unlikely ever to rival the major eastern core or to include the south or the northwest within its tributary area. They also have their own rapidly developing urban-commercial-industrial centers.

The North American essence

When all this has been said, what else seems distinctive about North America, as one tries to see it in perspective? Surely it is a racial and cultural mixture of which its inhabitants can be proud and from which North America has drawn much of its modern strength. North America is unique. It has developed as a joint creation of the culture, genius, and hard work of the people of every country in Europe, of originally African Negroes, and of smaller numbers of representatives from most of the rest of the world. Modern North American civilization is the result of that vigorous blend. It has borrowed and adapted most of its techniques, standards, ideas, and even its citizens from the richly varied stocks of the Old World. It has fused them into a mixture that is much more than the sum of its parts, and built not only an economy but also a society from which the Old World can learn. For all its glaring shortcomings, North America offers a challenging example to a world divided by culture and race. Its society and economy continue also to challenge Americans as a still imperfect ideal. To the social and political institutions North America largely inherited from its British parent, it has given special meaning, even though it is still far from complete. In the sphere of social democracy, this meaning acquires new significance in the light of the widely varied origins of North Americans as a whole.

In Canada differences of origin and culture have a marked spatial form. In Quebec, Canada is mainly French in language, religion, customs, and associated social and political attitudes. The rest of Canada remains dominantly British in origin, but now has the same varied mixture as the United States. The great influx of displaced eastern and southern Europeans into Canada since 1945 is still increasing the mixture of Canada's population. Previously it had been less diverse than that of the United States. The culture of French Canada was conservative from its inception, and in many ways tended to preserve the culture of the *ancien régime* of seventeenth-century France. The people of French Canada have spread into the rest of Canada very little, and have often resisted or resented the interference of the central government. They have self-consciously maintained themselves spatially and numerically, as a distinct group, while the rest of Canada has grown in numbers and cultural variety. The division of Canada into French and British-mixed has posed some awkward political problems. But compromise and cooperation within a huge and varied cultural complex have been strong enough so far to hold Canada together on a basis of mutual agreement. In recent years change and political participation within Canada have increasingly affected French Canada. It is now playing a larger, although still distinctive, role in national affairs. Cultural mixtures are far more spatially scattered in the United States, although the more recent arrivals tend to concentrate in the cities.

Nearly all North Americans share an enormously mixed heritage; it is their greatest strength. For all the economic achievement of North America since

1800, by the end of the twentieth century the perspectives of geography and history may well suggest that North America's most important contribution to world civilization was its noble experiment in cultural and racial mixture. The final chapter of that story, which appears to be in its opening stages, has yet to be acted.

Questions for further study and discussion

1. How and to what extent were the Appalachians related to the fact that European settlement in North America remained largely limited to the coastal plain until about 1800? What other factors must be considered?

2. What differences would there have been in the settlement pattern of North America if the continent had been discovered and settled first by Europeans on the west coast instead of the east? Would settlement have grown as fast? Would the pattern by, say, 1978 have been markedly different? By 1900?

3. What minerals are already being taken out of the arctic and subarctic areas of North America? Why?

4. Approximately how much of the United States economy, and how much of its employment, is directly dependent on foreign trade? In this light, how should the country's tariff structure be evaluated? Why has this structure developed as it has? What are the possibilities, and what would be the results, of reducing or abolishing it?

5. Why should communication, commercialization, and industrialization tend to favor regional concentration? What changes, if any, may be expected in the present concentration pattern in North America? Why?

25 Regions and contemporary problems of North America

Expressways leading to Detroit, Michigan. *(Photo from City of Detroit.)*

The United States and Canada are two of the world's largest countries in area, and each contains a very wide range of physical environments. Yet each is also remarkable for its cultural homogeneity, by comparison with most other countries, including many only a small fraction of their size. This relative homogeneity, of course, is the result of the recent settlement of North America by outsiders. Also, settlement took place in an originally very sparsely occupied continent on the eve of a technological revolution that greatly reduced the friction of distance. Had it not been for these circumstances, several states and different cultures would probably have evolved, even given the more or less uniform culture and origins of most of the early settlers. As it was, France, Spain, Mexico, and England all had long-standing claims to parts of North America. They settled in different areas, and fought wars with one another and with the government of the United States over territorial sovereignty. These conflicting claims were not finally settled until the age of the railway. For the first time the railroad made the notion of single sovereignties conceivable and workable for large areas such as Canada and the United States. The revolution in transport and communications, then, made it possible to superimpose an increasingly homogeneous culture and a single set of political forms over the early regional cultural diversity.

The only political unit approaching Canadian or American size that exhibits equal or greater cultural homogeneity is Australia, where the history of outside settlement is very similar. Like Australia, North America also owes its high degree of uniformity to isolation. Native Indian groups north of Mexico were small, fragmented, and technically undeveloped. They did not offer rival cultural influences or impose their own highly diversified regional patterns on the evolution of immigrant settlement. With the single exception of Mexico, there were no neighboring states (as opposed to tribal units) and no effective bases from which diverse influences might have come. This situation was the reverse of Europe's, for example, and opposite also to India's with its long and complex history of invasion and cultural diffusion through a great variety of outside forces. North America was shielded from outside contact by its vast surrounding oceans. Outside contact began only in the age of European expansion so the people and influences that did come in were, broadly speaking, very similar. They were more varied than those that acted on Australia, and there is, accordingly, a greater degree of cultural diversity in North America. But the transport and communications revolution and the rise of two instead of many strong central states have combined to develop a relatively uniform pattern by comparison with most areas of similar size elsewhere.

Aspects of homogeneity

Perhaps the most revealing of these uniformities, apart from the political units, is language. Although uniformity of language is not absolutely complete (in parts of southwestern United States and eastern Canada, Spanish and French, respectively, are the first languages), English is understood and spoken everywhere. There are regional speech differences, but none that prevent understanding by most people, such as the many mutually unintelligible Indian languages and dialects that existed before the Europeans came. Such uniformity contrasts even with states the size of France. Indeed language variety within England itself is far greater, reflecting the regional speech patterns of an earlier age of more restricted communication and interaction. Language patterns, here as elsewhere, are both cause and effect of cultural patterns. They indicate and also help to produce or perpetuate cultural uniformities and differences. The existing language patterns of North America are more significant than most North Americans realize. They tend to take language for granted, especially if it is compared with what existed in 1492, or with contemporary patterns in most other parts of the world, especially in Europe, Africa, or India.

The commercial, industrial, and urban revolutions have played a major role in this, together with the associated revolution in transport and communications. One of North America's most striking uniformities (overlooking a few, dwindling pockets) is the increasing sameness of its urban areas. They involve a rising majority of the population in similar life-styles, activities, occupations—and problems. Whether living in large urban areas or not, practically all of the population, including the rapidly shrinking group of farmers, is involved in a single commercial system. Goods and services are produced almost exclusively for sale within a national market, and nearly all of their consumption is from a uniform set of commodities and services produced by someone else who may be hundreds or even thousands of miles away in another part of the same commercial system. Most Americans and Canadians read the same news, the same advertisements, the same magazines. They buy the same goods, attend the same kinds of schools, watch the same television programs, and listen to the same music. They wear the same kinds of clothes, eat the same limited range of food, and in general terms share a common life-style, values, and attitudes. (The political structure, which overrepresents rural and small-town areas, does not always reflect this, and may mask both variety and uniformity.)

The most obvious exception, the shamefully large group of the poor (especially in the United States), both black and white, suggests that the spectacular commercial-industrial growth of North America has ignored or failed to cope with human values and social justice. But so far at least, poverty itself has not produced a separate culture, only a second-class subculture, that stands enviously and bitterly on the margins hoping to enter the prosperous world of the majority rather than create a separate cultural world of its own. Some black Americans understandably feel differently, and are attempting to legitimize a separate culture that would reflect black identity. It is easy for whites to argue against this in the name of interracial cooperation, but actions, past and present, speak louder than words. The white United States in particular still has a long way to go before it can offer its black citizens an attractive and equitable partnership. Such a goal seems very desirable, but it is not yet clear whether it can be achieved soon enough to prevent increasing black separatism. Whatever the outcome, however, regionally based separatism seems out of the question. All U.S. cities (although to differing degrees) and many rural areas contain large, disadvantaged groups of black Americans. This is one aspect of urban uniformity that has grown especially in the past two or three decades. Blacks (and poor whites) are also becoming mobile and shifting their occupations and residences over great distances, primarily into urban areas in every part of the country, in search of economic opportunity. But it has often proved marginal at best in a white-dominated society.

Even considering these critical problems, one

MAP 111
REGIONS OF NORTH AMERICA

of the striking things about North American development is the cultural uniformity that has been superimposed on a huge and highly diverse physical landscape. The United States alone ranges from subtropical to arctic climates, from perpetually wet to desert conditions. It varies from broad coastal and interior plains to extensive areas of high and rugged mountains, from rich to poor soil areas, from tundra to taiga to steppe to mixed forest to Mediterranean scrub to savanna. The effect of these widely differing environments on regional economies and cultures has been significant, as discussed in more detail later. But the recently arrived, technically powerful, white, European settlers have been remarkably effective in welding together and evening out different environments within a single economic, political, and cultural whole.

Environmental differences are most clearly reflected in the distribution of population (see Map 109, page 469). Canada is larger than the United States, but contains about one-tenth of its population. Over half of Canada's population is found in less than 5 percent of the total land area, in a thin strip along the St. Lawrence and the northern shores of Lakes Erie and Ontario where Canada reaches farthest south. Over half of the population of the United States lives in about a seventh of its area, roughly the manufacturing belt (see Map 106, page 466). But it also includes a large share of the most productive agricultural land with a permissive climate and fertile plains. The most pronounced regional variations in culture are associated with the most distinctive environments, especially those that differ sharply from the chief core in the northeast: the far south, notably Florida and Louisiana, a subtropical area where traces of earlier French and Spanish influences remain; the southwest, a desert where Spanish and Mexican influences are still more important; northern New England, with its glaciated landscape, harsh climate, and offside location; or French-speaking Canada, like northern New England in its relative barrenness and isolation. But even in palm-lined Los Angeles, icy Edmonton, adobe-built Santa Fe, and moss-draped Charleston, a homogenized culture is evident. Some areas have preserved their identity better than others, but there is far more uniformity in urban and rural character, problems, economic patterns, and life-styles than the enormous disparities in environment might suggest. Hamburgers, drive-ins, neon signs, motels, disc jockeys, traffic and housing problems, pollution, the commercial economy, schools, clubs, churches, the generation gap, sports, mass attitudes on most things, and other ingredients of the so-called American way of life differ surprisingly little from one end of North America to the other. Foreign visitors are almost invariably impressed with this uniformity that contrasts with the much greater regional variety characteristic of most of the rest of the world over comparable amounts of territory. Residents of North America, understandably enough, are more aware of regional differences within their own domain, and the rest of this chapter will attempt to examine them.

Regionalization

As with any area, North America may be divided into regions in a variety of ways and on different scales. Perhaps the commonest division is based on physiographic units, for example, the southeastern coastal plain from Texas to Long Island, or the Canadian Shield, or the Rocky Mountain area. Such regionalization has its uses and its validity; physiographic models for it and lines distinguishing such regions are found in most atlases. But it is obvious that most physiographically based regions, including those just cited, include such a variety of settlement, land use, and other cultural characteristics that it may be misleading to lump them together under a single regional label. That might tend to imply at least a degree of homogeneity and coherence, as pointed out in Chapter 3. Much of the purpose and value of regionalization is lost if, for example, one classifies Louisiana or southeastern Virginia with New Jersey or Long Island merely because all of them occupy the southeastern coastal plain, or applies the same regional label to northern New Mexico and the upper Yukon because both are in the Rockies. Geography is concerned with the patterns and interactions resulting from all differences and similarities between areas, not simply with those having to do with landforms. Landforms are usually one factor in helping to account for other kinds of regional differences. But they do not act alone and are rarely dominant over all other factors to such an extent that they produce adequate regional homogeneity and coherence despite other influences, physical and human, toward diversity. The regionalization scheme used in the following discussion is based on regions that have a wide spread of characteristics and that tend to function as distinct areal units, including the attitudes and regional identity of their inhabitants. Human perception has a major effect on behavior, especially as regions are given meaning and function by the perception, attitudes, and behavior of their inhabitants—their sense of regional identity.

In these terms, North America may be divided into twelve regional units (see Map 111, page 480): the empty arctic and subarctic (by far the largest unit in square miles, but the least important humanly speaking); French Canada; New England and the Canadian maritime provinces; the middle Atlantic coast; Appalachia; the south; the middle west; the arid mountains and high plains of the west; the southwest; western California; the Pacific northwest (including southern coastal Alaska); and Hawaii. There are no sharp lines around any of these regions, except for Hawaii; many of them are surrounded by wide transition zones, or merge gradually into the adjoining regions. A few smaller areas do not fit entirely well into any of these regions, but are not large or distinct enough themselves to warrant separate identification on this size scale. Any scheme of regionalization is arbitrary to some extent, but these regions have been selected to represent discernible areas that tend to include coherent and distinctive characteristics and functions, and that are considered as separate regions by most of their inhabitants.

Taiga landscape—Yukon. This is near the taiga-tundra boundary; note the treeless plain across the river. But it shows well the thick coniferous (evergreen) forest of the taiga, mainly spruce and fir, with occasional birches. This is the same kind of forest that covers much of Siberia in a similar climate. *(Photo by J. Lundquist.)*

The far north

The arctic and subarctic areas extend from about 200 or 300 miles (320 or 480 km) north of the U.S.-Canadian border to the Arctic Ocean, except for a thin strip along the Pacific Coast where marine influences moderate the climate. Most of this vast extent of taiga and tundra landscape supports more moose and caribou than people. There are extremely few permanent settlements, associated mainly with mineral resources, prospecting, or scattered groups of Eskimos dependent on marine resources or caribou herds. Marginal agriculture based on wheat, small grains, and root crops is spreading slowly northward in a thin wedge in western Saskatchewan and Alberta as far north as the valley of the Peace River, although growth there is also related to petroleum finds (see the outlines of this pattern in Map 104, page 458). There is a much smaller, similar development around Anchorage, Fairbanks, and Dawson stimulated by local urban markets. But in the whole of this enormous region, over a third of North America, there is only one city (as opposed to towns), Anchorage, which owes its size more to its function as a transport center (air, sea, rail, and road) than to local resources. Today Alaska contains fewer people than during the gold rush at the turn of the century, and enjoys the distinction of being both the largest of the American states and the least populous, with its small total clustered along the coast. Interior northern Alaska is experiencing a small boom following the discovery of huge petroleum reserves under the frozen tundra, but this is unlikely to result in any very large population increases. (Southern coastal Alaska is discussed as part of the Pacific Northwest.)

The spruce-fir-aspen taiga forest and tundra landscape that covers the rest of this frigid third of North America is hardly interrupted by human settlement except for a few tiny and widely scattered dots. Apart from the short and isolated rail lines to Fairbanks from Anchorage and from coastal Skagway to Whitehorse, the northernmost rail connection from the main continental network reaches through the taiga and out into the tundra at Churchill on the west shore of Hudson Bay. For a few short summer months it may ship wheat from the prairie provinces

Alaskan tundra. This sparse, low complex of bushes, mosses, and lichens covers roughly the northern third of Canada and most of Alaska, north of the taiga zone. The tundra is too cold for trees. The soil is permanently frozen except for a foot or two at the surface, which thaws in the brief summer, when this picture was taken. *(Photo by J. Lundquist.)*

to Europe. Churchill's trade function declined with the opening of the Panama Canal, and almost disappeared with the opening of the St. Lawrence Seaway as Thunder Bay on the Canadian shore of Lake Superior replaced it. Another shorter rail line along the border between Quebec and Labrador in eastern Canada reaches north from the St. Lawrence to Schefferville to tap a rich deposit of iron ore. But the bulk of northern North America is used or seen only by its original wild inhabitants, except occasionally from the air.

French Canada

French Canada lines the St. Lawrence valley from eastern Ontario through Quebec to the sea. Montreal, its big city, the largest in Canada, is dominantly French-speaking. So is Quebec city, the cultural capital of French Canada, as well as the administrative center of the province. This is the oldest settled part of Canada, and its language reflects the early primacy of France in this area. Canada's political destiny was fixed in a British pattern by the battle of Quebec in 1759, when the French army under Montcalm was defeated by Wolfe's forces. Under cover of darkness, they scaled the heights on which the old city and fort were built. The St. Lawrence valley was the obvious focus of early settlement. It was easily accessible by ship from Europe and opened into a broad lowland of reasonably productive agricultural land above Quebec. This area remains second only to southern Ontario along the Great Lakes as the most densely settled in Canada, and includes nearly a third of Canada's population.

Traces of originally French land systems remain in the "long-lot" ownership pattern of farms running back from the river in long thin strips. This pattern also suggests the importance of the river as the only link for communications and trade until relatively recent times. The valley is French in culture and identity as well as language, and has successfully pushed for greater recognition of its distinctiveness within Canada. The Catholic church plays an important role, and the region is traditionally conservative and resistant to the pressures exerted by British Canada. Agriculture is still the chief economic base, with predominantly small holdings engaged in mixed farming and dairying. Quebec is relatively isolated not only from the rest of Canada but from the rest of the world, jealous of its separate identity, and wedded to its own traditions. Even as a big city, Quebec mirrors this regional self-image in its physical appearance, the only walled city in North America. Montreal belongs only in part to French Canada because it plays a national more than a regional role. It is Canada's largest port and rail center, and also is a major industrial base for a national and international market. Its size and continuing growth are related to these functions and to its location on the St. Lawrence close to the center of Canada's largest area of high productivity and dense settlement.

Canada and the United States offer a model of cooperation by sharing a long, open border that is not fortified militarily. But Canadians fear and resent the overwhelming economic power of their immense neighbor. U.S. corporations, manufactured goods, and mining interests play a dominant role in Canada. Like the U.S. government, these commercial interests sometimes behave as if Canada were a U.S. colony or junior subsidiary. Canadian resentment is understandable. In fact, Canada has its own distinctive culture (or rather two cultures—French and British-Canadian) and separate economic and political conditions and preferences. Its political tie with Britain has become a purely formal one and is of little use against U.S. influence. But close economic and cultural ties across the Atlantic help to preserve Canadian distinctiveness.

New England and the maritimes

Dominantly French settlement extends to the Gaspé Peninsula, at the mouth of the St. Lawrence, and occurs on a minority basis in parts of the maritime provinces (Newfoundland, Prince Edward Island, Nova Scotia, and New Brunswick), as well as in northern and central New England. The maritimes, however, are mainly English-speaking, and also closely related to New England in other respects. They share a good deal of common identity, especially with northern New England, with which there is considerable interaction. Both occupy climatically and economically marginal glaciated landscapes, and are isolated from the major centers of economic activity and growth. Their economic activities reveal these circumstances: lumbering, pulp, paper, fishing, some dairying and potatoes, and recreational tourism—this being the chief economic base of many areas.

New England in some ways is best divided between the north and the south, with Maine, New Hampshire, and Vermont linked to the maritimes, Massachusetts, Rhode Island, and Connecticut to the northern edge of the middle Atlantic region. Southern New England lies in the lengthening shadow of the vast New York area, and has long been closely integrated with the industrial-urban world in its own right. The Industrial Revolution in the New World began in southern New England in factories originally based on water power at sites such as Fall River, Haverhill, or Waterbury. The Connecticut River valley occupies a broad corridor of productive agricultural land through the center of southern New England, still intensively and profitably farmed. The eastern half of the south is, in effect, the northern extension of the coastal plain. Western and northern New England (including northeastern New York state), by contrast, is dominated by the northern reach of the Appalachians, contains almost no good agricultural land and no significant mineral resources, and has continued to lose population over the past several decades.

Most of New England, except for southern Connecticut in the New York commuter belt, thinks of itself as a separate and single region, bounded on the west by the Hudson River and the Lake Champlain trough. It shares common attitudes and behavior patterns, and looks toward Boston as its regional capital, the only major city in New England of national as well as regional importance. Boston incorporates a large urban area outside its own narrow political limits. It is highly diversified in both specialized manufacturing and services, including finance, education, medicine, and insurance. Some of these goods and services have a national market, but Boston is basically a regional city and cultural center, with New England as its hinterland. Heavy industry, including iron and steel, has avoided New England for the most part, despite local efforts to attract it. The region offers almost no heavy resource base, and is also offside with respect to the major markets. Its regional market is not large enough to make local production worthwhile. In southern New England manufacturing is highly developed in a large number of smaller cities. It concentrates on precision industries and on goods where raw materials and transport are not a large proportion of the total costs. Textiles, once the region's leading industry, have tended to migrate closer to raw material supplies and to lower-cost labor in the south. This pattern parallels to some extent the progressive abandonment of much of New England's original cultivated land during the nineteenth century as better land became available across the Appalachians. Having pioneered the development of textile manufacturing, New England saw its original advantages exceeded by other areas. But on the whole it has been able to fill the resulting economic gap with at least equally profitable precision industries. The abandoned farms that once covered most of the New England landscape have been replaced, especially north of Connecticut, by woods and ski slopes, both profitable sources of recreation income. But the regional flavor and character of New England remain distinctive.

The middle Atlantic

The middle Atlantic is the dominant region of North America in nearly every way. Here is the largest city, the national capital, and two of the other dozen largest American cities, Philadelphia and Baltimore. Here also is the largest single concentration of manufacturing, not only in North America but in the entire world, and the densest transport net. The region fades into New England north of New York City, stops relatively abruptly south of Washington, and reaches west into the Appalachians. There transport routes or local resources tie the mountains in with the coastal plain (for example, Scranton–Wilkes Barre, or the Mohawk corridor from Utica to Albany). This is the most densely settled region in North America, both rural and urban. What is almost one continuous city along the New Haven to Washington corridor dominates the landscape and the character of the entire area. The megalopolis, as it has been called, is still expanding. There are already relatively few rural gaps separating the major and satellite urban areas along a stretch of nearly 350 miles (560 km), or of about 500 miles (800 km) if one includes the somewhat more interrupted urban corridor north of New Haven through Hartford, Springfield, Providence, and Boston to the northern limits of the Boston metropolitan area near the New Hampshire border. The middle Atlantic region has been dominant, in the same respects, almost since the beginning of European settlement. New York and Philadelphia were long rivals as the country's biggest cities, and the national capital migrated between them until it was fixed on a new site at Washington. The region still includes some of the country's most productive agricultural land, with rich soils and a permissive climate on the edge of the mountains, plus the sandy coastal plain as an intensively worked truck garden area for the huge nearby urban markets.

New York is the national capital in many ways that Washington is not: for finance, publishing, foreign and domestic trade, fashions, music, theatre—roles played by the primate city in most countries, commonly also the political capital. Although New York is also the largest single industrial center, it does not support any basic iron and steel plant and its industrial structure is dominated by light manufacturing, assembly, precision goods, clothing, publishing and style-centered production. New York, of course, also epitomizes the galloping problems of urban overcrowding, air and water pollution, the spreading cancer of substandard housing, the deterioration of public services including transport, and the racial and economic class tensions common to other American cities on a smaller scale. But New York is still where the action is, in the minds of most Americans. That aspect of the city is almost exclusively concentrated in the lower two-thirds of the island of Manhattan, the only one of the five boroughs that most visitors see or are even aware of. The other boroughs cover a far greater urban area. The New York metropolitan area also spills over deep into New Jersey as far south as New Brunswick and includes New Jersey's largest city, Newark.

New Jersey, "backyard of two cities," is almost completely tributary to New York and Philadelphia. It sends a daily stream of commuters into each, markets its truck, dairy, and poultry products there, and in summer turns over its beaches and seaside resorts to Philadelphians and New Yorkers. It serves also as a transport corridor linking the two cities to each other and to Baltimore and Washington. New York's rise to primacy resulted in part from its median location within the densely settled zone of the east coast, in part from easy connections westward, and in part from its unequalled harbor. By the second decade of the nineteenth century New York's harbor already had attracted the largest share of ocean trade and pushed New York permanently ahead of Philadelphia and Boston in size. Still about twice the size of Chicago or Los Angeles, its nearest rivals, New York seems certain to retain its primacy despite its inability to solve the problems of overcrowding. Philadelphia, deeply in the shadow of New York, is more a regional than a national center. It has highly diversified manufacturing that includes a major iron and steel plant at Morrisville just above the city, and is also a major port, especially for petroleum moving in for refining and out for distribution to market. Baltimore is a somewhat smaller version of Philadelphia, but like it and like Boston, it preserves a highly parochial and inward-looking sense of identity. All are old cities, by American standards, and all are conscious of their past (New York is perennially a city of the present and the future). All are self-consciously aware of their distinctiveness and cultural character, however ethnically and economically mixed their populations have become over the past century.

Appalachia

Appalachia is beautiful but economically depressed, qualities it has in common with most mountain areas. Although it is surrounded by some of the most highly developed and rapidly growing parts of North America, most of it is an isolated area of cultural and economic backwardness. Surviving pockets of archaic speech patterns more closely related to Elizabethan England than to contemporary America, folk songs long forgotten elsewhere, a long history of losing struggle with the surrounding lowlands, and a more recent history of exploitation, all help to define Appalachia's regional character. This broad finger between the east and the midwest reaches from northern Georgia through the western half of the Carolinas and eastern Tennessee and Kentucky to West Virginia and western Pennsylvania–southeastern Ohio. In many respects it continues through eastern New York into western and northern New England, whose relatively unspoiled beauty, poverty, and isolation link it as much as does its geologic structure with the rest of Appalachia farther south.

Exploitation has been associated largely with coal, which occurs in some very large and rich near-surface deposits scattered throughout most of Appalachia south of New York state. The coal of Appalachia has generally been owned by and mined for the profit of outsiders. This has led to the same grim cycles of depression that characterize the mining districts of Wales (see Chapter 5). Pittsburgh, closely linked from the beginning with the middle west by the Ohio River, is the outstanding but qualified exception to this. Appalachia is one of the largest pockets of poverty in North America, and its meager agricultural resources have not been adequately supplemented either by coal mining or by more constructive or developmental investment. It would almost certainly have been better for most of its inhabitants if, like northern New England, it had lacked obviously exploitable resources and had been allowed to concentrate on preserving its natural beauty. The overcrowded eastern third of the United States needs unspoiled natural areas especially badly. Perhaps as belated federal and other outside attention is turned to the pressing human problems of Appalachia, it may be able to play this positive and important role more successfully.

The south

The south is the most distinctive region in North America, and one of the hardest to define precisely on the map. Appalachia thrusts a deep wedge into it and helps to make the areas east of the mountains different from those to their west. The mountains originally helped to limit the spread of settlement and delayed occupation west of them until about the beginning of the nineteenth century. The northern border of the south on both sides of the Appalachians is relatively clear, more or less coinciding with the Virginia state line on the east and the Ohio River on the west. (West Virginia, a relic of the Civil War, is a part of Appalachia.) There is a much wider transition zone on both sides of the Ohio than along the Virginia state line, which may seem puzzling until one remembers the integrating rather than dividing role most navigable rivers play. This is particularly true of the Ohio with its very heavy volume of traffic and the high degree of economic unity within its valley. For similar reasons, the Mississippi south of its junction with the Ohio cannot be used as the western border of the south since this would exclude Louisiana and southeastern Arkansas, which belong with the south. Southeastern Texas to some extent is transitional between the south and the southwest, although many Texans would maintain that Texas constitutes its own region. It may be acceptable to partition it for regionalization purposes by assigning the moister eastern lowland to the south and the drier and higher rest of the state to the southwest, although this is not fully satisfactory in either case.

Cultural attitudes and life-styles predominantly give regional identity to the south. Broadly speaking it is also an area of mild, moist climate that has developed an agricultural system in which cotton plays a major role and other specialty crops, such as tobacco (in good soil areas along the plains edge of the piedmont in the Carolinas and Virginia) and citrus fruits, are also important. Eastern Texas is a booming industrial area based on oil, gas, and sulphur. Rapidly growing Atlanta (nearly 2 million), a transport junction where rail lines meet around the southern end of the Appalachians, is a diversified manufacturing and service center for most of the old south, east of Mississippi and south of Kentucky. There is a major iron and steel complex at Birmingham, Alabama (its name was taken from Birmingham, England—see Chapter 5). It is based on local and nearby coal and ore and serves all of the southern market. Smaller cities, especially in the Carolinas, are

Mountain homestead—Tennessee. The cultural isolation and limited economic opportunities of the Appalachian region are apparent in this farm dwelling and its adjacent fields. *(Photo from Tennessee Valley Authority.)*

centers of the textile and cigarette industries. But most of the south is primarily an agricultural area, especially by comparison with most of the rest of the United States. This is closely related to the fact that, outside of Appalachia, the south as a whole has the lowest per capita income of any of the regions of North America, and the largest proportion of black inhabitants. The economic doldrums in which most of the south stagnated for much of the century following the Civil War have been partially relieved during the past two or three decades, as new industry has developed or old industry migrated southward. The Texas and Louisiana mineral boom has continued. The great TVA hydro-power project revitalized the area it serves, and military contracts have gone increasingly to southern firms. But the south still lags behind the rest of the country economically, and also in many ways politically, socially, and culturally. It is an area of conservatism, traditionalism, and emotional commitment to often out-dated ideas. This is not fair to some of the larger and vigorously growing cities, notably Atlanta, but most of the south is rural and tends to remain outside the mainstream of contemporary North American development.

The black population, of course, is the direct result of slavery, associated with the plantation agriculture of the old south. It is still largely concentrated in the better soil areas on the coastal plain and the lower Mississippi valley that could support soil-demanding crops such as cotton, sugar, indigo, or tobacco. The slopes and poorer soils of the piedmont and mountain areas never supported this kind of agriculture and have remained dominantly white. Both areas have been increasingly affected by emigration as blacks and poor whites have moved to northern cities in search of better economic opportunity, especially since the two world wars. Cotton is still the most important single commercial crop of the south. But over two centuries of its cultivation, in most areas as a single crop, have tended to exhaust many of the original cotton soils. Cotton has moved progressively westward, into Texas, the southwest, and California, where there are more productive soils, where the drier climate is also an asset, and especially where irrigation is available. Cotton is highly susceptible to moisture damage in its ripening stage, and the humid climate of the old south also helped spread the ravages of the boll weevil. Finally, with the development of mechanical cotton pickers, the cheap labor (almost entirely black) of the old south was no longer a strong advantage.

One more recent development has helped to make up for some of the ailing fortunes of cotton in the south: the planting of huge areas of poorer or exhausted soil with trees, predominantly pines, as the basis for a large and thriving pulp, paper, and lumber industry. This takes advantage of the adequate moisture and long growing season of the south to produce a marketable crop especially quickly. There has also been a considerable increase in dairying, another effective use of lower-quality soils. The south has changed and grown, and, as already pointed out, there has been impressive industrial growth in some areas. But by comparison with the rest of the country, the south as a whole has moved more slowly and is still backward in many parts. The old ports of Charleston and Savannah have declined with the plantation economy they served. They have been almost entirely replaced as the commercial centers of the south by upstart Atlanta, on the edge of the piedmont. Mobile, Alabama has kept pace a little better, though on the newer basis of oil and ore refining. Much larger New Orleans, which plays both a national and a regional role as a port, is matched only by Quebec as a living replica of another cultural age. It is notable, however, that Houston-Galveston (joined by a deep-water canal and really a single port) has become an even larger conurbation, and is beginning to rival New Orleans in the volume of its port business. Dallas, slightly larger than Houston-Galveston, is growing still faster; but even east Texas is really different from the rest of the south.

Probably the part of the south that fits the above generalizations least well is coastal Florida. In some respects it is merely an outpost resort for the north, mainly the middle Atlantic region. Among southern cities, Miami is between Dallas and Houston in size, although it is really not a southern city. Interior Florida also, with large areas occupied by swamps, sparsely covered sandy scrub, and subtropical vegetation, is distinct from most of the south. The principal crop on the small proportion of the total land area that is cultivated (see Map 104, page 458) is citrus fruit, which links it more with California than with its neighboring states.

The midwest

The middle west, second in size only to the arctic and subarctic north, stretches from the western edge of the Appalachians and the borders of French Canada, north of the Ohio River and the Ozark Plateau, to a wide boundary zone of transition running through central Kansas, Nebraska, the Dakotas, and western Saskatchewan. This western boundary is primarily climatic, and south of Canada coincides roughly with the 20-inch annual isohyet, the transition from reasonably well-watered to steppe conditions (see Map 107, page 468). This change is reflected in the rail pattern (Map 103, page 456), and in the closely related pattern of population (Map 109, page 469). In the Canadian prairie provinces the transition in all these respects comes somewhat farther west, in part as a result of lower evaporation rates in this cooler climate, hence of greater moisture availability.

The middle west is North America's premier agricultural region, but also includes a substantial share of its industrial complex and big cities. The name *middle west* is really no longer accurate; it is now an uncompromisingly eastern region: green, intensively cultivated, densely settled, industrial, and urban, although not quite to the high degree of the middle Atlantic. What distinguishes it from the middle Atlantic region is primarily cultural attitudes, conditioned in part by the greater prominence of agriculture and rural farm population, and in part by its land-locked isolation. Much of the middle west, especially outside the big cities, still sees itself as "America's heartland." It is proud of its freedom from supposedly contaminating influences from overseas or the more cosmopolitan east coast. As the cities continue to become increasingly dominant and acquire much the same circumstances and character as those on the east coast, with which they are closely tied economically, such distinctions break down. Understandably, the farther west one goes within this region, the greater the felt differences from the middle Atlantic. Population density shades off westward from Chicago, cities become smaller and more scattered, and agriculture becomes proportionally more important. The heart of middle west agriculture is the corn belt (discussed briefly in Chapter 24) whose relatively intensive grain-and-animal complex changes westward into more extensive grain cultivation with drought-tolerant wheat increasingly important. Finally, grazing predominates as the middle west gives way to the drier region of the high plains and mountains. In the cooler climates with more heavily glaciated, poorer soils, such as Wisconsin and Minnesota, dairying is prominent, serving the large urban markets of the upper midwest especially.

The biggest midwest cities are clustered along the shores of the Great Lakes, from Toronto and Buffalo through Cleveland and Detroit to Chicago and Milwaukee. St. Louis is the major western gateway and service center on the important north-south route of the Mississippi located near its confluence with the navigable Missouri. Minneapolis–St. Paul performs a similar function for the northwest at the principal head of navigation on the Mississippi. All of these cities are clearly transport-oriented, as is appropriate to a region that lies in the middle. It was settled largely in the age of the railroad, after early urban bases were established along major waterways. The opening of the St. Lawrence Seaway has given new importance to the Great Lakes as a transport link to the east coast and abroad. But they have also long been important as a connection with rail routes and for internal movement within the region. They have provided a source of water and a means of low-cost assembly and distribution for iron and steel making and other heavy industries that cluster along the southern and western shores of Lakes Ontario, Erie, and Michigan. Pittsburgh became the premier iron and steel center at the head of the Ohio River system where it could tap rich local coal deposits on the edge of the Appalachians. It could also draw ore from mines at the western end of Lake Superior by cheap water transport and a short rail haul. The industrial prominence of the middle west is related also to its mineral resource base, which includes some of the country's largest reserves of coal and iron ore (see Map 105, page 462), supplemented by the coal fields of Appalachia, and significant petroleum and natural gas pools.

Except for the southern peninsula of Ontario, the northern shores of the Great Lakes mark the southern edge of the Canadian Shield (which also extends into northern Wisconsin and Minnesota). Agriculture is almost nonexistent on the shield, which reaches as far west as Lake Winnipeg. Population is extremely sparse except in a few small spots where mineral resources are present, notably the rich nickel mines at Sudbury, Ontario. Ottawa lies on the eastern edge of the shield, purposely selected as the national capital because of its border position between French Canada and British Ontario and the west. Although the choice was made in the mid-nineteenth century, Ottawa, like Washington, but even more so, is still not far from the geographical center of national population.

Western mountains and plains

The region of mountains and high, arid plains and basins of the west is one of limited economic opportunity and of huge, relatively empty space, which gives it its regional character. These are the "wide open spaces," "the land of the big sky." The inhabitants value these qualities and take much of their regional identity from them. This is ranch and cowboy country, most of it too dry to support agriculture except where it can be irrigated or where drought-tolerant wheat and sorghums can be raised on the more humid eastern and northwestern margins. The region includes desert and near-desert conditions in much of the basin area between the Rockies and the Sierras-Cascades, where population reaches its lowest densities within the United States, apart from Alaska. There are growing spots of irrigated agriculture (see Map 104, page 458), and a few smaller dots of economic activity based on local minerals. But for the most part this remains grazing country or mountain and desert wasteland, whatever its visual or aesthetic attractiveness. Some of the mountain areas support profitable recreation centers. Winter skiing is concentrated in a few places where facilities have been developed, but dude ranching is more widespread and extends onto parts of the high plains in Colorado, Wyoming, and Montana.

The region supports only two sizable cities. Denver, the larger of the two, is primarily a transport junction, a way station on transcontinental east-west routes, apart from its central place functions. Its hinterland is huge, by far the largest in the United States, but also one of the least productive. Salt Lake City, the capital of Mormonism, is based primarily on the intensive irrigated agriculture of its immediate hinterland. But it also profits from a small iron and steel complex in nearby Provo, using coal and iron from Utah, and the fortunate presence of a large freshwater lake at Provo. Although this region stretches from the borders of Arizona and New Mexico and east of the Sierras and Cascades north into Canada to the edge of the subarctic, most of its characteristics change surprisingly little. Calgary, the Canadian equivalent of Denver in many ways, shares its hinterland with Edmonton and also with Regina farther east. Alberta too, despite its somewhat greater moisture availability and the presence of wheat farming, is also ranch country of great open spaces with a life-style and attitudes to match. A clear example of this regional distinction can be found within Washington and Oregon, where the Cascades draw a climatic and regional line north-south through both states, leaving their eastern parts far more a part of this region and sharing its characteristics than of the different physical and cultural world of the Pacific northwest across the mountains.

The southwest

The southwest includes Oklahoma, western Texas, and, as its heart, Arizona and New Mexico, with the southeastern sections of Nevada and California closely related. This is desert or semidesert country, but has some spots of irrigated agriculture as well as extensive marginal grazing. Tourism is a major part of the economic base in this sunniest part of North America. Except in the highland areas winter is brief and mild with only occasional frost. This is the most important single reason why the southwest for many years has been the fastest growing region in North America in terms of population increase. The high average incomes, leisure, and easy mobility characteristic of the United States, and the growing proportion of retired people who have settled permanently in the southwest have contributed to the growth of this area. It is still sparsely settled, although per capita incomes are high. Phoenix is the only city with over 1 million population. Like Miami, it is unusual among large North American cities in doing little manufacturing apart from local service industries. It and nearby Tucson are commercial service centers for a region of ranches, motels, resorts, and desert. Phoenix is also the Arizona capital, completing the parallel with Albuquerque and Santa Fe, smaller-scale versions of Phoenix and Tucson. Irrigated agriculture is expanding rapidly. The southwest is already a major producer of cotton and citrus fruits in west Texas as well as in Arizona, New Mexico, and southeastern California. The oil and gas of west Texas and Oklahoma are a major resource. They are almost entirely exported in crude form by pipeline, and contribute little to local urban or industrial growth. Much of the attractiveness of the southwest lies in its relative emptiness and the stark beauty of its dry steppe and desert landscape, marred only in a few places by commercialization.

This is also the most recently acquired part of the continental United States. Only a little over a century ago all of it was detached from Mexico: a small parcel along the present border by purchase, the rest annexed as a part of or the result of the war fought by land-hungry Americans (including Texans) against Mexico. Since the first white settlements in the southwest were established by the Spanish as long ago as the seventeenth century, this area has been part of the United States for only a small fraction of its modern existence. Spanish is still used as the first language in many areas, especially in New Mexico. Not surprisingly, it retains Spanish, Mexican, and local Indian characteristics to a marked degree, including strong architectural influences. Many buildings in Santa Fe, including new ones, are still built of adobe (sun-baked clay), following both a Spanish-Mexican and a local Indian tradition that continues to make sense in this hot, dry climate. In Santa Fe, the best as well as most consciously preserved Spanish-Mexican-Indian city, adobe buildings dominate the urban landscape. The Santa Fe area as a whole is a striking reminder that this was long someone else's country. Its original Indian inhabitants have fared shamefully less well than has their architectural style at the hands of their conquerors. They have been shunted onto inadequate reservations carved out of the poorest land and discriminated against in the efforts they have made to join the white system. Once probably the most highly developed of the Indian groups in North America, and especially accomplished in both dry-field and irrigated agriculture, they have been largely reduced to a miserable level of existence by the march of "progress," North American style.

California

Western California is appropriately distinguished from the eastern part of the state, which belongs with the southwest (Death Valley, for example, lies well within California). The difference is created largely by the Sierra Nevada range. It has helped to produce desert conditions as well as isolation in the rain shadow it casts eastward, and to mark the different region between it and the sea. Much of this area is also dry, but mainly in a Mediterranean climatic pattern (see Chapter 7), so properly selected agriculture is possible. The rainfall can also be supplemented on a very large scale by irrigation, drawing on water from the heavy rain and snowfall of the Sierras. In fact, the great majority of California's cultivated acreage is irrigated, and is highly productive, especially in fruits, vegetables, and cotton. They are all produced for the national market and shipped over great distances. California was part of the southwest area wrested from Mexico in 1848, just before the discovery of gold near Sacramento. The Spanish place names, characteristic of most California settlements south of San Francisco, remain, together with some muted traces of Spanish architecture, and a few gently decaying or specially preserved Spanish missions. But most evidence of the long Spanish occupation has been buried under the avalanche of North American development. To many Americans, California still symbolizes a promised land, filled with sunshine and flowing with milk and honey, an image that its economic growth during the past century has tended to support.

The rush of immigration that followed the discovery of gold in 1849 was maintained and increased after the gold was worked out by expanding economic opportunity based on a variety of more enduring resources, both agricultural and industrial. The central valley of California, between the Coast Ranges and the Sierras, with the help of irrigation, became the most productive agricultural area of comparable size anywhere in the world, measured in terms of dollars per acre. This development was later extended beyond the central valley into the coastal and inland parts of southern California, where the chief crop is citrus fruits. The interior desert basin of Imperial Valley, adjoining the Mexican border, was added to the irrigated area and became an immensely productive source of fruits and vegetables for national marketing. To top off these riches, oil was discovered in coastal and near-coastal southern California, in pools that extend out under the sea and are being tapped there as well. California has been an economic bonanza, and this plus its attractive climate has drawn many more migrants than any other part of North America. The little paradise that

Air pollution in Los Angeles. The bottom picture was taken on a day relatively free from smog; the top picture on a very smoggy day. The temperature inversion layer over coastal southern California traps pollutants and dramatizes this problem. All cities suffer from it, especially those that have many automobiles. *(Photos from Los Angeles County Pollution Control District.)*

California represented only a few decades ago is rapidly succumbing to the urban sprawl, pollution, overirrigation, and overcrowding that have already appeared in the middle Atlantic and much of the middle west.

California to a large extent has become the hinterland of two cities, San Francisco for the northern half and Los Angeles for the southern. Both are major ports and both serve as central places for the west as a whole. Although San Francisco with its magnificent bay harbor handles more long-distance shipping, total port volume is greater at Los Angeles, which is also a much larger city. Southern California also supports another million-population city, San Diego, only about 100 miles (160 km) from Los Angeles. This reflects the greater productivity, in nearly all respects, of southern California, as well as the great human attractiveness of its year-round warm, dry climate. The climate was also responsible in part for southern California's early lead in airplane manufacture, an operation requiring large amounts of space that could be done if necessary in the open air in this mild climate. The early growth of the film industry also profited from a mild outdoor climate and from the availability of a wide variety of scenery within easy range of Hollywood (a suburb of Los Angeles), from ocean to city to orange grove to desert to snow-covered mountains in about 50 miles (80 km). Both Los Angeles and San Francisco, however, are major industrial centers of a highly diversified nature, and both are among the ten leading areas in North America in terms of value added by manufacture (see Map 106, page 466). Los Angeles enjoys the further dubious distinction of being the smog capital, which provides some measure of its industrial importance, but results from the frequent temperature inversions that build up between the relatively high Coast Ranges and the cold, offshore currents, inhibiting the dissipation or outward flow of polluted air masses trapped close to the ground.

Pacific northwest

The Pacific northwest runs in a thin strip from northernmost California to south coastal Alaska, between the Cascades and the Coast Mountains of British Columbia and the sea. It is a cool, wet region of mild winters and often chilly summers. The climate is similar to that of the Low Countries or northern France, but the soils are largely acid and poorly drained and offer only a limited agricultural basis. In part for these reasons, as well as for its relative remoteness, it was the last of the western regions to be settled by whites and still retains something of a pioneer air. Its principal resources from the beginning were the dense forest stands of coniferous trees and the rich fish population along its coasts, notably salmon. Agriculture is still limited in extent and importance, and lumbering and fishing are still significant. But the great economic growth beginning especially since the Second World War has been industrial and commercial. The Pacific northwest contains almost no mineral deposits of consequence, but the combination of rainfall and mountains creates an enormous hydro-power potential. This has been developed especially in the past three or four decades, with a series of big dams feeding power into industrial operations centered in the region's three major cities, Portland, Seattle-Tacoma, and Vancouver in Canada. Seattle-Tacoma is the largest, and rivals southern California in aircraft production. Like the other two cities, however, it has a diversified industrial structure, including operations that profit from large supplies of relatively low-cost electric power.

All three cities are important ports. All three are even more important as service centers, not only for the Pacific northwest, but for much of the mountain and plain region across the Cascades for which they are both the nearest ports and the nearest big cities. They occupy the end of the line on transcontinental transport routes, and are take-off points for routes to East Asia. Although the immediate hinterland of each in the wet, coastal lowlands is relatively small and unproductive, their wider hinterlands are very large and include some major trade flows.

South coastal Alaska is a smaller and more restricted northern extension of the Pacific northwest region. It has no large cities or industrial structure, and its economy is centered on fishing, lumbering, and the processing of these primary products. From Puget Sound north, the coast is highly indented by innumerable fjords in the same pattern as the coast of Norway. North of Vancouver, the glaciated mountains extend to the sea and there is little basis for a productive economy except through power generation near the fjord heads, as in Norway, and transmission to better-favored areas for consumption. The pulp and paper industry of the Pacific northwest extends in a narrow strip along the Canadian and Alaskan coasts as far as Sitka. A few minerals are also mined, but otherwise the region north of Vancouver necessarily faces and depends on the sea. The salmon and halibut fisheries are productive, but cannot support a large population. Juneau, at the northern limit of the Pacific northwest, was chosen as the capital of Alaska, but has remained only a small town, much smaller than Anchorage, which has a much larger hinterland.

Hawaii

No argument is necessary to establish Hawaii as a separate and legitimate region. Until recently it was a part of Polynesia in cultural and ethnic terms, before it was overwhelmed by the rush and commercial acculturation of the United States. There are eight main islands, all mountainous and of volcanic origin, and all sharing a wet, tropical climate. As in the Caribbean, however, the leeward side of the mountains may be relatively dry. The population is heavily concentrated on the relatively small island of Oahu, which also contains Honolulu, the largest city, of about 685 thousand people. Native Hawaiians are a fast-decreasing minority and their culture is already almost entirely eroded, although the United States did not gain possession of the islands until 1899 as a result of the Spanish-American War. Americanization is a powerful force, and has been further concentrated on Hawaii since statehood in 1959. Immigrants from the American mainland are now the largest single group, followed by Japanese-Americans who moved there after 1899 to work in the sugar cane and pineapple industries, accompanied by smaller numbers of Chinese and other East Asians, Samoans, and Puerto Ricans. Intermarriage among these groups has been common and the population is very mixed. But continued immigration from the mainland keeps whites dominant.

The Hawaiian economy revolves around sugar,

pineapples, and tourists, with an important addition provided by the U.S. military, which maintains several large installations, including Pearl Harbor. Nearly all of the cane and pineapple plantations, which benefit from fertile volcanic soils, and the major economic interests of the islands as a whole are owned by whites, with the labor provided by Asian-Americans and Hawaiian-Americans. The great bulk of what is produced is marketed in the mainland United States. Statehood further intensified cultural and administrative ties. A small separate world, which early American whalers and missionaries described as an earthly paradise, has been almost wholly transformed and given an abrupt about-face. Outside the island of Oahu, relatively little of the landscape has been altered. But the city of Honolulu and its satellite areas, which now cover most of Oahu and clutter its magnificent beaches with luxury hotels, cast a lengthening shadow over all of Hawaii, a shadow in which there is both much profit and much loss.

A summing up

Having inherited a continent abundantly rich in resources, and having made impressive use of it in economic terms for the most part, Canadians and Americans still have to solve two basic and important problems: working out social and economic justice for all the people of both countries, and care of the environment to keep it both habitable and productive. Both countries, especially the United States, have been built to a large extent by the ambition of individuals and powerful groups. As pointed out in the preceding chapter, the rugged individualism that was an understandable philosophy for an expanding frontier and a booming economic growth may not be good for the whole society. In particular it may work social and economic injustice on less powerful or less fortunate individuals or groups. Unrestrained and unplanned development may also have destructive effects on the resource base and on vital aspects of the environment. Twentieth-century North America confronts a frightening series of problems coming from abuse of the land, soil erosion, and the poisoning and unrestrained polluting of the atmosphere and the rivers, lakes, and the sea. Already the air in most big cities is a health hazard, and may also be harmful to vegetation. Many rivers, lakes, and near-coastal waters are poisonously polluted, some so much so that they can no longer sustain life. Millions of acres of once productive land have been destroyed by overirrigation, salinization, or erosion. Realization of these problems has come late and is still far from complete. Public action, regulation, and efforts at restoration have yet to become even barely adequate. Disastrous soil erosion continues in areas where deforestation, overgrazing, or improper cultivation have been unwisely pursued. North Americans have been careless with their environment in part because it seemed inexhaustibly rich and uncluttered. That time has long since passed, and misuse of the environment has produced a genuine crisis. North America's 5 percent of world population gobbles up 35 percent of world resource use, and probably also creates about half of world environmental pollution. The inequity these figures represent is obvious and desperately serious in an increasingly interrelated global community.

The remainder of the present century may be too much time to allow for North America to produce workable and acceptable solutions to these problems. It is certainly possible that economic, social, and racial justice will not wait that long before other kinds of solutions are tried. It is possible also that environmental deterioration will continue to take place faster than efforts to control it. Rugged individualism and old-fashioned American distrust of public authority, or unwillingness to sacrifice individual or group interest to the public welfare continue to exact a very high price, one that may be too high for the existing system to bear without fundamental and rapid change.

The same considerations apply in our relations with the rest of the world. The United States has often used its immense power, won mainly as a result of the Second World War, to promote by force in distant places an American pattern for how other people shall manage their affairs. Unbridled power in the past history of society has often proved to be its own undoing. Perhaps this is the lesson Americans need to learn most if they are to succeed in building a genuine community at home and in serving as useful members of a world community abroad.

Questions for further study and discussion

1. How and why are language patterns both a cause and an effect of cultural patterns in other respects?

2. What was the extent and nature of native Indian cultural diversity in North America? What major regional patterns had evolved, and why?

3. Why do the United States rail patterns become so much less dense west of the hundredth meridian of longitude?

4. Drawing regional lines is an arbitrary and often difficult business, and there may be arguments about particular lines. Argue against the lines and the regionalization scheme in this chapter, and suggest better alternatives.

5. What is the basis and nature of the TVA project, and why was it begun? What effects has it had on the area it serves?

Glossary

Alluvial soils—fine-grained soils deposited by a river or lake, usually very fertile.

Autarky—self-sufficiency in national economy; a German doctrine influential in the country's national development.

Break-in-bulk points—places where products are changed from one type of carrier to another, requiring that they be unloaded and reloaded.

Complementary trade relationship—fills mutual lacks between areas or countries.

Contour lines—lines connecting points of equal elevation on relief maps of small areas.

Culture—the total way of life of a people.

Drang nach Osten—the German drive to control the countries lying to the east.

Dispersed settlements—where houses are far apart or isolated, as in farming country.

Emergent coastline—characterized by an even, sandy coastline of bars and beaches with shallow water close to shore.

Entrepôt—"between ports"; a collecting and distributing center for the trade of several other, usually smaller, ports lying around it.

Escarpment—a long, high, steep cliff.

Exotic river—a river that has its source in well-watered lands and crosses a desert on its way to the sea.

Fjord—a narrow waterway from the sea with steep rock sides or cliffs on each side, gouged out by glacial action. Found in Norway and western Alaska, among other areas.

Gestalt—a pattern of factors operating as a unit that is more than the simple sum of its parts.

Glacial moraine—sand, gravel, and rocks pushed up by glaciers at the edge of their advance and left there as the ice melted.

Halophytic crops—tolerant of salty soils.

Hinterland—the area surrounding a city and dependent on it for goods and services. The size of a hinterland depends on ease of access to the city, and the size of the city depends to a large extent on the hinterland.

Isohyet—a line on a map connecting areas of equal rainfall.

Latitude, parallels of—imaginary lines running east and west around the earth parallel to the equator, and used for indicating location.

Loess—fine-grained soil deposited by the wind.

Longitude, meridians of—imaginary lines running north and south around the earth through the poles, and used for indicating location.

Nucleated settlements—where houses are close together, as in villages or towns.

Orographic rainfall—rainfall caused by mountains forcing the air masses to rise.

Permafrost—ground permanently frozen a few inches or feet (meters) below the surface, limiting its use for crops and causing problems for buildings and roads.

Piedmont—the rolling or foothills area along the base of a mountain range.

Podzolic soil—soil found in cold climates where the decay of humus (organic matter) is slow or incomplete, building up an acid balance; not very fertile soil.

Primate city—a city at least twice the size of its nearest domestic rival.

Prime meridian—the base line of reference for longitude; the meridian that runs through the Royal Observatory at Greenwich, England.

Region—a definable area possessing a wide range of more or less homogeneously distributed and interrelated characteristics.

Savanna—a tropical or subtropical grassland with limited or seasonal rainfall and only scattered, drought-resistant trees.

Settlements—*see* Dispersed settlements; Nucleated settlements.

Site—the actual ground on which a city is located.

Situation—the location of a place relative to other places, principally involving accessibility rather than distance.

Spatial interaction—relations between places.

Submergence, coast of—a "drowned" coastline with deep water close to shore with many bays, harbors, and indentations.

Taiga—northern forests of Siberia, Europe, and North America dominated by coniferous trees (spruces and firs).

Tundra—an area with permanently frozen subsoil, generally occurring north of the taiga regions, and supporting low-growing shrubs and lichens.

Veld—a grassland on level or rolling uplands at different elevations, specifically in southern Africa.

Wadi—channel of a stream that is dry except during the rainy season or a single violent shower, when it may suddenly fill with dangerously large amounts of water.

Xerophytic crops—tolerant of drought conditions.

Statistical appendix

Anyone who has attempted to gather or to compile statistics knows that it is a sobering experience. The most painstakingly careful techniques of collection, the most thorough surveys, and the most conscientious selections ultimately confront the irreducible margin of human error. This statistical appendix has been prepared as carefully as possible from a variety of authoritative sources. But there is almost no statistical compendium dealing with all of the world and concerned with multiple factors that does not contain internal inconsistencies suggestive of uncertainty or error. External inconsistencies between different statistical sources, not surprisingly, are much greater. Fully accurate data simply do not exist for most of the categories covered in this appendix, and original inaccuracies have been compounded within each country at successive stages of compilation. Small inaccuracies rapidly mushroom when there is a multiplier of millions of people, tons, acres, or dollars. Figures dealing with or derived from the agricultural sector in each country are probably the most prone to error, but in almost every category, there is some unavoidable error.

Probably the most unfortunate aspect of this problem is that the printing of a figure, no matter how inaccurate the figure is, tends to imply that it is accurate and that the compiler has complete faith in it, or at least that the compiler knows without reservation what the quantitative truth of the matter is. This is rarely, if ever, the case. The reader is warned to use the figures that follow with caution, disregarding the air of finality and precision a printed figure tends to carry with it, remembering the imperfect nature of even the best sources, and treating the figures as no more than conscientious attempts at the best possible approximations.

City population figures are notoriously ambiguous. Many national censuses give population figures for large metropolitan areas, but each national census defines "metropolitan area" differently, and each city even within a single country presents a unique problem. Other national censuses give population figures only for the political limits of cities. In addition, there are a great many cities in several countries for which only crude estimates are available. The population figures for cities listed in this appendix accordingly are not directly comparable with each other, but are based on the best available data.

Metropolitan areas as defined by the United States Census tend to be much broader than metropolitan areas as defined in other national censuses and in many cases also produce misleading results, especially for comparing United States cities with each other. However, U.S. city populations here are based on metropolitan areas.

Infant mortality rate refers to deaths of infants under one year of age per 1,000 live births.

The percentage of cultivated land devoted to a crop refers to the sown acreage of the crop, which takes account of multiple cropping of a given area in one calendar year.

Figures for annual timber production refer to output of wood whether for use as fuel or for conversion into lumber, veneer, or other products. The figures represent running averages from 1969 to 1973. In countries for which running averages cannot be obtained because of a lack of adequate data, only the recent years (or the most recent year) are considered. This applies also to other running averages, as below. Figures for any one year may often be misleading; a more reliable pattern is shown by averaging figures, especially for production, over several successive years. Therefore, running averages of this sort are used here wherever possible and appropriate.

Figures for annual fish catch refer to running averages from 1969 to 1973.

Figures for total value of imports and exports refer to running averages from 1970 to 1974.

Figures for per capita foreign trade refer to running averages from 1970 to 1974.

Figures for per capita energy consumption refer to running averages from 1970 to 1974 in terms of total commercial energy consumed in kilograms of coal equivalent.

It is a truism that accurate statistical data do not exist for large parts of the world, and that data are often simply not obtainable, whatever their accuracy. There are accordingly blank spaces in many places in the tables that follow, reflecting not oversight but the unavailability of data from any source. It is also not possible to be completely consistent, since the statistical sources from various countries are not consistent, especially in their labeling of categories. Thus, ethnic groups are variously described (but rarely *defined*) in different data collections as "white," "European," "Caucasian," "mixed," "Asian," or "African"; religions as "animist," "local cults," "shamanist." In the absence of further definition, these differing labels used by the different censuses or surveys have had to be repeated here, however unsatisfactory they may be. Measures of petroleum and natural gas are also inconsistent, and not readily convertible to metric tons since "barrel" as a quantitative unit is defined differently in different countries, but often not specified in particular cases.

A number of independent countries have not been included because not enough statistical information about them is available to make possible adequate coverage, or adequate comparison with the countries that are listed here. The countries omitted for such reasons are Andorra, Bahrain, Barbados, Cape Verde Islands, Comoro Islands, Djibouti, Equatorial Guinea, Grenada, Republic of the Maldives, Monaco, Nauru, San Marino, Sao Tomé and Principe, Seychelles, South West Africa (Namibia), Tonga, Transkei, Vatican City, and Western Samoa.

All figures are the latest available as of mid-1977. There is nearly always a long time lag between the gathering of figures and their publication; thus most figures are a year or more old by the time they are released.

The following sources were used for the compilation of this appendix:

Background Notes. Washington, D.C.: U.S. Government Printing Office.

Countries of the World and Their Leaders. 2nd ed. Detroit, Michigan: Gale Research Company, 1975.

The Europa Yearbook: A World Survey. Volumes 1 and 2.

International Bank for Reconstruction and Development World Bank. *World Tables 1976.* Baltimore and London: The Johns Hopkins University Press, 1976.

International Labor Office. *Yearbook of Labour Statistics.*

National Basic Intelligence Fact Book. Washington, D.C.: U.S. Government Printing Office.

United Nations. *Demographic Yearbook.*

United Nations. *Monthly Bulletin of Statistics.*

United Nations. *Population and Vital Statistics Report.*

United Nations. *Production Yearbook.*

United Nations. *Statistical Yearbook.*

United Nations. *World Energy Supplies, 1950–1974.*

United Nations. *Yearbook for Asia and the Pacific.*

United Nations. *Yearbook of International Statistics.*

United Nations. *Yearbook of National Account Statistics.*

Abbreviations

BM	basic manufactures (include textiles, iron and steel, metal manufactures, pulp and paper, and similar industrial products)
FA	food and live animals
GDP	gross domestic product
km^2	square kilometers
m^2	square miles
MTE	machinery and transportation equipment
PP	petroleum products
PPP	petroleum and petroleum products

	Afghanistan	Albania	Algeria	Angola
Total area: m² / km²	250,000 / 647,500	11,010 / 28,749	950,000 / 2,460,500	481,000 / 1,245,790
Total population:				
Latest official figure	18,294,000	2,432,000	16,780,000	5,800,000
1977 estimate	19,811,000	2,499,000	17,558,000	6,246,000
Annual rate of population increase (%)	2.4	3.1	3.2	
Infant mortality rate / Life expectancy	— / 38	87 / 69	128 / 53	203 / 38
Urban population (%)	15	38	50	17
Population of major cities (capitals are capitalized)	KABUL 534,000	TIRANA 175,000	ALGIERS 1,200,000 Oran 330,000 Constantine 254,000	LUANDA 480,613
Major ethnic groups (% of total population)	Pashtuns 52; Tadzhiks 25	Albanian 96	Arab-Berber 99	African 93; European 5
Major religious groups (% of total population)	Sunni Muslim 87; Shia Muslim 12	Muslim 70; Albanian Orthodox 20	Sunni Muslim 98	Roman Catholic 38; local cults
Major languages (% of total population)	Pushtu 50; Dari (Persian) 35	Tosk (official); Gheg	Arabic (official); Berber; French	Portuguese (official)
Literacy (% of population over age 15)	8	70	25	25
Land use (% of total area):				
1. cultivated	13	19	3	1
2. permanent meadow and pasture	9	24	16	23
3. forested	3	43	1	58
4. built on, waste, and other	75	14	80	18
Total cultivated area ('000 hectares)	8,467	640	7,000	1,310
Major crops (% of cultivated area)	wheat 29; barley 6; corn 4; rice 3	corn 31; wheat 26; dry beans 10; potatoes 4	wheat 29; barley 7; grapes 5; citrus fruits	corn 40; coffee 38; cotton 7
Annual timber production (millions of cubic meters)	6.6	2.3		7.3
Annual fish catch ('000 metric tons)	1.5	3.6	26.5	434.3
Percent of GDP from: farming, forestry, and fishing / manufacturing (and mining)	51 / 12			
Percent of labor force in: farming, forestry, and fishing / manufacturing (and mining)	82 / 7	61 / 18	50 / 7	63 / —
Major industrial products ('000 metric tons)	cotton yarn 1,093; cement 144; processed foods	textiles; processed foods	fuel oil 3,009; nitrogenous fertilizers 50	cement 767; refined petroleum 538
Major mineral products ('000 metric tons; natural gas in millions of cubic meters)	coal 149; natural gas 3,200; salt 51	petroleum 1,657; coal 675; chromium 230	petroleum 47,192; iron ore 3,797; natural gas 4,745	petroleum 8,175; iron ore 6,052; diamonds 2,125 m.c.
Number of passenger cars in use (per 1,000 inhabitants)	2		11	22
Number of telephones in use (per 100 inhabitants)	0.1		1.4	0.6
Total value of: imports / exports (millions of U.S. $)	153 / 90	159 / 91	2,049 / 1,863	468 / 667
Major imports (% of total value)	minerals 15; textiles 10; sugar 10	MTE 50; fuels and minerals 17	capital goods 35; semifinished goods 28; foodstuffs 23	vehicles 14; iron and steel 9
Major exports (% of total value)	fruits 43; natural gas 12	fuels and minerals 54; foodstuffs 23	petroleum 87	petroleum 30; coffee 27; diamonds 10
Per capita foreign trade (U.S. $)	13	103	233	196
Per capita calories consumed per day	2,440	2,100	1,730	2,000
Per capita energy consumption (kg)	53	687	400	181
Per capita national income (U.S.$)	60	520	425	492

	Argentina	Australia	Austria	The Bahamas
Total area: m² / km²	1,070,000 / 2,771,300	2,970,000 / 7,692,300	32,400 / 83,916	4,400 / 11,396
Total population:				
Latest official figure	25,380,000	13,500,000	7,520,000	205,000
1977 estimate	25,887,000	13,808,000	7,542,000	213,000
Annual rate of population increase (%)	1.3	1.6	0.5	3.9
Infant mortality rate / Life expectancy	60 / 67	17 / 71	24 / 71	33 / 67
Urban population (%)	80	86	72	72
Population of major cities (capitals are capitalized)	BUENOS AIRES 8,925,000 Rosario 807,000 Cordoba 791,000	CANBERRA 156,100 Sydney 2,922,760 Melbourne 2,661,400 Brisbane 958,800 Adelaide 899,300	VIENNA 1,614,841 Graz 249,000 Linz 203,000	NASSAU 120,000
Major ethnic groups (% of total population)	White 85; Mestizo, Indian, and others 15	Caucasian 99; Aborigine and Asian 1	German 98	Negro 85; Caucasian 10
Major religious groups (% of total population)	Roman Catholic	Church of England 38; Roman Catholic 23	Roman Catholic 90	Baptist 29; Roman Catholic 23; Church of England 23
Major languages (% of total population)	Spanish	English	German 95	English (official)
Literacy (% of population over age 15)	93	99	99	89
Land use (% of total area):				
1. cultivated	12	6	19	1
2. permanent meadow and pasture	52	59	26	0
3. forested	22	5	39	23
4. built on, waste, and other	14	30	16	76
Total cultivated area ('000 hectares)	34,420	44,980	1,612	16
Major crops (% of cultivated area)	wheat 12; corn 12; rye 7	wheat 20; barley 4; oats 3	wheat and rye 25; barley 20; corn 9	truck crops
Annual timber production (millions of cubic meters)	12	14	12	—
Annual fish catch ('000 metric tons)	238	109		1.5
Percent of GDP from: farming, forestry, and fishing / manufacturing (and mining)	11 / 31	6 / 30	6 / 38	—
Percent of labor force in: farming, forestry, and fishing / manufacturing (and mining)	15 / 20	7 / 26	13 / 31	7 / 6
Major industrial products ('000 metric tons)	cement 5,185; cellulose paper 243; cotton fiber 109	steel 8,017; cotton yarn 24; chemicals	iron and steel 11,514; cement 6,435	tourism; cement; pharmaceuticals
Major mineral products ('000 metric tons; natural gas in millions of cubic meters)	coal 451; petroleum 24,000,000 cu. m.	coal 59,833; iron ore 57,801; petroleum 19,595; bauxite 18,545; gold 16,271; tin 10,599	iron ore 4,245; lignite 3,629; petroleum 2,238; natural gas 2,206	salt; oragonite
Number of passenger cars in use (per 1,000 inhabitants)	85	353	218	196
Number of telephones in use (per 100 inhabitants)	9	38	26	27
Total value of: imports / exports (millions of U.S. $)	2,272 / 2,530	6,309 / 7,233	5,820 / 4,471	800 / 534
Major imports (% of total value)	BM 24; machinery 22; chemicals 13	MTE 34; fabrics 10; PPP 6	MTE 28; BM 22; mineral fuels 12; chemicals 10	petroleum 84
Major exports (% of total value)	FA 35; vegetable products 29	metal ores 19; wool 17; meat (beef) 12; wheat 11; coal 5	BM 39; MTE 24	PPP 93
Per capita foreign trade (U.S. $)	189	1,003	1,368	6,507
Per capita calories consumed per day	3,060	3,280	3,310	—
Per capita energy consumption (kg)	1,815	5,753	3,694	6,186
Per capita national income (U.S.$)	1,922	5,884	3,916	2,300

	Bangladesh	Belgium	Benin (formerly Dahomey)	Bhutan
Total area: m² / km²	55,019 / 142,500	11,800 / 30,562	44,700 / 115,773	17,992 / 46,600
Total population:				
Latest official figure	71,317,000	9,790,000	3,110,000	1,035,000
1977 estimate	76,560,000	9,826,000	3,235,000	1,217,000
Annual rate of population increase (%)	2.4	0.4	3.1	2.1
Infant mortality rate / Life expectancy	132 / 49	17 / 71	185 / 41	
Urban population (%)	5	87	13	4
Population of major cities (capitals are capitalized)	DACCA 1,319,970 Chittagong 458,000 Khulna 436,000	BRUSSELS 1,054,970 Antwerp 678,000	PORTO-NOVO (official) 80,000 Cotonou (de facto capital) 140,000	THIMBU
Major ethnic groups (% of total population)	Bengali 98; Bihari	Fleming 55; Walloon 33; Mixed and other 12	Fons; Adjas; Baribas; Yorubas	Bhotia 60; Nepalese 25
Major religious groups (% of total population)	Islam 85; Hinduism 14	Roman Catholic 75	Animist 65; Christian 15; Muslim 13	Buddhist 75; Hindu 25
Major languages (% of total population)	Bengali (official); English	Dutch 56; French 32; Bilingual 11	French (official)	Dzongkha (official)
Literacy (% of population over age 15)	20	98	20	5
Land use (% of total area):				
1. cultivated	63	27	26	3
2. permanent meadow and pasture	4	24	4	0
3. forested	16	20	19	64
4. built on, waste, and other	17	29	51	33
Total cultivated area ('000 hectares)	9,112	824	2,950	150
Major crops (% of cultivated area)	rice 88; jute 11	wheat 22; barley 17; oats 10	corn; yams; cassava	rice; wheat; barley
Annual timber production (millions of cubic meters)	11	3	2	
Annual fish catch ('000 metric tons)	247	57	32	
Percent of GDP from: farming, forestry, and fishing / manufacturing (and mining)	56 / 9	4 / 34	31 / 10	
Percent of labor force in: farming, forestry, and fishing / manufacturing (and mining)	80 / 7	4 / 31	90 / 10	99 / 1
Major industrial products ('000 metric tons)	PP 761; jute textiles 444; steel 150; cement 143	iron and steel 29,251; coke 8,050; cement 1,467; textiles 257	palm oil; cotton textiles	handicrafts
Major mineral products ('000 metric tons; natural gas in millions of cubic meters)	natural gas	coal 8,111; iron ore 123	iron ore; petroleum	
Number of passenger cars in use (per 1,000 inhabitants)	0.5	256	5	0.7
Number of telephones in use (per 100 inhabitants)		27	0.3	
Total value of: imports / exports (millions of U.S. $)	425 / 333	18,359 / 18,253 (with Luxembourg)	92 / 42	1.4 / 1
Major imports (% of total value)	BM 38; foodstuffs 27; raw materials 27	BM 24; MTE 22; fuels 14	textiles 23; machinery 12; chemicals 9	kerosene; sugar; textiles
Major exports (% of total value)	jute textiles 55; raw jute 33	BM 42; MTE 19; chemicals 13 (all with Luxembourg)	cocoa beans 18; cotton lint 15; palm oils 14	rice; dolomite; handicrafts
Per capita foreign trade (U.S. $)	11	3,740	43	2
Per capita calories consumed per day	2,310	3,380	2,250	
Per capita energy consumption (kg)	31	6,020	46	
Per capita national income (U.S.$)	70	5,029	85	100

	Bolivia	Botswana	Brazil	Bulgaria
Total area: m² / km²	424,000 / 1,098,160	220,00 / 569,800	3,290,000 / 8,521,100	43,186 / 111,852
Total population:				
Latest official figure	5,630,000	690,000	107,661,000	8,720,000
1977 estimate	5,624,000	720,000	111,666,000	8,759,000
Annual rate of population increase (%)	2.6	3.4	2.8	0.6
Infant mortality rate / Life expectancy	77 / 50	175 / 44	170 / 61	26 / 71
Urban population (%)	34	4	58	56
Population of major cities (capitals are capitalized)	LA PAZ 697,480	GABORONE 21,000	BRASILIA 545,000 São Paulo 7,200,000 Rio de Janeiro 4,856,000 Belo Horizonte 1,557,000	SOFIA 946,305 Plovdiv 287,744
Major ethnic groups (% of total population)	Ayamra 25; Mestizo 30; Quechua 30	Tswana 94; Bushmen 5; European 1	White 60; Mixed 30; Black 8; Indian 2	Bulgarian 85; Turk 3; Gypsy 3
Major religious groups (% of total population)	Roman Catholic 95	Animist 85; Christian 15	Roman Catholic 93	Bulgarian Orthodox 85; Muslim 13
Major languages (% of total population)	Spanish (official); Aymara	English (official); Botswana	Portuguese	Bulgarian
Literacy (% of population over age 15)	40	20	81	95
Land use (% of total area):				
1. cultivated	12	1	4	40
2. permanent meadow and pasture	52	68	20	14
3. forested	22	2	60	34
4. built on, waste, and other	14	29	16	12
Total cultivated area ('000 hectares)	3,217	512	36,060	4,488
Major crops (% of cultivated area)	corn; sugar cane; potatoes	sorghum 18; corn 3; millet 3; cow peas 3	corn 29; rice 15; coffee 6; sugar 6; cotton 6	wheat 23; corn 15; barley 11
Annual timber production (millions of cubic meters)	4.2	0.7	162	5
Annual fish catch ('000 metric tons)			545	98
Percent of GDP from: farming, forestry, and fishing / manufacturing (and mining)	14 / 30	29 / 20	13 / 22	22 / 51
Percent of labor force in: farming, forestry, and fishing / manufacturing (and mining)	67 / 15	91 / 2	44 / 12	44 / 26
Major industrial products ('000 metric tons)	cement 202,000; sugar 169,000		cement 14,920; iron and steel 13,078; sugar 6,673; PP	cement 4,300; iron and steel 3,716
Major mineral products ('000 metric tons; natural gas in millions of cubic meters)	petroleum 2,600,000 cu. m.; zinc 39; tin 30	diamonds 2,400,000 c.	iron ore 55,019; coal 5,574; manganese 2,594; bauxite 849; petroleum 10,500,000 cu. m.	coal 26,459; iron ore 890; petroleum 190
Number of passenger cars in use (per 1,000 inhabitants)	5	5	34	
Number of telephones in use (per 100 inhabitants)	1	0.6	3	8
Total value of: imports / exports (millions of U.S. $)	192 / 282	176 / 106	6,496 / 4,757	2,822 / 2,790
Major imports (% of total value)	capital goods 61; raw materials 35	MTE 30; BM 23; FA 13	petroleum and minerals 25; MTE 20; metals 17; chemicals 13	MTE 41; fuels and minerals 29
Major exports (% of total value)	minerals 68; petroleum 15	minerals 45; meat products 40	sugar 16; coffee 11; minerals 10; textiles 9	MTE 40; foodstuffs 24
Per capita foreign trade (U.S. $)	84	409	105	644
Per capita calories consumed per day	1,900	2,040	2,620	3,000
Per capita energy consumption (kg)	9		561	4,126
Per capita national income (U.S.$)	299	230	723	2,020

	Burma	Burundi	Cambodia (Kampuchea)	Cameroon
Total area: m² / km²	262,000 / 678,600	11,000 / 28,490	70,000 / 181,300	113,960 / 295,154
Total population:				
Latest official figure	30,170,000	3,763,000	7,735,000	6,539,000
1977 estimate	31,501,000	3,897,000	7,887,000	6,588,000
Annual rate of population increase (%)	2.4	2.4	2.2	2
Infant mortality rate / Life expectancy	126 / 47	150 / 39	127 / 43	137 / 49
Urban population (%)	20	3	10	20
Population of major cities (capitals are capitalized)	RANGOON 3,186,886; Mandalay 417,300	BUJUMBURA 78,810	PHNOM PENH 403,500	YAOUNDE 274,000; Douala 486,000
Major ethnic groups (% of total population)	Burman 65; Shan 11; Karen 7; Kachin 6	Hutu (Bantu) 85; Tutsi (Hamitic) 14	Cambodian (Khmer) 85; Chinese 5; Vietnamese 3	200 tribes
Major religious groups (% of total population)	Buddhist 85	Catholic 50; Animist 40; Protestant 10	Therarada Buddhist 95	Animist 50; Christian 33; Muslim 17
Major languages (% of total population)	Burmese (official); English 2	Kirundi (official); French (official); Swahili	Khmer (Cambodian)	English (official); French (official)
Literacy (% of population over age 15)	70 (official claim)	Kirundi 55; French 10	41	25
Land use (% of total area):				
1. cultivated	15	45	17	15
2. permanent meadow and pasture	1	16	3	18
3. forested	67	3	74	63
4. built on, waste, and other	17	36	6	4
Total cultivated area ('000 hectares)	10,373	1,255	3,046	7,335
Major crops (% of cultivated area)	rice 50; sesame seeds 10; groundnuts 7; pulses 5; rubber	sweet potatoes; cassava; corn; bananas	rice 22; rubber	sorghum and millet 7; cassava; corn
Annual timber production (millions of cubic meters)	19		4.2	8
Annual fish catch ('000 metric tons)	441	11	75.3	71
Percent of GDP from: farming, forestry, and fishing / manufacturing (and mining)	38 / 12		41 / 17	31 / 15
Percent of labor force in: farming, forestry, and fishing / manufacturing (and mining)	69 / 8	95 / 1		80 / 10
Major industrial products ('000 metric tons)	fuel oils 370; cement 193; sugar 67; cotton yarn 8		cotton yarn 0.4; soap 0.4; rubber footwear	palm oil 58; aluminum (unwrought) 46; raw sugar 12
Major mineral products ('000 metric tons; natural gas in millions of cubic meters)	petroleum 968; lead 7; zinc 3	tin 0.1	salt (unrefined) 31	bauxite; petroleum
Number of passenger cars in use (per 1,000 inhabitants)	1	1	4	6
Number of telephones in use (per 100 inhabitants)	0.1	0.1	0.1	
Total value of: imports / exports (millions of U.S. $)	152 / 137	30 / 26	66 / 15	313 / 297
Major imports (% of total value)	MTE 30; metal products 14	BM 84	foodstuffs 54; textiles 13; minerals 12	MTE 30; PPP 8; iron and steel 6; cereals 4
Major exports (% of total value)	teak and hardwood 31; other farm products 25; rice 19	coffee 84; hides and skins 4; cotton 4	rubber 93	cocoa 27; coffee 21; timber 10
Per capita foreign trade (U.S. $)	10	11	10	93
Per capita calories consumed per day	2,160	2,040	2,220	2,230
Per capita energy consumption (kg)	59	15	30	93
Per capita national income (U.S.$)	80	60	75	304

	Canada	Central African Empire	Chad	Chile
Total area: m² / km²	3,850,000 / 9,971,500	242,000 / 626,780	496,000 / 1,284,640	286,000 / 740,740
Total population:				
Latest official figure	22,998,000	1,637,000	4,030,000	10,250,000
1977 estimate	23,314,000	1,846,000	4,155,000	10,543,000
Annual rate of population increase (%)	1.3	2.3	2.1	1.7
Infant mortality rate / Life expectancy	18 / 72	190 / 40	160 / 43	79 / 63
Urban population (%)	76	27	7	76
Population of major cities (capitals are capitalized)	OTTAWA 626,000 Montreal 2,798,000 Toronto 2,741,000 Vancouver 1,137,000	BANGUI 301,000	N'DJAMENA 179,000	SANTIAGO 3,069,000 Valparaiso 293,000
Major ethnic groups (% of total population)	British origin 40; French origin 30	80 tribal groups	12 major groups	Mestizo and European 95; Indian 3
Major religious groups (% of total population)	Roman Catholic 46; United Church 18; Church of England 12	Protestant 40; Catholic 28; Animist 24; Muslim 8	Muslim 50; Animist 45; Christian 5	Roman Catholic 89; Protestant 11
Major languages (% of total population)	English (official); French (official)	French (official); Sangho (national)	French (official); Chadian Arabic; Sara; Sangho	Spanish
Literacy (% of population over age 15)	100	18	10	88
Land use (% of total area):				
1. cultivated	4	10	5	7
2. permanent meadow and pasture	3	0	35	16
3. forested	32	12	13	27
4. built on, waste, and other	61	78	47	50
Total cultivated area ('000 hectares)	43,100	5,900	7,000	5,742
Major crops (% of cultivated area)	hay 28; wheat 20; barley 10; oats 5	cassava 4; cotton 2; groundnuts 2	millet and sorghum 13; cotton 4; pulses 3; groundnuts 2	wheat 12; corn 2; barley 2; potatoes 2
Annual timber production (millions of cubic meters)	122.2	2.4	3.4	8
Annual fish catch ('000 metric tons)	1,281	3.5	117	1,044
Percent of GDP from: farming, forestry, and fishing / manufacturing (and mining)	5 / 26	31 / 18		7 / 34
Percent of labor force in: farming, forestry, and fishing / manufacturing (and mining)	6 / 21	87 / —	96 / 4	19 / 25
Major industrial products ('000 metric tons)	MTE; chemicals; wood and paper products	textiles	sugar 15; beverages	iron and steel 1,112; fuel oils; sugar 802
Major mineral products ('000 metric tons; natural gas in millions of cubic meters)	petroleum 91,976; iron ore 52,216; coal 23,600; natural gas 1,364	diamonds 380,000 c.	petroleum; uranium	iron ore 9,416; coal 1,390; copper 743; petroleum 1,800,000 cu. m.
Number of passenger cars in use (per 1,000 inhabitants)	368	6	1	22
Number of telephones in use (per 100 inhabitants)	55	0.2	0.2	4.3
Total value of: imports / exports (millions of U.S. $)	20,673 / 22,490	40 / 37	72 / 34	1,173 / 1,352
Major imports (% of total value)	automobile and parts 22; petroleum 8	MTE 29; cotton textiles 8	PPP 16; MTE 13; sugar and confectionery 11	MTE 32; chemicals 12; vegetable products 12
Major exports (% of total value)	vehicles 18; forest products 15; crude oil 11; wheat 7	diamonds 23; cotton 22; coffee 21	raw cotton 63; livestock products 16	minerals 80; BM 12
Per capita foreign trade (U.S. $)	1,877	47	26	246
Per capita calories consumed per day	3,180	2,170	2,060	2,670
Per capita energy consumption (kg)	9,335		22	1,408
Per capita national income (U.S.$)	5,672	155	73	793

	China, People's Republic of	China, Republic of (Taiwan)	Colombia	Congo, People's Republic of
Total area: m² / km²	3,692,000 / 9,562,210	12,456 / 32,260	440,000 / 1,139,600	135,000 / 349,650
Total population:				
Latest official figure	900,000,000	16,172,000	24,720,000	1,300,020
1977 estimate	958,540,000	16,453,000	25,541,000	1,400,000
Annual rate of population increase (%)	1.7	1.9	3.2	3.1
Infant mortality rate / Life expectancy	55 / 57	18 / 70	70 / 61	180 / 44
Urban population (%)	15	52	63	30
Population of major cities (capitals are capitalized)	PEKING 7,600,000 Shanghai 10,820,000 Tientsin 4,280,000 Shenyang 3,000,000 Wuhan 2,700,000 Canton 2,200,000	TAIPEI 2,043,000 Kaohsiung 1,000,000	BOGOTA 2,856,000 Medellin 1,100,000 Cali 920,000	BRAZZAVILLE 289,700 Pointe Noire 141,700
Major ethnic groups (% of total population)	Han Chinese 94; Manchu; Mongol; Hui; Chuang	Taiwanese 87; Mainland Chinese 13	Mestizo 58; Caucasian 20; Mulatto 14	Kongo 48; Sangha 20; Teke 17; M'Bochi 12
Major religious groups (% of total population)		Buddhist, Taoist, and Confucian 93; Christian 5	Roman Catholic 95	Christian (nominal) 50; Animist 49
Major languages (% of total population)	Standard Chinese 95	Chinese (official); Taiwanese	Spanish	French (official); Lingala; Kikongo
Literacy (% of population over age 15)	55	93	73	20
Land use (% of total area):				
1. cultivated	14	24	5	2
2. permanent meadow and pasture	22	6	15	42
3. forested	15	55	68	45
4. built on, waste, and other	49	15	12	11
Total cultivated area ('000 hectares)	130,000	902	5,090	647
Major crops (% of cultivated area)	rice 27; wheat 19; potatoes 11; soy beans 7; cotton 5	rice 87; sugar cane; sweet potatoes	corn 20; coffee 19; sugar cane 7	cassava 24; groundnuts 5; sugar cane 3
Annual timber production (millions of cubic meters)	173	0.8	26.4	2.2
Annual fish catch ('000 metric tons)	7,017	779	66	21
Percent of GDP from: farming, forestry, and fishing / manufacturing (and mining)		18 / 31	27 / 23	
Percent of labor force in: farming, forestry, and fishing / manufacturing (and mining)	85 / —	37 / 23	39 / 16	70 / —
Major industrial products ('000 metric tons)	iron and steel 60,000; cement 35,000; paper 3,000; sugar 2,300; cotton yarn 2,000	cement 6,795; diesel oil 1,714; petrol 1,122; sugar 706	cement 3,431; steel 244; sugar 894	palm oil 3,500; cement 52; sugar 35
Major mineral products ('000 metric tons; natural gas in millions of cubic meters)	coal 420; iron ore 105; petroleum 80	coal 3,140; natural gas 1.5	petroleum 9,104; coal	petroleum 2,071; potash 442; natural gas 15.7
Number of passenger cars in use (per 1,000 inhabitants)			13	15
Number of telephones in use (per 100 inhabitants)			5	1
Total value of: imports / exports (millions of U.S. $)	7,400 / 6,900	4,850 / 4,900	1,006 / 960	78 / 45
Major imports (% of total value)	BM 53; chemicals and raw materials 27; foodstuffs 20	petroleum 11; cotton 3; foodstuffs	chemicals 24; MTE 19; metals 15; vehicles 11	machinery 18; iron and steel 10; vehicles 6
Major exports (% of total value)	BM 40; chemicals and raw materials 33; foodstuffs 27	clothing 11; television receivers 5; raw sugar 4; plastic footwear 4	coffee 46; cotton 7; fuel oil 6; sugar 5	timber 60; potash 26
Per capita foreign trade (U.S. $)	18	603	80	95
Per capita calories consumed per day	2,000	2,360	2,200	2,160
Per capita energy consumption (kg)	590	1,300 kwh per capita	629	205
Per capita national income (U.S.$)	243	700	445	267

	Costa Rica	Cuba	Cyprus	Czechoslovakia
Total area: m² / km²	19,700 / 51,000	44,200 / 114,478	3,572 / 9,251	49,400 / 127,946
Total population:				
Latest official figure	1,970,000	9,090,000	639,000	14,800,000
1977 estimate	2,041,000	9,567,000	650,000	14,989,000
Annual rate of population increase (%)	2.7	1.8	1.1	0.6
Infant mortality rate / Life expectancy	54 / 64	36 / 67	33 / 71	21 / 70
Urban population (%)	37	61	44	56
Population of major cities (capitals are capitalized)	SAN JOSE 418,000	HAVANA 1,755,000	NICOSIA 116,100	PRAGUE 1,161,000 Brno 355,661 Ostrava 292,000
Major ethnic groups (% of total population)	White and Mestizo 98; Negro 2	Mulatto 51; White 37; Negro 11	Greek 78; Turk 18	Czech 65; Slovak 30
Major religious groups (% of total population)	Roman Catholic 95	Roman Catholic 85	Greek Orthodox 78; Muslim 4	Roman Catholic 77; Protestant 20
Major languages (% of total population)	Spanish	Spanish	Greek; Turkish; English	Czech; Slovak; Hungarian
Literacy (% of population over age 15)	89	78	76	nearly 100
Land use (% of total area):				
1. cultivated	10	32	47	41
2. permanent meadow and pasture	31	24	10	14
3. forested	50	11	18	35
4. built on, waste, and other	9	33	25	10
Total cultivated area ('000 hectares)	501	3,720	432	5,290
Major crops (% of cultivated area)	coffee 19; corn 11; rice 10; sugar 7; bananas 4	sugar 40; rice 5; corn 3; tobacco 2	wheat; barley; grapes; citrus fruits	wheat 25; barley 17; potatoes 5; sugar beets 4
Annual timber production (millions of cubic meters)	3	2		14.4
Annual fish catch ('000 metric tons)	11	113		16
Percent of GDP from: farming, forestry, and fishing / manufacturing (and mining)	20 / 20		12 / 18	11 / 62
Percent of labor force in: farming, forestry, and fishing / manufacturing (and mining)	36 / 12	30 / 20	28 / 13	16 / 38
Major industrial products ('000 metric tons)	fuel oil 315; cement 272; raw sugar 176	raw sugar 4,388; cement 1,474; steel 145; fertilizers 620	cement 338; alcoholic beverages	iron and steel 30,860; coke 10,898; fuel oil 10,739; cement 8,967
Major mineral products ('000 metric tons; natural gas in millions of cubic meters)	bauxite; manganese	salt 102; chromium 37; nickel and cobalt 37	pyrites 283; copper concentrates 45; asbestos 36	coal 110,681; magnesite 2,815; iron ore 1,688
Number of passenger cars in use (per 1,000 inhabitants)	26	8	122	90
Number of telephones in use (per 100 inhabitants)	5	3	11	17
Total value of: imports / exports (millions of U.S. $)	436 / 302	1,373 / 1,009	333 / 136	5,207 / 5,195
Major imports (% of total value)	BM 59; MTE 18	capital goods 33; raw materials 33; petroleum 10	BM 29; MTE 22; PPP 12; chemicals 7	MTE 36; BM 15; FA 10; fuels 9
Major exports (% of total value)	coffee 29; bananas 22; cattle and meat 8; sugar 6	sugar and its products 80; nickel oxide and sulphide 10	fruits and vegetables 48; alcoholic beverages 11	MTE 46; BM 22; iron and steel 12
Per capita foreign trade (U.S. $)	375	262	734	703
Per capita calories consumed per day	2,610		2,670	3,100
Per capita energy consumption (kg)	482	1,140	1,606	6,761
Per capita national income (U.S.$)	799	382	1,273	3,820

	Denmark	Dominican Republic	Ecuador	Egypt
Total area: m² / km²	16,600 / 42,994	18,800 / 48,692	106,000 / 274,540	386,200 / 1,000,258
Total population:				
Latest official figure	5,150,000	4,697,000	6,730,000	37,230,000
1977 estimate	5,079,000	4,906,000	7,133,000	38,533,000
Annual rate of population increase (%)	0.6	2.9	3.4	2.2
Infant mortality rate / Life expectancy	14 / 73	50 / 58	77 / 53	103 / 54
Urban population (%)	80	40	39	43
Population of major cities (capitals are capitalized)	COPENHAGEN 1,383,000 Arhus 245,000	SANTO DOMINGO 980,000	QUITO 557,000 Guayaquil 860,600	CAIRO 6,588,000 Alexandria 2,032,000
Major ethnic groups (% of total population)	Scandinavian; German	Mulatto 73; Caucasian 16; Negro 11	Indian 40; Mestizo 40; Spanish 10	Eastern Hamitic stock 90; Greek, Italian, and Syro-Lebanese 10
Major religious groups (% of total population)	Evangelical Lutheran 97	Roman Catholic 95	Roman Catholic (nominal) 95	Muslim 94; Coptic and other 6
Major languages (% of total population)	Danish	Spanish	Spanish (official); Quechua	Arabic (official); English; French
Literacy (% of population over age 15)	100	68	68	38
Land use (% of total area):				
1. cultivated	62	20	15	3
2. permanent meadow and pasture	7	30	8	0
3. forested	11	23	62	0
4. built on, waste, and other	20	27	15	97
Total cultivated area ('000 hectares)	2,658	995	4,324	2,855
Major crops (% of cultivated area)	barley 53; root crops 11; oats 6; wheat 5	coffee 16; sugar cane 15; cacao 7; rice 7	corn 6; coffee 5; cacao 5; sugar 3; bananas 3	corn 22; cotton 21; wheat 17
Annual timber production (millions of cubic meters)	2.1	1.8	3.1	
Annual fish catch ('000 metric tons)	1,362	4.5		77
Percent of GDP from: farming, forestry, and fishing / manufacturing (and mining)	8 / 28	22 / 21	20 / 23	26 / 21
Percent of labor force in: farming, forestry, and fishing / manufacturing (and mining)	9 / 24	44 / 8	56 / 18	53 / 13
Major industrial products ('000 metric tons)	motor and fuel oil 7,099; processed foods 4,778	cement 642; fertilizers 164; sugar 79	PP 3,731; cement 484; sugar 249	fuel oil 5,169; cement 3,617; sugar 634; cotton yarn 183; cotton cloth 118
Major mineral products ('000 metric tons; natural gas in millions of cubic meters)	salt 423; coal 135; iron ore 6	bauxite 1,087; gypsum 247	petroleum 11,343; copper 0.3	petroleum 9,616; iron ore 656; super phosphates 419
Number of passenger cars in use (per 1,000 inhabitants)	245	13	5	5
Number of telephones in use (per 100 inhabitants)	43	2	3	1
Total value of: imports / exports (millions of U.S. $)	6,358 / 5,104	404 / 377	482 / 484	1,172 / 1,002
Major imports (% of total value)	MTE 24; BM 22; PPP 18; chemicals 10	foodstuffs; petroleum	MTE 33; raw materials 33	cereals 31; chemicals 13
Major exports (% of total value)	FA 32; MTE 27; BM 11	sugar; coffee; cocoa; bauxite	petroleum 59; bananas 11; cocoa 10; coffee 6; sugar 4	raw cotton 47; cotton yarn 11; rice 7
Per capita foreign trade (U.S. $)	2,225	166	144	58
Per capita calories consumed per day	3,240	2,220	2,010	2,500
Per capita energy consumption (kg)	5,471	391	151	303
Per capita national income (U.S.$)	5,402	396	474	240

	El Salvador	Ethiopia	Fiji	Finland
Total area: m² / km²	8,263 / 21,400	455,000 / 1,178,450	7,056 / 18,272	130,000 / 336,700
Total population:				
Latest official figure	4,010,000	27,950,000	564,000	4,710,000
1977 estimate	4,195,000	29,044,000	572,000	4,738,000
Annual rate of population increase (%)	3	2.5	1.9	0.4
Infant mortality rate / Life expectancy	53 / 58	181 / —	26 / 69	10 / 70
Urban population (%)	41	11	33	57
Population of major cities (capitals are capitalized)	SAN SALVADOR 600,000	ADDIS ABABA 912,000; Asmara 249,000	SUVA 96,000	HELSINKI 837,553
Major ethnic groups (% of total population)	Mestizo 89; Indian 10	Galla 40; Amhara 25; Tigre 12; Sidama 9	Indian 50; Fijian 42	homogeneous white population; small Lapp minority
Major religious groups (% of total population)	Roman Catholic 97	Christian (Ethiopian Orthodox) 40; Muslim 40; Animist 20	Christian; Hindu	Evangelical Lutheran 92
Major languages (% of total population)	Spanish	Amharic (official); English	English (official)	Finnish 94; Swedish 6
Literacy (% of population over age 15)	57	7	85	100
Land use (% of total area):				
1. cultivated	32	12	13	8
2. permanent meadow and pasture	32	53	4	0
3. forested	12	6	65	67
4. built on, waste, and other	24	29	18	25
Total cultivated area ('000 hectares)	673	13,730	79	2,654
Major crops (% of cultivated area)	corn 9; coffee 7; cotton 5; sugar	teff 16; barley 13; sorghum 9; wheat 8; coffee 5	sugar cane; coconuts; bananas	oats 21; barley 17; wheat 8
Annual timber production (millions of cubic meters)	2.4	24		44
Annual fish catch ('000 metric tons)		21		80
Percent of GDP from: farming, forestry, and fishing / manufacturing (and mining)	26 / 21	48 / 10	21 / 13	11 / 30
Percent of labor force in: farming, forestry, and fishing / manufacturing (and mining)	47 / 8	86 / 10	54 / 9	16 / 26
Major industrial products ('000 metric tons)	PP 467; cement 296; sugar 201	fuel oil 273; cement 182; sugar 108	sugar	iron and steel 2,488; cement 2,203
Major mineral products ('000 metric tons; natural gas in millions of cubic meters)		salt 210; potash	gold; copper	zinc 61; copper 37
Number of passenger cars in use (per 1,000 inhabitants)	5	1	35	199
Number of telephones in use (per 100 inhabitants)	1	0.2	5	36
Total value of: imports / exports (millions of U.S. $)	335 / 321	208 / 184	173 / 86	3,966 / 3,393
Major imports (% of total value)	BM 21; chemicals 21; MTE 16; PPP 7	MTE 30; chemicals 14; fuels 9; iron and steel 5	foodstuffs 19; MTE 16; fuel 16	MTE 28; BM 20; PPP 18; chemicals 10
Major exports (% of total value)	coffee 42; cotton 10; chemicals 5	coffee 27; pulses 18; oilseeds 17; hides and skins 8	sugar 70; coconut oil 12; unrefined gold 9	timber and paper products 50; MTE 17
Per capita foreign trade (U.S. $)	164	14	459	1,562
Per capita calories consumed per day	1,930	2,160		3,050
Per capita energy consumption (kg)	222	33	472	4,503
Per capita national income (U.S.$)	373	88	958	4,184

	France	Gabon Republic	Gambia	Germany, East
Total area: m^2 / km^2	213,000 / 551,670	102,000 / 264,180	4,000 / 10,360	41,800 / 108,262
Total population:				
Latest official figure	52,658,000	530,000	524,000	16,820,000
1977 estimate	53,096,000	557,000	544,000	16,778,000
Annual rate of population increase (%)	0.8	1.0	2.5	0.2
Infant mortality rate / Life expectancy	13 / 72	229 / 45	165 / 41	16 / 71
Urban population (%)	70	19	10	74
Population of major cities (capitals are capitalized)	PARIS 9,250,000 Lyon 1,075,000 Marseille 964,000 Lille 881,000 Bordeaux 555,000	LIBREVILLE 251,400	BANJUL 52,640	EAST BERLIN 1,090,000 Leipzig 573,000 Dresden 506,000
Major ethnic groups (% of total population)	Celtic 45; Latin; Teutonic	40 Bantu tribes	Mandigo 40; Fula 13; Wolof 12	German 99.7; Slavic and other 0.3
Major religious groups (% of total population)	Roman Catholic 83; Protestant 2	Christian 65; Animist 30	Muslim 85; Christian and Animist 15	Protestant 59; Roman Catholic 8
Major languages (% of total population)	French 100	French (official)	English (official)	German
Literacy (% of population over age 15)	97	20	10	99
Land use (% of total area):				
1. cultivated	34	1	32	45
2. permanent meadow and pasture	25	18	23	13
3. forested	27	80	6	27
4. built on, waste, and other	14	1	39	15
Total cultivated area ('000 hectares)	18,844	155	360	4,890
Major crops (% of cultivated area)	wheat 25; barley 16; corn 12; oats 4	cassava 42; cacao 8	peanuts 50; millet 12; rice 9	barley 17; wheat 16; rye 14; potatoes 14; corn 7; sugar beets 5
Annual timber production (millions of cubic meters)	34.4	3.2		8
Annual fish catch ('000 metric tons)	757	4		335
Percent of GDP from: farming, forestry, and fishing / manufacturing (and mining)	6 / 36	13 / 42		11 / 64
Percent of labor force in: farming, forestry, and fishing / manufacturing (and mining)	11 / 28	70 / —	85 / —	12 / 40
Major industrial products ('000 metric tons)	fuel oil 83,334; iron and steel 71,013; automobiles	PP 579	peanut products; clothing	cement 9,548; iron and steel 8,092; automobiles
Major mineral products ('000 metric tons; natural gas in millions of cubic meters)	iron ore 53,260; coal 26,797; bauxite 2,675	petroleum 10,202; manganese 2,119; natural gas 45; uranium concentrates 1.7		coal 246,998; natural gas 7,012; potash 2,556; iron ore 52
Number of passenger cars in use (per 1,000 inhabitants)	285	19	6	101
Number of telephones in use (per 100 inhabitants)	24	1	0.5	15
Total value of: imports / exports (millions of U.S. $)	31,558 / 29,291	167 / 343	28 / 23	6,647 / 5,195
Major imports (% of total value)	MTE 21; PPP 21; BM 20; chemicals 9; FA 8	MTE 37; base metals 19	textiles 24; FA 20; MTE 13	coal; iron ore; petroleum; grains
Major exports (% of total value)	MTE 30; BM 23; FA 14; chemicals 11	petroleum 41; timber 33; manganese ore 15	peanut products 81	MTE; chemicals; textiles
Per capita foreign trade (U.S. $)	1,156	962	97	704
Per capita calories consumed per day	3,210	2,210	2,370	3,000
Per capita energy consumption (kg)	4,221	952	63	6,479
Per capita national income (U.S.$)	4,486	1,248	125	4,170

	Germany, West	Ghana	Greece	Guatemala
Total area: m² / km²	96,000 / 248,640	92,000 / 238,280	51,200 / 132,608	42,000 / 108,800
Total population:				
Latest official figure	61,916,000	9,870,000	9,050,000	5,160,000
1977 estimate	62,590,000	10,281,000	9,123,000	6,099,000
Annual rate of population increase (%)	0.6	2.7	0.5	2.8
Infant mortality rate / Life expectancy	20 / 70	156 / 48	28 / 69	79 / 48
Urban population (%)	77	31	53	34
Population of major cities (capitals are capitalized)	BONN 283,000; West Berlin 2,048,000; Hamburg 1,752,000; Munich 1,337,000; Cologne 832,000; Frankfurt 663,000	ACCRA 848,800; Kumasi 345,000	ATHENS 2,540,000; Thessaloniki 557,000	GUATEMALA CITY 1,000,000
Major ethnic groups (% of total population)	German 99	Negro 99.8	Greek 96; Turkish 2	Indian 51; Mestizo 38; Ladino 10
Major religious groups (% of total population)	Protestant 49; Roman Catholic 45	Animist 45; Christian 43; Muslim 12	Greek Orthodox 98	Roman Catholic predominant; Traditional Mayan
Major languages (% of total population)	German	English (official)	Greek; English; French	Spanish; 10 local dialects
Literacy (% of population over age 15)	100	25	84	38
Land use (% of total area):				
1. cultivated	33	11	30	16
2. permanent meadow and pasture	21	45	40	8
3. forested	29	11	20	53
4. built on, waste, and other	17	33	10	23
Total cultivated area ('000 hectares)	8,080	2,710	3,905	1,700
Major crops (% of cultivated area)	barley 23; wheat 21; oats 12; rye 8; sugar beets 6	cacao 55; corn 17; millet 8; sorghum 7	olives 27; wheat 23; barley 11; cotton 3; tobacco 3	corn 52; coffee 16; cotton 5; sugar 4; bananas 4
Annual timber production (millions of cubic meters)	28	10	2.8	7
Annual fish catch ('000 metric tons)	533	198	81	
Percent of GDP from: farming, forestry, and fishing / manufacturing (and mining)	3 / 44	48 / 13	16 / 20	29 / 16
Percent of labor force in: farming, forestry, and fishing / manufacturing (and mining)	7 / 41	60 / 10	41 / 18	57 / 14
Major industrial products ('000 metric tons)	iron and steel 132,438; cement 35,977; automobiles	textiles; wood products; processed foods	cement 7,026; fertilizers 1,426; chemicals 1,165	cement 259; sugar 194
Major mineral products ('000 metric tons; natural gas in millions of cubic meters)	coal 220,920; iron 1,565	diamonds 2,400,000 c.; bauxite 357; manganese 252; gold	lignite 13,230; bauxite 2,741; iron ore 1,843; magnesite 1,087	nickel
Number of passenger cars in use (per 1,000 inhabitants)	273	4	150	10
Number of telephones in use (per 100 inhabitants)	30	0.6	21	1
Total value of: imports / exports (millions of U.S. $)	46,322 / 56,177	481 / 475	2,852 / 1,132	408 / 391
Major imports (% of total value)	BM 20; PPP 18; MTE 15; FA 12	foodstuffs 21; machinery 21; BM 20; chemicals 17	MTE 28; PPP 21; BM 18; FA 11	BM 24; MTE 22; chemicals 21
Major exports (% of total value)	MTE 42; BM 24; chemicals 14	cocoa 47; timber 18; minerals 13	BM 34; FA 19	coffee 30; cotton 12; sugar 9; bananas 4; meat 4
Per capita foreign trade (U.S. $)	1,655	97	440	155
Per capita calories consumed per day	3,220	2,200	3,190	2,130
Per capita energy consumption (kg)	5,625	167		235
Per capita national income (U.S.$)	5,461	380	2,061	470

	Guinea	Guinea-Bissau	Guyana	Haiti
Total area: m² / km²	95,000 / 246,050	14,000 / 36,260	83,000 / 214,970	10,700 / 27,713
Total population:				
Latest official figure	5,143,000	482,000	794,000	4,580,000
1977 estimate	4,584,000	518,000	818,000	4,682,000
Annual rate of population increase (%)	2.4	3	2.2	1.6
Infant mortality rate / Life expectancy	216 / 41	208 / 40	38 / 61	— / 48
Urban population (%)	11	18	35	12
Population of major cities (capitals are capitalized)	CONAKRY 525,700; Labé 418,650	BISSAU 80,000	GEORGETOWN 195,000	PORT-AU-PRINCE 550,000
Major ethnic groups (% of total population)	3 major tribes; 15 smaller tribes	Balanta 30; Fulani 20; Mandyako 14; Malinke 13	East Indian 55; African 35	African descent 95
Major religious groups (% of total population)	Muslim 75; Animist 25	Animist 69; Muslim 30	Christian 57; Hindu 33; Muslim 9	Roman Catholic 80
Major languages (% of total population)	French (official)	Portuguese (official)	English	French (official)
Literacy (% of population over age 15)	10	5	76	10
Land use (% of total area):				
1. cultivated	17	8	4	33
2. permanent meadow and pasture	12	35	11	20
3. forested	5	30	84	2
4. built on, waste, and other	66	27	1	45
Total cultivated area ('000 hectares)	4,150	285	845	905
Major crops (% of cultivated area)	rice 11; corn 10; coffee 7	peanuts 28; rice 13; coconuts; palm oil	rice 17; sugar cane 7	corn 26; sorghum 24; sugar cane 8; rice 5; coffee 3
Annual timber production (millions of cubic meters)	2.8			4
Annual fish catch ('000 metric tons)	5	1.4	17	
Percent of GDP from: farming, forestry, and fishing / manufacturing (and mining)			22 / 22	47 / 14
Percent of labor force in: farming, forestry, and fishing / manufacturing (and mining)			30 / 18	83 / 12
Major industrial products ('000 metric tons)	palm oil 450; wood products 90	palm oil; hides and skins; beverages	wood products	cement 130; processed foods
Major mineral products ('000 metric tons; natural gas in millions of cubic meters)	bauxite 7,650; iron ore 1,040; diamonds 80,000 c.; alumina 700		bauxite 2,316; alumina 264	bauxite; copper
Number of passenger cars in use (per 1,000 inhabitants)	2		31	3
Number of telephones in use (per 100 inhabitants)	0.2		2	0.2
Total value of: imports / exports (millions of U.S. $)	132 / 113	35 / 3	167 / 166	70 / 51
Major imports (% of total value)	textiles 16; PPP 16; rice 10	rice 16; textiles 8	BM 73	BM 26; MTE 18; FA 18
Major exports (% of total value)	alumina 72; pineapples 10; coffee 6; palm products 6	peanuts 72; palm kernels 5	bauxite and alumina 43; sugar 35; rice 8	coffee 34; handicrafts 29; bauxite 11; sugar 4
Per capita foreign trade (U.S. $)	47	79	419	26
Per capita calories consumed per day	2,040		2,390	1,850
Per capita energy consumption (kg)	97	52		32
Per capita national income (U.S.$)	80	150	359	158

	Honduras	Hong Kong	Hungary	Iceland
Total area: m^2 / km^2	43,300 / 112,150	404 / 1,046	35,900 / 92,981	39,750 / 102,952
Total population:				
Latest official figure	2,752,000	4,300,000	10,540,000	220,000
1977 estimate	2,872,000	4,800,000	10,634,000	221,000
Annual rate of population increase (%)	4.0	1.4	0.3	1.3
Infant mortality rate / Life expectancy	40 / 49	18 / 67	34 / 69	12 / 74
Urban population (%)	28	98	49	86
Population of major cities (capitals are capitalized)	TEGUCIGALPA 300,000	HONG KONG 4,370,000	BUDAPEST 2,056,000	REYKJAVIK 110,000
Major ethnic groups (% of total population)	Mestizo 90; Indian 7	Chinese 98	Magyar 93; German 3; Gypsy 2.4	homogeneous white population
Major religious groups (% of total population)	Roman Catholic 97	Christian 10	Roman Catholic 68; Lutheran 8; Calvinist 5	Evangelical Lutheran 95
Major languages (% of total population)	Spanish	English (official); Cantonese (official)	Magyar 98	Icelandic
Literacy (% of population over age 15)	45	75	98	100
Land use (% of total area):				
1. cultivated	8	13	59	0
2. permanent meadow and pasture	18	0	14	22
3. forested	63	11	17	1
4. built on, waste, and other	11	76	10	77
Total cultivated area ('000 hectares)	870		5,503	1
Major crops (% of cultivated area)	corn 33; coffee 14; beans 7; bananas 4	rice; truck crops	corn 27; wheat 24; barley 5; grapes 4	potatoes nearly 100; hay; turnips
Annual timber production (millions of cubic meters)	4.3		5.1	
Annual fish catch ('000 metric tons)			27.5	748
Percent of GDP from: farming, forestry, and fishing / manufacturing (and mining)	32 / 17		19 / 42	
Percent of labor force in: farming, forestry, and fishing / manufacturing (and mining)	57 / 12	4 / 48	23 / 36	11 / 14
Major industrial products ('000 metric tons)	cement; sugar	textiles; clothing; metal and electrical products	iron and steel 8,150; cement 3,437; fertilizers 2,939	
Major mineral products ('000 metric tons; natural gas in millions of cubic meters)	zinc 204; lead 181; silver	iron ore 167	coal 25,761; natural gas 5,101; bauxite 2,751; petroleum 1,997; iron ore 545	diatomite
Number of passenger cars in use (per 1,000 inhabitants)	6	29	47	290
Number of telephones in use (per 100 inhabitants)	0.5	23	10	40
Total value of: imports / exports (millions of U.S. $)	252 / 211	6,295 / 5,599	3,629 / 3,534	295 / 222
Major imports (% of total value)	MTE 27; fuels 17; chemicals 15	BM 29; FA 18; MTE 17	MTE 29; BM 20; chemicals 15; FA 10	MTE 34; BM 20; PPP 12
Major exports (% of total value)	bananas 32; coffee 18; lumber 17	clothing 45; MTE 16; BM 13	MTE 33; FA 21; BM 17	fish 38; aluminum products 9
Per capita foreign trade (U.S. $)	168	2,766	680	2,350
Per capita calories consumed per day	2,140	2,370	3,140	2,900
Per capita energy consumption (kg)	234	1,600 kwh	3,342	4,607
Per capita national income (U.S.$)	306	1,435	1,825	6,170

	India	Indonesia	Iran	Iraq
Total area: m^2 / km^2	1,211,000 / 3,136,500	736,000 / 1,906,240	636,000 / 1,647,240	172,000 / 445,480
Total population:				
Latest official figure	589,170,000	127,590,000	33,020,000	10,765,000
1977 estimate	634,993,000	135,913,000	34,455,000	11,577,000
Annual rate of population increase (%)	2.1	2.4	2.8	3.3
Infant mortality rate / Life expectancy	139 / 54	125 / 48	139 / 50	99 / 53
Urban population (%)	20	18	43	47
Population of major cities (capitals are capitalized)	NEW DELHI 4,066,000 Calcutta 7,000,000 Bombay 6,000,000 Madras 2,500,000 Hyderabad 1,796,000 Ahmadabad 1,741,000	DJAKARTA 5,849,000 Surabaya 1,556,000 Bandung 1,202,000	TEHRAN 2,800,000 Mashhad 562,000 Isfahan 540,000 Tabriz 480,000	BAGHDAD 2,800,000 Basra 371,000 Mosul 293,000
Major ethnic groups (% of total population)	Indo-Aryan 72; Dravidian 25; Mongoloid 3	Javanese 45; Sudanese 14; Madurese 8; Malay 8	Persian 76; Turkic 18	Arab 75; Kurd 20
Major religious groups (% of total population)	Hindu 84; Muslim 10	Muslim 85; Christian 9; Hindu 2	Shia Muslim 93; Sunni Muslim 2	Muslim 90; Christian 8
Major languages (% of total population)	Hindi (official); English (official)	Indonesian (official); English	Persian; Turki; Kurdish; Arabic	Arabic 75; Kurdish 20
Literacy (% of population over age 15)	28	60	23	26
Land use (% of total area):				
1. cultivated	50	10	10	12
2. permanent meadow and pasture	4	5	7	9
3. forested	20	64	11	4
4. built on, waste, and other	26	21	72	75
Total cultivated area ('000 hectares)	165,070	18,500	16,280	5,280
Major crops (% of cultivated area)	rice 23; wheat 12; sorghum and millet 18; cotton 5	rice 47; corn 15; cassava 7; rubber; coconuts	wheat 31; barley 8; rice 3	wheat 11; barley 4; cotton; dates
Annual timber production (millions of cubic meters)	113.2	114.6	6.4	
Annual fish catch ('000 metric tons)	1,762	1,251		
Percent of GDP from: farming, forestry, and fishing / manufacturing (and mining)	44 / 14	40 / 18	15 / 39	13 / 45
Percent of labor force in: farming, forestry, and fishing / manufacturing (and mining)	72 / 10	62 / 8	42 / 17	18 / 14
Major industrial products ('000 metric tons)	PP 19,116; iron and steel 12,129; jute products 1,032	cement 652; fertilizers 177; textiles	cement 4,500; sugar 700; vegetable ghee 247	cement; cotton textiles
Major mineral products ('000 metric tons; natural gas in millions of cubic meters)	coal 80,382; iron ore 35,401; petroleum 7,197; manganese ore 1,461; bauxite 1,285; natural gas 682	petroleum 71,194; bauxite 922; nickel 801; coal 206	petroleum 267,800; natural gas 49,993; coal 1,000; chromite 120; iron ore 98	petroleum 107,313
Number of passenger cars in use (per 1,000 inhabitants)	1	2	12	30
Number of telephones in use (per 100 inhabitants)	0.3	0.2	2	2
Total value of: imports / exports (millions of U.S. $)	2,955 / 2,658	2,051 / 2,961	3,003 / 7,938	1,019 / 2,729
Major imports (% of total value)	fuels 26; foodstuffs 19; chemicals 16; MTE 15	MTE 32; iron and steel 7; yarns 4; rice 3	MTE 32; iron and steel 17; cereals 8	machinery 17; sugar 5; automobiles 4
Major exports (% of total value)	foodstuffs 31; textiles 25; jute products 9	petroleum 69; timber 7; rubber 5	petroleum 97	petroleum 98; dates
Per capita foreign trade (U.S. $)	9	39	331	348
Per capita calories consumed per day	2,210	2,160	2,400	2,160
Per capita energy consumption (kg)	190	148	1,087	728
Per capita national income (U.S.$)	110	110	1,295	570

	Ireland	Israel	Italy	Ivory Coast
Total area: m² / km²	26,600 / 68,894	8,000 / 20,720	116,300 / 301,217	125,000 / 323,750
Total population:				
Latest official figure	3,130,000	3,410,000	55,810,000	6,673,000
1977 estimate	3,167,000	3,545,000	56,410,000	7,045,000
Annual rate of population increase (%)	1.1	3.2	0.8	2.5
Infant mortality rate / Life expectancy	18 / 71	22 / 71	28 / 71	164 / 44
Urban population (%)	52	84	48	21
Population of major cities (capitals are capitalized)	DUBLIN 680,000 Cork 128,000	JERUSALEM 355,500 Tel Aviv–Jaffa 1,150,300 Haifa 351,900	ROME 2,856,000 Milan 1,732,000 Naples 1,224,000 Turin 1,203,000	ABIDJAN 500,000
Major ethnic groups (% of total population)	Celtic nearly 100	Jewish 85; Arab 15	Italian predominant	over 60 ethnic groups
Major religious groups (% of total population)	Roman Catholic 94; Episcopalian 4	Jewish 89; Muslim 8; Christian and other 3	Roman Catholic nearly 100	Animist 63; Muslim 24; Christian 13
Major languages (% of total population)	English (official); Gaelic (official)	Hebrew (official); Arabic; English	Italian predominant	French; Tribal dialects
Literacy (% of population over age 15)	99	90	93	20
Land use (% of total area):				
1. cultivated	15	20	41	28
2. permanent meadow and pasture	54	40	17	25
3. forested	3	6	21	28
4. built on, waste, and other	28	34	21	19
Total cultivated area ('000 hectares)	1,054	430	12,288	9,100
Major crops (% of cultivated area)	barley 58; wheat 13; oats 10; sugar beets 6	field crops 64; truck crops 29	wheat 30; grapes 11; olives 9; corn 7	coffee 10; cacao 5; rice 4
Annual timber production (millions of cubic meters)			13.5	9.5
Annual fish catch ('000 metric tons)	80	26.2	390	63
Percent of GDP from: farming, forestry, and fishing / manufacturing (and mining)	15 / 29	5 / 20	8 / 34	29 / 13
Percent of labor force in: farming, forestry, and fishing / manufacturing (and mining)	25 / 21	6 / 25	15 / 31	86 / 1
Major industrial products ('000 metric tons)	processed foods	paper 72,898; cement 1,428	iron and steel 55,231; fuel oil 51,208; automobiles	fuel oil 800; cement 665; palm oil 118
Major mineral products ('000 metric tons; natural gas in millions of cubic meters)		petroleum 45,000,000 liters; natural gas 66; phosphate rock 1,026	pyrites 1,168; petroleum 1,024; iron ore 659	diamonds 350,000 m.c.; manganese 10
Number of passenger cars in use (per 1,000 inhabitants)	157	70	256	14
Number of telephones in use (per 100 inhabitants)	13	22	25	1
Total value of: imports / exports (millions of U.S. $)	2,420 / 1,745	2,458 / 1,174	23,791 / 19,877	583 / 709
Major imports (% of total value)	MTE 22; BM 22; PPP 13; chemicals 12	machinery 16; rough diamonds 11; iron and steel 8; vehicles 5	fuels 26; BM 16; MTE 16; FA 14	MTE 34; wheat 5
Major exports (% of total value)	FA 37; BM 18; MTE 10; chemicals 9	cut diamonds 35; fruits 11; textiles 9	MTE 32; BM 23; chemicals 10	wood 29; coffee 24; cocoa 16
Per capita foreign trade (U.S. $)	1,331	1,065	782	194
Per capita calories consumed per day	3,410	2,960	3,180	2,490
Per capita energy consumption (kg)	3,289	2,785	2,979	322
Per capita national income (U.S.$)	2,040	2,638	2,441	600

	Jamaica	Japan	Jordan	Kenya
Total area: m² / km²	4,410 / 11,422	143,000 / 370,370	37,100 / 96,089	225,000 / 582,750
Total population:				
Latest official figure	2,025,000	111,934,000	2,700,000	13,400,000
1977 estimate	2,093,000	113,462,000	2,833,000	14,113,000
Annual rate of population increase (%)	1.7	1.2	3.5	3.4
Infant mortality rate / Life expectancy	26 / 65	12 / 73	99 / 53	135 / 49
Urban population (%)	37	72	43	10
Population of major cities (capitals are capitalized)	KINGSTON 506,200	TOKYO 11,669,000 Osaka 2,778,000 Yokohama 2,621,000 Nagoya 2,079,000 Kyoto 1,400,000	AMMAN 520,700	NAIROBI 750,000 Mombasa 247,100
Major ethnic groups (% of total population)	Black 85; Mixed 10	Japanese 99.2	Arab 98	Kikuyu 20; Luo 15; Balhya 13; Kamba 11; Kisii 6; Meru 5
Major religious groups (% of total population)	Protestant predominant	Shintoist and Buddhist 83; Christian 1; Other 16	Sunni Muslim 92; Christian 8	Animist 38; Protestant 37; Roman Catholic 22
Major languages (% of total population)	English; Jamaican; Creole	Japanese nearly 100	Arabic (official)	English (official); Swahili (official)
Literacy (% of population over age 15)	82	100	40	25
Land use (% of total area):				
1. cultivated	24	15	14	3
2. permanent meadow and pasture	20	1	1	7
3. forested	45	67	1	3
4. built on, waste, and other	11	17	84	87
Total cultivated area ('000 hectares)	260	5,615	1,360	1,755
Major crops (% of cultivated area)	sugar 24; bananas 12; yams 5	rice 49; oranges 3; potatoes 2; wheat 1	wheat 9; barley 4; truck crops	corn 71; wheat 6; coffee 5; sisal 5; sugar 3
Annual timber production (millions of cubic meters)		47.3		11.3
Annual fish catch ('000 metric tons)		9,786		31
Percent of GDP from: farming, forestry, and fishing / manufacturing (and mining)	8 / 26	6 / 40	18 / 12	30 / 15
Percent of labor force in: farming, forestry, and fishing / manufacturing (and mining)	27 / 12	13 / 27	33 / —	31 / 13
Major industrial products ('000 metric tons)	fertilizers 55,509,000 long tons; cement 393,000 long tons; textiles	iron and steel 189,191; machinery; automobiles	cement 572	cement 856; refined oil 3,300,000 liters; sugar 164
Major mineral products ('000 metric tons; natural gas in millions of cubic meters)	bauxite 15,327	coal 18,999; iron pyrites 1,698; iron 602; zinc 254	phosphates 1,352	soda ash 205; salt 62; magnetite 12
Number of passenger cars in use (per 1,000 inhabitants)	43	142	10	10
Number of telephones in use (per 100 inhabitants)	5	36	2	1
Total value of: imports / exports (millions of U.S. $)	662 / 439	32,497 / 32,879	297 / 65	597 / 309
Major imports (% of total value)	BM 21; foodstuffs 21; fuels 21; MTE 17	fuels and minerals 44; PPP 36; FA 8	MTE 32; minerals 11; textiles 6; chemicals 6	petroleum 19; iron and steel 8; motor vehicles 6; chemicals
Major exports (% of total value)	alumina 53; bauxite 21; sugar 4; bananas 2	MTE 49; iron and steel 18; electronics 15; optical goods 8	phosphates 40; fruits and vegetables 19	coffee 26; PP 17; tea 11; sisal 10
Per capita foreign trade (U.S. $)	543	584	134	68
Per capita calories consumed per day	2,360	2,510	2,430	2,360
Per capita energy consumption (kg)	1,430	3,633		156
Per capita national income (U.S.$)	1,066	4,100	320	181

	Korea, North	Korea, South	Kuwait	Laos
Total area: m² / km²	49,000 / 126,900	38,000 / 98,400	6,200 / 16,058	91,430 / 236,804
Total population:				
Latest official figure	11,568,000	34,708,000	990,400	3,257,000
1977 estimate	17,294,000	35,538,000	1,101,000	3,455,000
Annual rate of population increase (%)	2.7	1.7	5.6	2.4
Infant mortality rate / Life expectancy	— / 58	60 / 62	44 / 67	123 / 48
Urban population (%)	17	41	56	15
Population of major cities (capitals are capitalized)	PYONGYANG 1,500,000 Hamhung 484,000	SEOUL 6,889,000 Pusan 2,454,000 Taegu 1,311,000	KUWAIT 560,000 Hawalli 106,500	VIENTIANE 176,000
Major ethnic groups (% of total population)	Korean nearly 100	Korean nearly 100	Arab 85; Iranian 13	Lao 47; Kha 25; Tai 14
Major religious groups (% of total population)		Buddhist 15; Christian 6; Shamanist	Muslim 99; Christian 1	Theravada Buddhist 50; Animist 50
Major languages (% of total population)	Korean nearly 100	Korean nearly 100	Arabic (official); English	Lao
Literacy (% of population over age 15)	90	90	55	15
Land use (% of total area):				
1. cultivated	12	19	0	4
2. permanent meadow and pasture	1	1	8	4
3. forested	78	70	0	63
4. built on, waste, and other	9	10	92	29
Total cultivated area ('000 hectares)	1,400	2,100	1	960
Major crops (% of cultivated area)	minor grains 36; potatoes 28; wheat 25	rice 53; barley 38		rice 69; corn 4
Annual timber production (millions of cubic meters)	5	10.3		3
Annual fish catch ('000 metric tons)	800	1,176		20
Percent of GDP from: farming, forestry, and fishing / manufacturing (and mining)		26 / 29	0.0 / 70	
Percent of labor force in: farming, forestry, and fishing / manufacturing (and mining)	50 / 50	46 / 17	2 / 16	85 / —
Major industrial products ('000 metric tons)	cement 6,400; iron and steel 5,900	cement 10,129; iron and steel 3,195	fuel oil 15,102	
Major mineral products ('000 metric tons; natural gas in millions of cubic meters)	coal 40,900; iron ore 9,400; magnetite 2,000	anthracite coal 17,585; iron ore 524	petroleum 92,456; natural gas 207	tin 0.5
Number of passenger cars in use (per 1,000 inhabitants)		2	167	43
Number of telephones in use (per 100 inhabitants)		4	11	0.2
Total value of: imports / exports (millions of U.S. $)	1,100 / 800	3,592 / 2,242	929 / 4,462	74 / 5
Major imports (% of total value)	MTE; petroleum; foodstuffs; coking coal	PPP 18; machinery 5; chemicals 5; wheat 4	MTE 36; BM 25; FA 15	minerals 26; vegetables 18; MTE 15
Major exports (% of total value)	minerals; chemicals; metal products	clothing 10; electrical machinery 9; textiles 7; fish 5; plywood 4	petroleum 96	timber 64; tin 30
Per capita foreign trade (U.S. $)	164	168	5,445	24
Per capita calories consumed per day		2,350		2,220
Per capita energy consumption (kg)	2,356	876	9,989	80
Per capita national income (U.S.$)	450	501	11,182	66

	Lebanon	Lesotho	Liberia	Libya
Total area: m² / km²	4,000 / 10,360	11,700 / 30,303	43,000 / 111,370	679,000 / 1,758,610
Total population:				
Latest official figure	3,100,000	1,040,000	1,501,000	2,440,000
1977 estimate	3,150,000	1,074,000	1,554,000	2,599,000
Annual rate of population increase (%)	3.0	2.2	2.3	4.2
Infant mortality rate / Life expectancy	59 / 63	181 / 46	159 / 45	130 / 52
Urban population (%)	60	1	10	27
Population of major cities (capitals are capitalized)	BEIRUT 1,000,000	MASERU 5,000	MONROVIA 171,000	TRIPOLI 247,000 Benghazi 170,000
Major ethnic groups (% of total population)	Arab 93; Armenian 6	Sotho 85; Nguni 15	16 tribes	Arab and Berber 97
Major religious groups (% of total population)	Christian 50; Muslim and Druze 50	Christian 70; Animist 30	Animist 80; Muslim 20	Muslim 97
Major languages (% of total population)	Arabic (official)	English (official); Sesotho (official)	English (official)	Arabic; English; Italian
Literacy (% of population over age 15)	86	40	10	22
Land use (% of total area):				
1. cultivated	34	12	4	2
2. permanent meadow and pasture	1	82	2	4
3. forested	8	0	22	0
4. built on, waste, and other	57	6	72	94
Total cultivated area ('000 hectares)	348	350	380	2,544
Major crops (% of cultivated area)	wheat 14; olives 9; grapes 5; apples 3	corn 35; sorghum 26; wheat 24	rice 45; cassava 17; coffee 4; rubber	barley 12; wheat 5; dates; olives
Annual timber production (millions of cubic meters)	0.5		1.6	0.4
Annual fish catch ('000 metric tons)			12	5.7
Percent of GDP from: farming, forestry, and fishing / manufacturing (and mining)	9 / 16	62 / 3	19 / 33	3 / 60
Percent of labor force in: farming, forestry, and fishing / manufacturing (and mining)	18 / 17	91 / 1	25 / —	37 / 10
Major industrial products ('000 metric tons)	fuel oil 2,438; cement 1,744; processed foods; textiles	carpets; candles; fertilizers; pottery	PP 595; cement 91	processed foods; cement; chemicals
Major mineral products ('000 metric tons; natural gas in millions of cubic meters)		diamonds 8,588 c.	iron ore 25,188; diamonds 636,000 c.	petroleum 72,800; natural gas 10,909
Number of passenger cars in use (per 1,000 inhabitants)	71	2	8	93
Number of telephones in use (per 100 inhabitants)	8	0.3	0.2	2
Total value of: imports / exports (millions of U.S. $)	857 / 348	50 / 10	195 / 279	1,373 / 4,051
Major imports (% of total value)	MTE 24; metals and precious stones 19; textiles 12	FA 23; clothing 16; MTE 10	MTE 30; fuels 20; BM 19	machinery 33; BM 28; foodstuffs 16
Major exports (% of total value)	metals and precious stones 21; MTE 20; textiles 11; vegetable products 9	wool 35; mohair 16; animals 16	iron ore 66; rubber 16; diamonds 7	petroleum 98
Per capita foreign trade (U.S. $)	389	58	316	2,223
Per capita calories consumed per day	2,280		2,040	2,570
Per capita energy consumption (kg)	924		420	702
Per capita national income (U.S.$)	730	100	333	2,400

	Luxembourg	Malagasy Republic (Madagascar)	Malawi	Malaysia
Total area: m² / km²	1,000 / 2,590	230,000 / 595,700	36,700 / 95,053	128,400 / 332,556
Total population:				
Latest official figure	360,000	7,928,000	5,040,000	11,930,000
1977 estimate	361,000	7,812,000	5,245,000	12,395,000
Annual rate of population increase (%)	0.2	2.5	2.5	2.7
Infant mortality rate / Life expectancy	16 / 73	170 / 38	148 / 41	75 / 65
Urban population (%)		14	6	28
Population of major cities (capitals are capitalized)	LUXEMBOURG 78,800	TANANARIVE 400,000	LILONGWE 70,000	KUALA LUMPUR 500,000 Penang 280,000
Major ethnic groups (% of total population)	Luxembourger 83; Italian descent 5	18 Malagasy tribes	African 99	Malay 50; Chinese 36; Indian 10
Major religious groups (% of total population)	Roman Catholic 97	Animist 52; Christian 41; Muslim 7	Animist majority; Christian; Muslim	Muslim 40; Buddhist 25; Christian 16; Hindu 10
Major languages (% of total population)	Luxembourgian; German; French	Malagasy (official); French (official)	Chichewa; English	Malay; Chinese; English; Tamil
Literacy (% of population over age 15)	98	40	22	50
Land use (% of total area):				
1. cultivated	24	5	24	22
2. permanent meadow and pasture	24	58	16	0
3. forested	36	21	20	63
4. built on, waste, and other	16	16	40	15
Total cultivated area ('000 hectares)	62	2,950	2,895	2,880
Major crops (% of cultivated area)	barley 27; oats 20; wheat 18; grapes 2	rice 35; cassava 8; coffee 7	tobacco; tea; sugar; cotton	rubber 53; rice 19; oil palm 13
Annual timber production (millions of cubic meters)		5.4	4.3	26
Annual fish catch ('000 metric tons)		46.4	36	372
Percent of GDP from: farming, forestry, and fishing / manufacturing (and mining)	4 / 43	29 / 14	47 / 13	31 / 22
Percent of labor force in: farming, forestry, and fishing / manufacturing (and mining)	8 / 34	80 / 2	85 / 15	59 / 4
Major industrial products ('000 metric tons)	iron and steel 11,917; chemicals; rubber products	sugar 114; cement 61; textiles	processed foods; tobacco; clothing	rubber sheets 138; cement 1,342
Major mineral products ('000 metric tons; natural gas in millions of cubic meters)	iron ore 2,686	chromium ore 65; salt 24; graphite 17		bauxite 692; iron ore 342; tin concentrates 63
Number of passenger cars in use (per 1,000 inhabitants)	355	7	2	36
Number of telephones in use (per 100 inhabitants)	40	0.4	0.4	2
Total value of: imports / exports (millions of U.S. $)	(see Belgium)	215 / 181	130 / 86	2,360 / 2,629
Major imports (% of total value)	plastics; machinery; textiles	petroleum 16; chemicals 14; metals 11	PP 7; textiles 5; fertilizers 3	MTE; chemicals
Major exports (% of total value)	steel 66; metal products	coffee 29; cloves 11; vanilla 10; PP 6	tobacco 44; tea 19; sugar 10	rubber 28; tin 15; timber 15; palm oil 11
Per capita foreign trade (U.S. $)	(see Belgium)	50	43	418
Per capita calories consumed per day	3,380	2,530	2,210	2,240
Per capita energy consumption (kg)	17,907	74	52	490
Per capita national income (U.S.$)	5,535		129	716

	Mali	Malta	Mauritania	Mauritius
Total area: m² / km²	465,000 / 1,204,350	121 / 313	419,000 / 1,085,210	720 / 1,856
Total population:				
Latest official figure	5,376,000	324,000	1,320,000	872,000
1977 estimate	5,815,000	340,000	1,371,000	900,000
Annual rate of population increase (%)	2.5		2.7	1.1
Infant mortality rate / Life expectancy	188 / 37	24 / 70	189 / 38	64 / 66
Urban population (%)		94	7	44
Population of major cities (capitals are capitalized)	BAMAKO 350,000	VALLETTA 15,000 Sliema 22,000	NOUAKCHOTT 80,000	PORT LOUIS 140,000
Major ethnic groups (% of total population)	Mande 50; Peul 17; Voltaic 12	Arab; Italian; British	Moor 33; Negro 33; Mixed Moor and Negro 33	Indo-Mauritian 67; Creole 28
Major religious groups (% of total population)	Muslim 90; Animist 9; Christian 1	Roman Catholic 99	Muslim nearly 100	Hindu 51; Christian 33; Muslim 16
Major languages (% of total population)	French (official); Bambara 80	Maltese; English	French (official); Hassaniya 80; Arabic	French; English Creole; Hindi; Urdu
Literacy (% of population over age 15)	10	88	5	61
Land use (% of total area):				
1. cultivated	10	44	1	57
2. permanent meadow and pasture	3	0	38	4
3. forested	4	0	15	31
4. built on, waste, and other	83	56	46	8
Total cultivated area ('000 hectares)	11,700	14	1,005	106
Major crops (% of cultivated area)	millet 11; peanuts 2; cotton	cereals and legumes 40; truck crops 40	millet 16; dates; gum arabic	sugar 75; tea
Annual timber production (millions of cubic meters)	2.8		0.5	
Annual fish catch ('000 metric tons)	90	30	25	
Percent of GDP from: farming, forestry, and fishing / manufacturing (and mining)		6 / 30		25 / 16
Percent of labor force in: farming, forestry, and fishing / manufacturing (and mining)	90 / 1	6 / 29	97 / —	28 / 12
Major industrial products ('000 metric tons)	cement 47; sugar 4	textiles; processed foods		sugar 696; molasses 172
Major mineral products ('000 metric tons; natural gas in millions of cubic meters)	unrefined salt 5	salt	iron ore 8,280; copper ore 200	
Number of passenger cars in use (per 1,000 inhabitants)	1		0.3	18
Number of telephones in use (per 100 inhabitants)	0.1			3
Total value of: imports / exports (millions of U.S. $)	81 / 39	218 / 77	96 / 124	152 / 137
Major imports (% of total value)	foodstuffs 34; MTE 24; PPP 10	MTE 15; foodstuffs 14; fuels 11; chemicals 7	MTE 40; iron and steel 10; PPP 7	MTE 17; rice 13; PPP 9
Major exports (% of total value)	cotton 46; livestock 23; peanuts 12	textiles 56; rubber goods 7	iron ore 68; copper 10; fish 10; gum arabic	sugar 86; clothing 3
Per capita foreign trade (U.S. $)	22	910	167	331
Per capita calories consumed per day	2,170	2,820	1,970	2,360
Per capita energy consumption (kg)	24	1,125	125	208
Per capita national income (U.S.$)	70	1,132	180	612

	Mexico	Mongolia	Morocco	Mozambique
Total area: m² / km²	764,000 / 1,978,800	597,150 / 1,546,619	158,000 / 409,200	303,769 / 786,762
Total population:				
Latest official figure	60,150,000	1,450,000	16,880,000	8,519,000
1977 estimate	63,192,000	1,513,000	18,090,000	9,294,000
Annual rate of population increase (%)	3.5	3.0	3.4	2.3
Infant mortality rate / Life expectancy	63 / 61	— / 61	149 / 53	165 / 41
Urban population (%)	61	46	37	7
Population of major cities (capitals are capitalized)	MEXICO CITY 8,591,000 Guadalajara 1,560,000 Monterrey 1,049,000	ULAN BATOR 310,000	RABAT 385,000 Casablanca 1,561,000	LOURENÇO MARQUES 384,000 Beira 115,000
Major ethnic groups (% of total population)	Mestizo 60; American Indian 30	Mongol 90; Kazakh 2; Chinese 2	Arab-Berber 99	Portuguese; African tribal groups
Major religious groups (% of total population)	Roman Catholic 97	Tibetan Buddhist predominant; Muslim 4	Muslim 99	Catholic; Protestant; Muslim; Animist
Major languages (% of total population)	Spanish	Khalkha Mongol 90	Arabic (official); Berber dialects; French	Portuguese (official); African languages
Literacy (% of population over age 15)	65	80	14	21
Land use (% of total area):				
1. cultivated	12	1	17	4
2. permanent meadow and pasture	40	89	28	56
3. forested	22	10	12	25
4. built on, waste, and other	26	0	43	15
Total cultivated area ('000 hectares)	27,390	801	7,630	3,080
Major crops (% of cultivated area)	corn 30; dry beans 5; wheat 3; sugar cane 2	wheat 38	wheat; barley; grapes; sugar beets	cotton 9; sugar 2; cashew nuts; coconuts; tea
Annual timber production (millions of cubic meters)	14	2.2	2.8	8.7
Annual fish catch ('000 metric tons)	421		271	9
Percent of GDP from: farming, forestry, and fishing / manufacturing (and mining)	11 / 29		27 / 25	
Percent of labor force in: farming, forestry, and fishing / manufacturing (and mining)	41 / 19		50 / 10	73 / 10
Major industrial products ('000 metric tons)	iron and steel 7,404; fertilizers 5,187; paper and cellulose 1,372	processed foods	PP 1,875; cement 1,619; sugar 473	PP 924; cement 465; sugar 265
Major mineral products ('000 metric tons; natural gas in millions of cubic meters)	iron ore 3,338; sulphur 2,322; zinc 288; lead 218; manganese 145; silver; petro- leum 46,786,000 cu. m.; natural gas 22,273	coal 1,997	phosphates 19,749; coal 574; iron ore 534	coal 425; iron ore
Number of passenger cars in use (per 1,000 inhabitants)	29		18	10
Number of telephones in use (per 100 inhabitants)	5	2	1	0.6
Total value of: imports / exports (millions of U.S. $)	3,694 / 2,186	171 / 138	1,029 / 855	382 / 202
Major imports (% of total value)	MTE 37; foodstuffs 8	consumer goods 33; MTE 26; fuels 14; foodstuffs 14	machinery; fuels; foodstuffs	MTE 41; metal products 12; textiles 9
Major exports (% of total value)	BM 52; foodstuffs 42; minerals 4; petroleum	raw materials 58; foodstuffs 29	citrus fruits; phosphates; canned fish	textiles 25; cashew nuts 22; cotton 20; sugar 10; tea 4
Per capita foreign trade (U.S. $)	98	213	112	69
Per capita calories consumed per day	3,110		2,220	2,050
Per capita energy consumption (kg)	1,145	961	223	150
Per capita national income (U.S.$)	910		212	230

	Nepal	The Netherlands	New Zealand	Nicaragua
Total area: m² / km²	54,600 / 141,400	13,100 / 33,929	103,580 / 268,276	57,100 / 147,900
Total population:				
Latest official figure	12,570,000	13,734,000	3,148,000	2,160,000
1977 estimate	13,006,000	13,868,000	3,203,000	2,263,000
Annual rate of population increase (%)	2.3	1.0	1.9	3.3
Infant mortality rate / Life expectancy	169 / 44	11 / 74	16 / 72	45 / 54
Urban population (%)	4	77	81	49
Population of major cities (capitals are capitalized)	KATMANDU 150,000	THE HAGUE 797,820 Amsterdam 1,323,695 Rotterdam 1,259,620	WELLINGTON 354,700 Auckland 796,700 Christchurch 326,410	MANAGUA 450,000
Major ethnic groups (% of total population)	Indo-Nepalese 80; Tibeto-Nepalese 20	Dutch 99	European 93; Maori 7	Mestizo 70; Caucasian 17; Negro 9; Indian 4
Major religious groups (% of total population)	Hindu 90; Buddhist	Protestant 41; Roman Catholic 19	Anglican 34; Presbyterian 22; Roman Catholic 15	Roman Catholic 95
Major languages (% of total population)	Nepali; Newari	Dutch	English; Maori	Spanish
Literacy (% of population over age 15)	12	100	98	50
Land use (% of total area):				
1. cultivated	14	23	3	7
2. permanent meadow and pasture	14	34	49	14
3. forested	32	8	25	47
4. built on, waste, and other	40	35	23	32
Total cultivated area ('000 hectares)	2,000	837	834	960
Major crops (% of cultivated area)	rice 62; corn 22; wheat 14	potatoes 17; wheat 16; sugar beets 13; barley 9	barley 10; wheat 9; peas 3	corn 20; cotton 15; coffee 8
Annual timber production (millions of cubic meters)	9	3.8	8.6	2.4
Annual fish catch ('000 metric tons)	2.1	327	60	11
Percent of GDP from: farming, forestry, and fishing / manufacturing (and mining)	69 / 10	5 / 31		26 / 21
Percent of labor force in: farming, forestry, and fishing / manufacturing (and mining)	94 / 1	8 / 30	12 / 26	46 / 13
Major industrial products ('000 metric tons)	jute 12; sugar 12	iron and steel 10,621; metal fabrication; electronics; chemicals; textiles	fertilizers 1,801; cement 1,074; sugar 148; pulp and paper	PP 446; processed foods; textiles
Major mineral products ('000 metric tons; natural gas in millions of cubic meters)	copper; iron ore	petroleum 1,461; coal 758	coal 2,468; iron sands 2,352; natural gas 484	copper concentrates 1.5; gold 2,371 kg
Number of passenger cars in use (per 1,000 inhabitants)	0.3	250	357	15
Number of telephones in use (per 100 inhabitants)	0.1	34	48	1
Total value of: imports / exports (millions of U.S. $)	171 / 86	20,449 / 19,871	1,989 / 1,882	302 / 252
Major imports (% of total value)	BM 49; FA 19	MTE 21; BM 20; petroleum 17	machinery 29; BM 23; chemicals 11	MTE 22; chemicals 12; iron and steel 7; foodstuffs 6
Major exports (% of total value)	foodstuffs 66; BM 11	MTE 17; chemicals 17; BM 17; foodstuffs 17	meat 27; dairy products 17; wool 15	cotton 23; coffee 16; meat 16; sugar 5
Per capita foreign trade (U.S. $)	20	2,937	1,230	256
Per capita calories consumed per day	2,210	3,220	3,500	2,450
Per capita energy consumption (kg)	13	5,683	3,147	428
Per capita national income (U.S.$)	87	4,694	4,043	620

	Niger	Nigeria	Norway	Oman
Total area: m² / km²	489,000 / 1,266,510	357,000 / 924,630	125,000 / 323,750	82,000 / 212,380
Total population:				
Latest official figure	4,600,000	62,925,000	4,017,000	750,000
1977 estimate	4,792,000	65,628,000	4,037,000	770,000
Annual rate of population increase (%)	2.7	2.7	0.7	3.1
Infant mortality rate / Life expectancy	200 / 38	180 / 39	11 / 74	138 / 46
Urban population (%)	8	23	45	5
Population of major cities (capitals are capitalized)	NIAMEY 102,000	LAGOS 1,500,000 Ibadan 1,300,000 Kano 357,000	OSLO 477,500 Bergen 182,000	MUSCAT 7,000 Matrah 14,000
Major ethnic groups (% of total population)	Hausa 50; Djerma 23; Fulani 15; Tuareg 10	250 tribal groups	homogeneously white; Lapp minority	Arab 98
Major religious groups (% of total population)	Muslim 80	Muslim 47; Christian 34; Animist 19	Evangelical Lutheran (State Church) 95	Ibadhi Muslim; Sunni Muslim
Major languages (% of total population)	French (official); Hausa; Djerma	English (official)	Norwegian (official); Lappish dialect	Arabic (official); English; Fars
Literacy (% of population over age 15)	5	25	100	5
Land use (% of total area):				
1. cultivated	12	27	2	0
2. permanent meadow and pasture	2	22	0	5
3. forested	9	34	26	0
4. built on, waste, and other	77	17	72	95
Total cultivated area ('000 hectares)	15,000	25,300	793	36
Major crops (% of cultivated area)	millet 15; dry beans 7; sorghum 4; peanuts 3	sorghum 22; millet 20; corn 6; cacao 3	barley 21; oats 13; potatoes 4	wheat 8; dates; alfalfa; coconuts
Annual timber production (millions of cubic meters)	2.3	58	8.4	
Annual fish catch ('000 metric tons)	12	611	2,937	
Percent of GDP from: farming, forestry, and fishing / manufacturing (and mining)	51 / 7	35 / 30	6 / 25	
Percent of labor force in: farming, forestry, and fishing / manufacturing (and mining)	90 / —	70 / 10	10 / 25	— / 13
Major industrial products ('000 metric tons)	textiles; cement; farm products	PP 2,001; cement 1,206; textiles	processed foods; pulp and paper; chemicals; metal products	cement 626
Major mineral products ('000 metric tons; natural gas in millions of cubic meters)	uranium 1.2	petroleum 102,644; hard coal 304; tin concentrates 5.4; iron ore	iron ore 4,752; petroleum 1,706; pyrites 659; coal 461	petroleum 17,040
Number of passenger cars in use (per 1,000 inhabitants)	2	2	222	
Number of telephones in use (per 100 inhabitants)	0.1	0.2	34	
Total value of: imports / exports (millions of U.S. $)	72 / 47	1,732 / 3,591	5,352 / 3,844	106 / 239
Major imports (% of total value)	MTE 21; PP 10; cotton fabrics 7	MTE 31; iron and steel 12	MTE 34; fuels 12; chemicals 9	MTE 40; BM 22; FA 11
Major exports (% of total value)	uranium 43; peanuts 23; animals 18	petroleum 93; cocoa 3	ships 18; PPP 13; base metals 13; pulp and paper 10; iron and steel 7	petroleum 99.8
Per capita foreign trade (U.S. $)	26	85	2,289	460
Per capita calories consumed per day	2,180	2,290	2,960	
Per capita energy consumption (kg)	29	73	4,845	234
Per capita national income (U.S.$)	88	210	6,150	355

	Pakistan	Panama	Papua–New Guinea	Paraguay
Total area: m² / km²	310,000 / 803,000	29,208 / 75,650	183,540 / 475,369	157,000 / 406,630
Total population:				
Latest official figure	70,260,000	1,667,000	2,760,000	2,572,000
1977 estimate	73,452,000	1,745,000	2,865,000	2,764,000
Annual rate of population increase (%)	3.3	3.3	1.6	3.4
Infant mortality rate / Life expectancy	132 / 51	38 / 59	159 / 47	39 / 59
Urban population (%)	27	49	11	39
Population of major cities (capitals are capitalized)	ISLAMABAD 250,000 Karachi 3,469,000 Lahore 2,148,000 Lyallpur 1,109,000	PANAMA 455,000	PORT MORESBY 91,000	ASUNCIÓN 550,000
Major ethnic groups (% of total population)	Punjabi; Sindhi; Pushtun; Baluchi	Mestizo 70; Antillean Negro 13; Caucasian 10; Indian 6	Papua–New Guinean predominant	Mestizo 95; White and Indian 5
Major religious groups (% of total population)	Muslim 96	Roman Catholic 93; Protestant 6	Christian (nominal) 60; Animist 40	Roman Catholic 97
Major languages (% of total population)	Urdu (official) 7; Punjabi 64; English	Spanish (official); English	750 local languages	Spanish; Guarani
Literacy (% of population over age 15)	17	78	13	74
Land use (% of total area):				
1. cultivated	24	7	1	2
2. permanent meadow and pasture	6	15	0	37
3. forested	3	55	79	51
4. built on, waste, and other	67	23	20	10
Total cultivated area ('000 hectares)	19,450	555	349	970
Major crops (% of cultivated area)	wheat 30; cotton 19; rice 8; sugar cane 3	bananas 38; rice 20; corn 14; sugar cane 6; coffee 4	coconut 30; cacao 16; rubber 4	corn 29; cotton 12; manioc 8; wheat 3
Annual timber production (millions of cubic meters)	12.4	1.5	5.3	3.5
Annual fish catch ('000 metric tons)	180	56	37	
Percent of GDP from: farming, forestry, and fishing / manufacturing (and mining)	34 / 16	23 / 17		34 / 18
Percent of labor force in: farming, forestry, and fishing / manufacturing (and mining)	57 / 12	38 / 8		49 / 14
Major industrial products ('000 metric tons)	cement 3,261; sugar 496; chemicals 278; textiles		timber products; processed foods	cement 102; sugar 70; wood products
Major mineral products ('000 metric tons; natural gas in millions of cubic meters)	natural gas 4,400; coal 1,160; petroleum 428	copper; salt	copper 0.5; gold	
Number of passenger cars in use (per 1,000 inhabitants)	3	38	7	63
Number of telephones in use (per 100 inhabitants)	0.3	7		1
Total value of: imports / exports (millions of U.S. $)	1,098 / 830	498 / 139	320 / 330	115 / 102
Major imports (% of total value)	capital goods; raw materials; foodstuffs	petroleum 37; BM 20; MTE 19	MTE 27; BM 17	MTE 34; fuels 31
Major exports (% of total value)	rice; raw cotton; textiles	PP 42; bananas 24; sugar 13	copper 55; cocoa 9; coffee 8; coconut 7	meat products 21; timber 14; oil seeds 12; cotton fiber 10
Per capita foreign trade (U.S. $)	27	382	236	84
Per capita calories consumed per day	2,310	2,580		2,580
Per capita energy consumption (kg)	148	821	197	
Per capita national income (U.S.$)	149	935	310	457

	Peru	Philippines	Poland	Portugal
Total area: m^2 / km^2	496,000 / 1,284,640	116,000 / 300,440	120,600 / 312,354	36,400 / 94,276
Total population:				
Latest official figure	15,870,000	42,760,000	34,020,000	9,449,000
1977 estimate	16,027,000	43,705,000	34,566,000	9,587,000
Annual rate of population increase (%)	3.2	3.0	0.9	0.2
Infant mortality rate / Life expectancy	73 / 54	78 / 58	45 / 70	50 / 68
Urban population (%)	55	32	53	23
Population of major cities (capitals are capitalized)	LIMA 3,158,000 Arequipa 304,000 Callao 296,000	QUEZON CITY 960,000 Manila 4,500,000 Davao 592,000	WARSAW 1,436,000 Lodz 798,300 Krakow 684,000 Wroclaw 575,000	LISBON 1,640,000 Oporto 1,315,000
Major ethnic groups (% of total population)	Indian 46; Caucasian 11; Mestizo 43	Malay 95; Chinese 2	Polish 98	Portuguese predominant
Major religious groups (% of total population)	Roman Catholic 90	Catholic 83; Protestant 9; Muslim 5	Roman Catholic 95	Roman Catholic 97
Major languages (% of total population)	Spanish (official)	Tagalog (official); English; Spanish	Polish nearly 100	Portuguese
Literacy (% of population over age 15)	61	83	98	65
Land use (% of total area):				
1. cultivated	2	34	49	41
2. permanent meadow and pasture	21	2	13	6
3. forested	58	42	27	41
4. built on, waste, and other	19	22	11	12
Total cultivated area ('000 hectares)	2,880	10,308	15,078	3,640
Major crops (% of cultivated area)	corn 13; potatoes 10; barley 7; wheat 5; cotton 5; sugar	rice 33; corn 27; copra 21; sugar 5	rye 19; potatoes 17; wheat 12; oats 9	wheat 13; corn 10; grapes; truck crops
Annual timber production (millions of cubic meters)	6	34	20	6.6
Annual fish catch ('000 metric tons)	7,906	1,080	504	462
Percent of GDP from: farming, forestry, and fishing / manufacturing (and mining)	16 / 30	28 / 20	19 / 51	14 / 34
Percent of labor force in: farming, forestry, and fishing / manufacturing (and mining)	41 / 14	51 / 12	35 / 30	30 / 22
Major industrial products ('000 metric tons)	PP 4,687; cement 1,428	cement 3,485; sugar 2,446; processed foods	iron and steel 34,017; PP 11,670; chemicals; textiles	cement 3,432; fertilizers 1,201; steel 743
Major mineral products ('000 metric tons; natural gas in millions of cubic meters)	iron ore 5,648; petroleum 3,482; zinc 413; copper 220	iron ore 1,351; chrome 520; copper 226; coal 105	coal 171,625; sulphur 39,865; natural gas 5,963; iron ore 1,192; petroleum 552	iron pyrites 462; anthracite 221
Number of passenger cars in use (per 1,000 inhabitants)	146	8	27	90
Number of telephones in use (per 100 inhabitants)	2	1	7	12
Total value of: imports / exports (millions of U.S. $)	944 / 1,088	1,815 / 1,604	6,255 / 3,534	2,679 / 1,492
Major imports (% of total value)	MTE 29; BM 11; chemicals 10	MTE 32; fuels 22	metals 18; MTE 13; chemicals 11	foodstuffs; electronic products; machinery
Major exports (% of total value)	copper 20; fish products 13; zinc 9; sugar 9	sugar 25; coconut products 19; copper concentrates 9; timber 8	MTE 17; coal 16; chemicals 9	textiles and clothing 28; metals and ores 12; MTE 10
Per capita foreign trade (U.S. $)	128	80	288	441
Per capita calories consumed per day	2,320	2,260	3,200	2,900
Per capita energy consumption (kg)	623	311	4,505	977
Per capita national income (U.S.$)	444	319	1,650	1,469

	Qatar	Rhodesia (Zimbabwe)	Rumania	Rwanda
Total area: m² / km²	4,000 / 10,360	151,000 / 391,090	91,700 / 237,503	10,000 / 25,900
Total population:				
Latest official figure	180,000	6,420,000	21,245,000	4,200,000
1977 estimate	190,000	6,648,000	21,555,000	4,397,000
Annual rate of population increase (%)	3.0	3.5	0.9	2.9
Infant mortality rate / Life expectancy	138 / 47	122 / 51	38 / 68	133 / 41
Urban population (%)	73	20	42	1
Population of major cities (capitals are capitalized)	DOHA 95,000	SALISBURY 569,000 Bulawayo 340,000	BUCHAREST 1,706,000 Cluj 208,100 Timisoara 200,000 Iasi 194,000	KIGALI 54,000
Major ethnic groups (% of total population)	Arab 56; Iranian 37; Pakistani 7	Africans 96; Europeans 3	Rumanian 88; Magyar 8.5; German 2	Hutu 80; Tutsi 10
Major religious groups (% of total population)	Muslim 98	Syncretic 51; Christian 24; Animist 24	Rumanian Orthodox 80; Roman Catholic 5	Christian 60; Animist 39; Muslim 1
Major languages (% of total population)	Arabic (official)	English (official)	Rumanian 85; Hungarian; German	French (official); Kinyarwanda (official); Kiswahili
Literacy (% of population over age 15)	20	30	98	10
Land use (% of total area):				
1. cultivated	0	6	44	33
2. permanent meadow and pasture	5	12	19	31
3. forested	0	61	27	12
4. built on, waste, and other	95	21	10	24
Total cultivated area ('000 hectares)	2	2,480	10,469	881
Major crops (% of cultivated area)	truck crops	corn 19; millet 16; tobacco; cotton; sugar; wheat	corn 28; wheat 23; sunflowers 5; barley 4	cassava 4; coffee 3; potatoes 2; bananas
Annual timber production (millions of cubic meters)		5.7	22.4	3.8
Annual fish catch ('000 metric tons)			72	1.3
Percent of GDP from: farming, forestry, and fishing / manufacturing (and mining)		14 / 32	22 / 57	
Percent of labor force in: farming, forestry, and fishing / manufacturing (and mining)	30 / —		42 / 28	95 / 5
Major industrial products ('000 metric tons)		MTE; ferrochrome; textiles; processed foods	iron and steel 14,921; cement 11,195; PP 8,882; fertilizers 1,410	processed foods
Major mineral products ('000 metric tons; natural gas in millions of cubic meters)	petroleum 25,050	coal 3,060; iron ore 320; chromium ore 272	coal 29,207; natural gas 24,217; petroleum 14,486; iron ore 3,265	tin 1.4; natural gas 1; tungsten 0.4
Number of passenger cars in use (per 1,000 inhabitants)		28		1
Number of telephones in use (per 100 inhabitants)		3	5	0.1
Total value of: imports / exports (millions of U.S. $)	200 / 1,500	421 / 474	3,058 / 3,025	37 / 27
Major imports (% of total value)	consumer goods 65; industrial equipment 35	MTE; PPP	MTE 13; iron ore and steel	MTE 23; PP 7; chemicals 7
Major exports (% of total value)	petroleum 98	tobacco; chrome; asbestos	MTE 8; foodstuffs; fuels; BM	coffee 62; tin 16; tea 7; tungsten 5
Per capita foreign trade (U.S. $)	9,444	139	286	15
Per capita calories consumed per day		2,660		1,960
Per capita energy consumption (kg)	19,330	723	3,311	12
Per capita national income (U.S.$)	4,000	750	1,630	68

	Saudi Arabia	Senegal	Sierra Leone	Singapore
Total area: m² / km²	900,000 / 2,331,000	76,000 / 196,840	27,900 / 72,261	225 / 583
Total population:				
Latest official figure	8,966,000	4,140,000	2,710,000	2,250,000
1977 estimate	7,517,000	5,147,000	2,810,000	2,297,000
Annual rate of population increase (%)	3	2.4	1.5	1.7
Infant mortality rate / Life expectancy	152 / 45	159 / 40	136 / 44	20 / 67
Urban population (%)	14	26	14	98
Population of major cities (capitals are capitalized)	RIYADH 500,000 Jidda 400,000 Mecca 250,000	DAKAR 581,000 Kaolack 96,000	FREETOWN 274,000	SINGAPORE 2,250,000
Major ethnic groups (% of total population)	Arab 90; Afro-Asian 10	Wolof 36; Peuhl 18; Serere 17	Temne 30; Mende 30	Chinese 74; Malay 14; Indian and Pakistani 8
Major religious groups (% of total population)	Sunni Muslim nearly 100	Muslim 80; Animist 15; Christian 5	Animist 70; Muslim 25; Christian 5	Buddhist; Taoist; Muslim; Hindu; Christian
Major languages (% of total population)	Arabic	French (official); Wolof; Peuhl	English (official); Mende; Temne	Malay; Chinese dialects; Tamil; Hindi; English
Literacy (% of population over age 15)	25	10	10	22
Land use (% of total area):				
1. cultivated	0	12	52	16
2. permanent meadow and pasture	40	29	31	0
3. forested	1	27	4	7
4. built on, waste, and other	59	32	13	77
Total cultivated area ('000 hectares)	775	2,300	3,740	9
Major crops (% of cultivated area)	sorghum 18; wheat 17; millet 17	peanuts 48; millet 46; cotton 2	rice 10; oil palm; cacao; coffee; cassava	fruits 29; rubber 26; coconuts 21
Annual timber production (millions of cubic meters)		2.5	2.7	
Annual fish catch ('000 metric tons)	30	241	38	17
Percent of GDP from: farming, forestry, and fishing / manufacturing (and mining)	4 / 68			2 / 27
Percent of labor force in: farming, forestry, and fishing / manufacturing (and mining)	75 / 10	74 / —	75 / 10	3 / 28
Major industrial products ('000 metric tons)	PP 31,516; cement 964	PP 714; cement 296; peanut oil; textiles	beverages; textiles	rubber products 1,004,994; PP
Major mineral products ('000 metric tons; natural gas in millions of cubic meters)	petroleum 407,900; natural gas 3,200	phosphate 1,877; salt 150	diamonds 1,670,000 m. c.; iron ore 1,982; bauxite 661	
Number of passenger cars in use (per 1,000 inhabitants)	7	11	8	86
Number of telephones in use (per 100 inhabitants)	1	1	0.3	13
Total value of: imports / exports (millions of U.S. $)	1,614 / 10,387	300 / 210	146 / 149	4,439 / 2,993
Major imports (% of total value)	MTE 32; foodstuffs 24	MTE 23; PPP 11	BM 23; machinery 20; foodstuffs 17	MTE 26; fuels 24; BM 18
Major exports (% of total value)	petroleum 99	peanut products 31; phosphate products 31	diamonds 61; foodstuffs 15; iron ore 10	MTE 23; fuels 27
Per capita foreign trade (U.S. $)	1,339	123	109	3,303
Per capita calories consumed per day	2,270	2,300	2,240	2,430
Per capita energy consumption (kg)	918	159	115	1,839
Per capita national income (U.S.$)	5,530	235	170	2,376

	Somalia	South Africa, Republic of	Spain	Sri Lanka (Ceylon)
Total area: m² / km²	246,000 / 637,140	472,000 / 1,222,480	195,000 / 505,050	25,290 / 65,500
Total population:				
Latest official figure	2,941,000	25,470,000	35,470,000	13,874,000
1977 estimate	3,265,000	26,493,000	36,161,000	14,081,000
Annual rate of population increase (%)	2.6	2.6	1.1	2.3
Infant mortality rate / Life expectancy	177 / 41	117 / 52	19 / 69	45 / 66
Urban population (%)	20	51	43	22
Population of major cities (capitals are capitalized)	MOGADISHU 400,000	CAPE TOWN 1,100,000 PRETORIA 570,000 BLOEMFONTEIN 181,000 Johannesburg 2,550,000 Durban 1,040,000	MADRID 3,580,000 Barcelona 3,165,000 Valencia 700,000	COLOMBO 2,700,000 Jaffna 106,856
Major ethnic groups (% of total population)	Somali 99	African 70; Caucasian 18; Cape Colored 10	Mediterranean mixed	Sinhalese 71; Tamil 21; Moor 2
Major religious groups (% of total population)	Muslim 99	Animist 40; Christian 15	Roman Catholic 99	Buddhist 64; Hindu 20; Christian 9; Muslim 6
Major languages (% of total population)	Somali; Arabic; English; Italian	English (official); Afrikaans (official)	Spanish (official); Catalan 17; Galician 7; Basque 2	Sinhala (official) 70; Tamil 20; English 10
Literacy (% of population over age 15)	5	55	97	90
Land use (% of total area):				
1. cultivated	2	12	41	30
2. permanent meadow and pasture	45	67	22	7
3. forested	14	4	29	44
4. built on, waste, and other	39	17	8	19
Total cultivated area ('000 hectares)	1,050	14,470	20,885	1,979
Major crops (% of cultivated area)	sorghum 24; corn 16	corn 31; wheat 10; sugar cane; tobacco	wheat 15; barley 14; citrus fruits	rice 32; tea 12; rubber 12; coconuts
Annual timber production (millions of cubic meters)	3	10	15	4.7
Annual fish catch ('000 metric tons)	5	1,406	1,562	104
Percent of GDP from: farming, forestry, and fishing / manufacturing (and mining)		10 / 36	12 / 28	32 / 11
Percent of labor force in: farming, forestry, and fishing / manufacturing (and mining)	60 / 3	28 / 21	23 / 27	41 / 8
Major industrial products ('000 metric tons)	canned foods 3,956; sugar 50; textiles	MTE; chemicals; textiles; processed foods	cement 21,750; iron and steel 17,960; textiles; petrochemicals	cement 460; sugar 10; textiles; chemicals
Major mineral products ('000 metric tons; natural gas in millions of cubic meters)		coal 65,018; iron ore 11,734; chrome 1,877; gold 0.7; diamonds 7,400,000 c.	coal 14,030; iron ore 11,475; potash 4,704; copper 3,757	salt 156; graphite 7
Number of passenger cars in use (per 1,000 inhabitants)	3	76	121	6
Number of telephones in use (per 100 inhabitants)	0.2	8	20	0.5
Total value of: imports / exports (millions of U.S. $)	72 / 39	4,692 / 3,088	8,247 / 4,270	434 / 378
Major imports (% of total value)	BM 25; MTE 24; FA 21	MTE 52; BM 25	fuels 26; MTE 22; BM 11; FA 11	rice 43; PP 20; flour 16
Major exports (% of total value)	FA 90	BM 26; foodstuffs 21; precious stones 20	MTE 26; BM 26; FA 17	tea 39; rubber 21; coconut products 12
Per capita foreign trade (U.S. $)	38	305	353	59
Per capita calories consumed per day	1,830	2,740	2,600	2,220
Per capita energy consumption (kg)	36		1,809	143
Per capita national income (U.S.$)	65	1,125	2,234	223

	Sudan	Surinam	Swaziland	Sweden
Total area: m² / km²	967,000 / 2,504,530	55,100 / 142,709	6,700 / 17,353	173,000 / 448,070
Total population:				
Latest official figure	17,760,000	405,000	494,000	8,209,000
1977 estimate	18,428,000	431,000	518,000	8,326,000
Annual rate of population increase (%)	2.5	2.6	3.1	0.4
Infant mortality rate / Life expectancy	141 / 49	141 / 57	149 / 44	11 / 74
Urban population (%)	13	13	5	81
Population of major cities (capitals are capitalized)	KHARTOUM 321,666 Omdurman 305,308 Port Sudan 132,631	PARAMARIBO 151,000	MBABANE 24,000	STOCKHOLM 1,300,000 Göteborg 686,000 Malmö 451,000
Major ethnic groups (% of total population)	African 52; Arab 39	Hindu 37; Creole 31; Javanese 15	African 96; European 3	Swedish nearly 100
Major religious groups (% of total population)	Sunni Muslim 73; Animist 23; Christian 4	Muslim; Hindu; Moravian	Christian 57; Animist 43	State Lutheran 95
Major languages (% of total population)	Arabic (official); English	Dutch (official); English	Swati (official); English; Zulu	Swedish
Literacy (% of population over age 15)	7	80	36	100
Land use (% of total area):				
1. cultivated	3	1	10	7
2. permanent meadow and pasture	10	0	77	2
3. forested	36	89	6	50
4. built on, waste, and other	51	10	7	41
Total cultivated area ('000 hectares)	7,195	45	175	3,015
Major crops (% of cultivated area)	sorghum 35; millet 17; peanuts 12; sesame seeds 12; cotton 7	rice 92; sugar cane 7; coffee	corn 46; sugar cane 10; cotton 7	barley 20; oats 15; wheat 10; oleiferous plants 5
Annual timber production (millions of cubic meters)	21	0.2	1.6	58
Annual fish catch ('000 metric tons)	22.3	2.8		253
Percent of GDP from: farming, forestry, and fishing / manufacturing (and mining)	35 / 12	13 / 26	33 / 22	4 / 28
Percent of labor force in: farming, forestry, and fishing / manufacturing (and mining)	86 / 6	25 / 16	85 / 11	7 / 29
Major industrial products ('000 metric tons)	PP 1,195; cement 200; sugar 93; pharmaceuticals	aluminum 57; sugar 9		iron and steel 13,029; pulp and paper 10,873; aluminum 77; machinery
Major mineral products ('000 metric tons; natural gas in millions of cubic meters)	petroleum 46; chromium ore 16; iron ore 10	bauxite 6,863; alumina 1,186; natural gas 4	iron ore 2,076; coal 116; asbestos 38	iron ore 31,281; zinc ore 197; silver and lead ore 100
Number of passenger cars in use (per 1,000 inhabitants)	2	53	13	321
Number of telephones in use (per 100 inhabitants)	0.3	3	1	63
Total value of: imports / exports (millions of U.S. $)	428 / 350	128 / 153	116 / 147	9,855 / 10,232
Major imports (% of total value)	MTE 16; textiles 15; sugar 12; iron and steel 5	BM 29; MTE 27; fuels 12; foodstuffs 12	MTE 24; BM 14; fuels 12	MTE 31; BM 20; PPP 16
Major exports (% of total value)	cotton 45; peanuts 23; sesame seeds 8; gum arabic 5	alumina 46; bauxite 27; aluminum 14	sugar 39; wood pulp 26; iron ore 10; asbestos 5	MTE 43; BM 26; raw materials 15
Per capita foreign trade (U.S. $)	44	694	532	2,447
Per capita calories consumed per day	2,160	2,450		2,810
Per capita energy consumption (kg)	125	2,590	250 kwh	6,122
Per capita national income (U.S.$)	143	970	190	6,150

	Switzerland	Syria	Tanzania	Thailand
Total area: m² / km²	16,000 / 41,440	72,000 / 186,480	362,800 / 939,652	198,000 / 512,820
Total population:				
Latest official figure	6,400,000	7,354,000	15,160,000	41,280,000
1977 estimate	6,415,000	7,722,000	15,772,000	43,736,000
Annual rate of population increase (%)	0.7	3.3	2.7	3.2
Infant mortality rate / Life expectancy	14 / 75	93 / 53	162 / 44	65 / 61
Urban population (%)	58	44	6	13
Population of major cities (capitals are capitalized)	BERN 285,800 Zurich 714,000 Basel 375,000 Geneva 323,000	DAMASCUS 923,000 Aleppo 639,000	DAR ES SALAAM 344,000 Dodoma 23,500	BANGKOK 5,000,000 Chiang Mai 100,000
Major ethnic groups (% of total population)	German 69; French 19; Italian 10	Arab 90; Kurd; Armenian	African 99	Thai 75; Chinese 14
Major religious groups (% of total population)	Protestant 53; Roman Catholic 45	Sunni Muslim 71; other Muslim sects 13; Christian 13	Animist 40; Christian 30; Muslim 30	Buddhist 96; Muslim 4
Major languages (% of total population)	German 69; French 19; Italian 10	Arabic (official); French	Swahili (official); English (official)	Thai (official); English
Literacy (% of population over age 15)	98	50	20	70
Land use (% of total area):				
1. cultivated	9	33	13	27
2. permanent meadow and pasture	39	35	42	1
3. forested	26	2	33	40
4. built on, waste, and other	26	30	12	32
Total cultivated area ('000 hectares)	384	6,027	12,330	14,100
Major crops (% of cultivated area)	wheat 22; barley 11; potatoes 6; truck crops	wheat 28; barley 17; cotton 3; grapes 3	cassava 6; corn 6; cotton 3; coffee; sisal	rice 62; corn 9; rubber; sugar
Annual timber production (millions of cubic meters)	4		32	19
Annual fish catch ('000 metric tons)			168	1,535
Percent of GDP from: farming, forestry, and fishing / manufacturing (and mining)		18 / 20	36 / 12	31 / 20
Percent of labor force in: farming, forestry, and fishing / manufacturing (and mining)	8 / 38	49 / 13	90 / —	72 / 8
Major industrial products ('000 metric tons)	aluminum 79; chocolate 58; watches and clocks; machinery; chemicals	cement 1,080; textiles 273; sugar 140	sisal twine 29; textiles; processed foods	PP 7,848; sugar 839; crude steel 40; textiles
Major mineral products ('000 metric tons; natural gas in millions of cubic meters)	natural gas 122	petroleum 9,144; natural gas	diamonds 100,000 grams	brown coal 361; iron ore 21; tin concentrates 21; petroleum 6
Number of passenger cars in use (per 1,000 inhabitants)	269	5	3	6
Number of telephones in use (per 100 inhabitants)	59	2	0.4	1
Total value of: imports / exports (millions of U.S. $)	9,637 / 7,847	637 / 366	436 / 305	1,851 / 1,328
Major imports (% of total value)	BM 23; MTE 23; chemicals 12; PPP 10	MTE 16; foodstuffs 16; fuels 6	MTE 34; petroleum 17; foodstuffs 16; iron and steel 5	machinery 32; BM 21; chemicals 16; fuels 10
Major exports (% of total value)	MTE 32; chemicals 27; BM 18	petroleum 76; cotton textiles 15	cotton 19; sisal 18; coffee 15; cashew nuts 8; diamonds 4	rice 12; sugar 12; corn 12; tin 7; rubber
Per capita foreign trade (U.S. $)	2,732	136	49	77
Per capita calories consumed per day	3,190	2,650	2,260	2,220
Per capita energy consumption (kg)	3,748	523	63	296
Per capita national income (U.S.$)	7,253	300	144	303

	Togo	Trinidad and Tobago	Tunisia	Turkey
Total area: m² / km²	22,000 / 56,980	1,980 / 5,128	63,400 / 164,206	296,000 / 766,640
Total population:				
Latest official figure	2,220,000	1,070,000	5,770,000	39,180,000
1977 estimate	2,304,000	1,033,000	5,975,000	41,516,000
Annual rate of population increase (%)	2.6	1.5	2.4	2.4
Infant mortality rate / Life expectancy	179 / 40	26 / 66	128	119 / —
Urban population (%)	13	50	44	39
Population of major cities (capitals are capitalized)	LOMÉ 200,000	PORT OF SPAIN 310,000 San Fernando 50,000	TUNIS 469,000	ANKARA 1,554,000 Istanbul 3,136,000 Izmir 1,660,000
Major ethnic groups (% of total population)	African 99	Negro 43; East Indian 40; Mixed 14	Arab 98; European 1	Turk 90; Kurd 7
Major religious groups (% of total population)	Animist 75; Christian 20; Muslim 5	Roman Catholic 31; Protestant 27; Hindu 23; Muslim 6	Muslim 95; Christian 4	Muslim 98
Major languages (% of total population)	French (official); many local dialects	English	Arabic (official); French	Turkish; Kurdish; Arabic
Literacy (% of population over age 15)	10	78	32	55
Land use (% of total area):				
1. cultivated	41	31	27	36
2. permanent meadow and pasture	4	2	20	35
3. forested	6	44	3	26
4. built on, waste, and other	49	23	50	3
Total cultivated area ('000 hectares)	2,285	157	4,360	27,895
Major crops (% of cultivated area)	cacao 13; millet 9; corn 6; cassava 5	sugar cane 18; cocoa 13; coffee 6; coconuts	wheat 24; barley 9; truck crops; olives	wheat 31; barley 9; cotton 3; tobacco
Annual timber production (millions of cubic meters)	1.2		1.7	17.7
Annual fish catch ('000 metric tons)	10		28	174
Percent of GDP from: farming, forestry, and fishing / manufacturing (and mining)	43 / 19	8 / 44	19 / 16	26 / 23
Percent of labor force in: farming, forestry, and fishing / manufacturing (and mining)	78 / 22	16 / 19	41 / 12	69 / 9
Major industrial products ('000 metric tons)	palm oil 242; textiles	cement 249; PP; processed foods	PP 1,006; cement 540; superphosphates 430	PP 10,815; iron and steel 2,647
Major mineral products ('000 metric tons; natural gas in millions of cubic meters)	phosphate rock 2,552	petroleum 9,253; asphalt 107	petroleum 4,146; phosphate 3,827; iron ore 818; natural gas 201	coal 10,485; petroleum 3,209; iron ore 2,245; chromite 694
Number of passenger cars in use (per 1,000 inhabitants)	6	83	17	8
Number of telephones in use (per 100 inhabitants)	0.3	6	2	2
Total value of: imports / exports (millions of U.S. $)	86 / 81	924 / 853	567 / 402	1,852 / 1,000
Major imports (% of total value)	MTE 21; cotton fabrics 12; PP 9; chemicals 8	fuels 72; BM 8; FA 7; MTE 6	MTE 22; PP 10; iron and steel 5; sugar 5	MTE 28; minerals 22; base metals 18; chemicals 12
Major exports (% of total value)	phosphates 76; cocoa beans 12; coffee 4	PP 90; FA 4; chemicals 3	petroleum 34; phosphate products 24; olive oil 18	cotton 22; fruits and nuts 19; tobacco 13
Per capita foreign trade (U.S. $)	75	1,661	168	73
Per capita calories consumed per day	2,160	2,380	2,250	3,250
Per capita energy consumption (kg)	75	4,245	326	564
Per capita national income (U.S.$)	170	1,380	569	893

	Uganda	U.S.S.R.	United Arab Emirates	United Kingdom
Total area: m² / km²	91,000 / 235,690	8,600,000 / 22,274,000	32,000 / 82,880	94,200 / 243,978
Total population:				
Latest official figure	11,550,000	254,380,000	656,000	55,962,000
1977 estimate	12,148,000	258,106,000	688,000	56,043,000
Annual rate of population increase (%)	3.3	0.9	3.1	0.3
Infant mortality rate / Life expectancy	160 / 47	28 / 70	138 / 47	17.5 / 72
Urban population (%)	10	59	57	78
Population of major cities (capitals are capitalized)	KAMPALA 332,000	MOSCOW 7,734,000; Leningrad 4,372,000; Kiev 2,013,000; Tashkent 1,643,000; Baku 1,403,000	ABU DHABI 85,000; Dubayy 66,000	LONDON 9,618,000; Manchester 2,900,000; Birmingham 2,893,000; Glasgow 1,865,000; Liverpool 1,715,000
Major ethnic groups (% of total population)	African 99	Russian 53; Ukrainian 17; Uzbek 4; Belorussian 4	Arab 72; Iranian; Pakistani; Indian	English 83; Scottish 9; Welsh 5; Irish 3
Major religious groups (% of total population)	Christian 50; Muslim 6	Atheist 70; Russian Orthodox 18; Muslim 9	Muslim 96	Church of England 50; Roman Catholic 10; Presbyterian 4
Major languages (% of total population)	English (official); Swahili; Luganda	Russian (official); Slavic 76; Altaic 11	Arabic	English
Literacy (% of population over age 15)	30	99	25	99
Land use (% of total area):				
1. cultivated	22	10	0	29
2. permanent meadow and pasture	21	17	2	47
3. forested	12	41	0	8
4. built on, waste, and other	45	32	98	16
Total cultivated area ('000 hectares)	5,130	4,901	13	7,154
Major crops (% of cultivated area)	corn 11; cotton 11; cassava 7; coffee 5	wheat 27; barley 14; oats 5; rye 3	truck crops	barley 34; wheat 15; potatoes 3; sugar beets 3
Annual timber production (millions of cubic meters)	14	382		2.4
Annual fish catch ('000 metric tons)	145	7,493		1,103
Percent of GDP from: farming, forestry, and fishing / manufacturing (and mining)	48 / 10	20 / 51		3 / 31
Percent of labor force in: farming, forestry, and fishing / manufacturing (and mining)	90 / 4	27 / 73		2.5 / 34
Major industrial products ('000 metric tons)	cement 153; raw sugar 40; fertilizers 14	iron and steel 244,000; paper 8,581; MTE; chemicals		PP 83,422; iron and steel 32,329; MTE
Major mineral products ('000 metric tons; natural gas in millions of cubic meters)	copper 14; phosphate rock 23	petroleum 491; hard coal 485; natural gas 289; iron ore 233; nickel 152	petroleum 82,300; natural gas 1,982	hard coal 128,626; natural gas 36,365; iron ore 4,490; petroleum 1,223
Number of passenger cars in use (per 1,000 inhabitants)	3			252
Number of telephones in use (per 100 inhabitants)	0.4	6	10	37
Total value of: imports / exports (millions of U.S. $)	133 / 272	17,252 / 18,167	1,912 / 4,275	33,274 / 27,046
Major imports (% of total value)	MTE 40; metals 11; paper products 4	MTE 34; BM; foodstuffs	MTE 52; BM 23; foodstuffs 7	MTE 19; BM 19; PPP 17; FA 16
Major exports (% of total value)	coffee 59; cotton 16; copper 5	MTE 19; PPP; BM	petroleum 96	MTE 41; BM 21; chemicals 11
Per capita foreign trade (U.S. $)	35	139	9,431	1,078
Per capita calories consumed per day	2,130			3,190
Per capita energy consumption (kg)	66	4,842	10,615	5,440
Per capita national income (U.S.$)	100	2,800	16,600	3,072

	United States	Upper Volta	Uruguay	Venezuela
Total area: m² / km²	3,615,207 / 9,363,396	106,000 / 274,540	72,200 / 186,998	352,000 / 911,680
Total population:				
Latest official figure	214,500,000	5,541,000	2,764,000	11,993,000
1977 estimate	215,966,000	6,300,000	2,788,000	12,366,000
Annual rate of population increase (%)	0.8	2.3	1.2	3.1
Infant mortality rate / Life expectancy	18 / 71	182 / 38	40 / 70	50
Urban population (%)	74	4	80	76
Population of major cities (capitals are capitalized)	WASHINGTON 3,190,000 New York 11,400,200 Los Angeles 6,986,000 Chicago 6,971,000 Philadelphia 4,810,000 Detroit 4,434,000	OUAGADOUGOU 115,000	MONTEVIDEO 1,229,748	CARACAS 2,400,000 Maracaibo 900,000
Major ethnic groups (% of total population)	White 87.2; Black 11.3; Others 1.5	over 50 African tribes	European 85–90; Mestizo 5–10	Mestizo 67; European 21; Negro 10
Major religious groups (% of total population)	Protestant 33; Roman Catholic 23; Jewish 3	Animist 75; Muslim 20; Christian 5	Catholic 66; Agnostic 30	Roman Catholic 96
Major languages (% of total population)	English nearly 100	French (official) Sudanic dialects 50	Spanish	Spanish
Literacy (% of population over age 15)	99	10	91	78
Land use (% of total area):				
1. cultivated	19	20	11	6
2. permanent meadow and pasture	27	50	77	19
3. forested	32	15	3	52
4. built on, waste, and other	22	15	9	23
Total cultivated area ('000 hectares)	206,921	5,377	1,862	5,179
Major crops (% of cultivated area)	wheat 21; corn 21; hay 19; soybeans 17; cotton 8	sorghum 22; millet 16; groundnuts 3	wheat 24; corn 8	corn 11; coffee 6; cacao; sugar; cotton
Annual timber production (millions of cubic meters)	341	4.3		7.2
Annual fish catch ('000 metric tons)	2,681	4.2	15.7	143
Percent of GDP from: farming, forestry, and fishing / manufacturing (and mining)	4 / 29	44 / 10	11 / 21	7 / 38
Percent of labor force in: farming, forestry, and fishing / manufacturing (and mining)	4 / 24	80 / 1	8 / 34	20 / 15
Major industrial products ('000 metric tons)	iron and steel 221,477; MTE; BM; chemicals	sugar 13; soap 3; cotton yarn 0.8	crude steel 13; processed foods; textiles	crude steel 1,100; petrochemicals; paper products
Major mineral products ('000 metric tons; natural gas in millions of cubic meters)	coal 535,130; petroleum 492,615; iron ore 77,081; natural gas 9,558	manganese; bauxite; gold		gold 1,249,000 c.; crude oil 111,774; natural gas 46,426; iron ore 26,420; coal 57
Number of passenger cars in use (per 1,000 inhabitants)	486	2	50	50
Number of telephones in use (per 100 inhabitants)	68	0.1	9	5
Total value of: imports / exports (millions of U.S. $)	63,264 / 60,483	69 / 23	289 / 271	2,380 / 4,894
Major imports (% of total value)	BM 27; MTE 24; PPP 24	metal products 52; foodstuffs 22; MTE 18	raw materials 43; PPP 33	BM 65; chemicals 10; foodstuffs 10
Major exports (% of total value)	MTE 39; BM 17; FA 14	livestock 41; cotton 22	meat 38; wool 18; textiles 9; hides 6	petroleum 94; iron ore 3; coffee; cocoa
Per capita foreign trade (U.S. $)	577	17	183	607
Per capita calories consumed per day	3,330	1,940	2,880	2,430
Per capita energy consumption (kg)	11,426	13		2,520
Per capita national income (U.S.$)	6,590	60	618	2,168

	Vietnam	Yemen Arab Republic	Yemen, People's Dem. Rep. of	Yugoslavia
Total area: m² / km²	127,300 / 329,707	75,000 / 194,250	111,000 / 287,490	98,800 / 255,892
Total population:				
Latest official figure	43,734,000	6,472,000	1,630,000	21,352,000
1977 estimate	47,101,000	6,791,000	1,764,000	21,650,000
Annual rate of population increase (%)	2.4	2.9	3.3	0.9
Infant mortality rate / Life expectancy	43 / 48	152 / 42	80 / 45	43 / 67
Urban population (%)	26	6	29	39
Population of major cities (capitals are capitalized)	HANOI 736,000; Ho Chi Minh City (Saigon) 1,825,000; Haiphong 182,490	SAN'A 150,000	ADEN 285,000	BELGRADE 1,209,000; Zagreb 668,000
Major ethnic groups (% of total population)	Vietnamese 90; Chinese 3	Arab 90; Afro-Arab 10	Arab nearly 100	Serb 41; Croat 23; Slovene 8; Macedonian 6; Muslim Slavs 5; Albanian 5; Montenegrin 3; Hungarian 3
Major religious groups (% of total population)	Buddhist 80; Catholic 8; Taoist	Muslim 100	Muslim 98	Serbian Orthodox 41; Roman Catholic 32; Muslim 12
Major languages (% of total population)	Vietnamese	Arabic	Arabic (official)	Serbo-Croatian; Slovene; Macedonian; Albanian
Literacy (% of population over age 15)	60	10	15	83
Land use (% of total area):				
1. cultivated	14	6	1	32
2. permanent meadow and pasture	—	36	32	25
3. forested	50	2	9	35
4. built on, waste, and other	—	56	58	8
Total cultivated area ('000 hectares)	5,500	1,240	334	8,065
Major crops (% of cultivated area)	rice; rubber; corn; sugar cane	millet and sorghum 65; barley 9; cotton 2; coffee 1	millet and sorghum 13; cotton 5; wheat 4	corn 35; wheat 20; potatoes 4
Annual timber production (millions of cubic meters)	18			15.4
Annual fish catch ('000 metric tons)	728		108	47
Percent of GDP from: farming, forestry, and fishing / manufacturing (and mining)	29 / 7	71 / 3	19 / 27	19 / 39
Percent of labor force in: farming, forestry, and fishing / manufacturing (and mining)	70 / 12	73 / —	— / 3	45 / 18
Major industrial products ('000 metric tons)	processed foods; textiles; cement	aluminum products 200,000; cotton textiles	petrochemicals 2,916; clothing	cement 7,066; iron and steel 5,000; petrochemicals
Major mineral products ('000 metric tons; natural gas in millions of cubic meters)			unrefined salt 75	coal 35,537; copper ore 14,576; iron ore 5,239; petroleum 3,692
Number of passenger cars in use (per 1,000 inhabitants)			8	62
Number of telephones in use (per 100 inhabitants)		0.1	1	5
Total value of: imports / exports (millions of U.S. $)	541 / 81	92 / 6	173 / 135	4,340 / 2,516
Major imports (% of total value)	PPP; BM; foodstuffs	foodstuffs 43; consumer goods 27; MTE 14	petroleum 37; BM 29; foodstuffs 16	MTE 34; BM 23; chemicals 11; PPP 10
Major exports (% of total value)		cotton 51; coffee 21; hides and skins 13	PPP 77; textiles 5; cotton 4	BM 29; MTE 28; chemicals 9; FA 9
Per capita foreign trade (U.S. $)	14	15	189	321
Per capita calories consumed per day	2,000	2,040	2,070	3,190
Per capita energy consumption (kg)	200	24	433	1,675
Per capita national income (U.S.$)	120	130	100	1,840

	Zaire	Zambia
Total area: m² / km²	905,000 / 2,343,950	288,000 / 745,920
Total population:		
Latest official figure	24,900,000	4,900,000
1977 estimate	25,960,000	5,099,000
Annual rate of population increase (%)	2.8	3.3
Infant mortality rate / Life expectancy	160 / 43	157 / 44
Urban population (%)	16	24
Population of major cities (capitals are capitalized)	KINSHASA 2,008,000 Kananga 601,000	LUSAKA 415,000
Major ethnic groups (% of total population)	African predominant	African (mainly Bantu) 98
Major religious groups (% of total population)	Christian 51; Animist 45	Animist 82; Christian 17
Major languages (% of total population)	French; English; Swahili; Kikongo; Lingala; Chiluba (all official)	English (official)
Literacy (% of population over age 15)	35	20
Land use (% of total area):		
1. cultivated	3	6
2. permanent meadow and pasture	11	40
3. forested	52	50
4. built on, waste, and other	34	4
Total cultivated area ('000 hectares)	7,750	4,985
Major crops (% of cultivated area)	cassava 14; corn 8; coffee 4; cotton 2	corn 5; tobacco; cotton; peanuts
Annual timber production (millions of cubic meters)	14.3	4.8
Annual fish catch ('000 metric tons)	134.4	40
Percent of GDP from: farming, forestry, and fishing / manufacturing (and mining)	26 / 27	9 / 41
Percent of labor force in: farming, forestry, and fishing / manufacturing (and mining)	80 / —	15 / 7
Major industrial products ('000 metric tons)	PP 689; cement 537; refined copper 322	refined copper 638; processed foods; chemicals; textiles
Major mineral products ('000 metric tons; natural gas in millions of cubic meters)	industrial diamonds 12,000,000 c.; copper 499; manganese ore 309; coal 96; cobalt ore 17	coal 940; copper ore 706; zinc ore 73; lead ore 25
Number of passenger cars in use (per 1,000 inhabitants)	3	18
Number of telephones in use (per 100 inhabitants)	0.1	2
Total value of: imports / exports (millions of U.S. $)	665 / 893	586 / 996
Major imports (% of total value)	consumer goods 29; capital goods 19	MTE 33; BM 26
Major exports (% of total value)	copper 63; foodstuffs 11; cobalt 5; diamonds 4	copper 93; zinc; lead; cobalt
Per capita foreign trade (U.S. $)	63	323
Per capita calories consumed per day	2,060	2,590
Per capita energy consumption (kg)	81	520
Per capita national income (U.S.$)	124	504

Index

A

abaca, 302
Abadan, 375
Abu Dhabi, 371
Accra, 397
Addis Ababa, 364–65
Adelaide, 322
Aden, 371
Adriatic Sea, 155
Afghanistan, 350, 376
Aframomum, 379
Africa, 378–412
Afrikaans, 406
Ainu, 267
Aksai Chin, 242
Alaska, 482, 493
Albania, 156
Alexander the Great, 203
Alexandria, 363
Algeria, 357–59
Algiers, 359
alpaca, 421
Alsace, 122
altiplano, 441–42
aluminum (alumina), 27, 126, 399, 457
Amazon River and basin, 443–47
Amsterdam, 111
Amur River, 168–69, 171
Anatolia, 366
Andes Mountains, 442
Angkor, 310
Angola, 402
animal products, 91, 94, 96, 100, 105, 109, 113, 176, 188–89, 225–26, 274, 316, 324–25, 327, 335, 336. *See also* hides and skins
Ankara, 367
Annamese, 308
Anshan, 243
Antilles, 433
Antwerp, 113
apartheid, 407–08
Apennines, 130, 134
Appalachia, 485–86
Aquitaine, 117
Arabia, 370–71
Arabs, 342–43, 350, 363
Argentina, 423, 447–51
Armenians, 366
Aryans, 201
asbestos, 383, 457
Asoka, 203
Assam, 207–08, 229
Aswan, 361
Atacama, 452
Atlas Mountains, 357
atoll, 328
Australia, 313–24
Austria, 156–57
autarky, 143
Az Zahran (Dhahran), 371
Aztecs, 420–22

B

Baghdad, 370
Balkans, 154–57
Baltic Sea, 101, 105
bananas, 424, 437, 438, 441
Bangkok, 300
Bangladesh, 211, 213–14, 225, 227
Barbary, 358
barley, 100, 102, 120, 175, 273, 285, 286, 339, 347, 357, 366, 367, 370, 371, 372, 374, 376, 412, 437, 443
basic-service ratio, 9
Basuto, Basutoland, 407
bauxite, 218, 316, 399, 435, 441
Bechuanaland, 407
Beirut, 369
Belfast, 95
Belgium, 106, 112–14
Benelux, 114
Bengal, 213–14, 217, 229
Berbers, 358
Bergen, 99
Berlin, 140
Bhakra Dam, 218, 223
Black Sea, 366
Boers, 386, 405–06
Bohemia, 152

Bolivia, 441–43
Bombay, 209, 218
Bordeaux, 117
Bosporus, 367
Boston (Massachusetts), 484
Botswana, 407
Brabant, 106
Brahmaputra River, 207
Brasilia, 447
Bratsk, 172
Brazil, 443–47
Brazzaville, 402
break-in-bulk, 6
Brisbane, 322
Britain, 80, 89–90
Bruges, 106
Brunei, 301
Brussels, 112
Buddhism, 203
Buenos Aires, 447–49
Bulgaria, 155
Burma, 308

C

cabbage, 100, 141, 164, 174
cacao, 74, 328, 381, 382, 384, 395, 396, 435, 441, 444
Cairo, 363–64
Calcutta, 213, 220, 229
Caledonia, 92
California, 491–93
camels, 337–38, 342
Canadian Shield, 472, 489
Canberra, 321
Canton, 237, 249
Cape Colony, 386
Cape Colored, 386, 406
Cape Province, 407
Cape Town, 386, 407
Cardiff, 92
Caribbean, 429–35
Carnegie, Andrew, 92
Carpathian Mountains, 155
Casablanca, 359
cassava. *See* manioc
caste, in India, 204
cattle, 91, 100, 105, 127, 188, 189, 225–26, 234, 254, 316, 322, 383, 410, 421, 424, 427, 440, 447, 450. *See also* livestock
Celts, 81, 90–95
central place theory, 9
cereal grains, 335, 336, 339, 357, 376, 423, 445, 447, 458
Ceylon. *See* Sri Lanka
Chaco, 447–50
cheese, 96, 100, 109, 111, 121, 127
chemicals, 28, 111, 113, 137, 138, 141, 143, 280, 282, 286, 463
chernozem, 163
Chicago, 470
Chile, 450–52
China, 38–39, 231–62
Chin Ling (Tsin Ling) Mountains, 246
Cholon, 300
Chota Nagpur, 217
chrome, 165, 215, 218, 367, 408, 457
citrus fruits, 127, 129, 322, 340, 357, 366, 372, 374, 376, 406, 407, 408, 435, 443, 452, 458, 473, 486, 488, 490, 491
city states, 123–26
climatic change, variation, 343–45
cloves, 411
Clyde River (Scotland), 92
coal, 27–28, 29, 30, 73, 74, 84, 85–87, 88, 89, 90, 92, 94, 95, 100, 102, 103, 104, 111, 113, 117, 118, 120, 121, 126, 130, 137, 138, 141, 143, 151, 152, 153, 163, 165–66, 167, 168, 191, 196, 207, 215, 217, 218, 220, 228, 243, 252, 257–59, 269–70, 272, 277, 280, 281, 285, 286, 302–03, 323, 326, 351, 359, 367, 376, 382, 401, 403, 404–05, 408, 427, 436, 445, 446, 449, 452, 457, 458, 461, 473, 475, 486, 489, 490
cobalt, 383, 400
Cochin China, 308
coconuts and coconut products, 209, 227, 228, 302, 305, 306, 327–28, 435, 441
coffee, 306–07, 371, 381, 384, 402, 410, 411, 412, 422, 424, 428, 435, 437, 438, 439, 441, 444, 445
coke, 27
Cologne (Köln), 140
Colombia, 439–40
Colombo, 228–29
Columbus, Christopher, 416
Common Market (European), 79, 90
complementarity, 6
Congo (Zaire) River and basin, 399–403
Constantinople (Istanbul), 367
Cook, James, 313
Copenhagen, 105
copper, 27, 75, 126, 165, 270, 345, 346, 376, 381, 383, 400–02, 424, 427, 443, 452, 457
core areas, 39–40, 475–76
cork, 357
corn, 359, 447, 454, 458, 463, 471, 475, 489. *See also* maize
Costa Rica, 438
cotton, 88, 143, 176, 196, 207, 208, 215, 220, 221, 228, 229, 250, 279, 339–40, 351, 357, 359, 365, 366, 367, 370, 374, 375, 380, 381, 384, 396, 400, 402, 409, 411, 437, 440, 443, 444, 445, 446, 449, 450, 458, 463, 473, 486, 487, 490, 491
Cuba, 433–35
Czechoslovakia, 144, 152–54

D

Dairen, 249
dairying, 89, 96, 100, 105, 107, 109, 111, 113, 176, 274, 316, 322, 325, 336, 410, 451, 473, 484, 485, 488, 489

Dakar, 392
Damascus, 369
Damodar River, 217
Danube River, 153–57
Danzig, 152
Dar es Salaam, 411
Dardanelles, 367
Darwin, 322
dates, 340, 351, 357, 365, 370, 373
Dead Sea, 372
Deccan, 208
Delhi, 204, 230
Denmark, 97, 105–06
Denver (Colorado), 490
desert, 336–40
Dhahran, 371
diamonds, 372, 381, 383, 395, 400, 402, 404–05, 424, 444, 445, 447
Djakarta (Batavia), 301
Djibouti, 364
Donegal Mountains, 95
Donets Basin, 165
Drang nach Osten, 143–44
Dravidian, 203, 210
Dublin, 95
Durban, 407
Dutch, 108–12; in Africa, 386

E

ecology, 31
Ecuador, 441
Edinburgh, 94
Edo, 267
Egypt, 345–47, 359–64
Elbe River, 136, 153
Elburz Mountains, 374
England, 80–90
English Channel, 80
entrepôt, 81
environmental determinism, 23
environmental influence, 21–33
Esfahan. *See* Isfahan
Estonia, 152
Ethiopia, 364–65, 411–12
Eurasia, 180
European Common Market. *See* Common Market

F

fertilizer, 141, 189, 214, 215, 218, 223, 224, 225–26, 254–55, 272, 273, 275, 276, 286, 287, 293, 463
figs, 366
Fiji, 327
Finland, 97, 103–05
fish and fishing, 81, 102, 246, 255, 273, 275–76, 284, 293, 327, 328, 380, 443, 455, 456, 457, 473, 484, 493
fjeld, 100
fjord, 99
Flanders, 106, 108, 112
flax, 449
Flemish, 112
Florida, 488
fodder crops, 100, 119, 120, 130, 176, 188, 226, 227, 274, 319, 447, 450, 453
Formosa. *See* Taiwan
France, 115–22
free trade, 85
Freetown (Sierra Leone), 396
French Canada, 476, 483–84
frontier, influence of, 464–65
functional regions, 41
Fushun, 243

G

Gaelic, 95
Ganges River and valley, 204, 207, 214, 217, 226–27
Genoa (Genova), 130–33
geopolitics, 146 n.
Germany, 136–46
Ghana, 393–99
Ghent, 106
Glasgow, 94
glass, 113
gold, 7, 316, 323, 379, 381, 383, 395, 403–05, 414, 418, 421, 423, 424, 432, 444, 445, 447, 457, 482, 491
Göteborg, 102
grapes, 120, 127, 129, 322, 340, 357, 366, 372, 373, 374, 407, 452
Great Wall, 240
Greece, 130
Green Revolution, 223–25, 254
Greenwich (England), 14
Guatemala, 422
Guianas (South America), 441

H

hacienda, 423
Hague, The, 111
Haifa, 373
Haiphong, 300
Haiti, 433
Hamburg, 140
Han dynasty (China), 237–39
handicrafts, 91, 221, 273, 376
Hanoi, 300
harbors, 98–99
Hawaii, 493–94
heavy industry, 27, 87, 88, 92, 94, 95, 102, 111, 113, 118, 120, 133, 137, 138, 141, 146, 151, 153, 163, 165, 167, 168, 169, 171, 172–73, 178, 207, 215, 217, 218, 220–21, 229–30, 256, 257–61, 270, 276–77, 280, 281, 282, 285, 286, 449, 457, 459, 484, 489, 493. *See also* machinery
Helsinki, 104
herding, 334, 345, 351, 374, 376, 412
hides and skins, 176, 336, 338, 340, 357, 367, 376, 381, 440, 449, 450. *See also* animal products
Hindi, 210
Hinduism, 204
Hindustan, 204, 207
hinterland. *See* tributary area
Hirakud, 223

Hiroshima, 281
Hokkaido, 267–69
Holland, 106–12
Hong Kong, 249–50
horses, 96, 189, 234, 334–35, 421, 427
Hsi River, 237, 244
Hsinkiang. *See* Sinkiang
Huang River. *See* Yellow River
Hungary, 154–55
hydroelectric power. *See* water power

I

Incas, 420–22
India, 199–230
Indochinese peninsula, 308–10
Indus River and valley, 201, 215
industrialization, 27–29, 84–87
Inland Sea (Japan), 267
insularity, 80–84
International Date Line, 14
Iran, 349–50, 374–76. *See also* Persia
Iraq, 369–70
Ireland, 90–91, 95–96
iron ore, 27, 28–29, 73, 74, 84, 85, 87, 89, 92, 94, 100, 102, 103, 104, 113, 114, 117, 118, 120, 121, 126, 130, 137, 143, 151, 153, 157, 163, 165, 167, 168, 169, 207, 217, 218, 220, 228, 237, 243, 258, 259, 270, 277, 280, 285, 286, 302, 316, 322, 323, 326, 351, 358, 363, 367, 376, 382, 395, 403, 404–05, 424, 425, 427, 436, 440, 443, 445, 446, 447, 449, 452, 457, 458, 473, 475, 483, 486, 490. *See also* steel
Isfahan, 375
Islam, 156, 205, 330, 342–43, 350
Israel, 371–74
Istanbul, 367
Italy, 122–35
ivory, 379, 395, 400
Izmir (Smyrna), 366

J

Jamaica, 433
Jamshedpur, 217
Japan, 263–83
Java, 289–91
Jerusalem, 373
Jidda, 370
Johannesburg, 403–04
Jordan, 369–70
Jordan River, 369, 372, 373
Juddah. *See* Jidda
jute, 196, 207, 213, 214, 220, 229
Jutland, 105

K

Kabul, 376
Kaiping, 259
Kalahari Desert, 402, 407
Kano, 391
Kanto Plain, 269
kapok, 196, 302, 306
Karachi, 214
Karaganda, 167
Kashmir, 207, 217
Kazakhstan, 175
Kenya, 409–11
Khabarovsk, 168
Khartoum, 364–65
Khmers, 310
Kiev, 171
Kikuyu, 409–10
Kinshasa, 399
Kiruna, 102
Kisangani, 399
Kitakyushu, 281
Kobe, 279
Korea, 283–87
Krakow, 151
Krivoi Rog, 165
Kuala Lumpur, 300
Kumasi, 397
Kuroshio Current, 267
Kustanay, 165
Kuwait, 371
Kuznetsk coal basin, 168
Kyoto, 267, 279
Kyushu, 270, 280–81

L

Ladinos (Latin America), 417
Lagos, 397
Lahore, 214
Latin America, 413–53
latitude, 14
Latvia, 152
lead, 376, 400, 436, 443, 457
Lebanon, 368–69
Lena River, 172
Leningrad, 165, 171
Lesotho, 407
Liao River, 243
Liberia, 393,
Limburg, 106
Lithuania, 152
Liverpool, 92
livestock, 331, 333, 338, 347, 357, 365, 366, 405, 406, 408, 423, 444, 445, 447, 450, 452, 489, 490. *See also* cattle, sheep
llama, 421, 440, 443
llanos, 440
loess, 235, 245
Loire River, 117
Lombardy, 130
London, 88
longitude, 14
Lorraine, 121
Los Angeles, 492
Low Countries, 106–14
lumber and lumbering, 484, 488, 493
Luxembourg, 114

M

machinery, 91, 280, 281, 282, 287, 457, 461, 463, 473. *See also* heavy industry
Madras, 209, 229
Magnitogorsk, 165
Magyars, 157
maize, 164, 176, 195, 295, 381, 396, 406, 408, 410, 421, 424, 437, 443, 446, 447. *See also* corn
Malawi, 408–09

Malaya, 298, 300
Malaysia, 300–01
Malmö, 102
Manchuria, 242–43
manganese, 357, 381, 383, 395, 396, 427, 436, 445, 457
Manila, 301
manioc (cassava), 295, 381
Maoris, 317
maps, 11–16
Massif Central (France), 117
Matadi, 399
Maurya Empire, 203
Mayas, 420–22
Mecca, 370
Mediterranean Basin, 122–30
Mediterranean climate, 126–27, 322–23, 407, 451–52, 491
Meiji restoration, 272
Melanesia, 326–27
Melbourne, 322
melons, 340
Mesopotamia, 345–47
mestizo, 417
Mexico, 435–38
Micronesia, 327–29
Middle America, 435–38
Middle Atlantic region, 484–85
Middle West region (U.S.), 489–90
Midi (France), 115–16
Milan (Milano), 133
mixed farming, 119, 188, 316, 451, 453
Mogul Empire, 206
Mombasa, 409
Mongolia, 240
Mongolian People's Republic, 240
Mongols, 238–39, 240, 335
Monroe Doctrine, 435
Monrovia, 397
monsoon, 181, 223
Montevideo, 450
Moravia, 152
Morocco, 357–59
Moscow, 165, 171
Mozambique, 409
Mukden (Shenyang), 243
multiple-factor region, 41–42
Muroran, 281
Muscat, 371

N

Nagasaki, 281
Nagoya, 280
Nairobi, 409
Namib Desert, 407
Namibia, 403, 407
Naples (Napoli), 134
Nara, 267
Narbada River, 209
Narvik, 102
Natal, 406–07
natural frontiers, 122–23
natural gas, 29, 73, 90, 100, 111, 126, 130, 135, 215, 357, 457, 475, 486, 489, 490
Negev Desert, 371–72
Netherlands, 106, 108–12
New England (region), 484
New Guinea, 317
New Jersey, 485
New York City, 485
New Zealand, 313–19, 324–26
nickel, 435, 452, 457
Niger River, 395
Nigeria, 393–99
night soil, 189
Nile River and valley, 346, 361, 365
nitrates, 424
nomadism, 176, 331–36
Normans, 81
North Sea, 81, 90
Norway, 97–101
Novgorod, 171
nucleated settlements, 5, 188
Nyasaland. *See* Malawi

O

oases, 439–40
oats, 90, 91, 100, 102, 104, 120
Ob River, 172
Oder-Neisse Line, 137
oil. *See* petroleum
olives, 127, 129, 340, 357, 366, 372, 374
Oman, 370–71
opium, 247
Organization of Petroleum Exporting Countries (O.P.E.C.), 355–57
Osaka, 279–80
Oslo, 99
Ottawa, 490
Ottoman Empire, 156, 349
Oyoshio Current, 267

P

Pakistan, 213–17
palm products, 302, 306, 381, 382, 384, 395, 396, 399, 400, 402
pampa (South America), 447–50
Panama, 432–33, 438
Papua, 327
Paraguay, 450
Paris, 117–18
Patagonia, 451
Pataliputra, 203
peanuts, 381, 395, 396, 402, 411
peasant society, 185–86
Peking, 237
permafrost, 174
Persia, 347, 349–50. *See also* Iran
Perth, 323
Peru, 441–43
petroleum, 27, 28, 29, 73, 75, 90, 100, 103, 104, 111, 126, 130, 135, 143, 151, 157, 165, 166–67, 168, 196, 215, 217–18, 228, 259–60, 269, 270–72, 276, 297, 301, 303, 304, 305, 326, 331, 351, 353, 355–57, 358, 363, 367, 369, 370–71, 374, 375, 376, 382, 383, 395, 402, 424, 425, 427, 435, 436, 440, 441, 442, 449, 450, 452, 457, 458, 473, 475, 482, 486, 489, 490, 491

Philadelphia, 485
phosphates, 286, 328, 358, 457
Pilsen. *See* Plzeň
pineapples, 493–94
Pittsburgh, 489
plantations, 228, 289, 305
platinum, 383, 404
Plzeň (Pilsen), 153
Poland, 151–52
polders, 108
Polynesia, 327–29
Port Said, 363
Port Sudan, 364
Port Talbot, 92
ports, 98–99
possibilism, 23
potash, 140, 151, 372, 457
potatoes, 90, 96, 100, 104, 141, 152, 164, 188, 195, 273–74, 295, 381, 421, 437, 443, 484
Pretoria, 404
primate city, 105
prime meridian, 14
projections, map, 12
Prussia, 145
Puerto Rico, 433–35
Punjab, 207, 215
purdah, 366
Pusan, 286
Pyongyang, 286
Pyrenees, 129
pyrethrum, 410

Q

qanats, 339, 375, 442–43
Queensland, 322

R

race, 201
Rawalpindi, 215
region, 35–43
regional planning, 36
regionalism, 37–43
resources, 25–31
Rhine River and region, 106–11, 122, 137, 140
Rhodesia (Zimbabwe), 403, 408–09
rice, 130, 184–85, 188, 201, 207, 208, 209, 224, 225, 227, 228, 229, 234, 250–51, 269, 272, 273, 275, 285, 286, 289, 295, 297, 298, 300, 305, 307–08, 310, 311, 339, 359, 374, 441, 443, 444
Rio de Janeiro, 446–47
Rome, 133
Rotterdam, 111
rubber and rubber products, 74, 143, 196, 209, 225, 227, 228, 300, 301, 302, 304, 305–06, 307, 309, 381, 384, 395, 399, 400, 402, 424, 444
Ruhr, 137–40
rum, 380, 435
Rumania, 148, 155
Russia, 159–79
rye, 90, 91, 100, 102, 152, 164–65

S

Saar, 137
Sabah, 301
Saigon, 300
Sakhalin Island, 270
salt, 7, 190, 272, 336, 343, 347, 369, 372, 457
San Francisco, 492
Sanskrit, 201
São Paulo, 444
Sarawak, 301
Sardinia, 134
Saudi Arabia, 370–71
scale, in maps, 12
Scandinavia, 97–106
Scheldt River, 113
Scotland, 90–94
sea power, 81–83
Seine River, 118
Seoul, 286
Shaba, 400–02
Shang dynasty (China), 235
Shanghai, 244, 249, 259
Shantung, 245
sheep, 22, 91, 96, 100, 127, 188, 189, 225, 234, 316, 322, 323, 338, 345, 357, 376, 410, 421, 443, 449, 451
Shenyang. *See* Mukden
shifting cultivation, 293–95
Sikhs, 215
Silesia, 137–41, 151
silk, 121, 196, 247, 249, 274, 278
silver, 414, 421, 423, 425, 432, 436, 443, 452, 457
Singapore, 300
single-factor region, 41–42
Sinkiang (Hsingkiang), 238, 240–42
sisal, 302, 410, 411, 437
site, 6
situation, 6
Sjaelland, 105
Skåne, 102
slavery, slave trade, 370–80, 423
Slovakia, 152–53
small-holder agriculture, 306
Somalia, 364
sorghum, 188, 207, 208, 250, 339, 371, 412, 461, 490
South Africa, Republic of, 403–07
Southeast Asia, 288–311
Southwest Africa (Namibia), 403, 407
Soviet Union, 159–79
soybeans, 286
Spain, 129–30, 414–16
Spanish Main, 433
spatial interaction, 5–10
spices, 7, 196, 208, 302, 306, 379, 395
Sri Lanka (Ceylon), 211, 223, 225, 227–28
Stalin, Josef, 178
Standard Metropolitan Area (U.S. Census), 467
steel, 26, 92, 94, 102, 114, 121, 135, 137, 151, 167, 168, 215, 217, 220, 243, 258, 259, 270,

276, 279, 280, 281, 282, 302, 322, 323, 326, 350, 367, 372, 404, 436, 437, 440, 445, 449, 452, 457, 475, 485, 486, 489. *See also* iron ore
Stockholm, 102–03
Sudan (area), 385–86
Sudan (country), 365
Suez, 363
sugar and sugar cane, 196, 289, 291, 302, 306, 322, 327, 328, 340, 359, 365, 380, 381, 384, 406, 407, 409, 410, 423, 424, 425, 432, 437, 440, 441, 443, 446, 488, 493–94
sugar beets, 120, 176, 427
Sukkur, 215
sulphur, 143, 270, 425, 436, 457, 486
Swansea, 92
Swazi, Swaziland, 407
Sweden, 97, 101–03
Switzerland, 146–47
Sydney, 313, 322
Syria, 368–69

T

Taebak Mountains, 286
Tahiti, 328
Taiwan (Formosa), 247–48
Tanzania, 409–11
tea, 196, 207, 208, 209, 225, 227, 228, 229, 247, 249, 251, 274, 302, 306, 409
Tehran, 374
Tel Aviv, 373
textiles, 88, 91, 94, 108, 111, 113, 133, 137, 141, 151, 165, 176, 197, 215, 217, 218–20, 221, 228, 229, 279–80, 281, 282, 287, 335, 359, 363, 367, 375, 437, 449, 484, 486, 487
Thais, 309–10
Tientsin, 246, 249
tierra caliente, 437
Tierra del Fuego, 451
tierra fria, 437
tierra templada, 437–38
Tigris River, 346, 370
timber, 100, 102, 103, 104, 138, 163, 172, 246, 297, 305, 402, 451, 458, 473, 493
Timbuktu, 6–7
tin, 75, 92, 196, 297, 300, 302, 303–04, 305, 309, 381, 382, 383, 395, 427, 443, 457
tobacco, 196, 229, 289, 357, 366, 374, 376, 380, 384, 406, 408, 409, 435, 437, 440, 446, 450, 458, 486
Tokyo, 67, 279
Tonkin, 308–09
Transkei, 408
tributary area (hinterland), 8–10
Trieste, 155
Trondheim, 100
tropical hardwoods, 302, 399, 402, 444
truck crops, 105, 109, 113, 120, 127, 273, 406, 407, 473, 485, 491
tsetse fly, 383, 411
Tsin Ling Mountains. *See* Chin Ling Mountains
Tsingtao, 249
Tsushima, 263
Tumen River, 283
tungsten, 443, 457
Tunis, 359
Tunisia, 357–59
Turin (Torino), 133
Turkestan, 176
Turkey, 365–68
Turks, 156, 349–50. *See also* Ottoman Empire
Tweed River, 91

U

Uganda, 409–11
Ukraine, 163–65, 171, 173, 176
Ulster, 95
Union of Soviet Socialist Republics (U.S.S.R.), 159–79
United Arab Emirates, 370–71
United Kingdom, 80, 96
Ural Mountains, 165–67
uranium, 30, 218, 381, 383, 400, 404
Uruguay, 447–50

V

veld, 387, 406
Venezuela, 438, 440
Vienna (Wien), 157
Vietnam, 298, 308
Vikings, 454
Vindhya Mountains, 209
Vladivostok, 169
Volga River, 169, 171
Volta River, 399

W

Wales, 90–92
Walloons, 112
Warsaw, 151
water power, 74, 87, 94, 100–01, 102, 104, 117, 120, 126–27, 130, 133, 138, 147, 151, 155, 157, 169, 172, 217, 218, 220, 243, 253, 260, 272, 277, 281, 286, 302, 368, 372, 376, 382–83, 427, 446, 452, 457, 488, 493
West Indies, 429–35
West River (China). *See* Hsi River
wheat, 100, 102, 120, 127, 152, 163, 164, 175, 176, 188, 196, 207, 208, 215, 223, 225, 229, 234, 243, 250, 251, 272, 273, 285, 286, 316, 319, 322, 324, 339, 347, 357, 359, 366, 370, 372, 374, 376, 381, 384, 405, 406, 407, 408, 410, 412, 424, 437, 447, 449, 452, 463, 471, 473, 482, 489, 490
whiskey, 91, 380
wine, 119, 120, 121, 357, 366, 405, 407, 452

Witwatersrand, 403–04
wood products, 100, 104, 274–75, 286, 336, 359, 444, 450, 458, 484, 488, 493
wool, 339–40, 357, 367, 449, 451, 452
Wuhan, 244, 259

Y

Yalu River, 283, 286
yams, 396
Yangtze River and basin, 237, 244, 246, 249, 260
Yellow (Huang) River, 235, 244–45
Yemen, 370–71
Yenisei River, 172
Yokohama, 279
Yucatan, 422, 437
Yugoslavia, 155–56
yurt, 338

Z

Zaire, 399–403
Zaire River. *See* Congo River
Zambia, 400–03
Zamindars, 224
Zanzibar, 411
Zimbabwe, 403, 408
zinc, 400, 436, 443, 457

Printed in U.S.A.